Lecture Notes in Comp  587

Commenced Publication in 1973
Founding and Former Series Editors:
Gerhard Goos, Juris Hartmanis, and Jan van Leeuwen

Jeffrey Xu Yu   Myoung Ho Kim
Rainer Unland (Eds.)

Database Systems for Advanced Applications

16th International Conference, DASFAA 2011
Hong Kong, China, April 22-25, 2011
Proceedings, Part I

 Springer

Volume Editors

Jeffrey Xu Yu
The Chinese University of Hong Kong
Department of Systems Engineering and Engineering Management
Shatin, N.T., Hong Kong, China
E-mail: yu@se.cuhk.edu.hk

Myoung Ho Kim
Korea Advanced Institute of Science and Technology (KAIST)
Department of Computer Science
291 Daehak-ro (373-1 Guseong-don), Yuseong-gu, Daejeon 305-701, Korea
E-mail: mhkim@dbserver.kaist.ac.kr

Rainer Unland
University of Duisburg-Essen
Institute for Computer Science and Business Information Systems (ICB)
Schützenbahn 70, 45117 Essen, Germany
E-mail: rainer.unland@icb.uni-due.de

ISSN 0302-9743   e-ISSN 1611-3349
ISBN 978-3-642-20148-6   e-ISBN 978-3-642-20149-3
DOI 10.1007/978-3-642-20149-3
Springer Heidelberg Dordrecht London New York

Library of Congress Control Number: 2011923553

CR Subject Classification (1998): H.2-5, C.2, J.1, J.3

LNCS Sublibrary: SL 3 – Information Systems and Application, incl. Internet/Web
and HCI

Typesetting: Camera-ready by author, data conversion by Scientific Publishing Services, Chennai, India

Printed on acid-free paper

Springer is part of Springer Science+Business Media (www.springer.com)

Preface

DASFAA is an annual international database conference, which showcases state-of-the-art R&D activities in database systems and their applications. It provides a forum for technical presentations and discussions among database researchers, developers and users from academia, business and industry. It is our great pleasure to present you the proceedings of the 16th International Conference on Database Systems for Advanced Applications (DASFAA 2011), which was held in Hong Kong, China, during April 22-25, 2011.

DASFAA 2011 received 225 research paper submissions from 32 countries / regions (based on the affiliation of the first author). After a thorough review process for each submission by the Program Committee and specialists recommended by Program Committee members, DASFAA 2011 accepted 53 full research papers and 12 short research papers (the acceptance rates were 24% and 5%, respectively). Other papers in this volume include four industrial papers selected by a committee chaired by Takahiro Hara (Osaka University), Tengjiao Wang (Peking University), and Xing Xie (Microsoft Research China), and eight demo papers selected by a committee chaired by Jidong Chen (EMC Research China), Lei Chen (The Hong Kong University of Science and Technology), and Kyoung-Gu Woo (Samsung Electronics).

This volume also includes two invited keynote papers, presented by leading experts in database research and advanced applications at DASFAA 2011, Josephine M. Cheng (IBM Research Almaden Lab) and Divy Agrawal (University of California at Santa Barbara), on the topics of "Smarter Planet: Empower People with Information Insights" and "Database Scalability, Elasticity, and Autonomy in the Could," respectively; one extended abstract for the DASFAA 2011 ten-year best paper on "What Have We Learnt from Deductive Object-Oriented Database Research?" by Mengchi Liu (Carleton University), Gillian Dobbie (University of Auckland), and Tok Wang Ling (National University of Singapore); three tutorial abstracts, selected by Tutorial Co-chairs Reynold Cheng (The University of Hong Kong), Ken Lee (University of Massachusetts Dartmouth), and Ee-Peng Lim (Singapore Management University), "Managing Social Image Tags: Methods and Applications" by Aixin Sun and Sourav S. Bhowmick, "Searching, Analyzing and Exploring Databases" by Yi Chen, Wei Wang and Ziyang Liu, and "Web Search and Browse Log Mining: Challenges, Methods, and Applications" by Daxin Jiang; and one panel abstract selected by Panel Co-chairs, Haibo Hu (Hong Kong Baptist University), Haixun Wang (Microsoft Research China), and Baihua Zheng (Singapore Management University). The conference program boasts conference proceedings that span two volumes in Springer's *Lecture Notes in Computer Science* series.

Beyond the main conference, six workshops, held in conjunction with DASFAA 2011, were selected by Workshop Co-chairs Jianliang Xu (Hong Kong

Baptist University), Ge Yu (Northeastern University), and Shuigeng Zhou (Fudan University). They are the First International Workshop on Graph-structured Data Bases (GDB 2011), the First International Workshop on Spatial Information Modeling, Management and Mining (SIM3), the International Workshop on Flash-Based Database Systems (FlashDB), the Second International Workshop on Social Networks and Social Media Mining on the Web (SNSMW), the First International Workshop on Data Management for Emerging Network Infrastructures (DaMEN), and the 4th International Workshop on Data Quality in Integration Systems (DQIS). The workshop papers are included in a separate volume of proceedings also published by Springer in its *Lecture Notes in Computer Science* series.

DASFAA 2011 was jointly organized by The Chinese University of Hong Kong, The Hong Kong University of Science and Technology, Hong Kong Baptist University, The University of Hong Kong, City University of Hong Kong, and The Hong Kong Polytechnic University. It received in-cooperation sponsorship from the China Computer Federation Database Technical Committee. We are grateful to the sponsors who contributed generously to making DASFAA 2011 successful. They are the Department of Systems Engineering and Engineering Management of The Chinese University of Hong Kong, Oracle, IBM, K.C. Wong Education Foundation, and Hong Kong Pei Hua Education Foundation.

The conference would not have been possible without the support of many colleagues. We would like to express our special thanks to Honorary Conference Co-chairs, Xingui He (Peking University), Shan Wang (Renmin University of China), and Kyu-Young Whang (KAIST) for their valuable advice on all aspects of organizing the conference. We thank Organizing Committee Chair Kam-Fai Wong (The Chinese University of Hong Kong), Publicity Co-chairs, Raymond Wong (The Hong Kong University of Science and Technology), Xiaochun Yang (Northeastern University), and Xiaofang Zhou (University of Queensland), Publication Chair Rainer Unland (University of Duisburg-Essen), Finance Chair Vincent Ng (The Hong Kong Polytechnic University), Local Arrangements Chair Hong-va Leong (The Hong Kong Polytechnic University), Sponsor Chair Joseph Ng (Hong Kong Baptist University), Best Award Committee Co-chairs Ming-Syan Chen (Academia Sinica, Taiwan and National Taiwan University) and Aoying Zhou (East China Normal University), and Demo Award Committee Co-chairs Ben Kao (The University of Hong Kong) and Lizhu Zhou (Tsinghua University). Our thanks go to all the committee members and other individuals involved in putting it all together, and all authors who submitted their papers to this conference.

July 2010

Dik Lun Lee
Wang-Chien Lee
Kamal Karlapalem
Jeffrey Xu Yu
Myoung Ho Kim

Organization

Honorary Conference Co-chairs

Xingui He — Peking University, China
Shan Wang — Renmin University of China, China
Kyu-Young Whang — Korea Advanced Institute of Science and Technology (KAIST), Korea

Conference General Co-chairs

Dik Lun Lee — The Hong Kong University of Science and Technology, China
Wang-Chien Lee — Penn State University, USA
Kamal Karlapalem — IIIT-Hyderabad, India

Program Committee Co-chairs

Jeffrey Xu Yu — The Chinese University of Hong Kong, China
Myoung Ho Kim — Korea Advanced Institute of Science and Technology (KAIST), Korea

Organizing Committee Chair

Kam-Fai Wong — The Chinese University of Hong Kong, China

Workshop Co-chairs

Jianliang Xu — Hong Kong Baptist University, China
Ge Yu — Northeastern University, China
Shuigeng Zhou — Fudan University, China

Industrial Co-chairs

Takahiro Hara — Osaka University, Japan
Tengjiao Wang — Peking University, China
Xing Xie — Microsoft Research China, China

Tutorial Co-chairs

Reynold Cheng — The University of Hong Kong, China
Ken Lee — University of Massachusetts Dartmouth, USA
Ee-Peng Lim — Singapore Management University, Singapore

Panel Co-chairs

Haibo Hu	Hong Kong Baptist University, China
Haixun Wang	Microsoft Research China, China
Baihua Zheng	Singapore Management University, Singapore

Demo Co-chairs

Jidong Chen	EMC Research China, China
Lei Chen	The Hong Kong University of Science and Technology, China
Kyoung-Gu Woo	Samsung Electronics, Korea

Publicity Co-chairs

Raymond Wong	The Hong Kong University of Science and Technology, China
Xiaochun Yang	Northeastern University, China
Xiaofang Zhou	University of Queensland, Australia

Local Arrangements Chair

Hong-va Leong	The Hong Kong Polytechnic University, China

Finance Chair

Vincent Ng	The Hong Kong Polytechnic University, China

Publication Chair

Rainer Unland	University of Duisburg-Essen, Germany

Web Chair

Hong Cheng	The Chinese University of Hong Kong, China

Demo Award Committee Co-chairs

Ben Kao	The University of Hong Kong, China
Lizhu Zhou	Tsinghua University, China

Best Paper Committee Co-chairs

Ming-Syan Chen	Academia Sinica, Taiwan and National Taiwan University, Taiwan
Aoying Zhou	East China Normal University, China

Steering Committee Liaison

Qing Li   City University of Hong Kong, China

Sponsor Chair

Joseph Ng   Hong Kong Baptist University, China

CCF DBTC Liaison

Xiaofeng Meng   Renmin University of China, China

DASFAA Awards Committee

Tok Wang Ling (Chair)   National University of Singapore, Singapore
Jianzhong Li   Harbin Institute of Technology, China
Krithi Ramamirtham   Indian Institute of Technology at Bombay,
                                     India
Kian-Lee Tan   National University Singapore, Singapore
Katsumi Tanaka   Kyoto University, Japan
Kyu-Young Whang   Korea Advanced Institute of Science and
                                     Technology (KAIST), Korea
Jeffrey Xu Yu   The Chinese University of Hong Kong, China

DASFAA Steering Committee

Katsumi Tanaka (Chair)   Kyoto University, Japan
Ramamohanarao Kotagiri
   (Vice Chair)   University of Melbourne, Australia
Kyu-Young Whang (Advisor)   Korea Advanced Institute of Science and
                                     Technology (KAIST), Korea
Yoshihiko Imai (Treasurer)   Matsushita Electric Industrial Co., Ltd., Japan
Kian Lee Tan (Secretary)   National University of Singapore (NUS),
                                     Singapore
Yoon Joon Lee   Korea Advanced Institute of Science and
                                     Technology (KAIST), Korea
Qing Li   City University of Hong Kong, China
Krithi Ramamritham   Indian Institute of Technology at Bombay,
                                     India
Ming-Syan Chen   National Taiwan University, Taiwan
Eui Kyeong Hong   Univerity of Seoul, Korea
Hiroyuki Kitagawa   University of Tsukuba, Japan
Li-Zhu Zhou   Tsinghua University, China
Jianzhong Li   Harbin Institute of Technology, China
BongHee Hong   Pusan National University, Korea

Program Committees

Research Track

Toshiyuki Amagasa	University of Tsukuba, Japan
Masayoshi Aritsugi	Kumamoto University, Japan
James Bailey	University of Melbourne, Australia
Ladjel Bellatreche	Poitiers University, France
Boualem Benatallah	University of New South Wales, Australia
Sourav S. Bhowmick	Nanyang Technological University, Singapore
Athman Bouguettaya	CSIRO, Australia
Chee Yong Chan	National University Singapore, Singapore
Jae Woo Chang	Chonbuk National University, Korea
Lei Chen	The Hong Kong University of Science and Technology, China
Ming-Syan Chen	National Taiwan University, Taiwan
Reynold Cheng	The University of Hong Kong, China
Hong Cheng	The Chinese University of Hong Kong, China
James Cheng	Nanyang Technological University, Singapore
Byron Choi	Hong Kong Baptist University, China
Yon Dohn Chung	Korea University, Korea
Gao Cong	Nanyang Technological University, Singapore
Bin Cui	Peking University, China
Alfredo Cuzzocrea	ICAR-CNR / University of Calabria, Italy
Gill Dobbie	University of Auckland, New Zealand
Xiaoyong Du	Renmin University of China, China
Jianhua Feng	Tsinghua University, China
Ling Feng	Tsinghua University, China
Sumit Ganguly	IIT Kanpur, India
Yunjun Gao	Zhejiang University, China
Vivek Gopalkrishnan	Nanyang Technological University, Singapore
Wook-Shin Han	Kyungpook National University, Korea
Takahiro Hara	Osaka University, Japan
Bingsheng He	Nanyang Technological University, Singapore
Wynne Hsu	National University Singapore, Singapore
Haibo Hu	Hong Kong Baptist University, China
Seung-won Hwang	POSTECH, Korea
Yoshiharu Ishikawa	Nagoya University, Japan
Mizuho Iwaihara	Waseda University, Japan
Adam Jatowt	Kyoto University, Japan,
Ruoming Jin	Kent State University, USA
Jaewoo Kang	Korea University, Korea

Vincent S. Tseng	National Cheng Kung University, Taiwan
Vasilis Vassalos	Athens University of Economics and Business, Greece
John Wang	Griffith University, Australia
Jianyong Wang	Tsinghua University, China
Guoren Wang	Northeastern University, China
Wei Wang	University of New South Wales, Australia
Raymond Wong	The Hong Kong University of Science and Technology, China
Xiaokui Xiao	Nanyang Technological University, Singapore
Jianliang Xu	Hong Kong Baptist University, China
Man-Lung Yiu	Hong Kong Polytechnic University, China
Haruo Yokota	Tokyo Institute of Technology, Japan
Jae Soo Yoo	Chungbuk National University, Korea
Ge Yu	Northeastern University, China
Aidong Zhang	University of Buffalo, SUNY, USA
Rui Zhang	University of Melbourne, Australia
Yanchun Zhang	Victoria University, Australia
Baihua Zheng	Singapore Management University, Singapore
Aoying Zhou	East China Normal University, China
Xiaofang Zhou	University of Queensland, Australia

Industrial Track

Wolf-Tilo Balke	University of Hannover, Germany
Edward Chang	Google, China and University of California Santa Barbara, USA
Bin Cui	Peking University, China
Dimitrios Georgakopoulos	CSIRO, Australia
Seung-Won Hwang	POSTECH, Korea
Marek Kowalkiewicz	SAP, Australia
Sanjay Kumar Madria	Missouri University of Science and Technology, USA
Mukesh Mohania	IBM Research India, India
Makoto Onizuka	NTT Corporation, Japan
Jilei Tian	Nokia Research China, China
Masashi Tsuchida	Hitach, Ltd., Japan
Jianyong Wang	Tsinghua University, China
Wei Wang	Fudan University, China
Yu Zheng	Microsoft Research Asia, China

Demo Track

Ilaria Bartolini	University of Bologna, Italy
Bin Cui	Peking University, China
Heasoo Hwang	Samsung Electronics, Korea

Jin-ho Kim	Kangwon National University, Korea
Changkyu Kim	Intel Labs, USA
Guoqiong Liao	Jiangxi University of Finance and Economics, China
Hongrae Lee	Google Research, USA
Jiaheng Lu	Renmin University of China, China
Peng Wang	Fudan University, China
Feng Yaokai	Kyushu University, Japan

External Reviewers

Eunus Ali	Selma Khouri	Miao Qiao
Parvin Asadzadeh	Henning Koehler	Hongda Ren
He Bai	Neila Ben Lakhal	Jong-Won Roh
Moshe Barukh	Dong-Ho Lee	Seung Ryu
Seyed-Mehdi-Reza Beheshti	Injoon Lee	Sherif Sakr
	Jongwuk Lee	Shuo Shang
Arnab Bhattacharya	Kyubum Lee	Jie Shao
Nick Bozovic	Mu-Woong Lee	Mohamed A. Sharaf
Xin Cao	Sanghoon Lee	Gao Shen
Wing Kwan Chan	Sunwon Lee	Wei Shen
Lijun Chang	Guoliang Li	Zhitao Shen
Muhammad Aamir Cheema	Jianxin Li	Wanita Sherchan
	Jing Li	Reza Sherkat
Jinchuan Chen	Jiang Li	Lei Shi
Jiefeng Cheng	Lin Li	Chihwan Song
Taewon Cho	Xian Li	Ha-Joo Song
Jaehoon Choi	Xiang Li	Shaoxu Song
Shumo Chu	Xiang Lian	Yehia Taher
Ke Deng	Wenxin Liang	Takayuki Tamura
Wei Feng	Lian Liu	Guanting Tang
Shen Ge	Wenting Liu	Yuan Tian
Haris Georgiadis	Xingjie Liu	Guoping Wang
Kazuo Goda	Jiangang Ma	Puwei Wang
Jian Gong	Hossein Maserrat	Yi Wang
Reza Hemayati	Takeshi Mishima	Yousuke Watanabe
He Hu	Surya Nepal	Ingo Weber
Hai Huang	Bo Ning	Chuan Xiao
Jun Huang	Wee Siong Ng	Hairuo Xie
Stéphane Jean	Junho Oh	Kexin Xie
Bin Jiang	Sai Tung On	Lin Xin
Lili Jiang	Jin-woo Park	Jiajie Xu
Yifan Jin	Yu Peng	Zhiqiang Xu
Akimitsu Kanzaki	Jianzhong Qi	Kefeng Xuan
Hideyuki Kawashima	Kun Qian	Yuan Xue

Qingyan Yang
Xuan Yang
Zenglu Yang
Peifeng Yin
Mingxuan Yuan
Henry Ye
Mao Ye
Pengjie Ye
Peifeng Yin
Tomoki Yoshihisa

Naoki Yoshinaga
Gae-won You
Li Yu
Qi Yu
Weiren Yu
Bin Zhang
Shiming Zhang
Peiwu Zhang
Song Zhang
Geng Zhao

Xiang Zhao
Ye Zhen
Kai Zheng
Yu Zheng
Bin Zhou
Guangtong Zhou
Gaoping Zhu
Andreas Zuefle

Table of Contents – Part I

Data Mining I

Data Mining II

Probability and Uncertainty

Stream Processing

Graph

XML

XML and Graph

Table of Contents – Part II

Query Processing I

Query Processing II

Indexing and High Performance

Industrial Papers

Demo Papers

Panel

Tutorials

Smarter Planet: Empower People with Information Insights

Josephine Cheng

IBM Research - Almaden
chengjm@us.ibm.com

We are all now connected economically, technically and socially. Our planet is becoming smarter. Infusing intelligence into the way the world literally works the systems and processes that enable physical goods to be developed, manufactured, bought and sold services to be delivered everything from people and money to oil, water and electrons to move and billions of people to work and live. All these become possible via information integration scattering in many different data sources: from the sensors, on the web, in our personal devices, in documents and in databases, or hidden within application programs. Information is exploding with large amount of data generated every second. It creates many challenges in securely storing, managing, integrating, cleansing, analyzing and governing the massive generated information besides the privacy issue. This can be a difficult or time consuming endeavor. This talk describes some information-intensive tasks, choosing examples from such areas as healthcare, science, the business world and our personal lives. I will discuss the barriers to getting information together, delivering it to the people that need it, in a form they can understand, analyzing the diverse spectrum of information, giving insights to the decision makers. I will review key research on information integration and information interaction, indicate how the combination may enable real progress, and illustrate where research challenges remain.

J.X. Yu, M.H. Kim, and R. Unland (Eds.): DASFAA 2011, Part I, LNCS 6587, p. 1, 2011.

Database Scalability, Elasticity, and Autonomy in the Cloud⋆

(Extended Abstract)

Divyakant Agrawal, Amr El Abbadi, Sudipto Das, and Aaron J. Elmore

Department of Computer Science
University of California at Santa Barbara
Santa Barbara, CA 93106, USA
{agrawal,amr,sudipto,aelmore}@cs.ucsb.edu
http://www.cs.ucsb.edu/~dsl

Abstract. Cloud computing has emerged as an extremely successful paradigm for deploying web applications. *Scalability, elasticity, pay-per-use* pricing, and *economies of scale* from large scale operations are the major reasons for the successful and widespread adoption of cloud infrastructures. Since a majority of cloud applications are data driven, database management systems (DBMSs) powering these applications form a critical component in the cloud software stack. In this article, we present an overview of our work on instilling these above mentioned "cloud features" in a database system designed to support a variety of applications deployed in the cloud: designing scalable database management architectures using the concepts of *data fission* and *data fusion*, enabling lightweight elasticity using low cost live database migration, and designing intelligent and autonomic controllers for system management without human intervention.

Keywords: Cloud computing, scalability, elasticity, autonomic systems.

1 Introduction

The proliferation of technology in the past two decades has created an interesting dichotomy for users. There is very little disagreement that an individual's life is significantly enriched as a result of easy access to information and services using a wide spectrum of computing platforms such as personal workstations, laptop computers, and handheld devices such as smart-phones, PDAs, and tablets (e.g., Apple's iPads). The technology enablers are indeed the advances in networking and the Web-based service paradigms that allow users to obtain information and data-rich services at any time blurring the geographic or physical distance between the end-user and the service. As network providers continue to improve the capability of their wireless and broadband infrastructures, this paradigm will continue to fuel the invention of new and imaginative services that simplify and enrich the professional and personal lives of end-users. However, some will argue that the same technologies that have enriched the lives of

⋆ This work is partly funded by NSF grants III 1018637 and CNS 1053594 and an NEC Labs America University relations award.

J.X. Yu, M.H. Kim, and R. Unland (Eds.): DASFAA 2011, Part I, LNCS 6587, pp. 2–15, 2011.

the users, have also given rise to some challenges and complexities both from a user's perspective as well as from the service provider or system perspective. From the user's point-of-view, the users have to navigate through a web of multiple compute and storage platforms to get their work done. A significant end-user challenge is to keep track of all the applications and information services on his/her multiple devices and keep them synchronized. A natural solution to overcome this complexity and simplify the computation- and data-rich life of an end-user is to push the management and administration of most applications and services to the network core. The justification being that as networking technologies mature, from a user's perspective accessing an application on his/her personal device will be indistinguishable from accessing the application over the broadband wired or wireless network. In summary, the current technology trend is to host user applications, services, and data in the network core which is metaphorically referred to as the *cloud*.

The above transformation that has resulted in user applications and services being migrated from the user devices to the cloud has given rise to unprecedented technological and research challenges. Earlier, an application or service disruption was typically confined to a small number of users. Now, any disruption has global consequences making the service unavailable to an entire user community. In particular, the challenge now is to develop server-centric application platforms that are available to a virtually unlimited number of users 24×7 over the Internet using a plethora of modern Web-based technologies. Experiences gained in the last decade from some of the technology leaders that provide services over the Internet (e.g., Google, Amazon, Ebay, etc.) indicate that application infrastructures in the cloud context should be highly *reliable*, *available*, and *scalable*. Reliability is a key requirement to ensure continuous access to a service and is defined as the probability that a given application or system will be functioning when needed as measured over a given period of time. Similarly, availability is the percentage of times that a given system will be functioning as required. The scalability requirement arises due to the constant load fluctuations that are common in the context of Web-based services. In fact these load fluctuations occur at varying frequencies: daily, weekly, and over longer periods. The other source of load variation is due to unpredictable growth (or decline) in usage. The need for scalable design is to ensure that the system capacity can be augmented by adding additional hardware resources whenever warranted by load fluctuations. Thus, scalability has emerged both as a critical requirement as well as a fundamental challenge in the context of cloud computing.

In the context of most cloud-based application and service deployments, *data* and therefore the *database management system* (DBMS) is an integral technology component in the overall service architecture. The reason for the proliferation of DBMS, in the cloud computing space is due to the success DBMSs and in particular Relational DBMSs have had in modeling a wide variety of applications. The key ingredients to this success are due to many features DBMSs offer: overall functionality (modeling diverse types of application using the relational model which is intuitive and relatively simple), consistency (dealing with concurrent workloads without worrying about data becoming out-of-sync), performance (both high-throughput, low-latency and more than 25 years of engineering), and reliability (ensuring safety and persistence of data in the presence of different types of failures). In spite of this success, during the past decade there has been a growing concern that DBMSs and RDBMSs are not *cloud-friendly*. This is because, unlike other technology components for cloud service such as the web-servers and application servers, which can easily scale from a few machines to hundreds or even thousands of machines), DBMSs cannot be scaled very easily. In fact, current DBMS technology fails to provide adequate tools and guidance if an existing database deployment needs to scale-out from a few machines to a large number of machines.

At the hardware infrastructure level, the need to host scalable systems has necessitated the emergence of large-scale data centers comprising thousands to hundreds of thousands of compute nodes. Technology leaders such as Google, Amazon, and Microsoft have demonstrated that data centers provide unprecedented economies-of-scale since multiple applications can share a common infrastructure. All three companies have taken this notion of *sharing* beyond their internal applications and provide frameworks such as Amazon's AWS, Google's AppEngine, and Microsoft Azure for hosting third-party applications in their respective data-center infrastructures (viz. the clouds). Furthermore, most of these technology leaders have abandoned the traditional DBMSs and instead have developed proprietary data management technologies referred to as *key-value stores*. The main distinction is that in traditional DBMSs, all data within a database is treated as a "whole" and it is the responsibility of the DBMS to guarantee the consistency of the entire data. In the context of key-value stores this relationship is completely severed into key-values where each entity is considered an independent unit of data or information and hence can be freely moved from one machine to the other. Furthermore, the atomicity of application and user accesses are guaranteed only at a single-key level. Key-value stores in conjunction with the cloud computing frameworks have worked extremely well and a large number of web applications have deployed the combination of this cloud computing technology. More recent technology leaders such as Facebook have also benefited from this paradigm in building complex applications that are highly scalable.

The requirement of making web-based applications *scalable* in cloud-computing platforms arises primarily to support virtually unlimited number of end-users. Another challenge in the cloud that is closely tied to the issue of scalability is to develop mechanism to respond to sudden load fluctuations on an application or a service due to demand surges or troughs from the end-users. Scalability of a system only provides us a guarantee that a system can be scaled up from a few machines to a larger number of machines. In cloud computing environments, we need to support additional property that such scalability can be provisioned dynamically without causing any interruption in the service. This type of dynamic provisioning where a system can be scaled-up dynamically by adding more nodes or can be scaled-down by removing nodes is referred to as *elasticity*. Key-value stores such as BigTable and PNUTS have been designed so that they can be elastic or can be dynamically provisioned in the presence of load fluctuations. Traditional database management systems, on the other hand, are in general intended for an enterprise infrastructure that is statically provisioned. Therefore, the primary goal for DBMSs is to realize the highest level of performance for a given hardware and server infrastructure. Another requirement that is closely related to scalability and elasticity of data management software is that of *autonomic management*. Traditionally, data administration is a highly manual task in an enterprise setting where a highly-trained engineering staff continually monitor the health of the overall system and take actions to ensure that the operational platform continues to perform efficiently and effectively. As we move to the cloud-computing arena which typically comprises data-centers with thousands of servers, the manual approach of database administration is no longer feasible. Instead, there is a growing need to make the underlying data management layer *autonomic* or *self-managing* especially when it comes to load redistribution, scalability, and elasticity. This issue becomes especially acute in the context of *pay-per-use* cloud-computing platforms hosting multi-tenant applications. In this model, the service provider is interested in minimizing its operational cost by consolidating multiple tenants on as few machines as possible during periods of low activity and distributing these tenants on a larger number of servers during peak usage.

Due to the above desirable properties of key-value stores in the context of cloud computing and large-scale data-centers, they are being widely used as the data management tier for cloud-enabled Web applications. Although it is claimed that atomicity at a single key is adequate in the context of many Web-oriented applications, evidence is emerging that indicates that in many application scenarios this is not enough. In such cases, the responsibility to ensure atomicity and consistency of multiple data entities falls on the application developers. This results in the duplication of *multi-entity synchronization mechanisms* many times in the application software. In addition, as it is widely recognized that concurrent programs are highly vulnerable to subtle bugs and errors, this approach impacts the application reliability adversely. The realization of providing atomicity beyond single entities is widely discussed in developer blogs [28]. Recently, this problem has also been recognized by the senior architects from Amazon [23] and Google [16], leading to systems like MegaStore that provide transactional guarantees on key-value stores [3].

Cloud computing and the notion of large-scale data-centers will become a pervasive technology in the coming years. There are two major technology hurdles that we confront in deploying applications on cloud computing infrastructures: DBMS scalability and DBMS security. In this paper, we will focus on the problem of making DBMS technology cloud-friendly. In fact, we will argue that the success of cloud computing is critically contingent on making DBMSs scalable, elastic, and autonomic, which is in addition to the other well-known properties of database management technologies: high-level functionality, consistency, performance, and reliability. This paper summarizes the current state-of-the-art as well as identifies areas where research progress is sorely needed.

2 Database Scalability in the Cloud

In this section, we first formally establish the notion of scalability. In the context of cloud-computing paradigms, there are two options for scaling the data management layer. The first option is to start with key-value stores, which have almost limitless scalability, and explore ways in which such systems can be enriched to provide higher-level database functionality especially when it comes to providing transactional access to multiple data and informational entities. The other option is to start with a conventional DBMS architecture and leverage from key-value store architectural design features to make the DBMS highly scalable. We now explore these two options in detail.

2.1   Scalability

Scalability is a desirable property of a system, which indicates its ability to either handle growing amounts of work in a graceful manner or its ability to improve throughput when additional resources (typically hardware) are added. A system whose performance improves after adding hardware, proportionally to the capacity added, is said to be a scalable system. Similarly, an algorithm is said to scale if it is suitably efficient and practical when applied to large situations (e.g. a large input data set or large number of participating nodes in the case of a distributed system). If the algorithm fails to perform when the resources increase then it does not scale.

There are typically two ways in which a system can scale by adding hardware resources. The first approach is when the system scales *vertically* and is referred to as *scale-up*. To scale vertically (or scale up) means to add resources to a single node in a system, typically involving the addition of processors or memory to a single computer. Such vertical scaling of existing systems also enables them to use virtualization technology more effectively, as it provides more resources for the hosted set of operating system and application modules to share. An example of taking advantage of such shared resources is by by increasing the number of Apache daemon processes running. The other approach of scaling a system is by adding hardware resources *horizontally* referred to as *scale-out*. To scale horizontally (or scale out) means to add more nodes to a system, such as adding a new computer to a distributed software application. An example might be scaling out from one web-server system to a system with three web-servers.

As computer prices drop and performance demand continue to increase, low cost "commodity" systems can be used for building shared computational infrastructures for deploying high-performance applications such as Web search and other web-based services. Hundreds of small computers may be configured in a cluster to obtain aggregate computing power which often exceeds that of single traditional RISC processor based supercomputers. This model has been further fueled by the availability of high performance interconnects. The scale-out model also creates an increased demand for shared data storage with very high I/O performance especially where processing of large amounts of data is required. In general, the scale-out paradigm has served as the fundamental design paradigm for the large-scale data-centers of today. The additional complexity introduced by the scale-out design is the overall complexity of maintaining and administering a large number of compute and storage nodes.

Note that the scalability of a system is closely related to the underlying algorithm or computation. In particular, given an algorithm if there is a fraction α that is inherently sequential then that means that the remainder $1 - \alpha$ is parallelizable and hence can benefit from multiple processors. The maximum *scaling* or *speedup* of such a system using N CPUs is bounded as specified by Amdahl's law [1]:

$$Speedup = \frac{1}{\alpha + \frac{1-\alpha}{N}}.$$

For example if only 70% of the computation is parallelizable then the speedup with 4 CPUs is 2.105 whereas with 8 processors it is only 2.581. The above bound on scaling clearly establishes the need for designing algorithms and mechanisms that are inherently scalable. Blindly adding hardware resources may not necessarily yield the desired scalability in the system.

2.2 Data Fusion: Multi-key Atomicity in Key-Value Stores

As outlined earlier in the prior section, although key-value stores provide almost infinite scalability in that each entity can (potentially) be handled by in independent node, new application requirements are emerging that require multiple entities (or equivalently keys) to be accessed atomically. Some of these applications are in the domain of cooperative work as well as in the context of multi-player games. This need has been recognized by companies such as Google who have expanded their application portfolio from Web-search to more elaborate applications such as Google documents and others. Given this need, the question arises as to how to support multi-key atomicity in key-value stores such as Google's Bigtable [7], Amazon's Dynamo [17], and Yahoo's PNUTS [9].

The various key-value stores differ in terms of data model, availability, and consistency guarantees, but the property common to all systems is the Key-Value abstraction where data is viewed as key-value pairs and atomic access is supported only at the granularity of *single keys*. This *single key* atomic access semantics naturally allows efficient horizontal data partitioning, and provides the basis for scalability and availability in these systems. Even though a majority of current web applications have *single key* access patterns [17], many current applications, and a large number of Web 2.0 applications (such as those based on collaboration) go beyond the semantics of *single key* access, and foray into the space of *multi key* accesses [2]. Present scalable data management systems therefore cannot directly cater to the requirements of these modern applications, and these applications either have to fall back to traditional databases, or to rely on various ad-hoc solutions.

In order to deal with this challenge, Google has designed a system called MegaStore [3] that builds on Bigtable as an underlying system and creates the notion of entity groups on top of it. The basic idea of MegaStore is to allow users to group multiple entities as a single collection and then uses write-ahead logging [22, 32] and two-phase commit [21] as the building blocks to support ACID transactions on *statically* defined *entity groups*. The designers also postulate that accesses across multiple entity groups are also supported, however, at a weaker or loose consistency level. Although Megastore allows entities to be arbitrarily distributed over multiple nodes, Megastore provides higher level of performance when the entity-group is co-located on a single node. On the other hand if the entity group is distributed across multiple nodes, in that case, the overall performance may suffer since more complex synchronization mechanisms such as two-phase commit or persistent queues may be necessary. We refer to this approach as a **Data Fusion** architecture for multi-key atomicity while ensuring scalability.

Google's MegaStore takes a step beyond *single key* access patterns by supporting transactional access for *entity groups*. However, since keys cannot be updated in place, once a key is created as a part of a group, it has to be in the group for the rest of its lifetime. This static nature of *entity groups*, in addition to the requirement that keys be contiguous in sort order, are in many cases insufficient and restrictive. For instance, in case of an online casino application where different users correspond to different key-value pairs, *multi key* access guarantees are needed only during the course of a game. Once a game terminates, different users can move to different game instances thereby requiring guarantees on dynamic groups of keys–a feature not currently supported by MegaStore.

To circumvent this disadvantage, we have designed **G-Store** [14], a scalable data store providing transactional *multi key* access guarantees over *dynamic, non-overlapping groups of keys* using a key-value store as an underlying substrate, and therefore inheriting its scalability, fault-tolerance, and high availability. The basic innovation that allows scalable *multi key* access is the Key Group abstraction which defines a granule of on-demand transactional access. The Key Grouping protocol uses the Key Group abstraction to transfer *ownership*—i.e. the exclusive read/write access to keys—for all keys in a group to a single node which then efficiently executes the operations on the Key Group. This design is suitable for applications that require transactional access to groups of keys that are transient in nature, but live long enough to amortize the cost of group formation. Our assumption is that the number of keys in a group is small enough to be *owned* by a single node. Considering the size and capacity of present commodity hardware, groups with thousands to hundreds of thousands of keys can be efficiently supported. Furthermore, the system can scale-out from tens to hundreds of commodity nodes to support millions of Key Groups. G-Store inherits the data model as well as the set of operations from the underlying Key-Value store; the only addition being that the notions of atomicity and consistency are extended from a single key to a group of keys.

A Key Group consists of a **leader** key and a set of follower keys. The leader is part of the group's identity, but from an applications perspective, the semantics of operations on the leader is no different from that on the followers. Once the application specifies the Key Group, the *group creation* phase of Key Grouping protocol transfers ownership of follower keys to the node currently hosting the leader key, such that transactions executing on the group can be executed locally. Intuitively, the goal of the proposed Key Grouping protocol is to transfer key ownership safely from the followers to the leader during group formation, and from the leader to the followers during group deletion. Conceptually, the follower keys are locked during the lifetime of the group. Safety or correctness requires that there should never be an instance where more than one node claims ownership of an item. Liveness, on the other hand, requires that in the absence of repeated failures, no data item is without an owner indefinitely. The Key Grouping protocol can tolerate message and node failures as well as message re-ordering, concurrent group creation requests as well as detect overlapping group create requests [14].

This data fusion approach provides the building block for designing scalable data systems with consistency guarantees on data granules of different sizes, supporting different application semantics. The two alternative designs have resulted in systems with different characteristics and behavior.

2.3 Data Fission: Database Partitioning Support in DBMS

Contrary to the approach of data fusion, where multiple small data granules are combined to provide stringent transactional guarantees on larger data granules at scale, another approach to scalability is to split a large database unit into relatively independent

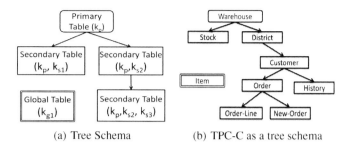

(a) Tree Schema   (b) TPC-C as a tree schema

Fig. 1. Schema level database partitioning

shards or partitions and provide transactional guarantees only on these shards. We refer to this approach as **Data Fission**. This approach of partitioning the database and scaling out with partitioning is popularly used for scaling web-applications. Since the inefficiencies resulting from distributed transactions are well known (see [11] for some performance numbers), the choice of a good partitioning technique is critical to support flexible functionality while limiting transactions to a single partition. Many modern systems therefore partition the schema in a way such that the need for distributed transactions is minimized–an approach referred to as *schema level partitioning*. Transactions accessing a single partition can be executed efficiently without any dependency and synchronization between the database servers serving the partitions, thus allowing high scalability and availability. Partitioning the database schema, instead of partitioning individual tables, allows supporting rich functionality even when limiting most transactions to a single partition. The rationale behind schema level partitioning is that in a large number of database schemas and applications, transactions only access a small number of related rows which can be potentially spread across a number of tables. This pattern can be used to group related data together in the same partition.

One popular example of partitioning arises when the schema is a "tree schema". Even though such a schema does not encompass the entire spectrum of OLTP applications, a survey of real applications within a commercial enterprise shows that a large number of applications either have such an inherent schema pattern or can be easily adapted to it [4]. Figure 1(a) provides an illustration of such a schema type. This schema supports three types of tables: **Primary Tables**, **Secondary Tables**, and **Global Tables**. The primary table forms the root of the tree; a schema has exactly one primary table whose primary key acts as the partitioning key. A schema can however have multiple secondary and global tables. Every secondary table in a database schema will have the primary table's key as a foreign key. Referring to Figure 1(a), the key k_p of the primary table appears as a foreign key in each of the secondary tables. This structure implies that corresponding to every row in the primary table, there are a group of related rows in the secondary tables, a structure called a **row group** [4]. All rows in the same row group are guaranteed to be co-located and a transaction can only access rows in a particular row group. A database partition is a collection of such row groups. This schema structure also allows efficient dynamic splitting and merging of partitions. In contrast to these two table types, global tables are look up tables that are mostly read-only. Since global tables are not updated frequently, these tables are replicated on all the nodes. In addition to accessing only one row group, an operation in a transaction can only read a global table. Figure 1(b) shows a representation of the TPC-C schema [29] as a tree schema. Such a schema forms the basis of the design of a number of systems such as MS SQL Azure [4], ElasTraS [12], and Relational Cloud [10]. The MS SQL Azure and Relational Cloud designs are based on the *shared nothing* storage model where each DBMS instance on a node is independent and an integrative layer is provided on the top for routing queries and transactions to an appropriate database server. The ElasTraS design on the other hand utilizes the *shared storage* model based on append-only distributed file-systems such as GFS [20] or HDFS [25]. The desirable feature of the ElasTraS design is that it supports elasticity of data in a much more integrated manner. In particular, both MS SQL Azure and Relational Cloud designs need to be augmented with database migration mechanisms to support elasticity where database partition migration involves moving both memory-resident database state and disk-resident data. ElasTraS, on the other hand, can support database elasticity for relocating database partitions by simply migrating the memory state of the database which is considerably simpler. In fact, well-known VM migration techniques [6, 8, 27] can be easily adopted in the case of ElasTraS [15].

This schema level partitioning splits large databases into smaller granules which can then be scaled out on a cluster of nodes. Our prototype system—named ElasTraS [12, 13]—uses this concept of data fission to scale-out database systems. ElasTraS is a culmination of two major design philosophies: traditional relational database systems (RDBMS) that allow efficient execution of OLTP workloads and provide ACID guarantees for small databases and the Key-Value stores that are elastic, scalable, and highly available allowing the system to scale-out. Effective resource sharing and the consolidation of multiple tenants on a single server allows the system to efficiently deal with tenants with small data and resource requirements, while advanced database partitioning and scale-out allows it to serve tenants that grow big, both in terms of data as well as load. ElasTraS operates at the granularity of these data granules called *partitions*. It extends techniques developed for Key-Value stores to scale to large numbers of partitions distributed over tens to hundreds of servers. On the other hand, each partition acts as a self contained database; ElasTraS uses technology developed for relational databases [22] to execute transactions efficiently on these partitions. The partitioning approach described here can be considered as static partitioning. There have been recent efforts to achieve database partitioning at run-time by analyzing the data access patterns of user queries and transactions on-the-fly [11].

3   Database Elasticity in the Cloud

One of the major factors for the success of the cloud as an IT infrastructure is its *pay per use* pricing model and *elasticity*. For a DBMS deployed on a pay-per-use cloud infrastructure, an added goal is to optimize the system's operating cost. *Elasticity*, i.e. the ability to deal with load variations by adding more resources during high load or consolidating the tenants to fewer nodes when the load decreases, all in a live system without service disruption, is therefore critical for these systems.

Even though elasticity is often associated with the scale of the system, a subtle difference exists between elasticity and scalability when used to express a system's behavior.

Scalability is a static property of the system that specifies its behavior on a static configuration. For instance, a system design might scale to hundreds or even to thousands of nodes. On the other hand, elasticity is dynamic property that allows the system's scale to be increased *on-demand* while the system is operational. For instance, a system design is elastic if it can scale from 10 servers to 20 servers (or vice-versa) on-demand. A system can have any combination of these two properties.

Elasticity is a desirable and important property of large scale systems. For a system deployed on a pay-per-use cloud service, such as the Infrastructure as a Service (IaaS) abstraction, elasticity is critical to minimize operating cost while ensuring good performance during high loads. It allows consolidation of the system to consume less resources and thus minimize the operating cost during periods of low load while allowing it to dynamically scale up its size as the load decreases. On the other hand, enterprise infrastructures are often statically provisioned. Elasticity is also desirable in such scenarios where it allows for realizing energy efficiency. Even though the infrastructure is statically provisioned, significant savings can be achieved by consolidating the system in a way that some servers can be powered down reducing the power usage and cooling costs. This, however, is an open research topic in its own merit, since powering down random servers does not necessarily reduce energy usage. Careful planning is needed to select servers to power down such that entire racks and alleys in a data-center are powered down so that significant savings in cooling can be achieved. One must also consider the impact of powering down on availability. For instance, consolidating the system to a set of servers all within a single point of failure (for instance a switch or a power supply unit) can result in an entire service outage resulting from a single failure. Furthermore, bringing up powered down servers is more expensive, so the penalty for a miss-predicted power down operation is higher.

In our context of a database system, migrating parts of a system while the system is operational is important to achieve on-demand elasticity—an operation called **live database migration**. While being elastic, the system must also guarantee the tenants' service level agreements (SLA). Therefore, to be effectively used for elasticity, live migration must have low impact—i.e. negligible effect on performance and minimal service interruption—on the tenant being migrated as well as other tenants co-located at the source and destination of migration.

Since migration is a necessary primitive for achieving elasticity, we focus our efforts on developing live migration for the two most common common cloud database architectures: shared disk and shared nothing. Shared disk architectures are utilized for their ability to abstract replication, fault-tolerance, consistency, fault tolerance, and independent scaling of the storage layer from the DBMS logic. Bigtable [7], HBase [24] and ElasTraS [12, 13] are examples of databases that use a shared disk architecture. On the other hand, a shared nothing multi-tenant architecture uses locally attached storage for storing the persistent database image. Live migration for a shared nothing architecture requires that all database components are migrated between nodes, including physical storage files. For ease of presentation, we use the term **partition** to represent a self-contained granule of the database that will be migrated for elasticity.

In a shared storage DBMS architecture the persistent image of the database is stored in a network attached storage (**NAS**). In the shared storage DBMS architecture, the persistent data of a partition is stored in the NAS and does not need migration. We have designed **Iterative Copy** for live database migration in a shared storage architecture. To minimize service interruption and to ensure low migration overhead, Iterative Copy focuses on transferring the main memory state of the partition so that the partition starts "warm" at the destination node resulting in minimal impact on transactions at the destination, allowing transactions active during migration to continue execution at the destination, and minimizing the tenant's unavailability window. The main-memory state of a partition consists of the cached database state (DB state), and the transaction execution state (Transaction state). For most common database engines [22], the DB state includes the cached database pages or some variant of this. For a two phase locking (**2PL**) based scheduler [22], the transaction state consists of the lock table; for an Optimistic Concurrency Control (**OCC**) [26] scheduler, this state consists of the read and write sets of active transactions and a subset of committed transactions. Iterative Copy guarantees serializability for transactions active during migration and ensures correctness during failures. A detailed analysis of this technique, optimizations, and a detailed evaluation can be found in [15].

In the shared nothing architecture, the persistent image of the database must also be migrated, which is typically much larger than the database cache migrated in the shared disk architecture. As a result, an approach different from Iterative Copy is needed. We have designed **Zephyr**, a technique for live migration in a shared nothing transactional database architecture [19]. Zephyr minimizes service interruption for the tenant being migrated by introducing a synchronized phase that allows both the source and destination to simultaneously execute transactions for the tenant. Using a combination of on-demand pull and asynchronous push of data, Zephyr allows the source node to complete the execution of active transactions, while allowing the destination to execute new transactions. Lightweight synchronization between the source and the destination, only during the short mode of synchronized operation, guarantees serializability, while obviating the need for two phase commit [21]. Zephyr guarantees no service disruption for other tenants, no system downtime, minimizes data transferred between the nodes, guarantees *safe* migration in the presence of failures, and ensures the strongest level of transaction isolation. It uses standard tree based indices and lock based concurrency control, thus allowing it to be used in a variety of DBMS implementations. Zephyr does not rely on replication in the database layer, thus providing greater flexibility in selecting the destination for migration, which might or might not have the tenant's replica. However, considerable performance improvement is possible in the presence of replication when a tenant is migrated to one of the replicas.

4 Database Autonomy in the Cloud

Managing large systems poses significant challenges in monitoring, management, and system operation. Moreover, to reduce the operating cost, considerable autonomy is needed in the administration of such systems. In the context of database systems, the responsibilities of this autonomic controller include monitoring the behavior and performance of the system, elastic scaling and load balancing based on dynamic usage patterns, modeling behavior to forecast workload spikes and take pro-active measures to handle such spikes. An autonomous and intelligent system controller is essential to properly manage such large systems.

Modeling the behavior of a database system and performance tuning has been an active area of research over the last couple of decades. A large body of work focuses on tuning the appropriate parameters for optimizing database performance [18, 31], primarily in the context of a single database server. Another line of work has focused on resource prediction, provisioning, and placement in large distributed systems [5, 30].

To enable autonomy in a cloud database, an intelligent system controller must also consider various additional aspects, specifically in the case when the database system is deployed on a pay-per-use cloud infrastructure while serving multiple application tenant instances, i.e., a multitenant cloud database system. In such a multitenant system, each tenant pays for the service provided and different tenants in the system can have competing goals. On the other hand, the service provider must share resources amongst the tenants, wherever possible, to minimize the operating cost to maximize profits. A controller for such a system must be able to model the dynamic characteristics and resource requirements of the different application tenants to allow elastic scaling while ensuring good tenant performance and ensuring that the tenants' service level agreements (**SLAs**) are met. An autonomic controller consists of two logical components: the *static* component and the *dynamic* component.

The static component is responsible for modeling the behavior of the tenants and their resource usage to determine tenant placement to co-locate tenants with complementary resource requirements. The goal of this tenant placement algorithm is to minimize the total resource utilization and hence minimize operating cost while ensuring that the tenant SLAs are met. Our current work uses a combination of machine learning techniques to classify tenant behavior followed by tenant placement algorithms to determine optimal tenant co-location and consolidation. This model assumes that once the behavior of a tenant is modeled and a tenant placement determined, the system will continue to behave the way in which the workload was modeled, and hence is called the static component. The dynamic component complements this static model by detecting dynamic change in load and resource usage behavior, modeling the overall system's behavior to determine the opportune moment for elastic load balancing, selecting the minimal changes in tenant placement needed to counter the dynamic behavior, and use the live database migration techniques to re-balance the tenants. In addition to modeling tenant behavior, it is also important to predict the migration cost such that a migration to minimize the operating cost does not violate a tenant's SLA. Again, we use machine learning models to predict the migration cost of tenants and the re-placement model accounts for this cost when determining *which* tenant to migrate, *when* to migrate, and *where* to migrate [15].

5   Concluding Remarks

Database systems deployed on a cloud computing infrastructure face many new challenges such as dealing with large scale operations, lightweight elasticity, and autonomic control to minimize the operating cost. These challenges are in addition to making the systems fault-tolerant and highly available. In this article, we presented an overview of some of our current research activities to address the above-mentioned challenges in designing a scalable data management layer in the cloud.

References

1. Amdahl, G.: Validity of the single processor approach to achieving large-scale computing capabilities. In: AFIPS Conference, pp. 483–485 (1967)
2. Amer-Yahia, S., Markl, V., Halevy, A., Doan, A., Alonso, G., Kossmann, D., Weikum, G.: Databases and Web 2.0 panel at VLDB 2007. SIGMOD Rec. 37(1), 49–52 (2008)
3. Baker, J., Bond, C., Corbett, J., Furman, J., Khorlin, A., Larson, J., Leon, J.M., Li, Y., Lloyd, A., Yushprakh, V.: Megastore: Providing Scalable, Highly Available Storage for Interactive Services. In: CIDR, pp. 223–234 (2011)
4. Bernstein, P.A., Cseri, I., Dani, N., Ellis, N., Kalhan, A., Kakivaya, G., Lomet, D.B., Manner, R., Novik, L., Talius, T.: Adapting Microsoft SQL Server for Cloud Computing. In: ICDE (2011)
5. Bodík, P., Goldszmidt, M., Fox, A.: Hilighter: Automatically building robust signatures of performance behavior for small- and large-scale systems. In: SysML (2008)
6. Bradford, R., Kotsovinos, E., Feldmann, A., Schiöberg, H.: Live wide-area migration of virtual machines including local persistent state. In: VEE, pp. 169–179 (2007)
7. Chang, F., Dean, J., Ghemawat, S., Hsieh, W.C., Wallach, D.A., Burrows, M., Chandra, T., Fikes, A., Gruber, R.E.: Bigtable: A Distributed Storage System for Structured Data. In: OSDI, pp. 205–218 (2006)
8. Clark, C., Fraser, K., Hand, S., Hansen, J.G., Jul, E., Limpach, C., Pratt, I., Warfield, A.: Live migration of virtual machines. In: NSDI, pp. 273–286 (2005)
9. Cooper, B.F., Ramakrishnan, R., Srivastava, U., Silberstein, A., Bohannon, P., Jacobsen, H.A., Puz, N., Weaver, D., Yerneni, R.: PNUTS: Yahoo!'s hosted data serving platform. In: Proc. VLDB Endow., vol. 1(2), pp. 1277–1288 (2008)
10. Curino, C., Jones, E., Popa, R., Malviya, N., Wu, E., Madden, S., Balakrishnan, H., Zeldovich, N.: Relational Cloud: A Database Service for the Cloud. In: CIDR, pp. 235–240 (2011)
11. Curino, C., Zhang, Y., Jones, E.P.C., Madden, S.: Schism: a workload-driven approach to database replication and partitioning. PVLDB 3(1), 48–57 (2010)
12. Das, S., Agarwal, S., Agrawal, D., El Abbadi, A.: ElasTraS: An Elastic, Scalable, and Self Managing Transactional Database for the Cloud. Tech. Rep. 2010-04, CS, UCSB (2010)
13. Das, S., Agrawal, D., El Abbadi, A.: ElasTraS: An Elastic Transactional Data Store in the Cloud. In: USENIX HotCloud (2009)
14. Das, S., Agrawal, D., El Abbadi, A.: G-Store: A Scalable Data Store for Transactional Multi key Access in the Cloud. In: ACM SoCC, pp. 163–174 (2010)
15. Das, S., Nishimura, S., Agrawal, D., El Abbadi, A.: Live Database Migration for Elasticity in a Multitenant Database for Cloud Platforms. Tech. Rep. 2010-09, CS, UCSB (2010)
16. Dean, J.: Talk at the Google Faculty Summit (2010)
17. DeCandia, G., Hastorun, D., Jampani, M., Kakulapati, G., Lakshman, A., Pilchin, A., Sivasubramanian, S., Vosshall, P., Vogels, W.: Dynamo: Amazon's highly available key-value store. In: SOSP, pp. 205–220 (2007)
18. Duan, S., Thummala, V., Babu, S.: Tuning database configuration parameters with ituned. Proc. VLDB Endow. 2, 1246–1257 (2009)
19. Elmore, A., Das, S., Agrawal, D., El Abbadi, A.: Zephyr: Live Database Migration for Lightweight Elasticity in Multitenant Cloud Platforms. under submission for review

20. Ghemawat, S., Gobioff, H., Leung, S.T.: The Google file system. In: SOSP, pp. 29–43 (2003)
21. Gray, J.: Notes on data base operating systems. In: Flynn, M.J., Jones, A.K., Opderbeck, H., Randell, B., Wiehle, H.R., Gray, J.N., Lagally, K., Popek, G.J., Saltzer, J.H. (eds.) Operating Systems. LNCS, vol. 60, pp. 393–481. Springer, Heidelberg (1978)
22. Gray, J., Reuter, A.: Transaction Processing: Concepts and Techniques. Morgan Kaufmann Publishers Inc., San Francisco (1992)
23. Hamilton, J.: I love eventual consistency but... (April 2010),
    `http://bit.ly/hamilton-eventual`
24. HBase: Bigtable-like structured storage for Hadoop HDFS (2010),
    `http://hadoop.apache.org/hbase/`
25. HDFS: A distributed file system that provides high throughput access to application data (2010), `http://hadoop.apache.org/hdfs/`
26. Kung, H.T., Robinson, J.T.: On optimistic methods for concurrency control. ACM Trans. Database Syst. 6(2), 213–226 (1981)
27. Liu, H., et al.: Live migration of virtual machine based on full system trace and replay. In: HPDC, pp. 101–110 (2009)
28. Obasanjo, D.: When databases lie: Consistency vs. availability in distributed systems (2009), `http://bit.ly/4J0Zm`
29. The Transaction Processing Performance Council: TPC-C benchmark (Version 5.10.1) (2009)
30. Urgaonkar, B., Rosenberg, A.L., Shenoy, P.J.: Application placement on a cluster of servers. Int. J. Found. Comput. Sci. 18(5), 1023–1041 (2007)
31. Weikum, G., Moenkeberg, A., Hasse, C., Zabback, P.: Self-tuning database technology and information services: from wishful thinking to viable engineering. In: VLDB, pp. 20–31 (2002)
32. Weikum, G., Vossen, G.: Transactional information systems: theory, algorithms, and the practice of concurrency control and recovery. Morgan Kaufmann Publishers Inc., San Francisco (2001)

What Have We Learnt from Deductive Object-Oriented Database Research?

Mengchi Liu[1], Gillian Dobbie[2], and Tok Wang Ling[3]

[1] State Key Lab of Software Engineering, Wuhan University, China
mengchi@sklse.org
[2] Department of Computer Science, University of Auckland, New Zealand
g.dobbie@auckland.ac.nz
[3] School of Computing, National University of Singapore, Singapore
lingtw@comp.nus.edu.sg

Deductive databases and object-oriented databases (DOOD) are two important extensions of the traditional relational database technology.

Deductive databases provide a rule-based language called Datalog¬ (Datalog with negation) that uses function-free Horn clauses with negation to express deductive rules [1], and is a simplified version of the logic programming language Prolog [2]. A deductive database consists of an extensional database and an intensional database. The extensional database (EDB) consists of the relations stored in a relational database whereas the intensional database (IDB) consists of a Datalog¬ program that is a set of deductive rules used to derive relations that are the logical consequences of the program and the extensional database. Datalog¬ is more expressive than pure relational query languages such as relational algebra and relational calculus as it supports recursive deductive rules and recursive queries. Moreover, deductive databases have a firm logical foundation that consists of both model-theoretic semantics in terms of the minimal model [3], the stable model [4], and the well-founded model [5], and proof-theoretic semantics in terms of bottom-up fixpoint semantics [2].

Object-oriented databases provide richer data modeling mechanisms such as object identity, property/attribute, method, complex object, encapsulation, class hierarchy, non-monotonic multiple inheritance, overloading, late binding, and polymorphism, but without much research on good DOOD database design, and the accompanying theoretical progress.

The objective of deductive object-oriented databases is to combine the best features of deductive databases and object-oriented databases, namely to combine the expressive power of rule-based language and logical foundation of the deductive approach with various data modeling mechanisms of the object-oriented approach. In the late 80s and 90s, a large number of deductive object-oriented database languages were proposed, such as O-logic, revised O-logic, C-logic, F-logic, IQL, LOGRES, LLO, COMPLEX, ORLOG, LIVING IN LATTICE, Datalog^{meth}, CORAL++, Noodle, DTL, Gulog, Rock & Roll, ROL, Datalog^{++}, ROL2, Chimera, and DO2. These proposals can be roughly classified into two kinds: loosely-coupled and tightly coupled. The first kind mainly uses or extends

J.X. Yu, M.H. Kim, and R. Unland (Eds.): DASFAA 2011, Part I, LNCS 6587, pp. 16–21, 2011.

Datalog⁻-like languages as a query language for object-oriented databases. This is not a satisfactory approach as the resulting system consists of two clearly distinct parts with no unifying semantics. Typical examples of this kind are: IQL, Rock & Roll, CORAL++, and Chimera. The other approach is more fundamental in which new unifying logics are proposed to formalize the notions underlying object-oriented databases. It started with a small set of simple object-oriented features, and more and more powerful features were gradually incorporated into successive languages. Typical examples of this kind are: O-logic, revised O-logic, C-logic, F-logic, ORLOG, ROL, and ROL2.

However, a clean logical semantics that could naturally account for all the important object-oriented features was still missing. Such a semantics played an important role in database research. In particular, two key object-oriented features that were not addressed logically were encapsulation of rule-based methods in classes, and non-monotonic structural and behavioral inheritance with overriding, conflict resolution and blocking.

Method Encapsulation

In object-oriented programming languages and data models, methods are defined using functions or procedures and are encapsulated in class definitions. They are invoked through instances of the classes. In deductive databases, we use deductive rules based on Horn clauses with negation instead of functions and procedures. By analogy, methods in deductive object-oriented databases can be defined using deductive rules and encapsulated in class definitions. Such methods should be invoked through instances of the classes as well. However, earlier deductive object-oriented database languages did not allow rule-based methods to be encapsulated in the class definitions. The main reason is that the logical semantics of deductive databases is based on programs that are sets of rules as in logic programming. If rules are encapsulated into classes and classes can non-monotonically inherit methods, then it is not clear how to define logical semantics directly.

Inheritance Conflict Resolution

Non-monotonic multiple structural, and behavioral inheritance is a fundamental feature of object-oriented databases. Users can explicitly redefine (or override) the inherited attributes or methods and stop (or block) the inheritance of attributes or methods from the superclasses. Ambiguities may arise when an attribute or method is defined in two or more superclasses, and the conflicts need to be handled (or resolved). Most systems use the superclass ordering to resolve the conflicts. Unfortunately, a logical semantics for multiple inheritance with overriding, blocking and conflict resolution had not been defined directly. The main difficulty is that the inherited instances of a superclass may not be well typed with respect to its type definition because of overriding and blocking. Most deductive object-oriented database languages only allow monotonic multiple structural inheritance or non-monotonic single inheritance, which is not powerful enough.

Main Contribution

Our DASFAA 2001 paper successfully solved these two problems. It provides a direct well-defined declarative semantics for a deductive object-oriented database language with (1) encapsulated rule-based methods and (2) non-monotonic structural and behavioral inheritance with overriding, conflict resolution, and blocking. In the language, methods are declared in the class definitions, and the methods are invoked through instances of the classes. We introduce a special class, none, to indicate that the inheritance of an attribute or method in a subclass is blocked; that is, it won't be inherited from its superclasses. We provide a very flexible approach to inheritance conflict resolution. Our mechanism consists of two ways. The first and default is similar to the method used in Orion, namely a subclass inherits from the classes in the order they are declared in the class definition. The second allows the explicit naming of the class the attribute or method is to be inherited from. Therefore, a subclass can inherit an attribute or a method from any superclass. We then define a class of databases, called well-defined databases that have an intuitive meaning and develop a direct logical semantics for this class of databases. The semantics naturally accounts for method encapsulation, multiple structural and behavioral inheritance, overriding, blocking, and conflict resolution, and is based on the well-founded semantics from logic programming. However, our semantics differs from the well-founded semantics in the following ways. We are concerned with a typed language with methods rather than an untyped language with predicates. We introduce a well-typed concept and take typing into account when deducing new facts from methods. The definition of satisfaction of expressions is more complex in our definition because we define the truth values for our many kinds of expressions. Our definition reflects the fact that our model effectively has two parts, an extensional database (EDB) that models oid membership and attribute expressions, and an intensional database (IDB) that models method expressions. The EDB is a 2-valued model, in which oid membership and attribute expressions are true if they are in the model; otherwise, they are false. The IDB is a 3-valued model, in which method expressions are true if they are in the model, false if their complement belongs to the model; otherwise, they are undefined. When a method expression is undefined, either the method isn't defined on the invoking object, or it isn't possible to assign a truth value to that expression. The reason we use a 3-valued model for IDB is that we can infer both positive and negative method expressions using method rules. On the other hand, EDB only contains positive oid membership and attribute expressions so we just use a 2-valued model. In the well-founded semantics, a program may have a partial model. This is not the case in our definition, in fact we prove that every well-defined program has a minimal model. We define a transformation that has a fixpoint, I^* for well-defined databases, and prove that if I^* is defined, then it is a minimal model of the database.

Our work differs from the work of others in many ways. Most existing deductive object-oriented database languages do not allow rule-based methods to be encapsulated in the class definitions. Those that do, do not address the issue directly. In contrast, we have provided a direct semantics for methods encapsulated in class definitions. Also, most existing deductive object-oriented database languages do not allow non-monotonic multiple structural and behavioral inheritance.

By providing a direct logical semantics to the two most difficult object-oriented features, our work has shown that the object-oriented features that were believed to be difficult to address, can indeed be completely captured logically.

Applications of Deductive Object-Oriented Database Techniques

There was not a lot more work undertaken into the semantics of object-oriented databases after 2001, as most important semantic issues have been successfully solved. However, the logical semantics developed in our work and also deductive object-oriented database techniques in general have been used in several new research areas.

In [6], a Datalog-like language is proposed to define derivation rules in object-oriented conceptual modeling language. As pointed out in [6], the language is quite similar to our work, which shows that our work can be used for conceptual modeling.

One is the development of knowledge based applications for information extraction and text classification [7]. The language developed for this purpose was DLV+. A later development was OntoDLV which supports a powerful interoperability mechanism with OWL, allowing the user to retrieve information from OWL ontologies, and build rule-based reasoning on top of OWL ontologies [8]. There was further work into providing a formal approach to ontology representation that was mainly based on our deductive object-oriented database semantics [9].

Our logical semantics has been used in new languages for extending the power of XML schema definition languages, for example adding non-monotonic inheritance to XML schema definition languages [10,11,12].

Another XML development is the active deductive XML database system ADM [13], which extends XML with logical variables, logical procedures and event-condition-action (ECA) rules.

Also, W3C has created a new working group called Rule Interchange Format (RIF) to produce a recommendation for a standardized rule language that can be used as an interchange format for various rule-based systems [14]. It will support features in deductive object-oriented databases in an XML syntax in order to exchange and merge rules from different sources

Three perhaps surprising areas where the deductive object-oriented database technique has been used are source code analysis, metamodel translation and conceptual modeling in the biological domain.

There is one project in Source Code Analysis that used an object-oriented query language (.QL) [15,16]. .QL can be used to assess software quality, namely to find bugs, to compute metrics and to enforce coding conventions.

Metamodels are used for performing translations of schemas and databases from one model to another. Typical object-oriented constructs and Datalog provide significant advantages in the language independent model [17].

It is recognized that the biological domain has some specific requirements when it comes to conceptual modeling for data integration and database design [18]. Some of the concepts in the biological domain are similar to object-oriented concepts.

References

1. Ceri, S., Gottlob, G., Tanca, T.: Logic Programming and Databases. Springer, Heidelberg (1990)
2. van Emden, M.H., Kowalski, R.A.: The Semantics of Predicate Logic as a Programming Language. Journal of ACM 23, 733–742 (1976)
3. Apt, K.R., Blair, H.A., Walker, A.: Towards a theory of declarative knowledge. In: Minker, J. (ed.) Foundation of Deductive Databases and Logic Programming, pp. 89–148. Morgan Kaufmann, Los Altos (1988)
4. Gelfond, M., Lifschitz, V.: The Stable Model Semantics for Logic Programming. In: Proceedings of the International Conference and Symposium on Logic Programming (ICLP/SLP 1988), pp. 1070–1080 (1988)
5. Gelder, A.V., Ross, K.A., Schlipf, J.S.: The Well-Founded Semantics for General Logic Programs. Journal of ACM 38, 620–650 (1991)
6. Olivé, A.: Derivation Rules in Object-Oriented Conceptual Modeling Languages. In: Eder, J., Missikoff, M. (eds.) CAiSE 2003. LNCS, vol. 2681, pp. 404–420. Springer, Heidelberg (2003)
7. Ricca, F., Leone, N.: Disjunctive logic programming with types and objects: The DLV+ system. Journal of Applied Logic 5, 545–573 (2007)
8. Ricca, F., Gallucci, L., Schindlauer, R., Dell'Armi, T., Grasso, G., Leone, N.: OntoDLV: An ASP-based System for Enterprise Ontologies. Journal of Logic and Computation 19, 643–670 (2009)
9. Sun, Y., Sui, Y.: A logical foundation for ontology representation in NKI. In: Proceedings of IEEE International Conference on Natural Language Processing and Knowledge Engineering (NLPKE 2005), pp. 342–347 (2005)
10. Wang, G., Han, D., Qiao, B., Wang, B.: Extending XML Schema with Object-Oriented Features. Information Technology Journal 4, 44–45 (2005)
11. Wang, G., Liu, M.: Extending XML with Nonmonotonic Multiple Inheritance. In: Proceedings of Database Systems for Advanced Applications (DASFAA 2005) (2005)
12. Liu, M.: DTD schema: a simple but powerful XML schema language. International Journal of Web Information System 4, 465–483 (2008)
13. Olmedo-Aguirre, O., Escobar-Vzquez, K., Alor-Hernndez, G., Morales-Luna, G.: ADM: An Active Deductive XML Database System. In: Monroy, R., Arroyo-Figueroa, G., Sucar, L.E., Sossa, H. (eds.) MICAI 2004. LNCS (LNAI), vol. 2972, pp. 139–148. Springer, Heidelberg (2004)
14. Rule Interchange Format: W3C Working Group (2005),
    http://www.w3.org/2005/rules
15. de Moor, O., Verbaere, M., Hajiyev, E., Avgustinov, P., Ekman, T., Ongkingco, N., Sereni, D., Tibble, J.: Keynote Address: .QL for Source Code Analysis. In: Proceedings of the IEEE International Workshop on Source Code Analysis and Manipulation, pp. 3–16 (2007)

16. de Moor, O., Sereni, D., Verbaere, M., Hajiyev, E., Avgustinov, P., Ekman, T., Ongkingco, N., Tibble, J.: QL: Object-Oriented Queries Made Easy. In: Lämmel, R., Visser, J., Saraiva, J. (eds.) Generative and Transformational Techniques in Software Engineering II. LNCS, vol. 5235, Springer, Heidelberg (2008)
17. Atzeni, P., Gianforme, G., Toti, D.: Polymorphism in Datalog and Inheritance in a Metamodel. In: Proceedings of the 6th International Symposium on Foundations of Information and Knowledge Systems, FoIKS 2010 (2010)
18. de Macedo, J.A.F., Porto, F., Lifschitz, S., Picouet, P.: Dealing with Some Conceptual Data Model Requirements for Biological Domains. In: Proceedings of the International Conference on Advanced Information Networking an Applications Workshops, AINAW 2007 (2007)

ECODE: Event-Based Community Detection from Social Networks

Xiao-Li Li[1], Aloysius Tan[1], Philip S. Yu[2], and See-Kiong Ng[1]

[1] Institute for Infocomm Research, 1 Fusionopolis Way #21-01 Connexis Singapore 138632
[2] Department of Computer Science, University of Illinois at Chicago, IL 60607-7053
{xlli,skng}@i2r.a-star.edu.sg, aloysius_tan@hotmail.com,
psyu@cs.uic.edu

Abstract. People regularly attend various social events to interact with other community members. For example, researchers attend conferences to present their work and to network with other researchers. In this paper, we propose an *Event*-based COmmunity DEtection algorithm ECODE to mine the underlying community substructures of social networks from event information. Unlike conventional approaches, ECODE makes use of content similarity-based *virtual links* which are found to be more useful for community detection than the physical links. By performing partial computation between an event and its candidate relevant set instead of computing pair-wise similarities between all the events, ECODE is able to achieve significant computational speedup. Extensive experimental results and comparisons with other existing methods showed that our ECODE algorithm is both efficient and effective in detecting communities from social networks.

Keywords: social network mining, community detection, virtual links.

1 Introduction

In recent years, many real world networks, such as worldwide web [1], social networks [2] [3], biological networks [4] [5] [6] [7] [8] [9], citation networks [10], communication networks [11] etc, have become available for data mining. A key task of mining these networks is to unravel the underlying community substructures. Community detection can reveal important functional information about the real-world networks. For example, communities in the biological networks usually correspond to functional modules or biological pathways that are useful for understanding the causes of various diseases [7]. In the social networks, knowledge about the underlying community substructures can be used for searching for potential collaborators, devising strategies to optimize the social relationships, identifying key persons in the various communities, etc.

Qualitatively, detecting communities from networks involves dividing the vertices into groups such that there is a higher density of links within groups than between them [12] [13]. Numerous algorithms have been proposed to detect communities from various networks in recent years. However, detecting community substructures from *large scale networks* is still a challenging issue [12]. First of all, these algorithms are

J.X. Yu, M.H. Kim, and R. Unland (Eds.): DASFAA 2011, Part I, LNCS 6587, pp. 22–37, 2011.
© Springer-Verlag Berlin Heidelberg 2011

not very efficient as they either compute pair-wise similarities between all the entities or cliques (agglomerative methods), or iteratively calculate the cutting edges (divisive methods) based on the values of some measures, e.g. betweenness scores. Secondly, it is common in practice that the social entities only interact with a limited subset of community members. As such, there exist communities which do not have very dense connections among all its members. This will make existing algorithms, most of which are density-based, suffer.

We observe that people regularly attend various social events to interact with other community members. Many communities are formed and strengthened during such events as the members are able to effectively interact *en masse* in addition to traditional one-on-one interactions with one another. For example, in the academic domain, researchers often attend conferences, seminars and workshops to network with other community members whom they may not yet have direct working relationships with, but who have common research background and interests with them. In such events, old links are strengthened while new links are formed as the community members present their work, talk about the possible technical solutions for specific problems, provide feedbacks and suggestions to their peers' work, discuss the possibility of future research direction and the collaboration topics, etc, during the formal programs as well as the informal tea breaks, lunches, and dinners. Similarly, in business domain, professionals also often attend business meetings and trade exhibitions to find potential collaborators, discuss with their business plans, exchange ideas on the issues regarding the economic situation, and find commercial opportunities in current and/or emerging markets. Event information can thus be quite useful for inferring communities from social networks.

In this paper, we have proposed a novel ECODE algorithm which detects community substructures from events. ECODE stands for <u>E</u>vent-based <u>CO</u>mmunity <u>De</u>tection. In ECODE, similar events are merged using hierarchical clustering to form bigger communities. We summarize the main contributions in this paper as follows:

- For the first time, the *event*-based community detection problem is formally defined. This will facilitate the use of *event* data for better detection of community substructures in social networks.

- Our proposed ECODE algorithm uses *events* instead of single persons or cliques as the basic unit to perform hierarchical clustering efficiently. In fact, ECODE only computes the similarity scores from a part of the selected potentially similar events, which further speeds up our algorithm.

- A novel idea termed as *virtual links* has been proposed to enhance the connectivity among members within same communities. The *virtual links,* which are content-based associations, can be used to enrich the potentially sparse connections amongst the community members, resulting in effective community detection.

- Experimental results showed that our method that can effectively address the challenging problems in the community detection, namely, the issues of low efficiency and low connectivity within community. ECODE not only significantly outperformed the existing state-of-the-art community detection methods, but it also detects the hierarchical substructures of communities in the social networks, which can provide more insights on community formations. Our algorithm also allows the communities discovered to have overlapping structures.

2 Related Work

Detecting communities or modules from networks has attracted considerable attention in recent years [14]. The current research on community detection can be divided into two main thrusts, namely, agglomerative methods and divisive methods [13] [15].

Agglomerative methods adopt bottom-up strategies to build a tree where the leaves can be either a single node or dense graphs [16] [12] [17] [18] [4] [19] [20]. The agglomerative methods proposed in [12], [16], [17] and [18] evaluate the pair-wise similarities or closeness $s(i, j)$ for every pair of nodes i and j in the network. Starting from individual nodes as initial groups, the process involves iteratively merging the two most similar groups into larger and larger communities. A tree which represents the whole network is built from the bottom up to the root. In comparison, the agglomerative methods proposed in [4], [19] and [20] detect dense graphs, such as the maximal cliques [4] [19] and k-core [20], as the initial leaves. They then repeatedly join together the two most similar dense graphs to larger communities.

Divisive methods, on the other hand, construct a tree in the reversed order [21] [22] [23] [15] [24] [25]. They start from the root, which represents the whole network, and divide the network progressively into smaller and smaller disconnected sub-networks which can correspond to the communities. The fundamental idea of the divisive methods is to select links that are inter-cluster links and not intra-cluster links to be cut. A well-known divisive algorithm has been proposed by Girvan and Newman [26]. The algorithm selects the links to be cut based on the values of the "edge betweenness" — a link's betweenness score is defined as the number of shortest paths between all pairs of nodes in the network that pass along it. Links with large betweenness score are thus "bridge"-like edges (or inter-cluster links) linking densely connected clusters, since many shortest paths between the different clusters will have to pass through these edges. Spectral graph partitioning methods have also been employed to detect the groups by identifying an approximately minimal set of links from the given graph [27] [28]. The block modeling method can be considered as a classical Social Network Analysis (SNA) method for this problem [29].

Many interesting problems have been explored recently by taking time factors into consideration. The work in [30] investigated communities that grow rapidly and explored how the overlaps between pairs of communities change over time. The work in [31] showed to discover what the "normal" growth patterns in social, technological and information networks are. A tractable model for information diffusion in social networks was proposed in [32], while the work in [33] studied how communities evolve over time in dynamic multi-mode networks.

3 The Proposed Technique

In this section, we present our proposed ECODE algorithm. In Subsection 3.1, we provide the problem definition of event-based community detection. Then, in Subsection 3.2, we introduce a content-based *virtual link* method. Next, in Subsection 3.3, we describe three different similarity measures. We present our ECODE algorithm in Subsection 3.4. Finally, we assign people to corresponding communities.

3.1 Problem Definition

Let event set $E = \{\ \varphi_i\ |\ \varphi_i$ is a event, $i = 1, 2, ..., n\}$. Each event φ_i can be represented as a graph $\varphi_i = \{V_i, E_i\}$ where $V_i = \{v_j\ |\ v_j$ is an individual entity who attended the event $\varphi_i\}$, $E_i = \{(v_j, v_k)\ |\ v_j$ and v_k are two individual entities who have certain relationships, v_j, $v_k \in V_i\}$. Each link (v_j, v_k) in E_i could be v_j and v_k work together (*physical* links).

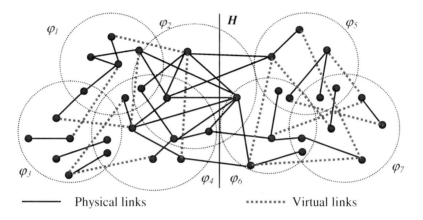

Fig. 1. Illustration of community detection

Given an event set E, our objective is to detect those communities $\{C_1, C_2, ..., C_p\}$ where each individual community $C_i (i=1, 2, ..., p)$ has much more intra-links (links within C_i) but relatively less inter-links cross different communities (links between C_i and $C_j, j = 1, 2, ..., p, i \neq j$) (*link* perspective). In addition, there should be relatively small number of vertices that participate in two communities C_i and C_j ($i \neq j$). Fig. 1 illustrates an event community detection problem where the nodes are individual entities (represented by colored circles) and there are two different types of links, i.e. physical links (represented by solid lines) and virtual links (represented by dotted lines). Virtual links connect a pair of entities from different events who do not have direct interactions but work on some similar topics. Fig. 1 depicts 7 events $\varphi_1, ..., \varphi_7$ (circled), and 2 main communities existing separated by H (community 1: $\varphi_1, \varphi_2, \varphi_3$, φ_4 and community 2: $\varphi_5, \varphi_6, \varphi_7$). Note that some people attend multiple events and they are thus located in the intersections of these communities. If there are many common participants in two separate events (vertex overlapping), then these events are probably related and those people in the two events should belong to the same community. The virtual links enhance the connectivity across different events within the same community, which are useful to merge events to form bigger communities.

3.2 Virtual Links between Events

Given a vertex v_i, we consider its associated content in various events d_i : for the researchers' social networks, these could be v_i's research papers, presentation slides, project descriptions, curriculum vitae, etc to profile v_i's interests. For a pair of vertices v_j and v_k from different events ($v_j \in \varphi_j = \{V_j, E_j\}$, $v_k \in \varphi_k = \{V_k, E_k\}, j \neq k$), we evaluate if there is a virtual link between v_j and v_k by computing whether their content similarity $consim(v_j,v_k)$ is big enough, i.e. $consim(v_j,v_k)>\delta$, where δ is a threshold which can be computed by averaging the similarities among the non-connected entities within randomly selected events $\varphi_R \subset E$, i.e.

$$\delta = \frac{1}{|\varphi_R|} \sum_{\varphi \in \varphi_R} \frac{consim\,(v_j,v_k)}{|\varphi_i|}, v_j, v_k \in \varphi_i \tag{1}$$

where (v_j,v_k) is a pair of non-connected entities in event φ_i, $|\varphi_i|$ is the number of all the non-physical link pairs in φ_i, and φ_R is the event set selected from event set E.

The content similarity v_j and v_k, $consim(v_j, v_k)$ in equation (1) can be defined as

$$consim\,(v_j,v_k) = \frac{1}{K_{con}} \frac{|f(v_j) \cap f(v_k)|}{|f(v_j)|*|f(v_k)|} \tag{2}$$

In equation (2), $f(v_j)$ represents the feature set of vertex v_j after eliminating the stop words; K_{con} is a normalization constant and $K_{con} = \max_{a,b}(consim\,(v_a,v_b))$. $consim(v_j, v_k)$ (ranged from 0 to 1) will be bigger if two vertices shared a lot of common features. While the people within same community have a higher chance to interact with each other, each individual entity typically still only interacts with a limited number of his or her community members. In community detection, it is thus quite important to enrich the social network by linking those entities with common interests together. Here, we propose virtual links to connect those people from different events whose content similarity is equal to or higher than the average feature similarity between people within randomly selected events.

3.3 Similarity Measures

Communities can consist of the people from different *events*. It is thus necessary to combine the smaller events together to form those bigger communities. We evaluate the similarities between events by the following three different similarity measures.

Given two graphs $\varphi_i = \{V_i, E_i\}$ and $\varphi_j = \{V_j, E_j\}$, the vertex similarity between two events φ_i and φ_j is defined as

$$Vex\_sim(\varphi_i,\varphi_j) = \frac{|V_i \cap V_j|}{|V_i \cup V_j|} / K_{vex} \tag{3}$$

where K_{Vex} is a normalization constant and $K_{Vex} = \max_{a,b}(Vex\_sim(\varphi_a,\varphi_b))$. According to equation (3), if two events share a high proportion of members, then they are considered to be events for the same community.

The physical similarity between two events φ_i and φ_j is defined as

$$PL\_sim(\varphi_i,\varphi_j) = \frac{|\{(v_i,v_j)|(v_i,v_j) \in \varphi_k, k \neq i, k \neq j, v_i \in V_i \setminus V_j, v_j \in V_j \setminus V_i\}|}{|V_i \setminus V_j|*|V_j \setminus V_i|} / K_{PL} \tag{4}$$

where K_{PL} is a normalization constant and $K_{PL} = \max_{a,b}(PL\_sim(\varphi_i,\varphi_j))$. Equation (4) basically evaluates how closely the members from different events interact with each other. If there are a lot of physical inter-links (involved in different events, such as φ_k) between the members from two events φ_i and φ_j, then the events are highly likely to be events for the same community.

In the same way, we define the virtual link similarity between two events φ_i and φ_j:

$$VL\_sim(\varphi_i,\varphi_j) = \frac{|\{(v_i,v_j)\,|\,v_i \in \varphi_i, v_j \in \varphi_j, consim(v_i,v_j) > \delta, v_i \in V_i \setminus V_j, v_j \in V_j \setminus V_i\}|}{|V_i \setminus V_j|*|V_j \setminus V_i|}/K_{VL} \qquad (5)$$

where K_{VL} is a normalization constant and $K_{VL} = \max_{a,b}(VL\_sim(\varphi_i,\varphi_j))$. Note that the virtual link similarity in Equation 5 is similar to the physical link similarity in Equation 4 – the only difference is that we use the virtual links replace the physical inter-links between the members from the two events. It may appear that the virtual links are not very useful as the virtual links between entities not involving in the same events merely indicate that the individuals are doing something similar but they do not have any physical interactions. However, we will show that the virtual links are actually more useful for community detection than the physical interaction links.

3.4 ECODE Algorithm

We adopt an agglomerative clustering approach for our community detection algorithm ECODE (Fig. 2). The objective is to detect similar events in terms of overlapping vertices and virtual links, and then merging them to form bigger communities. The algorithm terminates when the quality of the detected communities in the merging process have become maximal. In Fig. 2, ECODE algorithm starts with the members from each event forming an initial community. Although we could employ standard hierarchical clustering, the need to compute pair-wise similarities between all initial leaf nodes is too time-consuming for large networks. To improve the efficiency of our technique, in Step 2 of our algorithm, we only select those event pairs which are potentially similar to compute their similarities.

1. **For** each event $\varphi_i = \{V_i, E_i\}$ ($i = 1, 2, \ldots\ldots, n$), $\varphi_i \in E$
2. Find its candidate relevant set $E\varphi_i$ where the members from φ_i also frequently participated in the each event in $E\varphi_i$
3. Compute the similarities between φ_i and each event φ_{ip} in $E\varphi_i$
4. **While** (quality of current-level of tree increases)
5. Find the most similar events φ_i and φ_j and merge them into a new event φ_{new}
6. Construct a candidate set $E\varphi_{new}$ for φ_{new} from its children's candidate sets $E\varphi_i$ and $E\varphi_j$
7. Compute the similarities between the new event φ_{new} and each event in $E\varphi_{new}$
8. Compute the quality of current level of the tree

Fig. 2. ECODE algorithm for community mining

Given an event φ_i, we want to find its candidate relevant set $E\varphi_i$, which consists of potential similar events that φ_i's entities/members have also participated in. To do this, we first construct event transaction set $T\varphi_i$ where each record includes an entity and the various events that he/she is involved in. We want to detect candidate relevant event set $E\varphi_i$ for event φ_i where those events in $E\varphi_i$ have high *support* in $T\varphi_i$, *i.e.*

$$E\varphi_i = \{\varphi_j | \, support_{T\varphi_i}(\varphi_i, \varphi_j) > \alpha, \varphi_i, \varphi_j \in E, j \neq i\} \qquad (6)$$

where α is a parameter to control the size of candidate relevant set $T\varphi_i$. The problem to find high *support* associated events can be modeled as mining frequent item sets problem — there exist many efficient algorithms for this problem in the data mining domain [34]. We are thus able to compute $E\varphi_i$ rapidly.

For each event φ_i and its candidate relevant set $E\varphi_i$, Step 3 computes the similarities $sim(\varphi_i, \varphi_j)$ between φ_i and each event φ_{ip} in $E\varphi_i$ which is defined as the linear combination of vertex similarity $Vex\_sim(\varphi_i, \varphi_{ip})$ and virtual link similarity $VL\_sim(\varphi_i, \varphi_{ip})$:

$$sim(\varphi_i, \varphi_{ip}) = \lambda * Vex\_sim(\varphi_i, \varphi_{ip}) + (1 - \lambda) * VL\_sim(\varphi_i, \varphi_{ip}), \qquad (7)$$

where λ $(0 \leq \lambda \leq 1)$ is a parameter to adjust the weighs for the importance of vertex similarity and virtual link similarity. If $\lambda = 1$ ($\lambda = 0$), then we only consider the vertex similarity (virtual similarity). In our experiments, we will test λ's sensitivity.

Note that according to Equation (5), obtaining $VL\_sim(\varphi_i, \varphi_j)$ will incur significant computational costs because of the computation of feature similarities between all the pair-wise events φ_i and φ_j. In order to speed up its computation, we adopt a sampling strategy which randomly selects some entities, i.e. φ_{ip} and φ_{jp} from φ_i and φ_j respectively and reduce $VL\_sim(\varphi_i, \varphi_j)$ to a manageable $VL\_sim(\varphi_{ip}, \varphi_{jp})$.

Steps 4 to 8 perform the hierarchical clustering process. In Step 5, the most similar events φ_i and φ_j are merged together into a new event φ_{new}. We then construct candidate relevant set $E\varphi_{new}$ for φ_{new} (merge $E\varphi_i$ and $E\varphi_j$ to get the events whose support is larger than α) and compute the similarities between φ_{new} and each event φ_k in $E\varphi_{new}$ based on their children's similarities, *i.e.*

$$sim(\varphi_{new}, \varphi_k) = sim(\varphi_i, \varphi_k) + sim(\varphi_j, \varphi_k) \qquad (8)$$

Finally, we compute the quality of the current level of the tree. Note that our hierarchical clustering may not necessarily result in a tree since we are not building one big community – we will stop the merging process if the current merging step does not improve the quality of the current level of tree. Newman has proposed a quality function Q (modularity) to evaluate the goodness of a partition [15]:

$$Q = \sum_i (e_{ii} - a_i^2) \qquad (9)$$

where e_{ii} is the number of edges in the same group/community connecting the vertices (intralinks) and a_i^2 is the sum of edges from the vertices in group i to another group j (interlinks). Since we have observed that there are many interactions across different communities, instead of using the physical links, we use the content/feature-based approach. We represent each event using a TFIDF representation, and then use cosine similarity to compute the intra-similarities and inter-similarities. The quality equation in (9) can be rewritten into Equation (10),

$$Q = \sum_i (cos\,sim\,(i,i) - \sum_j cos\,sim\,(i,j)^2) \qquad (10)$$

Basically, using Equation 10 favors a community substructure which has in overall bigger intra-similarity and less inter-similarity in terms of their topics and content. Our ECODE algorithm stops at a level of tree with the maximal Q value.

3.5 Assign People to Corresponding Communities

We note that each entity may occur in multiple communities. For each entity, we discover the core communities in which the entity is highly involved in. If a_i is a member of community set $C=\{C_1, C_2, \ldots, C_p\}$, we compute the community attachment scores of a_i to C_j (j=1, 2, ..., p) as follows

$$s(a_i, C_j) = \sum \frac{int(a_i, a_k)}{int(a_i)} \tag{11}$$

where $int(a_i, a_k)$ is the number of links between a_i and other members in community C_j and $int(a_i)$ is the total number of links of a_i, i.e. a_i's degree.

If a_i is not a member of C_j but it can be connected to C_j through intermediate connectors (indirect neighbors), its community attachment scores can be computed as

$$s(a_i, C_j) = \sum_{\substack{a_k \in C_m, a_l \in C_j \\ m \neq j}} \frac{int(a_i, a_k)}{int(a_i)} * \frac{int(a_k, a_l)}{int(a_k)} \tag{12}$$

Note if $s(a_i, C_k) > \sum_{i=1}^{p} s(a_i, C_i) / p$, then C_k will be regarded as a_i's core community.

4   Experimental Results

We evaluate the proposed ECODE algorithm by using it to mine communities from a large researcher social network built by using bibliography data. The datasets that we have used for our experiment are publication data from the Digital Bibliography and Library Project (DBLP). The DBLP database provides bibliography information on major computer science journals and conferences (http://www.informatik.uni-trier.de/~ley/db/). DBLP currently lists more than one million articles; each article record contains the author names, paper title, conference or journal name, and year of publication, as well as other bibliographic information. For our work, we used only the information on the author names, paper titles and conference names. In our experiments, each conference will be regarded as one event.

We have selected 6 domains in computer science, namely *database*, *data mining*, *machine learning*, *multimedia*, *bioinformatics*, and *natural language processing*, which represent different communities in computer science. For each community, 3 events (in this case, top conferences) were selected and a total of 28,998 papers (from 1970 to 2008) were retrieved from DBLP, including 31,122 authors/entities and 127,238 links (physical links between every two co-authors). The link density of the network is quite small, which is equal to 127238/(31122*31121/2)= 0.00026274, indicating that each researcher will only interact with a very limited subset of community members which results the low connectivity issue in the network.

Table 1. Communities, events and community core members

Domain/Communities	Events	#PC members
Database (DB)	SIGMOD, VLDB, ICDE	557
Data Mining (DM)	KDD, ICDM, SDM	738
Machine Learning (ML)	ICML, NIPS, ECML	1,007
Multimedia (MM)	CVPR, ICCV, ACM MM	802
Bioinformatics (BI)	RECOMB, ISMB, CSB	951
Natural Language Processing (NLP)	ACL, COLING, EACL	187

Table 1 summarizes the communities and the corresponding events (column 1 and 2). To evaluate the quality of the detected communities, we also manually construct gold standard community data sets consisting of community core members, namely, the technical program committee (PC) members, for each event (note that there is no existing gold standard for evaluating the communities in social networks). The third column lists the number of PC members for these top conferences from 2000 to 2007.

Next, we describe the experimental setting. In our ECODE algorithm, for each event, we will find its "*candidate relevant set*" which consists of its potential similar events where α is used to control the size of candidate relevant set (Equation 6). In our experiments, α is set as 4, but we have also tested the sensitivity how α affects our algorithm later on (Fig. 4). In order to compute the virtual links between two events, we randomly selected 10 events/conferences and compute the average similarity of non-connected community members as the virtual link threshold δ (Equation 1). In addition, in order to speed up the computation, we randomly selected 100 members in each event to compute the virtual links between them. We also tested how the number of members affects the performance of our technique (Fig. 5). For Equations 2 and 10, we only used the paper titles as the associated documents since they are readily available. In ECODE algorithm, we combined the vertex similarity and virtual link similarity (equation 7) where λ is used to weight the two similarities. In our experiments, λ is fixed as 0.9, and we also test λ's sensitivity in Fig. 6. Note all our experiments were run with a standard Intel Core 2 2.40 GHz desktop with 2GB RAM.

Let us now present the experimental results. Table 2 lists the results using two recently published techniques CONGO [24], EAGLE [19] (they have performed better than state-of-the-art techniques), as well as our proposed techniques with different similarity measures, such as vertex similarity (ECODE_Vex), physical links (ECODE_PL), virtual links (ECODE_VL), combined vertex and virtual links (ECODE). The table lists the performance of various techniques in terms of Recall_BM, which is obtained by computing the best match of discovered communities to gold standard communities with one to one mapping. To do so, we find all the similarity scores (Equation 3 was used to compute the scores) between the discovered communities and the gold standard communities. Then, we find the first best match pair with the biggest similarity score to match a discovered community with a gold standard community. We continue this process for the remaining gold standard communities until all the gold standard communities have found their best match discovered communities, or that no discovered community can be matched to the gold standard communities. Recall_BM is defined as the number of the members in gold standard communities retrieved by discovered communities divided by the total number of members in gold standard communities.

Table 2. Overall performances of various techniques

Methods	CONGO	EAGLE	ECODE_Vex	ECODE_PL	ECODE_VL	ECODE
Performance	13.7%	27.6%	63.5%	53.3%	64.1%	69.9%

Table 2 shows that ECODE produces the best results, achieving a Recall_BM score of 69.9%, which is 42.3%, and 56.2% higher than the Recall_BM of the two existing techniques CONGO and EAGLE respectively. Compared with only using physical link, virtual link, and vertex similarities, ECODE also generated better results, illustrating that integrating the vertex links and virtual links improves the effectiveness of detecting communities in the social networks.

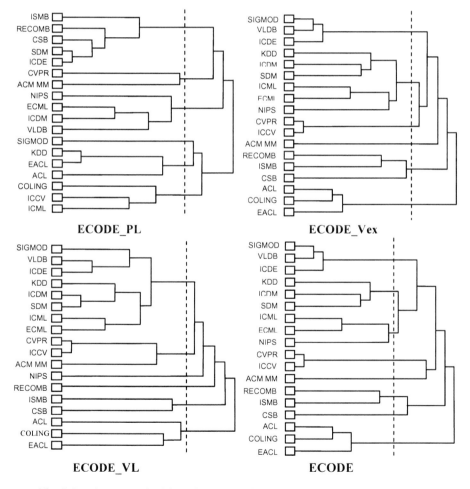

Fig. 3. Dendrograms of DCODE for communities using different similarity measures

We further checked the four dendrograms of our ECODE algorithms with different similarity settings, which are shown in Fig. 3. We observe that the dendrogram of ECODE is much more meaningful than ECODE_PL, ECODE_Vex as well as ECODE_VL. From the results of ECODE_PL, we can see that it is almost like random clustering. This is because researchers tend to have collaborations with those from different communities. As such, the physical links are rather misleading for forming community substructures. ECODE_Vex showed a more meaningful result; however, there are still some faults in the clustering process. For example, CVPR and ICCV (multimedia conferences) were grouped with data mining and machine learning communities first before being grouped with ACM MM although ACM MM, CVPR and ICCV are multimedia-related conferences. As for ECODE_VL, while their results are better than ECODE_PL in terms of its dendrogram, conferences belonging to the same community such as ICML, ECML and NIPS (machine learning conferences) are not grouped together. NIPS was grouped with ICML and ECML after both have been clustered to the database, data mining, and multimedia community.

In comparison, by weighting and combining different similarity measures (vertex and virtual links), ECODE algorithm was able to categorize the right conferences to the right communities. Our ECODE algorithm discovered 8 communities where all the merging steps are correct. This shows that ECODE's integrating of the virtual links and vertex overlapping was effectively used to detect the community substructures. In addition, our clustering cut-off (Equation 10)—the dotted line in Fig. 3—is also very accurate, showing that ECODE was able to early-stop the hierarchical clustering and detect meaningful communities.

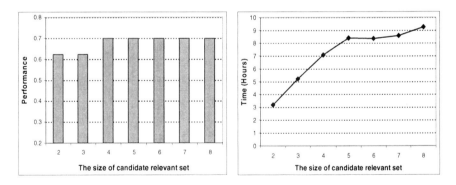

Fig. 4. Performance of ECODE with different size of candidate relevant set

Fig. 4 shows the performance of ECODE with different sizes of candidate relevant set, from 2 to 8 with Step 1. The performance of ECODE increases as the size of candidate relevant size increases from 2, 3 to 4, but it does not change after that. The plot on the right portion of Fig. 4 shows the actual running time against the size of candidate relevant set. As expected, more computations were needed when the size of candidate relevant set increases. However, by considering both performance and running time plots together, Fig. 4 indicates that after the size of the candidate relevant set has increased to a certain degree — in this case, 4 — more computation is no longer useful for community detection as it only increases the computational time without increasing the performance. It also shows that our candidate relevant set has effectively captured the more related events so that it can save a large amount of computational time, as compared with computing all the pair-wise similarities which is typically used for hierarchical clustering.

Recall that we also selected a subset of authors to compute the virtual links among two events in order to improve the efficiency of our algorithm. To study the sensitivity of the number of authors selected, we performed a series of experiments using different numbers of authors, from 50 to 200 with a step of 50. The results are shown in Fig. 5. While the results of using 100, 150 and 200 authors are better than using only 50, there are no significant improvements. This means that our ECODE algorithm with small number of authors can perform reasonably well even when we select only 50 or 100 authors from each event. On the other hand, in terms of efficiency, our algorithm will perform much fast when we use less authors for computing virtual links, as shown in the right part of Fig. 5.

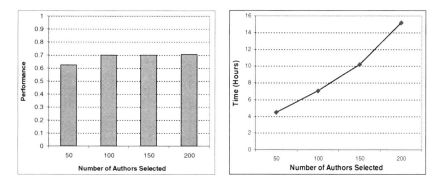

Fig. 5. Performance of ECODE with different number of selected authors

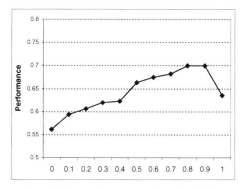

Fig. 6. The performance of ecode with λ

In equation 7, we have used λ to weight the importance of vertex similarity and virtual link similarity. Fig. 6 shows how the values of λ affect the performance of ECODE algorithm. In Fig. 6, when λ increases, the performance of ECODE also increases until λ reaches 0.9. Fig. 6 shows that combining the vertex similarity and virtual link similarity can get consistent better results when $\lambda \in [0.5, 0.9]$ than using vertex similarity and virtual link similarity individually.

Finally, in Fig. 7, we show the 20 top representative researchers with the most publications in our discovered communities. We observe that these top researchers are typically shared by two or more communities (e.g. Tao Jiang is shared by four, i.e. bioinformatics, database, data mining and machine learning). In fact, all the data

Fig. 7. The top researchers in our discovered communities

mining researchers are shared by two or more communities (no single data mining circle in Fig. 7), indicating that data mining is highly related to other domains, and data mining researchers are always doing applications or research in other domains.

Note that out of the total 3,488 PC members in Table 1, there were about a third of them (1,118) who were not assigned any community because these community (PC) members did not have any publications in the conferences listed. We have searched all their publications in DBLP (they published in other related conferences anyway). By incorporating their publication data into current publication data sets, we were able to assign 782 of them to one or more correct communities through indirect neighbors with an accuracy of 77.75% (using Equation 12). After assigning all these authors to their respective communities, the Recall_BM score for ECODE algorithm has a significant further improvement from 69.9% to 87.1%, as shown in Table 3. As such, the Equation 12 (assigning entities to community based on connectivity information) can be useful to effectively infer the underlying community belongings.

Table 3. Performance before and after assigning unpublished authors for various communities

Communities	Before assigning unpublished authors	After assigning unpublished authors
Bioinformatics (BI)	49.7%	77.6%
Database (DB)	87.1%	97.3%
Data Mining (DM)	59.4%	80.1%
Multimedia (MM)	62.6%	85.9%
Machine Learning (ML)	70.5%	87.5%
Natural Language Processing (NLP)	89.8%	94.1%
Average	**69.9%**	**87.1%**

5 Conclusions

Communities are often formed and strengthened during various social events attended by individuals to interact with other members of the community. Event information can thus be quite useful for inferring communities from social networks. In this paper, we have therefore proposed an *Event*-based COmmunity DEtection algorithm ECODE to mine the underlying community substructures of social networks from event information. Unlike conventional approaches, ECODE makes use of content similarity-based *virtual links* in the social networks. The virtual links are found to be more useful for community detection than the physical links. By performing computation between an event and its candidate relevant set instead of computing pair-wise similarities between all the events, ECODE was able to achieve significant computational speedup. We have performed extensive experimental results on the events and social network data of Computer Science researchers. Comparisons with other existing methods showed that our ECODE algorithm is both efficient and effective in detecting communities from social networks.

We have so far focused on the social networks for our approach in this work. In our future work, we plan to generalize our current approach to mine other networks. For example, we aim to mine protein complexes from protein interaction networks where proteins are vertices and protein interactions between two proteins are the links [9]. Each protein in protein interaction networks will have various biological evidences (similar to content profiling data in social networks) such as sequences, protein domains, motifs, molecular functions, cellular components as well as other protein's physico-chemical properties etc. In this scenario, virtual links will connect two proteins if they have overall bigger similarities in terms of sequence similarity, functional similarity, location similarity etc. We will leave this as our future work.

References

1. Albert, R., Jeong, H., Barabási, A.-L.: Diameter of the world-wide web. Nature 401, 130–131 (1999)
2. Wasserman, S., Faust, K.: Social Network Analysis. Cambridge University Press, Cambridge (1994)

3. Li, X.-L., et al.: Searching for Rising Stars in Bibliography Networks. In: DASFAA (2009)
4. Palla, G., et al.: Uncovering the overlapping community structure of complex networks in nature and society. Nature 435, 814–818 (2005)
5. Li, X.-L., et al.: Interaction Graph Mining for Protein Complexes Using Local Clique Merging. Genome Informatics 16(2) (2005)
6. Li, X.-L., Foo, C.-S., Ng, S.-K.: Discovering Protein Complexes in Dense Reliable Neighborhoods of Protein Interaction Networks. In: CSB (2007)
7. Steinhaeuser, K., Chawla, N.: A Network-Based Approach to Understanding and Predicting Diseases. Springer, Heidelberg (2009)
8. Wu, M., et al.: A Core-Attachment based Method to Detect Protein Complexes in PPI Networks. BMC Bioinformatics 10(169) (2009)
9. Li, X.-L., et al.: Computational approaches for detecting protein complexes from protein interaction networks: a survey. BMC Genomics 11(Suppl 1:S3) (2010)
10. Redner, S.: How popular is your paper? An Empirical Study of the Citation Distribution. Eur. Phys. J. B(4), 131–138 (1998)
11. Nisheeth, S., Anirban, M., Rastogi, R.: Mining (Social) Network Graphs to Detect Random Link Attacks. In: ICDE (2008)
12. Clauset, A., Newman, M.E.J., Moore, C.: Finding community structure in very large networks. Phys. Rev. E 70, 066111 (2004)
13. Radicchi, F., et al.: Defining and identifying communities in networks. PNAS 101(9), 2658–2663 (2004)
14. Fortunato, S.: Community detection in graphs. Physics Reports 486, 75–174 (2010)
15. Newman, M.E.J.: Modularity and community structure in networks. PNAS 103(23), 8577–8582 (2006)
16. Ravasz, E., et al.: Hierarchical Organization of Modularity in Metabolic Networks. Science 297, 1551–1555 (2002)
17. Clauset, A.: Finding local community structure in networks. Phys. Rev. E 72 (2005)
18. Boccaletti, S., et al.: Detection of Complex Networks Modularity by Dynamical Clustering. Physical Review E, 75 (2007)
19. Shen, H., et al.: Detect overlapping and hierarchical community structure in networks. CoRR abs/0810.3093 (2008)
20. Seidman, S.B.: Network structure and minimum degree. Social Networks 5, 269–287 (1983)
21. Holme, P., Huss, M., Jeong, H.: Subnetwork hierarchies of biochemical pathways. Bioinformatics 19(4), 532–538 (2003)
22. Gleiser, P., Danon, L.: Community structure in jazz. Advances in Complex Systems 6, 565 (2003)
23. Tyler, J.R., Wilkinson, D.M., Huberman, B.A.: Email as Spectroscopy: Automated Discovery of Community Structure within Organizations. Communities and Technologies, 81–96 (2003)
24. Gregory, S.: A fast algorithm to find overlapping communities in networks. In: Daelemans, W., Goethals, B., Morik, K. (eds.) ECML PKDD 2008, Part I. LNCS (LNAI), vol. 5211, pp. 408–423. Springer, Heidelberg (2008)
25. Bie, T.D., Cristianini, N.: Fast SDP relaxations of graph cut clustering, transduction, and other combinatorial problems. Journal of Machine Learning Research 7, 1409–1436 (2006)
26. Girvan, M., Newman, M.E.J.: Community structure in social and biological networks. PNAS 99(12), 7821–7826 (2002)
27. Newman, M.E.J.: Detecting community structure in networks. European Physical Journal B 38, 321–330 (2004)

28. Ding, C., He, X., Zha, H.: A Spectral Method to Separate Disconnected and Nearly-disconnected Web Graph Components. In: KDD (2001)
29. Wasserman, S., Faust, K.: Social network analysis: methods and applications. Cambridge University Press, Cambridge (1994)
30. Backstrom, L., et al.: Group Formation in Large Social Networks: Membership, Growth, and Evolution. In: Proceedings of the Twelfth ACM SIGKDD International Conference on Knowledge Discovery and Data Mining (KDD 2006), Philadelphia, USA (2006)
31. Leskovec, J., Kleinberg, J., Faloutsos, C.: Graphs over Time: Densification Laws, Shrinking Diameters and Possible Explanations. In: ACM SIGKDD International Conference on Knowledge Discovery and Data Mining, KDD (2005)
32. Kimura, M., Saito, K.: Tractable Models for Information Diffusion in Social Networks. In: ECML/PKDD (2006)
33. Tang, L., et al.: Community Evolution in Dynamic Multi-Mode Networks. In: SIGKDD (2008)
34. Agrawal, R., Imielinski, T., Swami, A.: Mining Association Rules Between Sets of Items in Large Databases. In: SIGMOD Conference (1993)

A User Similarity Calculation Based on the Location for Social Network Services

Min-Joong Lee and Chin-Wan Chung

Department of Computer Science,
Korea Advanced Institute of Science and Technology(KAIST)
335 Gwahangno, Yuseong-gu, Daejeon, Republic of Korea
mjlee@islab.kaist.ac.kr, chungcw@kaist.edu

Abstract. The online social network services have been growing rapidly over the past few years, and the social network services can easily obtain the locations of users with the recent increasing popularity of the GPS enabled mobile device. In the social network, calculating the similarity between users is an important issue. The user similarity has significant impacts to users, communities and service providers by helping them acquire suitable information effectively.

There are numerous factors such as the location, the interest and the gender to calculate the user similarity. The location becomes a very important factor among them, since nowadays the social network services are highly coupled with the mobile device which the user holds all the time. There have been several researches on calculating the user similarity. However, most of them did not consider the location. Even if some methods consider the location, they only consider the physical location of the user which cannot be used for capturing the user's intention.

We propose an effective method to calculate the user similarity using the semantics of the location. By using the semantics of the location, we can capture the user's intention and interest. Moreover, we can calculate the similarity between different locations using the hierarchical location category. To the best of our knowledge, this is the first research that uses the semantics of the location in order to calculate the user similarity. We evaluate the proposed method with a real-world use case: finding the most similar user of a user. We collected more than 251,000 visited locations over 591 users from foursquare. The experimental results show that the proposed method outperforms a popular existing method calculating the user similarity.

Keywords: User similarity, Social network, Location based service.

1 Introduction

Over the past few years, the online social network services such as facebook, twitter and foursquare have been rapidly increasing their territory in the Internet world. With the help of recently growing prevalence of mobile devices, the online social network services have naturally permeated into the mobile devices

J.X. Yu, M.H. Kim, and R. Unland (Eds.): DASFAA 2011, Part I, LNCS 6587, pp. 38–52, 2011.
© Springer-Verlag Berlin Heidelberg 2011

such as smartphones. Nowadays most of the smartphone users create, share and communicate with other users by using the online social network services at anytime, anywhere.

Moreover, the emergence of the Global Positioning System (GPS) enabled smartphone brings a great opportunity to the online social networks services. The GPS-enabled smartphones are able to acquire their current position through the GPS sensor and tag the acquired location of a device to user generated contents. For instance, when a user writes a post or takes a photo, a smartphone can tag the location of the user on the post or the photo automatically. Especially, if the post is regarding the user's current location or the photo is a landscape of the user's location, the location information will be a huge asset to the online social network. For example, for other online social network users trying to get information about the specific location, this tagged location information can be used to increase the quality of search results.

As the GPS-enabled smartphones become more and more popular, the location based social network service is getting into the spotlight as a new type of the online social network service. For instance, foursquare belongs to this category. foursquare lets a user record a place of one's current location and tell friends where he/she is and leave a short commentary about the place. We will describe the details of foursquare in section 5.1.

In social network services, finding similar users is a very important issue since we can recommend similar users as friends to a new user and recommend a lot of things such as products, search results and experiences. However there are numerous factors to calculate the user similarity. Among the various factors, we utilize the user's location to calculate the user similarity. Since the users carry mobile devices most of the day, especially smartphones, the location of a smartphone has more meaning then just a specific point of the earth. The location of a smartphone is the user's current position and it implies user's interest and life style.

When calculating the user similarity by using the user's location information, the physical location cannot capture the user's real intention why the user visits there. In the real world, one specific physical location is related to many places such as a coffee shop and a theater, and we cannot determine the exact place by using the physical location. This problem is worsened if the user is in a building in a downtown. Therefore, we use the semantics of the location such as the name of the place, and the type of the place to determine the exact place and capture the user's intention. For example, if a user visits a theater frequently, it is reasonable to infer that the user likes watching movies and if a user visits a university regularly, we can infer the user is a student or a faculty member of the university.

In this paper, we propose a new method to calculate the user similarity by using the semantics of the location. We only consider the top-k visited locations of each user. Infrequently visited locations incur incorrect results since people occasionally visit some locations against their will. To take advantage of using the semantics of the location we utilize a location category hierarchy. Also, we devise the human sense imitated similarity calculation which is able to calculate the interest for another location by using a user's current interest in a certain location.

The contributions of this paper are as follows:

- We address an importance of the semantics of the location than the physical location, and use the semantics of the location to calculate the user similarity. This is the first research that uses the semantics of the location in order to calculate the user similarity.
- We devise an efficient method to calculate the user similarity by using the semantics of the location. In our method, locations and their categories form a hierarchical graph structure. By considering only relevant nodes and computing the similarity at necessary nodes, the proposed method generates the result quickly.
- Our proposed method can also be used to efficiently calculate the similarity between the two objects other than users when the object can be associated with hierarchical categories of elements where each category has a weight.
- We experimentally evaluate the proposed method with a real-world use case: finding the most similar user of a user. Experimental results show that our proposed method is 84% higher in precision, 61% in recall and 72% in f-measure than Jaccard index.

The rest of the paper is organized as follows. In Section 2, we discuss the related works on user similarity calculation. In Section 3, we explain basic concepts and derive basic equations of our proposed method. In Section 4, we describe the details of our proposed method. The experimental results are shown in Section 5. Finally, in Section 6, we make conclusions.

2   Related Work

There have been numerous efforts to calculate the user similarity for different objectives. Recommending people is one of the popular objectives. Guy et al. [4] proposed a method based on the various aggregated information about people relationships but it focused on the people that the user is already familiar with. Therefore, this method cannot be used for calculating the similarity with an unknown user and finding a new friend in the online social network. Terveen et al. [9] proposed a framework called *social matching*. The *social matching* framework aims to match people mainly using the physical locations of people, while we focus on the semantics of the location.

Some methods recommend experts. McDonald et al. [7] proposed an expert locating system that recommends people for possible collaboration within a work place. Also an expert search engine is described in [2]. The expert search engine finds relevance people according to query keywords. Those approaches are useful to find co-workers or experts but their life style can be varying since the authors focused on a domain to find experts of that domain. Therefore, this approach can not used for finding similar users in general.

Nisgav et al. [8] proposed a method to find the user similarity in the social network. They utilize the user's typed queries to calculate the user similarity. However, since the location based social network is mainly accessed by using the smartphone, typed queries are not much used. In addition, considering the importance of the location information in the location based social network, using the user's queries is not suitable for the location based social network.

The increasing pervasiveness of the location-acquisition technologies such as GPS and WiFi has produced a large amount of location data, and there have been numerous attempts to utilize these location data. Several researchers manipulated and extracted valuable information by using the individual's location data. Chen et al. [1] proposed the raw-GPS trajectory simplification method for the location based social network service. They considered both the shape skeleton and the semantic meanings of a GPS trajectory but they did not report about the semantic meanings of a GPS trajectory.

Also, multiple users' locations have been used to extract meaningful information by several researchers. Krumm et al. [5] described a method that uses a history of the GPS driving data to predict the destination as a trip progresses, and Gonotti et al. [3] developed an extension of the sequential pattern mining paradigm that analyzes the trajectories of moving objects. In contrast to these techniques, Li et al. [6] proposed a framework for mining the user similarity based on the location history. Li et al. extended the paradigm of mining multiple users' location histories, from exploring user's behaviors to exploring a correlation between two users.

The purpose of [6] is similar to ours as finding the similarity between users by using the location history. First, they identified the stay points from the GPS-trajectories and clustered stay points. Then, they matched clustered sequences of two users. The higher user similarity in their framework means two users are physically close such as a family, roommates and lovers, since their framework is based on physical locations. In general, people do not think roommates are similar to each other. On the other hands, people intuitively think two users are similar to each other, if their life styles are alike such as two users both often go coffee shops even if coffee shops are different. Our proposed method efficiently finds the similarity of two users who have similar life styles, since we use the semantics of the location.

3   Preliminary

We explain basic concepts and derive basic equations.

3.1   Location Category

As we mentioned in previous sections, we are using the semantics of the location instead of the physical location. For the use of the semantics of the location, we construct and utilize the location category hierarchy graph. We extract the location category from foursquare since we use a foursquare dataset. A part of the location category hierarchy graph is shown in Fig. 1.

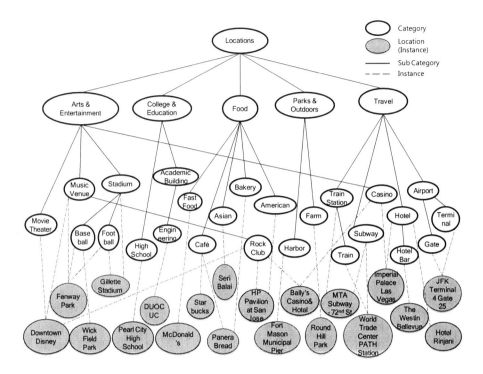

Fig. 1. The location category hierarchy graph (selected)

The location category hierarchy graph consists of two kinds of nodes, *location nodes* and *category nodes*. A *location node* represents the corresponding unique location such as Downtown Disney, Hotel Rinjanis and Gillette Stadium. A *category node* represents a location category such as a movie theater, a hotel and a stadium.

3.2 Significant Score

$SigS_n(u)$ denotes the significant score of node n of user u and it is calculated as follows:

$$SigS_n(u) = \frac{Visit_n(u)}{TotalVisit(u)} \qquad (1)$$

where $Visit_n(u)$ is the number of visits at *location node* n of user u, $TotalVisit(u)$ is the total number of visits of user u.

We denote a user's most frequently visited k locations as the *top − k locations* of the user. There are many locations that users visit, but users visit only a few locations frequently. We can consider that the location visited more by a user represents the user's characteristic better than the location visited less, and we experimentally show that the visits to *top − k locations* take up great part of total visits in Section 5.4. To avoid a time-consuming process, we consider only *top − k locations* of a user to calculate the similarity.

3.3   Similarity Score

If two users have their own significant scores at the same *node*, we can compute a similarity score at that node (denote as a *match node*). Let $SimS_n(u,v)$ be the similarity score between user u and user v at *match node* n. To compute $SimS_n(u,v)$, we take the minimum significant score of user u and user v at *match node* n as follows:

$$SimS_n(u,v) = min(SigS_n(u), SigS_n(v)) \qquad (2)$$

We take the minimum value of two users' significant scores since the minimum value intuitively represents two users' common interest at the *match node*.

3.4   Significant Score Propagation

We should take into account miss-matching nodes between two users to get more accurate similarity. For example, consider that two users like watching movies, where one user often goes to 'theater A' and the other user often goes to 'theater B'. In such a case, people intuitively think two users are similar. Furthermore, we can infer that their hobby is watching movies. The human intuition also tells us two users in different theaters are less similar than two users in the same theater.

To imitate this human intuition, we design a method to give the similarity score at the common nearest ancestor node when two nodes are different in the location category hierarchy graph. For instance, 'theater A' and 'theater B' belong to a movie theater category node.

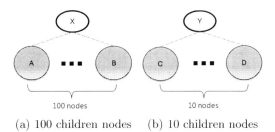

(a) 100 children nodes   (b) 10 children nodes

Fig. 2. Different propagation rate according to the number of children nodes

Consider the following two cases depicted in Fig 2. First, the similarity between two users when one user visits location A, corresponding to node A, and the other user visits location B, corresponding to node B, where node A and node B are children of node X, represents category X, which has 100 children (Fig. 2 (a)). Second, the similarity between two users when one user visits location C, corresponding to node C, and the other user visits location D, corresponding to node D, where node C and node D are children of node Y, represents category Y, which has 10 children (Fig. 2 (b)). The first case is more similar than the second case from a probabilistic perspective. The difference between location A and location B in the first case is less than that between location C and location D

in the second case because location A and location B are among 100 locations, while location C and location D are among 10 locations. Therefore, the similarity between two users in the first case is probabilistically higher than that in the second case.

Since each category has various numbers of children, we introduce the logarithm to lessen the effect of various numbers of children nodes. Let $PR(n)$ be the propagation rate of node n. It is calculated as follows:

$$PR(n) = \frac{\log(|Sibling(n)| + 1)}{\log(totalNumberofNodes)} \tag{3}$$

where $Sibling(n)$ is the node n's sibling node set, including node n. We add one to $|Sibling(n)|$ to prevent a case which a dividend becomes zero. $totalNumberof$ $Nodes$ is used because $PR(n)$ should be a small number, and $totalNumberof$ $Nodes$ is always much bigger than $|Sibiling(n)|$ and easy to obtain. The significant score at node n is multiplied by the propagation rate $(PR(n))$ when node n's significant score propagates to the parent node.

4   User Similarity Calculation

In this section, we first overview our proposed method, and then explain the details.

4.1   Overall Process

The procedure of our proposed method is as follows:

1. Compute the significant score (Equation 1) of each visited location of user u and user v
2. Find $top - k\ locations$ of user u and user v, and construct a top-k significant score table for each user.
3. Construct a location category hierarchy graph by using only visited *location nodes* of two user and visited *location nodes*' ancestor nodes.
4. Find the *match nodes* and its calculation order by using algorithm **MatchNodeOrder()** (Fig. 4).
5. Calculate the user similarity between user u and v by using algorithm **Similarity()** (Fig. 5).

The details of **MatchNodeOrder()** will be discussed in Section 4.2, and the details of **Similarity()** will be discussed in Section 4.3.

4.2   Order of Match Nodes

There are two difficulties to calculate the user similarity. First, as we showed in Fig. 1, the structure of *category nodes* is a tree structure. However, some of *location nodes* have multiple parent nodes since some locations belong to more than one category. If a *location node* has multiple parent nodes, we should select one of the parent nodes to propagate the significant score. Second, the diverse

depths of location nodes make it hard to find match nodes at which two users'
significant scores are propagated to come across each other.

Regarding to first difficultly, to overcome the multiple parent nodes problem,
we split a *location node* to the number of parent nodes and also the significant
score is divided equally among split nodes. This step does not require too much
workload since only some of *location nodes* belong to more than one parent node
as shown in Fig. 1. Regarding the second difficulty, to efficiently calculate the
similarity score at a match node, we find match nodes which need to calculate the
similarity score, and the calculation order of match nodes. Without this match
node order, we should calculate the similarity score recursively. The algorithm
of splitting *location nodes* and finding the match nodes and the order of the
match nodes is as follows:

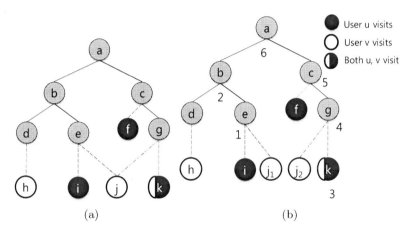

Fig. 3. Example of finding match nodes and its calculation order

Fig. 3 (b) shows an example of finding match nodes and its calculation order.
Black colored nodes correspond to user u's visited locations and white colored
nodes correspond to user v's corresponding visited locations. The nodes which
have a number at the bottom are match nodes and the numbers indicate the
calculation order which is the output of the algorithm. We explain the details of
the algorithm with an example case in Fig. 3.

As shown in Fig. 3 (a), node j has two parent nodes, e and g. We split node j
to j_1, j_2 and make them the children of node e and g, respectively in Fig. 3 (b)
(Line(4) - Line(6)). Then, we equally divide the initial significant score of node j
into split node j_1, j_2 and store split nodes in the ST_u table (Line(7) - Line(10)).
We construct empty sets for each user, *ancestorU* and *ancestorV* (Line(13)).
For user u, add all ancestor nodes of all visited nodes (a, b, c, e and g) and all
visited nodes (f, i and k) to *ancestorU*. And for user v, ancestor nodes (a, b, c,
d, e and g) and visited nodes (h, j_1, j_2 and k) are added to *ancestorV* (Line(14)
- Line(19)).

After that, let *MatchNodeSet* be the intersection of *ancestorU* and *ancestorV*
(Line(20)). In this example, *MatchNodeSet* is $\{a, b, c, e, g, k\}$ node elements

```
Algorithm MatchNodeOrder()
Input location category hierarchy graph G
        user u's top-k significant score table ST_u
        user v's top-k significant score table ST_v
Output match node order list m
begin
1.    Let list m be an empty node list
2.    Let ST_u[n] be a significant score of user u at node n
      /* to handle multiple parent nodes problem */
3.    foreach leafnode l in graph G
4.        if location node l has more than one parent node
5.            pNum := location node l's parent nodes count
6.            Split node l to l_1, l_2, ..., l_{pNum} as each parent's child node
              /* Do following if step on ST_u */
7.            if ST_u contains score for node l
8.                Add l_1, l_2, ...., l_{pNum} to ST_u with (ST_u[l])/(pNum) score
9.                Delete node l and its significant score from ST_u table
10.           endif
11.       endif
12.   endforeach
13.   Let MatchNodeSet, ancestorU and ancestorV be empty node sets
14.   foreach node n in ST_u
15.       Add node n's ancestor nodes to ancestorU
16.   endforeach
17.   foreach node n in ST_v
18.       Add node n's ancestor nodes to ancestorV
19.   endforeach
20.   MatchNodeSet := ancestorU ∩ ancestorV
21.   m := Sort MatchNodeSet in post-order by using graph G structure
22.   return m
end
```

Fig. 4. Algorithm of finding match nodes order

which is intersection of $\{a, b, c, e, g, i, k, f\}$ and $\{a, b, c, d, e, g, j_1, j_2, h, k\}$. Finally, a post-order list of $MatchNodeSet$ elements are assigned in a list m and returned as an output of this algorithm (Line(21) - Line(22)). As shown as numbers in Fig. 3 (b), the calculation order list is $<e, b, k, g, c, a>$.

4.3   User Similarity Calculation

After we determine the match nodes and its calculation order, we can efficiently calculate the user similarity. The algorithm of calculating the user similarity is as follows:

For efficiency, we devise the multiple propagation rate based on Equation 3 for propagating a significant score of a node to its ancestor node through multiple levels. $MPR(n, v)$ denotes the multiple propagation rate, from node v to ancestor node n, and it is calculated as follows:

$$MPR(n,v) = \frac{\log(|Sibling(k_1)|+1) \times \log(|Sibling(k_2)|+1) \times ... \times \log(|Sibling(k_n)|+1)}{depthDiff \times \log(totalNumber of Nodes)}$$

(4)

where $depthDiff$ is the depth difference between node n and node v, $Sibling(n)$ is the number of node n's sibling nodes. k_1 is node v, k_2 is the parent node of v, ... , and node $k_{depthDiff}$ is the node n since $depthDiff$ is the depth difference between node n and node v.

Algorithm Similarity()
Input location category hierarchy graph G
            user u's top-k significant score table ST_u
            user v's top-k significant score table ST_v
            match node order list m
Output User similarity score $SimScore$
begin
1.   Let $ST_u[n]$ be a significant score of user u at node n
2. $SimScore := 0.0$
       /* Calculate user similarity at each node in list m in order*/
3.   **foreach** node n **in** list m
4. $descendants :=$ the set of descendant of node n (use graph G structure)
5.   **foreach** node d **in** $descendants$
             /* Do following **if** step on ST_v */
6.   **if** ST_u has node v
7.   **if** ST_u does not have node n
8.   Add node n to ST_u
9. $ST_u[n] := 0.0$
10.   **endif**
                /* propagate all descendants significant score to node n */
11. $ST_u[n] := ST_u[n] + ST_u[d] * MPR(n, d)$ (Equation 4)
12.   Delete node d from ST_u table
13.   **endif**
14.   **endforeach**
15. $SimScore := SimScore + SimS_n(u, v)$ (Equation 2)
16. $ST_u[n] := ST_u[n] - SimS_n(u, v)$
17. $ST_v[n] := ST_u[n] - SimS_n(u, v)$
18.   **endforeach**
19.   **return** $SimScore$
end

Fig. 5. Algorithm of the user similarity calculation

The user similarity calculation algorithm (Fig. 5) utilizes match node order list m (output of Fig. 4 algorithm) as one of the inputs. At the beginning of the algorithm, we initialize $SimScore$ with zero (Line (2)) and enumerate each node n in list m (Line (3) - Line (18)). At the beginning of enumeration steps, let $descendants$ be the descendant node set of node n (Line (4)). For user u, if ST_u has node v (Line (6)), and if ST_u does not have node n we make a empty table entry for node n to store the propagated score of descendant nodes (Line(7) - (10)). Then we propagate user u's significant scores at node v to the match node n using Equation 4 (Line (11) - Line(12)). We calculate the similarity score at node n and add the similarity score to $SimScore$ (Line (15)), and we subtract similarity score $SimS_n(u, v)$ from two user's significant scores since the similarity score $SimS_n(u, v)$ is already added to the user similarity score $SimScore$. (Line (16) - Line (17)). Finally, the algorithm returns $SimScore$ as the output (Line(19)).

5   Experiment

We evaluate the proposed method with a real-world use case. We use foursquare user's data as a dataset. Before discussing about the experimental results, we briefly introduce our dataset.

5.1   Dataset

foursquare is one of the most spotlighted location based social network services. We briefly introduce foursquare since we collect data from foursquare for evaluating our method.

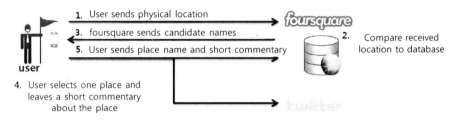

Fig. 6. Diagram of foursquare service

Fig. 6 shows a diagram of foursquare's 'check-in'[1], feature. When a user tries to 'check-in' to a certain place, the user sends an exact physical location of the user (Step 1). Then foursquare compares the received location with their huge venue database (Step 2) and suggests a few names of places in distance order (Step 3). After that, the user selects one place name for his/her current locations, also the user can make a short commentary about the place (Step 4). After that, the user sends the selected place name and a commentary to foursquare and twitter (Step 5). Therefore, by using foursquare APIs[2] (Step 1-3), we can convert the physical location to the semantics of the location.

We collect users' visited locations through users' twitter pages since users also post 'check-in' information to their twitter accounts (Step 5 in Fig. 6). At first, we collected 1,358,287 visiting locations over 17,863 users. However, most of users are not active enough to use their recoeds as a dataset. Therefore, we select 591 users based on their activity. The selected users visited 251,053 locations and they are distributed around the world.

5.2   Finding a Similar User

Fig. 7 shows an example of two similar users which are selected by our method. User A and user B live in very different locations, but they are similar because they are both students and they like to go shopping. By only considering the physical location of two users, the similarity score between user A and user B is close to zero. However, our proposed method finds the similarity between them since our method utilizes the semantics of the location.

5.3   Performance of Proposed Method

We experimentally evaluate the proposed method with a real-world use case; finding the most similar user of a user. We compare our method to Jaccard

---

[1] Records a user's place and able to leave a short commentary about the place.

[2] http://groups.google.com/group/foursquare-api/web/api-documentation

# of visits	Location name	Category	Region
34	Universitas Kristen Petra – Gedung T	Academic Building, University, Math	East Java
29	Petra Christian University	University	East Java
18	CITO (City of Tomorrow)	Mall	East Java
13	Alfamidi	Other – Shopping	East Java
13	Pakuwon	Mall	Indonesia

(a) User A

# of visits	Location name	Category	Region
49	Pearl City High School	High School	HI
37	Waimalu Plaza	Mall	HI
35	Pearl City Cultural Center	Concert Hall, Event Space	Hawaii
32	Foodland	Other – Shopping, Bakery, Seafood	Hi
31	Westridge Shipping Center	Mall	Hi

(b) User B

Fig. 7. Example of two similar users selected by our method

index which is a popular method to calculate the similarity. As we utilize the semantics of the location to calculate the user similarity for the first time, there is no existing method to be compared with.

To find the most similar user of a user, firstly, we calculate pairwise the user similarity between 591 users. Then, we select the most similar user to each of 591 users. After that, we compare a user's visited locations with the most similar user's visited locations for every user.

In order to measure the accuracy of two methods, we compute the precision, recall and F-measure by comparing his/her visited locations with the visited locations of the most similar user recommended by each method.

Let $f(u)$ returns the set of categories of the top-k locations of user u. The precision is calculated as follows:

$$Precision = \frac{|f(u) \bigcap f(u_r)|}{|f(u_r)|} \qquad (5)$$

where u_r is the most similar user selected by a method, which can be our method or Jaccard index.

The recall is calculated as follows:

$$Recall = \frac{|f(u) \bigcap f(u_r)|}{|f(u)|} \qquad (6)$$

The F- measure is calculated as follows:

$$F - measure = \frac{2 \times Precision \times Recall}{Precision + Recall} \qquad (7)$$

Then, we average the precisions, recalls and F-measures for all users.

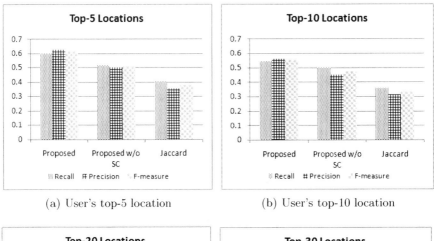

(a) User's top-5 location   (b) User's top-10 location

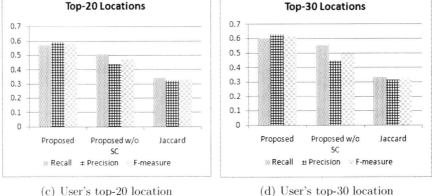

(c) User's top-20 location   (d) User's top-30 location

Fig. 8. Experimental results with various top-k locations

As shown in Fig. 8, our proposed method outperforms the Jaccard index for every different top-k location setting. Our method is 84% higher in the precision, 61% in the recall and 72% in the f-measure than the Jaccard index on the average. Since a frequency of visits to a location is represented by a binary value in Jaccard index, we make our proposed method not to use significant scores and compare with Jaccard index. This modified version of our method is labeled as *Proposed w/o SC* in Fig. 8. However, the result that our method shows higher performance than the Jaccard index remains unchanged.

5.4   Top-k Location

Since we use only the *top − k locations* of a user to calculate the similarity, we experimentally show that considering only the *top − k locations* of a user results is better than considering all the visited location of a user.

To show the visits to $top - k\ locations$ are greater part of total visits, we devise $TopKCover(k)$ which shows the ratio of $top - k\ locations$ visits to total visits.

$$TopKCover(k) = \frac{\sum_{u \in U} \left(\frac{TopkVisit_k(u)}{TotalVisit(u)} \right)}{|U|} \qquad (8)$$

where U is the set of all users, $TopkVisit_k(u)$ is the user u's number of visits to $top - k\ locations$ of user u, $TotalVisit(u)$ is the total number of visits of user u.

From the result of Fig. 9 (a), we can consider that using more than 100 visited locations to characterize a user is meaningless, also the small number of top-k locations covers a large part of visits. The top-5 locations cover 32%, top-10 cover 43%, top-20 cover 55% and top-30 cover 62% of visits.

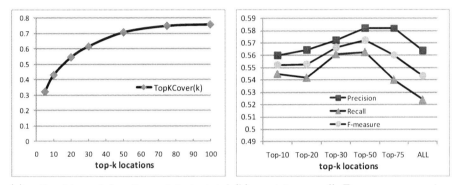

(a) ratio of $top - k\ locations$ visits to total visits

(b) precision, recall, F-measure on various top-k settings

Fig. 9. Two experimental results to select proper top-k

Fig. 9 (b) shows that considering only the $top - k\ locations$ of a user is better than considering a large number or all of the user's visits. All the three measure start dropping after top-50 locations.

6   Conclusion

In this paper, we proposed an accurate and efficient user similarity calculation method. Our method utilizes the semantics of the location, while the other existing previous researches have been focused on only the physical location. We also utilize the location category hierarchy to semantically match locations, and the experimental results show that the proposed method outperforms the popular Jaccard index. We also experimentally show how many numbers of the locations has the meaning to a user, and it helps us to understand the user's behavior. As a future work, we would aggregate the semantics of the location with some other information such as user generated tags to get more accurate results.

Acknowledgments. This work was supported by the National Research Foundation of Korea(NRF) grant funded by the Korea government (MEST) (No. 2010-0000863).

References

1. Chen, Y., Jiang, K., Zheng, Y., Li, C., Yu, N.: Trajectory simplification method for location-based social networking services. In: International Workshop on Location Based Social Networks, pp. 33–40 (2009)
2. Ehrlich, K., Lin, C.Y., Griffiths-Fisher, V.: Searching for experts in the enterprise: combining text and social network analysis. In: International ACM SIGGROUP Conference on Supporting Group Work, pp. 117–126 (2007)
3. Giannotti, F., Nanni, M., Pinelli, F., Pedreschi, D.: Trajectory pattern mining. In: 13th ACM SIGKDD International Conference on Knowledge Discovery and Data Mining, pp. 330–339 (2007)
4. Guy, I., Ronen, I., Wilcox, E.: Do you know?: recommending people to invite into your social network. In: International Conference on Intelligent User Interfaces, pp. 77–86 (2009)
5. Krumm, J., Horvitz, E.: Predestination: Inferring destinations from partial trajectories. In: 8th International Conference on Ubiquitous Computing, pp. 243–260 (2006)
6. Li, Q., Zheng, Y., Xie, X., Chen, Y., Liu, W., Ma, W.Y.: Mining user similarity based on location history. In: 16th ACM SIGSPATIAL International Symposium on Advances in Geographic Information Systems, p. 34 (2008)
7. McDonald, D.W.: Recommending collaboration with social networks: a comparative evaluation. In: Conference on Human Factors in Computing Systems, pp. 593–600 (2003)
8. Nisgav, A., Patt-Shamir, B.: Finding similar users in social networks: extended abstract. In: 21st Annual ACM Symposium on Parallel Algorithms and Architectures, pp. 169–177 (2009)
9. Terveen, L.G., McDonald, D.W.: Social matching: A framework and research agenda. ACM Trans. Comput. -Hum. Interact, 401–434 (2005)

Modeling User Expertise in Folksonomies by Fusing Multi-type Features

Junjie Yao[1], Bin Cui[1], Qiaosha Han[1], Ce Zhang[2], and Yanhong Zhou[3]

[1] Department of Computer Science & Key Laboratory of High Confidence Software
Technologies (Ministry of Education), Peking University
{junjie.yao,bin.cui,qiaoshahan}@pku.edu.cn
[2] Department of Computer Science, University of Wisconsin-Madison
czhang@cs.wisc.edu
[3] Yahoo! Global R&D Center, Beijing
zhouyh@yahoo-inc.com

Abstract. The folksonomy refers to the online collaborative tagging system which offers a new open platform for content annotation with uncontrolled vocabulary. As folksonomies are gaining in popularity, the expert search and spammer detection in folksonomies attract more and more attention. However, most of previous work are limited on some folksonomy features. In this paper, we introduce a generic and flexible user expertise model for expert search and spammer detection. We first investigate a comprehensive set of expertise evidences related to users, objects and tags in folksonomies. Then we discuss the rich interactions between them and propose a unified Continuous CRF model to integrate these features and interactions. This model's applications for expert recommendation and spammer detection are also exploited. Extensive experiments are conducted on a real tagging dataset and demonstrate the model's advantages over previous methods, both in performance and coverage.

1 Introduction

Collaborative tagging is an emerging method for online content organization and management. By annotation using uncontrolled vocabulary, collaborative tagging systems provide better experience of resource sharing as well as organization. There are many sites assisted by collaborative tagging. For example, Delicious (http://delicious.com) for web page bookmarking, YouTube (http://www.youtube.com) for video sharing, Flickr (http://www.flickr.com) for photo sharing, and Twitter (http://www.twitter.com) with hashtag. This collaborative organization approach is also called *folksonomy*.

Along with the developments of these tagging systems, many research problems have been studied to improve folksonomy. For example, personalized recommendation is discussed in [4], an improved tag based content retrieval is presented in [13], and one of our previous work in [2] exploits novel features fusion methods for tagged resources.

As tagging systems gain in popularity, experts and spammers flow into the tagging sites at the same time. User interaction becomes difficult, and finding appropriate information is urgent. In this paper, we study the problem of modeling users' expertise in collaborative tagging communities. That is to discover the user's expertise with respect to a given topic (or a tag), of which can be made use to clearly distinguish the experts and spammers.

J.X. Yu, M.H. Kim, and R. Unland (Eds.): DASFAA 2011, Part I, LNCS 6587, pp. 53–67, 2011.

The expert search problem, which has already caught our eyes in enterprise corpora [1] and recently social networks [6], is also very meaningful to current tagging communities. Nevertheless, without the ability to combat spammers, the system will suffer the misleading influences of them. For example, an expertise model is helpful in the case that a user wants to find top experts on a specific topic and then follow their activities. With a suitable expertise model, we can also directly recommend a user the experts of certain topics which may be interesting to him/her. However, the expert list recommended by the system for the user will be filled with some useless spammers if we cannot eliminate them accurately. What is more, a suitable expertise model can also improve tag qualities. On one side, tagging systems can directly push the resources with imprecise tags to the expert users and let them tag. On the other side, we can avoid the misleading influences of spammers on tag quality calculation, by simultaneously distinguishing spammers from users. We believe that, with the help of a reliable expertise model to search experts and combat spammers, a lot of applications as those we mention above can be supported in such communities to improve user satisfaction significantly.

The most crucial difference between folksonomy and traditional classic enterprise corpus or social network is that the former has a comprehensive set of features, e.g., users, tags and objects. There also exist rich and meaningful interactions among them. For example, it is not surprising to see that the expertise of a certain user on tag t is determined not only by his/her tagging behavior on t, but also on tags similar to t, or the user's social network involved. Such multi-type features can not only help us get more reasonable expert ranking, but also better protect our system from malevolent attacks of spammers.

To the best of our knowledge, there do not exist many solutions to deal with the expertise modeling problem in folksonomies. Current methods usually utilize a part of information about users, objects and tags [15,5,11]. However, no work have explored all features related to users, objects and tags in folksonomies. We believe this could result in better representation models, and further make folksonomies more accurate in expert finding and more resistant to spammers.

To fully utilize existing multi-type features, we propose a novel expertise model for collaborative tagging communities. We extract several expertise evidences/features hidden in folksonomies. An expertise model is used to combine those expertise evidences and generate users' expertise over topics/tags. Experiments demonstrate the advantage of this integrated model.

We outline the contributions of this paper as below:

1. We extract a comprehensive set of expertise evidences/features hidden in such tagging communities. Considering the fact that collaborative tagging communities basically consist of three parts: tags, users, and objects, those evidences can be classified into three categories similarly: 1) tag-related evidences, 2) user-related evidences and 3) object-related evidences.
2. The expertise model based on Continuous Conditional Random Fields (CRF) [12] is introduced to automatically integrate those expertise evidences and generate user's expertise over topics/tags as a result. This model is inspired by the successful applications of CRF technique [8] to model interactions between different items in an undirected graph. As for our expertise modeling problem, the proposed model can keep the balance among different kind of features and also make full use of them.

3. Our experiments conducted on the expertise ranking problem in a real tagging dataset show clearly that our CRF-based expertise model is obtaining much higher precision on searching experts and more resistant to spamming activities than any other baselines.

The rest of this paper proceeds as follows. In Section 2 we formulate the problem and explore the evidences inspiring our expertise model. Section 3 presents the CRF-based expertise model. We discuss the modeling and learning of this model in details. Quantitative experiments are shown in Section 4. After Section 5 reviews the related work, we conclude this paper in Section 6.

2 Expertise Evidence in Folksonomy

In this section, we carry out a thorough analysis of the folksonomies and investigate all evidences which are helpful to our expertise model. We begin with the problem formulation of expertise modeling task and correlated structure of features in folksonomy. Then a comprehensive set of expertise evidences are studied.

2.1 Feature Correlation

Let $X = \{O, U, T, R\}$ denote all the observations we have in a folksonomy including all the objects $O = \{o_1, o_2, \ldots, o_M\}$, users $U = \{u_1, u_2, \ldots, u_N\}$, tags $T = \{t_1, t_2, \ldots, t_L\}$ and their relationships $R = \{(o_i, u_j, t_k)\}$. Also, define a matrix $\mathbf{E} = \{e_{ij}\}$, where e_{ij} denotes the expertise score of user u_i on tag t_j. With higher value of e_{ij}, user u_i is more likely to be an expert on tag t_j.

The expertise modeling problem is to determine the expertise score matrix $\mathbf{E}$ given all the observations X in a folksonomy. We need to infer reasonable expertise scores based on all observations. Hence, our task can be modeled to find $\mathbf{E}$ that maximizes the appearance probability of $\mathbf{E}$ given current observations X, i.e.,

$$\mathbf{E} = \arg\max_{\mathbf{E}} P_\theta(\mathbf{E}|X) = \arg\max_{\mathbf{E}} P_\theta(\mathbf{E}|O, U, T, R), \qquad (1)$$

where θ denotes the model parameter. However, the computing $P_\theta(\mathbf{E}|O, U, T, R)$ is not a trivial task, as discussed in the above section.

To model the interactions among users, objects and tags and show the influences of the interactions on expertise, we introduce a graph structure, in Figure 1. Shown in this graph, three core elements: users, objects and tags are represented by grey nodes and expertise scores by blank nodes in the upper part. Here we present relations among the same type of nodes as well as the cross-type relations. For example, the edge between user u_i and u_j stands for a subscribe/as-a-friend relation in the tagging system. And for the cross-type relation, the use of tag t_j by user u_i is represented by an edge between the corresponding nodes of t_j and u_i. Edges in this graph are weighted. In this scenario, the more frequently user u_i uses tag t_j, the larger the edge weight between node u_i and node t_j is.

Besides those observations $X = \{O, U, T, R\}$, our expertise scores, which are denoted in the top layer $\mathbf{E}$, are the output of our expertise model. Each node e_{ij} in this

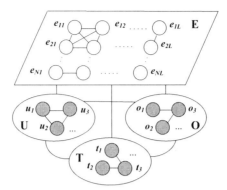

Fig. 1. Graph Structure of Folksonomy

layer represents the expertise of user u_i on tag t_j. In this graph the expertise nodes are interlinked. This aligns with our intuition that there are hidden interactions among the expertise scores of different users on different tags.

To present these mutual influences, each node in the expertise layer is connected to the nodes corresponding to observations in folksonomy, which implies that a user's expertise in folksonomy is influenced by all items, i.e., users, objects and tags. The intuitive meanings of these edges will be clear after explaining the *Expertise Evidences* we find.

2.2   Expertise Evidences

Only the structure of folksonomy can not give us enough information about the user expertise, here we illustrate more facets and evidences in folksonomy for more inspirations to estimate expertise reasonably.

The *Expertise Evidences* refer to the information which can help discover experts and tell spammers in a folksonomy system. Some state-of-the-art work have already proposed some indications, e.g., user authority [15], tag reliability [5] and post date [11]. However, they usually fail to conduct a comprehensive study of feature interactions, which are actually the most crucial difference between folksonomy and classic web environment or the enterprise corpus. In contrast, we first investigate an overall set of *Expertise Evidences*. Considering the fact that users, objects, and tags are three basic foundations of folksonomy, we observe the system from these three perspectives and explore the corresponding evidences.

We categorize our discovered evidences into three types: 1) tag-related evidences, 2) object-related evidences and 3) user-related evidences.

Tag-related Evidences.   Two tag-related expertise evidences are applied in our model.

The first evidence suggests that, if a user often agrees with others on the choice of tag to label on an object, there is a great chance that he/she is an expert. This is consistent with our intuition [15,5], however, this feature alone is very vulnerable to attacks conducted by spammers as shown in our following experiments. *(TE-1)*

The second one is that an expert on tag t should have high expertise on similar tag t', too. Imagine an expert on topic *Web2.0*, even though he/she may not tag many tags on similar topics, e.g., *Delicious*, we can trust these tags because that his/her knowledge has been reflected on a very similar topic. This evidence may also help us handle the different personal customs on tagging. For example, tell an expert on *Puppy* in spite of his preference to use *Dog* when tagging. *(TE-2)*

User-related Evidences. The user-related evidence comes from the subscribe relation in folksonomy. Two metrics are used to depict user's characteristics: the expertise on certain tag and the ability to find an expert on certain tag.

In the directed subscribe relation, the difference and interaction between the two metrics are important. The expert user tends to be subscribed by users who are good at discovering experts, and the user with high ability of discovering experts is more likely to subscribe many experts. This is reasonable for us, given that in practice experts are often subscribed by others because they often provide useful knowledge. Also a user subscribing lots of experts is good at finding experts because his behavior is a good sign of the ability to distinguish the usefulness of provided knowledge. These two sub-rules together show the mechanism of the interactive influence between user's expertise and user's ability of finding experts. *(UE-1)*

Object-related Evidences. It is always not easy to analyze the content of objects, and hence we will not use object's content information in this work. Instead, we should notice that the user's expertise on tag t should be increased if he/she uses tag t to annotate the object o which later becomes a popular object.

This evidence combines the popularity of an object and the post time information. It satisfies the assumption in [11] that an expert should have the ability of discovering popular objects instead of just following others. However, compared to [11], we model this feature in a different way to integrate it into our model seamlessly. *(OE-1)*

For any folksonomy system, the three elements, i.e., users, objects and tags, are pillars of the pyramid, and all the facts and evidences we are considering have fully covered the three main foundations of the folksonomy. The evidences and rules directly resulting from these facts are laconic, easy to obtain, reasonable intuitively at the same time. It cannot be denied that other features can also be drawn from the folksonomy graph structure we presented above, too. For example, one can write *ad hoc* evidences specific to application scenario or user preference. But we need to point out that the four basic evidences we pick now can fully cover all three facets of a folksonomy system and as shown in our latter experiment, our model integrating barely these four evidences can still beat other state-of-the-art techniques and perform satisfactorily. What is more important, any other evidence can be easily integrated into our model. As a result, in this paper, we can mainly focus on these evidences and show how they can be integrated into a unified model.

3   CRF Based Expertise Model in Folksonomy

In this section, we present our CRF based expertise model in detail. The CRF based expertise model is proposed to fuse multi-type features in folksonomy and to finally generate expertise scores of users on specific tags. We formulate the relations in this unified model and also discuss the parameter learning methods.

3.1 Model Formulation

In order to well fuse the expertise evidences discussed in Section 2, a Continuous Conditional Random Fields [12] based expertise model is proposed here to model the user expertise in folksonomy. Compared to other fusion models or heuristic methods, this model is powerful in automatic feature weighting and interaction combination. To cope with requirements in this problem setting, we also discuss the improvement over basic CRF model.

Recall the problem defined above, we aim at estimating the probability of $\mathbf{E}$ given the observation $X = \{O, U, T, R\}$, and hence we can maximize this probability to obtain an optimal $\mathbf{E} = \arg\max_{\mathbf{E}} \{P_\theta(\mathbf{E}|X)\}$. Continuous CRF provides a way to estimate such probability:

$$P_\theta(\mathbf{E}|X) = \frac{1}{Z(X)} \exp\left\{\sum_{i=1}^{k} \left(\boldsymbol{\lambda_o} \cdot \boldsymbol{F_o}(c_i) + \boldsymbol{\lambda_u} \cdot \boldsymbol{F_u}(c_i) + \boldsymbol{\lambda_t} \cdot \boldsymbol{F_t}(c_i)\right)\right\}, \quad (2)$$

where $\{c_i | i \in [1, k]\}$ is a set of k cliques in our graph. For example, in Figure 1, $c = \{u_1, t_1, t_2, e_{11}, e_{12}\}$ can be taken as a clique if the corresponding nodes of these elements in the set are fully connected. In the above equation, $\boldsymbol{F_o}$ denotes a function vector consisting of *feature functions* designed for the object-related evidences and $\boldsymbol{\lambda_o}$ represents a weight vector of those features or evidences. $\boldsymbol{F_u}, \boldsymbol{\lambda_u}, \boldsymbol{F_t}$ and $\boldsymbol{\lambda_t}$ all have similar definitions. The variable θ stands for the parameter set of this model: $\theta = \{\boldsymbol{\lambda_o}, \boldsymbol{\lambda_u}, \boldsymbol{\lambda_t}\}$ which satisfies $\sum_i \lambda_o^i + \sum_j \lambda_u^j + \sum_k \lambda_t^k = 1$ and $Z(X)$ is a normalization factor defined as

$$Z(X) = \int_{\mathbf{E}} \exp\left\{\sum_{i=1}^{k} \left(\boldsymbol{\lambda_o} \cdot \boldsymbol{F_o}(c_i) + \boldsymbol{\lambda_u} \cdot \boldsymbol{F_u}(c_i) + \boldsymbol{\lambda_t} \cdot \boldsymbol{F_t}(c_i)\right)\right\}. \quad (3)$$

We need to define different *feature functions* for the three kinds of evidences to integrate them into this model. Here the *feature function* refers to a function defined on clique c to measure the fitness of nodes in c to appear together. Specifically, four feature functions are designed with respect to the four expertise evidences stated above. Before we explain the detailed *feature functions*, we first define variables extracted to describe information in folksonomy.

- **Tag similarity matrix** $\mathbf{S}_{tag}$: each entry $\mathbf{S}_{tag}(t_i, t_j)$ equals the similarity between tag t_i and t_j, which is calculated by tag co-occurrence in our implementation.
- **User subscription matrix** $\mathbf{Sub}$: $\mathbf{Sub}(u_i, u_j) = 1$ iff. user u_j subscribes user u_i.
- **User temporal similarity matrix** $\mathbf{S}_T$: $\mathbf{S}_T(u_i, u_j | t_k)$ is the similarity between u_i and u_j tagging behaviors computed based on the average number of users who follow u_i and u_j on objects tagged by t_k.
- **Expertise matrix E:** each entry e_{ij} represents the expertise score of user u_i on tag t_j.
- **Expert finding ability matrix E′:** each entry e'_{ij} denotes the ability of user u_i to find experts on tag t_j.

We then write feature functions according to the suggested evidences above. Note that, these feature functions only have non-zero values to certain types of cliques and are automatically set to zero for cliques of other kinds.

- **TE-1:** We define f_t^1 on clique c like $\{e_{ij}, u_i, t_j\}$ as

$$f_t^1(e_{ij}, u_i, t_j) = -\left(e_{ij} - \mathcal{N}\left(\sum_{o_{t_j}} CoIn(u_i, t_j)\right)\right)^2, \tag{4}$$

where o_{t_j} enumerates the objects tagged by u_i with tag t_j, and $CoIn(u_i, t_j)$ is the number of users who agree with user u_i to apply tag t_j to object o_{t_j} for tag t_j. Function $\mathcal{N}(.)$ is introduced to normalize the input variable to $[0, 1]$. This feature represents the evidence that if one user agrees with more other users on a certain tag, his/her expertise on this tag should be higher.

- **TE-2:** We define f_t^2 on clique c like $\{e_{ij}, e_{ik}, t_j, t_k\}$ as

$$f_t^2(e_{ij}, e_{ik}, t_j, t_k) = -\frac{1}{2(|T|-1)}\mathbf{S}_{tag}(t_j, t_k) \times (e_{ij} - e_{ik})^2, \tag{5}$$

where $|T|$ is the number of all tags in the system. By this feature function, the user u_i's expertise scores on similar tags t_j and t_k would be close.

- **UE-1:** We define f_u^1 on clique c like $\{e_{ij}, e'_{kj}, u_i, u_k\}$ as

$$f_u^1(e_{ij}, e'_{kj}, u_i, u_k) = -\mathbf{Sub}(u_i, u_k) \times (e_{ij} - e'_{kj})^2, \tag{6}$$

where $e'_{kj} = \sum_{i=1}^{|U|} \left(\mathbf{Sub}(u_i, u_k) \times e_{ij}\right)$ and $|U|$ is the number of all users in folksonomy. This user-related feature function encodes the two-side interactions between expertise score and finding expert ability into a unified framework.

- **OE-1:** We define f_o^1 on clique c like $\{e_{ij}, e_{kj}, u_i, u_k, t_j\}$ as

$$f_o^1(e_{ij}, e_{kj}, u_i, u_k, t_j) = -\frac{1}{2}\mathbf{S}_T(u_i, u_k|t_j) \times (e_{ij} - e_{kj})^2. \tag{7}$$

This means it is better for two users u_i and u_k to own similar high expertise on tag t_j if they both label a popular object with tag t_j and they discover the object earlier than most other users.

3.2   Parameter Learning

The learning process of our expertise model is to obtain parameters $\theta = \{\lambda_o, \lambda_u, \lambda_t\}$, given a training dataset $D = (X, \mathbf{E})$, X includes objects O, users U, tags T and their relations R. In matrix $\mathbf{E} = \{e_{ij}\}$, each entry e_{ij} represents the expertise score of user u_i on tag t_j. We normalize the expertise scores to $[0, 1]$.

One traditional technique for parameter learning is to train a model which can maximize log-likelihood of training dataset D's appearance. There exist lots of discussions about how to learn the optimal parameters in CRF framework, e.g., *Gibbs Sampling* from [14]. However, it may not optimize the desired objective function, i.e., the average precision of expertise ranking problem in our case. In contrast, direct optimization aiming at the evaluation metric is better in some scenarios [10]. In this paper, we use the methodology applied in [10]. Specifically speaking, we enumerate the combination of parameter θ and select parameter which makes the model obtain the maximal average precision of expert ranking task.

In our problem setting, we are only interested in ranking users by their expertise, so the inference process can be simplified. The $Z(X)$ will influence only the absolute expertise scores, not the ranking positions. Under this occasion, the ranking score matrix of users on certain tags is denoted by $\mathbf{E}'$.

$$\mathbf{E}' \propto \arg\max_{\mathbf{E}} \exp \left\{ \sum_{i=1}^{k} \left(\boldsymbol{\lambda_o} \cdot \boldsymbol{F_o}(c_i) + \boldsymbol{\lambda_u} \cdot \boldsymbol{F_u}(c_i) + \boldsymbol{\lambda_t} \cdot \boldsymbol{F_t}(c_i) \right) \right\}$$

$$\propto \arg\max_{\mathbf{E}} \sum_{i=1}^{k} \left(\boldsymbol{\lambda_o} \cdot \boldsymbol{F_o}(c_i) + \boldsymbol{\lambda_u} \cdot \boldsymbol{F_u}(c_i) + \boldsymbol{\lambda_t} \cdot \boldsymbol{F_t}(c_i) \right)$$

After substituting the detailed feature functions, i.e., tag related functions: $f_t^1(.)$, $f_t^2(.)$, user related function: $f_u^1(.)$, as well as object related function: $f_o^1(.)$, into this equation, we can generate the solution of $\mathbf{E}'$ by using standard Lagrange multiplier methods.

4   Empirical Study

In this section, we evaluate our expertise model by the expertise ranking problem in folksonomies. Specifically, we conduct the evaluations on *expert ranking* and *spammer ranking* to answer the following two questions respectively:

Q1: How exactly is the performance of our expertise model on searching experts for specific tags?

Q2: Is the proposed expertise model robust enough to resist the spammers' attacks?

4.1   Experimental Setup

Experiments are conducted on a real tagging dataset, collected from Delicious (http://delicious.com/) website. These tags range from Jan. to Jun. 2010. The dataset contains $10, 800, 690$ web page urls, $197, 783$ users, $1, 928, 677$ tags. We also fetch subscription relations between users.

The distribution of tag frequency is shown in Table 1. It is easy to tell that, less tags show in the dataset with higher frequency. In our experiments, we mainly focus on the tags ranging from level-3 to level-5. In truth, more people will be interested in experts on tags such as "asp.net" in level-4 than those like "vibes" in level-1. Since users' interests mainly focus on popular tags, these tags deserve more attention. And the improvement in experts search on these tags can dramatically enhance user satisfaction.

Training Set and Testing Set. To construct our testing query set, we randomly select 10 tags for each frequency level from level-3 to level-5. This is the base to run the model and to evaluate its performance on *expert ranking* and *spammer ranking*.

For *expert ranking* part, parameter learning is crucial to our expertise model as illustrated before. To learn the parameter set $\theta = \{\lambda_o, \lambda_u, \lambda_t\}$, a small training set is manually annotated by two annotators. Annotators are asked to assign the binary expertise score (expert or not) to randomly selected users to some randomly selected tags. With the annotation result as ground truth, we adjust the parameter to achieve higher expert recommendation precision.

Table 1. Statistics of tag frequency

Level-ID	Frequency Interval	#Tags
0	[1,9]	1,694,768
1	[10, 99]	199,563
2	[100,999]	28,557
3	[1000,9999]	4,921
4	[10000,99999]	780
5	[100000,999999]	88

Turning to *spammer ranking* part, in order to measure how resistant the expertise model is to the spammers' attacks, we follow the method used in [11]. Three types of spammers are randomly inserted, i.e., *Flooders*, *Promoters* and *Trojans*. *Flooders* refer to users who tag a extremely large number of tags, while *Promoters* always tag their own web pages and pay little attention to objects provided by other users. Much more crafty, *Trojans* tag a lot for their own pages but at the same time conceal their malicious intentions by acting like regular users. More details can be found in [11]. Specifically, each query tag has 20 spammers of each type.

Evaluation Metrics. We apply the Precision@N as our main evaluation metric, which represents the percentage of answers that are "correct" in all N candidate answers retrieved, taking the manually annotated experts list and inserted spammers list as ground truth.

In particular, in *expert ranking* part, the retrieved user is "correct" if he/she is labeled as expert by annotators. As for *spammer ranking* task, the retrieved user is "correct" if he/she is a simulated spammer we inserted. Hence, the higher Precision@N for expert ranking, the more reliable and suitable the model. In contrast, a higher Precision@N for spammer ranking task is an indication that the model is more vulnerable to spammers' attack activities.

In addition to the Precision@N, we use another metric, i.e., Average Ranking Position, to measure the difference in model's ability to demote spammers in the expert list by giving spammers lower scores than true experts. The higher the metric, the more resistant the expertise model to spammers' attacks.

Baseline Methods. We compare our method with three state-of-the-art approaches for both expert ranking and spammer ranking.

- Baseline 1: *HITS* [15]. It applies HITS algorithm to determine the user expertise by assuming that there exit reciprocal reinforcements between user expertise and tag quality.
- Baseline 2: *CoIn* [5]. It uses the coincidence between users as the expertise of users.
- Baseline 3: *SPEAR* [11]. It assumes the users tagging objects of more popularity or tagging objects earlier deserve higher expertise scores.

4.2   Quantitative Result

We report our performance study on two tasks, i.e., *expert ranking* and *spammer ranking*.

Expert Ranking. We present our experimental results to answer how exactly the performance of our expertise model on searching experts is. Table 2 shows the results of different approaches, including our method named Multi-type Feature Fusion (MFF) and three baselines, i.e., SPEAR, CoIn and HITS. We have tried different popular levels of tags, however they obtain similar results so we do not report them separately.

As seen in this table, in respect of Precision@1, our MMF is tied with other baselines. Also, our MFF approach obtains the best performance in top-5 precision, however, it meets a slightly decrease in performance for larger N. From the overall perspective HITS obtains the best performance but our approach is the second best.

Table 2. Average Precision for Expert Ranking Task

Average Precision	SPEAR	CoIn	HITS	MFF
$P@1$	**1**	**1**	**1**	**1**
$P@5$	0.80	0.96	0.96	**1**
$P@10$	0.68	0.90	**0.92**	0.88
$P@15$	0.64	0.85	**0.91**	0.85
$P@20$	0.61	0.87	**0.91**	0.88

Several facts can interpret this result. First of all, due to the limited time, when the annotators determine whether the retrieved user is an expert on the query tag, they usually take the tag frequency of users as the most important factor, but true users in a collaborative tagging system will consider more in fact. Hence, the annotated results may be more inclined to HITS method. Secondly, as reported by the annotators, they can not easily determine whether some users with tremendous tags are spammers or not. Generally, they just take those users as experts instead of spammers.

Despite having slightly worse overall performance than HITS when measured by annotated results and assuming all manually annotated experts are accurate, our model can improve user satisfaction empirically and also, the top 5 retrieved users of our method for all query tags are all annotated as experts. We believe in fact, our algorithm can averagely achieve satisfactory results, and can be better or at least comparable to most state-of-the-art approaches.

Spammer Ranking. This experiment is conducted to test how our expertise model's performance is when confronting malicious spammers.

When measuring with the metric of Precision@N, Figure 2, 3 and 4 present the performance of *spammer ranking* for different models here. The overall performance of whole query set is shown in Figure 2, while Figure 3 and 4 give more detailed information about the model's performance concerning the discrepancies of frequency levels in tags. The result for the level-4 query tags is quite similar to level-5, so we do not show it.

In addition, we show the evaluation results in Figure 5 for different expertise models with respect to various types of spammers. The y-axis represents the average ranking position of specified type spammers in top $10,000$ retrieved users. In Figure 2 and Figure 3, for the *spammer ranking* task, the lower value in Precision@N suggests there are less spammers in the first N experts, which means the expertise model is more

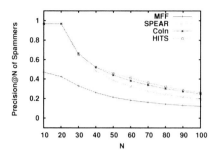

Fig. 2. Average Precision of Spammer Ranking, from level-3 to level-5

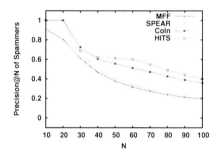

Fig. 3. Average Precision of Spammer Ranking, level-3

resistant to spammers' attacks. In contrast, in Figure 5, the lower value in y-axis serves as an indication that inserted spammers are decided to be experts by the model with high expertise scores, showing the system's vulnerability to spammers' attacks.

In Figure 2, the average Precision@20 almost equals 1 for three baselines, but only approximately 0.4 for our approach. With the increase of the recommended expert number, the advantage of our model shrinks. However, even at Precision@100, our approach is still better than the other three. As discussed later, it is the seamless fuse of different types of expertise information that makes our method most resistant to spammers on average.

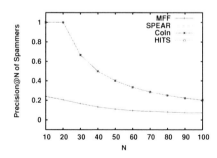

Fig. 4. Average Precision of Spammer Ranking, level-5

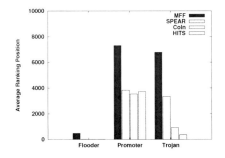

Fig. 5. Average Spammer Ranking Positions

According to Figure 3 and Figure 4, we can find that all the expertise models can obtain better performance when dealing with tags of higher popularity. One possible reason might be that for the tags with less popularity, the spammers are easier to "beat" regular users and become the "experts" on certain topics. Figure 5 suggests that no matter what kind the spammers belong to, our expertise model MMF shows great improvement in the ability to demote all the spammers from the top of the expert list. To be more specific, all three baselines show nearly no resistance to *Flooders* while MMF goes a big step further than them. As for *Promoters* and *Trojans*, these spammers will be demoted twice or more in the expert list by MMF than by any other method in the three.

Result Analysis and Discussion. From the above expert ranking and spammer ranking experiments, we can see that our MFF approach outperforms the baseline models. And we should point out that the good performance of MMF approach results from its ability to integrate multi-type information extracted from the folksonomy system.

When distinguishing experts, we have more clues and features to inspire experts search. So we can get accurate experts list especially when most top experts are needed. While telling spammers, CoIn and HITS both cannot well separate *Trojans* from regular users. This failure is a result of their too much dependence on the tag frequency information. Although when added to the temporal information, SPEAR has great improvements in telling *Trojans* and basically performs better than CoIn and HITS, it still cannot identify the *Flooders* like MMF because *Flooders* tag too many resources to be distinguished by temporal information easily. However, our MMF model integrates information of a wider range, e.g., subscription network among users, yielding more satisfactory results both on the average spammers ranking and spammers demotion of special kinds.

Considering the performance of our expertise model on *expert ranking* and *spammer ranking*, the model is believed to suitable for practical applications in real world collaborative tagging systems. First, although the model is not best when recommending large scale of experts, but in our daily life, only the top experts recommended are interesting to users. Too much patience are needed for a user to browse 10 or more recommended experts everyday. Second, the outstanding performance on *spammer ranking* means our recommended expert list will not be filled with spammers, especially *Flooders*. Hence, compared to other methods, such as HITS, our expertise model is good at demoting crafty spammers like *Flooders*, who will waste user's energy to follow and reduce user satisfaction significantly. Given all these reasons, by accurate top experts ranking without misleading spammers in the top positions, our expertise model will provide satisfactory services in real world folksonomies.

5   Related Work

Our work in this paper is broadly related to several areas. We review some in this section.

Social Media Management: With the recent startling increase of social media application, the uncontrolled vocabulary annotation is becoming popular. Researchers have discussed various directions of tagging systems. A structured tag recommendation approached was discussed in [4], and [13] presented an improved retrieval algorithm based on sequential tags. One of our previous work [2] proposed a feature fusion approach for social media retrieval and recommendation. Work in this paper focuses on a unified model for both spammer detection and expert recommendation.

Spammer Detection: Another line of related work is spammer detection which aims to detect the spammers in collaborative tagging systems or other similar systems. Here we do not pay much attention to explicitly illustrate the details in spammer detection work, instead, we focus more on the information utilized in these work. In [9], the classification methods were utilized to differentiate spammers from regular users. The features used in those machine learning method mainly focused on the content of resources. Another example was [7], in which co-occurrence information of tags and resources was used to detect spammers. In detail, the manually annotated spammer scores were propagated through a user graph, the edges of which were generated from the co-tagging, co-resource and co-tag-resource relations among different users. However, the system suffered from the problem of human labeled training data, which limited the use of the system in large scale data. Also, in [15], Xu et al. applied HITS algorithm on the bipartite graph of users and tags to implement the mutually reinforcement between tag qualities and user authorities. In addition, when dealing with tag recommendation problem in [5], the authors measured user's reliability by the frequency of the user's tags agreeing with other users' postings.

Expert Recommendation: With the widespread use of social communities in our daily life, online user modeling and expert recommendation or expert search show its importance. Researchers have made great efforts towards this direction. For example, [16] explored expertise networks in online systems, user interest and expertise modeling in social search engine was discussed in [6]. Usually, content based and structure based methods are used in user profiling. Expert search task in enterprise corpora is always of interest for many researchers. There exist two seminar models applied, i.e., document based model [3] and profile based model [1]. In one of our previous work, we combine the profile and structure based method together for community expert recommendation [17].

To tell experts in folksonomies, Noll et al. focused on structure property and proposed a HITS based algorithm on the bipartite graph among users and objects graph to extract users' expertise information in [11].

Different from all existing methods for expert ranking in tagging systems, we introduce a new expertise model to integrate a comprehensive set of expertise evidences among users, tags, and objects. With this fusion framework, our method can obtain better performance than those state-of-the-art approaches, both in combating spammer and expert ranking.

6   Conclusion

In this paper, we have addressed the problem of modeling users' expertise in folksonomies by fusing multi-type features. Compared to state-of-the-art methods, we highlighted coding the multiple interactive evidences into a unified framework by employing Continuous Conditional Random Fields techniques.

We examined the performance of our method in large scale real-world tagging data both about expert ranking and spammer ranking. According to our experiments, we find our proposed model obtains high precision in expert ranking problem in folksonomies and is also far more resistant to the spamming attacks than those state-of-the-art approaches.

We plan to extend our expertise model in two aspects. First, we will further investigate more evidences from real world folksonomies while considering the characteristics of different social sites. Second, we will employ our expertise model to facilitate other applications in folksonomies, e.g., tag-based retrieval or tag ranking.

Acknowledgements. This research was supported by the National Natural Science foundation of China under Grant No. 61073019, 60933004 and 60811120098.

References

1. Balog, K., Azzopardi, L., de Rijke, M.: A language modeling framework for expert finding. Information Processing and Management 45(1), 1–19 (2009)
2. Cui, B., Tung, A., Zhang, C., Zhao, Z.: Multiple feature fusion for social media applications. In: Proc. of ACM SIGMOD, pp. 435–446 (2010)
3. Deng, H., King, I., Lyu, M.R.: Enhancing expertise retrieval using community-aware strategies. In: Proc. of ACM CIKM, pp. 1733–1736 (2009)
4. Guan, Z., Bu, J., Mei, Q., Chen, C., Wang, C.: Personalized tag recommendation using graph-based ranking on multi-type interrelated objects. In: Proc. of ACM SIGIR, pp. 540–547 (2009)
5. Heymann, P., Koutrika, G., Garcia-Molina, H.: Fighting spam on social web sites: A survey of approaches and future challenges. IEEE Internet Computing 11(6), 36–45 (2007)
6. Horowitz, D., Kamvar, S.: The anatomy of a large-scale social search engine. In: Proc. of WWW, pp. 431–440 (2010)
7. Krestel, R., Chen, L.: Using co-occurence of tags and resources to identify spammers. In: Proc. of ECML PKDD Discovery Challenge Workshop, pp. 38–46 (2008)
8. Lafferty, J., McCallum, A., Pereira, F.: Conditional random fields: probabilistic models for segmenting and labeling sequence data. In: Proc. of ICML, pp. 282–289 (2001)

9. Madkour, A., Hefni, T., Hefny, A., Refaat, K.S.: Using semantic features to detect spamming in social bookmarking systems. In: Proc. of ECML PKDD Discovery Challenge Workshop, pp. 55–62 (2008)
10. Metzler, D., Croft, W.B.: A markov random field model for term dependencies. In: Proc. of ACM SIGIR, pp. 472–479 (2005)
11. Noll, M.G., Yeung, A.: et al. Telling experts from spammers: expertise ranking in folksonomies. In: Proc. of ACM SIGIR, pp. 612–619 (2009)
12. Qin, T., Liu, T., Zhang, X., Wang, D., Li, H.: Global ranking using continuous conditional random fields. In: Proc. of NIPS, pp. 1281–1288 (2008)
13. Sarkas, N., Das, G., Koudas, N.: Improved Search for Socially Annotated Data. PVLDB 2(1), 778–789 (2009)
14. Xin, X., King, I., Deng, H., Lyu, M.R.: A social recommendation framework based on multiscale continuous conditional random fields. In: Proc. of ACM CIKM, pp. 1247–1256 (2009)
15. Xu, Z., Fu, Y., Mao, J., Su, D.: Towards the semantic web: collaborative tag suggestions. In: Proc. of WWW Collaborative Web Tagging Workshop (2006)
16. Zhang, J., Ackerman, M.S., Adamic, L.: Expertise networks in online communities: structure and algorithms. In: Proc. of WWW, pp. 221–230 (2007)
17. Zhou, Y.H., Cong, G., Cui, B., Jensen, C.S., Yao, J.J.: Routing Questions to the Right Users in Online Communities. In: Proc. of ICDE, pp. 700–711 (2009)

Identifying Topic Experts and Topic Communities in the Blogspace

Xiaoling Liu, Yitong Wang[*], Yujia Li, and Baile Shi

School of Computer Science, Fudan University, Shanghai 200433, China
{Xiaolingl,yitongw,bshi}@fudan.edu.cn,
Liyujia2008@gmail.com

Abstract. Blogs have become an important media of self-expression recently. Millions of people write blog posts, share their interests, give suggestions and form groups in blogspace. An important way to understand the development of blogspace is to identify topic experts as well as blog communities and to further find how they interact with each other. Topic experts are influential bloggers who usually publish "authoritative" opinions on a specific topic and influence their followers. Here we first discuss the challenge of efficient identifying topic experts and then propose a novel model to quantify topic experts. Based on the topic experts identified, we further propose a new approach to identify the related blog communities on that topic. Experiments are conducted and the results demonstrate that our approaches are very effective and efficient.

Keywords: topic expert; blog community; identify; blogspace.

1 Introduction

Blogs have become an important media within the last decade, and millions of people adopt it as a tool to express themselves. These people (called bloggers), might be not acquainted with each other at all but they share their opinions on hot political events, their pets, their lovely children etc. through blogs. Until August 2006, the size of blogspace was two orders of magnitude larger than three years ago [8]; most bloggers are young people in the age group of 13-29, and they generate 91% blog activities.

Bloggers are connected by hyperlinks, comments, and blogroll (list of friends). The whole blogspace can be modelled as a huge graph by considering bloggers as nodes and connections as edges. For example, bloggers talked about SARS could produce many connections (edges) and made a community (dense sub graph) about this topic. We make the following contributions:

(1) Present a novel approach to identify topic experts. Considering topic and many other factors, our approach can identify the true experts efficiently.
(2) Present a novel approach to identify blog communities on a specific topic effectively. Comparing with other approaches, our approach could produce more stable, more cohesive, and larger topic-related communities.

---

[*] Corresponding author.

J.X. Yu, M.H. Kim, and R. Unland (Eds.): DASFAA 2011, Part I, LNCS 6587, pp. 68–77, 2011.
© Springer-Verlag Berlin Heidelberg 2011

(3)   Thorough experiments are conducted and the results demonstrate that our approach can extract meaningful topic experts and communities effectively.

This paper is organized as follows. In section 2, we briefly review related work in recent years. We present topic expert identification algorithm and blog community identification algorithm in section 3 and section 4. Experiments are introduced in section 5. In section 6, we conclude the paper with some discussion and future work.

2   Related Work

Topic expert identification and blog community identification are two hot topics of research about blogs. Song [3] tried to find opinion leaders in the blogspace, and he regard importance and novelty as two considerable factors to identify opinion leaders. Agarwal proposed a method to identify influential bloggers in a blog community [2], some factors such as post citations, post length and comment count were considered. However, he did not consider the quality and main topic of posts or comments, which we think are rather important factors to make the blogger to be considered as a topic expert. Unlike Agrawal, we focus on identifying topic experts. We will integrate a series of factors, such as topic similarity, length of a post or a comment, and the number of posts a blogger published on the topic etc. to identify topic experts.

Lin's method [9] based on blogs' mutual awareness, and the interaction of blogs was considered as an important factor. Belle's approach [10] was first ranking blogs by their citations, and then discovered communities composed of important blogs. Gruhl [11] studied the information propagation through blogspace, they characterize and model the data at two levels, one is macroscopic characterization which formalizing the notion of long-running "chatter" topics sensitising restively of "spike" topics generated by real world events; the other is a microscopic characterization of propagation among individuals; then proposed an algorithm to induce the propagation network from a sequence of posts. Bulters presented a method for discovering blog communities that incorporates both topology and content analysis. [12].

In Kumar's algorithm [1], blogs were organized as a time graph, in which timestamps were used to analyze bursty events through Kleinberg's method [13]. His work is most similar to our method of identifying blog communities. In his opinion, a dense sub-graph of time graph is a signature of a blog community. However, his method has some weaknesses, such as low topic relativity, bad stability, fragmentary communities, and low cohesiveness of community members. The method we presented overcomes these weaknesses to identify high quality blog communities.

3   Experts on a Specific Topic

We regard a blogger as a topic expert if he/she: (1) publish many high quality posts on that topic; (2) give many high quality comments to others on that topic; (3) is commented positively by many other bloggers; (4) is acquainted by many other high quality bloggers. The length of a post, similarity between a post and a given topic, the number of posts published, the number of comments replied, relevance of a comment to a post, and commenter's quality etc. are considered to evaluate the blog quality.

3.1 Clusters of Keywords

Keywords in a cluster appear together in many posts and we look on these keywords as a hot topic. Gabriel [7] and Bansal [4] developed different algorithms to identify keyword clusters. Gabriel developed a probability model to identify keyword clusters, which represent hot events at a specific temporal interval. Bansal produced a keyword graph using all pairs of keywords in posts, and then find bi-connected components of the keyword graph. We extend Bansal's approach and take average frequency of a keyword similarity between a post (or a comment) and a cluster into account.

3.2 Approach of Identifying Experts

Formula (1) is a keyword vector and formula (2) is a keyword frequency vector. Note that $freq_i$ in $freq$ is the average appearing times of f_i in related posts (posts include at least one keyword in f). Keyword frequency vector is used to compute topic similarity between a post (comment) and the given topic. Suppose p is a post, and $freq_p$ is the keyword frequency of p; the topic similarity between p and f can be calculated using Cosine Similarity as formula (3) shows. Given a comment c and the keyword frequency of c, we can compute $sim(c)$ in the same way.

$$f = \{f_1, f_2, ..., f_n\} \tag{1}$$

$$freq = \{freq_1, freq_2, ..., freq_n\} \tag{2}$$

$$sim(p) = \frac{\sum_{k=1}^{n} freq_k \times freq_{pk}}{\sqrt{\sum_{k=1}^{n} freq_k^2} \times \sqrt{\sum_{k=1}^{n} freq_{pk}^2}} \tag{3}$$

We identify a topic expert in the blogspace by his/her quality score, which is composed of three parts: scores of post quality, scores of comment quality and scores from commenters. We will explain the three parts in later section respectively.

For a post p, we denote the quality score of p as $quality(p)$, which can be calculated by formula (4). In formula (4), $length(p)$ is the size of p; the number "1000" is a scale factor. Formula (4) ensures that a larger post similar to a specific topic has a larger quality score. For a comment c, $quality(c)$ can be calculated in the same way.

$$quality(p) = \frac{length(p)}{1000} \times sim(p) \tag{4}$$

Fig. 1 shows the relationship of comments and posts. Ellipses represent bloggers; squares and black points represent posts and comments respectively; the numbers attached to arcs are comment quality scores and the numbers attached to the squares are post quality scores. The dashed line (c31, p12) means that c31 has nothing to do with the given topic (sim(c31)= 0), but it also plays a part in the quality score calculation of b1, so we give it a smaller basic quality score 0.1(called "fame quality score").

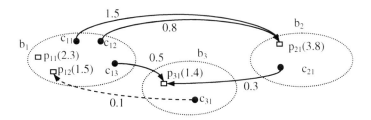

Fig. 1. Relationship of blogs

For a comment c with $sim(c)=0$, we calculate its fame quality score by formula (5). Note that this kind of comment can not take part in the computation of the second part quality score. Constant *fame_sim* represents the fame similarity and can be adjusted.

$$fame\_quality(c) = \frac{length(c)}{1000} \times fame\_sim \qquad (5)$$

There are three parts of a blog's quality score, and the first two parts are fixed: total quality score of posts and total quality score of comments. The fixed quality score is defined in formula (6), where n is the number of posts of b_i, and m is the number of comments of b_i. Note that if $sim(p_{ij})=0$ or $sim(c_{ij})=0$, the corresponding quality is equal to 0. w_p and w_c are the weights used to adjust the contribution of the two parts.

$$fqs(b_i) = w_p \sum_{j=1}^{n} quality(p_{ij}) + w_c \sum_{k=1}^{m} quality(c_{ik}) \qquad (6)$$

The third part of blog quality score is obtained from other blogs. We call it mutable quality score (*mqs*) because it is obtained by iteratively computation. Suppose the number of comments a blogger published is *num_c*, so the *mqs* of that blog can be divided into *num_c* parts which are given to the bloggers he/she commented on. *mqs* can be computed by formula (7), where o is the number of comments with zero similarity. The first part is necessary because some blogs' *fqs*=0, but their comments play a part in the *mqs* calculation of the blogs they commented on.

$$mqs(b_i) = \sum_{j=1}^{o} fame\_quality(c_{ij}) + fqs(b_i) \qquad (7)$$

For a blogger b_i, assume the total quality score of all his/her comments is *sum_cs*, the contribution ratio of a comment c_{ij} on other blog can be defined as:

$$ratio(c_{ij}) = \begin{cases} \dfrac{quality(c_{ij})}{sum\_cs} & sim(c_{ij}) > 0 \\[3mm] \dfrac{fame\_quality(c_{ij})}{sum\_cs} & sim(c_{ij}) = 0 \end{cases} \qquad (8)$$

Now calculate the new third part quality score of blog b_i using formula (9), where *commenter_count(b_i)* is the number of blogs that have given some comments to b_i, and *num(c_k_to_b_i)* is the number of comments that blog b_k gives to b_i. Formula (9) can be iteratively calculated and the procedure will terminate when the mutable quality scores of all blogs reach stable, then the third part quality score of a blog is obtained.

$$mqs(b_i) = \sum_{k=1}^{commenter\_count(b_i)} (mqs(b_k) \times \sum_{r=1}^{num(c_k\_to\_b_i)} ratio(c_{kr})) \tag{9}$$

4   Blog Communities on a Specific Topic

4.1   Some Symbols and Definitions about Blog Graph

Table 1 lists some useful symbols for topic communities.

Table 1. Symbols

Symbol	Definition and Description
$G = \{B, E\}$	G: blog graph; B: vertex set; E: edge set
$B = \{b_1, b_2, ..., b_n\}$	b_i ($1 <= i <= n$): blog; B: collective of b_i
$E = \{ele = (b_i, b_j) \wedge 1 <= i, j <= n$ $\wedge i! = j \wedge b_i, b_j \in B\}$	E: edge between b_i and b_j if there exist a comment on a given post.
$P = \{p_1, p_2, ..., p_n\}$	p_i is the set of posts posted by b_i
$C = \{b_1, ..., b_m \mid 3 <= m <= n\}$	C is a community composed by some blogs in B

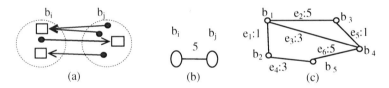

Fig. 2. The undirected graph of blogs

Fig. 2 (a) and (b) represent two blogs and their connections. In Fig. 2 (a), b_i and b_j represent two blogs; squares and black points represent posts and comments. We don't consider the blogger's comments on his/her self's posts, so there are no lines between nodes inside a blog. Fig. 2 (b) simplifies the connection between b_i and b_j, the undirected edge represents their "contacts"; the edge's weight is their contact times. A huge blog graph shown in Fig. 2 (c) can be formed using this structure.

Definition 1. We denote the vertex incident degree between a blog b_i and a community as $degreeVtoC(b_i)$. Suppose the adjacent vertex set of b_i is $Adjacent(b_i)$, then for b_i and a Community C, $degreeVtoC(b_i) = count(Adjacent(b_i) \cap C)$. We consider only one community at the same time, so for b_i, $degreeVtoC(b_i)$ is changing with the forming of community. Note that iff there exists at least one edge $e = (b_i, b_j)$, subject to $b_j C$ and $b_i C$, $degreeVtoC(b_i) > 0$, or else $degreeVtoC(b_i) = 0$. Fig. 3 shows the changing of $degreeVtoC(b_5)$; grey vertices are the members of community C. In Fig. 3(a), $degreeVtoC(b_5)$ is initialized to zero and $C = \{\}$; using the rule mentioned, we can get a number list $\{0, 1, 1, 2, 0\}$ of $degreeVtoC(b_5)$.

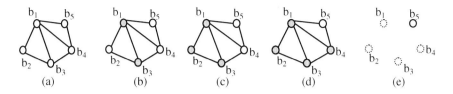

Fig. 3. The value evolvement of degreeVtoC

Definition 2. For an edge e, we denote the multiplicity of e as *edgeMultiplicity(e)*. Edge multiplicity represent the contact times of two blogs.

4.2 Identifying Topic Blog Communities

We give every vertex of the blog graph a property *QS* (Quality Score) which is obtained through the expert identification algorithm in section 3 and denote the *QS* value of blog b_i as $QS(b_i)$. Unlike Kumar's algorithm, we choose candidate vertices mainly by their *QS*, and then by the *degreeVtoC* and the *edgeMultiplicity*.

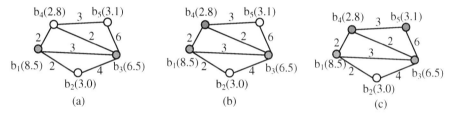

Fig. 4. Identifying expert communities on a specific topic

In Fig. 4 (a), we firstly choose (b_1, b_3) as a seed because $QS(b_1)+QS(b_2)=15$, the largest value among all edges. According to Kumar, b_2 and b_4 are two candidates that can join in $C=\{b_1, b_3\}$; because $QS(b_2)=3.0$ and $QS(b_4)=2.8$, at this point, b_2 is a better choice. However, because $degree(b_4)>degree(b_2)$, so b_4 will brings more candidates to C. To make a good choice, we give a threshold $t=|max(QS)-c|$, where c is a constant and $max(QS)$ is the max *QS* among all candidates. For a candidate b_i, if $QS(b_i)>=t$, then add it to the candidate list. If a blog in this list have the largest degree, then choose it as the best one; if there are many best candidates, choose the one with the largest *edgeMultiplicity* value. Suppose $c=0.3$, so based on the rules, b_4 is chosen and the current community is $\{b_1, b_3, b_4\}$. Now $max(QS)=3.1$, so the candidate list is $\{b_2, b_5\}$. The average *edgeMultiplicity* of edges between b_5 and C is $(3+6)/2=4.5$, between b_2 and C is $(2+4)/2=3$, so b_5 is chosen and the final community $C=\{b_1, b_3, b_4, b_5\}$.

5 Experiments

In this section, we will discuss our experiments based on the real dataset collected from "www.sbnation.com". We collected the data from Nov. 26, to Dec. 25, 2008, which includes 9408 bloggers, 6804 posts, and 197,933 comments.

5.1 Identifying Experts on a Specific Topic

We adopt Bansal's method [4] and make a change by calculating the average appearance times of keywords to identify topics. Two keyword clusters f_1 (about baseball) and f_2 (about football) are selected from the results. f_1 has 19 elements and each element appears 1.41 times in a related post on average; f_2 has 10 elements and each element appears 1.29 times in a related post on average.

Table 2. Top 5 experts of topic one

Rank	Name	TQS	PQS	CQS	VQS	In(Out)
1	King Billy Royal	145.44	105.38	0.55	39.52	817(43)
2	the pinstripes	111.99	87.13	0.12	24.74	748(16)
3	Zonis	94.37	74.22	0	20.15	1581(0)
4	TwistNHook	64.90	33.02	1.58	30.30	2263(11)
5	Dewey Finn	36.85	9.19	0.95	26.72	422(8)

Table 3. Top 5 experts of topic two

Rank	Name	TQS	PQS	CQS	VQS	In(Out)
1	King Billy Royal	166.59	103.61	0.46	62.52	817(34)
2	the pinstripes	117.81	82.54	0	35.27	743(14)
3	Dewey Finn	47.16	9.43	0.20	37.53	301(5)
4	marcello	32.83	6.13	0.10	26.61	454(2)
5	Zonis	25.71	14.01	0	11.71	814(0)

Table 2 and Table 3 show the top 5 experts using f_1 and f_2 respectively. *TQS* is the total quality score of three parts, *PQS* is the quality score from this blogger's posts, *CQS* is the quality score from this blogger's comments, *VQS* is vote quality score from others and *In (Out)* is the indegree (outdegree). In Table 2, the best expert is "King Billy Royal"(*KBR*) with the highest *TQS*, and the main contribution to his/her *TQS* is *PQS* (105.38), which means that *KBR* has published many topic-related posts. However, the low *CQS* shows that *KBR*'s comments have a weaker contribution. The indegree of *KBR* is 817, which is very lower than that of "Zonis"(*Z*) and "TwistNHook"(*T*), but his VQS is still larger than that of *Z* and *T*, that's because his commenter, "the pinstripes"(*TP*), has high quality. *T* can't be listed in Table 3 because of low *TQS* (4.49). Obviously, he/she has little interests in topic two.

Table 4. The important posts' information of expert blogs

Rank	Post1				Post2				Post3			
	pqs	len	cn	cbqs	pqs	len	Cn	cbqs	pqs	len	cn	cbqs
1	4.68	4909	55	5.40	4.56	4938	12	10.74	4.55	4886	30	94.75
2	4.35	4716	13	5.28	4.35	4716	13	2.55	4.30	4657	32	5.49
3	8.22	9116	21	4.71	8.21	9106	12	0.72	3.27	3608	37	0.72
4	4.09	11087	210	1.57	3.89	8248	338	1.77	3.27	9809	315	1.89
5	5.10	6814	156	114.48	2.72	4865	162	115.02	0.76	1967	112	108.49

In table 4, we analyze the top 3 best posts of every expert in table 2 to illustrate why they are experts. *pqs* is the post quality score, *len* is the post length, *cn* is the number of comments on that post, and *cbqs* is the total quality score of the commenters. The first row shows *KBR*'s three best posts. We can see that the *cbqs* of *Post3* is much larger than that of *Post1* and *Post2*, that's because *TP* gives him/her a high quality comment. Because of *KBR*'s small outdegree and large *TQS*, his contributions to others are huge. For "Dewey Finn"(*DF*), all his/her three posts have high *cbqs* because of *KBR*' comments, so the blog's *VQS* depends on not only the number of the comments but also the quality and outdegree of the commenters.

Now we consider the results generated from different *fame_sim*. In the above experiments, *fame_sim*=0.1; now we will analyze the change using different *fame_sim*. Table 5 shows the results when *fame_sim*=0.1, 0.2, and 0.02. *VQS_0.1*, *VQS_0.2*, and *VQS_0.02* represent the *VQS* adopting different *fame_sim* values; *CR_0.2* and *CR_0.02* are the change ratios relative to *VQS_0.1*. For every expert in Table 2, Fig. 5 (a) shows the ratio of the number of high quality comments ($sim(c_{ij})>0$) to low quality comments ($sim(c_{ij})=0$), and Fig. 5 (b) shows the values of *CR_0.2* and *CR_0.02*.

Table 5. Vote quality score of different weights

Rank	Blog	VQS_0.1	VQS_0.2(CR_0.2)	VQS_0.02(CR_0.02)	In(Out)
1	King Billy Royal	39.52	49.83(26.0%)	31.96(19.1%)	817(43)
2	the pinstripes	24.74	32.01(29.4%)	18.05(27.0%)	748(16)
3	Zonis	20.15	25.47(26.4%)	16.52(18.0%)	1581(0)
4	TwistNHook	30.30	38.98(28.6%)	23.25(23.3%)	2263(11)
5	Dewey Finn	26.72	33.81(26.5%)	22.13(17.2%)	422(8)

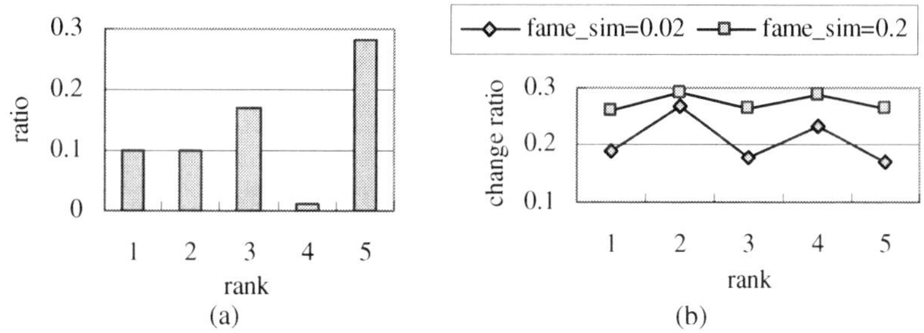

Fig. 5. The ratios of different kinds of comments and the ratios of change

From Table 5 and Fig. 5, we know that when *fame_sim*=0.2 or 0.02, *TP*'s and *T*'s change ratio are always larger than that of others. From Table 4, we know that the three *cbqs* values of *T* are all small, and the three *cn* values of *T* are all relatively large; in Fig. 5 (a), *T* has the lowest ratio; so it is no doubt that *T*'s change ratio is unstable. That is because the main contribution of *T*'s *VQS* is obtained from plenty of low quality comments. However, the question is that the first and the second expert have similar ratios in Fig. 5 (a), but have different stabilities in Fig. 5 (b). To interpret this, we calculated the sum of all their commenters' *TQS*(*sum_cq(rank)*) respectively, and found that *sum_cq(1)*=124.58, *sum_cq(2)*=129.82. Further analysis shows that

KBR is the main contributor of *TP* and *TP* is the main contributor of *KBR*; however, *KBR* gives *TP* his/her 16% *mqs* (about 17) by three comments, while *TP* gives *KBR* his/her 32% *mqs* (about 28) by two comments. The final results suggest that the *VQS* of *KBR* is more stable than that of *TP*.

5.2 Identifying Topic Blog Communities

To prove that our community extraction approach is better than Kumar's, we make a vector *vec100* that consists of top 100 experts. For every community *C* obtained, we calculate the intersection of *vec100* and *C*. Fig. 6 shows the properties of top 5 (sorted by size) such intersections. In Fig. 6, x-axis is the intersection's index; left y-axis represents the average quality score of all community members; right y-axis represents the size of a community or an intersection. The first intersection obtained by our algorithm has 8 elements; and the according community size and average quality score are 9 and 39.37 respectively. The first intersection obtained by Kumar's algorithm has 6 elements; and the according community size and average quality score are 11 and 31.28 respectively. The result is shown in Fig. 6, and this result means that our approach can generate communities with higher quality and purer members.

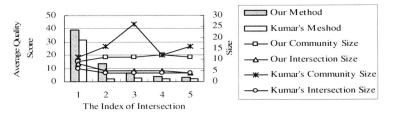

Fig. 6. The properties of the intersection of top 100 experts and every community

Fig. 7 shows the top 100 experts' distribution; x-axis represents expert rank and y-axis represents the generating order of communities. Obviously, when adopting our method, these experts are almost all distributed in the first 50 communities. That's because we only choose the blog with the highest quality as candidate when extracting communities. This ensures the community's purity and high quality.

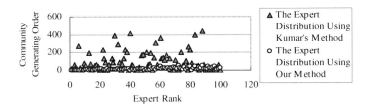

Fig. 7. The distribution of the top 100 experts when extracting communities

6   Conclusion

Blogs have spread fast all over the world and form all kinds of topics and virtual blog communities. In this paper, we are focusing on identifying topic experts and topic communities. These two issues are very important not only for understanding the development of blogspace, but can also provide people the facility of sales, advertisements, etc. We propose several algorithms to tackle these two issues, and experimental results demonstrate that our algorithms are very effective and efficient.

We focus on identifying experts on a given topic and identifying topic communities based on the results of topic expert identification. Our algorithms could produce purer, high quality, and highly topic-related communities. We plan to further extend our research in several directions. First, the algorithm proposed is only based on comments and not combined with link analysis, and we plan to incorporate link analysis into our framework. Second, we plan to analyse the evolution of topic experts and topic communities over a period of time by tracing their activities and understanding the interplay between them.

Acknowledgement

This research is supported by NSF China with Grant No. 90818023.

References

1. Kumar, R., Novak, J., Raghavan, P., Tomkins, A.: On the Bursty Evolution of Blogspace. In: Proc. WWW, pp. 568–576. ACM Press, New York (2003)
2. Agarwal, N., Liu, H., Tang, L., Yu, P.S.: Identifying the Influential Bloggers in a Community. In: Proc. of WSDM, pp. 207–218. ACM Press, New York (2008)
3. Song, X., Chi, Y., Hino, K., Tseng, B.L.: Identifying Opinion Leaders in the Blogosphere. In: Proc. of CIKM, pp. 971–974. ACM Press, New York (2007)
4. Bansal, N., Chiang, F., Koudas, N., Wm, F.: Tompa: Seeking Stable Clusters in the Blogosphere. In: Proc. of VLDB, pp. 806–817. VLDB Endowment (2007)
5. Kumar, R., Raghavan, P., et al.: Trawling the Web for Emerging Cyber-Communities. In: Proc.of WWW, New York, pp. 1481–1493 (1999)
6. Cormen, T.H., Leiserson, C.E., Rivest, R.L.: Introduction to Algorithms. McGraw Hill and MIT Press (1990)
7. Pui, G., Fung, C., et al.: Parameter Free Bursty Events Detection in TextStreams. In: Proc. of VLDB, pp. 181–192. VLDB Endowment (2005)
8. State of the Blogosphere – (August 2006), http://www.sifry.com/alerts/archives/000436.html
9. Lin, Y., Sundaram, H., Chi, Y., Tatemura, J., Tseng, B.: Blog Community Discovery and Evolution Based on Mutual Awareness Expansion. In: Proceedings of the IEEE/WIC/ACM International Conference on Web Intelligence, pp. 48–56. IEEE Computer Society, Washington (2007)
10. Tseng, B., Tatemura, J., Wu, Y.: Tomographic Clustering to Visualize Blog Communities as Mountain Views. In: Proc. of the World Wide Web (2005)
11. Gruhl, D., Guha, R., Liben-Nowell, D., Tomkins, A.: Information Diffusion Through Blogspace. In: Proc. of the World Wide Web, pp. 43–52. ACM Press, New York (2004)
12. Bulters, J., de Pijke, M.: Discovering Weblog Communities. In: International AAAI Conference on Weblogs and Social Media Boulder, pp. 211–214 (2007)
13. Kleinberg, J.: Bursty and Hierarchical Structure in Streams. In: Proc. 8th ACM SIGKDD Intl. Conf. on Knowledge Discovery and Data Mining, pp. 373–397. Kluwer Academic Publishers, HingHam (2003)

Utility-Oriented K-Anonymization on Social Networks

Yazhe Wang[1], Long Xie[1], Baihua Zheng[1], and Ken C.K. Lee[2]

[1] Singapore Management University
{yazhe.wang.2008,longxie,bhzheng}@smu.edu.sg
[2] University of Massachusetts Dartmouth
ken.ck.lee@umassd.edu

Abstract. "Identity disclosure" problem on publishing social network data has gained intensive focus from academia. Existing k-anonymization algorithms on social network may result in nontrivial utility loss. The reason is that the number of the edges modified when anonymizing the social network is the only metric to evaluate utility loss, not considering the fact that different edge modifications have different impact on the network structure. To tackle this issue, we propose a novel *utility-oriented social network anonymization* scheme to achieve privacy protection with relatively low utility loss. First, a proper utility evaluation model is proposed. It focuses on the changes on social network topological feature, but not purely the number of edge modifications. Second, an efficient algorithm is designed to anonymize a given social network with relatively low utility loss. Experimental evaluation shows that our approach effectively generates anonymized social network with high utility.

Keywords: social networks, privacy,k-anonymity, utility, HRG.

1 Introduction

With the rapid growing of social network applications and the proliferation of the social network data in recent years, social network data privacy has attracted more and more attentions from academia [1–4]. Among various privacy problems on social networks, *identity disclosure* [1] in publishing social network data is most concerned. Usually, a social network is modeled as a complex graph. Given a social network G, a published social network G^* has identity disclosure problem if there is a vertex v in G^* that can be mapped to an original entity t in G with a high probability. It has been demonstrated that even after removing all identifiable personal information (e.g., names and identity card numbers), an attacker is still able to identify an original entity in a published social network with high confidence based on the knowledge of the topological structure around the entity, such as degree, neighborhood and subgraph.

To tackle this issue, various anonymization models have been proposed based on the principle of *k-anonymity*. They all have to make changes to the original social networks in order to protect the privacy. Generally, from privacy protection

J.X. Yu, M.H. Kim, and R. Unland (Eds.): DASFAA 2011, Part I, LNCS 6587, pp. 78–92, 2011.

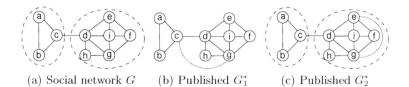

(a) Social network G (b) Published G_1^* (c) Published G_2^*

Fig. 1. An example of the impact of adding edges to achieve 2-degree anonymity

point of view, more changes on the original social network are preferred. However, it will greatly affect the *utility* of the social network. Ideally, we prefer that a modified social network does not disclose the true identity of each vertex, and meanwhile it still provides comparable level of accuracy with the original data for the corresponding mining and analysis activities. The trade-off between privacy and utility in publishing tabular data has been well studied [5], however, it is still new in the field of social network publishing.

To the best of our knowledge, most of previous works use the total number of modified edges to measure the social network utility loss. In other words, they try to achieve anonymity with minimum number of edge modifications. However, this measurement is not effective as it assumes each edge modification has an equal impact on the original social network properties. For example, a social network G is given in Fig. 1(a). Its vertices are naturally divided into two communities, as indicated by the dash circles. The vertices within the same community are strongly connected, while connections between the vertices of different communities are weak. Assume there are two corresponding social networks G_1^* and G_2^* published based on G, as illustrated in Fig. 1(b) and Fig. 1(c) respectively. In terms of privacy, both G_1^* and G_2^* satisfy 2-degree anonymity, that is for any given vertex, there is at least one other vertex sharing the same degree. In terms of utility, they are same as they both only add one edge to the original social network. However, the change that G_1^* makes (i.e., adding edge between vertex g and vertex c) is more significant, compared with the change made by G_2^* (i.e., adding edge between vertex g and vertex e), as G_2^* remains the two-community structure of G, while G_1^* blurs the boundary of the communities.

Based on the above observations, we believe that the number of edge modifications alone is not a good measurement of the utility loss and hence the existing k-anonymization algorithms based on this measurement have nature flaws in providing high-utility anonymized social network data. To address this concern, we propose a novel *utility-oriented social network anonymization* approach in this paper to achieve high privacy protection and low utility loss. First, a proper utility model is proposed based on the *hierarchical community structure* of the social network, to measure the utility loss of a published social network. It focuses on social network topological feature changes instead of purely the number of edge modifications. Second, an efficient k-anonymization algorithm is designed to modify a given social network G to G^*, where G^* satisfies the privacy requirement (e.g., k-degree anonymity) with relatively low utility loss.

The rest of the paper is organized as follows. Section 2 presents some background knowledge and reviews related works about social network privacy protection. Section 3 details the new utility model based on the hierarchical community structure of the social network. Section 4 presents the k-anonymization algorithm based on the proposed utility model. Section 5 reports the experiment results. Finally, section 6 concludes the paper.

2 Preliminaries and Related Work

We first present the terminology that will be used in this paper. Similar as other works, we model the social network as an undirected graph $G(V, E)$, where vertex set V represents the entities (e.g., persons, organizations, et al), and edge set E represents the relationships between two entities (e.g., friendship, collaboration, et al). An edge between vertex v_i and v_j is denoted as $e(v_i, v_j) \in E$.[1]

2.1 Structural Re-identification Attack and K-Anonymity

Social network data publishing faces various privacy challenges, and we only focus on the identity privacy problem in this work. We assume the entities' true identities in the original social network G are sensitive, and hence they are eliminated in the released social network G'. An attacker tries to locate a target entity in G' based on her background knowledge about the target. We use F to denote the type of background knowledge that an attacker uses and $F(t)$ to represent the evaluated value of F for a target t. If F is based on the structure of the graph, such as degree, neighborhood and subgraph, this attack is called *structural re-identification attack (SRA)* [6], as defined in Definition 1.[2]

Definition 1 (Structural Re-identification Attack (SRA)). *Given a social network $G(V, E)$, its published graph $G'(V', E')$, a target entity $t \in V$ and the attacker's background knowledge $F(t)$, the attacker performs the structural re-identification attack by searching for all the vertices in G' that could be mapped to t, i.e., $V_{F(t)} = \{v \in V' | F(v) = F(t)\}$. If $|V_{F(t)}| << |V'|$,[3] then t has a high probability to be re-identified.*

K-anonymity is a widely adopted principle to prevent the SRA on social networks [1–4], formally defined in Definition 2. Please note that we need to specify the type of background knowledge F (e.g., degree, neighborhood) that an attacker has in order to formally define k-anonymity. However, when the context of F is clear, we use k-anonymity in this paper for the brevity of presentation.

Definition 2 (K-Anonymity). *Given a graph $G(V, E)$, and a type of attacker's background knowledge F, G satisfies k-anonymity against F, iff for each $v \in V$, there are at least $(k - 1)$ other vertices in V with the same F value of $F(v)$.*

---

[1] For ease of presentation, we use "graph" and "social network" interchangeably.
[2] The attacker could possess some non-structural information as well (e.g., the vertex(edge) labels), but we only consider the structural information in this paper.
[3] Notation $|V|$ refers to the cardinality of a set V.

Different approaches have been proposed to convert a given graph into a k-anonymized graph. In this work, we only focus on edge modification, that is to modify a graph into a k-anonymized graph only via inserting and/or deleting edges. It is expected that the topological structure of a graph will be changed by modifications and the published graph is expected to lose some utility of the original one. Consequently, social network publishing should take both the privacy and utility into consideration. Ideally a published social network G' should satisfy k-anonymity and meanwhile cause a utility loss as small as possible.

2.2   Related Work

Structural re-identification is one of the major privacy concerns in social network publishing [7]. The initial study demonstrates that simply removing the identification information of the entities is not sufficient to protect privacy as the true identities of the vertices can be inferred due to the structural uniqueness of some embedded small subgraphs (i.e., SRA). Various classes of SRAs are thereafter proposed based on the types of attackers' background knowledge, including vertex refinement queries, subgraph queries and hub-fingerprint queries [6].

   To counter the SRAs, various protection schemes were proposed [1–4, 6, 8]. For example, the random permutation approach protects privacy by randomly inserting and deleting edges [8], which is simple but may significantly affect the graph utility. Then graph generalization based approaches abstract an original graph into a super graph by grouping the vertices into small blocks represented by super nodes and linking super nodes via edges if the corresponding blocks are connected [6]. The super graph introduces great uncertainty in the released data, thus increasing the difficulties of using the data.

   Recently, researchers start to apply the principle of k-anonymity [9] to protect the social network privacy. Based on the types of attacker's background knowledge, various k-anonymity schemes and algorithms have been proposed. For example, k-degree anonymity scheme is to against the attackers with knowledge of entity degree [1]; scheme proposed in [3] considers the attackers with the information about the vertices' neighborhood; k-automorphism and k-symmetry schemes can resist multiple structural attacks. k-automorphism modifies the graph such that for each vertex, there are at least $(k-1)$ other structurally equivalent vertices [4]; and k-symmetry utilizes the symmetry property of the social network to modify the graph [2]. All these algorithms anonymize graphs based on edge/node modification operations (i.e., addition and/or deletion), and try to preserve the utility of the released graphs. However, most of them employ the number of edge/node changes as the only measurement to quantify the utility loss, which is *not* effective, as demonstrated in Section 1. The new model we propose, as will be detailed in next section, actually gives a better measurement.

3   Graph Utility Measurement

In order to support utility-oriented k-anonymization, the first issue to address is how to measure the utility loss of a published social network, compared with

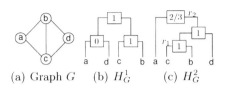

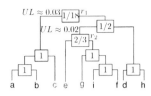

Fig. 2. A graph G and its corresponding HRGs

Fig. 3. An example of HRG and utility loss

the original social network. As pointed out in previous sections, the number of edge modifications, the most common utility loss measurement, is not effective as it treats all the edges equally. In our work, we aim at developing a new measurement that reflects the different impacts of various edge operations on the social network structure. Given the fact that social network is a complex graph, there are many aspects of its topological properties, such as transitivity, eigenvector, and community structure. Among them, the community structure is a central organizing principle of complex graph and it is a core graph topological feature which has a strong correlation with other important features (e.g., transitivity and betweenness). Consequently, we decide to use the community structure to represent the topological features of social networks as it provides a simple representative to reflect the influence of the edge modification on social network structure in a micro perspective. In this section, we first introduce the *Hierarchical Random Graph* (HRG) [10] for modeling the community structure, then present its construction algorithm, and finally introduce a novel hierarchical community entropy to quantify the graph structural features (i.e., utility).

3.1   Hierarchical Random Graph

The community structure of a graph is a nature grouping of its vertices. The vertices have dense connections within the groups, but sparse connections between groups. Recent studies suggest that the communities of social networks often exhibit hierarchical organization (i.e. the large communities further contain small communities). Consequently, we adopt a *Hierarchical Random Graph* (HRG) model to capture this hierarchical organization of communities [10].

Given a graph $G(V, E)$, the HRG is a binary tree, denoted as H_G. The leaf nodes of H_G correspond to vertices of G, and each internal node r is associated with a *connection probability* p_r. Given a sub-tree T_r that is rooted at node r, p_r is the probability that a vertex in the left subtree T_r^L has an edge with a vertex in the right subtree T_r^R. It reflects the connection strength between the vertices in the left and right subtrees. The larger the p_r is, the stronger the connection is. Mathematically, connection probability p_r is defined in Equation (1).

$$p_r = |E_r|/(|T_r^L| \cdot |T_r^R|) , \tag{1}$$

where $|E_r|$ is the number of edges $e(v_i, v_j) \in E$ with $v_i \in T_r^L$ and $v_j \in T_r^R$, and $|T_r^L|$ ($|T_r^R|$) is the number of vertices in r's left (right) subtree. A graph G and its two possible HRGs are depicted in Fig. 2. Naturally, the vertices in sub-tree T_r rooted at node r are regarded as a community C_r.

3.2   Constructing HRG

As mentioned above, the tree-structure of HRG organizes the underlying social network hierarchically. However, for a given social network G, there are multiple possible HRGs. How to construct an HRG that captures the topological structure of a given social network G best is a key issue we have to address in order to use HRG to model social network. In the literature, a likelihood function $\mathcal{L}$ has been developed to evaluate the fitness of a given HRG H_G to G, with $\mathcal{L}(H_G) = \prod_{r \in H_G} \left[p_r^{p_r} (1 - p_r)^{1-p_r} \right]^{|T_r^L| \cdot |T_r^{lt}|}$ [10]. Accordingly, a representative HRG construction algorithm uses Markov chain Monte Carlo method to sample the space of all possible HRGs with probability proportional to $\mathcal{L}$ and returns the sampled HRG having the maximum $\mathcal{L}$ value.

Essentially, $\mathcal{L}(H_G)$ is the posterior probability that the model H_G generates G. However, in fact, a model H_G generates G with high probability does not necessarily mean that H_G is a good model of the hierarchical community structure of G. We use example depicted in Fig. 2 to support our statement. We observe that the partition of H_G^2 is more meaningful than H_G^1 in terms of the community structure. This is because H_G^1 groups a and d into the same community which is improper as there is even no edge between them. However, $\mathcal{L}(H_G^1) = 1 >> \mathcal{L}(H_G^2) = 0.148$. Consequently, maximize the likelihood value $\mathcal{L}(H_G)$ does not necessarily reflect a good community organization of a social network G. Thus, the Monte Carlo sample algorithm which is developed based on $\mathcal{L}(H_G)$ cannot return the best HRG that preserves most, if not all, the topological structure properties of a given social network G as we expect, not to mention its extremely high construction cost.

To overcome the shortcomings of the existing HRG construction approach, we propose a simple greedy bottom-up construction algorithm. Initially, the algorithm forms each vertex of G as one community (i.e. the leaf nodes of H_G). Thereafter, communities (i.e. subtrees) with strong connections are merged from bottom to up until one unified community is achieved. The connection strength of two community C_i and C_j is again evaluated by the connection probability $p_{C_i,C_j} = |E_{ij}|/(|C_i||C_j|)$, with $|E_{ij}|$ the number of edges connecting vertices of C_i with vertices of C_j, and $|C_i|$ ($|C_j|$) the number of vertices of C_i (C_j). Due to the limitation of the space, we omit the detailed HRG construction algorithm.

3.3   Hierarchical Community Entropy

As mentioned above, we use an HRG H_G to represent the topological features of a given social network G. In this subsection, we introduce an information entropy based utility function to quantify the information (i.e. utility) of G reflected by H_G. In the literature, there are various graph entropy definitions available, based on different focuses. For example, entropy of the degree distribution, target entropy and road entropy [11]. However, none of the above entropy definitions considers the graph's hierarchical community information. Consequently, we propose a new *Hierarchical Community Entropy* (HCE) to represent the information embedded in the graph community structure.

HCE is defined based on the edge grouping. Given a graph $G(V, E)$ and its community structure represented by H_G, there are $|V| - 1$ internal nodes in H_G as H_G is a complete binary tree with $|V|$ leaf nodes. Each internal node r in H_G roots a subtree corresponding to a group of crossing edges E_r of G. Given the numbers of vertices in left subtree and right subtree represented by $|T_r^L|$ and $|T_r^R|$ respectively, $|E_r| = |T_r^L| \cdot |T_r^R| \cdot p_r$. The HCE of a given H_G of a social network G, denoted as $HCE(G, H_G)$ is defined in Equation (2), with p_t represents the connection probability. For example, the graph depicted in Fig. 1(a), the HCE of its H_G shown in Fig. 3 is 2.807.

$$HCE(G, H_G) = - \sum_{t=1}^{|V|-1} \frac{|T_t^L| \cdot |T_t^R| \cdot p_t}{|E|} \log(\frac{|T_t^L| \cdot |T_t^R| \cdot p_t}{|E|}) \, . \qquad (2)$$

When we insert/delete an edge on a graph G, the modification will be reflected by the connection probability change of an internal node on H_G, thus changing the HCE value. Continue our example. When we add a new edge $e(v_g, v_c)$ to G in Fig. 1(a), the connection probability of the lowest common ancestor of v_g and v_c (i.e. the root) is changed from $\frac{1}{18}$ to $\frac{1}{9}$, with the new HCE value of the modified graph being 2.840. Similarly, if we add a new edge $e(v_g, v_e)$ to G, its HCE value will be 2.790. The utility loss caused by the edge operation is evaluated by the change of the HCE value, as defined in Equation 3.

$$UL(G, G') = |HCE(G, H_G) - HCE(G', H_G')| \, , \qquad (3)$$

where G' is the modified graph, and H_G' is the corresponding HRG derived from H_G with updated connection probabilities. The main goal of this work is to anonymize the social network while making the utility loss as small as possible. Continue above example. As adding edge $e(v_g, v_c)$ causes the utility loss of $|2.840 - 2.807| = 0.033$, and adding edge $e(v_g, v_e)$ causes utility loss of 0.017, the second modification has a less significant impact on the graph structure and hence is preferred. It also confirms our observation in Fig. 1.

4 HRG Based *K*-Anonymization

After introducing the HRG model and the information entropy based utility measurement, we are ready to present HRG-based k-anonymization algorithm that tries to anonymize a given social network via edge operations with the utility loss as small as possible. In the following, we first present the basic idea of HRG-based k-anonymization and then detail its main components individually. Notice that although we only focus on k-degree anonymity in this section, our approach is general and it is applicable to other k-anonymity based privacy protection schemes on social networks (e.g. k-neighborhood anonymity).

Algorithm 1. HRG based k-anonymization algorithm

    Input: Graph $G(V, E)$, H_G, F, and k
    Output: K-anonymized graph G'
1 $G'(V', E') = G(V, E)$;
2 $D^* = \textbf{estimate}(G, F, k)$;
3 **while** G' *is not k-anonymized* **do**
4     $Set_{op} = \textbf{findcandidateOp}(G', D^*, H_G)$;
5   **while** $Set_{op} \neq \emptyset$ **do**
6   operation $p = Set_{op}.min\_op()$;
7        $execute(p, G', H_G)$;
8        $Set_{op} = \textbf{findcandidateOp}(G', D^*, H_G)$;
9   **if** G' *is not k-anonymized* **then** $D^*=\textbf{refine}(D^*, G')$;
10 **return** G';

4.1 Basic Idea and Algorithm Framework

The optimal k-anonymization problem (i.e. k-anonymization with minimum utility loss) on social networks is NP-hard.[4] To simplify the problem, we assume the utility loss is affected by the number of edge operations performed and the utility loss caused by each edge operation. In other words, we try to solve the problem by reducing the number of edge operations and meanwhile always performing the edge operations that cause smaller utility loss first. A greedy algorithm is designed accordingly.

The basic idea of our algorithm is as follows. Given a graph G, the attack model F and the privacy requirement k, we perform edge operations one at a time on G to achieve k-anonymity. To restrain the utility loss, we perform the edge operation that directs the current G towards its "nearest" k-anonymized graph and meanwhile causes the smallest utility loss. Here, "nearest" k-anonymized graph refers to the graph that satisfies k-anonymity with the smallest number of edge operations, which is denoted as G^* to facilitate our explanation. The knowledge of G^* is essential for our algorithm. However, G^* is unknown in advance and it is hard to locate. Given that forming G^* directly is not always possible, we try to estimate the local structure information of the vertices of G^* (e.g., the degrees and/or the degrees of the neighbors of each vertex) which, based on the given G, F and k, is possible. Then, according to the local structure information, a set of candidate edge operations are generated to lead G towards G^*.

Algorithm 1 sketches a high-level outline of our HRG based k-anonymization algorithm. It takes a graph G, its HRG H_G, attacker's background knowledge F and privacy parameter k as inputs, and outputs a modified graph G' that is k-anonymized and meanwhile has small utility loss. Initially, the algorithm sets G' to G, and sets D^* as an estimation of G^* based on G, F and k (lines 1-2). Thereafter, it generates a set of candidate edge operations, maintained by a set Set_{op} with the utility loss caused by each edge operation (line 4).

---

[4] The NP-hardness is proved by reducing the traditional set-packing problem [12] to the optimal k-anonymization problem.

At each step, it gets the edge operation which causes the smallest utility loss, performs that edge operation on G', at the same time, updates the corresponding connection probability on H_G, and then re-generates the candidate set based on the updated G' (lines 6-8). This process continues until Set_{op} becomes empty (lines 5-8). After performing all the identified candidate edge operations, there are two possible outcomes, i.e., the current G' is k-anonymized or not. In case G' still does not satisfy the privacy requirement, it means the k-anonymized graph which has the local structure information D^* is not achievable by the current executed operation sequence and we need to refine D^* via small adjustments and continue previous process (line 9). We would like to point out that when refining D^*, we only consider additive adjustment, i.e. adjust the graph via adding edges. Thus, in the worst case, G' will be modified towards a complete graph, which always satisfies the privacy requirement. Therefore, our algorithm is convergent.

As highlighted in Algorithm 1, there are three key components, i.e., estimation of local structure information, generation of candidate edge operations, and refinement of D^*. Each of them will be detailed in following subsections.

4.2   Estimating Local Structure Information

As pointed out earlier, we only focus on k-degree anonymization for presentation simplicity. In the following, we explain how to find a good estimation of the k-anonymized graph with smallest number of edge operations, i.e., G^*. Our approach is to perform the estimation on the local structure information based on *degree sequence*. Degree sequence D of a graph $G(V, E)$ is a vector of size $|V|$ with each element $D[i] \in D$ representing the degree of vertex v_i in G. We further assume the degree sequence is sorted by the decreasing order of its elements.

Given a graph G, its degree sequence D and k, we want to estimate the degree sequence D^* of its "nearest" k-degree anonymized graph G^*. We list some pre-knowledge that can guide the estimation. First, D^* shares equal size with D, because we only consider graph modification via edge insertion/deletion but not vertex insertion/deletion. Second, D^* must be k-anonymized since D^* is the degree sequence of a k-degree anonymized graph of G. In other word, for each element $D^*[i] \in D^*$, there are at least $(k-1)$ other elements sharing the same value as $D^*[i]$. Third, because that D^* is the degree sequence of the "nearest" k-anonymized graph of G, the L_1 distance between D^* and D should be minimized. Based on the above knowledge, we can employ the dynamic programming method proposed in [1] to find D^*. We ignore the detail due to space limitation.

4.3   Generating Candidate Edge Operation Set

Once D^* that represents the target local structure information is ready, we need to find candidate edge operations that convert G' to a k-anonymized graph with its degree sequence represented by D^*. Before we introduce the detailed algorithm, we first define three basic edge operations, i.e., *edge insertion*, *edge deletion*, and *edge shift*, denoted as $ins(v_i, v_j)$, $del(v_i, v_j)$, and $shift((v_i, v_j), v_k)$. As suggested by their names, $ins(v_i, v_j)$ is to insert a new edge that links vertex

Fig. 4. $shift((v_c, v_d), v_e)$ **Fig. 5.** HRG based 2-degree anonymization

v_i to vertex v_j and $del(v_i, v_j)$ is to remove the edge between v_i and v_j. Operation $shift((v_i, v_j), v_k)$ is to replace the edge $e(v_i, v_j)$ with edge $e(v_i, v_k)$. It is motivated by the observation that the HCE value is only sensitive to the number of the crossing edges between two communities. For example, as shown in Fig. 1(a), G is partitioned into two main communities as demonstrated by the dash circles. Edge $e(v_c, v_d)$ is the crossing edge connecting those two communities, and their lowest common ancestor is the root (based on H_G shown in Fig. 3). If we shift the end point v_d of the edge $e(v_c, v_d)$ to v_e (i.e., $shift((v_c, v_d), v_e)$) as illustrated in Fig. 4, it will not affect the connection probability of the root in H_G and hence HCE value, as the number of the crossing edge is not changed. Therefore, edge shift operation should receive a higher priority when modifying the graph to achieve k-anonymity. Definition 3 gives formal definition of this operation.

Definition 3 (Edge Shift). *Given a graph $G(V, E)$, the corresponding HRG H_G, an edge $e(v_i, v_j) \in E$, and a vertex $v_k \in V$ such that $e(v_i, v_k) \notin E$, let r be the lowest common ancestor of v_j and v_k on H_G, and assume v_i is not in the subtree of r. Edge shift $shift((v_i, v_j), v_k)$ is to replace $e(v_i, v_j)$ with $e(v_i, v_k)$.*

The goal of the edge operations is to modify the graph such that its degree sequence D' matches the target degree sequence D^*. Consequently, the difference sequence $\delta = (D^* - D')$ can give some guidance. For each element $\delta[i] \in \delta$ with $\delta[i] > 0$ (i.e. $D'[i] < D^*[i]$), it means a vertex in G' with degree $D'[i]$ needs to increase its degree, i.e., it should have more edges connected to. We maintain $D'[i]$ with $\delta[i] > 0$ via set DS^+ and maintain all vertices $v \in G'$ that have degree of $D'[i]$ via set VS^+ which includes all the vertices that may require edge insertion operation. Similarly, for each element $\delta[j] \in \delta$ with $\delta[j] < 0$ (i.e. $D'[j] < D^*[j]$), it means a vertex in G' with degree $D'[i]$ needs to decrease its degree, i.e., it should have less edges connected to. We maintain $D'[i]$ with $\delta[j] < 0$ via set DS^- and maintain all the vertices $v \in G'$ that have degree of $D'[j]$ via set VS^- which includes all vertices that may require edge deletion operation. Notice that the degree value of $D'[i]$ or $D'[j]$ may correspond to multiple vertices in G' and we treat them equally in our work. In addition, if the degree $D'[i]$ ($D'[j]$) only appears once in DS^+ (DS^-), we cannot perform edge insertion (deletion) to connect (disconnect) two vertices v_l, v_m both with original degree of $D'[i]$ ($D'[j]$) and hence we mark these vertices *mutual exclusive*, denoted as $EX(v_l, v_m) = True$.

Back to the graph G depicted in Fig. 1(a). Its degree sequence D and the target 2-degree anonymized degree sequence D^* are shown in Fig. 5. Based on $\delta = (D^* - D)$, we find $\delta[2] = \delta[4] = 1 > 0$ and hence $D[2]$ ($= 4$) and $D[4]$ ($= 3$)

are inserted into DS^+. Consequently, all the vertices in G with degree being 4 or 3 are inserted into $VS^+ = \{v_i, v_g, v_c, v_e, v_f\}$. Notice that all pair of vertices among of $\{v_i, v_g\}$ and among of $\{v_c, v_e, v_f\}$ are marked mutual exclusive. As there is no element of δ with its value smaller than 0, $DS^- = VS^- = \emptyset$.

The reason that we form VS^+ set and VS^- set is to facilitate the generation of candidate edge operations. As VS^+ set contains those vertices that need larger degree, a new edge connecting v_i to v_j (i.e., $ins(v_i, v_j)$) is a candidate, if $v_i, v_j (i \neq j) \in VS^+ \wedge e(v_i, v_j) \notin E' \wedge EX(v_i, v_j) \neq True$. We can enumerate all the candidate edge insertion operations based on VS^+ and preserve them in set Op^{ins}. Similarly, removing edge $e(v_i, v_j)$ (i.e., $del(v_i, v_j)$) forms an edge deletion operation, if $e(v_i, v_j) \in E' \wedge v_i, v_j (i \neq j) \in VS^- \wedge EX(v_i, v_j) \neq True$. Again, we explore all the candidate edge deletion operations and preserve them in set Op^{del}. We also consider the candidate edge shift operation. For a pair of vertices (v_j, v_k) with $v_j \in VS^- \wedge v_k \in VS^+ \wedge (j \neq k)$, if there is a vertex $v_i, (i \neq j, k)$ such that $e(v_i, v_j) \in E' \wedge e(v_i, v_k) \notin E' \wedge v_i$ is not in the subtree of v_j and v_k's lowest common ancestor on the HRG, $shift((v_i, v_j), v_k)$ is a candidate. All possible edge shift operations form another set Op^{shift}. We continue the above example shown in Fig. 5. As $VS^- = \emptyset$, we only need to consider possible edge insertion operations, i.e., $Op^{del} = Op^{shift} = \emptyset$. Based on $VS^+ = \{v_i, v_g, v_c, v_e, v_f\}$, we have $Op^{ins} = \{ins(v_g, v_e), ins(v_i, v_c), ins(v_g, v_c)\}$.

Given all the candidate edge operations maintained in the operation sets Op^{ins}, Op^{del}, and Op^{shift} respectively, we can insert them into the candidate operation set Set_{op} that is used by HRG-based k-anonymization algorithm (i.e., Algorithm 1). We sort Set_{op} by the increasing order of the HCE value changes caused by each operations, so that the edge operation that causes smaller utility loss will be performed earlier. Based on the HRG in Fig. 3, the corresponding Set_{op} is set to $\{\langle ins(v_g, v_e), 0.017\rangle, \langle ins(v_i, v_c), 0.033\rangle, \langle ins(v_g, v_c), 0.033\rangle\}$. The whole process of finding candidate operations is summarized in Algorithm 2.

4.4   Refining Target Local Structure Information

As mentioned above, our HRG-based k-anonymization algorithm generates D^* that estimates the local structure information of the "nearest" k-anonymized graph as the target, and performs edge operation to change graph towards D^*. However, it is possible that k-anonymized graph with degree sequence D^* is not achievable by the current executed operation sequence. If this happens, we need to refine D^* and start another round of attempt. To ensure the convergence of our algorithm, we only consider additive adjustment and we prefer that the new target degree sequence is close to that of the original D^*. The basic idea is to find a point on D^* to make adjustment and hopefully, after the adjustment, we can find executable candidate operations on G'.

In our work, we take VS^+ as a candidate set for the adjustable points. It contains the vertices that have not been k-anonymized and need to increase their degrees. For each $v_i \in VS^+$, we find $v_j \in V, (i \neq j)$, such that $e(v_i, v_j) \notin E'$, and $EX(v_i, v_j) \neq True$, and preserve $ins(v_i, v_j)$ via an operation set Op. Within Op, we choose the $ins(v_i, v_j)$ that causes smallest utility loss. Notice

---

Algorithm 2. findcandidateOp algorithm

Input: $G'(V, E')$, D^*, H_G

Output: Candidate operation set Set_{op}

1   $D' =$ degree sequence of G';

2   $\delta = (D^* - D')$;

3   $DS^+ = \{D'[i] \mid \delta[i] > 0, 1 \leq i \leq |D'|\}$;

4   $DS^- = \{D'[i] \mid \delta[i] < 0, 1 \leq i \leq |D'|\}$;

5   $VS^+ = VS^- = \emptyset$;

6   **foreach** $d \in DS^+$ **do**

7     $\lfloor \; VS^+ = VS^+ \cup \{v_i | v_i \in G', v_i.degree = d\}$;

8   **foreach** $d \in DS^-$ **do**

9     $\lfloor \; VS^- = VS^- \cup \{v_i | v_i \in G', v_i.degree = d\}$;

10   $Op^{ins} = getOp(VS^+, VS^+)$, $Op^{del} = getOp(VS^-, VS^-)$,

    $Op^{shift} = getOp(VS^+, VS^-)$;

11   calculate the cost of each operation in Op^{ins}, Op^{del}, and Op^{shift};

12   $Set_{op}.insert(Op^{ins}, Op^{del}, Op^{shift})$;

13   **return** Set_{op};

---

that this operation changes degree of v_j even if the degree of v_j does not request adjustment. One simple method to address this issue is to change v_j's degree in D^* but it breaks the k-anonymity of D^*. Therefore, we increase the degree of v_j in the original G (i.e., the corresponding element in the degree sequence D is changed), and re-generate D^* based on the updated D. As the changes made to D are very small, the new D^* should be very similar as the old one.

We consider VS^+ first because we want to make additive change on D^*. However, if VS^+ is empty, we have to use VS^- that contains the vertices having not been k-anonymized and need to decrease their degrees. We can decrease the degrees of those vertices of VS^-, but it is against our goal of only making additive change. Alternatively, for a vertex $v_i \in VS^-$, we increase the degree of another vertex v_j, whose degree value is close to v_i according to G'. The rationale is that because v_i and v_j have similar degrees, they are very likely to have the same degree in the anonymized graph. Increasing the degree of v_j will cause the degree of v_i and v_j in the anonymized graph to be increased. In this case, v_i will not need to decrease its degree anymore. Still, we increase the degree of v_j on the seed degree sequence D instead of D^* by the same reason mentioned above.

5   Experimental Evaluation

In this section, we compare the utility loss of our HRG-based k-anonymization algorithm, referred as HRG, with the existing k-anonymization approaches that only consider minimizing the number of edge modifications. We choose k-degree anonymity as the privacy requirement, and use two existing k-degree anonymization algorithms proposed in [1] as competitors, namely *probing method* that only considers edge addition operations, and *greedy swap method* that considers both edge addition and deletion operations. We refer them as Prob. and Swap respectively. We implemented all the evaluated algorithms in C++, running on a PC having an Intel Duo 2.13GHz processor and 2GB RAM.

We first examine the utility loss by measuring the HCE value change. In addition, we use some common graph structural properties to further evaluate the utility loss of different algorithms, such as clustering coefficient (**CC**), average path length (**APL**), and average betweenness (**BTN**) (see [11] for more information). We use function $PCR = |P - P'|/|P|$ to measure the property change ratio, where P and P' are the property value (i.e., HCE, CC, APL, or BTN) of the original graph G and the corresponding k-anonymized G' respectively.

Two real datasets are used in our tests, namely **dblp** and **dogster**. The former is extracted from dblp (dblp.uni-trier.de/xml), and the latter is crawled from a dog-theme online social network (www.dogster.com). We sampled subgraphs from these datasets with the size changing from 500 to 3000 respectively.

5.1   Utility Loss v.s. Graph Size

In our first set of experiments, we evaluate the impact of graph size in terms of number of vertices on the graph utility loss (i.e., HCE and other graph properties changes) under different k-anonymization methods. We set $k = 25$.

Fig. 6 shows the change ratio of different graph properties with different graph size of two datasets. Generally, our HRG method is most effective in terms of preserving graph properties. Take HCE value as an example. As depicted in Fig. 6(a) and Fig. 6(e), the change ratio of our method (i.e., HRG) is around 0.1% for both **dblp** and **dogster**, while that under Prob. is 0.5% for **dblp** and 3% for **dogster**, and that under Swap is 60% for **dblp** and 14% for **dogster**. The other example is APL value. As depicted in Fig. 6(c) and Fig. 6(g), as number of vertices increases from 500 to 3000, our HRG method causes around 0.4% and 0.07% utility loss on average for **dblp** and **dogster**, respectively. On contrary, Prob. causes 4% and 7% utility loss on average, and Swap causes around 11% and 1.6% utility loss for **dblp** and **dogster** datasets. All these observations verify that HRG model does successfully capture most, if not all, core features of the social network, as our HRG method which employs HRG model to represent graph feature is most effective.

5.2   Utility Loss v.s. k

In the second set of experiments, we evaluate the impact of k on the graph property change ratio of different k-anonymity methods. Here, the size of the graph is fixed to 2000 vertices.

Fig. 7 presents the results. We can observe that, in most cases, our HRG approach outperforms the others. As privacy requirement increases (i.e., k value increases), the utility loss under HRG and Prob. becomes more significant. This is because more edge operations are needed to achieve k-anonymity with large k

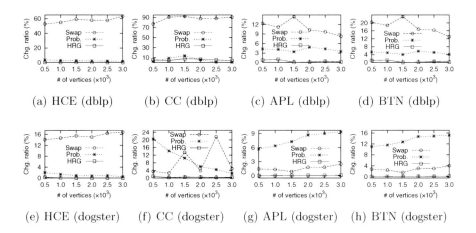

(a) HCE (dblp)   (b) CC (dblp)   (c) APL (dblp)   (d) BTN (dblp)

(e) HCE (dogster)   (f) CC (dogster)   (g) APL (dogster)   (h) BTN (dogster)

Fig. 6. Graph property change ratio v.s. the graph size

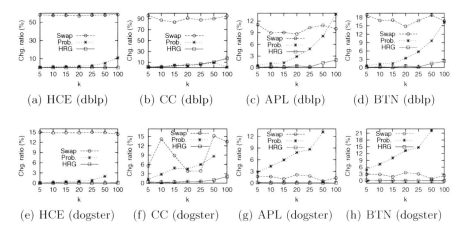

(a) HCE (dblp)   (b) CC (dblp)   (c) APL (dblp)   (d) BTN (dblp)

(e) HCE (dogster)   (f) CC (dogster)   (g) APL (dogster)   (h) BTN (dogster)

Fig. 7. Graph property change ratio v.s. k

under both methods. On the other hand, the utility loss caused by Swap algorithm is not affected by the change of k value that much. This is because, Swap method has to perform a large number of edge operations even for a small k. When k increases, the number of edge operations does not change much.[5]

To sum up, our experiments use different graph properties to evaluate the utility loss, although our HRG method is developed based on HCE values. The experimental results clearly verify that our approach can generate anonymized social networks with much lower utility loss.

---

[5] Due to the extremely long converge time of the Prob. method, its results on the **dogster** graph with $k = 100$ were missing. However, it should not affect our observations of the experimental trend.

6   Conclusion

Privacy and utility are two main components of a good privacy protection scheme. Existing k-anonymization approaches on social networks provide good protection for entities' identity privacy, but fail to give an effective utility measurement, thus are unable to generate anonymized data with high utility. Motivated by this issue, in this paper, we propose a novel utility-oriented social network anonymization approach to achieve high privacy protection with low utility loss. We define a new utility measurement HCE based on the HRG model, then design an efficient k-anonymization algorithm to generate anonymized social network with low utility loss. Experimental evaluation on real datasets shows our approach outperforms the existing approaches in terms of the utility with the same privacy requirment.

Acknowledgment. This study was funded through a research grant from the Office of Research, Singapore Management University.

References

1. Liu, K., Terzi, E.: Towards Identity Anonymization on Graphs. In: SIGMOD 2008, pp. 93–106 (2008)
2. Wu, W., Xiao, Y., Wang, W., He, Z., Wang, Z.: K-Symmetry Model for Identity Anonymization in Social Networks. In: EDBT 2010, pp. 111–122 (2010)
3. Zhou, B., Pei, J.: Preserving Privacy in Social Networks Against Neighborhood Attacks. In: ICDE 2008, pp. 506–515 (2008)
4. Zou, L., Chen, L., Özsu, M.: K-Automorphism: A General Framework for Privacy Preserving Network Publication. VLDB Endowment 2(1), 946–957 (2009)
5. Li, T., Li, N.: On the Tradeoff Between Privacy and Utility in Data Publishing. In: SIGKDD 2009, pp. 517–525 (2009)
6. Hay, M., Miklau, G., Jensen, D., Towsley, D., Weis, P.: Resisting Structural Re-identification in Anonymized Social Networks. VLDB Endowment 1(1), 102–114 (2008)
7. Backstrom, L., Dwork, C., Kleinberg, J.: Wherefore Art Thou R3579X? Anonymized Social Networks, Hidden Patterns, and Structural Steganography. In: WWW 2007, pp. 181–190 (2007)
8. Hay, M., Miklau, G., Jensen, D.: Anonymizing Social Networks. Technical report, UMass Amberst (2007)
9. Sweeney, L.: K-anonymity: A Model for Protecting Privacy. IJUFKS 10(5), 557–570 (2002)
10. Clauset, A., Moore, C., Newman, M.E.J.: Hierarchical Structure and The Prediction of Missing Links in Networks.. Nature 453(7191), 98–101 (2008)
11. Costa, L.D.F., Rodrigues, F.A., Travieso, G., Boas, P.R.V.: Characterization of Complex Networks: A Survey of Measurements. Advances in Physics 56(1), 167–242 (2007)
12. Karp, R.M.: Reducibility among combinatorial problems. In: Miller, R.E., Thatcher, J.W. (eds.) Complexity of Computer Computations, pp. 85–103. Plenum Press, New York (1972)

Distributed Privacy Preserving Data Collection

Mingqiang Xue[1], Panagiotis Papadimitriou[2], Chedy Raïssi[3],
Panos Kalnis[4], and Hung Keng Pung[1]

[1] Computer Science Department, National University of Singapore
[2] Stanford University
[3] INRIA Nancy
[4] King Abdullah University of Science and Technology

Abstract. We study the distributed privacy preserving data collection problem:
an untrusted data collector (e.g., a medical research institute) wishes to collect
data (e.g., medical records) from a group of respondents (e.g., patients). Each respondent owns a multi-attributed record which contains both non-sensitive (e.g.,
quasi-identifiers) and sensitive information (e.g., a particular disease), and submits it to the data collector. Assuming T is the table formed by all the respondent
data records, we say that the data collection process is privacy preserving if it
allows the data collector to obtain a k-anonymized or l-diversified version of T
without revealing the original records to the adversary.

We propose a distributed data collection protocol that outputs an anonymized
table by generalization of quasi-identifier attributes. The protocol employs cryptographic techniques such as homomorphic encryption, private information retrieval and secure multiparty computation to ensure the privacy goal in the process
of data collection. Meanwhile, the protocol is designed to leak limited but non-
critical information to achieve practicability and efficiency. Experiments show
that the utility of the anonymized table derived by our protocol is in par with the
utility achieved by traditional anonymization techniques .

1 Introduction

In the data collection problem a third party collects data from a set of individuals who
concern about their privacy. Specifically, we consider a setting in which there is a set of
data *respondents*, each of whom has a row of a table, and a data *collector*, who wants
to collect all the rows of the table. For example, a medical researcher may request from
some patients that each of them provides him with a health record that consists of three
attributes: ⟨*age, weight, disease*⟩. Figure 1(a) shows the table of the patients' records.

Although the health record contains no explicit identifiers such as name and phone
numbers, an adversarial medical researcher may be able to retrieve a patient's identity
using the combination of *age* and *weight* with external information. For instance, in
the data records of Figure 1(a), we see that there is only one patient with age 45 and
weight 60 and this patient suffers from Gastritis (the third row). If the researcher knows
a particular patient with the same age and weight values, after collecting all the data
records he learns that this patient suffers from Gastritis. In this case the attributes *age*
and *weight* serve as a quasi-identifier. Generally, the patients feel comfortable to provide
the researcher with medical records only if there is a guarantee that the researcher can

J.X. Yu, M.H. Kim, and R. Unland (Eds.): DASFAA 2011, Part I, LNCS 6587, pp. 93–107, 2011.
© Springer-Verlag Berlin Heidelberg 2011

only form an anonymized table with their records. In k-anonymity [16], each record has at least $k - 1$ other records whose values are indistinct over the quasi-identifier attributes [16]. l-diversity [10] further requires that there are at least l well represented sensitive values for records in the same *equivalence class*. The patients may achieve k-anonymity or l-diversity by generalizing the values that correspond to the quasi-identifiers [14]. In Figure 1(b), observe that if each patient discloses only some appropriate range of his age and weight instead of the actual values, then the medical researcher sees a 4-anonymous and 3-diverse table. In this case, the researcher can only determine with probability at most 1/2 the disease of the 45-year old patient.

In the privacy preserving data collection the data respondents look for the minimum possible generalization of the quasi-identifier values so that the collector receives an anonymized table. The constraint of the problem is that although the respondents can communicate with each other and with the collector, no single participant can leak any information to the others except from his final anonymous record. Traditional table anonymization techniques [16] are not applicable to our problem, as they assume that there is a single trusted party that has access to all the table records. If the trusted party is compromised then the privacy of all respondents is compromised as well. In our approach, each respondent owns his own record and does not convey its information to any other party prior to its anonymization.

Our setting is similar to the distributed data collection scenario studied by Zhong et al [19]. The difference is that in their work the respondents create a k-anonymous table for the collector by suppressing quasi-identifier attribute values. We use generalization instead of suppression, which makes the problem not only more general but also much more practical. Our problem is more general because suppression is considered as a special case of generalization: a suppressed attribute value is equivalent to the value generalization to the higher level of abstraction. The problem is also more practical because generalized attribute values have greater utility than suppressed values, since they convey more information to the data collector without compromising the respondents' privacy. Moreover, our solution not only achieves k-anonymity, but also l-diversity. Our contributions are the following:

- We formally define the problem of distributed privacy preserving data collection with respondents that can generalize attribute values.
- We present an efficient and privacy-preserving protocol for k-anonymous or l-diverse data collection.
- We show theoretically the information leakage that our protocol yields.
- We evaluate our protocol experimentally to show that it achieves similar utility preservation as the state-of-the art non-distributed anonymization algorithm [6].

2   Related Work

In [19], the authors proposed a distributed, privacy-preserving version of the MW [11], which is an $O(k \log k)$ approximation to optimal k-anonymity based on entry suppression; in contrast, our algorithm supports generalization. Similar to our scheme, in order to achieve efficient distributed anonymization the distributed MW algorithm reveals information about the relative distance between different data record pairs. In [19], the

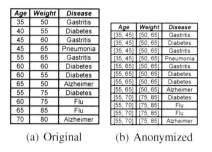

Age	Weight	Disease
35	50	Gastritis
40	55	Diabetes
45	60	Gastritis
45	65	Pneumonia
55	65	Gastritis
60	60	Diabetes
60	55	Diabetes
65	50	Alzheimer
55	75	Diabetes
60	75	Flu
65	85	Flu
70	80	Alzheimer

Age	Weight	Disease
[35, 45]	[50, 65]	Gastritis
[35, 45]	[50, 65]	Diabetes
[35, 45]	[50, 65]	Gastritis
[35, 45]	[50, 65]	Pneumonia
[55, 65]	[50, 65]	Gastritis
[55, 65]	[50, 65]	Diabetes
[55, 65]	[50, 65]	Diabetes
[55, 65]	[50, 65]	Alzheimer
[55, 70]	[75, 85]	Diabetes
[55, 70]	[75, 85]	Flu
[55, 70]	[75, 85]	Flu
[55, 70]	[75, 85]	Alzheimer

(a) Original   (b) Anonymized

Fig. 1. Distributed medical records table

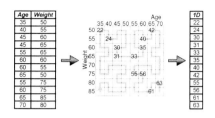

Fig. 2. Mapping 2D to 1D points using Hilbert curve

distance between two records is the *number* of differences in the attribute values. For example, in Figure 1(a), the distance between the first two records is 2, since age 35 is different from age 40 and weight 50 is different from weight 55. In our approach, the distance between two records depends on the *distance* between the corresponding attribute values, which is more difficult to evaluate securely. In [18] the authors proposed another k-anonymous data collection scheme. Opposed to the work in [19], this scheme has eliminated the need for unidentified communication channel by the data respondent. However, this scheme is still based on entry suppression and it is unclear whether the scheme can be generated to l-diversity. A similar approach appears in [7]. They considered distributed data collection problem based on a suppression based k-anonymity algorithm Mondrian and they only consider the k-anonymity case. Different from the above, [3] and [17] considered the anonymity-preserving data collection problem. Although, the setting of the problem is similar to ours, their protocols aims to allow the miner to collect original data from the respondents without linking the data to individuals. When the data collected contains identifiable information as in our case, their solutions are not applicable.

The anonymization algorithm that we present in this paper is based on the the Fast data Anonymization with Low information Loss (*FALL*)*et al* [6]. In this work, efficient anonymization is achieved in two steps. The first step includes the transformation of u-dimensional to 1-dimensional data, in which a multi-attributed data record is converted to an integer using a space filling curve (e.g. Hilbert curve [12]). For example, Figure 2 shows a Hilbert walk that visits each cell in the two dimensional space (*Weight* × *Age*) and assigns each cell with an integer in increasing order along the walk. In the second step, an optimal 1D k-anonymization is performed over the set of integers obtained in the first step using an efficient algorithm based on dynamic programming. The same partitions will be used for forming the *equivalence classes* of data records. Similarly, efficient l-diversity can be achieved using heuristics in a similar manner as k-anonymity.

3   Problem Formulation

3.1   The System and the Adversaries

The system employs the Client-Server architecture. Each respondent runs a client. There is an untrusted server that facilitates the communication and computation in the system on behalf of the collector. We assume that all messages are encrypted, and secure communication channels exist between any pair of communicating parties. By the end of the protocol execution, an anonymized table, generalized from the data records of the respondents, is created at the server side (i.e., the collector).

The adversaries can either be the respondents or the server. We assume that the adversaries follow the semi-honest model, which means that they always correctly follow the protocol but are curious in gaining additional information during the execution of the protocol. In addition, we assume that the adversarial respondents can collaborate with each other to gain additional information. We assume there can be up to $t_{ss} - 1$ adversaries among the respondents, where t_{ss} is a security parameter.

3.2 Notion of Privacy

Initially, there are x number of respondents each running an instance of the client. We denote the set of non-sensitive attributes of the data records $A = \{a_1, a_2, \ldots, a_u\}$ and the sensitive attribute s^i. The data record for the ith respondent is represented as $t_i = \{a_1^i, \ldots, a_u^i, s^i\}$ and $T = \{t_1, t_2, \ldots, t_x\}$ is the table formed by the original data records of the respondents. $t_i.A$ represents the non-sensitive attribute values for the data record t_i. Similarly, $T.A$ represents the non-sensitive attribute columns of table T. Let $\mathcal{K}(T)$ denote the final output of the protocol, which is an anonymized table generalized from T. Let $\mathcal{L}_i$ and $\mathcal{L}_{svr}$ denote the amount of information leaked in the process of protocol execution to the respondent i and the server, respectively.

During the execution of the protocol, the view of a party uniquely consists of four objects: (i) the data owned by the party, (ii) the assigned key shares, (iii) the set of received messages and (iv) all the random coin flips picked by this party. Let $\mathsf{view}_i(T)$ (respectively $\mathsf{view}_{svr}(T)$) denote the view of the respondent i (respectively the view of server). We adopt a similar privacy notion as in [19]:

Definition 1. *A protocol for k-anonymous data collection leaks only $\mathcal{L}_i$ for the respondent i and $\mathcal{L}_{svr}$ for the server if there exist probabilistic polynomial-time simulators M_{svr} and $M_1, M_2, \ldots, M_x$ such that:*

$$\{M_{svr}(keys_{svr}, \mathcal{K}(T), \mathcal{L}_{svr})\}_T \equiv_c \{\mathsf{view}_{svr}(T)\}_T \tag{1}$$

and for each $i \in [1, x]$,

$$\{M_i(keys_i, \mathcal{K}(T), \mathcal{L}_i)\}_T \equiv_c \{\mathsf{view}_i(T)\}_T \tag{2}$$

The contents of $\mathcal{L}_{svr}$ and $\mathcal{L}_i$ are statistical information about the respondent's quasi-identifiers. Later in this paper, we prove that the execution of our proposed protocol respects the previous definition by only leaking $\mathcal{L}_{svr}$ and $\mathcal{L}_i$ for each i.

3.3 Using Secret Sharing

To conquer up to $t_{ss} - 1$ collaborating adversaries among the respondents, we initially assume that there is a global private key SK shared by all the respondents and the server using a $(t_{ss}, x + 1)$ threshold secret sharing scheme [15]. The shares owned by the respondents and the server are denoted as $sk_1, sk_2, \ldots, sk_x$, and sk_{svr}, respectively. With a $(t_{ss}, x + 1)$ secret sharing scheme, t_{ss} or more key shares are necessary in order to successfully reconstruct the decryption function with the secret key SK, while less than t_{ss} key shares give absolutely no information about SK. The corresponding public key of the private key SK is denoted as PK. The public key encryption algorithm that we use in this paper is the Paillier's cryptosystem [13] because of its useful additive homomorphic property. To support threshold secret sharing, we use a threshold version of Paillier's encryption as described in [8] based on Asmuth-Bloom secret sharing [1].

4   Towards the Solution

4.1   A Sketch of the Solution

Preparation stage. The main goal of this stage is to map the uD records to 1D integers. In this stage, each respondent independently performs uD to 1D mapping using a space filling curve, e.g., the Hilbert curve. Symbolically, the mapping for $t_i.A$ is denoted as $c_i = \mathcal{S}(t_i.A)$. Each integer c_i is in the range $[1, c_{max}]$, where c_{max} denotes the maximum possible value that the mapping function can yield. The set of mapped values is denoted as $S = \{c_1, c_2, \ldots, c_x\}$. Without loss of generality, we assume that the values in S are already sorted in ascending order for the ease of subsequent discussion.

Stage 1. The goal of this stage is to achieve *p-probabilistic locality preserving* mapping. Symbolically, the ith respondent maps the secret integer c_i to a real number $r_{c_i}^+$ using function $\mathcal{F}()$, i.e. $r_{c_i}^+ = \mathcal{F}(c_i)$. Note that, although the encryption algorithm that we use do not support encryption of real numbers directly, we can use integers in the chosen field to simulate a fixed point real number which is sufficient for our purpose. The set of mapped values for all the respondents is represented as $\mathcal{F}(S) = \{r_{c_1}^+, r_{c_2}^+, \ldots, r_{c_x}^+\}$. We require that the mapping from each c_i to $r_{c_i}^+$ by $\mathcal{F}()$ preserves certain order and distance relations for the integers in S for utility efficient anonymization, which is known as *p-probabilistic locality preserving* and is defined as follows:

Definition 2. *Given any two pre-images c_{i_1}, c_{i_2}, a mapping function $\mathcal{F}()$ is order preserving if:*

$$c_{i_1} \leq c_{i_2} \Rightarrow \mathcal{F}(c_{i_1}) \leq \mathcal{F}(c_{i_2}) \tag{3}$$

Given any three pre-images $c_{i_1}, c_{i_2}, c_{i_3}$, and the distances $dist_1 = |c_{i_1} - c_{i_2}|$, $dist_2 = |c_{i_2} - c_{i_3}|$, a mapping function $\mathcal{F}()$ is p-probabilistic distance preserving if:

$$dist_1 \leq dist_2 \Rightarrow \Pr(fdist_1 \leq fdist_2) \geq p \tag{4}$$

and it increases with $dist_2$, where $fdist_1 = |\mathcal{F}(c_{i_1}) - \mathcal{F}(c_{i_2})|$, $fdist_2 = |\mathcal{F}(c_{i_2}) - \mathcal{F}(c_{i_3})|$, and p is a parameter in the $[0, 1]$.

A mapping function $\mathcal{F}()$ is *p-probabilistic locality preserving* if it is both *order preserving* and *p-probabilistic distance preserving*. In addition, we also require that the mapping from c_i to $r_{c_i}^+$ reveals limited information about c_i, which is to be γ-concealing:

Definition 3. *Given the pre-image c_i and $r_{c_i}^+ = \mathcal{F}(c_i)$, the function $\mathcal{F}()$ is γ-concealing if $\Pr(c_{mle} = c_i | r_{c_i}^+) \leq 1 - \gamma$ for the Maximum Likelihood Estimation (MLE) c_{mle} of c_i.*

Stage 2. The goal of this stage is to determine a set of partitions of respondents based on the set of values in $\mathcal{F}(S)$ using 1D optimal k-anonymization algorithm or the l-diversity heuristics as proposed in *FALL*.

Stage 3. The goal of this stage is to privately anonymize the respondent data records based on the partitions from *Stage 2*, which involves secure computation of *equivalence classes* for the respondents in the same partition. As $\mathcal{F}(S)$ is p-probabilistic locality preserving, if we use the same partitions created on $\mathcal{F}(S)$ to anonymize T, we expect that the anonymized table $\mathcal{K}(T)$ preserves the utility well.

4.2   Technical Details

Stage 1. Probabilistic Locality Preserving Mapping. The challenge of performing p-*probabilistic locality preserving* mapping in this application is that all the data values in S are distributed, and we must ensure the secrecy of c_i for respondent i in the mapping process. In our approach we build an encrypted index $E(R+) = \{E(r_1^+), \ldots, E(r_{c_{max}}^+)\}$ on the server side containing c_{max} randomly generated numbers that correspond to all integers in the range $[1, c_{max}]$ of the mapping function $\mathcal{S}$. Later, each respondent i retrieves then the c_i^{th} item in the encrypted index, i.e., the item $E(r_{c_i}^+)$, in a private manner and can jointly and safely decrypt it with other respondents in order to build the anonymized data.

Essentially, four steps are needed in order to achieve p-*probabilistic locality preserving mapping*: *Step 1.* Two sets of encrypted real numbers are created at the server side, i.e. $E(R_{init})$ and $E(R_p)$. *Step 2.* The set of encrypted real numbers $E(R+)$ is created in a recursive way using the two sets of encrypted real numbers from *Step 1*: the set $E(R_{init})$ is used to define the value of the first encrypted number $E(r_1^+)$ and the set $E(R_p)$ is used to define number $E(r_i^+)$ in terms of $E(r_{i-1}^+)$. *Step 3.* Respondent i retrieves the c_i^{th} item from index $E(R+)$ created in *Step 2* using a *private information retrieval* scheme. *Step 4:* The retrieved encrypted item is jointly decrypted by t_{ss} parties, and uploaded to the server. Its plaintext is defined as $r_{c_i}^+$, i.e., the image of c_i under $\mathcal{F}()$. In the following, we describe the above four steps in detail. In *Step 1*, we first describe how to create one encrypted random real number whose plaintext value is not known by any parties. The creation of two sets of encrypted real numbers is just a simple repetition of this process.

In order to hide the value of a random number, each of these is jointly created by both a respondent and the server. To compute an encrypted joint random number $E(r)$, the respondent randomly selects a real number r_{dr} from a uniform distribution in the interval $[\rho_{min}, \rho_{max}]$. Then the respondent sends the encrypted number $E(r_{dr})$ to the server. The server independently chooses another random real number r_{svr} from the same interval $[\rho_{min}, \rho_{max}]$ and encrypts it to obtain $E(r_{svr})$. The join of the two encrypted real numbers is computed as $E(r) = E(r_{dr}) \cdot E(r_{svr}) = E(r_{dr} + r_{svr})$ by the additive homomorphic property of the Paillier's encryption (assuming a large modulus N is used so that round up does not take place). We are aware that both the respondent i and the server knows the range information about r. We denote such range knowledge about the joint random numbers for respondent i and the server as $\mathcal{RG}_i$ and $\mathcal{RG}_{svr}$, respectively. Recall that $\mathcal{L}_{svr}$ and $\mathcal{L}_i$ are the information leakage for the server and the data respectively. Therefore, we have that $\mathcal{RG}_{svr} \in \mathcal{L}_{svr}$ and $\mathcal{RG}_i \in \mathcal{L}_i$.

With the above technique, the first encrypted set of joint random numbers that we create is $E(R_{init}) = \{E(\iota_1), E(\iota_2), E(\ldots), E(\iota_b)\}$, where the size b is a security parameter of the system. Each of the encrypted joint random numbers is created by the server and a randomly selected respondent. The second set of encrypted joint random numbers that we create on the server side is $E(R_p) = \{E(r_1), E(r_2), \ldots, E(r_{c_{max}})\}$. To create $E(R_p)$, each respondent needs to generate $\lfloor \frac{c_{max}}{x} \rfloor$ or $\lceil \frac{c_{max}}{x} \rceil$ encrypted joint random numbers with the server, if we distribute this task evenly.

In *Step 2*, to build an encrypted set of real numbers $E(R^+) = \{E(r_1^+), E(r_2^+), \ldots, E(r_{c_{max}}^+)\}$ whose plaintexts values are in ascending order based on $E(R_{init})$ and $E(R_p)$, we once again use the additive homomorphic property of Paillier's encryption:

$$\begin{cases} E(r_i^+) = E(r_i) \cdot \prod_{j=1}^{b} E(\iota_j) & i = 1 \\ E(r_i^+) = E(r_{i-1}^+) \cdot E(r_i) & i = 2, \ldots, c_{max} \end{cases} \quad (5)$$

In *Step 3*, $E(r_{c_i}^+)$ is retrieved from the server by the respondent i who owns the secret c_i using Private Information Retrieval (*PIR*) scheme. We adopt the single database *PIR* scheme developed in [5] which supports the retrieval of a block of bits with constant communication rate. This *PIR* scheme is proven to be secure based on a simple variant of the Φ-hiding assumption. To hide the complexity of the *PIR* communications, we use the $\mathcal{PIR}(c_i, E(R^+))$ to represent the sub-protocol that privately retrieves the $c_i{}^{th}$ item in the set $E(R^+)$ by the ith respondent, and the result of retrieval is $E(r_{c_i}^+)$.

In *Step 4*, after the respondent i has retrieved $E(r_{c_i}^+)$, he partially decrypts $E(r_{c_i}^+)$ and sends the partially decrypted cipher to $t_{ss} - 2$ other respondents for further decryption. The last partial decryption is done by the server, after which the server obtains the plaintext $r_{c_i}^+$. Note that the server cannot identify the value of c_i by re-encrypting the $r_{c_i}^+$ and search through $E(R^+)$, as the Paillier's encryption is a randomized algorithm in which the output ciphers are different for the same plaintext with different random inputs. Finally, we have achieved the mapping from the c_i to $r_{c_i}^+$. The server obtains the set $\mathcal{F}(S)$ by the end of this step.

We illustrate these four steps in the Figure 3. The first column describes the respondents 1D data. The second column represents the 33rd to 40th entries in $E(R^+)$. The third column represents 33rd to 40th entries in $E(R_p)$. The ith entry of $E(R^+)$ is computed based on the product of the $(i-1)^{th}$ entry of $E(R^+)$ and the ith entry of $E(R_p)$. For example, $E(r_{34}^+) = E(r_{33}^+) \cdot E(r_{34})$ by the additive homomorphic property, $E(r_{34}^+) = E(r_{33}^+ + r_{34})$ which translated in terms of real values gives $E(304.7) = E(293.5) \cdot E(11.2) = E(293.5 + 11.2)$.

Theorem 1. *The mapping function $\mathcal{F}()$ is $\frac{1}{2}$-probabilistic locality preserving.*

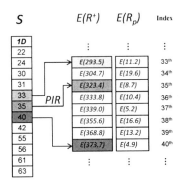

Fig. 3. Example of the probabilistic locality preserving mapping construction

Proof. Since R_p^+ is a set of ascending real numbers, we have $r_{c_{i_1}}^+ \leq r_{c_{i_2}}^+$, if $c_{i_1} \leq c_{i_2}$. Therefore, $\mathcal{F}()$ is order preserving by Equation 3. To prove that it is also $\frac{1}{2}$-*probabilistic distance preserving*, let $c_{i_1}, c_{i_2}, c_{i_3}$ be any randomly selected pre-images, and $dist_1$, $dist_2$, $fdist_1$ and $fdist_2$ follow the definitions in Definition 2 Equation 4. Assume that $c_{i_1} \leq c_{i_2} \leq c_{i_3}$ and $dist_1 \leq dist_2$. The exact form of the distributions of $fdist_1$ and $fdist_2$ are difficult to estimate. However, since $fdist_1$ ($fdist_2$ resp.) is the sum of $dist_1$ ($dist_2$ resp.) number of joint random numbers, where each joint random number is the sum of two random uniformly selected real numbers in the interval $[\rho_{min}, \rho_{max}]$, $fdist_1$ and $fdist_2$ can be unbiasedly approximated by continuous normal distribution according to the *central limit theorem*. Let $\mu = \frac{\rho_{min}+\rho_{max}}{2}$ and $\sigma^2 = \frac{(\rho_{min}-\rho_{max})^2}{12}$ be the mean and variance of the uniform distribution respectively, and without ambiguity, $fdist_1$ and $fdist_2$ be the continuous random variables. From the *central limit theorem*, we have $fdist_1 \sim N(dist_1 \cdot 2\mu, dist_1 \cdot 2\sigma^2)$ and $fdist_2 \sim N(dist_2 \cdot 2\mu, dist_2 \cdot 2\sigma^2)$. Therefore, $fdist_1 - fdist_2 \sim N((dist_1 - dist_2) \cdot 2\mu, (dist_1 + dist_2) \cdot 2\sigma^2)$. From the property of continuous normal distribution, $\Pr(fdist_1 - fdist_2 \leq 0) = \Pr(fdist_1 \leq fdist_2) \geq \frac{1}{2}$ when $dist_1 \leq dist_2$ and it increases with $dist_2$. Hence, by Equation 4, $\mathcal{F}()$ is also $\frac{1}{2}$-*probabilistic distance preserving*.

Stage 2. Anonymization in the mapped space. Suppose the anonymization algorithm in *FALL*(i.e. the 1D optimal k-anonymization or l-diversity heuristics) is used by the server for determining the partitions. Let $Z = \{z_1, z_2, \ldots, z_\pi\}$ be the result of anonymization, where the ith element in Z is the ending index of the ith partition of respondents and there are π number of partitions. Without losing generality, we assume the indices in Z are sorted in ascending order.

Stage 3. Secure computation of equivalence classes. In this stage, the quasi-identifiers of respondents in the same partition defined by Z form an equivalence class in $\mathcal{K}(T)$. Consider the ith partition defined by Z, which is formed by the $z_{i+1} - z_i$ number of respondents with IDs $z_i, z_i + 1, \ldots, z_{i+1} - 1$, where $k \leq z_{i+1} - z_i \leq 2k - 1$. Note that each non-sensitive attribute in the partition will be generalized to an interval in the $\mathcal{K}(T)$. Moreover, the interval for a particular attribute is the same for all the data records in this partition. We use $lep(a_j, i)$ and $rep(a_j, i)$ to represent the left endpoint and right endpoint of the interval, for the attribute a_j ($1 \leq j \leq u$) in the ith partition in the $\mathcal{K}(T)$, respectively. From the anonymization algorithm, we have:

$$
\begin{aligned}
lep(a_j, i) &= \min(a_j^{z_i}, a_j^{z_i+1}, \ldots, a_j^{z_{i+1}-1}) \\
rep(a_j, i) &= \max(a_j^{z_i}, a_j^{z_i+1}, \ldots, a_j^{z_{i+1}-1})
\end{aligned}
\tag{6}
$$

To find the minimum and maximum values of the set $\{ a_j^{z_i}, a_j^{z_i+1}, \ldots, a_j^{z_{i+1}-1} \}$ by the $z_{i+1} - z_i$ respondents, we employ the unconditionally secure constant-rounds *SMPC* scheme in [4]. This *SMPC* scheme provides a set of protocols that compute the shares of a function of the shared values.

Based on the result of [4], we can define a primitive comparison function $\overset{?}{<} : \mathbb{F}_\delta \times \mathbb{F}_\delta \to \mathbb{F}_\delta$ for some prime δ, such that $(\alpha \overset{?}{<} \beta) \in \{0, 1\}$ and $(\alpha \overset{?}{<} \beta) = 1$ iff $\alpha < \beta$. This function securely compares two numbers α and β, and outputs if α is less than β. With this function, the maximum and minimum numbers in a set are easily found based on a series of pairwise comparisons. We omit the details of the implementation the comparison function where the readers can find in [4].

The sub-protocol that uses the primitive comparison function $\overset{?}{<}$ to find the maximum and minimum values for the attribute a_j in the ith partition is called $\mathcal{M}(a_j, i)$ with the output $< lep(a_j, i), rep(a_j, i) >$. This sub-protocol is described as follows: first, each value in this set is shared using Shamir's (t_{ss}, t_{ss}) secret sharing. The shares are distributed via an *anonymous protocol* so that the identities of the shares' owners are not revealed. Second, with the shares, the pairwise comparison of values based on $\overset{?}{<}$ can be successfully constructed. The maximum and minimum values in $\{a_j^{z_i}, a_j^{z_i+1}, \ldots, a_j^{z_{i+1}-1}\}$ can be found with maximally $\left\lceil \frac{3 \cdot (z_{i+1}-z_i)}{2} \right\rceil - 2$ number of pairwise comparisons. Finally, the owners of the maximum value and minimum value publish their values of a_j anonymously and each respondent in the partition assigns the values of $lep(a_j, i)$ and $rep(a_j, i)$ accordingly.

For each non-sensitive attribute a_j ($1 \leq j \leq u$) and each partition i ($1 \leq i \leq \pi$), $\mathcal{M}(a_j, i)$ is run once. Therefore, the $\mathcal{M}$ sub-protocol runs for $\pi \cdot u$ rounds. Since the $\mathcal{M}$ sub-protocol runs independently within each partition, the sub-protocol can run simultaneously for each partition. By the end, the respondent j in the ith partition submits the anonymized data record $\mathcal{K}(t_j)=\{[lep(a_1, i), rep(a_1, i)], \ldots, [lep(a_u, i), rep(a_u, i)], s_1, \ldots, s_v\}$ to the server. After collecting $\mathcal{K}(t_1), \mathcal{K}(t_2), \ldots, \mathcal{K}(t_x)$ from all x respondents, the final anonymized table $\mathcal{K}(T)$ is created and is returned to the collector.

5   Analysis

5.1   Information Leakage

Theorem 2. *The privacy preserving data collection protocol only leaks $\mathcal{L}_{svr}$ for the server and $\mathcal{L}_i$ for the respondent i, where $\mathcal{L}_{svr} = \{\mathcal{RG}_{svr}, \mathcal{F}(S)\}$ and $\mathcal{L}_i = \{\mathcal{RG}_i\}$.*

Proof. We first construct the simulator M_{svr} for the server. In step 1 in stage 1, the knowledge of the server is described by $\mathcal{RG}_{svr}$, in which the server knows the range of each of the random numbers in $E(R)$ and $E(R_{init})$. Each joint encrypted random number in $E(R)$ and $E(R_{init})$ in the view of the server can be simulated by M_{svr} by multiplying an encrypted random number in the range of $[\rho_{min}, \rho_{max}]$ to the encrypted random number contributed by the server. In step 2, the $E(R^+)$ is constructed based on $E(R)$ and $E(R_{init})$, where no information is leaked during the computation based on the semantic security of the Paillier's encryption. Therefore, M_{svr} simulates $E(R^+)$ based on the simulations of $E(R)$ and $E(R_{init})$. In step 3, the server gains no information about the retrieved item which is guaranteed by the property of $\mathcal{PIR}()$ function. The decrypted value in step 4 is $\mathcal{F}(S)$, which is part of the knowledge of the server. In stage 2, the input is based on $\mathcal{F}(S)$, therefore the server does not gain any additional information. In stage 3, the server receives the anonymized tuples from the respondents, the received data are equivalent to the knowledge of the server $\mathcal{K}(T)$.

Now, we construct the simulator M_i for the respondent i. In stage step 1 in stage 1, the knowledge of respondent i is described by $\mathcal{RG}_i$, in which he knows the range of joint random numbers which are jointly created by him and the server. The respondent is not participating in step 2. In step 3, M_i simulates the retrieved ciphertext by a random ciphertext. In step 4, M_i simulates the partially decrypted message by partially decrypted the random ciphertext. The respondent is not participating in stage 2. In stage 3, the secret shares and messages can be simulated by M_i using random ciphers, guaranteed by the function sharing algorithm in [4]. The output is equivalent to the knowledge of the respondent $\mathcal{K}(T)$.

5.2 γ-Concealing Property

A property explaining how well the mapped value $r_{c_i}^+$ hides the value c_i is described by the notion of γ-concealing. The value of $1 - \gamma$ (the probability the adversary can guess c_i correctly based on $r_{c_i}^+$) can be approximated as follows: with $r_{c_i}^+$, the Maximum Likelihood Estimation of c_i is $c_{mle} = \left\| r_{c_i}^+ / \mu \right\| - b$ (i.e. $c_{mle} =$ roundup$(r_{c_i}^+ / \mu) - b$). As the condition for $c_i = \left\| r_{c_i}^+ / \mu \right\| - b$ is equivalent to the condition for $r_{c_i}^+$ to be in the range of $[(c_i - \frac{1}{2} - b)\mu, (c_i + \frac{1}{2} - b)\mu]$, we can establish the following equivalence:

$$\Pr(c_{mle} = c_i | r_{c_i}^+) = \Pr(r_{c_i}^+ \in [(c_i - \frac{1}{2} - b)\mu, (c_i + \frac{1}{2} - b)\mu]) \tag{7}$$

The probability value on the r.h.s of the above equation can be approximated using the *central limit theorem*. According to the *central limit theorem*, $r_{c_i}^+$ is approximately normally distributed with $r_{c_i}^+ \sim N((c_i + b)\mu, (c_i + b)\sigma^2)$. Thus, the following approximation holds:

$$1 - \gamma \approx \begin{array}{l} \Phi_{(c_i + b)\mu, (c_i + b)\sigma^2}[(c_i + \frac{1}{2} - b)\mu] \\ - \Phi_{(c_i + b)\mu, (c_i + b)\sigma^2}[(c_i - \frac{1}{2} - b)\mu] \end{array} \tag{8}$$

In the above equation, $\Phi_{(c_i + b)\mu, (c_i + b)\sigma^2}$ is the distribution function of a normal distribution with mean $(c_i + b)\mu$, and variance $(c_i + b)\sigma^2$. The equation shows that, the value of $1 - \gamma$ relies on the values of μ, σ^2, b and c_i. Particularly, the protocol tends to be secure when large σ^2, b, and c_i values, and small μ value are used.

6   Experimental Evaluation

In this section, we carry out several experiments to evaluate the performance of the proposed privacy preserving data collection protocol. The experiments are divided into four parts: in the first part, we evaluate the γ-concealing property of the proposed protocol. In the second part, we evaluate the *probabilistic distance preserving* property in the proposed protocol due to its importance in utility preservation. In the third part, we evaluated the performance of the protocol in utility preservation. In order to compare with *FALL* $-$ the k-anonymization algorithm that the proposed protocol is based on, we employ the utility metric *GCP* [6]. Lastly, we evaluate the running time of the protocol to show the practicality.

The dataset that we use for the experiments is from the website of Minnesota Population Center (MPC)[1], which provides census data over various locations through different time periods. For the experiments, we have extracted 1% sample USA population records with attributes *age, sex, marital status, occupation* and *salary* for the year 2000. The dataset contains $2,808,457$ number of data records, however, we only use a subset of these records. Among the five attributes, the age is numerical data while others are categorical data. For the categorical data, we can use taxonomy trees (e.g. [2,9]) to convert a categorical data to numerical data for generalization purposes. Among all the seven attributes, the *salary* is considered as the sensitive attribute, while others are non-sensitive and are considered as quasi-identifiers. The domain sizes for *age, sex, marital status* and *occupation* are 80, 2, 6 and 50, respectively. The programs for the experiments are implemented in Java and run on Windows XP PC with $4.00\,GB$ memory and Intel(R) Core(TM)2 Duo CPU each at $2.53\,GHz$.

6.1   Evaluation of γ-Concealing Property

In this part of experiments, we compute some real values of $1-\gamma$ with predefined parameters based on the formulas in Equation 8, to show that the proposed protocol is privacy preserving. The Figure 4(a) shows the result of how the value of $1-\gamma$ changes with the value of μ and σ^2. In the first three rows of the table, we keep the value of μ constant ($\mu = 150$) while increasing the value of σ^2. Notice that the value of $1-\gamma$ decreases with increasing σ^2. In the last three rows of the table, we keep the values of σ^2 constant ($\sigma^2 = 7,500$) instead, and increase the values of μ. Notice in this case that the value of $1-\gamma$ increases with increasing μ. In Figure 4(b), we experimented how the value of $1-\gamma$ changes with the value of b and c_i. In the first three rows of the table, we keep the value of b constant ($b = 200$) and increase the value of c_i. We find that the value of $1-\gamma$ decreases with increasing c_i. In the last three rows of the table, we keep the value of c_i constant ($c_i = 0$) and increase b. It is true that the value of $1-\gamma$ decreases with increasing b. Since the minimum c_i is 0, the last three rows of the table shows the maximum values of $1-\gamma$ under different values of b. The values of $1-\gamma$ are all below 0.1 which supports the level of privacy that a respondent can hide his quasi-identifiers with probability at least 90% in the process of data collection. For stronger privacy, we can further lower the value of $1-\gamma$, by either decreasing μ or increasing σ^2 or b.

---

[1] http://www.ipums.org/

ρ_{min}	ρ_{max}	μ	σ^2	$1-\gamma$
100	200	150	833.333	0.119235
50	250	150	3333.33	0.0597853
0	300	150	7500	0.0398776
100	400	250	7500	0.0664135
150	450	300	7500	0.0796557
200	500	350	7500	0.0928758

b	c_i	$1-\gamma$
200	100	0.0796557
200	200	0.0690126
200	300	0.0617421
200	0	0.0974767
300	0	0.0796557
400	0	0.0690126

ρ_{min}	ρ_{max}	μ	σ^2	DPR
100	200	150	833.333	0.999525
50	250	150	3333.33	0.999174
0	300	150	7500	0.998411
100	400	250	7500	0.999223
150	450	300	7500	0.999438
200	500	350	7500	0.999536

(a) $b = 200$, $c_i = 100$ (b) $\rho_{min} = 100$, $\rho_{max} = 300$ $b = 200$

Fig. 4. γ-Concealing and relative distance preserving

6.2   Evaluation of *Distance Preserving* Mapping

In this part of experiments, we show that the proposed mapping function $\mathcal{F}()$ can quite well preserve the *relative distance*. For this purpose, we propose the *Distance Preserving Ratio* (DPR) metric. Given a set of pre-images $\{c_1, c_2, \ldots, c_x\}$, and the set of images $\{\mathcal{F}(c_1), \mathcal{F}(c_2), \ldots, \mathcal{F}(c_x)\}$. A *relative distance preserving triple* (*RDPT*), is a combination of three pre-images $< c_{i_1}, c_{i_2}, c_{i_3} >$ whose images $< \mathcal{F}(c_{i_1}), \mathcal{F}(c_{i_2}), \mathcal{F}(c_{i_1}) >$ preserve their relative distances. The DPR is defined as follows:

$$DPR = \frac{\text{total no. of } RDPT < c_{i_1}, c_{i_2}, c_{i_3} >}{\text{total no. of triples } C(x, 3)} \tag{9}$$

The DPR describes the ratio between the number of triples of pre-images whose mapping preserve relative distances and the total number of triples in the set of pre-images. In the experiments, we randomly select 2,000 data records from the dataset. We convert the non-sensitive attributes of selected data records into a set of integers using Hilbert curve, and input it to $\mathcal{F}()$ as the set of pre-images. The set of parameters used is the same as the one used in the experiments for γ-concealing property. In Figure 4(c), we see that when μ is fixed to 150, the value of DPR decreases with increasing of σ^2. On the other hand, when we fix the value of σ^2 to be 7,500, the value of DPR increases with μ. In other words, large μ and small σ^2 has positive impacts on *relative distance preserving*. In all cases, the values of DPR are extremely high (almost close to 1), which clearly show that the mapping function $\mathcal{F}()$ achieves excellent *relative distance preserving*.

6.3   Evaluation of Utility Preservation

Lastly, we evaluate the utility preservation property of the proposed protocol by measuring the utility loss (the *GCP* metric) against several parameters. The set of data records used in the first three experiments is the same set of 2,000 data records used in the last part of the experiments.

In the first experiment, we measure the *GCP* value against increasing k. The parameters that we use are $b = 200$, $\rho_{min} = 200$ and $\rho_{max} = 500$. Figure 5(a) shows that the value of *GCP* increases with increasing k (as expected). Moreover, the *GCP* value computed based on table created by *FALL* (as labeled) and the proposed protocol (labeled as *Distr.*) are almost the same, showing that our approach can achieve almost the same level of utility preservation as the *FALL*. A naive method (labeled as *Order only*), which only sorts the respondents in 1D space and group every consecutive k respondents, results in much higher *GCP* values compared to *FALL* and our approach. Figure 5(b) shows the utility loss for both *FALL* and the proposed protocol with increasing σ^2. Though from Figure 5(a), the curve of utility loss for *FALL* and the proposed protocol appear to be overlapping, when we focus the *GCP* values in the interval of [0.55, 0.6] in Figure 5(b), we indeed observe that the performance of the proposed protocol in utility preservation is slightly less optimal compare to *FALL*. Moreover, the Figure 5(b) shows that the *GCP* value based on the proposed approach increases with increasing σ^2 at relatively slow rate. Similarly, Figure 5(c) shows that increasing μ value helps to reduce the *GCP* value. In Figure 5(d), in order to evaluate how the *GCP* value changes with the data size, we increase the data size from 10,000 to 50,000. It shows that the *GCP* value for both *FALL* and the proposed approach decreases at similar rate with increasing data size. The decreasing of *GCP* value is due to the fact that when data size increases, the density of data also increases. To conclude this part, these experiments show that with appropriate parameters, the proposed approach achieves almost the same utility preservation performance as *FALL*. Figures 5(e)(f)(g)(h) show the utilities in the anonymized table based on the l-diversity heuristics in *FALL*. The pattern is similar to the experiments for k-anonymity except that the utilities for l-diversity is lower than the k-anonymity as expected.

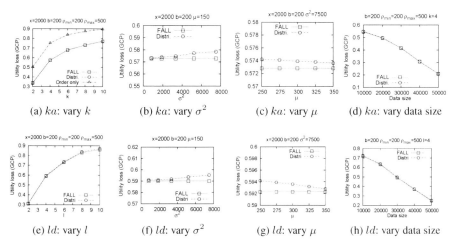

(a) ka: vary k (b) ka: vary σ^2 (c) ka: vary μ (d) ka: vary data size

(e) ld: vary l (f) ld: vary σ^2 (g) ld: vary μ (h) ld: vary data size

Fig. 5. Utility preservation evaluation

6.4   Evaluation of System Time

In this experiment, we show the practicality of our protocol using experiment implemented with java BigInteger class and Security package. We aim to verify the practicality of the solution rather than comparing its efficiency against the requirement for real time applications. The experiments results match our expectation for privacy preserving data collection application.

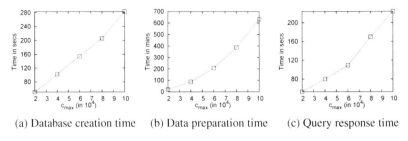

(a) Database creation time   (b) Data preparation time   (c) Query response time

Fig. 6. System time

The time evaluation focuses on the server database generation and PIR part which are the main time components of the protocol for both respondent side and server side. On the server side, the first time component is the creation of the encrypted database of size c_{max} that is equivalent to the size of domain of T as in step 1 and 2 in stage 1 of the protocol. Since c_{max} is usually large, we vary this value from 10,000 to 100,000. For the parameters setting, we choose $b = 200$, $\rho_{min} = 0$ and $\rho_{max} = 300$. In the Paillier's encryption, the modulus is set to 512 bits which creates blocks of size 512 bits on the server. Figure 6(a) shows the time (in seconds) needed by the server to create the joint encrypted random variables and encrypted database by using Paillier's encryption. Experiment shows that this step is efficient as the number of encryptions is linear to c_{max}. The second time component for the server is database preparation time. The PIR scheme in [5] requires the database to be prepared as a big integer using Chinese Remainder Theorem so as to answer PIR query. Figure 6(b) shows the database preparation time (in mins) and suggests that the time needed for this step could last for about 10.5 hours for $c_{max} = 100,000$. However, as this step is taken independently on the server for once only and requires no interaction with the data respondents, we still consider it as practical. The third server time component is the response time to the respondent's PIR query. The parameters for PIR are determined by database size and block size. In each query, the respondent privately retrieves 512 bits from the server. Figure 6(c) shows the server response time which is in minutes. Compare to the server, the client is lightly loaded in cryptographic operations. We assume 1,000 respondents are participating and measure the sum of joint random number generation time as in step 1 in stage 1 of the protocol, the query generation time and answer extraction time. Our experiments show that the PIR query generation and answer extraction at the respondent side is less sensitive to the size of database on the server side and the number of participating data respondents, and the total time needed by the respondents are less than 5 seconds. From both complexity analysis and experiment, we show that our protocol is practical. The efficiency of the our protocol could be further improved by optimizing the memory or CPU usage, or using dedicated hardware circuits for cryptographic operations.

7   Conclusions

We proposed a privacy preserving data collection protocol under the assumption that the data collector is not trustworthy. With our protocol, the collector receives an anonymized (either k-anonymized or l-diverse) table generalized from the data records of the respondents. We show that the privacy threat caused by the information leakage remains limited. Lastly, we show with experiments that the protocol is scalable, practical and that the data utility is almost as good as in the case of a trustworthy collector.

References

1. Asmuth, C., Bloom, J.: A modular approach to key safeguarding. IEEE Trans. Information Theory 29(2), 208–210 (1983)
2. Bayardo, R., Agrawal, R.: Data privacy through optimal k-anonymization. In: Proc. of ICDE, pp. 217–228 (2005)
3. Brickell, J., Shmatikov, V.: Efficient anonymity-preserving data collection. In: KDD 2006, pp. 76–85. ACM, New York (2006)
4. Damgard, I., Fitzi, M., Kiltz, E., Nielsen, J., Toft, T.: Unconditionally secure constant-rounds multi-party computation for equality, comparison, bits and exponentiation, pp. 285–304 (2006)
5. Gentry, C., Ramzan, Z.: Single-database private information retrieval with constant communication rate, pp. 803–815 (2005)
6. Ghinita, G., Karras, P., Kalnis, P., Mamoulis, N.: Fast data anonymization with low information loss. In: Proc. of VLDB, pp. 758–769 (2007)
7. Jurczyk, P., Xiong, L.: Privacy-preserving data publishing for horizontally partitioned databases. In: CIKM 2008: Proceeding of the 17th ACM Conference on Information and Knowledge Mmanagement, pp. 1321–1322. ACM, New York (2008)
8. Kaya, K., Selçuk, A.A.: Threshold cryptography based on asmuth-bloom secret sharing. Inf. Sci. 177(19), 4148–4160 (2007)
9. LeFevre, K., DeWitt, D.J., Ramakrishnan, R.: Incognito: Efficient full-domain k-anonymity. In: Proc. of ACM SIGMOD, pp. 49–60 (2005)
10. Machanavajjhala, A., Gehrke, J., Kifer, D., Venkitasubramaniam, M.: l-diversity: Privacy beyond k-anonymity. In: Proc. of ICDE (2006)
11. Meyerson, A., Williams, R.: On the complexity of optimal k-anonymity. In: PODS 2004, pp. 223–228. ACM, New York (2004)
12. Moon, B., Jagadish, H.v., Faloutsos, C., Saltz, J.H.: Analysis of the clustering properties of the hilbert space-filling curve. IEEE TKDE 13(1), 124–141 (2001)
13. Paillier, P.: Public-key cryptosystems based on composite degree residuosity classes, pp. 223–238 (1999)
14. Samarati, P., Sweeney, L.: Generalizing data to provide anonymity when disclosing information (abstract). In: Proc. of ACM PODS, p. 188 (1998)
15. Shamir, A.: How to share a secret. Commun. ACM 22(11), 612–613 (1979)
16. Sweeney, L.: k-anonymity: A model for protecting privacy. Int. J. of Uncertainty, Fuzziness and Knowledge-Based Systems 10(5), 557–570 (2002)
17. Yang, Z., Zhong, S., Wright, R.N.: Anonymity-preserving data collection. In: KDD 2005, pp. 334–343. ACM, New York (2005)
18. Zhong, S., Yang, Z., Chen, T.: k-anonymous data collection. Inf. Sci. 179(17), 2948–2963 (2009)
19. Zhong, S., Yang, Z., Wright, R.N.: Privacy-enhancing k-anonymization of customer data. In: PODS 2005, pp. 139–147. ACM, New York (2005)

Privacy Preserving Query Processing on Secret Share Based Data Storage

XiuXia Tian[1,2], ChaoFeng Sha[1], XiaoLing Wang[3], and AoYing Zhou[3]

[1] School of Computer Science, Fudan University, Shanghai 200433, China
[2] School of Computer and Information Engineering, Shanghai University of Electric Power,
Shanghai 200090, China
[3] Institute of Massive Computing, East China Normal University, Shanghai 200062, China
{xxtian,cfsha,wxling,ayzhou}@fudan.edu.cn

Abstract. Database as a Service(DaaS) is a paradigm for data management in which the Database Service Provider(DSP), usually a professional third party for data management, can host the database as a service. Many security and query problems are brought about because of the possible untrusted or malicious DSP in this context. Most of the proposed papers are concentrated on using symmetric encryption to guarantee the confidentiality of the delegated data, and using partition based index to help execute the privacy preserving range query. However, encryption and decryption operations on large volume of data are time consuming, and query results always consist of many irrelevant data tuples. Different from encryption based scheme, in this paper, we present a secret share based scheme to guarantee the confidentiality of delegated data. And what is more important, we construct a privacy preserving index to accelerate query and to help return the exactly required data tuples. Finally we analyze the security properties and demonstrate the efficiency and query response time of our approach through empirical data.

1 Introduction

With the development of practical cloud computing applications, many IT enterprises have designed service products of their own, such as EC2 and S3 of Amazon, AppEngine of Google, in which EC2 can provide clients with scalable servers service, S3 can provide low cost and convenient network storage service, and AppEngine can provide network application platform service. Database as a Service(DaaS) emerges and conforms to this trend, which is more attractive for small enterprises who couldn't afford the expensive hardware and software, especially the expensive costs for hiring data management expertise. Despite of advantages of the DaaS paradigm, it also introduces security and query efficiency problems due to the possible untrusted or malicious DSP. Most of the proposed papers[2] and [3] are focused on using encryption, especially symmetric encryption algorithm, to guarantee the confidentiality and authorized access of the delegated data, and using partition based index to execute privacy preserving range query[8], [9].

However, data encryption and decryption operations are time consuming, and the query results always consist of many irrelevant data tuples. In this paper, we present a

J.X. Yu, M.H. Kim, and R. Unland (Eds.): DASFAA 2011, Part I, LNCS 6587, pp. 108–122, 2011.
© Springer-Verlag Berlin Heidelberg 2011

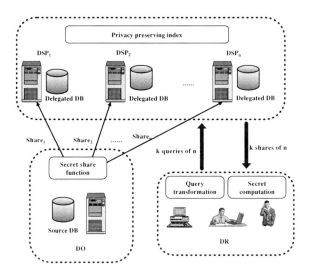

Fig. 1. Secret share based system architecture

secret share based scheme to guarantee the confidentiality of delegated data. And what is more important, we construct a privacy preserving index to accelerate query and to help return the exactly required data tuples. We illustrate our secret share based system architecture in Fig. 1.

- As the general DaaS paradigm, there are three types of entities in our architecture, Data Owner(DO), Data Requestor(DR), and Database Service Provider(DSP). However, in our approach there is a set of DSPs, not one, who cooperate to guarantee the confidentiality of delegated data.
- DO delegates his/her source database(Source DB) to n DSPs by using the secret share function, in which DO divides a private data in source database into n $shares$ by using the secret share function, and each $share$ is delegated to one different DSP, such as $share_1$ for DSP_1, $share_2$ for DSP_2, and $share_n$ for DSP_n respectively.
- DR transforms a query into the privacy preserving form by using the query transformation function, and derives the real query results by using the secret computation function in terms of the returned results from any k different DSPs.

The contributions of our paper are in the following:

- This paper introduces efficient secret share scheme[1] to guarantee the confidentiality of delegated data, in which a private data is divided into n $shares$(n is the number of DSPs), and then each $share$ of them is delegated to one different DSP.
- We construct an ordered privacy preserving index through combining the partition and encryption techniques, which can be used to accelerate data access for DRs online without disclosing data privacy.
- In order to guarantee the privacy of query from DRs in client, the submitted query is transformed into a private form in terms of the proposed privacy preserving index.
- We evaluate our approach with empirical data from two different aspects, security analysis and efficiency evaluation respectively.

The rest of the paper is organized as follows. Section 2 first introduces the preliminary knowledge about the secret share scheme. Then we describe the secret share based storage model and give the general concept of privacy preserving index in section 3. We elaborate our proposed privacy preserving index in section 4. Section 5 describes the query transformation and processing in terms of the proposed privacy preserving index. Then comes the experiment evaluation in section 6. Finally we give the related work and conclusion in section 7 and section 8 respectively.

2 Preliminary

In this section, we first introduce the secret share scheme and its related properties, then describe how to divide and delegate a private data to n DSPs by using the secret share scheme, and finally describe how DR reconstructs the required private data from any k different *shares* of n DSPs.

2.1 Secret Share Scheme

In cryptography community, everyone knows that encryption or decryption algorithm can be publicly known, but the secret or private key must be kept secret. That is to say, the security of secret key is critical to the security of encrypted data. Once the secret key is leaked, the encrypted data will be in danger or leaked to the malicious attackers. Shamir[1] proposed a secret share scheme to protect the security of secret key, where a secret key can be divided into n *shares*, and each *share* is distributed to one participant, only the designated number of participants like k or more participants like t, $t > k$ together can reconstruct the secret key. The scheme is called (k, n) threshold scheme if and only if it satisfies the following two properties:

- Knowledge of any k or more than k of n *shares* makes the secret key easily computable.
- Knowledge of any $k-1$ or fewer than $k-1$ of n *shares* makes the secret completely undetermined.

In order to understand our approach clearly, we give the definition of *share* used throughout the paper as follows:

Definition 1. *A share is the result value y by computing the following polynomial on inputting a known x.*

$$y = f(x) = a_{k-1}x^{k-1} + a_{k-2}x^{k-2} + ... + a_1x + a_0$$

where $a_{k-1}, a_{k-2}, ..., a_0 \in \mathcal{F}_q$, q is a large prime, $\mathcal{F}_q$ is a finite domain on q, a_0 is the secret value.

From the definition above we know that n *shares* are $y_1, y_2, ..., y_n$ computed from known $x_1, x_2, ..., x_n$ respectively, and the polynomial $y = f(x)$ can be reconstructed from any k known pairs $(x_{i1}, y_{i1}), (x_{i2}, y_{i2}), ..., (x_{ik}, y_{ik})$ of n pairs $(x_1, y_1), ..., (x_n, y_n)$.

2.2   Data Division at DO

Assume there are one DO, n DSPs, and many DRs. DO wants to use the database service provided by DSPs to relieve him/her from expensive database management. In order to prevent the delegated database from reading by the possible untrusted or malicious DSPs and other attackers, DO first uses the secret share function in Fig. 1 to deal with the private data in source database and then delegates them to n DSPs. After that DRs in client can access the delegated database from n DSPs conveniently and efficiently without revealing any data privacy. For simplicity, assume there is only one table *Employees* in DO's source database, and there are three attributes *empno, name*, and *salary* respectively in Table 1. The unit of *salary* is thousand, for example, 20 is equivalent to 20000. In order to protect the private value of attribute *salary*, DO uses the secret share function to divide each value for attribute *salary* to n *shares* and then distributes each of them to one different DSP.

Table 1. Employees

empno	name	salary
20060019	Mary	20
20060011	John	35
20050012	Kate	40
20050001	Mike	50
20040018	Henry	75

Example 1. Assume the scheme is a $(3, 5)$ threshold scheme created on finite domain $\mathcal{F}_{103}$, then DO divides each numeric value v_{sal} for *salary* attribute in Table 1 into 5 *shares*, and stores each share $share_i$ into *share* table $Employees_i$ at $DSP_i, 1 \leq i \leq 5$, such as five $share_1$s into *share* table $Employees_1$ at $DSP_1, \ldots$, five $share_5$s into *share* table $Employees_5$ at DSP_5 respectively. In terms of the secret share properties above, we know that any knowledge of less than 3 *shares* can't reconstruct the private value. The concrete private value division processing is as follows:

DO randomly chooses a polynomial on finite domain $\mathcal{F}_{103}$, such as

$$y = f_{20}(x) = 20 + 18x + 3x^2,$$

and selects the minimal generator $g = 5$ of prime $q = 103$. The following computation is how to divide the private *salary* value $v_{sal} = 20$ in Table 1 into five *shares* on finite domain $\mathcal{F}_{103}$. The notation $< x >_q = y$ means that the result y is the remainder of prime q divided by numeric value x, $x \equiv y \pmod{q}$.

$$< 5^1 >_{103} = 5, < 5^2 >_{103} = 25, < 5^3 >_{103} = 22, < 5^4 >_{103} = 7, < 5^5 >_{103} = 35$$

Assume $X = \{5, 25, 22, 7, 35\}$ is a set of chosen $x_i, x_i \in X$ such as $x_1 = 5, x_2 = 25$, $\ldots, x_5 = 35$, then the five *shares* are computed as follows:

$$share_{(20,1)} = f_{20}(< 5^1 >_{103}) = < 20 + 18 * 5 + 3 * 5^2 >_{103} = 82$$

$$share_{(20,2)} = f_{20}(< 5^2 >_{103}) = < 20 + 18 * 25 + 3 * 25^2 >_{103} = 79$$

Computed as above $share_{(20,3)} = 14$, $share_{(20,4)} = 87$, $share_{(20,5)} = 102$. For simplicity, we omit the notation $<>$, and the subscript 103, then 5^1 is equal to $< 5^1 >_{103}$, $5^1 = 5,...,5^5 = 35$. After the computation above, DO sends share $share_{(20,i)}$ to DSP_i, $1 \leq i \leq 5$. Therefore each DSP_i couldn't gain the private *salary* value 20, because he/she only knows one *share* of the private value. In fact even if more than 3 DSPs collude they couldn't derive the private salary value 20, since only the DO and the DR belongs to DO who know the secret set X.

2.3 Private Data Reconstruction at DR

The DRs can access the delegated database from n DSPs, each of whom stores one *share* table of the source database. When a DR wants to query from DSPs, he/she needs to transform the query into at least 3 equal privacy preserving queries for any 3 of 5 DSPs. Assume $share_{(20,1)}$, $share_{(20,2)}$, and $share_{(20,3)}$ are the three *shares*, then by using Lagrange interpolation the reconstructed polynomial is in the following:

$$y = 82 \cdot \frac{x - 5^2}{5 - 5^2} \frac{x - 5^3}{5 - 5^3} + 79 \cdot \frac{x - 5}{5^2 - 5} \cdot \frac{x - 5^3}{5^2 - 5^3} + 14 \cdot \frac{x - 5}{5^3 - 5} \cdot \frac{x - 5^2}{5^3 - 5^2} = 20 + 18x + 3x^2$$

Setting $x = 0$ and computing $f_{20}(0)$ the DR can get the private salary value 20. All the computation is implemented on the finite domain $\mathcal{F}_{103}$. The division of private value 35, 40, 50, 75 is the same as above.

In order to protect the private data against the untrusted or malicious DSPs, the values for *salary* attribute and the corresponding random polynomials can't be stored at DSPs, each of them only stores one *share* for each private salary value. Otherwise the private value will be leaked to the untrusted DSPs.

In fact by division, $share_i$s for all salary values stored in $share_i$ table at DSP_i are unordered even if the values of *salary* attribute in source database are ordered, $1 \leq i \leq 5$. Therefore all the queries submitted by DR should be returned all *shares* from any k DSPs, which will in turn result in the large communication and bandwidth costs between the DR and any k of n DSPs, as well as large computation costs in DR. However, in general, the DR has limit computation capability and lower bandwidth, therefore the simple application of secret share scheme to guarantee the confidentiality of the delegated data is not practical. In section 3 we will introduce the efficient secret share based storage model and present the concept of privacy preserving index.

3   Storage Model and Privacy Preserving Index

We first introduce storage model by using the secret share based data division described above, and then introduce the general definition of privacy preserving index.

3.1 Storage Model

In our approach, any relation in the following

$$R(A_1, A_2, ..., A_m),$$

where $m \geq 1$, in source database at DO, will be stored into n DSPs in the form of the following relation

$$R_i(key_1, share_i(A_1, i), ..., key_m, share_i(A_m, i))$$

where $1 \leq i \leq m$, relation R_i is delegated to DSP_i. $key_1, ..., key_m$ is the extra introduced key attribute for attributes $A_1, ..., A_m$ respectively, and each key is ordered in terms of the corresponding attribute values, such as key_1 is ordered in terms of attribute A_1. How to generate the keys for different attributes will be introduced in section 4. Each attribute key_i can be used to construct the index corresponding to attribute A_i, and the constructed index will be used to accelerate the query processing at each DSP_i.

Example 2. Table 1 is divided into five tables by using the secret share scheme, in which Table 2 is the delegated *share* table $Employees_3$ for DSP_3 to manage and maintain. In Table 2 the attribute name starting from character k represents the added key attribute for the corresponding attribute in source database. For instance k_{sal} is the key attribute for attribute *salary*. The attribute name starting from character s represents the *share* attribute for the corresponding attribute in source database. For instance s_{sal} is the *share* attribute for attribute *salary*.

Table 2. $Employees_3$

k_{no}	s_{no}	k_{name}	s_{name}	k_{sal}	s_{sal}
no214	20	name303	99	sal100	14
no211	38	name301	101	sal101	62
no212	65	name302	57	sal102	39
no210	18	name304	79	sal103	56
no213	34	name300	33	sal104	91

As a matter of fact, not all attributes k_{no}, k_{name} and k_{sal} should be added, we only add key attributes for the attributes which are accessed frequently, such as k_{sal}. Therefore in practical application, the keys added may be much less than those are showed in R_i.

3.2   Privacy Preserving Index

Definition 2. *Privacy preserving index is an index which can make the untrusted DSP to evaluate queries correctly with minimal information leakage.*

A partition based index is proposed in [3], in which an attribute domain is partitioned into a set of buckets, and each bucket is identified by a tag. These bucket tags are maintained as an index and are used by the untrusted DSP to accelerate queries. The bucket tags leak less information about the source data, that is to say, the DSP can't derive the private data from the delegated bucket tags. For example, assume the salary attribute in Table 1 is partitioned into three buckets, so the range of each bucket is $[0, 40)$, $[40, 80)$, $[80, 120)$ respectively. Each bucket is given a tag by using a map function, such as 4, 2, 7 respectively, and the bucket tags are sent to the untrusted DSP as an index.

Therefore for an exact query or obfuscated rang query from DR, the DSP returns a superset of tuples in terms of the partition based index, which include many irrelevant tuples.

However, the returned irrelevant tuples will result in large costs of communication, and large extra computation for the DR to decrypt them, and moreover lead to more than k times of that in one DSP scenario. So in the subsequent section 4 we will introduce our proposed privacy preserving index, which is expected to only retrieve the required or a small superset of tuples from any k of n DSPs.

4   Proposed Privacy Preserving Index

In order to give an efficient solution to the problems resulted by query on unordered data and minimize the returned irrelevant tuples, we construct a privacy preserving B+ tree index on the k_{sal} attribute for n DSPs to share. The constructed B+ tree index can accelerate query without leaking data privacy, because the values of k_{sal} attribute and the *shares* for *salary* attribute are irrelevant. So even if an attacker or a malicious DSP knows the values of k_{sal} attribute and the corresponding *shares* of *salary* attribute, he/she couldn't derive the real values of *salary* attribute. In the following subsections, we will present the construction of privacy preserving B+ tree index from two functions, key generation function and index creation function.

4.1   Key Generation Function $key\_generation$

We define an algorithm named as $key\_generation$ in Fig. 2 to generate keys for private attribute values, such as *salary* attribute in Table 1. In steps $1 - 4$, we first divide a specific domain different from *salary* domain into N uniform buckets $bucket\_ids$. Then in steps $5 - 6$ of the for loop the algorithm generates a random key to encrypt the corresponding *salary* value to the *encrypted_sal*. Finally in step 7 the algorithm concatenates $bucket\_id$ and $encrypted\_sal$ and takes the concatenating result as key_sal. After the for loop all values of key_sals are in order in terms of the corresponding $bucket\_ids$ and stored into the delegated Table 2 as values of attribute k_{sal}.

```
key_generation(domain_start, domain_end, int N)        // executed by DO
// domain_start and domain_end are the first and the last number in the domain
// N is the number of private value for some private attribute, such as salary
1. Divide the finite domain into at least N uniform buckets
2. For each bucket range from bucket_satrt to bucket_end
3.    Generate a random seed
4.    bucket_id=generator(seed)%(bucket_end-bucket_start)
5.    Generate a random secret_key
6.    encrypted_sal=E(the value of salary, secret_key)
7.    Concatenate the bucket_id and  encrypted_sal to form the key_sal
8.    Save the seed and the key_sal as the metadata for the subsequent query
//key_sal is ordered in terms of the bucket_id
//E is a symmetric algorithm
```

Fig. 2. The key_sal generating algorithm for salary attribute

Example 3. The *salary* attribute is usually used in the where clause for queries submitted by DRs, but it is usually private. In order to accelerate the query as well as protect the private value of *salary* attribute against the malicious DSPs or attackers, we adopt the secret share based scheme to implement the confidentiality of private salary and use $key\_generation$ algorithm to generate key_sals for each salary value. For instance, the key_sal for salary value 20 is generated in the following steps:

- Through steps $1 - 4$ in Fig. 2, the $bucket\_id$ for value 20 is 37, $bucket\_id = 37$.
- Choosing a symmetric algorithm DES and a random key in steps $5 - 6$ to generate the encrypted value $i89uhgnk$ for value 20, $encrypted\_sal = i89uhgnk$.
- Concatenating the results from the steps above obtains the key_sal for value 20, that is $key\_sal = bucket\_id||encrypted\_sal = 37||i89uhgnk = 37i89uhgnk$.

key_sal	37i89uhgnk	128h8jbka8g	279sudvyjl3	333ij9kyfbh	471rfyhujl2
bucket_id	37	128	279	333	471
encrypted_sal	i89uhgnk	h8jbka8g	sudvyjl3	ij9kyfbh	rfyhujl2
salary	20	35	40	50	75

Fig. 3. The key_sal for each salary value

Fig. 3 lists all key_sals for each value of *salary* attribute in Table 1 by using $key\_generation$. The security of encrypted value is determined by the security of the secret key chosen.

4.2   Index Creation Function $index\_creation$

We present an algorithm $index\_creation$ in Fig. 4 to construct a B+ tree index based on the key_sals generated above. In our approach, we sort the key_sals in term of its former part $bucket\_id$. And then construct the B+ tree index like the standard B+ tree algorithm. The proposed index satisfies the following properties:

- The proposed B+ tree index meets with the definition of privacy preserving index in section 3.2. The former part of key_sal is irrelevant with the *salary* value, and the latter part is the encrypted value of *salary* value. So DSP can't derive any information useful to breach the private salary except for the order of salary.
- We introduce another bucket named as $bucket\_pointer$ for each pointer field in the leaf node on the B+ tree index. The $bucket\_pointer$ consists of n rows, and each row consists of two fields, one is the identifier field for DSP_i, and the other is the pointer field which points to the record in the delegated database at DSP_i with key value equaling to the value of the data field on the leaf node.
- n DSPs can share the same B+ tree index without leaking any private information. Each DSP_i locates the corresponding record with the designated query key in term of the id of DSP_i in the $bucket\_pointer$. This property makes the query efficient and save much storage for each DSP_i.

```
index_creation(Relation R)        // executed by DO
// R is any relation in the datatabe of DO,such as relation Employees
1. For each key_sal computed from the algorithm key_generation
2.   Sort all key_sals in term of the former part in the key_sal
3. Create B+ tree in terms of the sorted key_sals
// The data field is filled with the key_sal not the former part
// The pointer is the same as the pointer in the standard B+ tree
```

Fig. 4. The index creation algorithm for key_sal

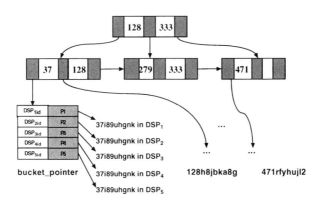

Fig. 5. The B+ tree for key_sal

- The proposed B+ tree index supports the exact query and range query efficiently and securely. This property makes the DSPs minimize the returned results and relieve DRs from computing many results irrelevant to the submitted query.

Example 4. Fig. 5 is an B+ tree index instance for key_sals generated above. For simplicity, we only put the former part $bucket\_id$ of key_sal into the data field such as 37, 128, however, in the practical B+ tree the concatenating results $key\_sal = bucket\_id\|encrypted\_sal$ is filled in the data field. For example, 37 is 37i89uhgnk, 128 is 128h8jbka8g in fact. In Fig. 5 we list the $bucket\_pointer$ for pointer field of key value 37i89uhgnk, which consists of five rows, one row for each DSP_i, $1 \le i \le 5$, in which DSP_{iid} is the identifier for DSP_i, Pi is the pointer pointing to the record with key value 37i89uhgnk at DSP_i.

In our approach the same value of salary will be mapped into different key_sals. That is to make the former part and the latter part both different by using different secret keys to encrypt the same value of salary. In order to describe how our proposed B+ tree index support all kind of queries efficiently, we will give the concrete query processing in next section. But due to the space limitation, we only focus on the exact and range query processing.

5 Query Processing

In this section we describe how to process queries submitted by DRs on the architecture in Fig. 1. For simplicity, we will take Table 1 and one of its corresponding secret share based Table 2 as relations. For any query, DR first needs to do the following steps to transform the submitted query into a private form:

- Encrypting the salary value in where clause into its corresponding encrypted value under the DR's secret key.
- Comparing the encrypted value with the latter part of the key_sal in the metadata, until find the correct key_sal.
- Rewriting the query in terms of the value computed above.

Example 5. Suppose the exact query submitted by DR is as follows

SELECT name FROM Employees WHERE salary=35

it is transformed into the following private form in term of the three steps above:

SELECT s_{namei} FROM $Employees_i$ WHERE $k_{sal} = 128h8jbka8g$

where $1 \leq i \leq k$, k is the at least number of DSPs. Then DR submits all k query $shares$ of any k of n DSPs. Each DSP_i searches the share $share_{(name,i)}$ for attribute $name$ in his/her own delegated database under the condition $key\_sal = 128h8jbka8g$, that is to say, $k_{sal} = 128h8jbka8g$. When searching to the leaf node, DSP_i finds the record pointer in the $bucket\_pointer$ on the condition that DSP_{iid} equals to the id proposed by the corresponding DSP_i.

Example 6. Suppose the range query submitted by DR is in the following

SELECT name FROM Employees WHERE $salary \geq 35$ and $salary \leq 50$

it is transformed into the following private form in term of the three steps above:

SELECT s_{namei} FROM $Employees_i$ WHERE $F(k_{sal}) \geq 128$ and $F(k_{sal}) \leq 333$

F is a defined function by DO, and used to extract the former part of k_{sal}. The salary range above has the corresponding start value and end value in the delegated database. However, not all query in which the start value or the end value for salary is in the predefined delegated database. Therefore if DR submits query in the following form:

SELECT name FROM Employees WHERE $salary \geq 30$ and $salary \leq 60$

where neither the start value 30 nor the end value 60 for salary are in the delegated database. Therefore DR first executes a binary search to find the maximal value which is less than or equal to 30 and the minimal value which is larger than or equal to 60. These can be implemented easily because we construct an order preserving $bucket\_id$. After this, DR computes the corresponding key key_max for the minimum and key key_min for the maximum. Finally, each DSP_i searches the s_{namei} for attribute $name$ in his/her own delegated database under the condition $F(k_{sal}) \geq F(key\_max)$ and

$F(k_{sal}) \leq F(key\_min)$. Since we construct a B+ tree index based on the order preserving $bucket\_id$ which keeps the same order with the practical salary value, each DSP_i can only return the required tuples. After receiving all $shares$ $s_{name_i}s$ from each DSP_i, DR or DO can obtain the practical results by reconstructing the polynomials for *name* attribute values.

6   Experiments Evaluation

Due to the space limitation we only analyze the security and efficiency of our approach through the following experiments.

6.1   Security Analysis

The blue line in Fig. 6(a) is based on the AMD sempron(tm) processor 1.6GHz PC with RAM 512M, 80G hard disks, VC++6.0 as the integrated development environment. The pink line is based on the Intel(R) Core(TM)2 Duo CPU, 2.33GHz PC with RAM 2G and 160GB hard disks, VS.NET 2008 as the integrated development environment, coding in C++ with Framework 3.5.

Aggarwal et al[7] proposed a distributed architecture, which took use of two different DSPs to implement the privacy guarantee for sensitive data. The query in proposed distributed architecture is efficient, however, the data privacy is completely determined by the condition that the two DSPs are unable to communicate directly with each other. In fact they need not even be aware of each other's existence. Once one of the provider is compromised, the important private information such as the credit card number or salary will be leaked. Compared to theirs, the privacy guarantee in our approach doesn't depend on the DSPs, but determined by the trusted DO and DRs. And the DSPs in our approach can communicate or collude freely with each other, which couldn't endanger the privacy of sensitive data. Only the DR colludes with at least k DSPs, does the privacy for the very DR be leaked.

In the information theoretically secure aspect, from Fig. 6(a) we know that if an attacker or the DSP doesn't know the prime number used by DO, he/she must take more time or exploit large scale computation on multi-computers to gain the appropriate prime number, because the computing time is sharply increasing with the decimal digits of prime over 8. It is shown in Fig. 6(a) that the computing time is sharply increasing to more than 1500 seconds(almost half an hour) in the blue line, even reach to 2200 seconds in the pink line.

In conclusion, the proposed approach is information theoretically secure and can be proved as secure as other common cryptographic primitives.

6.2   Efficiency Evaluation

We experiment on the Intel(R) Core(TM)2 Duo CPU, 2.33GHz PC with RAM 2G and 160GB hard disks, VS.NET 2008 as the integrated development environment, coding in $C\sharp$ with Framework 3.5, and SQL Server 2005 as the database server. We use five synthetic relations R_i, $1 \leq i \leq 5$, each contains two independent attributes att_1 and

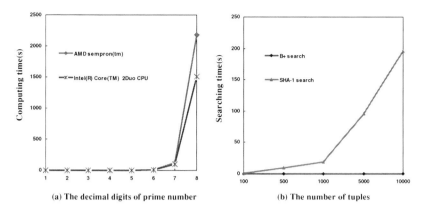

(a) The decimal digits of prime number   (b) The number of tuples

Fig. 6. The prime computation in (a) and search time comparison in (b)

att_2. The size of the tuples are $1 \cdot 10^4$, $5 \cdot 10^4$, $10 \cdot 10^4$, $50 \cdot 10^4$, and $100 \cdot 10^4$ respectively. In addition, attribute att_1 is the primary key for all five relations. We choose SHA-1 as the hash function and DES as the encryption algorithm.

Time Comparison Between Hash Based Searching and Our Index Based Searching
From Fig. 6(b) we know that the searching time based on the B+ tree(red line) is almost 0 seconds, but the searching time based on the SHA-1(pink line) grows fast when the number of the tuples are more than 1000 tuples. The SHA-1 approach proposed by [4] is to change the key in source database into the corresponding digest in delegated database. The digest can protect the privacy of data when the join happens among multi-tables from different n DSPs. For example, the searching time is 0.031 seconds in red line, however the searching time is 95.797 seconds in pink line, which is almost three thousands times that in the red line. So the proposed approach is much efficient than that in SHA-1 based approach. The quick speed in the red line mainly depends on the B+ tree index constructed on the key_sal. In contrast, the SHA-1 approach should search the whole database under the condition that the searching digest equals to the digest stored in the delegated database. Our approach can be applied to the range query well, however, the SHA-1 approach works bad because of the unordered key resulted by using the hash function.

The Comparison Between Encryption and Polynomial Computation
From Fig. 7 we know that the encryption time in green line and the polynomial computation time in blue line are almost the same with the number of tuples less than $5 \cdot 10^4$, but the encryption time is growing faster than that of the polynomial computation with the number of tuples more than $10 \cdot 10^4$.

The data extension in data encryption is much more than that in polynomial computation, for example, if the size of original relation is 0.172MB, then the corresponding encrypted relation is 0.398MB, but the corresponding polynomial relation is 0.172MB, Table 3 lists all the five relations with their corresponding data extension. The delegated relation in the polynomial computation has the same size with the original

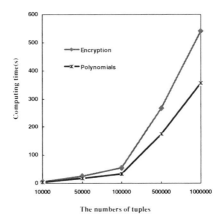

Fig. 7. The time for encryption and polynomial computation

Table 3. Data extension vs. tuple size

$Relation(MB)$	0.172	0.828	1.648	8.211	16.414
Encryption	0.398	2.016	4.039	28.445	58.961
Polynomial	0.172	0.828	1.648	8.211	16.414

relation, while the delegated relation in the encryption has more than three times that in the original relation, and the data extension even more larger with the size increasing of the original relation. So the communication and bandwidth costs between the DO and the DSP in the encryption must be more higher than that in the polynomial computation resulted by the different data extension.

In conclusion, our proposed approach is secure and efficient compared to the encryption or hash based approach. The constructed index based on the constructed key improves the query efficiency greatly. And the information leakage does not depend on DSPs, but the DR and the DO. That is to say, even if the DSPs collude, they couldn't leak private information delegated by DO.

7   Related Work

The concept of DaaS is first proposed by Hacigumus et al[2], in which a prototype system NetDB2 was developed. NetDB2 resolved two important challenges about data privacy and performance in DaaS scenario. From then on a lot of related papers are proposed[3], [10], [11], [5], but most of which used the encryption algorithm to protect the confidentiality of delegated data, and used asymmetric encryption algorithm to implement the integrity and completeness.

However, encryption on data makes the usability of the delegated encrypted database worse. In order to improve the usability of the delegated encrypted database most current proposed approaches[9], [11], are based on exploiting index information, which is stored together with the delegated encrypted database. This is used to help the service provider select the returned data for the query without the need of decrypting the data. Especially the order preserving encryption function proposed in [8] can support range queries and is adopted widely in many subsequent schemes. However, because of the low speed of the software-based encryption[13], [14] experiments database encryption efficiency through three different dimensions by adopting software encryption, hardware encryption, and hybrid encryption respectively. Finally, they concluded that the hybird encryption on database-level is most efficient and can prevent theft of critical data and protect against threats such as the storage theft and storage attacks.

Almost all of the schemes proposed above are implemented by using encryption to protect the data privacy, However, as we all know that the encryption and decryption on large amount of data are time consuming[13], [14], and the query results always consist of many irrelevant data tuples besides the required data tuples[9], [10]. In order to overcome the inefficiency problems of the encryption for cinfidentiality guarantee, [4], [5], [6] proposed a new way to protect data confidentiality, that is the secret share based, not encryption based, data storage. However, the schemes proposed by [4], [5], [6] couldn't give an efficient privacy preserving index to accelerate the query speed. In this paper, on one hand we present an information theoretic secure secret share[1] based approach, not encryption based, to guarantee the confidentiality of the delegated data at n DSPs , and on the other hand we construct a privacy preserving index to accelerate the query speed and help return the exactly required data tuples. Finally we present the query transformation techniques at DRs to access the delegated database from different DSPs.

8   Conclusions

Due to the paper limitation, we couldn't list all my algorithms and query processing proposed and couldn't provide all kinds of experiment evaluation from different aspects either. We will give the unwritten part in another full paper. In this paper we proposed a privacy preserving index for n DSPs on secret share based data storage, in which the index is constructed on the added key generated through the concatenation between the bucket id and the encrypted private value. Finally we analyze the security and efficiency of the proposed approach from two different angles.

Our future research interest is to focus on implementing the integrity verification based on the secret share based data storage. Then we will design a secure middleware as a middle interface among the different three parties, the DO, DRs and DSPs in DaaS scenario. The middleware can deal with the integrity, confidentiality and privacy preserving security. Finally we will develop a client tool to guarantee the correctness of the returned results from the n DSPs.

Acknowledgments

This work was supported by the NSFC grant (No.60903014), and the China National Funds for Distinguished Young Scientists grant (No.60925008).

References

1. Shamir, A.: How to share a secret. Communications of the ACM 22(11), 612–613 (1979)
2. Hacigumus, H., Iyer, B., Mehrotra, S.: Providing database as a service. In: Proc. of the 18th ICDE Conf., pp. 29–38 (2002)
3. Hacigumus, H., Iyer, B., Mehrotra, S., Li, C.: Executing SQL over encrypted data in the database-service-provider model. In: Proc. of the ACM SIGMOD Conf., pp. 216–227 (2002)
4. Emekci, F., Agrawal, D., Abbadi, A.E.: Abacus: A distributed middleware for privacy preserving data sharing across private data warehouses. In: Alonso, G. (ed.) Middleware 2005. LNCS, vol. 3790, pp. 21–41. Springer, Heidelberg (2005)
5. Emekci, F., Agrawal, D., Abbadi, A.E., Gulbeden, A.: Privacy preserving query processing using third parties. In: Proc. of 22th ICDE Conf. (2006)
6. Agrawal, D., Abbadi, A.E., Emekci, F., Metwally, A.: Datamanagement as a service:challenges and opportunities. Keynotes. In: Proc. of the 25th ICDE Conf. (2009)
7. Aggarwal, G., Bawa, M., Ganesan, P., Garcia-Molina, H., Kenthapadi, K., Motwani, R., Srivastava, U., Thomas, D., Xu, Y.: Two can keep a secret: A distributed architecture for secure database services. In: Proc. of CIDR Conf., Asilomar, CA, pp. 186–199 (2005)
8. Agrawal, R., Kiernan, J., Srikant, R., Xu, Y.: Order preserving encryption for numeric data. In: Proc. of the ACM SIGMOD Conf., pp. 563–574 (2004)
9. Hore, B., Mehrotra, S., Tsudik, G.: A privacy-preserving index for range queries. In: Proc. of the 30th VLDB Conf., Toronto, Canada, pp. 720–731 (2004)
10. Li, J., Omiecinski, R.: Efficiency, security trade-off in supporting range queries on encrypted databases. In: Jajodia, S., Wijesekera, D. (eds.) Data and Applications Security 2005. LNCS, vol. 3654, pp. 69–83. Springer, Heidelberg (2005)
11. Shmueli, E., Waisenberg, R., Elovici, Y., Gudes, E.: Designing secure indexes for encrypted databases. In: Proc. of the IFIP Conf. on Database and Application Security (2005)
12. Anciaux, N., Benzine, M., Bouganim, L., Pucheral, P., Shasha, D.: Ghostdb:querying visible and hidden data without leaks. In: Proc. of the ACM SIGMOD Conf., pp. 677–688 (2007)
13. Schneier, B.: Applied Cryptography. John Wiley Sons, Chichester (1996)
14. Mattsson, U.: Database encryption-how to balance security with performance. Protegrity Corp. (2005)

Node Protection in Weighted Social Networks

Mingxuan Yuan and Lei Chen

Hong Kong University of Science and Technology, Hong Kong
{csyuan,leichen}@cse.ust.hk

Abstract. Weighted social network has a broad usage in the data mining fields, such as collaborative filtering, influence analysis, phone log analysis, etc. However, current privacy models which prevent node re-identification for the social network only dealt with unweighted graphs. In this paper, we make use of the special characteristic of edge weights to define a novel k-weighted-degree anonymous model. While keeping the weight utilities, this model helps prevent node re-identification in the weighted graph based on three distance functions which measure the nodes' difference. We also design corresponding algorithms for each distance to achieve anonymity. Some experiments on real datasets show the effectiveness of our methods.

1 Introduction

With the popularity of social network websites, such as Facebook, LinkedIn, and Twitter, more and more social network data are available for the public to analyze social connections and user online behaviors. However, social network users might not want to be found in the social network due to their privacy concerns. As a result, those social network websites have to be careful when they publish their user data or provide API to third parties to develop new applications.

Lots of research has been done to generate anonymized social graphs. The published graph should make sure an attacker could not re-identify any user in the graph using the structural information, such as a node's degree [9], the neighborhood graph [14], the whole graph [8][16], etc. For example, given a graph G, for any node u, if there are another $k - 1$ nodes that have the same degree as u, we call G k-degree-anonymous graph. For this graph, if an attacker uses a node's degree information to re-identify it, he will always get at least k candidates. These pioneering works all treat the social graph as an un-weighted graph. However, in reality, many social networks are weighted graphs [10][7]. How to publish a privacy preserving weighted graph becomes an important problem. Two works [10][7] considered how to protect the edge weights in weighted graphs. But no work has been done on how to prevent node re-identification when the edge weights also work as part of an attacker's background knowledge. For example, Figure 1(a) is already a 2-degree-anonymous graph. Besides a node u's degree, if an attacker also knows the weights on the edges adjacent to u, he can still successfully re-identify u. Suppose an attacker knows there are two edges connect to Bob and the weights on these two edges are 1 and 2 respectively, the attacker can immediately re-identify node A is Bob in Figure 1(a). Target on this problem, in this paper, we study how to prevent node re-identification in weighted graphs.

J.X. Yu, M.H. Kim, and R. Unland (Eds.): DASFAA 2011, Part I, LNCS 6587, pp. 123–137, 2011.

The naive way to anonymize a weighted graph is to make the nodes with the same structure (e.g. the same degree) also have the same weights on the corresponding edges. We call a graph G which satisfies the above constraint as the strict weighted anonymous graph. For k-degree-anonymous model, we should make sure for any node u, there are at least $k-1$ other nodes which have the same degree as u and the weights on the edges adjacent to these nodes are also the same as u. One way is to generalize the weight to a certain range. Figure 1(a) is a 2-degree-anonymous graph. In it, M, N have the same degree and A, B, C have the same degree. Figure 1(b) is a strict weighted 2-anonymous graph by generalizing the weights. The shortcoming of this solution is quite obvious. To mine such an anonymized graph, it needs to sample the graph (assign a concrete weight for each edge within this edge's weight range) and it is unclear how many graphs have to be sampled in order to achieve the required accuracy. Furthermore, since each edge is adjacent to two endpoints, its weight influences two nodes. This makes the final weight range of each edge become very large. Instead of generalizing the weights, we can also use a new weight value to replace each weight. If we apply strict anonymous to Figure 1(a) to let each edge have one weight value, we can get an anonymized graph as shown in Figure 1(c). Since the weight on each edge influences this edge's two endpoints, we might change most or all weights to be the same (as shown in Figure 1(c)).

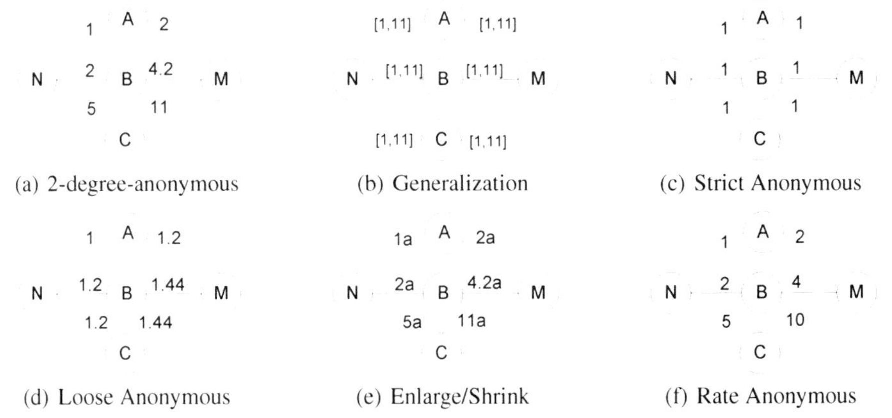

(a) 2-degree-anonymous   (b) Generalization   (c) Strict Anonymous

(d) Loose Anonymous   (e) Enlarge/Shrink   (f) Rate Anonymous

Fig. 1. Anonymous Examples

The strict weighted anonymous model helps to avoid node re-identification in a weighted graph. However, weight is a special label in a social network. It is only useful when the weights in the graph have different values. Otherwise, the data mining technique could not mine the useful results. As a consequence, when anonymizing a weighted social graph, we could not adjust all weights to the same value or a group of values. Let's consider the special characteristic of edge weights. Weight is a numerical number which is converted from some properties of social network. Although all the converting algorithms try to represent the same fact, they may use different values. The attacker could not get the complete information about what algorithm the publisher uses to construct the graph. Therefore, the attacker is impossible to get the exact number. He can only estimate a number around the exact value based on the fact he knows. It is not necessary to keep the weights in the same group identical. If we assume weights with less than 20% difference could be considered as equal, then we can have a loose anonymous version of the graph (as shown in Figure 1(d)).

Moreover, different extraction algorithms may use different metrics to represent the social network. Thus the edge weights may have different values using different graph generation algorithms. For example, the transactions between two companies can either be represented by dollars based on billions, or by euro based on millions. We can either enlarge or shrink the weight with the same ratio on the whole graph when publishing it. For example, we can multiply an a in the graph and get the same graph properties as shown in Figure 1(e). In this case, an attacker could only know the relative weight of a certain user in the social graph. For example, an attacker knows that Alice cooperated with 2 other researchers in the past and the ratio of these two collaboration weights is around 2. He does not know what metric the published graph uses. Thus all the nodes whose degree is 2 are candidates if their two weights' ratio is also around 2. Therefore, it is more important to protect the ratio between weights other than the real value. We can have Figure 1(f) as the published graph which is very similar to Figure 1(a).

In this paper, we assume the attacker may know one node's degree information and the weight information on the edges adjacent to this node. We propose a k-weighted-degree anonymous model to prevent the node re-identification using this background knowledge. Moreover, we make use of the numerical characteristic of the edge weights which improves the utilities of the published graph. The k-weighted-degree anonymous model is defined on three different distances which an attacker might use to re-identify a user. We propose graph construction algorithms for each distance definition in order to publish an anonymized graph.

The rest of the paper is organized as follows: Section 2 defines the problem. Section 3 defines three different distances and Section 4 describes the graph construction algorithms. We report the experiment results in section 5. Comparison of our work with related previous works is given in Section 6 followed by the conclusion in Section 7.

2   Problem Description

In this paper, we focus on preventing node re-identification in a weighted graph. For a node u, the weights on the edges adjacent to it can be represented as a ***sorted*** number sequence W, which is called node u's ***weight sequence***. For example, node N's weight sequence in Figure 1(a) is $(1, 2, 5)$. Then, for any two nodes u and v that already have the same degree, the difference between u and v can be measured by the distance between their corresponding weight sequences. (See Section 3 for the details of the distance between weight sequences). Based on the distance between weight sequences, we can give the definition of k-weighted-degree anonymous on a weighted graph:

Definition 1. *k-weighted-degree anonymous: for a weighted graph G, it is k-weighted-degree anonymous if and only if for each node u, there are at least $k - 1$ other nodes having the same degree as u and the distance between the weight sequences of u and anyone in these $k - 1$ nodes is within a threshold t.*

The threshold t is the users' equal expectation [7] on two weight sequences. The problem we solve in this paper is:

Problem 1. k-weighted-degree anonymous problem: given a weighted graph G, an integer k and a threshold t, convert G to a k-weighted-degree anonymous graph G' with the minimum cost of edge change.

Here, the cost of "edge change" contains two parts 1) The number of edges added/deleted; 2) The total weight change on edges.

3   Distance between Weight Sequences

In this part, we define three distances between weight sequences. The first two are based on the absolute values of weights and the last one is based on the ratio between weights.

3.1   Absolute Distance

In [11], a distance based on EMD (Earth Mover Distance) is defined to compare two distributions. The weight sequence can be transformed to a distribution on real numbers (weights) that appear in the sequence. For example, Table 1 shows the distribution of node N's weight sequence in Figure 1(f). Thus, the distance between two weight sequences can be represented as the EMD between the two corresponding distributions on real numbers. For two distributions on the numerical values $\{v_1, v_2, ..., v_m\}$ with ordered distance (the distance between two values is based on the number of values between them in the total order), the EMD distance between them is computed by:

$$d = \frac{1}{m-1} \sum_{i=1}^{m} |\sum_{j=1}^{i} r_i| (r_i = d_i - d'_i) \tag{1}$$

where d_i and d'_i are the corresponding values on v_i in two distributions respectively. Since the distributions of two weight sequences may be on two different real number sets, we need to combine these two real number sets together before computing the EMD distance. For two ordered real number sets $V_1 = (v_1, v_2, ..., v_p)$ and $V_2 = (v'_1, v'_2, ..., v'_q)$, we can compute a new real number set $N = (n_1, n_2, ..., n_m)$ where $n_1 = min(v_1, v'_1)$, $n_m = max(v_p, v_q)$, $V_1 \subseteq N$, $V_2 \subseteq N$ and for $\forall i < m$, $n_{i+1} - n_i = constant$. After getting this new real number set, the two distributions can be put on this new set and the EMD between them can be computed using Equation 1.

For example, the distributions of nodes M and N's weight sequences in Figure 1(f) are shown in Table 1 & 2, their new distributions on the combined numerical set are shown in Table 3. The distance between M and $N = (0.33 + 0.33 + 0.33 + 0 + 0.33 + 0.33 + 0.33 + 0.33 + 0)/9 = 0.293$. We give Lemma 1 to compute this distance on weight sequences directly without transforming them to distributions.

Table 1. Node N's weight sequence distribution **Table 2.** Node M's weight sequence distribution

weight sequence	1	2	5
$distribution_N$	0.33	0.33	0.33

weight sequence	2	4	10
$distribution_M$	0.33	0.33	0.33

Lemma 1. *For two weight sequences* $(x_1, x_2, ..., x_n)$ *and* $(y_1, y_2, ..., y_n)$, *if the corresponding distributions of* X, Y *are* DWS_X *and* DWS_Y, $EMD(DWS_X, DWS_Y) = \frac{1}{n(max(x_n, y_n) - min(x_1, y_1))} \sum_{i=1}^{n} |X_i - Y_i|$.

Due to the space limitation, we ignore the proof here. For node M and N in Figure1(f), $EMD(M, N) = \frac{1+2+5}{3*(10-1)} = 0.293$.

Table 3. Combined distributions of node M and N's weight sequences

	1	2	3	4	5	6	7	8	9	10
$distribution'_N$	0.33	0.33	0	0	0.33	0	0	0	0	0
$distribution'_M$	0	0.33	0	0.33	0	0	0	0	0	0.33

Table 4. weight sequence example

S_1	0.1	0.2	0.3	0.4
S_2	0.3	0.4	0.5	0.6
S'_1	1000.1	1000.2	1000.3	1000.4
S'_2	1000.3	1000.4	1000.5	1000.6

We name EMD as Absolute Distance and use Dis_{abs} to represent this distance for the rest part of our paper. The value of Absolute Distance is between [0, 1].

3.2 Relative Distance

The Absolute Distance does not consider the relativity between the distance and the original sequences. For example, in Table 4, $Dis_{abs}(S_1, S_2) = Dis_{abs}(S'_1, S'_2)$. However, comparing with S_1 and S_2, S'_1 and S'_2 are almost the same. In order to handle this problem, we give another distance definition:

Definition 2. *Relative Distance: the relative distance between two weight sequences* $A(a_1, a_2, ..., a_n)$ *and* $B(b_1, b_2, ..., b_n)$ *equals* $max(\frac{2|a_i - b_i|}{a_i + b_i})$.

This definition counts the maximum relative change from one weight sequence to the other comparing with the mean value on each position. By doing this, the distance relative to the original values of the two sequences can be represented. We use Dis_{rel} to represent the Relative Distance. The value of Dis_{rel} is between [0, 2).

3.3 Rate Distance

In many situations, an attacker can only use the relative weights among the neighbors of one node instead of using the absolute values of them. For example, company A has two cooperating companies and one of them has 3-5 times flow with A than the other. The attacker cannot know the exact metric that the data publisher uses to represent the flow. The flow may be based on thousand dollars, or based on million dollars. The weight sequences such as (1, 3.5), (10, 35) and (0.1, 0.35) can all be used to show node u's one edge has 3.5 times weight than the other. Therefore, we define another distance based on the rate between one item and the smallest item in the same weight sequence.

Definition 3. *Rate Distance: the rate distance between two weight sequences* $A(a_1, a_2, ..., a_n)$ *and* $B(b_1, b_2, ..., b_n)$ *is defined as* $max(max(\frac{a_i b_1}{a_1 b_i}, \frac{a_1 b_i}{a_i b_1}))$.

The Rate Distance computes the maximum ratio between the relative weights appearing in the same position in two weight sequences. We call $max(\frac{a_i b_1}{a_1 b_i}, \frac{a_1 b_i}{a_i b_1})$ as the rate at position i. The minimum distance (the distance between two same weight sequences) is 1. When using this distance, weight sequences such as (1, 3.5), (10, 35) and (0.1, 0.35) can be seen as the same. (1000, 4000) and (0.01, 0.04) also have the same distance to (1, 3.5), (10, 35) and (0.1, 0.35). We use Dis_{rate} to present this distance for the rest part of our paper. The value of Dis_{rate} is between $[1, +\infty)$.

In this paper, we also use $Distance(u_1, u_2)$ to directly represent the distance between two nodes u_1 and u_2. We use the words "cluster" and "group" interchangeably.

4   Graph Construction Algorithm

Before presenting the details of our k-weighted-degree anonymous graph constructing algorithms, we first show the hardness of Problem 1.

Lemma 2. *k-weighted-degree anonymous problem is NP-Hard.*

We prove the k-weighted-degree anonymous problem is NP-Hard from the r-Gather Problem [1]. Due to the space limitation, we ignore the proof here.

4.1   Algorithm Structure

We design a four step algorithm to construct a k-weighted-degree anonymous graph:

1. Construct a k-degree-anonymous [9] graph G' from G which minimizes the change of the edge number;
2. Assign weights on each new added edges;
3. Classify the nodes into groups with the same degree, where each group has size $\geq k$ and the diameter of each group is as small as possible;
4. Adjust weights on edges to ensure the diameter of each group $\leq t$.

where the diameter of a cluster is the maximum distance between any two nodes in it.

Firstly, we use the algorithm in [9] to construct a k-degree-anonymous graph with the minimum edge change. Through the algorithm in [9], we can obtain a k-degree-anonymous graph by adding edges. Next, we introduce how to implement steps (2)-(4).

4.2   Assign New Weights

When a group of new edges are added into the graph G to achieve a k-degree-anonymous graph G', we should set weights on the new created edges. For a new created edge e, we want to set the edge weight as similar as its "neighbors". Here "neighbors" means the edges sharing the same endpoints with e. So we compute the mean edge weight of all the "neighbors" of e and set it as e's weight.

4.3   Clustering

In the k-degree-anonymous graph G', for each degree, there are at least k nodes. Sometimes, for a certain degree, there are more nodes than k. So in step (3), we classify the

---

Algorithm 1. Clustering

---

1 Set superGroups = new Set;	11 **for** *Each Set sgn in superGroups* **do**
2 **for** *Each node u in G′* **do**	12 Sort sgn by group size;
3 Group *gn* = new Group(*u*);	13 **while** *sgn[0].size < k* **do**
4 **if** *superGroups.Contains(u.degree)* **then**	14 Group *gn* = sgn[0];
5 Set sgn =	15 Remove sgn[0];
superGroups.getSet(*u*.degree);	16 **for** *Each group gn′ in sgn* **do**
6 sgn.addGroup(*gn*);	17 *gn′.merge_cost* = cost of
7 **else**	*gn ∪ gn′*;
8 Set sgn = new Set(*u*.degree);	18 Sort sgn by the *merge_cost*;
9 sgn.addGroup(*gn*);	19 merge sgn[0] with gn;
10 superGroups.add(sgn);	20 Sort sgn by group size;

---

nodes with the same degree to groups so that each group has size bigger than or equal to k with the minimum group diameter. Then, in step (4), we can adjust the diameter of all the groups less than t with the minimum weight change.

The problem that needs to be solved in this step can be formulated as:

$$\text{objective} : min \quad \sum_i (\text{cost that makes Diameter}(Cluster_i) \leq t)$$

$$\text{subject to} : \quad \forall i, \ |Cluster_i| \geq k;$$

$$\forall i, \ \forall n_i, n_j \in Cluster_i, Degree(n_i) = Degree(n_j);$$

We design an incremental algorithm as shown in Algorithm 1 to cluster the nodes to groups. The algorithm first creates one group for each single node and stores the groups with the same degree into the same set. Then, for each degree, we sort its corresponding group set by group size. If the minimum group size is less than k, we merge this group with another group (the same degree) with the minimum merging cost. The merging cost is estimated by the diameter of the merged group. We repeat this process until all the groups have size $\geq k$. In the worst case, the algorithm merges all nodes into a single group with $|V|$ merging operations. Each merging operation computes distances between at most $|V|^2$ pair of nodes. So the time complexity of this algorithm is $O(|V|^3)$.

4.4 Weight Adjustment

After the nodes are divided into groups, the diameter of each group should be adjusted to a value $\leq t$. The objective is to minimize the change of weights on all edges. So, the problem that needs to be solved in this step can be formulated as:

$$\text{objective} : min \quad \sum_i^{|E|} (|w_i' - w_i|) \quad \vee \quad \sum_i^{|E|} (max(\frac{w_i'}{w_i}, \frac{w_i}{w_i'}))$$

$$\text{subject to} : \quad \forall j, \ Diameter(Cluster_j) \leq t;$$

$$\forall i, \ w_i' \in [min_W, max_W];$$

Where min_W and max_W are the allowed minimum and maximum edge weights. For the Absolute Distance and Relative Distance, the objective is to minimize the sum of weight changes on all edges (min $\sum_i^{|E|} |w_i' - w_i|$)). For the Rate Distance, since the weight rate becomes the main concern, the sum of weight rate change becomes the objective (min $\sum_i^{|E|} max(\frac{w_i'}{w_i}, \frac{w_i}{w_i'})$)).

We design our adjusting algorithm based on the local optimized adjustment. That is iteratively adjusting each group's diameter $\leq t$ with the minimum weight change of the edges in this group until all groups have diameters $\leq t$. Since the weight on one edge may appear in two nodes' weight sequences, if these two nodes are clustered into two different groups, this weight appears in two groups. In one group, the local optimal adjustment may require decreasing this weight. While at the same time, in the other group, the local optimal adjustment may need to increase this weight. This phenomenon makes the local optimization vibrate some weights and fall in infinite loop. Target on this problem, in our algorithm, we only allow each weight be adjusted in one direction: only increasing or only decreasing. The adjusting algorithm is shown in Algorithm 2.

Algorithm 2. Adjusting Framework

---

1 **while** ¬(*All group in Groups has diameter* $\leq t$) **do**
2   **for** *Each group g in Groups* **do**
3   **if** *g's diameter* $> t$ **then**
4   **for** *each pair of nodes* (n_1, n_2) *in g* **do**
5   **if** *Distance*$(n_1, n_2) > t$ **then**
6   Adjusting to *Distance*$(n_1, n_2) = t$;

The algorithm iteratively adjusts each group that has diameter bigger than t until all the groups have diameters $\leq t$. For a group which needs to be adjusted, if a pair of nodes (u_1, u_2) in it has distance bigger than t, we adjust the weights on the edges adjacent to u_1 and u_2 to make their distance equal to t. The key step of Algorithm 2 is to make $Distance(u_1, u_2) = t$ (line 6 in Algorithm 2) with the minimum weight change. Next we introduce how to implement this for the three distances defined in Section 3.

Adjusting for Absolute Distance. The weight adjusting algorithm for Absolute Distance is described in Algorithm 3. For a pair of nodes (u_1, u_2) with $Dis_{abs}(u_1, u_2) > t$, we do the following adjustment:

– Step 1: Find the maximum and second maximum value gaps on each position of the weight sequences (max($|S_1[i] - S_2[i]|$), suppose these two values are max_1 and max_2 respectively and there are num_{max_1} positions have max_1 gap;
– Step 2: Compute the total weight change $change_{need}$ that needs to be reduced to let $Dis_{abs}(n_1, n_2) = t$;[1]

---

[1] $Dis_{abs}(S_1, S_2) = \frac{\sum_{i=1}^{n} |S_1[i] - S_2[i]|}{n(max(S_1[n], S_2[n]) - min(S_1[1], S_2[1]))}$ ($n = |S_1|/|S_2|$ and S_1'/S_2' are the new weight sequences), $t = \frac{\sum_{i=1}^{n} |S_1'[i] - S_2'[i]|}{n(max(S_1[n], S_2[n]) - min(S_1[1], S_2[1]))}$, so $change_{need} = \sum_{i=1}^{n}(|S_1[i] - S_2[i]| - |S_1'[i] - S_2'[i]|) = (Dis_{abs}(S_1, S_2) - t) \times n(max(S_1[n], S_2[n]) - min(S_1[1], S_2[1]))$.

Algorithm 3. Adjusting for Absolute Distance

1 $S_1 = u_1$'s weight sequence; $S_2 = u_2$'s weight sequence; $n = |S_1|$;
2 **while** $Dis_{abs}(S_1, S_2) > t$ **do**
3      $max_1 = \max(|S_1[i] - S_2[i]|)$;
4   num_{max_1} = No. of pairs with ($|S_1[i] - S_2[i]| = max_1$);
5      $max_2 = \max(|S_1[i] - S_2[i]| < max_1)$;
6      $change_{need} = (Dis_{abs}(S_1, S_2) - t) \times n(max(S_1[n], S_2[n]) - min(S_1[1], S_2[1]))$;
7   **if** $((max_1 - max_2) \times num_{max_1}) \geq change_{need}$ **then**
8   **for** *each* $|S_1[i] - S_2[i]| = max_1$ **do**
9   Adjust corresponding weights, let $|S_1[i] - S_2[i]| = max_2$;

10   **else**
11          $target = max_1 - change_{need}/num_{max_1}$;
12   **for** *each* $|S_1[i] - S_2[i]| = max_1$ **do**
13   Adjust corresponding weights, let $|S_1[i] - S_2[i]| = target$;

- Step 3: If reducing all the max_1 gap to max_2 can cover $change_{need}$ ($change_{need} \leq$ $(max_1 - max_2) \times num_{max_1}$), we change the weights appear in the positions of max_1 to let all max_1 be reduced to a new value $target$: $target = max_1 - change_{need}/num_{max_1}$. By doing this, the $Dis_{abs}(n_1, n_2)$ is reduced to t. If changing all max_1 to max_2 cannot cover $change_{need}$, we reduce all max_1 to max_2, go to Step 1 and repeat this process until $Dis_{abs}(n_1, n_2) \leq t$.

Suppose the degree of the two adjusting nodes is d, then the time to find the maximum and second maximum value gap is $O(d)$. In the worst case, all the gaps are adjusted and the procedure repeats d times. So the time complexity of Algorithm 3 is $O(d^2)$.

Adjusting for Relative Distance. Algorithm 4 is the weight adjusting algorithm for Relative Distance. For a pair of nodes (u_1, u_2) with $Dis_{rel}(u_1, u_2) > t$, we find a position with $\frac{2|S_1[i] - S_2[i]|}{S_1[i] + S_2[i]} > t$. Suppose this position is $adjust_i$, we need to adjust the edge weight at $adjust_i$ to ensure $\frac{2|S_1[adjust_i] - S_2[adjust_i]|}{S_1[adjust_i] + S_2[adjust_i]} = t$. For weight decreasing only, suppose v_{min} is $min(S_1[adjust_i], s_2[adjust_i])$ and v_{max} is $max(S_1[adjust_i], s_2[adjust_i])$, we should reduce v_{max} to x that satisfies: $\frac{2(x - v_{min})}{x + v_{min}} = t$, then $x = \frac{(2+t)v_{min}}{2-t}$. For weight increasing only, similar computation can be deduced. In the worst case, every position should be adjusted. So the running time of Algorithm 4 is $O(d)$.

Adjusting for Rate Distance. The definition of Rate Distance is $max(max(\frac{a_i b_1}{a_1 b_i}, \frac{a_1 b_i}{a_i b_1}))$, so there are two options for adjustment, adjusting the minimum weight in a weight sequence or adjusting the weight at position i.

If we choose the first option, the rate at every position is changed. This makes the problem be a non-linear optimization problem. In order to solve this problem, we model the optimal weight adjusting problem between two nodes as *a Geometric Programming (GP) problem* and use a GP solver to find the proper weights on edges.

Algorithm 4. Adjusting for Relative Distance

---

1 $S_1 = u_1$'s weight sequence; $S_2 = u_2$'s weight sequence;

2 **while** $Dis_{rel}(S_1, S_2) > t$ **do**

3      $adjust_i = i$ such that $\frac{2|S_1[i] - S_2[i]|}{S_1[i] + S_2[i]} > t$;

4   **if** *Decrease* **then**

5          $v_{min} = min(S_1[adjust_i], S_2[adjust_i])$;

6          $v_{max} = max(S_1[adjust_i], S_2[adjust_i])$;

7          $v_{max} = \frac{(2+t)v_{min}}{2-t}$;

8   **else**

9          $v_{max} = max(S_1[adjust_i], S_2[adjust_i])$;

10         $v_{min} = min(S_1[adjust_i], S_2[adjust_i])$;

11         $v_{min} = \frac{(2-t)v_{max}}{2+t}$;

We model the local optimization problem for Rate Distance adjustment as a GP problem and iteratively invoke a GP solver to do the adjustment. A GP problem is an optimization problem that has the following form:

$$objective : \quad min \quad f_0(x)$$

$$subject\ to : \quad f_i(x) \leq 1; \quad h_i(x) = 1;$$

where f_i is a posynomial and h_i is a monomial. A monomial is a function which is defined as: $f(x) = cx_1^{a_1} x_2^{a_2} ... x_n^{a_n}$, where $c > 0$ and $a_i \in R$ (Real number). A posynomial is the summation of monomials. If x_is are real numbers, GP problem can be solved in polynomial time.

We use the weight decreasing adjustment to demonstrate how GP model can be built. In the case using weight increasing, similar model can be built. For two nodes u_1 and u_2, the weight sequences of these two nodes are $S_1(v_1, v_2, ..., v_n)$ and $S_2(v'_1, v'_2, ..., v'_n)$ respectively. Suppose the edge sequences correspond to these two weight sequences are $E_1(e_1, e_2, ..., e_n)$ and $E_2(e'_1, e'_2, ..., e'_n)$ (note there may exist one edge (u_1, u_2) which appears both in E_1 and E_2), we can build the following GP model:

The ***constants*** used in the GP model:

- The original weights on the edges in E_1 and E_2: $v_1, v_2, ..., v_n, v'_1, v'_2, ..., v'_n$.

The ***variables*** used in the GP model:

- nv_i : the new weight of edge e_i;
- nv'_i : the new weight of edge e'_i.

The ***objective*** of GP model:

- If there's no overlapping between E_1 and E_2: $min \sum_{i=1}^{n} \frac{v_i}{nv_i} + \sum_{i=1}^{n} \frac{v'_i}{nv'_i}$
- If there's one edge which appears both in E_1 and E_2, suppose e_p in E_1 and e_q in E_2 are the same edge (we should only count e_p/e_q one time):

$min \sum_{i=1}^{n} \frac{v_i}{nv_i} + \sum_{i=1}^{q-1} \frac{v'_i}{nv'_i} + \sum_{i=q+1}^{n} \frac{v'_i}{nv'_i}$

The *constraints* used in the GP model:

- The Rate Distance between new S_1 and S_2 should be less than t:
  $$\forall i \quad \frac{nv_1 \cdot nv_i'}{nv_i \cdot nv_1'} \le t \quad \wedge \quad \forall i \quad \frac{nv_1' \cdot nv_i}{nv_i' \cdot nv_1} \le t$$
- Only weight decreasing is allowed: $\forall i \quad \frac{nv_i}{v_i} \le 1 \quad \wedge \quad \forall i \quad \frac{nv_i'}{v_i'} \le 1$
- The new weight on each edge should be bigger than the allowed minimum weight min_W: $\forall i \quad \frac{min_W}{nv_i} \le 1 \quad \wedge \quad \forall i \quad \frac{min_W}{nv_i'} \le 1$
- If there's an edge appearing both in E_1 and E_2, suppose e_p in E_1 and e_q in E_2 are the same edge, the weights of them should be the same: $\frac{nv_i}{nv_i'} = 1$

We use the optimization software Mosek (www.mosek.com) solve our GP models. To build the GP model, we should scan all the edges adjacent to the two adjusting nodes, so the time complexity of this adjustment is $O(d + t_{GP})$ (d is the degree of the two adjusting nodes). Here t_{GP} is the running time of the GP solver. Since all the values in the GP model are real numbers, the GP solver can get the solution in polynomial time. t_{GP} has polynomial relation with d.

It is obvious that our adjusting algorithms will always find a solution. The worst case of an adjusting algorithm for weight decreasing adjustment is that all the edges have the equal weight min_W. The worst case for weight increasing adjustment is that all the edges have the equal weight max_W, so the adjusting algorithm will terminate with a solution. The worst case of the k-weighted-degree anonymous algorithms are to generate a full connected graph with the same weight on all edges.

5   Experiments

5.1   DataSets

In this section, we test the effectiveness of our methods on two real datasets.

Arnet. (www.arnetminer.org) is a researcher social network collected by Tsinghua University. It contains information extracted from crawled web pages of computer scientists. The data also contains the co-authorship graph among these people. The weight on each edge represents the number of papers coauthored by two researchers. In this experiment, we extract a subgraph which contains 6000 nodes and 37848 edges.

ArXiv Data. (arXiv.org) is an e-print service system in Physics, Mathematics, Computer Science, etc. We extract a subset of co-author graph in Computer Science, which contains 19835 nodes and 40221 edges. Each node denotes an author, and each edge has a weight which represents the number of papers co-authored by the two endpoints.

5.2   Utilities

Since our k-weighted-degree anonymous model targets on the special characteristic of the edge weight, we use the relative edge weight distribution's change as our utilities. Suppose the minimum edge weight in the graph is min, then for any edge e, its relative edge weight is $\frac{e.weight}{min}$. We define two utilities based on the relative weights in our experiment. The reason we choose relative weight is that the information represented by this graph is not changed after multiplying or dividing all edge weights by a parameter. The two utilities we tested are:

- Standard deviation change ratio of the relative edge weights (SDC)
  Standard deviation is a widely used measure of the variability or dispersion of data. Here we use it to measure the distribution of the relative edge weights. For a graph $G(V, E)$ (V is the node set and E is the edge set), the standard deviation of G: G_{SD} is computed as: $SD_G = \sqrt{\frac{\sum_{e \in E}(\frac{e.weight}{min} - mean)^2}{|V|-1}}$ with $mean = \frac{\sum_{e \in E}\frac{e.weight}{min}}{|V|}$.

  The SDC of a published graph G_a compared with the original Graph G_o is: $SDC = \frac{|SD_{G_a} - SD_{G_o}|}{SD_{G_o}}$. The smaller this value is, the better utility G_a gets.
- Average relative edge weights change ratio (CR)
  For a graph G_o and its corresponding k-weighted-degree anonymous graph G_a, the relative weight change ratio of all the edges can be estimated by: $CR = \frac{\sum_{e \in G_o, e' \in G_a, e.x = e'.x \wedge e.y = e'.y} \frac{|e.weight - e'.weight|}{e.weight}}{|E_o|}$. The smaller this value is, the better utility G_a gets.

5.3   Results

In order to show the effects of our privacy model, we also generate the strict k-weighted-degree anonymous graph. The algorithm to generate the strict k-weighted-degree anonymous graph from a k-degree-anonymous graph is: For any set of edges that need to have the same weight, if there exist two edges whose weights' difference is bigger than a threshold t, we set all the edge weights in this set to the mean weight of the set. After this adjustment, if there's no edge set needs to be adjusted, finish; otherwise, we repeat this process. We set $t = 0.0005$ in this experiment.

We generate the 3-weighted-degree anonymous graphs for Arnet and ArXiv datasets respectively using different distance definitions and thresholds. Figure 2 shows the utilities of the 3-weighted-degree anonymous graphs with the Absolute Distance, where x axis is the threshold and the y axis is the corresponding utility. "A. Graph" stands for the 3-weighted-degree anonymous graph and "S.A.Graph" stands for the strict 3-weighted-degree anonymous graph. From the result we can see, if we use the strict weight anonymous, the utilities are very bad (The SD is close to 1 and CR is bigger than 0.3). Given a distance threshold, an anonymized graph with much better utilities can be generated based on Absolute Distance. The utilities become better when the threshold increases.

Figure 3 and Figure 4 show the utilities of the 3-weighted-degree anonymous graphs with the Relative Distance and Rate Distance respectively. Similar results are observed as Figure 2. Especially, for Rate Distance case, even with threshold = 1, it gets much better utilities than the strict k-weighted-degree anonymous. These results demonstrate that our privacy model helps keep the utilities of weighted graphs. The Rate Distance gets the best effect on preserving the edge weights of the original graph.

We also test the protection effect and efficiency of our algorithms. We do the attack using the distances defined in Section 3 with corresponding thresholds, no node can be

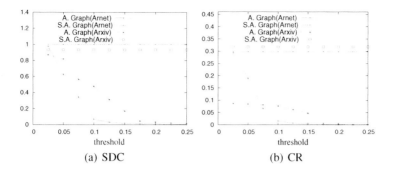

(a) SDC   (b) CR

Fig. 2. Utilities using Absolute Distance

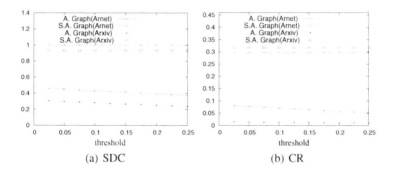

(a) SDC   (b) CR

Fig. 3. Utilities using Relative Distance

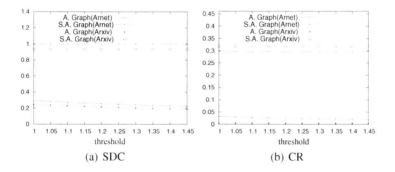

(a) SDC   (b) CR

Fig. 4. Utilities using Rate Distance

re-identified. Our algorithms also have good time performance. We run our algorithms on a notepad with $1GB$ memory and $1.66GHz$ CPU. For the ArXiv data (19835 nodes) with the smallest t (0.05 for the Absolute/Relative Distance and 1 for the Rate Distance), the computer takes 13.4 seconds, 3.1 seconds and 69.8 seconds for Absolute Distance, Relative Distance and Rate Distance respectively to finish the computation. Due to the space limitation, we do not show all the results here.

6   Related Works

Simply removing the identifiers in social networks does not guarantee privacy. The unique patterns such as node degree, subgraph, or distance to special nodes can be used to re-identify the nodes/links [8]. The attack that only uses certain background knowledge and doesn't "actively" change the graph is called passive attack, and the one "actively" changes the graph when social networks are collecting data is called active attack (An attacker will find the "actively" changed parts in the anonymized graph and follows them to attack other nodes). Most current works can be categorized into two classes: to prevent passive attack [9,8,5,15,4,13,14] and to prevent active attack [12].

The basic methods to prevent the passive attack include clustering and edge editing. Clustering is to cluster "similar" nodes together and publish a super node instead of the original nodes. Hay[8] proposed a heuristic clustering algorithm to prevent privacy leakage using vertex refinement, subgraph, and hub-print attacks. Zheleva[15] developed a clustering method to prevent the sensitive link leakage. Campan[4] discussed how to implement clustering when considering both node labels and structure information. Cormode[5][3] introduced (k, l)-groupings for bipartite graph and social networks to do the protection respectively. The edge editing-based approach tries to add or delete edges in order to make the graph satisfy certain constraints. Liu[9] defined and implemented k-degree-anonymous model on network structure, that is for published network, for any node, there exists at least $k - 1$ other nodes have the same degree as this node. Zhou[14] considered a stricter model: for every node, there exist at least $k - 1$ other nodes share isomorphic neighborhoods when taking node labels into account. Zou[16] proposed a k-Automorphism protection model: A graph is k-Automorphism if and only if for every node there exist at least $k - 1$ other nodes do not have any structure difference with it. Cheng[6] designed a k-isomorphism model to protect both nodes and links: a graph is k-isomorphism if this graph consists k disjoint isomorphic subgraphs. Ying[13] studied how random deleting and swapping edges change graph properties and proposed an eigenvalues oriented random graph change algorithm.

One method to prevent active attack is to recognize the fake nodes added by attackers and remove them before publishing the data. Shrivastava[12] proposed an algorithm that can identify fake nodes based on the triangle probability difference between normal nodes and fake nodes.

All the previous works that prevent node re-identification only considered unweighted graphs. For the privacy protection in weighted graphs, Liu[10] treated weights on the edges as sensitive labels and proposed a method to preserve shortest paths between most pairs of nodes in the graph. Das[7] proposed a Linear Programming based method to protect the edge weights while preserving the path of shortest paths. Both these two works treated the edge weights as sensitive information and considered how to protect them. Our work is different since we treat the edge weights as the background knowledge and study how to prevent node re-identification in a weighted graph.

7   Conclusion

In this paper, we study the problem about how to prevent node re-identification in weighted graphs. We define the k-weighted-degree anonymous model by considering the special characteristic of edge weights. Our model takes benefit from the numerical characteristic of edge weights, which helps to remain the weight diversity of the graph. Our experiments on the real datasets confirm that it is possible to preserve user's privacy while keeping weights in the social graph.

Acknowledgement

Funding for this work was provided by Hong Kong RGC under project no. N_HKUST61 2/09 and NSFC Project Grants 60736013, 60873022, 60903053.

References

1. Aggarwal, G., Feder, T., Kenthapadi, K., Khuller, S., Panigrahy, R., Thomas, D., Zhu, A.: Achieving Anonymity via Clustering. In: PODS 2006, pp. 153–162 (2006)
2. Backstrom, L., Dwork, C., Kleinberg, J.: Wherefore art thou r3579x?: anonymized social networks, hidden patterns, and structural steganography. In: WWW 2007, pp. 181–190 (2007)
3. Bhagat, S., Cormode, G., Krishnamurthy, B., Srivastava, D.: Class-based graph anonymization for social network data. Proc. VLDB Endow. 2(1), 766–777 (2009)
4. Campan, A., Truta, T.M.: A clustering approach for data and structural anonymity in social networks. In: Bonchi, F., Ferrari, E., Jiang, W., Malin, B. (eds.) PinKDD 2008. LNCS, vol. 5456. Springer, Heidelberg (2008)
5. Cormode, G., Srivastava, D., Yu, T., Zhang, Q.: Anonymizing bipartite graph data using safe groupings. Proc. VLDB Endow. 1(1), 833–844 (2008)
6. Cheng, J., Fu, A., Liu, J.: K-Isomorphism: Privacy Preserving Network Publication against Structural Attacks. In: SIGMOD 2010, pp. 459–470 (2010)
7. Das, S., Egecioglu, O., Abbadi, A.: Privacy Preserving in Weighted Social Network. In: ICDE 2010, pp. 904–907 (2010)
8. Hay, M., Miklau, G., Jensen, D., Towsley, D., Weis, P.: Resisting structural re-identification in anonymized social networks. Proc. VLDB Endow. 1(1), 102–114 (2008)
9. Liu, K., Terzi, E.: Towards identity anonymization on graphs. In: SIGMOD 2008, pp. 93–106 (2008)
10. Liu, L., Wang, J., Liu, J., Zhang, J.: Privacy preserving in social networks against sensitive edge disclosure. Technical Report CMIDA-HiPSCCS 006-08 (2008)
11. Li, N., Li, T.: t-closeness: Privacy beyond k-anonymity and l-diversity. In: ICDE 2007, pp. 106–115 (2007)
12. Shrivastava, N., Majumder, A., Rastogi, R.: Mining (social) network graphs to detect random link attacks. In: ICDE 2008, pp. 486–495 (2008)
13. Ying, X., Wu, X.: Randomizing social networks: a spectrum preserving approach. In: Jonker, W., Petković, M. (eds.) SDM 2008. LNCS, vol. 5159, pp. 739–750. Springer, Heidelberg (2008)
14. Zhou, B., Pei, J.: Preserving privacy in social networks against neighborhood attacks. In: ICDE 2008, pp. 506–515 (2008)
15. Zheleva, E., Getoor, L.: Preserving the privacy of sensitive relationships in graph data. In: Bonchi, F., Malin, B., Saygın, Y. (eds.) PInKDD 2007. LNCS, vol. 4890, pp. 153–171. Springer, Heidelberg (2008)
16. Zou, L., Chen, L., Özsu, M.T.: k-automorphism: a general framework for privacy preserving network publication. Proc. VLDB Endow. 2(1), 946–957 (2009)

An Unbiased Distance-Based Outlier Detection Approach for High-Dimensional Data

Hoang Vu Nguyen[1], Vivekanand Gopalkrishnan[1], and Ira Assent[2]

[1] School of Computer Engineering, Nanyang Technological University, Singapore
{ng0001vu,asvivek}@ntu.edu.sg
[2] Department of Computer Science, Aarhus University, Denmark
ira@cs.au.dk

Abstract. Traditional outlier detection techniques usually fail to work efficiently on high-dimensional data due to the curse of dimensionality. This work proposes a novel method for subspace outlier detection, that specifically deals with multidimensional spaces where feature relevance is a local rather than a global property. Different from existing approaches, it is not grid-based and dimensionality unbiased. Thus, its performance is impervious to grid resolution as well as the curse of dimensionality. In addition, our approach ranks the outliers, allowing users to select the number of desired outliers, thus mitigating the issue of high false alarm rate. Extensive empirical studies on real datasets show that our approach efficiently and effectively detects outliers, even in high-dimensional spaces.

1 Introduction

Popular techniques for outlier detection, especially density-based [1] and distance-based [2] ones, usually rely on the notion of distance functions defining the (dis)similarity between data points. However, since they take full-dimensional spaces into account, their performance is impacted by noisy or even irrelevant features. This issue was addressed in [3], which asserts that in such spaces, the concept of nearest neighbors becomes meaningless since nearest and farthest neighbors are alike. Even employing global dimension reduction techniques does not resolve this problem, because feature irrelevance is a local rather than a global property [3]. Therefore, in recent years, researchers have switched to subspace anomaly detection [3,4,5]. This paradigm shift is feasible as outliers though may be difficult to find in full-dimensional space, where they are hidden by irrelevant/noisy features, they nevertheless can be found completely in subspaces [3]. In addition, because subspaces are typically much fewer dimensions than the entire problem space, detection algorithms are able to overcome the curse of dimensionality. However, this approach opens new challenges:

Unavoidable Exploration of all Subspaces to Mine Full Result Set. Since the monotonicity property does not hold in the case of outliers, one cannot apply apriori-like heuristic [6] (as used in mining frequent itemsets) for mining outliers.

J.X. Yu, M.H. Kim, and R. Unland (Eds.): DASFAA 2011, Part I, LNCS 6587, pp. 138–152, 2011.
© Springer-Verlag Berlin Heidelberg 2011

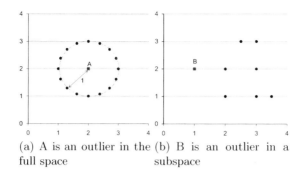

(a) A is an outlier in the (b) B is an outlier in a
full space   subspace

Fig. 1. Non-monotonicity of subspace outliers

In other words, if a data point p does not show any anomalous behavior in some subspace S, it may still be an outlier in some lower-dimensional projection(s) of S (and this is also the reason why we find outliers in subspaces). On the other hand, if p is a normal data point in all projections of S, it can still be an outlier in S. Figure 1 provides two examples on synthetic datasets illustrating our point. Though A is not an outlier on any 1-dimensional projection of the dataset, it is an outlier in the 2-dimensional space. Conversely, B is not an outlier in the 2-dimensional space, but it shows anomalous behavior when the dataset is mapped to the x-axis. Therefore, exploring all subspaces is unavoidable in order to find all outliers. However, algorithms that explore all subspaces are infeasible on high-dimensional data, because the number of available subspaces grows exponentially with the number of data dimensions.

Difficulty in Devising an Outlier Notion. Defining what an outlier should be in high-dimensional data is not straightforward. Popular distance-based outlier definitions (e.g., r-Distance [2], CumNN [7]) have been successfully used in full-dimensional space outlier detection. However, when they are applied for mining outliers in subspaces, they suffer the issue of *dimensionality bias*. In particular, they assign data points higher outlier scores in high-dimensional subspaces than they do in lower-dimensional ones. This leads to the loss of outliers located in low-dimensional subspaces. Moreover, it is so far unclear how these metrics should be extended to subspaces. Current notions specifically developed for mining outliers in subspaces simply consider outliers as by-products of the clustering process [8], or are arbitrary, i.e., they work with grids whose resolution cannot be easily determined [3,5], or use cutoff thresholds without clear details on how the values may be set [4]. On the other hand, it is not easy to devise a subspace outlier notion whose parameters can be decided intuitively.

Exposure to High False Alarm Rate. Because typical mining approaches make a binary decision on each data point (normal or outlier) in each subspace, they flag too many points as outliers as the number of possible subspaces to examine is large. This not only causes high false alarm rates, but also requires additional effort in analyzing the results. This problem can be avoided by employing a ranking-based algorithm [9], which would allow users to limit such large result sets. However, such algorithms are difficult to design, mainly because of the lack of appropriate scoring functions. In the context of subspace mining, a score function permitting ranking must be able to produce scores that can be compared to each other across subspaces, i.e., it should be dimensionality-unbiased.

Problem Statement. Our goal is to resolve the above challenges, and build an efficient technique for mining outliers in subspaces. It should: (a) avoid expensive scan of all subspaces while still yielding high detection accuracy, (b) include an outlier notion that eases the task of parameter setting, and facilitates the design of pruning heuristics to speed up the detection process, and (c) provide a ranking of outliers across subspaces. We achieve this goal by presenting *High-dimensional Distance-based Outlier Detection* (HighDOD), a novel technique for outlier detection in feature subspaces. Overcoming the aforementioned difficulties, High-DOD provides a distance-based approach [10] towards detecting outliers in very high-dimensional datasets. Though being distance-based, the notion of outliers here is unbiased w.r.t. the dimensionality of different subspaces. Furthermore, HighDOD produces a ranking of outliers using a direct, integrated nested-loop algorithm [10], which helps to reduce the overall runtime cost. HighDOD is also not grid-based (i.e., it does not require division of data attributes into ranges) and hence, is not dependent on grid resolution. Similar to other subspace outlier detection approaches [3,4], HighDOD explores subspaces of dimensionality up to some threshold. However, it is able to yield very high accuracy (c.f., Section 4). Our experimental results on real datasets demonstrate that it works efficiently and effectively to meet our purpose.

The rest of this paper is organized as follows. We provide a formal background of the problem and review related work in the next section. Then we present the HighDOD approach in in Section 3 and empirically compare with other existing techniques in Section 4. Finally, we conclude our paper in Section 5 with directions for future work.

2   Background and Literature Review

Consider a dataset DS with N data points in d dimensions. Each of the dimensions is normalized such that they all have the same scale (without loss of generality, we assume the range is $[0, 1]$). The distance between any two data points $p = (p_1, p_2, \cdots, p_d)$ and $q = (q_1, q_2, \cdots, q_d)$ in subspace $S = \{s_1, s_2, \cdots, s_{dim(S)}\} \subset \{1, 2, \cdots, d\}$ is defined as $D(p^S, q^S) = (\sum_{i \in S} |p_i - q_i|^l)^{1/l}$ where l is a positive integer. In other words, we restrict the distance function to the L-norm class.

Work in distance-based outlier detection was pioneered by Knorr and Ng in 1998 [2]. According to their proposal, outliers are points from which there are fewer than k other points within distance r. While this definition requires the specification of k and r (which is not easy) and produces only binary results (outlier or non-outlier), Angiulli *et al.* [7] proceed further by defining a data point's total distances to its k nearest neighbors to be its outlier score. This notion allows the design of ranking-based methods.

Ranking-based techniques in general have many advantages over threshold-based ones (for more details on the classification of a detection method into either ranking-based or threshold-based, please refer to [9]). First, as pointed out in [8], binary decision on whether or not a data point is an outlier is not practical in many applications and hard to parameterize. As for ranking-based methods, the difficulty one would face in setting the cutoff threshold is lifted off. Instead, users may specify how many outliers they want to see. Subsequently, the respective technique will produce a list of anomalies sorted in ascending/descending order of outlier scores and whose cardinality is equal to the user input parameter. This is of great convenience because users can avoid analyzing excessively large amount of outputs. Moreover, they are provided with an intuition on the degree of outlier-ness of output points.

Subspace mining has been studied extensively in the field of data clustering, which aims to group similar data points together. Typical clustering approaches are ineffective on high-dimensional data, because irrelevant features hide some underlying structures, and also distance functions utilizing all attributes are meaningless in such spaces [3]. Reducing data dimensionality using global dimension reduction techniques like PCA is also ineffective because feature irrelevance is a local rather than a global property, i.e., a feature may be irrelevant in one subspace but required in another subspace. This issue has been addressed by many subspace clustering methods, which efficiently explore the subspaces by employing the monotonicity property of clusters. In other words, if a cluster is present in a subspace, it is reflected in some projection of the latter, so apriori-like algorithms can be designed to avoid irrelevant subspaces.

Unfortunately, as the monotonicity property does not hold in our case, designing efficient outlier detection techniques in subspaces becomes very challenging. This problem was first addressed by Aggarwal *et al.* in their proposal HighOut [3], which defines a data point to be an outlier if it is present in a region of abnormally low density in some lower-dimensional projection. HighOut performs a grid *discretization* of the data by dividing each data attribute into ϕ equi-depth ranges, and then employs a genetic approach to mine hypercubes (of dimensionality up to m), with the smallest densities. There are a few issues with HighOut. First, its notion of outliers is grid-based whereas the grid resolution cannot be easily determined. Second, it suffers the intrinsic problems of evolutionary approaches - its accuracy is unstable and varies depending on the selection of initial population size as well as the crossover and mutation probabilities. Finally, it may suffer high false alarm rates, since it only produces a ranking of hypercubes whose total cardinality may be high while actual the number of outliers is small. In order to improve HighOut, the authors later introduced DenSamp, a non-grid-based subspace detection approach [4]. Though designed to work with uncertain data, it is also applicable on normal data. Similar to HighOut, DenSamp also mines outliers in subspaces of dimensionality up to m. However, it flags a data point p as a (δ, η)-outlier if the η-probability of p in some subspace is less than δ. Here, the η-probability of p is defined as the probability that p lies in a region with data density of at least η. One major drawback of DenSamp is that the two important parameters δ and η are not easy to set. Furthermore, δ is dimensionality biased. In particular, with increasing dimensionality, distances between points grow and densities drop [11]. Thus, in high-dimensional subspaces, the η-probability of p tends to be less than that in lower-dimensional ones. Besides, DenSamp does not provide a ranking of outliers, i.e., the number of outputs may be very high making post-analysis difficult.

Recently Ye *et al.* presented PODM [5], an apriori-like method for mining outliers in subspaces. Based on the idea of HighOut, PODM discretizes each dimension into ranges. It then assigns each subspace an anomaly score based on Gini-entropy, designed such that an apriori-like pruning rule can be applied. Consequently, PODM claims to explore all subspaces efficiently. PODM discards irrelevant subspaces, and then for the remaining ones, it calculates each hypercube's outlying degree and outputs those with highest values. This approach has several limitations. First, its performance is dependent on the grid resolution which is not easy to determine. Second, and perhaps more vital, is that it discards potentially useful subspaces (by implicitly assuming monotonicity), which causes loss of knowledge. In addition, similar to HighOut, PODM only ranks hypercubes, so it cannot provide a ranking of outliers. Finally, PODM lacks intuition on how to choose the cutoff thresholds utilized in its subspace pruning process. Our experiments (c.f., Section 4), show that PODM yields unsatisfactory accuracy for subspace outlier detection.

Motivated by the need of a ranking-based and dimensionality unbiased detection technique, Müller *et al.* [8] proposed the OutRank approach for ranking outliers in high-dimensional data. In order to overcome the curse of dimensionality, OutRank first performs subspace clustering and then assigns each object an outlier score that is an aggregation of its presence in different subspace clusters. This nonetheless smooths out the density deviations that we are trying to detect. Hence, OutRank does not account for great deviations in each individual subspace. Furthermore, the aggregation nature of outlier score in OutRank fails to address the issue of local feature irrelevance that we are interested in studying in this paper.

3   Approach

In this section, we first introduce a novel dimensionality unbiased notion of subspace outliers. Based on that, we proceed to present HighDOD and then theoretically explain why HighDOD works well to meet our purpose.

3.1   Outlier Score Function

Formally, we make the following assertion for subspace outliers:

Property 1. [NON-MONOTONICITY PROPERTY] Consider a data point p in the dataset DS. Even if p is not anomalous in subspace S of DS, it may be an outlier in some projection(s) of S. Even if p is a normal data point in all projections of S, it may be an outlier in S.

Among the available notions of distance-based outliers, the proposal by Angiulli *et al.* [7] is the most efficient and has been applied in many works, e.g. [10]. Denoting the set of k nearest neighbors of a data point p in DS as kNN_p, we can present their outlier score function as follows.

Definition 1. [OUTLIER SCORE FUNCTION: F_{out} [7]] *The dissimilarity of a point p with respect to its k nearest neighbors is known by its cumulative neighborhood distance. This is defined as the total distance from p to its k nearest neighbors in DS. In other words, we have: $F_{out}(p) = \sum_{m \in kNN_p} D(p, m)$.*

This function is dimensionality biased and violates Property 1. In particular, it is easy to recognize that if S is a subspace of T, then we have: $D(p^S, q^S) \le D(p^T, q^T)$. Thus, data points in higher-dimensional subspaces will have larger outlier scores than in lower-dimensional ones, i.e., if p is not an outlier in T, p will not be an outlier in S when a ranking-based technique is in play. This is obviously unanticipated. Let us denote the set of k nearest neighbors of a data point $p \in DS$ in subspace S as $kNN_p(S)$. In order to ensure Property 1 is not violated, we redefine the outlier score function as below.

Definition 2. [SUBSPACE OUTLIER SCORE FUNCTION: FS_{out}] *The dissimilarity of a point p with respect to its k nearest neighbors in a subspace S of dimensionality $dim(S)$, is known by its cumulative neighborhood distance. This is defined as the total distance from p to its k nearest neighbors in DS (projected onto S), normalized by $dim(S)$. In other words, we have:*

$$FS_{out}(p, S) = \frac{1}{[dim(S)]^{1/l}} \sum_{m \in kNN_p(S)} D(p^S, m^S),$$

where q^S is the projection of a data point $q \in DS$ onto S.

Besides assigning multiple outlier scores (one per subspace) to each data point, FS_{out} is also *dimensionality unbiased* and *globally comparable*.

We illustrate the dimensionality unbiased property of FS_{out} by revisiting the examples in Figure 1. Let us set $k = 1$ and $l = 2$ (i.e., using Euclidean distance). In Figure 1(a), A's outlier score in the 2-dimensional space is $1/\sqrt{2}$ which is the largest across all subspaces. In Figure 1(b), the outlier score of B when projected on the subspace of the x-axis is 1, which is also the largest in all subspaces. Hence, FS_{out} flags A and B (in their respective datasets) as outliers.

The globally comparable property of FS_{out} is established by the following lemmas.

Lemma 1. [RANGE OF DISTANCE] *In each subspace S of DS, the distance between any arbitrary data points p and q is bounded by $(dim(S))^{1/l}$. Mathematically:*

$$D(p^S, q^S) \le (dim(S))^{1/l}$$

Proof. From the definition of distance function D, we have:

$$D(p^S, q^S) = \left(\sum_{i \in S} |p_i - q_i|^l \right)^{1/l}$$

a. When $l < \infty$: Since $p_i, q_i \in [0, 1]$, it holds that $|p_i - q_i| \leq 1$. Thus, $|p_i - q_i|^l \leq 1$. As a result:

$$D(p^S, q^S) \leq \left(\sum_{i \in S} 1 \right)^{1/l} = (dim(S))^{1/l}$$

b. When $l \to \infty$: By definition of the Minkowski distance for $l \to \infty$:

$$D(p^S, q^S) = \lim_{l \to \infty} \left(\sum_{i \in S} |p_i - q_i|^l \right)^{1/l} = \max_{i \in S} |p_i - q_i|$$

Thus, $D(p^S, q^S) \leq 1$ (1). As $\lim_{l \to \infty} (1/l) = 0$ and $1 \leq dim(S) < \infty$, we have $\lim_{l \to \infty} (dim(S))^{1/l} = (dim(S))^{\lim_{l \to \infty} (1/l)} = (dim(S))^0 = 1$ (2). From (1) and (2), we conclude that as $l \to \infty, D(p^S, q^S) \leq (dim(S))^{1/l}$. $\qquad\square$

Lemma 2. [RANGE OF OUTLIER SCORE] *For an arbitrary data point p and any subspace S, we have $0 \leq FS_{out}(p, S) \leq k$.*

Proof. By definition, we get:

$$FS_{out}(p, S) = \frac{1}{(dim(S))^{1/l}} \sum_{m \in kNN_p(S)} D(p^S, m^S)$$

Following Lemma 1: $D(p^S, m^S) \leq (dim(S))^{1/l}$. Thus, it holds that:

$$FS_{out}(p, S) \leq \frac{1}{(dim(S))^{1/l}} \sum_{m \in kNN_p(S)} (dim(S))^{1/l}$$

i.e.,

$$FS_{out}(p, S) \leq k \qquad\qquad\square$$

From Lemma 2, it can be seen outlier scores of all points across different subspaces have the same scale. Therefore, they are comparable to each other. This is of great advantage towards designing a technique for mining ranking-based subspace outliers. Having obtained a desirable score function for outliers in subspaces, now we can formally define the mining problem based upon this score function.

Definition 3. [SUBSPACE OUTLIER DETECTION PROBLEM] *Given two positive integers k and n, mine the top n distinct anomalies whose outlier scores (in any subspace) are largest.*

By using the novel FS_{out}, we are able to reformulate the problem of detecting subspace outliers to facilitate the design of a ranking-based method. Notice that both the input parameters (k and n) can be easily decided as has been solidly founded in previous works on distance-based outliers [12]. In other words, in practical applications, it is much easier to tune the dependent parameters of the solution to our problem, than it is to determine the two threshold parameters of DenSamp, or the grid resolution of HighOut and PODM.

3.2   The HighDOD Method

We now present our solution towards the subspace outlier detection problem as mentioned in Definition 3. Our approach, HighDOD, is described in Algorithms 1, 2, and 3, and explained below.

OutlierDetection. Property 1 highlights that *in order to mine all outliers, it is unavoidable to explore all subspaces.* This poses a great burden towards designing a subspace detection method. For addressing the issue, Aggarwal *et al.* [3,4] suggested to examine subspaces of dimensionality up to some threshold m. Though this might cause some loss in accuracy, it is efficient in terms of runtime cost. The same article shows that the accuracy loss is not that severe as long as m is about $O(logN)$. Recent work by Ailon *et al.* [13] also mentions that a dataset's properties can be preserved after dimensionality reduction as long as the number of features extracted is $O(logN)$. Thus, we choose to pursue this direction in HighDOD.

First we call OutlierDetection (Algorithm 1) to carry out a bottom-up exploration of all subspaces of up to a dimensionality of m, where m is an input parameter. Therefore, 1-dimensional subspaces are examined first, then 2-dimensional ones, and so on. The top n outliers found so far are maintained in $TopOut$. In addition, the cutoff threshold c equal to the score of the nth outlier found so far is also maintained. It plays exactly the same role as in traditional nest-loop algorithms [10]. For each investigated subspace S, we first extract some candidate anomalies by calling CandidateExtraction (Algorithm 2), and then update $TopOut$ with those candidates by invoking SubspaceMining (Algorithm 3). After exhausting all i-dimensional subspaces, we proceed to the $(i+1)$-dimensional subspaces, and stop when the maximum dimensionality m has been reached, or there are no more subspaces to explore.

CandidateExtraction. This procedure is used for extracting the top βn ($\beta \geq 1$) potential candidate outliers in any subspace S. Without it, we would have to perform the traditional nested-loop algorithm in each individual subspaces, which is too expensive.

The main idea here is to estimate the data points' local densities by using a kernel density estimator, and choose βn data points with the lowest estimates as potential candidates. This comes from the intuition that outliers are rare events and not surrounded by many objects, i.e., their densities are expected to be very low. Note that in practice, though we only need to mine top n outliers, we may

---

Algorithm 1. OUTLIERDETECTION

Input: k: number of nearest neighbors; n: number of outliers to mine; m: maximum dimensionality; x: number of kernel centers; DS: the underlying dataset

Output: $TopOut$: the set of outliers

1 Set $c = 0$
2 Set $TopOut = \emptyset$
3 Set C_1 = the set of all 1-dimensional subspaces
4 Set $i = 1$
5 **while** $C_i \neq \emptyset$ and $i \leq m$ **do**
6   **foreach** *subspace* $S \in C_i$ **do**
7   Set $CandOut = CandidateExtraction(n, x, DS, S)$
8   Call $SubspaceMining(k, n, DS, S, TopOut, c)$
9   Set C_{i+1} = the set of distinct $(i + 1)$-dimensional subspaces created by combining C_i with C_1
10   Set $i = i + 1$

Algorithm 2. CANDIDATEEXTRACTION

Input: n: number of outliers to mine; x: number of kernel centers; DS: the underlying dataset; S: the considered subspace

Output: $CandOut$: the set of candidate outliers

1 Set $Ctrs$ = randomly sample x data points from DS
2 Construct x clusters $C_1, C_2, \cdots, C_x$ of DS on subspace S whose centroids are from $Ctrs$
3 Compute kernel bandwidths h_i on subspace S
4 Set $CandOut = \emptyset$
5 **foreach** *data point* $p \in DS$ **do**
6   Set $f(p, S) = (1/N) \cdot \sum_{j=1}^{x} |C_j| \cdot K(p^S - ctr(C_j)^S)$
7   Set $CandOut$ = extract $2n$ points from $CandOut \cup \{p\}$ with smallest density values

---

need to extract many more candidates to account for any error caused by the density estimator, so $\beta > 1$. Empirically, we find that $\beta = 2$ is sufficient. In other words, we extract $2n$ candidate outliers in each explored subspace.

Outlier detection by kernel density estimation has been studied before by Kollios *et al.* [14] and Aggarwal *et al.* [4]. Here, we follow the technique proposed in [4] though others like [14] are also applicable. We initially cluster the data incrementally with fixed centroids (initially chosen randomly from DS) to obtain a compact representation of the underlying dataset. Then we use those centroids across all subspaces for density estimation. However, since feature relevance varies among subspaces, the one-size-fits-all clustering centroids are unsuitable for our purpose. Hence, we suggest to perform clustering in each individual subspace to account for such variance. As analyzed later in this section, the additional runtime overhead incurred is not so high as one may be concerned. We employ the Gaussian kernel function, whose 1-dimensional form is expressed as $K_h(p - ctr) = (1/(h\sqrt{2\pi})) \cdot e^{-(p-ctr)^2/2h^2}$, where ctr and h are the kernel center and bandwidth, respectively. We choose the bandwidth h to be $1.06 \cdot \sigma \cdot N^{-1/5}$, where σ is the standard deviation of N data points [15]. The s-dimensional kernel function is the product of s identical kernels $K_{h_i}(\cdot)$, where $h_i = 1.06 \cdot \sigma_i \cdot N^{-1/5}$ with σ_i being the standard deviation along the ith dimension. On a subspace S, the density of a data point p can be approximated as $f(p, S) = (1/N) \cdot \sum_{j=1}^{x} |C_j| \cdot K(p^S - ctr(C_j)^S)$. More details for the reasoning of the approximation method are given in [4].

SubspaceMining. This procedure is used to update the set of outliers $TopOut$ with $2n$ candidate outliers extracted from a subspace S. Since outlier scores across subspaces have the same scale (c.f., Lemma 2), we can maintain one global cutoff threshold c and design a nested-loop-like algorithm for the update process.

Note that in high-dimensional data, an outlier may spread its anomalous behavior in more than one subspace. Thus, if we simply replace the nth outlier found so far with a new data point whose score is larger, we may end up with duplicate outliers. To prevent this from occurring, we only maintain a version of each outlier in $TopOut$. More specifically, if a data point is already in $TopOut$ before we use its newly computed score in some subspace for our updating purpose, we replace its score with the new score if it is higher, hence avoiding removal of any point maintained in $TopOut$. Otherwise, we update $TopOut$ with the new data point as in traditional nested-loop methods. The cutoff c is adjusted along the way to ensure good pruning efficiency.

3.3   Theoretical Analysis

Analysis of Parameters Used. As suggested in [3,13], setting the maximum dimensionality m as logarithmic to the size of the dataset N is sufficient, so we suggest to select $m = \lfloor log_{10}N \rfloor$[1]. This is equivalent to dividing each dimension of the original dataset into $\phi = 10$ ranges and choosing m such that the dataset is not very sparse w.r.t. m. The number of m-dimensional hypercubes is 10^m. To ensure that the sparsity condition is met, the average number of data points falling into each m-dimensional hypercube should be ≥ 1, i.e., $N/10^m \geq 1$. Solving the latter inequality, we arrive at $m = \lfloor log_{10}N \rfloor$.

The number of kernel centers x represents the level of data summarization. Following [4], we fix x to 140. We set the two remaining parameters, the number of nearest neighbors k and the number of outliers to detect n, based on many solid works on distance-based outlier mining [12].

Time Complexity. For each subspace S, the cost of extracting candidates includes: (a) clustering cost, (b) bandwidth computation cost, and (c) density estimation cost. The cost of clustering is $O(N \cdot x \cdot dim(S))$. To compute the

---

[1] We here consider cases where d is very high so that $N \ll 10^d$.

Algorithm 3. SUBSPACEMINING

Input: k: number of nearest neighbors; n: number of outliers to mine; DS: the underlying dataset; S: the underlying subspace; $CandOut$: the set of candidate outliers; $TopOut$: the set of outliers; c: cutoff threshold

1   **foreach** *data point* $p \in CandOut$ **do**
2   **foreach** *data point* $q \in DS$ **do**
3   **if** $q \neq p$ **then**
4   Update p's k nearest neighbors in subspace S using q
5   **if** $|kNN_p(S)| = k$ and $FS_{out}(p, S) < c$ **then**
6   Mark p as non-outlier
7   Process next data point in $CandOut$

8   /* p is not marked as non-outlier, so it is used to update $TopOut$ */
9   **if** $TopOut$ *contains* p **then**
10   **if** $FS_{out}(p, S) >$ *outlier score of* p *stored in* $TopOut$ **then**
11   Set p's score in $TopOut = FS_{out}(p, S)$

12   **else**
13   Set $TopOut =$ extract top n outliers from $TopOut \cup \{p\}$

14   **if** $Min(TopOut) > c$ **then**
15   Set $c = Min(TopOut)$

---

bandwidths, we need to compute the data's mean and standard deviation vectors on all dimensions of S. This incurs a cost of $O(N \cdot dim(S))$. For each data point $p \in DS$, the cost of density estimation and maintaining the set of $2n$ candidates is $O(N \cdot x \cdot dim(S) + 2 \cdot n \cdot N)$, which can be reduced to $O(N \cdot x \cdot dim(S))$ as $n \ll x \cdot N$. Hence, the total cost of executing CandidateExtraction is $O(N \cdot x \cdot dim(S))$. The cost of executing the SubspaceMining procedure is $O(2 \cdot n \cdot N \cdot dim(S))$. As a result, the overall cost of exploring the subspace S is $O((x + n) \cdot N \cdot dim(S))$.

Given an integer $r < d$, the number of subspaces of dimensionality r is given by C_d^r. Since we only examine subspaces of dimensionality up to m, the total runtime cost of HighDOD is: $\sum_{i=1}^{m} C_d^i \cdot O((x+n) \cdot N \cdot i) = O((x+n) \cdot N) \cdot \sum_{i=1}^{m} i \cdot C_d^i$.

In order to understand the efficiency of our approach, let us consider a dataset with 100,000 data points (i.e., $m = 5$) and $d = 20$. In this dataset, though N is large, it is very sparse w.r.t. d. As $\sum_{i=1}^{d} i \cdot C_d^i = d \cdot 2^{d-1}$, the reduction one obtains by using HighDOD instead of exploring all subspaces is $(d \cdot 2^{d-1})/(\sum_{i=1}^{m} i \cdot C_d^i) = (20 \cdot 2^{19})/(\sum_{i=1}^{5} i \cdot C_{20}^i) > 100$. In other words, HighDOD leads to a reduction of more than 100 times in execution time. Notice that the analysis above does not take into account the pruning rule used in Algorithm 3. The result is only an upper bound and our experiments (c.f., Section 4) show that the runtime cost of HighDOD is much lesser.

Benefits of HighDOD. Compared to HighOut and DenSamp, HighDOD utilizes a ranking-based outlier notion which allows easier parametrization. Furthermore, our proposed definition of outliers is derived from a popular distance-based notion [7] which has already been verified to be very suitable and intuitive for practical applications. As for OutRank, in the worst case, i.e., for poor parameterization of the subspace clustering or for rather homogeneous data, it will cluster almost all subspaces and can then only start to compute any scores, i.e., its execution time will be high. HighDOD's ranking-based algorithm is done in a nested-loop fashion, so it can avoid such costly clustering process. Different from HighOut and PODM, HighDOD is non-grid-based and hence not susceptible to the issues of grid resolution and position. The data compression in HighDOD is performed on every subspace which helps it better tune to feature local relevance than DenSamp. All of these points give HighDOD advantages over existing methods and make it be very applicable to outlier detection in high-dimensional spaces.

4   Empirical Results and Analyses

In this section, we compare the performance of HighDOD with DenSamp, HighOut, PODM, and LOF [1] (the best-known detection technique using full-dimensional space) by performing empirical studies on real datasets taken from the UCI Repository. As mentioned above, OutRank requires a clustering phase before starting the detection process. Further, OutRank is a "global" outlier detection approach that aggregates scores from different subspaces to come up with a global value, and hence, does not account for great deviations in each individual subspace. Thus, we decide not to include OutRank in our experiments.

Detection Accuracy. This experiment aims to assess the effectiveness of each method in terms of detection accuracy using four real datasets whose descriptions and setup procedures are given in Table 1. It is noted that the chosen datasets' dimensionality conforms to that of related work in the field for high-dimensional data [11,3,4]. We measure the quality of results by constructing the Precision-Recall tradeoff curve that is used widely in data classification as well as in outlier detection [3,5]. We build this curve by varying: (a) the number of outliers to detect n for HighDOD and LOF, (b) the number of hypercubes with lowest densities/highest outlier scores to mine for HighOut/PODM, and (c) the parameter η for DenSamp. For HighDOD, we set k to $0.05\% \cdot N$ following established work on distance-based outliers [12]. Parameter settings for other methods follow their respective papers. The results shown in Figure 2, indicate that in all test cases HighDOD yields the best accuracy. Among the remaining subspace detectors, DenSamp produces better accuracy than HighOut, while PODM has the worst results. The superior performance of HighDOD compared to DenSamp stems from the fact that it constructs kernel centers separately for each examined subspace. This helps HighDOD to better adapt to the local change in feature relevance. Though PODM explores all subspaces, its notion of anomalies fails to capture Property 1 which causes its detection quality to become unsatisfactory (the margin with HighOut is quite pronounced). While HighOut performs better than PODM, it is less accurate than the two non-grid-based methods, HighDOD and DenSamp. As for LOF, it performs relatively

Table 1. Characteristics of datasets used for measuring accuracy of techniques

Dataset	Description	Outlier	Normal
Ann-thyroid 1	21 features, 3428 instances	class 1	class 3
Ann-thyroid 2	21 features, 3428 instances	class 2	class 3
Breast Cancer (WSBC)[2]	32 features, 569 instances	'Malignant' class	'Benign' class
Musk (Version 2)[3]	168 features, 6598 instances	'Musk' class	'Non-musk' class
Arrythmia[4]	279 features, 452 instances	class 7, 8, 9, 14, 15	class 1

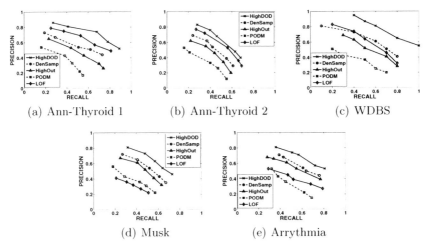

(a) Ann-Thyroid 1   (b) Ann-Thyroid 2   (c) WDBS

(d) Musk   (e) Arrythmia

Fig. 2. Detection accuracy of HighDOD, DenSamp, HighOut, PODM, and LOF

well when the number of dimensions is low (in Ann-Thyroid 1, Ann-Thyroid 2, and WDBS datasets). However, its accuracy deteriorates greatly with higher dimensionality where there are more noisy/irrelevant features.

Scalability. We also evaluate the scalability of HighDOD with respect to the dataset's size N, and the dataset's dimensionality d. Since PODM yields very unsatisfactory accuracy, we choose not to include it in this experiment. LOF is a full-dimensional detector and its accuracy on high-dimensional data (particularly those with more than 100 dimensions) is very poor. This points out that LOF is not suitable for high-dimensional outlier detection. Hence, the study of its scalability is also not of our interest. Thus, we only compare the efficiency of HighDOD against DenSamp and HighOut. Parameters settings for DenSamp and HighOut follow their respective papers. For HighDOD, n is set to the maximum number of outliers detected by DenSamp and HighOut while k is kept at $0.05\% \cdot N$. In this experiment, we test with the CorelHistogram (CH) dataset

---

[2] We randomly extract 10 'Malignant' records as outliers. We discard the record ID and label (i.e., 'Benign' or 'Malignant'), and use the remaining 30 real-valued features.

[3] The test set has 166 features (the first two symbolic attributes are excluded).

[4] We consider instances in classes whose cardinality less than 10 to be outliers.

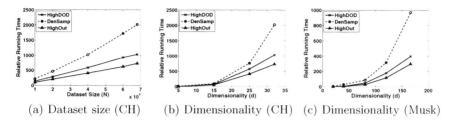

(a) Dataset size (CH)   (b) Dimensionality (CH)   (c) Dimensionality (Musk)

Fig. 3. Scalability of HighDOD, DenSamp, and HighOut

consisting of 68040 points in 32-dimensional space. CH contains 68040 records corresponding to 68040 photo images taken from various categories. However this dataset cannot be used to measure accuracy, because there is no clear concept of outlier among those images. Instead, it is often used for measuring the scalability of detection methods [7,10]. To better illustrate the efficiency of our method against high dimensionality, we include the Musk dataset in the experiment. Since its cardinality is not large enough, we choose not to test techniques' scalability against its size. The Arrythmia dataset though has large number of attributes is not selected due to its too small number of instances.

As in [3,4], we scale the running times obtained and present the relative running time of the three methods. Figure 3(a) shows that w.r.t. N, HighDOD scales better than DenSamp and worse than HighOut. The genetic-based searching process of HighOut prunes much of the search space, giving it the best scalability. Even though DenSamp only identifies the kernel centers once, it suffers highest execution time because its process of calculating data points' η-probability is costly, and moreover, it lacks pruning rules. On the contrary, the pruning rule in HighDOD's nested-loop approach helps to offset the cost of computing kernel centers in each explored subspace. Therefore, HighDOD yields better scalability than DenSamp. Figures 3(b) and 3(c) point out that the three algorithms scale super-linear with increasing data dimensionality with the same order: HighOut produces the best performance, next is HighDOD, and finally DenSamp. In addition, HighDOD's running time is just slightly worse than that of HighOut. From the empirical studies carried out, our proposed approach, HighDOD, obtains a better tradeoff between time and accuracy than existing methods.

5   Conclusions

This work proposes a new subspace outlier scoring scheme which is dimensionality unbiased. It extends the well-established distance-based anomaly detection to subspace analysis. Our notion of subspace outliers not only eases the parameter setting task but also facilitates the design of ranking-based algorithms. Utilizing this score, we introduced HighDOD, a novel ranking-based technique for subspace outlier mining. In brief, HighDOD detects outliers in a nested-loop fashion allowing it to effectively prune the search space. Empirical studies carried out on real datasets demonstrate HighDOD's efficiency as well as efficacy compared to other existing methods in the field. As future work, we are exploring possible ways to further reduce HighDOD's running time. We are also studying how to use our novel notion of subspace outliers to effectively explore all subspaces at low cost. The availability of such a technique would help us to mine all outliers in all subspaces, and hence, to further increase the detection accuracy.

References

1. Breunig, M.M., Kriegel, H.P., Ng, R.T., Sander, J.: LOF: Identifying density-based local outliers. In: SIGMOD Conference, pp. 93–104 (2000)
2. Knorr, E.M., Ng, R.T.: Algorithms for mining distance-based outliers in large datasets. In: VLDB, pp. 392–403 (1998)
3. Aggarwal, C.C., Yu, P.S.: An effective and efficient algorithm for high-dimensional outlier detection. VLDB J 14(2), 211–221 (2005)
4. Aggarwal, C.C., Yu, P.S.: Outlier detection with uncertain data. In: SDM, pp. 483–493 (2008)
5. Ye, M., Li, X., Orlowska, M.E.: Projected outlier detection in high-dimensional mixed-attributes data set. Expert Syst. Appl. 36(3), 7104–7113 (2009)
6. Agrawal, R., Srikant, R.: Fast algorithms for mining association rules in large databases. In: VLDB, pp. 487–499 (1994)
7. Angiulli, F., Pizzuti, C.: Outlier mining in large high-dimensional data sets. IEEE Trans. Knowl. Data Eng. 17(2), 203–215 (2005)
8. Müller, E., Assent, I., Steinhausen, U., Seidl, T.: OutRank: ranking outliers in high dimensional data. In: ICDE Workshops, pp. 600–603 (2008)
9. Nguyen, H.V., Ang, H.H., Gopalkrishnan, V.: Mining outliers with ensemble of heterogeneous detectors on random subspaces. In: Kitagawa, H., Ishikawa, Y., Li, Q., Watanabe, C. (eds.) DASFAA 2010, Part I. LNCS, vol. 5981, pp. 368–383. Springer, Heidelberg (2010)
10. Bay, S.D., Schwabacher, M.: Mining distance-based outliers in near linear time with randomization and a simple pruning rule. In: KDD, pp. 29–38 (2003)
11. Assent, I., Krieger, R., Müller, E., Seidl, T.: DUSC: Dimensionality unbiased subspace clustering. In: ICDM, pp. 409–414 (2007)
12. Tao, Y., Xiao, X., Zhou, S.: Mining distance-based outliers from large databases in any metric space. In: KDD, pp. 394–403 (2006)
13. Ailon, N., Chazelle, B.: Faster dimension reduction. Commun. CACM 53(2), 97–104 (2010)
14. Kollios, G., Gunopulos, D., Koudas, N., Berchtold, S.: Efficient biased sampling for approximate clustering and outlier detection in large data sets. IEEE Trans. Knowl. Data Eng. 15(5), 1170–1187 (2003)
15. Silverman, B.W.: Density Estimation for Statistics and Data Analysis. Chapman and Hall, Boca Raton (1986)

A Relational View of Pattern Discovery

Arnaud Giacometti, Patrick Marcel, and Arnaud Soulet

Université François Rabelais Tours, LI
3 place Jean Jaurès
F-41029 Blois France
`forename.surname@univ-tours.fr`

Abstract. The elegant integration of pattern mining techniques into database remains an open issue. In particular, no language is able to manipulate data and patterns without introducing opaque operators or loop-like statement. In this paper, we cope with this problem using relational algebra to formulate pattern mining queries. We introduce several operators based on the notion of cover allowing to express a wide range of queries like the mining of frequent patterns. Beyond modeling aspects, we show how to reason on queries for characterizing and rewriting them for optimization purpose. Thus, we algebraically reformulate the principle of the levelwise algorithm.

1 Introduction

Pattern discovery is a significant field of Knowledge Discovery in Databases (KDD). A broad spectrum of powerful techniques for producing local patterns has been developed over the two last decades [3–5]. But, it is widely agreed that the need of theoretical fusion between database and data mining still remains a crucial issue [14, 18, 23, 24]. We would force the pattern mining methods to fit in the relational model [1] which is the main database theory. Unlike most of the proposals [6, 10, 14, 16, 20, 23, 28, 33, 34], we desire to only address the pattern mining that we distinguish from the construction of global models [17] like decision trees.

Let us consider the popular task of frequent pattern mining [3] as a motivating example. Most works treat this task as a "black box" which input parameters are defined by the user [6, 7, 14, 16, 20, 28, 32, 34]. Instead of only specifying the minimal frequency threshold and the dataset, we think that the user query should fully formalize the notion of frequent patterns (e.g., it should describe how the frequency of a pattern is computed starting from the dataset). Ideally, we would like to express the frequent pattern mining query in the relational algebra in order to manipulate both the data and the patterns. As declarative aspects should be promoted on physical ones, a pattern discovery process has to be fully specified without considering algorithmic points. For this purpose, loop-like operators [10, 23, 33] are not relevant for us. Furthermore, the improvement of query performances mainly rests on physical optimizations in the field of pattern mining. Typically, the frequent pattern mining is efficiently performed

J.X. Yu, M.H. Kim, and R. Unland (Eds.): DASFAA 2011, Part I, LNCS 6587, pp. 153–167, 2011.
© Springer-Verlag Berlin Heidelberg 2011

by an adequate implementation [3–5, 25]. Such algorithmic optimizations (even specified at a higher level [10, 23, 33]) reduce the opportunity of integrating other optimizations. We prefer to favor logical reasoning for optimizing query performances. For instance, the rewriting of the naive frequent pattern mining query should enable us to algebraically formulate the levelwise pruning [25].

The main goal of this paper is to propose an algebraic framework for pattern discovery for expressing a wide range of queries without introducing opaque operators or loop-like statements. Our framework brings two meaningful contributions: expressive modeling and logical reasoning. First, it allows a large set of queries manipulating relations which contain both data and patterns. We add to the relational algebra several specific operators, like the cover operator ◁, to coherently and easily join such relations. We also define a new operator Δ for generating a language starting from a relation. Typically, the query $\sigma_{freq \geq f}(\gamma_{patt, \texttt{COUNT}(trans) \rightarrow freq}(\Delta(L) \triangleleft D))$ returns the patterns of language L frequent in dataset D. Second, the pattern-oriented relational algebra enables to characterize and rewrite queries in order to optimize their performance. In particular, we formalize the notions of syntactic constraint [9] and global constraint [12] by characterizing the degree of dependence between a query and a relation. Besides, we not only benefit from usual query rewriting methods stemming from the relational model, but we also algebraically reformulate the levelwise pruning.

This paper is organized in the following way. Section 2 introduces basic notions about the relational algebra and the pattern discovery. Section 3 defines the cover-like and domain operators which are at the core of our algebra. We then study the properties of downward closure and independence in Section 4. We rewrite queries satisfying such properties for optimization purpose in Section 5. Finally, Section 6 provides a related work.

2   Basic Notions

2.1   Relational Algebra

We enumerate here our notations for the relational algebra mainly inspired from [1]. Let **att** be a set of distinct literals, named *attributes*, **dom**(A) denotes the finite *domain* of the attribute $A \in$ **att**. The *relation schema* (or relation for brevity) $R[U]$ denotes a relation named by R where $U \subset$ **att**. An *instance* of R is a subset of **dom**$(U) = \times_{A \in U}$ **dom**(A). Given a relation $R[A_1, \ldots, A_n]$, R' renames the attributes $A_1, \ldots, A_n$ into $A'_1, \ldots, A'_n$. A *database schema* is a nonempty finite set $\mathbf{R} = \{R_1[U_1], \ldots, R_n[U_n]\}$ of relations. A *database instance* of $\mathbf{R}$ is a set $\mathbf{I} = \{I_1, \ldots, I_n\}$ such that I_i is an instance of the relation R_i. Finally, a *query* q maps a database instance to an instance of a relation. The set of attributes of this relation is denoted by **sch**(q). A query q' is *equivalent* to q, denoted by $q' \equiv q$, iff for any database instance $\mathbf{I}$, one has $q'(\mathbf{I}) = q(\mathbf{I})$.

Let I be an instance of R and J be an instance of S. The relations can be manipulated by means of set operators including Cartesian product $R \times S$ where $I \times J = \{(t, u) | t \in I \wedge u \in J\}$. If R and S are relations which have the same schema, then $R \cup S$, $R \cap S$ and $R - S$ are respectively the union, the intersection and the difference of R and S. *Selection:* $\sigma_f(I) = \{t | t \in I \wedge f(t)\}$ selects the tuples of I satisfying the logical formula f where f is built from (i) the logical operators ($\wedge$, $\vee$ or $\neg$), (ii) the arithmetic relational operators and (iii) operands based on attributes and constants. *Extended projection:* $\pi_{A_1,\ldots,A_n}(I) = \{t[A_1,\ldots,A_n] | t \in I\}$ only preserves the attributes $A_1,\ldots,A_n$ of R. Besides, the projection also permits to extend the relation by arithmetic expressions and to (re)name expressions. For instance, $\pi_{A+B \to B', C \to C'}(R)$ creates a new instance where the first attribute named B' results from the arithmetic expression $A + B$ and the second attribute corresponds to C, renamed C'. *Grouping:* $\gamma_{A_1,\ldots,A_n,\mathtt{AGG}(B)}(I) = \{(a_1,\ldots,a_n, \mathtt{AGG}(\pi_B(\sigma_{A_1=a_1 \wedge \cdots \wedge A_n=a_n}(I)))) \,|\, (a_1,\ldots,a_n) \in \pi_{A_1,\ldots,A_n}(I)\}$ groups tuples of I by attributes $A_1,\ldots,A_n$ and applies an aggregate function $\mathtt{AGG}$ on B.

2.2   Pattern Discovery

We provide here an overview of pattern discovery based on [25, 32] focusing on the main proposals of the field. A *language* $\mathcal{L}$ is a set of patterns: itemsets $\mathcal{L}_\mathcal{I}$ [3], sequences $\mathcal{L}_\mathcal{S}$ [4] and so on [5]. A *specialization relation* $\preceq$ of a language $\mathcal{L}$ is a partial order relation on $\mathcal{L}$ [25, 27]. Given a specialization relation $\preceq$ on $\mathcal{L}$, $l \preceq l'$ means that l is more general than l', and l' is more specific than l. For instance, the set inclusion is a specialization relation for the itemsets. Given two posets $(\mathcal{L}_1, \preceq_1)$ and $(\mathcal{L}_2, \preceq_2)$, a *cover relation* is a binary relation $\lhd \subseteq \mathcal{L}_1 \times \mathcal{L}_2$ iff when $l_1 \lhd l_2$, one has $l'_1 \lhd l_2$ (resp. $l_1 \lhd l'_2$) for any pattern $l'_1 \preceq_1 l_1$ (resp. $l_2 \preceq_2 l'_2$). The relation $l_1 \lhd l_2$ means that l_1 covers l_2, and l_2 is covered by l_1. The cover relation is useful to relate different languages together (e.g., for linking patterns to data). Note that a specialization relation on $\mathcal{L}$ is also a cover relation on $\mathcal{L}$ (e.g., the set inclusion is a cover relation for the itemsets).

The pattern can be manipulated by means of three kinds of operators non exhaustively illustrated hereafter. 1) *Pattern mining operators* produce patterns starting from a dataset: theory [25], MINERULE [26] and so on. More precisely, the *theory* denoted by $Th(\mathcal{L}, q, \mathcal{D})$ returns all the patterns of a language $\mathcal{L}$ satisfying a predicate q in the dataset $\mathcal{D}$ [25]. Typically, the minimal frequency constraint selects the patterns which occur in at least f transactions [3, 4]: $freq(\varphi, D) > f$. As mentioned in introduction, we notice that the query $Th(\mathcal{L}, freq(\varphi, \mathcal{D}) \geq f, \mathcal{D})$ does not make explicit how the frequency of a pattern is computed from the dataset. Other approaches find the k patterns maximizing a measure m in the dataset $\mathcal{D}$ [12, 15]. 2) *Pattern set reducing operators* compress a collection of patterns. For instance, the minimal and maximal operator denoted by $\mathcal{M}in(\mathcal{S})$ and $\mathcal{M}ax(\mathcal{S})$, return respectively the most general and specific patterns of $\mathcal{S}$ w.r.t. a specialization relation $\preceq$ [25]. The notion of negative and positive borders [25] is very similar. 3) *Pattern applying operators* cross patterns and data. For instance, the data covering operator $\theta_d(P, \mathcal{D}) = \{d \in \mathcal{D} | \exists p \in P : p \lhd d\}$ returns the data of $\mathcal{D}$ covered by at least one pattern of P [32]. Dually, the pattern covering operator $\theta_p(P, \mathcal{D})$ returns the patterns of P covering at least one element of $\mathcal{D}$ [32].

The next sections aim at stating an algebra based on the relational model to simultaneously and homogeneously handle data and patterns. In particular, all the manipulations of patterns described here will be expressed in our algebra.

3   Pattern-Oriented Relational Algebra

3.1   Pattern-Oriented Attributes

The pattern-oriented relational algebra pays attention to the attributes describing patterns, named *pattern-oriented attributes*. Indeed, several operations are specifically designed to handle such attributes which the domain corresponds to a pattern language together with a specialization relation.

Definition 1 (Pattern-oriented attributes). *The pattern-oriented attributes* **patt** *is a subset of the attributes:* **patt** $\subseteq$ **att** *such that for every* $A \in$ **patt**, **dom**(A) *is a poset. Let* $U \sqsubseteq$ **att** *be a set of attributes, the pattern-oriented attributes of* U *is denoted by* $\tilde{U}$.

For example, Table 1 provides instances of relations D, L and P containing pattern-oriented attributes. The relations $D[trans]$ and $L[patt]$ respectively describe a transactional dataset and the corresponding language in the context of (a) itemsets and (b) sequences. The relation $P[item, type, price]$ gives the item identifier, the type and the price of products. We consider that *trans*, *patt* and *item* are pattern-oriented attributes where **dom**$(item) = \mathcal{I}$ and **dom**$(trans) =$ **dom**$(patt) = \mathcal{L}_\mathcal{I}$ for itemsets (or $= \mathcal{L}_\mathcal{S}$ for sequences). Thereafter, the proposed queries can address instances where the domain of *patt* differs from that of *trans*.

Of course, the relations can be handled with relational operators. For instance, the query $\sigma_{patt \preceq \varphi}(L)$ returns all the patterns of L being more general than the pattern φ. The formula $patt \preceq \varphi$ is allowed because $\sigma_{patt \preceq \varphi}(L) \equiv \pi_{patt}(\sigma_{patt = left \wedge right = \varphi}(L \times C))$ where the relation $C[left, right]$ extensively enumerates in its instance the tuples (l, r) such that $l \preceq r$. On the contrary, the query $\sigma_{freq(patt, D) \geq f}(L)$ is not correct for computing the frequent patterns

Table 1. Instances for pattern discovery

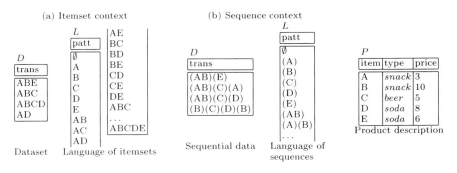

because the formula $freq(patt, D)$ requires a relation D and it is not allowed in a selection (see Section 2.1). Besides, we desire to make the computation of frequency explicit. The next section explains how to compute it with the relational algebra.

3.2   Cover, Semi-cover and Anti-cover Operators

We now indicate how to formulate the frequent pattern mining query (fpm query in brief) in the relational algebra which illustrates the need of the cover-like operators. Assume that $L[patt]$ and $D[trans]$ are two relations that respectively contain the language and the dataset as proposed in Table 1. The main challenge is to compute the frequency of each pattern of L. The Cartesian product of L by D gathers all the patterns of L with all the transactions of D. Of course, we only select the relevant tuples such that the pattern covers the transaction: $\sigma_{patt \lhd trans}(L \times D)$. Finally, we count for each pattern how many transactions it covers and we select the frequent ones: $\sigma_{freq \geq s}(\gamma_{patt, \texttt{COUNT}(trans) \rightarrow freq}(\sigma_{patt \lhd trans}(L \times D)))$. As the notion of cover relation plays a central role to relate pattern-oriented attributes, we introduce three operators based on this notion. The cover operator for the pattern discovery is as important as the join operator for classical data manipulations.

Cover operator. The result of a cover operation gathers all the combinations of tuples in R and S that have comparable pattern-oriented attributes.

Definition 2 (Cover operation). *The cover of a relation $R[U]$ for a relation $S[V]$ w.r.t. a cover relation[1] $\lhd \subseteq \mathbf{dom}(\widetilde{U}) \times \mathbf{dom}(\widetilde{V})$ is $R \lhd S = \sigma_{\widetilde{U} \lhd \widetilde{V}}(R \times S)$, i.e. for any instances I of R and J of S, $I \lhd J = \{(t, u) | t \in I \wedge u \in J \wedge t[\widetilde{U}] \lhd u[\widetilde{V}]\}$.*

As θ-join is a shortcut of $\sigma_f(R \times S)$, the cover operator is derived from primitive operations defined in Section 2.1. In fact, $R \lhd S$ is equivalentl to $\sigma_{\widetilde{U} \lhd \widetilde{V}}(R \times S)$ where the formula $\widetilde{U} \lhd \widetilde{V}$ can be expressed with usual relational operators as done above with $patt \preceq \varphi$. Then, as semi-cover and anti-cover defined below, the cover operator does not increase the expressive power of the relational algebra. However, such operators bring two main advantages. First, algebraic properties of cover-like operators can be formulated, in order to be used by a query optimizer (see Section 5). Second, specialized and efficient query evaluation methods for these operators could be developed.

Let us illustrate the cover operation on several examples of pattern manipulations. Given a dataset $D[trans]$ and a language $L[patt]$, the frequent patterns (with their frequency) correspond to the following query:

$$F = \sigma_{freq \geq f}(\gamma_{patt, \texttt{COUNT}(trans) \rightarrow freq}(L \lhd D))$$

This fpm query fulfills our modeling objective by explicitly and declaratively describing how the frequency is computed. Given the instances of L and D

---

[1] Definitions 2 to 4 consider that the binary relation $\lhd$ is a cover relation w.r.t. the specialization relations $\preceq_{\widetilde{U}}$ and $\preceq_{\widetilde{V}}$ respectively defined on $\mathbf{dom}(\widetilde{U})$ and $\mathbf{dom}(\widetilde{V})$.

Table 2. Instances containing mined patterns of instance D

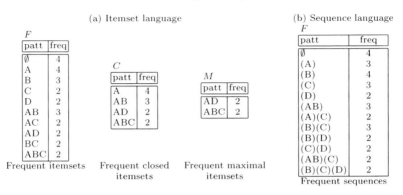

(a) Itemset language

F

patt	freq
∅	4
A	4
B	3
C	2
D	2
AB	3
AC	2
AD	2
BC	2
ABC	2

Frequent itemsets

C

patt	freq
A	4
AB	3
AD	2
ABC	2

Frequent closed itemsets

M

patt	freq
AD	2
ABC	2

Frequent maximal itemsets

(b) Sequence language

F

patt	freq
∅	4
(A)	3
(B)	4
(C)	3
(D)	2
(AB)	3
(A)(C)	2
(B)(C)	3
(B)(D)	2
(C)(D)	2
(AB)(C)	2
(B)(C)(D)	2

Frequent sequences

provided by Table 1 and $f = 2$, it exactly returns the instance of F (see Table 2). In the fpm query, the relation $\lhd \subseteq \mathbf{dom}(patt) \times \mathbf{dom}(trans)$ is a cover relation w.r.t. $\preceq_{patt}$ and $\preceq_{trans}$ (e.g., the inclusion for itemsets [3] or sequences [4]).

As mentioned earlier, a specialization relation is a particular kind of cover relation. Thereby, it can be exactly used as a cover operator. For instance, starting from the frequent patterns F, the frequent closed patterns of D [5] are computed as follows: $C = \pi_{patt,freq}(\sigma_{freq>max}(\gamma_{patt,freq,\texttt{MAX}(freq')\to max}(F \prec F')))$ (we recall that F' renames the attributes $patt$ and $freq$ into $patt'$ and $freq'$). Table 2 illustrates this query applied to a particular instance of F in the case of itemsets. Furthermore, the query $\gamma_{patt,\texttt{MAX}(freq')\to freq}(L \preceq C')$ regenerates the instance F.

Semi-cover operator. The semi-cover operator returns all the tuples of a relation covering at least one tuple of the other relation:

Definition 3 (Semi-cover operation). *The semi-cover of a relation $R[U]$ for a relation $S[V]$ w.r.t. a cover relation $\lhd \subseteq \mathbf{dom}(\widetilde{U}) \times \mathbf{dom}(\widetilde{V})$ is $R \lhd_{\ltimes} S = \pi_U(R \lhd S)$.*

Definition 3 implicitly means that $R \rhd_{\ltimes} S$ returns all the tuples of R covered by at least one tuple of S. Indeed, $R \rhd_{\ltimes} S$ has a sense because if the binary relation $\lhd$ is a cover relation on $\mathbf{dom}(\widetilde{U}) \times \mathbf{dom}(\widetilde{V})$ w.r.t. $\preceq_{\widetilde{U}}$ and $\preceq_{\widetilde{V}}$, then $\rhd$ is also a cover relation on $\mathbf{dom}(\widetilde{U}) \times \mathbf{dom}(\widetilde{V})$ w.r.t. $\succeq_{\widetilde{U}}$ and $\succeq_{\widetilde{V}}$. Table 3 illustrates Definition 3 by showing semi-cover operation of L for D which is the whole set of patterns occurring at least once in D: $L \lhd_{\ltimes} D$. Then, $\sigma_{patt \preceq \varphi}(L \lhd_{\ltimes} D)$ returns the patterns being more general than φ and present in D.

Let us come back to the data and pattern covering operators [32] presented in Section 2.2. The operation $\theta_p(P, D)$ which gives the tuples of P covering at least one tuple of D, is equivalent to $P \lhd_{\ltimes} D$. Dually, $\theta_d(P, D) = D \rhd_{\ltimes} P$ returns the tuples of D covered by at least one tuple of P.

Anti-cover operator. The anti-cover operator returns all the tuples of a relation not covering any tuple of the other relation:

Table 3. The semi-cover and anti-cover of L for D

$L \lhd_{\ltimes} D$	AE		$L \lhd_{\neg} D$	ACDE
patt	BC		patt	BCDE
$\emptyset$	BD		CE	ABCDE
A	BE		DE	
B	CD		ACE	
C	ABC		ADE	
D	ABD		BCE	
E	ABE		BDE	
AB	ACD		CDE	
AC	BCD		ABCE	
AD	ABCD		ABDE	

Definition 4 (Anti-cover operation). *The anti-cover of a relation $R[U]$ for a relation $S[V]$ w.r.t. a cover relation $\lhd \subseteq \mathbf{dom}(\tilde{U}) \times \mathbf{dom}(\tilde{V})$ is $R \lhd_{\neg} S = R - R \lhd_{\ltimes} S$.*

As for the semi-cover relation, $R \rhd_{\neg} S$ has a sense and returns all the tuples of R not covered by any tuple of S. Table 3 gives the patterns of L that do not occur in D by means of the anti-cover of L for D: $L \lhd_{\neg} D$. The anti-cover operator enables us to easily express the minimal and maximal pattern operators [25] (see Section 2.2): $\mathcal{M}in(R) = R \succ_{\neg} R$ and $\mathcal{M}ax(R) = R \prec_{\neg} R$. For instance, the frequent maximal itemsets are the frequent itemsets having no more specific frequent itemset: $M = F \prec_{\neg} F$ (see Table 2). A pattern of L is either present in D (i.e., in $L \lhd_{\ltimes} D$) or absent from D (i.e., in $L \lhd_{\neg} D$). Then, we obtain that $L = L \lhd_{\ltimes} D \cup L \lhd_{\neg} D$ (see Table 3). More generally, the semi-cover and anti-cover operator are complementary by definition (see Definitions 3 and 4): $R = R \lhd_{\ltimes} S \cup R \lhd_{\neg} S$ for any relations R and S.

3.3   Domain Operator

Let us come back to the query $\sigma_{freq \geq f}(\gamma_{patt, \texttt{COUNT}(trans) \to freq}(L \lhd D))$ that can be applied to any instance of relation L. However, in a practical pattern discovery task, the instance of L has to gather all the existing patterns of $\mathbf{dom}(patt)$ (as given by Table 1). To cope with this problem, we introduce a new operator that outputs the domain of the schema for a given relation.

Definition 5 (Domain operation). *The domain of a relation $R[U]$ is $\Delta(R)$ where for any instance I of R, $\Delta(I) = \mathbf{dom}(U)$.*

As the domain of each attribute is finite, the instance $\Delta(I)$ is finite. Assume that $I = \emptyset$ is an instance of $L[patt]$, $\Delta(I)$ returns the instance depicted by Table 1. The domain operator enables us to complete the frequent pattern mining query: $\sigma_{freq \geq f}(\gamma_{patt, \texttt{COUNT}(trans) \to freq}(\Delta(L) \lhd D))$. Other practical queries require the use of a language of patterns. For instance, negative border of R [25] can now be formulated: $\mathcal{B}d^-(R) = (\Delta(R) - R) \succ_{\neg} (\Delta(R) - R)$. Similarly, the downward and upward closure operators of R are respectively expressed by $\Delta(R) \preceq_{\ltimes} R$ and $\Delta(R) \succeq_{\ltimes} R$.

3.4 Scope of the Pattern-Oriented Relational Algebra

The *pattern-oriented relational algebra* which refers to the relational algebra plus the cover-like operators plus the domain operator, is strictly more expressive than the relational algebra. As aforementioned, the cover-like operators do not increase the expressive power of the relational algebra. In contrast, the domain operator cannot be expressed with relational operators because it induces domain dependent queries [1]. Let us note that [10] has already demonstrated that the frequent pattern mining query cannot be formulated in terms of the relational algebra.

From a practical point of view, the large number of query examples illustrating the previous sections (partially reported in Table 4 with q_1-q_5) highlights the generality of the pattern-oriented relational algebra. The other queries of Table 4 complete this overview by giving examples about the top-k frequent pattern mining with q_6 [15], the syntactic pattern mining q_7 [9], the utility-based pattern mining q_8 or the association rule mining q_9 [3]. Note that $\in$ is a cover relation on $\mathbf{dom}(item) \times \mathbf{dom}(patt)$ that relates one item with an itemset or a sequence. The query q_7 returns the patterns of L occurring in D and not containing a product of type 'snack'. q_8 returns the patterns of L occurring in D such that the sum of product prices is less than a threshold t.

Table 4. Examples of pattern-oriented queries and their properties

Pattern-oriented query		DC	Local	Global
q_1	$\sigma_{freq \geq f}(\gamma_{patt, \text{COUNT}(trans) \to freq}(L \lhd D))$	L	L	D
q_2	$\pi_{patt, freq}(\sigma_{freq > max}(\gamma_{patt, freq, \text{MAX}(freq') \to max}(F \prec F')))$			F
q_3	$\sigma_{patt \prec \varphi}(L)$	L	L	
q_4	$\sigma_{patt \prec \varphi}(L \lhd_\bowtie D)$	L	L, D	
q_5	$F \prec_\neg F$			F
q_6	$\sigma_{rank \leq k}(\gamma_{patt, freq, \text{COUNT}(patt') \to rank}(\sigma_{supp < supp'}(F \times F')))$	F		F
q_7	$(L \lhd_\bowtie D) \ni_\neg \sigma_{type=snack}(P)$	L	L, D, P	
q_8	$\sigma_{total < t}(\gamma_{patt, \text{SUM}(price) \to total}(P \in (L \lhd_\bowtie D)))$	L	L, D, P	
q_9	$\pi_{patt' \to head, patt \backslash patt' \to body, freq, freq/freq' \to conf}(F' \prec F)$			F

Most of these typical queries are difficult to evaluate because the handled instances may be very large especially when the domain operator is used for generating the language. The following sections explain how to rewrite queries for optimization purpose.

4   Characterizing Pattern-Oriented Queries

In the field of pattern mining, it is well known that some properties are useful to reduce the computation time (e.g., anti-monotone constraint or pre/post-processing ability). This section aims at characterizing such properties in the pattern-oriented relational algebra. More precisely, we first study the structuration of the instance resulting from a query w.r.t. the initial instance. Then, we analyze three levels of dependency between a query and a relation.

Thereafter we assume that q is a query formulated with the pattern-oriented relational algebra and the database schema $\{R_1[U_1], \ldots, R_{n-1}[U_{n-1}], R[U]\}$. Then, this query q is often applied to the database instance $\mathbf{I} = \{I_1, \ldots, I_{n-1}, I\}$.

4.1 Downward Closed Query

Intuitively, the notion of downward closed query expresses that of anti-monotone constraints [25] in the pattern-oriented relational algebra. A query q is downward closed in R if for any instance I of $R[U]$, any tuple of I more general than at least one tuple of $\pi_U(q(\mathbf{I}))$ also belongs to $\pi_U(q(\mathbf{I}))$.

Definition 6 (Downward closed queries). *A query q is downward closed in $R[U]$ w.r.t. $\preceq$ iff $U \subseteq \mathbf{sch}(q)$ and $(R \preceq_\ltimes q) \equiv \pi_U(q)$.*

Definition 6 means that if a tuple t of R is more general than at least one tuple of the answer of q, t is also present in this answer. The downward closed property is very interesting for pruning an instance (more details are given in Section 5.2). The query $\sigma_{freq \geq f}(\gamma_{patt, \texttt{COUNT}(trans) \rightarrow freq}(L \triangleleft D))$ is downward closed in L w.r.t. $\preceq$. Indeed, all the generalizations of a frequent pattern are frequent (e.g., ABC is frequent and then, A, B, C, AB and so on are also frequent, see Table 1). Similarly, the top-k frequent pattern query q_6 is also downward closed in F w.r.t. $\preceq$. The column 'DC' of Table 4 indicates the relations in which the query is downward closed w.r.t. $\preceq$.

4.2 Local and Global Dependent Queries

A query is dependent on the relation R whenever its result varies with the instance of R. Whereas the query $\sigma_{patt \preceq \varphi}(L)$ is independent of D, $\sigma_{patt \preceq \varphi}(L \triangleleft_\ltimes D)$ depends on D because it only returns the tuples of $\sigma_{patt \preceq \varphi}(L)$ that cover at least one tuple of the instance of D. Definition 7 formalizes the notion of total independence (or independence in brief):

Definition 7 (Total independence). *A query q is totally independent of R iff for any instances I, J of R, one has $q(\{I_1, \ldots, I_{n-1}, I\}) = q(\{I_1, \ldots, I_{n-1}, J\})$.*

In other words, a query which is independent of R is equivalent to another query not involving R. Note that the queries which are totally independent of D correspond to syntactical constraints [9].

We now refine this notion of dependence by introducing the *global independence*. Both queries $\sigma_{patt \preceq \varphi}(L \triangleleft_\ltimes D)$ and $\sigma_{freq \geq f}(\gamma_{patt, \texttt{COUNT}(trans) \rightarrow freq}(L \triangleleft D))$ are dependent on D. But, the dependency of the second query on D is stronger than that of the first query. Indeed, the computation of the frequency for a tuple of L requires to simultaneously take into account several tuples of D.

Definition 8 (Local/global dependence). *A query q is globally independent of R iff for any instances I, J of R, one has $q(\{I_1, \ldots, I_{n-1}, I \cup J\}) = q(\{I_1, \ldots, I_{n-1}, I\}) \cup q(\{I_1, \ldots, I_{n-1}, J\})$. A query being globally independent of R but dependent on R is said to be locally dependent on R.*

Definition 8 formalizes the notion of global constraints [12] which compare several patterns together to check whether the constraint is satisfied or not. The queries (like q_2, q_5, q_6 or q_9) which are globally dependent on L or F correspond to such global constraints. Besides, the query q_1 globally depends on D and locally depends on L. It means that q_1 can be evaluated by considering separately each tuple of the instance of L. Conversely, it is impossible to consider individually each tuple of the instance of D. Thus, the higher the overall number of global dependencies, the harder the evaluation of the query. The columns 'Local' and 'Global' of Table 4 indicates the local/global dependent relations for each query. As expected, the queries q_1, q_4, q_7 and q_8 depend on D because they benefit from the dataset to select the right patterns. We also observe that the queries q_2, q_5, q_6 and q_9 globally depend on F as they postprocess the frequent patterns by comparing them.

5   Rewriting Pattern-Oriented Queries

This section examines algebraic equivalences to rewrite queries into forms that may be implemented more efficiently.

5.1   Algebraic Laws Involving Cover-Like Operators

Let us consider the query q_4: $\sigma_{patt \preceq \varphi}(L \triangleleft_\ltimes D)$. As the predicate $patt \preceq \varphi$ is highly selective, it is preferable to first apply it for reducing the language. Thereby, the equivalent query $\sigma_{patt \preceq \varphi}(L) \triangleleft_\ltimes D$ may be more efficient than $\sigma_{patt \preceq \varphi}(L \triangleleft_\ltimes D)$. The property below enumerates equivalences:

Property 1 (Laws involving cover-like operators). *Let $R[U]$ and $S[V]$ be two relation schemas. Let f and g be two predicates respectively on R and S. Let A and B be two sets of attributes such that $\tilde{U} \subseteq A \subseteq U$ and $\tilde{V} \subseteq B \subseteq V$. One has the following equivalences:*

1. $\sigma_{f \wedge g}(R \triangleleft S) \equiv \sigma_f(R) \triangleleft \sigma_g(S)$ $\pi_{A \cup B}(R \triangleleft S) \equiv \pi_A(R) \triangleleft \pi_B(S)$
2. $\sigma_f(R \triangleleft_\ltimes S) \equiv \sigma_f(R) \triangleleft_\ltimes S$ $\pi_A(R \triangleleft_\ltimes S) \equiv \pi_A(R) \triangleleft_\ltimes S$
3. $\sigma_f(R \triangleleft_\neg S) \equiv \sigma_f(R) \triangleleft_\neg S$ $\pi_A(R \triangleleft_\neg S) \equiv \pi_A(R) \triangleleft_\neg S$
4. $R \triangleleft_\ltimes S \equiv R \triangleleft_\ltimes (S \prec_\neg S)$ $R \triangleleft_\neg S \equiv R \triangleleft_\neg (S \prec_\neg S)$

Intuitively, the right hand side of each equivalence listed in Property 1 (proofs are omitted due to lack of space) may lead to optimize the query. Indeed, Lines 1 to 3 "pushes down" the selection and projection operators to reduce the size of the operands before applying a cover-like operator. This technique is successfully exploited in database with Cartesian product or join operator [1]. Besides, Line 4 benefits from the maximal tuples of S (i.e., $S \prec_\neg S$) as done in pattern mining [25]. If a tuple t of the instance of R covers a tuple of the instance J of

S, then t also covers a tuple of $J \prec_\neg J$. As $|J \prec_\neg J| \leq |J|$, the rewritten query $R \bowtie_\ltimes (S \prec_\neg S)$ may be less costly than $R \bowtie_\ltimes S$ provided $J \prec_\neg J$ is not too costly.

5.2 Algebraic Reformulation of the Levelwise Algorithm

We now take into account the downward closed and the global independence properties for reformulating queries. For instance, assume that the instance of L is now equal to $\pi_{patt}(F)$. A new computation of q_1 again returns F: $F = \sigma_{freq \geq 2}(\gamma_{patt, \text{COUNT}(trans) \to freq}(\pi_{patt}(F) \triangleleft D))$. Of course, this query is faster to compute than the original fpm query because the instance of F is very small compared to $\Delta(L)$. We generalize this observation:

Property 2. *Let q be a downward closed query in $R[U]$ w.r.t. $\preceq$ and globally independent of R such that $U \subseteq \mathbf{sch}(q)$, one has $q(\mathbf{I}) = q(\mathbf{J})$ for any instances $\mathbf{I} = \{I_1, \ldots, I_{n-1}, I\}$ and $\mathbf{J} = \{I_1, \ldots, I_{n-1}, J\}$ such that $\pi_U(q(\mathbf{J})) \subseteq I \subseteq J$.*

Given a downward closed and independent query q, Property 2 demonstrates that $q(\mathbf{I}) = q(\mathbf{J})$ when I is an instance of R such that $\pi_U(q(\mathbf{J})) \subseteq I \subseteq J$. As $I \subseteq J$ and then $|I| \leq |J|$, we suppose that evaluating $q(\mathbf{I})$ is less costly than evaluating $q(\mathbf{J})$ because the cost generally decreases with the cardinality of the instance. Thus, in order to reduce the cost of the evaluation of $q(\mathbf{I})$, we aim at turning I into the smallest instance of R including $q(\mathbf{J})$. Such an approach can be seen as a *pruning* of the instance of R.

Table 5. Levelwise computation of the fpm query (level 2)

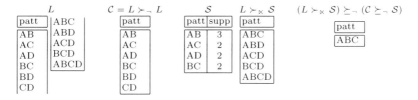

Table 5 illustrates how to prune the instance L for evaluating the fpm query q_1. As q_1 is globally independent of L, we first divide L into two parts: the most general tuples of L denoted by $\mathcal{C} = L \succ_\neg L$ (i.e., the candidates of the level 2 of APRIORI [3]) and others, i.e. $L \succ_\ltimes L$. We then apply the fpm query to $\mathcal{C}$ for computing $\mathcal{S}$: the frequent patterns of $\mathcal{C}$ and their frequency. Finally, we benefit from $\mathcal{S}$ for pruning $L \succ_\ltimes L$ using the downward closed property of q_1 in L w.r.t. $\preceq$ (see Definition 6). We only preserve the tuples which are more specific than at least one frequent tuple of $\mathcal{S}$: $L \succ_\ltimes \mathcal{S}$. Finally, we filter out the tuples having a non-frequent generalization: $(L \succ_\ltimes \mathcal{S}) \succeq_\neg (\mathcal{C} \succeq_\neg \mathcal{S})$. As the cardinality of this instance is smaller than $|L \succ_\ltimes L|$, we have achieved our goal.

This principle is generalized with this theorem:

Theorem 1 (Levelwise equivalence). *Let q be a downward closed query w.r.t. $\preceq$ and globally independent of R, one has the below equality for any database instance $\mathbf{I} = \{I_1, \ldots, I_{n-1}, I\}$:*

$$q(\mathbf{I}) = q(\{I_1, \ldots, I_{n-1}, \underbrace{I \succ_\neg I}_{\mathcal{C}=}\}) \cup q(\{I_1, \ldots, I_{n-1}, (I \succ_\ltimes \mathcal{S}) \succeq_\neg (\mathcal{C} \succeq_\neg \mathcal{S})\})$$

$$\underbrace{\phantom{q(\{I_1, \ldots, I_{n-1}, I \succ_\neg I\})}}_{\mathcal{S}=}$$

Proof. Let q be a downward closed query w.r.t. $\preceq$ and globally independent of R. To alleviate the notations, $q(I)$ refers to $q(\{I_1, \ldots, I_{n-1}, I\})$ where I is any instance of R. Besides, we fix that $\mathcal{C} = I \succ_\neg I$ and $\mathcal{S} = q(I \succ_\neg I) = q(\mathcal{C})$:

$$q(I) = q(I \succ_\neg I \cup I \succ_\ltimes I) = q(\mathcal{C} \cup I \succ_\ltimes I) \tag{1}$$
$$= q(\mathcal{C}) \cup q(I \succ_\ltimes I) \tag{2}$$
$$= q(\mathcal{C}) \cup q(I \succ_\ltimes q(\mathcal{C})) = q(\mathcal{C}) \cup q(I \succ_\ltimes \mathcal{S}) \tag{3}$$
$$= q(\mathcal{C}) \cup q((I \succ_\ltimes \mathcal{S}) \succeq_\neg (\mathcal{C} \succeq_\neg \mathcal{S})) \tag{4}$$

Line 1 stems from the complementary property: $R = R \triangleleft_\ltimes S \cup R \triangleleft_\neg S$. Line 2 is allowed because q is globally independent of R. Line 3-4 are due to the downward closed property in R (see Definition 6). □

Theorem 1 can be used for rewriting queries by considering two important points. Firstly, the redundant subqueries as candidate tuples $\mathcal{C} = I \succ_\neg I$ and satisfied tuples $\mathcal{S} = q(\{I_1, \ldots, I_{n-1}, I \succ_\neg I\})$ have to be evaluated only once. Secondly, the practical evaluation of q requires to recursively apply the equality proposed in Theorem 1. Indeed, the subquery $q(\{I_1, \ldots, I_{n-1}, (I \succ_\ltimes \mathcal{S}) \succeq_\neg (\mathcal{C} \succeq_\neg \mathcal{S})\})$ can also be rewritten by a query plan optimizer using the same identity. Therefore, Theorem 1 leads to algebraically reformulate the levelwise algorithm [3, 4, 25]. This algorithm repeats this equality for computing which candidate patterns satisfy the predicate and then, generating those of the next level. Other efficient pruning strategies like depth-first search techniques [5] could also be expressed in pattern-oriented relational algebra. Finally, as observed in [12, 15], we cannot apply Theorem 1 to q_6 because it globally depends on F.

6   Related Work

Inductive databases [18, 24] aims at tightly integrating databases with data mining. Our approach is less ambitious because it is "only" restricted to the pattern mining. Obviously, many proposals provide an environment merging a RDBMS with pattern mining tools: Quest [2], ConQueSt [7], DBminer [16], Sindbad [34] and many other prototypes [6]. In such a context, there are many extensions of the SQL language [31] like DMX or MINERULE [26]. There are also extended relational model [13] like 3W model [20]. However, such methods don't fuse the theoretical concepts stemming from both the relational model and the pattern discovery. For instance, the query optimizer of DBMS is isolated from pattern mining algorithms. Indeed, most of the approaches consider a pattern mining query as the result of a "black box". Only few works [10, 23, 33] express pattern mining operators by benefiting from the relational algebra. Such approaches add a loop statement for implementing the levelwise algorithm. On the contrary, our proposal extends the relational algebra by still using a declarative approach.

Many frameworks inspired from relational and logical databases, but created from scratch, are proposed during the last decade: constraint-based pattern mining [9, 25], distance-based framework [14], rule-base [19], tuple relational calculus [28], logical database [29], pattern-base [32] and so on. Other directions are suggested in [24] like probabilistic approach or data compression. Besides, constraint programming is another promising way for expressing and mining patterns [21, 30]. Such frameworks are less convenient for handling data (which are often initially stored in relational databases). Besides, they suffer from a lack of simple and powerful languages like the relational algebra (in particular, the manipulation of patterns is frequently separated from that of data).

From a more general point of view, many works add new operators to the relational algebra in order to express more sophisticated queries. Even if such new operators don't necessary increase the expressive power of the relational algebra, most of the time they facilitate the formulation of user queries and provide specific optimizations. Typically, several operators are introduced for comparing tuples with each other, as does a specialization relation with patterns. For instance, the winnow operator is specifically dedicated to handle preferences [11]. Several operators are dedicated for selecting the best tuples by means of relational dominant queries [8] or relational top-k queries [22]. The cover-like operators are very closed to such operators. But, they enable to compare tuples based on different languages, as does a cover relation with patterns. Finally, the domain operator enables us to manipulate values not initially present in the relations. The same concept is used in [13] for generating tables containing patterns.

7   Conclusion

In this paper, we have proposed a new and general framework for pattern discovery by only adding cover-like and domain operators to the relational algebra. The pattern-oriented relational algebra interestingly inherits good properties from the relational algebra as closure or declarativity. This framework deals with any language of patterns for expressing a wide spectrum of queries including constraint-based pattern mining, condensed representations and so on. We identify crucial aspects of queries as the downward closed and independence properties. We then benefit from such properties to algebraically reformulate the levelwise algorithm. We think that our algebraisation is an important step towards the elegant integration of pattern discovery in database systems.

Further work addresses the implementation of a complete system based on the pattern-oriented relational algebra. As done in the database field, we project to implement the physical cover operators and to design a query plan optimizer taking advantage of our proposed algebraic laws. We also study the test of local and global dependence between a query and a relation.

References

1. Abiteboul, S., Hull, R., Vianu, V.: Foundations of Databases. Addison-Wesley, Reading (1995)
2. Agrawal, R., Mehta, M., Shafer, J.C., Srikant, R., Arning, A., Bollinger, T.: The quest data mining system. In: KDD, pp. 244–249 (1996)
3. Agrawal, R., Srikant, R.: Fast algorithms for mining association rules in large databases. In: Bocca, J.B., Jarke, M., Zaniolo, C. (eds.) VLDB, pp. 487–499. Morgan Kaufmann, San Francisco (1994)
4. Agrawal, R., Srikant, R.: Mining sequential patterns. In: Yu, P.S., Chen, A.L.P. (eds.) ICDE, pp. 3–14. IEEE Computer Society, Los Alamitos (1995)
5. Arimura, H., Uno, T.: Polynomial-delay and polynomial-space algorithms for mining closed sequences, graphs, and pictures in accessible set systems. In: SDM, pp. 1087–1098. SIAM, Philadelphia (2009)
6. Blockeel, H., Calders, T., Fromont, É., Goethals, B., Prado, A., Robardet, C.: An inductive database prototype based on virtual mining views. In: KDD, pp. 1061–1064. ACM, New York (2008)
7. Bonchi, F., Giannotti, F., Lucchese, C., Orlando, S., Perego, R., Trasarti, R.: ConQueSt: a constraint-based querying system for exploratory pattern discovery. In: ICDE, p. 159. IEEE Computer Society, Los Alamitos (2006)
8. Börzsönyi, S., Kossmann, D., Stocker, K.: The skyline operator. In: ICDE, pp. 421–430. IEEE Computer Society, Los Alamitos (2001)
9. Boulicaut, J.F., Jeudy, B.: Constraint-based data mining. In: Maimon, O., Rokach, L. (eds.) The Data Mining and Knowledge Discovery Handbook, pp. 399–416. Springer, Heidelberg (2005)
10. Calders, T., Lakshmanan, L.V.S., Ng, R.T., Paredaens, J.: Expressive power of an algebra for data mining. ACM Trans. Database Syst. 31(4), 1169–1214 (2006)
11. Chomicki, J.: Querying with intrinsic preferences. In: Jensen, C.S., Jeffery, K., Pokorný, J., Šaltenis, S., Hwang, J., Böhm, K., Jarke, M. (eds.) EDBT 2002. LNCS, vol. 2287, pp. 34–51. Springer, Heidelberg (2002)
12. Crémilleux, B., Soulet, A.: Discovering knowledge from local patterns with global constraints. In: Gervasi, O., Murgante, B., Laganà, A., Taniar, D., Mun, Y., Gavrilova, M.L. (eds.) ICCSA 2008, Part II. LNCS, vol. 5073, pp. 1242–1257. Springer, Heidelberg (2008)
13. Diop, C.T., Giacometti, A., Laurent, D., Spyratos, N.: Composition of mining contexts for efficient extraction of association rules. In: Jensen, C.S., Jeffery, K., Pokorný, J., Šaltenis, S., Hwang, J., Böhm, K., Jarke, M. (eds.) EDBT 2002. LNCS, vol. 2287, pp. 106–123. Springer, Heidelberg (2002)
14. Dzeroski, S.: Towards a general framework for data mining. In: Džeroski, S., Struyf, J. (eds.) KDID 2006. LNCS, vol. 4747, pp. 259–300. Springer, Heidelberg (2007)
15. Fu, A.W.C., Kwong, R.W., Tang, J.: Mining n-most interesting itemsets. In: Ohsuga, S., Raś, Z.W. (eds.) ISMIS 2000. LNCS (LNAI), vol. 1932, pp. 59–67. Springer, Heidelberg (2000)
16. Han, J., Fu, Y., Wang, W., Chiang, J., Gong, W., Koperski, K., Li, D., Lu, Y., Rajan, A., Stefanovic, N., Xia, B., Zaïane, O.R.: DBMiner: a system for mining knowledge in large relational databases. In: KDD, pp. 250–255 (1996)

17. Hand, D.J.: Pattern detection and discovery. In: Hand, D.J., Adams, N.M., Bolton, R.J. (eds.) Pattern Detection and Discovery. LNCS (LNAI), vol. 2447, pp. 1–12. Springer, Heidelberg (2002)
18. Imielinski, T., Mannila, H.: A database perspective on knowledge discovery. Commun. ACM 39(11), 58–64 (1996)
19. Imielinski, T., Virmani, A.: MSQL: a query language for database mining. Data Min. Knowl. Discov. 3(4), 373–408 (1999)
20. Johnson, T., Lakshmanan, L.V.S., Ng, R.T.: The 3W model and algebra for unified data mining. In: Abbadi, A.E., Brodie, M.L., Chakravarthy, S., Dayal, U., Kamel, N., Schlageter, G., Whang, K.Y. (eds.) VLDB, pp. 21–32. Morgan Kaufmann, San Francisco (2000)
21. Khiari, M., Boizumault, P., Crémilleux, B.: Combining CSP and constraint-based mining for pattern discovery. In: Taniar, D., Gervasi, O., Murgante, B., Pardede, E., Apduhan, B.O. (eds.) ICCSA 2010. LNCS, vol. 6017, pp. 432–447. Springer, Heidelberg (2010)
22. Li, C., Chang, K.C.C., Ilyas, I.F., Song, S.: RankSQL: query algebra and optimization for relational top-k queries. In: Özcan, F. (ed.) SIGMOD Conference, pp. 131–142. ACM Press, New York (2005)
23. Liu, H.C., Ghose, A., Zeleznikow, J.: Towards an algebraic framework for querying inductive databases. In: Kitagawa, H., Ishikawa, Y., Li, Q., Watanabe, C. (eds.) DASFAA 2010. LNCS, vol. 5982, pp. 306–312. Springer, Heidelberg (2010)
24. Mannila, H.: Theoretical frameworks for data mining. SIGKDD Explorations 1(2), 30–32 (2000)
25. Mannila, H., Toivonen, H.: Levelwise search and borders of theories in knowledge discovery. Data Min. Knowl. Discov. 1(3), 241–258 (1997)
26. Meo, R., Psaila, G., Ceri, S.: A new SQL-like operator for mining association rules. In: Vijayaraman, T.M., Buchmann, A.P., Mohan, C., Sarda, N.L. (eds.) VLDB, pp. 122–133. Morgan Kaufmann, San Francisco (1996)
27. Mitchell, T.M.: Generalization as search. Artif. Intell. 18(2), 203–226 (1982)
28. Nijssen, S., Raedt, L.D.: IQL: a proposal for an inductive query language. In: Dzeroski, S., Struyf, J. (eds.) KDID 2006. LNCS, vol. 4747, pp. 189–207. Springer, Heidelberg (2007)
29. Raedt, L.D.: A logical database mining query language. In: Cussens, J., Frisch, A.M. (eds.) ILP 2000. LNCS (LNAI), vol. 1866, pp. 78–92. Springer, Heidelberg (2000)
30. Raedt, L.D., Guns, T., Nijssen, S.: Constraint programming for itemset mining. In: KDD, pp. 204–212. ACM, New York (2008)
31. Romei, A., Turini, F.: Inductive database languages: requirements and examples. Knowledge and Information Systems 1–34 (2010), http://dx.doi.org/10.1007/s10115-009-0281-4
32. Terrovitis, M., Vassiliadis, P., Skiadopoulos, S., Bertino, E., Catania, B., Maddalena, A., Rizzi, S.: Modeling and language support for the management of pattern-bases. Data Knowl. Eng. 62(2), 368–397 (2007)
33. Wang, H., Zaniolo, C.: ATLaS: a native extension of SQL for data mining. In: Barbará, D., Kamath, C. (eds.) SDM. SIAM, Philadelphia (2003)
34. Wicker, J., Richter, L., Kessler, K., Kramer, S.: SINDBAD and SiQL: an inductive database and query language in the relational model. In: Daelemans, W., Goethals, B., Morik, K. (eds.) ECML PKDD 2008, Part II. LNCS (LNAI), vol. 5212, pp. 690–694. Springer, Heidelberg (2008)

Efficient Incremental Mining of Frequent Sequence Generators[*]

Yukai He[1], Jianyong Wang[2], and Lizhu Zhou[2]

[1] Tsinghua National Laboratory for Information Science and Technology
[2] Department of Computer Science and Technology, Tsinghua University,
Beijing 100084, China
heyk05@mails.tsinghua.edu.cn,
{jianyong,dcszlz}@tsinghua.edu.cn

Abstract. Recently, mining sequential patterns, especially closed sequential patterns and generator patterns, has attracted much attention from both academic and industrial communities. In recent years, incremental mining of all sequential patterns (all closed sequential patterns) has been widely studied. However, to our best knowledge, there has not been any study for incremental mining of sequence generators. In this paper, by carefully examining the existing expansion strategies for mining sequential databases, we design a GenTree structure to keep track of the relevant mining information, and propose an efficient algorithm, IncGen, for incremental generator mining. We have conducted thorough experiment evaluation and the experimental results show that the IncGen algorithm outperforms state-of-the-art generator-mining method FEAT significantly.

1 Introduction

Sequential pattern mining is an important task in the data mining community. The purpose is to mine all frequent subsequences in a sequence database. Since the first work in [1], sequence mining has attracted much attention in the data mining community. Typical sequential pattern mining algorithms include GSP [6], SPADE [7], PrefixSpan [8], and SPAM [9]. Various applications benefit from this problem, such as classifying sequential data [2], detecting erroneous sentences [3], identifying comparative sentences from Web forum posting and product reviews [4], and Web log data analysis [5].

Due to the famous downward closure property of frequent patterns, if the minimum support is low or the database is dense, the complete set of frequent subsequences will grow exponentially. Recently, people are more interested in mining compact form of sequential patterns: closed subsequences, maximal

---

[*] This work was supported in part by National Natural Science Foundation of China under grant No. 60833003 and 60873171, and the Program of State Education Ministry of China for New Century Excellent Talents in University under Grant No. NCET-07-0491.

J.X. Yu, M.H. Kim, and R. Unland (Eds.): DASFAA 2011, Part I, LNCS 6587, pp. 168–182, 2011.

subsequences, and sequence generators. CloSpan [10] and BIDE [11] are two well-known closed sequential pattern mining algorithms. FEAT [12] and Gen-Miner [13] were proposed to mine sequence generators. These algorithms make full use of the search space pruning techniques and greatly improve the mining efficiency, and they are more efficient than the sequential pattern mining algorithms.

However, as presented in [14], databases are updated incrementally in many domains, such as user behavior analysis, DNA sequence analysis, web click stream mining, etc. Among all the **incremental mining algorithms** on an updating database, IncSpan [14] and IncSpan+ [15] are designed for mining the complete set of frequent subsequences, and GSP+/MFS+ [16] are for maximal subsequences. In addition, IMCS [17] can be used to mine closed subsequences. But to the best of our knowledge, there is no study on incremental mining of sequence generators.

To address this problem, in this paper, we study how to incrementally mine sequence generators. When a database is updated, the generators may change to non-generators, and non-generators may become generators too. Meanwhile, newly appended items to existing sequences or newly inserted sequences may cause previous infrequent subsequences become frequent. In addition, the increased minimum support may make previous frequent subsequences become infrequent again.

To deal with the above difficulties and solve the incremental generator mining problem, we propose a $GenTree$ (Generating Tree) structure to store the useful information of the original database, and design an efficient algorithm $IncGen$ (Incremental Generator Miner) to mine the updated database and maintain the $GenTree$ structure (for further updating). We have implemented our methods and the experimental results show that the $IncGen$ algorithm outperforms the traditional generator mining algorithm $FEAT$, especially on non-sparse databases.

The remainder of the paper is organized as follows. Section 2 describes the problem formulation. Section 3 presents the $GenTree$ structure and its properties. The $IncGen$ algorithm is introduced in Section 4. Performance study is shown in Section 5. Section 6 introduces the related works. Finally, Section 7 concludes the paper.

2 Problem Formulation

2.1 Sequences and Generators

Let $I = \{i_1, i_2, \cdots, i_n\}$ be a set of distinct items. A **sequence** S is defined as a list of items in I, where each item can occur multiple times, denoted by $S = \{e_1, e_2, \cdots, e_m\}$, or $S = e_1 e_2 \cdots e_m$ for short. A sequence $S_a = a_1 a_2 \cdots a_n$ is **contain**ed by (or a **subsequence** of) another sequence $S_b = b_1 b_2 \cdots b_m$ iff there exists integers $1 \leq i_1 < i_2 < \cdots < i_n \leq m$ such as $a_1 = b_{i_1}, a_2 = b_{i_2}, \cdots, a_n = b_{i_n}$. We use $S_a \sqsubseteq S_b$ to denote that S_a is a **subsequence** of S_b

(S_b is a **supersequence** of S_a vice versa). $S_a \sqsubset S_b$ denotes that $S_a \sqsubseteq S_b$ and $S_a \neq S_b$. Besides, $X \diamond Y$ is used to indicate the concatenation of two subsequences or items X and Y.

A **sequence database** SDB is a set of tuples, where each tuple is in the form of $\langle ID, sequence \rangle$. See Table 1 for an example. The number of the tuples is called the **base size** of SDB, denoted by $|SDB|$. The **absolute support** of a subsequence S_a in a sequence database SDB is the number of sequences in SDB containing S_a, denoted by $sup^{SDB}(S_a)$, or $sup(S_a)$ if there is no ambiguous; the **relative support** is the percentage of sequences in SDB containing S_a, that is, $sup^{SDB}(S_a)/|SDB|$. In the rest of the paper, they are used exchangeably if it is clear in the context.

Table 1. An example sequence database SDB

ID	Sequence
1	C A A B C
2	A B C B
3	C A B
4	A B B

Definition 1. *(Generator). A subsequence S_a is a generator if and only if $\nexists S_b$ that $sup(S_a) = sup(S_b)$ and $S_b \sqsubset S_a$.*

Given a **minimum support** threshold min_sup, a subsequence S_a is **frequent** on SDB if $sup^{SDB}(S_a) \geq min\_sup$. If a generator is frequent, it is called a frequent generator. In this paper, we focus on incrementally mining frequent generators.

Example 1. Table 1 shows an example of a sequence database. The database has three different distinct items and four input sequences, that is, $|SDB| = 4$. Supposing $min\_sup = 2$, there are seven generators: A:4, AC:2, B:4, BB:2, BC:2, C:3, and CA:2. (The numbers after the colons are the supports of the subsequences.) The remaining four frequent subsequences are not generators, because for each of them, there is a proper subsequence with the same support. For instance, CB is not a generator because $C \sqsubset CB$ and $sup(C) = sup(CB)$.

2.2   Incremental Generator Mining

When a sequence database is updated, it can be updated in two main ways: *INSERT* and *APPEND*. *INSERT* means inserting new sequences into the database, and *APPEND* is appending new items to some of the existing sequences. Rarely we may encounter the third way of updating: *MIXTURE*, which means the combination between *INSERT* and *APPEND*. See Tables 2(a), 2(b) and 2(c) for an example.

Table 2. Three updating manners of the SDB in Table 1

(a) $INSERT$			(b) $APPEND$			(c) $MIXTURE$	
ID	**Sequence**		**ID**	**Sequence** (New)		**ID**	**Sequence** (New)
1	CAABC		1	CAABC		1	CAABC
2	ABCB		2	ABCB		2	ABCB
3	CAB		3	CAB	**DC**	3	CAB	**DC**
4	ABB		4	ABB	**CA**	4	ABB	**CA**
5 (New)	**BDCBA**					5 (New)	**BDCBA**

In the rest of the paper, we use the notation $OriDB$ for the original database, $AppDB$ for the appended items of the database, and $InsDB$ for the inserted sequences. In the meantime, we also use D for $OriDB$, Δ for the updating parts of $OriDB$, and D' for the updated database.

Now we formalize the problem of incremental mining of generators.

Incremental Sequence Generator Mining Problem: Given a sequence database D, an updating database Δ, and a minimum support min_sup, the **incremental sequence generator mining problem** is to mine all *frequent* sequence generators on the updated database D'.

3   GenTree: The Generating Tree

In this section, we introduce the concept of the generating tree ($GenTree$), then discuss the construction and the node type switching rules of $GenTree$.

3.1   The Concept of GenTree

The $GenTree$ structure is designed for storing useful information of D's mining results. Each node represents an item of a frequent subsequence of D. For the three types of expansion manner: $APPEND$, $INSERT$ or $MIXTURE$, we use the same $GenTree$.

In a $GenTree$, each node s_n has four fields:

1. **Item.** The *Item* field stores an item. The items on the path from the root to the node s_n represent a subsequence S_n.
2. **Support.** The support of the node. Only frequent nodes are kept.
3. **Type.** The node type. There are three different types of nodes, which are:
    - **Generator Node:** If a subsequence S_n is a generator, the corresponding node s_n is called a generator node.
    - **Stub Node:** If a subsequence S_n is unpromising to be extended for generators, its corresponding node is called a stub node.
    - **Branch Node:** Except for the two types above, the rest nodes are branch nodes.
4. **Children.** Pointers to children's nodes.

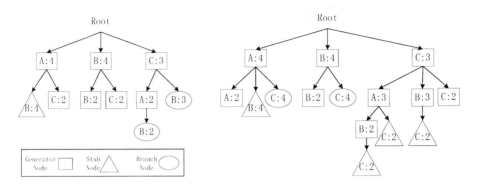

Fig. 1. *GenTree* for the database in Table 1

Fig. 2. Updated *GenTree* for database in Table 2(b)

Figure 1 is the corresponding *GenTree* of the database in Table 1. The generator nodes, stub nodes and branch nodes are shown in rectangles, triangles, and ellipses, respectively.

3.2   The Construction of GenTree

Before introducing the construction of the *GenTree*, we first give several definitions for ease of presentation.

Definition 2. *(Projected Sequence). Given an input sequence S, a prefix subsequence S_p, the projected sequence of S with respect to S_p is defined as the subsequence of S after the first appearance of S_p.*

Definition 3. *(Projected Database). Given an input sequence database SDB, a prefix sequence S_p, the projected database SDB_{S_p} of SDB with respect to S_p is defined as the complete set of projected sequences in SDB with respect to S_p.*

Definition 4. *(The i-th item missing subsequence) For sequence $S = e_1 e_2 \ldots e_n$, we define the i-th item missing subsequence of S as the subsequence derived from S by removing its i-th item, denoted by $S^{(i)} = e_1 e_2 \ldots e_{i-1} e_{i+1} \ldots e_n$.*

For example, let $S=CAABC$ and $S_p=CA$. The projected sequence of S with respect to S_p is ABC. The projected database of prefix sequence A in our example database is $\{ABC, BCB, B, BB\}$. And, the 4-th item missing subsequence of S is $S^{(4)} = CAAC$.

Here we first introduce the basis of a generator checking theorem and its proof. A generators is often checked using its definition and the concept of the projected database. The similar idea was first proposed in CloSpan [10] and also adopted in our previous *FEAT* algorithm [12].

Theorem 1. *(**Generator Checking Theorem**) Given two subsequences S_{p1} and S_{p2}, if $S_{p1} \sqsubset S_{p2}$ and $SDB_{S_{p1}} = SDB_{S_{p2}}$, then any extension to S_{p2} cannot be a generator.*

Proof. Assume there exists any subsequence S, which can be used to grow S_{p2} to get a subsequence $S'=S_{p2} \diamond S$, we can always use S to grow S_{p1} and get another subsequence $S''=S_{p1} \diamond S$. Since $S_{p1} \sqsubseteq S_{p2}$ and $SDB_{S_{p1}}= SDB_{S_{p2}}$ hold, we can get $sup(S') = sup(S'')$ and $S'' \sqsubseteq S'$, thus S' cannot be a generator.   □

The $FEAT$ algorithm proposed a useful pruning method: *backward prune*, and it is also appropriate for our $IncGen$ algorithm. The following is the slightly revised theorem from the $FEAT$ algorithm, added with the proof.

Theorem 2. (Backward Pruning Theorem) *Given* $S_p=e_1e_2 \ldots e_n$, *if* $\exists i \in \{1,2,\ldots,n-1\}$ *such that* $SDB_{S_p^{(i)}} = SDB_{S_p}$, *then* S_p *can be safely pruned.*

Proof. Because $SDB_{S_p^{(i)}}=SDB_{S_p}$ and $S_p^{(i)} \sqsubset S_p$ both hold, according to Theorem 1 we can safely prune S_p.   □

In the $GenTree$, if a node s_n is in depth k, the subsequence it represents is denoted by $S_n = \langle \alpha_1\alpha_2...\alpha_k \rangle$. We can judge two things about S_n: one is "is it a generator", and the other is "can it be pruned (impossible to be extended to generators)". We use $isGenerator$ and $canPrune$ to denote them respectively. In our implementation, with the procedure $backwardCheck()$, the status $isGenerator$ and $canPrune$ can be determined simultaneously. With the above definitions and theorems, we give the Algorithm 1 for constructing a GenTree.

It is a recursive algorithm. Let $\emptyset_{seq}$ be an empty sequence, then, to get the whole $GenTree$ of a sequence database SDB, we can invoke the algorithm: $ConstructGenTree(rootNode, \emptyset_{seq}, SDB, min\_sup)$. In each call of the algorithm, we list all the local frequent items in the projected database, then for each of the items: first we allocate a child node and use $childNode$ pointing to it; second, some basic initialization for $childNode$ is performed; third, $backwardCheck()$ procedure is invoked and the $canPrune$ and $isGenerator$ status are updated; fourth, among the four combinations of two boolean indicator $\langle isGenerator, canPrune \rangle$, only three are possible, that is, $\langle false, false \rangle$, $\langle false, true \rangle$, and $\langle true, false \rangle$, corresponding to the three types of the node: $BRANCH$, $STUB$, and $GENERATOR$; finally, if the pruning conditions cannot be met, we construct the subtree of the $childNode$ recursively.

3.3   The Node Type Switching of GenTree

A $GenTree$ has several nice properties for our incremental mining algorithm. When a database is updated, the switching rules of the nodes are made clear by the following theorems. In the following, we assume the original database is D, the updated database is D', and we use the corresponding subsequence S to represent the current tree node.

Theorem 3. (Branch node switching rule.) *No matter how D expands, and whether min_sup increases or not, a branch node may hold, or be changed to a generator node. It never becomes a stub node.*

Algorithm 1. $ConstructGenTree(curNode, S_p, SDB_{S_p}, min\_sup)$

Input : Current node pointer $curNode$, prefix sequence S_p, S_p's projected
database SDB_{S_p}, minimum support min_sup

Output: The subtree represented by current node pointer $curNode$

1 **begin**
2   **foreach** i in $localFrequentItems(SDB_{S_p}, min\_sup)$ **do**
3          $S_p^i \leftarrow S_p \diamond i$;
4          $SDB_{S_p^i} \leftarrow projectedDatabase(SDB_{S_p}, i)$;
5          $newSup \leftarrow |SDB_{S_p^i}|$;
6          $childNode \leftarrow new\_allocated\_tree\_node$;
7          $curNode.children.add(childNode)$;
8          $curNode.item \leftarrow i$;
9          $curNode.support \leftarrow newSup$;
10         $backwardCheck(S_p^i, SDB_{S_p^i}, newSup, canPrune, isGenerator)$;
11   **if** $isGenerator$ **then**
12 | $childNode.type \leftarrow GENERATOR$;
13   **else if** $canPrune$ **then**
14 | $childNode.type \leftarrow STUB$;
15   **else**
16 | $childNode.type \leftarrow BRANCH$;
17   **end**
18   **if** not $canPrune$ **then**
19 | $ConstructGenTree(childNode, S_p^i, SDB_{S_p^i}, newSup)$;
20   **end**
21   **end**
22 **end**

Proof. For branch nodes:

"$isGenerator=false$" may or may not change. S is not a generator, then we have $\exists k \in \{1, 2, \cdots, n\}$ that $sup^D(S^{(k)}) = sup^D(S)$. When D expands, since $S^{(k)} \sqsubset S$, $sup(S^{(k)})$ will not increase slower than $sup(S)$, so $sup^{D'}(S^{(k)}) > sup^{D'}(S)$ or $sup^{D'}(S^{(k)}) = sup^{D'}(S)$.

"$canPrune=false$" will remain unchanged. Here let $Seq_{<L>}^D$ be the L-th sequence in D. S cannot be pruned, then $\forall i \in \{1, 2, \cdots, n\}$, $\exists L_i \in \{l | Seq_{<l>}^D \sqsupseteq S\}$, making $SDB_{S^{(i)}}^D$ in the L_i-th line appear earlier than SDB_S^D. When D expands in any way, for $\forall i$, in the same L_i-th line of D', $SDB_{S^{(i)}}^D$ will still appear earlier than SDB_S^D. So, S cannot be pruned. $\square$

Theorem 4. (Stub node switching rule.) *No matter how D expands, and whether min_sup increases or not, a stub node may be switched to any type of node.*

Proof. For stub nodes:

"$isGenerator=false$" may or may not change. (The reason is the same as the above.)

"$canPrune=true$" may or may not change. S can be pruned means: $\exists i$ that $SDB_{S^{(i)}} = SDB_S$. When D expands, $S^{(i)}$ may occur in some new sequences of D' which do not contain S. So, in that case, we cannot prune S.   □

Theorem 5. (Generator node switching rule.) *If a database is updated in the INSERT way (see Subsection 2.2), the status to be a generator will remain unchanged. Otherwise, the generator node will remain unchanged if the new absolute min_sup remains unchanged, or be switched to a branch node if the min_sup increases. In both cases, it never becomes a stub node.*

Proof. For generator nodes:

In $INSERT$ expansion manner: "$isGenerator=true$" will remain unchanged. Since S is a generator before updating, we have: for $\forall i \in \{1, 2, \cdots, n\}$, $sup^D(S^{(i)}) > sup^D(S)$. In the insertion part $InsDB$, as $S^{(i)} \sqsubseteq S$, we get $sup^{InsDB}(S^{(i)}) \geqslant sup^{InsDB}(S)$. So, adding the two, we have: $\forall i \in \{1, 2, \cdots, n\}$, $sup^{D'}(S^{(i)}) > sup^{D'}(S)$, which means in the database D', S is still a generator.

In $MIXTURE$ or $APPEND$ expansion manner: "$isGenerator=true$" may or may not change. If $sup(S)$ increases, it will be possible that: $\exists k$, for some certain sequences in D containing $S^{(k)}$ but not containing S, appended items make them contain both $S^{(k)}$ and S. In this case, the generator's definition may be violated.

Meanwhile, "$canPrune=false$" will always remain unchanged. The reason is the same as the second-part proof of Theorem 3.   □

Here, Table 3 summarizes the above three rules.

Table 3. A summarization for the switching rules

Type	isGenerator	canPrune	INSERT	Switching_Rules
Branch	$false$	$false$	$\{yes, no\}$	$B \rightarrow \{B, G\}$
Stub	$false$	$true$	$\{yes, no\}$	$S \rightarrow \{S, G, B\}$
Generator	$true$	$false$	yes	$G \rightarrow G$
Generator	$true$	$false$	no	$G \rightarrow \{G, B\}$

Example 2. Figure 2 gives an updated $GenTree$ representing the Table 2(b). The three nodes in Depth 1 are always considered generators. Among the other seven non-root nodes in Figure 1, four of them change their types, and the left three keep their types unchanged.

4   The IncGen Algorithm

In this section we introduce the IncGen algorithm for incremental mining of sequence generators. There are three implementations of IncGen, corresponding to the three ways in which a database expands. They are: IncGen-Ins for $INSERT$, IncGen-App for $APPEND$, and IncGen-Mix for $MIXTURE$. The

---

Algorithm 2. $IncGen\_App(GenTreeFile, SDB, min\_sup)$

Input   : old $GenTreeFile$, updated SDB, minimum support min_sup
Output: all generators in D'

1 **begin**
2 | $rootNode \leftarrow loadFromDisk(GenTreeFile)$;
3 | $SuspSet \leftarrow \emptyset$;
4 | $IncGenAppDFS(rootNode, \emptyset_{seq}, SDB, min\_sup, SuspSet)$;
5 | **foreach** $S_i \in SuspSet$ **do**
6 | | $ReCheck(rootNode, S_i)$;
7 | | update the node type corresponding to S_i;
8 | **end**
9 | $outputGen(rootNode)$;
10 | $saveToDisk(rootNode, GenTreeFile)$;
11 **end**

---

three implementations are very similar. They all follow the Depth-First enumeration framework [8]. In the algorithm, D' is scanned to update the $GenTree$ (partially). Meantime, information in the old $GenTree$ is sufficiently utilized to avoid unnecessary operations. Then, some suspected generators are re-checked and the corresponding tree nodes are updated.

4.1   The IncGen-App Algorithm

Algorithm 2 gives the framework of IncGen-App. We first load the old $GenTree$ from the disk (Line 2), then invoke the procedure $IncGenAppDFS()$ to get the partially updated $GenTree$ of the expanded database D', with the suspected generators stored in the set $SuspSet$ (Line 4). Afterwards, each S in the $SuspSet$ is examined using the support information on $GenTree$ nodes, and $GenTree$ is updated in the meantime (Lines 5-8). The re-checking procedure checks all S's subsequences one by one. If any subsequence S_k of S satisfying $sup(S_k) = sup(S)$, S is not a generator. Otherwise, S is a generator. Finally, we can output all generators according to the $GenTree$ (Line 9) and save the $GenTree$ into the disk for future incremental mining (Line 10).

In the recursive $IncGenAppDFS()$ procedure, each time $IncGenAppDFS$ $(curNode, S_p, SDB_{S_p}, min\_sup, SuspSet)$ is invoked, the subtree corresponding to the parameter S_p is partially updated, leaving the suspects in the $SuspSet$. In each invoking, after getting the locally frequent item set (Line 2), for each item i we do the following things. First, some initializations are made (Lines 3-5). Second, we find whether or not the subsequence S_p^i ($S_p \diamond i$) is in the children of the current node. If it exists, let $childNode$ point to it; otherwise, we allocate a new tree node, let $childNode$ point to it, and add it to the $curNode$'s children list. In both cases, after this step, $childNode$ points to the very node for S_p^i (Lines 6-12). Third, the three types of $childNode$ are judged and processed (one of Case 1 to 3, Lines 13-20).

Procedure. IncGenAppDFS($curNode$, S_p, SDB_{S_p}, min_sup, $SuspSet$)

Input   : current node pointer $curNode$ to represent $GenTree$, prefix sequence
S_p, S_p's projected database SDB_{S_p}, minimum support min_sup, the
set $SuspSet$ for suspected generators

Output: updated $GenTree$ pointed by $curNode$

1 **begin**
2   **foreach** i in $localFrequentItems(SDB_{S_p}, min\_sup)$ **do**
3          $S_p^i \leftarrow S_p \diamond i$;
4          $SDB_{S_p^i} \leftarrow projectedDatabase(SDB_{S_p}, S_p^i)$;
5          $newSup \leftarrow |SDB_{S_p^i}|$;
6   **if** S_p^i exists in $curNode.children$ **then**
7 | $childNode \leftarrow the\_found\_child$;
8   **else**
9 | $childNode \leftarrow new\_allocated\_tree\_node$;
10 | $childNode.item \leftarrow i$;
11 | $curNode.children.add(childNode)$;
12   **end**
13   **if** $childNode.isNew = true$ OR $childNode.type = STUB$ **then**
14 | $backwardCheck(S_p^i, SDB_{S_p^i}, newSup, canPrune, isGenerator)$;
15 | update $childNode.type$;
16   **else if** $childNode.type = BRANCH$ **then**
17 | $SuspSet \leftarrow SuspSet \cup S_p^i$;
18   **else if** $newSup > childNode.support$ **then**
19 | $SuspSet \leftarrow SuspSet \cup S_p^i$;
20   **end**
21         $childNode.support \leftarrow newSup$;
22   **if** $canPrune = false$ **then**
23 | $IncGenAppDFS(childNode, S_p^i, SDB_{S_p^i}, newSup, SuspSet)$;
24   **end**
25   **end**
26 **end**

Case 1. $childNode$ is new or is a $STUB$, there is little information for us.
When it is a stub node, according to Table 3, it can be switched to any type.
So, here we invoke the $backwardCheck()$ (based on Theorem 2) procedure to
determined $canPrune$ and $isGenerator$. (Lines 13-15)

Case 2. $childNode$ is a $BRANCH$, then it never becomes a $STUB$. So, besides
extending it, all we have to do is to determine whether it is a generator. This
means it will be a waste of time to invoke the $backwardCheck()$ procedure. That
is why we abandon the procedure and leave it to be post-processed. (Lines 16-17)

Case 3. $childNode$ is a $GENERATOR$, and $sup(S_p^i)$ increases, which implies
it may become a $BRANCH$. The processing operations are the same as in Case
2. (Lines 18-19)

Case 4. $childNode$ is a $GENERATOR$, but $sup(S_p^i)$ remains unchanged.
That means it is still a generator node and no action is need.

Finally, after updating *childNode*'s support, we recursively invoke the same procedure for *childNode* if it is necessary. (Lines 21-24)

4.2   The IncGen-Ins and IncGen-Mix Algorithms

The three versions of IncGen are similar to each other. Here, based on the IncGen-App algorithm just introduced, we talk about the other two briefly.

The IncGen-Mix Algorithm

If the database expands in the $MIXTURE$ manner, new sequences will be inserted and the absolute minimum support may increase. In this case, some nodes of the loaded *GenTree* may become infrequent, but the information they take keeps correct and useful. We just need to delete the infrequent nodes before we save *GenTree* to the disk. Thus, we get the IncGen-Mix algorithm by inserting only one piece of code "*deleteInfrequentNodes(curNode)*" between Line 9 and Line 10 in IncGen-App (Algorithm 2).

The IncGen-Ins Algorithm

From Theorem 5, we know that in the $INSERT$ expansion manner, a generator will hold in every cases. Thus, based on the IncGen-Mix algorithm, we only need to delete Lines 18-19 in the $IncGenAppDFS()$ procedure to get the IncGen-Ins algorithm.

5   Performance Study

In this section, a thorough evaluation of the IncGen algorithm is performed on both realistic and synthetic datasets.

5.1   Test Environment and Datasets

To test the performance of *IncGen*, we used three datasets. The first dataset, *Gazelle*, is a sparse web click-stream dataset. The other two datasets are from the well-known IBM Data Generator. One is C10S10T2.5I0.03, containing only 30 distinct items, which is a small but dense dataset, referred as *IBM-Dense*. The other is C10S10T2.5I0.3, which is sparser than *IBM-Dense* but denser than *Gazelle*, referred as *IBM-Medial*. Table 4 summarizes the characteristics of the three datasets, including the number of sequences, number of distinct items, and the average sequence length.

Table 4. Characteristics of the datasets

Dataset	#sequence	#items	#avg_seq_len
Gazelle	29,369	1,423	3
IBM-Medial	10,000	300	60
IBM-Dense	500	30	50

The algorithms were implemented in standard C++ and compiled by Microsoft Visual Studio 2008. We conducted the performance study on a computer with Intel Core Duo 2 E6300 CPU and 2GB memory running Windows 7.

5.2   Performance Evaluation

Since our algorithm *IncGen* is *the first algorithm* designed for incremental mining of sequence generators, we can only compare *IncGen* with the non-incremental algorithm *FEAT* [12], which is also designed for mining generators. For each test, the 90% of the original dataset is randomly selected as the "base" dataset, to ensure that up-to 10% of the incremental part can be added.

We use the parameter hr for the horizontal ratio when a dataset is inserted, use vr_{seq} for the ratio regarding how many sequences are updated, and use vr_{item} for the average ratio regarding how many items will be appended to the updated sequences. For example, in Table 2(a), $hr = 25\%$; in Table 2(b), $vr_{seq} = 50\%$ and $vr_{item} = 66.7\%$.

Figure 3 shows the results when the sparse dataset *Gazelle* is inserted. It gives the running time of the two algorithms when the horizontal incremental ratio hr varies from 2% to 10% and min_sup is set to 0.022%. Note that when $hr = 10\%$, the running time drops rapidly because with the insertion process, the absolute support may increase. Figure 4 shows the result when a sparse dataset meets higher min_sup. Since less and less pruning operations are needed, the *IncGen* will gradually lose its efficiency.

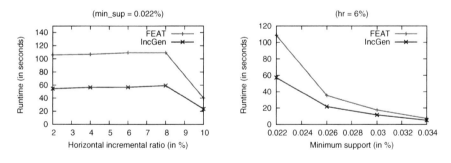

Fig. 3. *INSERT* to *Gazelle*   **Fig. 4.** *INSERT* to *Gazelle*

Figure 5 and Figure 6 compare the appending and inserting to the same dataset *IBM-Medial*. The two figures here indicate that: with the same incremental ratio (2% to 10%), the *APPEND* updating is more time-consuming than *INSERT*.

Figure 7 shows the insertion in the *IBM-Dense* dataset, and it gives the situation when min_sup varies with $hr = 6\%$. Note that with the incensing of the min_sup, the running time of *IncGen* varies from about 1/5 to nearly 3/10 of *FEAT*'s time.

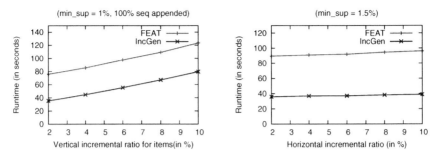

Fig. 5. *APPEND* to *IBM-Medial*   **Fig. 6.** *INSERT* to *IBM-Medial*

Although it is rarely seen, Figure 8 shows the situation when a dataset is updated in the $MIXTURE$ manner. Suppose $vr_{seq} \equiv 20\%$, each percentage point of expansion is divided into two parts: one for growing some existing sequences and the other for inserting new sequences. Again, we get satisfied results.

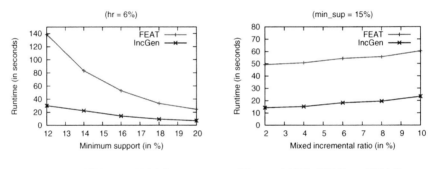

Fig. 7. *INSERT* to *IBM-Dense*   **Fig. 8.** *MIXTURE* to *IBM-Dense*

The memory cost of $IncGen$ is usually more than the non-incremental algorithm $FEAT$. But, since the nodes in $GenTree$ are all frequent and the pruned parts are not included in the tree, the memory usage of $IncGen$ is acceptable for most circumstances. For example, even on the denser dataset in Figure 7 with the lower $min\_sup = 12\%$ (usually implying more memory requirement), $IncGen$ only used approximate 62MB memory to mine 781,936 generators.

To sum up, the denser is a dataset or the lower is min_sup, the better our $IncGen$ algorithm will perform, because more time on pruning operations could be saved owing to the information from $GenTree$.

6   Related Works

The sequential pattern mining problem was first proposed by Agrawal and Srikant in [1], and an improved algorithm, called Generalized Sequential Patterns (GSP) [6], was later proposed. Then, improved SPADE [7], PrefixSpan [8],

SPAM [9] and PLWAP [18] were developed. These algorithms mine the complete set of frequent sequences. Later, two representative closed sequential pattern mining algorithms, CloSpan [10] and BIDE [11], were proposed. While the opposite concept – the "sequence generator" appears too. FEAT [12] and GenMiner [13] are two algorithms for it. Usually a well-designed closed sequence mining or generator mining algorithm can remove some redundant patterns, and can be more efficient in many cases by pruning some unpromising parts of search space.

Often an incremental mining algorithm is based on an algorithm introduced above. Based on GSP, algorithms ISE [19], IncSP [20] and GSP+/MFS+ [16] were proposed. They all follow the candidate-generation-and-test paradigm. This kind of algorithms have common disadvantages: they need to scan the database many times, and the candidate set may be very huge to process. The algorithm ISM [21] is based on SPADE. It is an interactive and incremental algorithm using vertical format data representation. It stores a sequence lattice in memory to save time, but consumes too much memory. IncSpan [14] and its revised version IncSpan+ [15] use the "semi-frequent" concept. It studies the switching rules between frequent, infrequent and semi-frequent sequences. Also, it may suffer the memory consuming problem. BSPinc [22] is another algorithm for incremental mining of frequent subsequences. Unlike others, it uses a backward mining strategy and can be 2.5 times faster than IncSpan. Algorithm PL2UP [23] is an incremental version of PLWAP. In PL2UP, a PLWAP tree is used to avoid scanning the database repeatedly. The experimental results show that PL2UP outperforms GSP+, MFS+ and IncSpan.

IMCS [17] is the first incremental algorithm for closed sequential patterns. With a CSTree, it stores the previous closed sequence information. When the database is updated, the CSTree is updated too and closed subsequences of the new database are also obtained.

7   Conclusions

In this paper, we have studied the problem of incremental mining of frequent sequence generators. We proposed a new structure GenTree to keep the useful information. Several properties and switching rules were studied, and based on these techniques we devised an incremental mining algorithm IncGen. Extensive experiments on both realistic and synthetic datasets were conducted and experimental results prove the efficiency of the IncGen algorithm.

References

1. Agrawal, R., Srikant, R.: Mining sequential patterns. In: ICDE, pp. 3–14 (1995)
2. She, R., Chen, F., Wang, K., Ester, M., Gardy, J.L., Brinkman, F.S.L.: Frequent-subsequence-based prediction of outer membrane proteins. In: KDD, pp. 436–445 (2003)

3. Sun, G., Liu, X., Cong, G., Zhou, M., Xiong, Z., Lee, J., Lin, C.Y.: Detecting Erroneous Sentences using Automatically Mined Sequential Patterns. ACL, 81–88 (2007)
4. Jindal, N., Liu, B.: Identifying comparative sentences in text documents. In: SIGIR, pp. 244–251 (2006)
5. Chen, J., Cook, T.: Mining contiguous sequential patterns from web logs. In: WWW, pp. 177–1178 (2007)
6. Srikant, R., Agrawal, R.: Mining Sequential Patterns: Generalizations and Performance Improvements. In: EDBT (1996)
7. Zaki, M.J.: SPADE: An Efficient Algorithm for Mining Frequent Sequences. ML, 31–60 (2001)
8. Pei, J., Han, J., Mortazavi-Asl, B., Pinto, H., Chen, Q., Dayal, U., Hsu, M.C.: PrefixSpan: Mining Sequential Patterns Efficiently by Prefix-Projected Pattern. In: ICDE (2001)
9. Ayres, J., Flannick, J., Gehrke, J., Yiu, T.: Sequential PAttern mining using a bitmap representation. In: KDD, pp. 429–435 (2002)
10. Yan, X., Han, J., Afshar, R.: CloSpan: Mining closed sequential patterns in large datasets. In: SDM, pp. 166–177 (2003)
11. Wang, J., Han, J.: BIDE: efficient mining of frequent closed sequences. In: ICDE, pp. 79–90 (2004)
12. Gao, C., Wang, J., He, Y., Zhou, L.: Efficient mining of frequent sequence generators. In: WWW, pp. 1051–1052 (2008)
13. Lo, D., Khoo, S.C., Li, J.: Mining and ranking generators of sequential patterns. In: SDM, pp. 553–564 (2008)
14. Cheng, H., Yan, X., Han, J.: IncSpan: incremental mining of sequential patterns in large database. In: KDD, pp. 527–532 (2004)
15. Nguyen, S.N., Sun, X., Orlowska, M.E.: Improvements of incSpan: Incremental mining of sequential patterns in large database. In: Ho, T.-B., Cheung, D., Liu, H. (eds.) PAKDD 2005. LNCS (LNAI), vol. 3518, pp. 442–451. Springer, Heidelberg (2005)
16. Kao, B., Zhang, M., Yip, C.L., Cheung, D.W., Fayyad, U.: Efficient algorithms for mining and incremental update of maximal frequent sequences. In: DMKD, pp. 87–116 (2005)
17. Chang, L., Wang, T., Yang, D., Luan, H., Tang, S.: Efficient algorithms for incremental maintenance of closed sequential patterns in large databases. In: DKE, pp. 68–106 (2009)
18. Ezeife, C.I., Lu, Y.: Mining web log sequential patterns with position coded preorder linked wap-tree. In: DMKD, pp. 5–38 (2005)
19. Masseglia, F., Poncelet, P., Teisseire, M.: Incremental mining of sequential patterns in large databases. In: DKE, pp. 97–121 (2003)
20. Lin, M.Y., Lee, S.Y.: Incremental update on sequential patterns in large databases by implicit merging and efficient counting. In: IS, pp. 385–404 (2004)
21. Parthasarathy, S., Zaki, M.J., Ogihara, M., Dwarkadas, S.: Incremental and interactive sequence mining. In: CIKM, pp. 251–258 (1999)
22. Lin, M.Y., Hsueh, S.C., Chan, C.C.: Incremental Discovery of Sequential Patterns Using a Backward Mining Approach. In: CSE, pp. 64–70 (2009)
23. Ezeife, C.I., Liu, Y.: Fast incremental mining of web sequential patterns with PLWAP tree. In: DMKD, pp. 376–416 (2009)

An Alternative Interestingness Measure for Mining Periodic-Frequent Patterns

R. Uday Kiran and P. Krishna Reddy

Center for Data Engineering,
International Institute of Information Technology-Hyderabad,
Hyderabad, Andhra Pradesh, India - 500032
uday_rage@research.iiit.ac.in, pkreddy@iiit.ac.in

Abstract. Periodic-frequent patterns are a class of user-interest-based frequent patterns that exist in a transactional database. A frequent pattern can be said *periodic-frequent* if it appears at a regular user-specified interval in a database. In the literature, an approach has been proposed to extract periodic-frequent patterns that occur periodically throughout the database. However, it is generally difficult for a frequent pattern to appear periodically throughout the database without any interruption in many real-world applications. In this paper, we propose an improved approach by introducing a new interestingness measure to discover periodic-frequent patterns that occur almost periodically in the database. A pattern-growth algorithm has been proposed to discover the complete set of periodic-frequent patterns. Experimental results show that the proposed model is effective.

Keywords: Data mining, knowledge discovery, frequent patterns and periodic-frequent pattern.

1 Introduction

Periodic-frequent pattern mining is an important model in data mining. Periodic-frequent patterns can provide useful information regarding the patterns that occur both frequently and periodically in a transactional database. The basic model of periodic-frequent patterns is as follows [4].

Let $I = \{i_1, i_2, \cdots, i_n\}$ be a set of items. A set $X \subseteq I$ is called an itemset (or a **pattern**). A pattern containing 'k' number of items is called a k-**pattern**. A transaction $t = (tid, Y)$ is a tuple, where tid represents a transaction-id (or a timestamp) and Y is a pattern. A transactional database T over I is a set of transactions, $T = \{t_1, \cdots, t_m\}$, $m = |T|$, where $|T|$ is the size of T in total number of transactions. If $X \subseteq Y$, it is said that t contains X or X occurs in t and such tid is denoted as t_j^X, $j \in [1, m]$. Let $T^X = \{t_k^X, \cdots, t_l^X\} \subseteq T$, where $k \leq l$ and $k, l \in [1, m]$ be the ordered set of transactions in which pattern X has occurred. Let t_q^X and t_r^X, where $k \leq q < r \leq l$ be the two consecutive transactions in T^X. The number of transactions or time difference between t_r^X and t_q^X can be defined as a **period** of X, say p_a^X. That is, $p_a^X = t_r^X - t_q^X$. Let $P^X = \{p_1^X, p_2^X, \cdots, p_r^X\}$, be the set of periods for pattern X. The **periodicity** of X, denoted as $Per(X) = maximum(p_1^X, p_2^X, \cdots, p_r^X)$. The **support** of X is denoted as $S(X) = |T^X|$. The pattern X is periodic-frequent if

J.X. Yu, M.H. Kim, and R. Unland (Eds.): DASFAA 2011, Part I, LNCS 6587, pp. 183–192, 2011.
© Springer-Verlag Berlin Heidelberg 2011

$S(X) \geq minSup$ and $Per(X) \leq maxPrd$. $MinSup$ and $maxPrd$ are the user-specified minimum support and maximum periodicity constraints. Both *periodicity* and *support* of a pattern can be described in percentage of $|T|$.

Table 1. Transactional database

ID	Items	ID	Items	ID	Items	ID	Items	ID	Items
1	a, b	3	a, b, e, f	5	a, b, c, d	7	c, d	9	c, d, e, f
2	c, d	4	b, e	6	e, f	8	a, b	10	a, b

Example 1. Consider the transactional database shown in Table 1. Each transaction in this database is uniquely identifiable with a *tid*. We consider that the *tid* of a transaction also represents a time stamp. Consider the first five transactions in Table 1. The set of items, $I = \{a, b, c, d, e, f\}$. The set of 'a' and 'b' i.e., $\{a, b\}$ is a pattern. It is a 2-pattern. The pattern 'ab' occurs in *tids* 1, 3 and 5. Therefore, $T^{ab} = \{1, 3, 5\}$ and $S(ab) = |T^{ab}| = 3$. The periods for this pattern are 1 $(= 1 - t_i)$, 2 $(= 3 - 1)$, 2 $(= 5 - 3)$ and 0 $(= t_l - 5)$, where $t_i = 0$ represents the initial transaction and $t_l = 5$ represents the last transaction in the sub-transactional database. The *periodicity* of ab, $Per(ab) = maximum(1, 2, 2, 0) = 2$. If the user-specified $minSup = 2$ and $maxPrd = 2$, then ab is a periodic-frequent pattern because $S(ab) \geq minSup$ and $Per(ab) \leq maxPrd$.

A pattern-growth algorithm using a tree structure known as periodic-frequent tree (PF-tree) has also been discussed in [4] to discover the complete set of periodic-frequent patterns in a database. In [3], the basic model was extended to multiple *minSups* and multiple *maxPrds* framework.

The basic model of periodic-frequent patterns mines only those frequent patterns that are occurring periodically throughout the database. However, in many real-world applications, it is difficult for the frequent patterns to appear periodically throughout the database (without any interruption), because items' occurrence behavior can vary over time causing periodically occurring patterns to be non-periodic and/or vice-versa. As a result, the existing model of periodic-frequent patterns misses the knowledge pertaining to the interesting frequent patterns that are appearing **almost periodically** throughout the database. However, depending upon the user and/or application requirements such patterns can be considered periodically interesting.

Example 2. In the entire transactional database shown in Table 1, $T^{ab} = \{1, 3, 5, 8, 10\}$, $P^{ab} = \{1, 2, 2, 3, 2, 0\}$, $S(ab) = 5$ and $Per(ab) = 3$. If the user-defined $minSup$ and $maxPrd$ values are respectively 3 and 2, we miss the frequent pattern 'ab' as a periodic-frequent pattern. However, this pattern can still be periodically interesting to the users as it has failed to appear periodically only once throughout the transactional database.

In this paper, we propose an approach to extract the frequent patterns that occur almost periodically throughout the database. The proposed approach evaluates the periodic interestingness of a frequent pattern based on the proportion of its periodic occurrences in a database. As a result, the proposed approach is able to generate the same set of periodic-frequent patterns as in the basic approach. In other words, the proposed approach generalizes the existing approach of periodic-frequent patterns. The proposed approach do not satisfy *downward closure property*. However, by exploiting the relationship between the *"support"* and *periods* of a pattern, we propose two pruning techniques for reducing the search space. A pattern-growth algorithm is also proposed to discover the complete set of periodic-frequent patterns. Experimental results on both synthetic and real-world datasets demonstrate that both the proposed model and algorithm are efficient.

The rest of the paper is organized as follows. The proposed model along with the pruning techniques are presented in Section 2. The pattern-growth algorithm is presented in Section 3. Experimental results are provided in Section 4. Conclusions are provided in the last section.

2   The Proposed Model and Pruning Techniques

2.1   Proposed Model

To extract frequent patterns that appear almost periodically in a database, we propose an interestingness measure, called *periodic ratio*. We use the notions *support*, *period* and *set of periods* which are defined in Section 1.

Definition 1. *Periodic ratio of a pattern X ($Pr(X)$): Let $IP^X \subseteq P^X$ be the set of periods such that $\forall p_a^X \in IP^X$, $p_a^X \leq maxPeriod$. The variable maxPeriod is the user-defined maximum period threshold value that determines the periodic interestingness of a pattern. The periodic ratio of the pattern X, $Pr(X)$, is equal to $\frac{|IP^X|}{|P^X|}$.*

The measure Pr captures the proportion of periods that satisfy the user-defined *maxPeriod* value. For a pattern X, $Pr(X) \in [0,1]$. If $Pr(X) = 0$, it means X has not appeared periodically anywhere in the transactional database. If $Pr(X) = 1$, it means X has appeared periodically throughout the database without any interruption.

Definition 2. *Periodic-frequent pattern: The pattern X can be said* periodic-frequent *if $S(X) \geq minSup$ and $Pr(X) \geq minPr$. MinSup and minPr are the user-defined minimum support* and *minimum periodic ratio values.*

For a periodic-frequent pattern X, if $S(X) = a$ and $Pr(X) = b$ then it is represented as shown in Equation 1.

$$X\ [support = a, Pr = b] \tag{1}$$

Example 3. Continuing with Example 2, if the user-specified *maxPeriod* = 2, then $IP^{ab} = \{1,2,2,2,0\}$. Therefore, the *periodic ratio* of the pattern 'ab' i.e., $Pr(ab) = \frac{|IP^{ab}|}{|P^{ab}|} = \frac{5}{6} = 0.83\ (= 83\%)$. If the user-specified *minPr* = 0.8 and *minSup* = 3, then 'ab' is a periodic-frequent pattern and is described as: $ab\ [support = 5,\ Pr = 0.83]$.

Problem definition: Given a transactional database T, minimum support (*minSup*), maximum period (*maxPeriod*) and minimum periodic ratio (*minPr*) constraints, the objective is to discover the complete set of periodic-frequent patterns in T that have *support* and Pr no less than the user-specified *minSup* and *minPr*, respectively.

2.2 Pruning Techniques

Let the time duration of the database be $[t_i, t_l]$, where t_i and t_l respectively denote the timestamps of the first and last transactions of a database. Support of the pattern X, $S(X)$, indicates the number of transactions containing X. The time difference between any two transactions constitute a period. The number of periods come to $S(X) - 1$. In addition, two more periods are to be added: one is from t_i to the first occurrence of X and another is from the last occurrence of X to t_l. Total number of periods become equal to $S(X) - 1 + 2 = S(X) + 1$.

Property 1. The total number of periods for a pattern X i.e., $|P^X| = S(X) + 1$.

Property 2. Let X and Y be the two patterns in a transactional database. If $X \subset Y$, then $|P^X| \geq |P^Y|$ and $|IP^X| \geq |IP^Y|$ because $T^X \supseteq T^Y$.

The measure *periodic ratio* captures the ratio of interesting periods to the total number of periods for a given pattern X. As a result, if a given pattern X satisfies the *periodic ratio* threshold value, its subset $Y \subset X$ or superset $Z \supset X$ may not satisfy the *periodic ratio* threshold value. So, the periodic-frequent patterns discovered with the proposed model do not satisfy *downward closure property* (see Lemma 1). That is, not all non-empty subsets of a periodic-frequent pattern may be periodic-frequent.

Lemma 1. *The periodic-frequent patterns discovered with the proposed model do not satisfy* downward closure property.

Proof. Let $Y = \{i_a, \cdots, i_b\}$, where $1 \leq a \leq b \leq n$ be a periodic-frequent pattern with $S(Y) = minSup$ and $Pr(Y) = minPr$. Let $X \subset Y$ be another pattern. From Property 2, we derive $|P^X| \geq |P^Y|$ and $|IP^X| \geq |IP^Y|$. Considering the scenario where $|P^X| > |P^Y|$ and $|IP^X| = |IP^Y|$, we derive $\frac{|IP^X|}{|P^X|} < \frac{|IP^Y|}{|P^Y|}$ (Property 2). In other words, $Pr(X) < Pr(Y)$. Since, $Pr(Y) = minPr$, we get $Pr(X) < minPr$. Therefore, X is not a periodic-frequent pattern.

However, by exploiting the relationship between the support and number of periods of a pattern (Property 1), there exists a scope to reduce the search space. It can be noted that every periodic-frequent pattern must have the support value greater than or equal to user-defined *minsup*, which implies that every pattern will have the minimum number periods. If the ratio of total number of interesting periods to minimum number of periods is not satisfying the user-defined *minPr* value, it can be pruned. Also, its supersets can be pruned as they cannot generate any periodic-frequent pattern. The following pruning techniques are proposed to reduce the search space.

i. The *minSup* constraint follows *downward closure property* [1]. Therefore, if a pattern X fails to satisfy *minSup*, then X can be eliminated as X and its supersets cannot generate any periodic-frequent pattern.

ii. For a pattern X, if $\frac{|IP^X|}{(minSup+1)} < minPr$, then X can be eliminated as X and its supersets cannot generate any periodic-frequent pattern (see Theorem 1). If $\frac{|IP^X|}{(minSup+1)} \geq minPr$ and $S(X) \geq minSup$, then X is called a **potential pattern**. A potential pattern need not necessarily be a periodic-frequent pattern. However, only potential patterns can generate periodic-frequent patterns. A potential pattern containing only one item (1-pattern) is called a **potential item**.

Theorem 1. *Let X and Y be the two patterns such that $Y \supset X$. For the pattern X, if $\frac{|IP^X|}{(minSup+1)} < minPr$, then X and Y cannot be periodic-frequent patterns.*

Proof. Every periodic-frequent pattern will have support greater than or equal to $minSup$. Hence, every periodic-frequent pattern will have at least $(minSup + 1)$ number of periods (Property 1). For a frequent pattern X, $\frac{|IP^X|}{(minSup+1)} \geq \frac{|IP^X|}{|P^X|}(= Pr(X))$ because $|P^X| \geq (minSup + 1)$. If $\frac{|IP^X|}{(minSup+1)} < minPr$, then X cannot be a periodic-frequent pattern as $\frac{|IP^X|}{|P^X|} \leq \frac{|IP^X|}{(minSup+1)} < minPr$. In addition, if $Y \supset X$, then Y cannot be a periodic-frequent pattern as $\frac{|IP^Y|}{|P^Y|} \leq \frac{|IP^Y|}{(minSup+1)} \leq \frac{|IP^X|}{(minSup+1)} < minPr$ (Property 2).

3   Proposed Algorithm

The PF-tree structure discussed in [4] cannot be used directly for mining periodic-frequent patterns with the proposed model. The reasons are as follows: (*i*) Periodic-frequent patterns discovered with the proposed model do not satisfy *downward closure property* and (*ii*) The *tree* structure must capture different interestingness measures, i.e., *maxPeriod* and *minPr*. To mine periodic-frequent patterns with the proposed model, we need to modify both the PF-tree and PF-growth algorithms. In this paper, we call the modified PF-tree and PF-growth as Extended PF-tree (ExPF-tree) and Extended PF-growth (ExPF-growth), respectively.

3.1   ExPF-tree: Structure and Construction

Structure of ExPF-tree. The ExPF-tree consists of two components: ExPF-list and a prefix-tree. An ExPF-list is a list with three fields: item (i), support or frequency (s) and number of interesting periods (ip). The node structure of prefix-tree in ExPF-tree is same as the prefix-tree in PF-tree [4], which is as follows.

The prefix-tree in ExPF-tree explicitly maintains the occurrence information for each transaction in the tree structure by keeping an occurrence *tid* list, called *tid-list*, only at the last node of every transaction. The ExPF-tree maintains two types of nodes: ordinary node and *tail*-node. The ordinary node is similar to the nodes used in FP-tree, whereas the latter is the node that represents the last item of any sorted transaction. The structure of a *tail*-node is $N[t_1, t_2, \cdots, t_n]$, where N is the node's item name and $t_i, i \in [1, m]$ is a transaction-id where item N is the last item. Like the FP-tree [2], each node in ExPF-tree maintains parent, children, and node traversal pointers. However, irrespective of the node type, no node in ExPF-tree maintains support count value in it.

Construction of ExPF-list. Let id_l be a temporary array that explicitly records the *tids* of the last occurring transactions of all items in the ExPF-list. Let t_{cur} be the *tid* of current transaction. The ExPF-list is maintained according to the process given in Fig. 2.

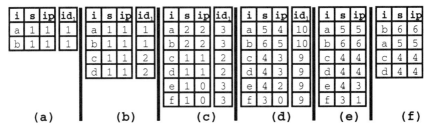

Fig. 1. ExPF-list. (a) After scanning first transaction (b) After scanning second transaction (c) After scanning third transaction (d) After scanning entire transactional database (e) Reflecting correct number of interesting periods (f) compact ExPF-list containing only potential items.

In Fig. 1, we show how the ExPF-list is populated for the transactional database shown in Table 1. With the scan of the first transaction $\{a,b\}$ (i.e., $t_{cur} = 1$), the items 'a' and 'b' in the list are initialized as shown in Fig. 1(a) (lines 4 to 6 in Algorithm 2). The scan on the next transaction $\{c,d\}$ with $t_{cur} = 2$ initializes the items 'c' and 'd' in ExPF-list as shown in Fig. 1(b). The scan on next transaction $\{a,b,e,f\}$ with $t_{cur} = 3$ initializes ExPF-list entries for the items 'e' and 'f' with $id_l = 3$, $s = 1$ and $ip = 0$ because $t_{cur} > maxPeriod$ (line 6 in Algorithm 2). Also, the $\{s;ip\}$ and id_l values for the items 'a' and 'b' are updated to $\{2;2\}$ and 3, respectively (lines 8 to 12 in Algorithm 2). Fig. 1(c) shows the ExPF-list generated after scanning third transaction. Fig. 1(d) shows the ExPF-list after scanning all ten transactions. To reflect the correct number of interesting periods for each item in the ExPF-list, the whole ExPF-list is refreshed as mentioned from lines 15 to 19 in Algorithm 2. The resultant ExPF-list is shown in Fig. 1(e). Based on the above discussed ideas, the items 'e' and 'f' are pruned from the ExPF-list because $\frac{|IP^e|}{(minSup+1)} < minPr$ and $\frac{|IP^f|}{(minSup+1)} < minPr$ (lines 21 to 23 in Algorithm 2). The items 'a', 'b', 'c' and 'd' are generated as ExPF-patterns (lines 24 to 26 in Algorithm 2). The items which are not pruned are sorted in descending order of their support values (line 29 in Algorithm 2). The resultant ExPF-list is shown in Fig. 1(f). Let PI be the set of potential items that exist in ExPF-list.

Construction of ExPF-tree. With the second database scan, we construct ExPF-tree in such a way that it only contains nodes for items in ExPF-list.

Continuing with the ongoing example, using the FP-tree [2] construction technique, only the items in ExPF-list take part in the construction of ExPF-tree. The tree construction starts with inserting the first transaction $\{a,b\}$ according to the ExPF-list order i.e., $\{b,a\}$, as shown in Fig. 3(a). The *tail*-node "$a:1$" carries the *tid* of the transaction. Fig. 3(b) and (c) respectively show the ExPF-tree generated in the similar procedure after scanning the second and every transaction in the database. For simplicity of figures, we do not show the node traversal pointers in trees; however, they are maintained as in the construction process of FP-tree.

It is to be noted that ExPF-tree is memory efficient. It has been shown in [4] that the tree achieves memory efficiency by keeping such transaction information only at the *tail*-nodes and avoiding the support count value at each node. Furthermore, ExPF-tree avoids the *complicate combinatorial problem of candidate generation* as in Apriori-like algorithms [1].

Algorithm 1 ExPF-list (T: Transactional database, I: set of items, $minsup$: minimum support, $maxperiod$: maximum period, $minPr$: minimum Pr)

1: Let id_l be a temporary array that explicitly records the $tids$ of the last occurring transactions of all items in the ExPF-list. Let t_{cur} be the tid of current transaction.
2: **for** each transaction $t_{cur} \in T$ **do**
3:    **for** each item i in t_{cur} **do**
4:       **if** t_{cur} is i's first occurrence **then**
5:          $s = 1, id_l = t_{cur}$;
6:          $ip = (t_{cur} \leq maxperiod)?1 : 0$;
7:       **else**
8:          **if** $(t_{cur} - id_l) \leq maxperiod$ **then**
9:             $ip++$;
10:          **end if**
11:          $s++, id_l = t_{cur}$;
12:       **end if**
13:    **end for**
14: **end for**
15: **for** each item i in ExPF-list **do**
16:    **if** $(|T| - id_l) \leq maxperiod$ **then**
17:       $ip++$;
18:    **end if**
19: **end for**
20: **for** each item i in ExPF-list **do**
21:    **if** $(s < minsup) || (\frac{ip}{(minsup+1)} < minPr)$ **then**
22:       remove i from ExPF-list;
23:    **else**
24:       **if** $\frac{ip}{(s+1)} \geq minPr$ **then**
25:          output i as ExPF-pattern.
26:       **end if**
27:    **end if**
28: **end for**
29: Sort the remaining items in the ExPF-list in descending order of their support values. Let PI be the set of potential items.

Fig. 2. Algorithm for constructing ExPF-list

3.2 Mining ExPF-tree

The basic operations in mining ExPF-tree are as follows: (i) counting length-1 potential items, (ii) constructing the prefix-tree for each potential pattern, and (iii) constructing the conditional tree from each prefix-tree. The ExPF-list provides the length-1 potential items. Before discussing the prefix-tree construction process, we explore the following important property and lemma concerning to ExPF-tree.

Property 3. A tail-node in an ExPF-tree maintains the occurrence information for all the nodes in the path (from that *tail*-node to the root) at least in the transactions in its tid-list.

Lemma 2. *Let $B = \{b_1, b_2, \cdots, b_n\}$ be a branch in ExPF-tree where node b_n is the tail-node carrying the tid-list of the path. If the tid-list is pushed-up to node b_{n-1}, then b_{n-1}*

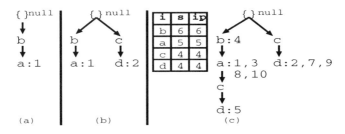

Fig. 3. ExPF-tree. (a) After scanning first transaction (b) After scanning second transaction and (c) After scanning complete transactional database.

maintains the occurrence information of the path $B' = \{b_1, b_2, \cdots, b_{n-1}\}$ for the same set of transactions in the tid-list without any loss [4].

Using the feature revealed by the above property and lemma, we proceed to construct prefix-tree starting from the bottom-most item, say i, of the ExPF-list. Only the prefix sub-paths of nodes labeled i in the ExPF-tree are accumulated as the prefix-tree for i, say PT_i. Since i is the bottom-most item in the ExPF-list, each node labeled i in the ExPF-tree must be a *tail*-node. While constructing the PT_i, based on Property 3 we map the *tid*-list of every node of i to all items in the respective path explicitly in a temporary array (one for each item). It facilitates the support and number of interesting periods' calculation for each item in the ExPF-list of PT_i. Moreover, to enable the construction of the prefix-tree for the next item in the ExPF-list, based on Lemma 2 the *tid*-lists are pushed-up to respective parent nodes in the original ExPF-tree and in PT_i as well. All nodes of i in the ExPF-tree and i's entry in the ExPF-list are deleted thereafter. Fig. 4(a) shows the status of the ExPF-tree of Fig. 3(c) after removing the bottom-most item 'd'. Besides, the prefix-tree for 'd', PT_d is shown in Fig. 4(b).

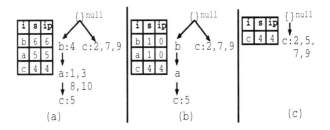

Fig. 4. Prefix-tree and conditional tree construction with ExPF-tree. (a) ExPF-tree after removing item 'd' (b) Prefix-tree for 'd' and (c) Conditional tree for 'd'.

The conditional tree CT_i for PT_i is constructed by removing the items whose support is less than $minSup$ or $\frac{|IP|}{(minSup+1)} < minPr$. If the deleted node is a *tail*-node, its *tid*-list is pushed-up to its parent node. Fig. 4(c), for instance, shows the conditional tree for 'd', CT_d constructed from the PT_d of Fig. 4(b). The contents of the temporary array for the bottom item 'j' in the ExPF-list of CT_i represent T^{ij} (i.e., the set of all *tids* where items i and j are occurring together). Therefore, it is a rather simple calculation to compute

$S(ij)$, $\frac{|IP^{ij}|}{(minSup+1)}$ and $Pr(ij)$ from T^{ij} by generating P^{ij}. If $S(ij) \geq minSup$ and $Pr(ij) \geq minPr$, then the pattern "ij" is generated as a periodic-frequent pattern with support and Pr values of $S(ij)$ and $Pr(ij)$, respectively. The same process of creating prefix-tree and its corresponding conditional tree is repeated for further extensions of "ij". Else, if $\frac{|IP^{ij}|}{(minSup+1)} \geq minPr$ and $S(ij) \geq minSup$, then the above process is still repeated for further extensions of "ij" even though "ij" is not a periodic-frequent pattern. The whole process is repeated until $ExPF\text{-}list \neq \emptyset$.

4 Experimental Results

The ExPF-growth algorithm is written in C++ and run with Ubuntu operating system on a 2.66 GHz machine with 1 GB memory. We pursued experiments on widely used synthetic ($T10I4D100K$) and real-world datasets ($Retail$ and $Mushroom$). $T10I4D100K$ is a sparse dataset containing 100,000 transactions and 870 items. $Retail$ is another sparse dataset containing 88,162 transactions and 16,470 items. $Mushroom$ is a dense dataset containing 8,124 transactions and 119 items. We have considered the transactions in these datasets as an ordered set based on a particular time stamp.

Table 2. Periodic-frequent patterns generated in different datasets

Database	minSup	maxPeriod$_1$ = 0.1%						maxPeriod$_2$ = 0.5%					
		minPr=0.5		minPr=0.75		minPr=1		minPr=0.5		minPr=0.75		minPr=1	
		A	B	A	B	A	B	A	B	A	B	A	B
T10I4D100k	0.1%	624	5	272	1	0	0	20748	10	6360	8	229	2
	1.0%	385	3	272	1	0	0	385	3	385	3	229	2
Retail	0.2%	643	5	205	5	4	2	2691	5	1749	5	15	3
	1.0%	159	4	102	4	4	2	159	4	159	4	15	3
Mushroom	10%	574,431	16	570,929	16	15	4	574,431	16	574,431	16	135	6
	20%	53,583	15	53,583	15	15	4	53,583	15	53,583	15	135	6

The periodic-frequent patterns discovered with the proposed model in different databases at various $minSup$, $maxPeriod$ and $minPr$ values are reported in Table 2. At different $minSup$, $maxPeriod$ and $minPr$ values, the column "A" shows the number of periodic-frequent patterns mined and the column "B" shows the maximal length of the periodic-frequent pattern(s) discovered. The columns with $minPr = 1$ indicates the performance of the basic model that extracts the periodic-frequent patterns that appears in the entire transactional database. It can be observed that the increase in $minSup$ or $minPr$ (keeping other constraints fixed) decreases the number of periodic-frequent patterns because many items (or patterns) fail to satisfy the increased threshold values. Also, the increase in $maxPeriod$ increases the number of periodic-frequent patterns. It is because of the increased interval range in which a pattern should reappear. More important, it can be observed that very few patterns, mostly of shorter lengths, are being generated as periodic-frequent patterns when $minPr = 1$. Overall, the experimental results show that the proposed model provides more flexibility and improves performance by extracting more number of periodic-frequent patterns of longer length over the basic model.

The runtime taken by ExPF-growth for generating periodic-frequent patterns at different *minSup*, *maxPeriod* and *minPr* values on various datasets is shown in Table 3. The runtime encompasses all phases of ExPF-list and ExPF-tree constructions, and the corresponding mining operation. It can be observed that the runtime taken by the proposed algorithm depends on the number of periodic-frequent patterns being generated.

Table 3. Runtime requirements for the ExPF-tree. Runtime is measured in seconds.

Dataset	*minSup*	*maxPeriod*$_1$ = 0.1%			*maxPeriod*$_2$ = 0.5%		
		minPr=0.5	*minPr*=0.75	*minPr*=1	*minPr*=0.5	*minPr*=0.75	*minPr*=1
T10I4D100k	0.1%	120.533	124.766	128.057	105.951	110.085	113.957
	1.0%	109.257	102.138	83.009	112.400	107.000	112.604
Retail	0.2%	29.638	25.479	22.529	43.254	40.178	34.305
	1.0%	15.441	15.182	14.352	15.780	15.909	15.894
Mushroom	10%	20.580	20.180	16.510	20.320	20.270	18.860
	20%	2.920	2.790	2.780	2.930	3.000	2.740

5 Conclusions

In this paper, we have proposed an improved approach to extract periodic-frequent patterns in a transactional database. A new interestingness measure, called *periodic ratio*, has been proposed for mining frequent patterns that occur almost periodically in a database. Two pruning techniques are proposed to improve the efficiency of proposed algorithm as the periodic-frequent patterns under the proposed model do not satisfy *downward closure property*. Also, a pattern-growth algorithm has been proposed to discover periodic-frequent patterns. The effectiveness of the proposed model and algorithm are shown practically by conducting experiments on various datasets.

References

1. Agrawal, R., Imieliński, T., Swami, A.: Mining association rules between sets of items in large databases. In: SIGMOD 1993: Proceedings of the 1993 ACM SIGMOD International Conference on Management of Data, pp. 207–216. ACM, New York (1993)
2. Han, J., Pei, J., Yin, Y., Mao, R.: Mining frequent patterns without candidate generation: A frequent-pattern tree approach. Data Min. Knowl. Discov. 8(1), 53–87 (2004)
3. Kiran, R.U., Reddy, P.K.: Towards efficient mining of periodic-frequent patterns in transactional databases. In: DEXA, vol. (2), pp. 194–208 (2010)
4. Tanbeer, S.K., Ahmed, C.F., Jeong, B.-S., Lee, Y.-K.: Discovering periodic-frequent patterns in transactional databases. In: Theeramunkong, T., Kijsirikul, B., Cercone, N., Ho, T.-B. (eds.) PAKDD 2009. LNCS, vol. 5476, pp. 242–253. Springer, Heidelberg (2009)

A Framework of Mining Semantic Regions from Trajectories

Chun-Ta Lu[1], Po-Ruey Lei[2], Wen-Chih Peng[1], and Ing-Jiunn Su[2]

[1] National Chiao Tung University, Hsinchu, Taiwan, ROC
{lucangel,wcpeng}@gmail.com
[2] Chung Cheng Institute of Technology, National Defense University,
Taoyuan, Taiwan, ROC
{kdboy1225,suhanson}@gmail.com

Abstract. With the pervasive use of mobile devices with location sensing and positioning functions, such as Wi-Fi and GPS, people now are able to acquire present locations and collect their movement. As the availability of trajectory data prospers, mining activities hidden in raw trajectories becomes a hot research problem. Given a set of trajectories, prior works either explore density-based approaches to extract regions with high density of GPS data points or utilize time thresholds to identify users' stay points. However, users may have different activities along with trajectories. Prior works only can extract one kind of activity by specifying thresholds, such as spatial density or temporal time threshold. In this paper, we explore both spatial and temporal relationships among data points of trajectories to extract semantic regions that refer to regions in where users are likely to have some kinds of activities. In order to extract semantic regions, we propose a sequential clustering approach to discover clusters as the semantic regions from individual trajectory according to the spatial-temporal density. Based on semantic region discovery, we develop a shared nearest neighbor (SNN) based clustering algorithm to discover the frequent semantic region where the moving object often stay, which consists of a group of similar semantic regions from multiple trajectories. Experimental results demonstrate that our techniques are more accurate than existing clustering schemes.

Keywords: Trajectory pattern mining, sequential clustering and spatial-temporal mining.

1 Introduction

Knowledge discovery from spatial-temporal data has risen as an active research because of the large amount of trajectory data produced by mobile devices. A trajectory is a sequence of spatial-temporal points which records the movement of a moving object. Each point specifies a moving location in space at a certain instant of time. The semantic knowledge may contain in some re-appear trajectories and can be applied in many applications, such as trajectory pattern mining

J.X. Yu, M.H. Kim, and R. Unland (Eds.): DASFAA 2011, Part I, LNCS 6587, pp. 193–207, 2011.

for movement behaviors [6,19,8], predicting user location [10,18], and location-based activity discovery [13,14,7]. Unfortunately, locations may not be repeated exactly in similar trajectories. The common preceding task for the above works is to discover the regions for replacing the exact locations where moving objects often pass by or stay. Such a region summarizes a set of location points from different trajectories that are close enough in the spatial space. Then, the relation between regions can be extracted for knowledge analysis. Intuitively, the quality of regions directly affects the analysis result of trajectory data. Thus, in this paper, we focus on effectively and precisely discovering regions from trajectory data where can imply the potential of users are likely to have some kinds of activities, called semantic regions.

Traditionally, regions are extracted from trajectory points by density-based clustering methods (e.g., DBSCAN [4]). Given the definition of distance (i.e., measure of dissimilarity) between any two points, regions with higher density are extracted in terms of clustering similar data points in the spatial domain.

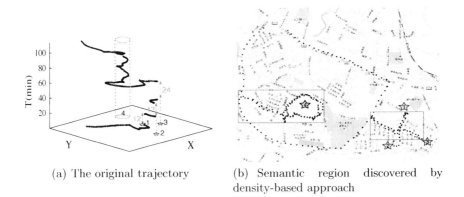

(a) The original trajectory

(b) Semantic region discovered by density-based approach

Fig. 1. An example of discovering semantic regions from a trajectory

However, such regions, extracted by clustering spatial points without considering sequential relation, only focus on the geometrical properties of trajectories. Consider an example in Figure 1, where there are four activities involved in this trajectory. Each region associated with one activity is marked with a star in Figure 1(b). We observed that, from the original trajectory based on spatial-temporal domain, there should be three indoor regions (region 1, 2 and 3) because of the appearance of temporal transition gaps which are labeled with stay durations in Figure 1(a). The temporal transition gap between sampled location points is generated due to the loss of satellite signal when GPS-embedded location recorder is inside a building (e.g., restaurant, home or office). In additional, in Figure 1(b), there is an outdoor activity (i,e., in region 4) where the user is walking around a lake. Two regions where represented by minimum bounding rectangles (MBRs) are discovered by a spatial density based clustering algorithm, DBSCAN ($Minpts = 7$, $Eps = 50$ meters) . There are three problems in this example. First, some semantic regions are missing. By verifying with the ground truth (i.e., four regions with stars in Figure 1(b)), only two regions are detected by DBSCAN and region 2 is missing. As shown in Figure 1(b), while this user stays in the region 2 for 17 minutes, DBSCAN cannot discover region 2. This is because that region 2 does not have a sufficient amount of GPS data points to form a cluster. Second, granularity problem causes the indistinguishability between region 1 and 3. Third, road-sections and intersections, where an object often passes but carries non-semantic meaning to the user, are included in both discovered regions. The above example indicates that only exploring density-based approaches in the spatial domain of data points in trajectories cannot discover semantic regions.

Recently, the authors in [21] proposed the concept of stay point detection to discover the stay regions. Unlike density based clustering, stay point is detected when the consecutive points of a examined point do not exceed the predefined distance threshold during the specified period of time threshold. The authors claimed that a stay point can stand for a geographic region and carry a particular semantic meaning. However, a trajectory usually contains more than one activity, such as driving, walking, sightseeing, staying and so on. Each activity has different distance density and speed. In other words, the density of trajectory points varies from different activities. Thus, the traditional density clustering approach or stay point detection, which using universal parameter to detect the clusters only for a certain density, cannot discover all semantic regions. Figure 2 shows regions discovered by the stay points approach, where the time threshold is fixed to 10 minutes and three distance thresholds are set to 100 meters, 200 meters and 250 meters. When distance threshold is set to 100 meters, in Figure 2(a), there are three stay points mapping to three semantic regions, but the semantic region 4 (lake), a much larger area with an activity of walking, cannot be detected. The regions are not detected completely until the distance threshold is larger than 250 meters. On the other hand, the other three regions have been mixed and their coverage have been overlapped shown in Figure 2(b) and Figure 2(c). As such, the stay point approach considers both the temporal and spatial thresholds for detecting regions. However, the stay point approach is highly dependent to thresholds. Consequently, to detect regions with a variety of activities, the stay point approach may need to have different settings of thresholds.

In this paper, we first propose a sequential density clustering approach to extract candidate semantic regions based on both the spatial and the temporal domains for GPS data points in trajectories. The density is measured by cost function to analyze the density distribution of a trajectory. The cost function reflects the local configuration of the trajectory points in spatial-temporal data space. In light of candidate semantic regions, we further propose shared nearest neighbor (SNN) clustering to extract frequent semantic regions from a set of candidate semantic regions. Our approach is nonexclusive to be applied in many different activity scenarios, not being to one single application. Our

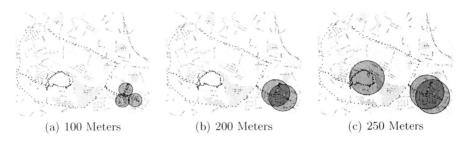

| (a) 100 Meters | (b) 200 Meters | (c) 250 Meters |

Fig. 2. An example of regions discovered by the stay point approach

experiments demonstrate that semantic regions can be extracted precisely as well as efficiently. The main contributions of this paper are summarized below.

– The scheme of region extraction is proposed for effectively and precisely semantic region discovery.
– We propose a sequential density based clustering method to discover semantic regions from a trajectory. The clustering method takes both spatial and temporal domain into account.
– We define the similarity between semantic regions and develop a shared nearest neighbor based clustering algorithm to discover frequent semantic regions from trajectory dataset.
– We present comprehensive experimental results over various real datasets. The results demonstrate that our techniques are more accurate than existing clustering schemes.

The remainder of this paper is organized as follows. Section 2 reviews the related literature. Our framework of mining semantic regions is proposed in section 3. In section 4, we evaluate our framework by real trajectory datasets. Finally, section 5 concludes this paper.

2   Related Work

Hot region detection has been widely used in the field of trajectory data analysis such as trajectory pattern mining [17,2,6,11,8], moving objects' location prediction [10,18], location-based activity discovery [14,13,21] and so on. Most of proposed methods employ density based clustering techniques to group a set of trajectory points into a cluster as a region, such as DBSCAN [4] and OPTICS [1]. In density based clustering, clusters are regions of high density separated by regions of low density. Based on density based clustering algorithms [6,11,10,18], the regions are extracted only according to the density in spatial domain without considering the density in temporal domain. Giannotti et. al. [6] adopted grid density clustering to discover popular regions as ROIs where dense cells in space are detected and merged if they are neighbors. It implies that popular regions can be extremely large. Thus, they have to give additional constrains to select significant and limited regions to represent ROIs. The authors in [11] extracted frequent regions by applying the clustering method DBSCAN. In spite of advantage of DBSCAN that clusters in arbitrary shape can be detected, they have to decompose a cluster when it is too large to describe correlations between frequent regions. The hybrid location prediction model proposed by [10] that divides a trajectory into several periodic sub-trajectories. Then, frequent regions of the same time offset are extracted by using DBSCAN to cluster locations from sub-trajectories.

For the purpose of knowledge discovery of the ROIs which contains activity-related meaning to users, few existing works [12,20,21] aimed to applying sequential constraint to a single sequence. The authors in [12,20,21] proposed a stay point and claimed that can stand for a geographic region and carry a particular semantic meaning. A stay point is the mean point of a sub-sequence where the consecutive points of a examined point do not exceed the distance threshold during the period of time threshold. Each stay point contains information about mean coordinates, arrival time and leaving time. In addition, the authors in [21] proposed stay regions extracted from stay points via grid based clustering algorithm.

All of the above techniques have some deficiencies for discovery ROIs from trajectory data. First, traditional clustering approach only considers similarity in one domain, i.e.,the spatial domain only. They have focused on geometric properties of trajectories, without considering the temporal information or sequential relation. The region extraction for semantic analysis has to consider both spatial and temporal domains. Second, applying a universal density threshold for cluster discovery may either miss regions with different density or merge non-related regions. In this paper, the challenge is that trajectories may consist of different activities and each activity has different distance density and speed distribution. We want to extract significant and precise regions with semantic meaning from trajectories and these regions can imply certain activities of moving objects by spatial-temporal clustering approach.

3   A Framework of Mining Semantic Regions

3.1   Overview

We propose an effective and precise algorithm to discover semantic regions from trajectory data based on spatial-temporal density model and sequential density clustering, and we develop a shared nearest neighbor based clustering method to discover frequent semantic regions from multiple trajectories. Figure 3 outlines the framework for semantic region discovery. On the process of semantic region discovery, each trajectory is first partitioned into a set of trajectory segments. The spatial-temporal density of each segment is computed by cost function. Then, the sequential density clustering is applied to sequentially group the segments with similar density. The region where users may have some kinds of activities locates in the cluster with local maxima density. Finally, while each trajectory is transformed into a sequence of semantic regions, a set of similar semantic regions is clustered to indicate the major frequent semantic regions from multiple trajectories.

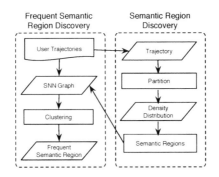

Fig. 3. Overview of extracting semantic regions

3.2   Problem Formulation

Given a trajectory dataset of a moving object, our algorithm generates a set of clusters as semantic regions of each trajectory and a set of frequent semantic regions from the trajectory dataset. An object's trajectory is represented as a sequence of points $\{p_1, p_2, ..., p_i, ..., p_n\}$. Each point $p_i (1 \leq i \leq n)$ contains location (x_i, y_i) and timestamp (t_i). A trajectory can be partitioned into continuous segments $\{s_1, s_2, ..., s_l, ..., s_m, ...\}$ according to user-defined parameter T. Let T be an integer called period of activity that is the minimum duration of activity proceeding time we are interested in. For example, T can be set to 30 minutes for sightseeing at an interesting spot or 2 hours for exercising at the gym.

A semantic region is a spatial-temporal based cluster and is denoted as SR. The cluster C_k is a set of trajectory segments $\{s_l, s_{l+1}, s_{l+2}, ..., s_m\}$, where $m \geq l$. The cluster C_k is a semantic region if (1) the stay duration of each segment in C_k is not less than T (i.e. $|t_j - t_i| \geq T$) (2) and spatial-temporal density of C_k is higher than that of its adjacent clusters (C_{k-1} and C_{k+1}) by a predefined threshold ξ. A frequent semantic region is a representative region which indicates that this region appears in a sufficient number of trajectories. Such a sufficient number is defined as MinSR.

The spatial-temporal density of a segment mentioned above is defined to reflect the local configuration of the points in the spatial-temporal data space and a cost function is used as a density measurement. Generally, the cost function is designed to represent the penalty of dissimilarity of the points within a segment. Previous work [15] defines the cost of a segment as the sum of squared Euclidean distance between points and its spatial centroid, where the cost is also called the variance of the segment. Without loss of generality, the squared Euclidean distance function is adopted as given below to measure the dissimilarity between two points.

$$D_{E^2}(p_i, p_j) = (x_i - x_j)^2 + (y_i - y_j)^2. \tag{1}$$

However, it only counts the spatial dissimilarity without considering the temporal feature such as the duration of a moving object staying in a location or lingering around some places. Our main idea of this research is to extract the region with semantic information where involves some activities of user. Because a trajectory does not involve only one activity in real world, the distance between location points can vary with different activities in spatial domain and the temporal interval from a point p_i to its succeeding point p_{i+1} can vary from seconds to hours. Furthermore, most location-acquisition technologies cannot localize and record current location under some condition. For example, when a GPS-embedded object enters a building or a cave, the GPS tracking device will lose satellite for a time interval until coming back outside and few points are recorded on such place. If we directly measure the spatial dissimilarity of the segment around this area, we cannot detect its significance. It implies that both spatial and temporal feature can affect the result of semantic region discovery. Thus, we take temporal feature as a weight and compounded with spatial relation to measure the dissimilar cost of a segment, i.e, the spatial-temporal density of a segment.

Given a segment $s_l = p_i, p_{i+1}, ...p_j$, the definition of weighted cost function is stated as follows.

$$Cost(s_l) = \frac{\sum_{k=i}^{j} w_k * D_{E^2}(p_k, c)}{\sum_{k=i}^{j} w_k}, \tag{2}$$

$$c = (\frac{\sum_{k=i}^{j} w_k * x_k}{\sum_{k=i}^{j} w_k}, \frac{\sum_{k=i}^{j} w_k * y_k}{\sum_{k=i}^{j} w_k}), \tag{3}$$

$$w_k = \frac{(t_k - t_{k-1}) + (t_{k+1} - t_k)}{2} \tag{4}$$

where w_k is the weight of point p_k, c is the weighted centroid of segment s_l, respectively. Because there are different activities processing in a trajectory, the $Cost(s_l)$ can vary in a wide range. To normalize the density of clusters with different activities, the density function is measured as the logarithm of one over the cost. The definition of density function is stated as follows.

$$Density(s_l) = \log_e(1 + \frac{1}{Cost(s_l) + \gamma}), \tag{5}$$

where the $Density(s_l)$ is in the boundary of $[0, log_e(1 + \frac{1}{\gamma})]$ and γ is a constant (in order to keep the maximum density equals to 1, γ is given as $\frac{1}{e-1}$ in this paper).

3.3   Discovering Semantic Regions

3.3.1   Trajectory Partition

Given an object's trajectory $\{p_1, p_2, ..., p_i, ..., p_n\}$, we aim to analyze its spatial-temporal density distribution to extract the region where the trajectory movement is more dense than the neighboring regions, i.e. the density in this region is a local maximum in the trajectory density distribution. Unlike the problem in [15], we are not pursuing to partition a trajectory such that the total cost of partitioned segments is minimized. Instead, we partition the trajectory in order to compare the density variance between sequent segments in an efficient way. To simplify the description of the spatial-temporal density distribution of a trajectory, each trajectory is periodically partitioned into $\lfloor \frac{p_n.t - p_1.t}{T} \rfloor$ trajectory segments, where T is a period of activity, i.e., a minimum duration of activity we are interested in and the density of each sequential segment is computed. We assume such a sequential density set can be used to describe the density distribution of the trajectory.

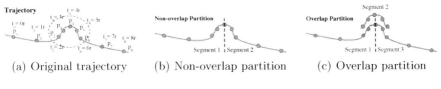

(a) Original trajectory   (b) Non-overlap partition   (c) Overlap partition

Fig. 4. Overlap partition

However, while the density distribution of a trajectory can be represented by the density distribution of a sequence of segments, it may occur a partition loss that a dense region of a trajectory is lost because of partition. A dense region may be separated into several segments because of partitioning. Under this condition, the density of each segment may become smaller than original density of the dense region. As shown in Figure 4(a), the dense region of a trajectory is in the center, marked within a circle. Given the time interval of each point to its neighbor point is r and the period T is set as $4r$, the partitioned segments are S_1 and S_2 shown in Figure 4(b). As a result, the dense region in the center of this trajectory is split into two segments and the dense region cannot be detected.

To solve this problem, overlapping partition is implemented to smooth the region-split property when partitioning the trajectory. The time interval of the trajectory in Figure 4(c) is set as $[0, 4r), [2r, 6r), [4r, 8r)$ corresponding to segment S_1, S_2, S_3, respectively. The time interval function of overlapping partition is given as follows. $[t_{start_k}, t_{end_k}) = [\frac{(k-1)*T}{fold}, \frac{(k-1)*T}{fold} + T)$, where $fold$ is a parameter to smooth the partition. In this paper, $fold$ can be set as a fixed integer and our experiment shows the result change slightly when $fold \geqslant 3$.

3.3.2   Sequential Density Clustering Algorithm

We now present our sequential density clustering algorithm for semantic region discovery. Given a set of sequential trajectory segments S, our algorithm generates a set of clusters as semantic regions. We define a cluster as a sequential density-connected set. It requires a parameter ξ, a density threshold for similarity measurement. Before clustering, each density D_k of partitioned segment S_k is calculated by spatial-temporal cost. In a trajectory density distribution, a segment with a local maximum can correspond to a dense region of a trajectory and the segments with similar density are grouped into a cluster if they are adjacent to each other. Finally, the boundary of a semantic region, i.e. a cluster, is extracted at where the density dramatically change. Thus, the semantic region discovery involves grouping the segments (if they belong to the same dense region) and setting boundary of dense region.

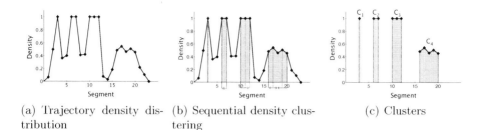

(a) Trajectory density distribution

(b) Sequential density clustering

(c) Clusters

Fig. 5. Sequential density clustering over trajectory density distribution

---

Algorithm 1. Sequential Density Clustering

Input: A set of trajectory segments S, a density threshold ξ
Output: A set of sequential density cluster SDC

1 Compute $D = Density(S)$ for each segment in S
2 Sequentially record the local max and local min from D to an array E
3 **foreach** *local max E_i in E* **do**
4   Take nearby local min and E_i as a group $G : \{E_{i-1}, E_i, E_{i+1}\}$ and take the local max of last group in GroupSet as E_{last}
5   **if** $|E_i - E_{i-1} \leqslant \xi|$ *and* $|E_{last} - E_{i-1} \leqslant \xi|$ **then**
6   Merge G with last group in GroupSet
7   **end**
8   **else**
9   Add G into GroupSet
10   **end**
11 **end**
12 **foreach** *group G in GroupSet* **do**
13   **foreach** *D_j from local max to local min in G* **do**
14   **if** $|D_j - D_{j-1}| \leqslant \xi$ *and* $(|D_j - D_{j+1}| \leqslant \xi)$ **then**
15   Add S_j into of a density cluster C
16   **end**
17   **end**
18   **if** $boundary(C) \neq boundary(G)$ **then**
19   Add cluster C into SDC
20   **end**
21 **end**
22 **return** SDC

For instance, we set $T = 10$, $fold = 2$ to partition the trajectory in Figure 1(a) into 23 segments and compute the density for each segment. The density distribution D of the periodically two-fold-partitioned trajectory is shown in Figure 5(a). Such a sequence of density D is the input of the algorithm. Algorithm 1 shows the sequential density clustering to extract semantic regions from the density distribution. Initially, the local extremes (maxima and minima) are identified and recorded as a set of group G. Each G is a group of local maxima E_i and its nearby local minima E_{i-1} and E_{i+1}, i.e., $G : \{E_{i-1}, E_i, E_{i+1}\}$.

The algorithm consists two steps. In the first step (Line 3-11), the algorithm computes the density similarity between two adjacent groups. If density difference between two adjacent groups is equal to or smaller than density threshold ξ, these groups are sequentially similar. The algorithm performs the clustering to merge them into a new group. For example, there are two connected groups $G_1 : \{D_{14}, D_{17}, D_{18}\}$ and $G_2 : \{D_{18}, D_{19}, D_{23}\}$ in Figure 5(b). Given $\xi = 0.1$, G_1 and G_2 are similar ($|D_{17} - D_{18}| \leqslant \xi$ and $|D_{19} - D_{18}| \leqslant \xi$) and can be merged into a new group $G' : \{D_{14}, D_{17}, D_{23}\}$. The clustering results are added to $GroupSet$ as a sequence of groups. In the second step (Line 12-21), the boundary of a cluster is extracted from each group G. The precise boundary of a cluster C is extended from the local maximum in G to its nearby local minima until the density difference between two continuous segment is more than ξ. The cluster $C_4 : \{S_{16}, S_{17}, S_{18}, S_{19}, S_{20}\}$ is extracted from group $\{D_{14}, D_{17}, D_{23}\}$ as shown in Figure 5(c). Only regions with significant change in density are taken as semantic regions. If there are no continuous density changes more than ξ inside a group, this implies the region enclosed in the group can be viewed as an non-semantic area.

3.4   Mining Frequent Semantic Regions

While semantic regions represent the location where a moving object proceeds with some kind of high dense activities in duration of time from a trajectory, it does not imply that those semantic regions are an object's "frequently" appearing at. Thus, given a set of trajectory data, we want to find out the region where an object frequently stays or lingers around for a certain activity, i.e, a frequent semantic region. A frequent semantic region is a summary of a set of similar semantic regions from different trajectories. To define the similarity between semantic regions and discover the frequent semantic regions, we adopt the definition of shared nearest neighbor (SNN) [9] and SNN density-based clustering [3]. That is, the similarity between a pair of points is measured by the number of their shared nearest neighbors. In graph terms, a link is created between a pair of nodes if both have each other in their K nearest neighbor (KNN) lists and an SNN similarity graph is created. Clusters are simply the connected components of the SNN graph. The discovery of frequent semantic regions is similar to find clusters. For each semantic region, it can be viewed as a node in SNN graph. However, if nodes are not close enough, they do not stay in the same region apparently. When applying SNN density based clustering to discover frequent semantic regions, we constrain the searching range of nearest neighbors is a radius D_h around the examined nodes. We define a semantic region is a frequent semantic region if each semantic region of which contains at least $MinSR$ number of neighbors in the distance radius D_h. The nodes without $MinSR$ nearest neighbors are viewed as non-frequent regions and discarded. All the connected components in the resulting graph are clusters finally. These clusters can be considered as frequent semantic regions where an object often visits for certain activities.

Algorithm 2. Frequent Semantic Region Discovery Algorithm

Input: A set of nodes, distance threshold D_h, minimum support $MinSR$
Output: a set of clusters

1 Find the $MinSR$-nearst neighbors in D_h of all nodes.
2 Construct the shared nearest neighbor similarity graph.
3 For every node in the graph, calculate the number of links.
4 Identify core nodes which has more or equal to $MinSR$ links.
5 Identify noise nodes which is neither a core node nor linked to a core node and remove them.
6 Take connected components of nodes to form clusters.
7 **return** the union of all clusters

We develop a frequent semantic region discovery algorithm (Algorithm 2) based on the property described in new SNN clustering algorithm [3]. The nodes that have at least $MinSR$ connectivity in the SNN graph are candidates for core nodes since they tend to be located well inside the natural cluster, and the nodes with connectivity lower than $MinSR$ and not connected to any core node are identified as noise nodes. As a result, a cluster is detected if there exists a connected component in SNN graph. The cluster is regarded as a frequent semantic region. For each semantic region which has at least $MinSR$ similar semantic regions, it will be included in a frequent semantic region. Notice that the number of clusters is not considered as a parameter. Depending on the nature of the data, the algorithm finds the nature number of clusters for given set of parameters, $MinSR$ and D_h.

4   Experiments

The experiments in this study are designed for two objectives. First, we compare the semantic region coverage of our method, Sequential Density Clustering (SDC), with Stay Point (SP) that is the method considering the sequential constraint in literature. Second, we verify the accuracy of frequent semantic region discovery. We conduct experiments on our prototype which was implemented in the python language on CarWeb [16], a traffic data collection platform on Ubuntu 9.10 operating system.

Table 1. Dataset of each activity in California

Activity	# Trajectory	# Photo
Hiking	3839	33065
Road Biking	5032	11968
Walking	955	4685

We evaluate the experiment with real dataset from EveryTrail [5] in California. Each data includes an labelled activity trail (a trajectory) and a set of photos with geographic information where are taken by user. We assume the ground truth that location with photo is where the activity happen at. Each photo represents a interesting of the user (photo taker) and each region containing the photos can be considered as a interesting (semantic) region. Three kind of activity (Hiking, Road Biking, Walking) in California are selected. The major difference between each activity is the average speed (Road Biking > Walking > Hiking). Table 1 shows the total number of trajectories and photos for each activity.

4.1 Evaluation of Semantic Regions

In order to evaluate the effectiveness of semantic region discovery, we compare the semantic region coverage of SDC with that of SP under varying conditions. A semantic region coverage is measured as the hit ratio of the photos enclosed by discovered region to total photos for each activity. We set SDC parameters as follows: the partition smoothing parameter $Fold = 3$ and the density threshold $\xi = 0.02$ for all datasets. There are two parameters setting for SP: distance and time thresholds. For comparison with SP fairly, the dynamic size of a sematic region is constrained as a fixed size of SP. Thus, We set various distance thresholds (100, 200, 300 meters) of stay point as the radius of the region around stay point and also as the radius around mean point discovered via SDC. In additional, we compare above regions of fixed size with the regions of dynamic size discovered via SDC. The time threshold of SP is set as the period of activity for SDC and varied from 5 minutes to 30 minutes.

For each activity, in Figure 6, the hit ratio of our method is much higher than that of SP. As expected, SDC shows the coverage of discovered region with dynamic size is better than that with fixed size while the average size (the size number marked with SDC curve in Figure 6) is smaller than the fixed size, especially in datasets of slow-speed activity (Hiking and Walking). It implies that our method is adaptive to various shape and size of discovered region. Besides, the hit ratio is much lower in high-speed activity than in low-speed activity, since the semantic region is much harder to be obtained when the activity has higher average speed and has many sudden changes of direction or speed. These results prove that using SDC for semantic region discovery is obviously more precise than using SP under different average speed. Another observation found in the results demonstrates that hit ratio decreases when the period of activity (time threshold) increases. Because the period of activity is a user-defined parameter which indicates the minimum duration of an activity, the semantic regions with the period of activity which is shorter than minimum requirement will be ignored when the activity is expected to keep running longer.

4.2 Accuracy of Frequent Semantic Regions

To show the accuracy of frequent semantic region discovery, we obtained a user's trajectory over one week and labeled the top five frequent semantic regions. We then generated 1000 different trajectory dataset each have 100 similar trajectories to the original trajectory. For each trajectory, we set the period of activity

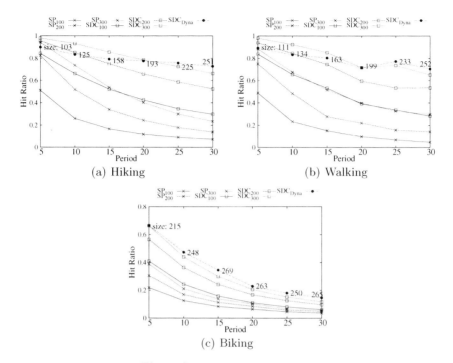

Fig. 6. Semantic region coverage

$T = 10$ minutes to discover semantic regions. We take the semantic regions as nodes in a $5 * 5$ map. Frequency and radius of each frequent semantic regions are stated in Table 2.

Table 2. Dataset of frequent semantic regions

# Region	Frequency	Radius
2	50%	0.5
2	80%	0.1
1	30%	0.8

We take F-measure to analyze the accuracy of discovered frequent semantic regions. Precision is defined as the overlapped area discovered in labelled regions divided by the total discovered area, and recall is defined as the overlapped area discovered in labelled regions divided by the total area of existing labelled regions. The definition of F-measure is the harmonic mean of precision and recall:

$$F = 2 * \frac{precision * recall}{precision + recall}$$

A higher precision score means the higher representative of discovered regions while a higher recall score means the higher coverage of labelled regions. Although a larger region can cover more labelled regions and obtain high recall, it

Table 3. Impact of minimum support

$MinSR$	Precision	Recall	F-measure
10%	0.856 ± 0.068	0.995 ± 0.013	0.919 ± 0.04
20%	0.903 ± 0.065	0.884 ± 0.005	0.894 ± 0.036
30%	0.916 ± 0.066	0.441 ± 0.074	0.592 ± 0.068

is hard to distinguish these labelled regions and results in low precision. In Table 3, we fix the radius D_h as 0.5 and report the performances of our model under different minimum support ($MinSR$) requirement for a frequent semantic region. The entry value in Table 3 denotes the mean and standard deviation of precision, recall and F-measure. As shown in the table, our method can achieve high precision under different $MinSR$. However, when the requirement of $MinSR$ increases, it is much harder to find regions with low frequency in a large radius.

5   Conclusion

In this paper, we propose the concept of semantic region that indicates regions along with trajectories where users may proceed with some activities. First, spatial-temporal cost is introduced to model the density distribution of a trajectory. Then, we adopt a sequential density clustering algorithm to extract the semantic regions. Based on semantic region discovery, we define the similarity between semantic regions and devise a SNN based clustering algorithm to discover frequent semantic regions from multiple trajectories. Finally, to show the preciseness and effectiveness of our framework, we present comprehensive experimental results over various real datasets. The results demonstrate that our framework is able to accurately extract semantic regions.

Acknowledgments. Wen-Chih Peng was supported in part by ITRI-NCTU JRC program (No. 99EC17A05010626), Microsoft, D-Link, National Science Council (No. NSC-97-2221-E-009-053-MY3) and Taiwan MoE ATU Program.

References

1. Ankerst, M., Breunig, M.M., Kriegel, H.P., Sander, J.: OPTICS: Ordering Points To Identify the Clustering Structure. In: SIGMOD, pp. 49–60 (1999)
2. Cao, H., Mamoulis, N., Cheung, D.W.: Mining Frequent Spatio-Temporal Sequential Patterns. In: ICDM, pp. 82–89 (2005)
3. Ertoz, L., Steinbach, M., Kumar, V.: A New Shared Nearest Neighbor Clustering Algorithm and its Applications. In: 2nd SIAM International Conference on Data Mining (2002)
4. Ester, M., Kriegel, H.P., Sander, J., Xu, X.: A Density-Based Algorithm for Discovering Clusters in Large Spatial Databases with Noise. In: KDD, pp. 226–231 (1996)
5. Everytrail – gps travel community, http://www.everytrail.com

6. Giannotti, F., Nanni, M., Pinelli, F., Pedreschi, D.: Trajectory Pattern Mining. In: KDD, pp. 330–339 (2007)
7. Hung, C.C., Chang, C.W., Peng, W.C.: Mining Trajectory Profiles for Discovering User Communities. In: GIS-LBSN, pp. 1–8 (2009)
8. Hung, C.C., Peng, W.C.: Clustering Object Moving Patterns for Prediction-Based Object Tracking Sensor Networks. In: CIKM, pp. 1633–1636 (2009)
9. Jarvis, R.A., Patrick, E.A.: Clustering Using a Similarity Measure Based on Shared Near Neighbors. IEEE Trans. Comput. 22(11), 1025–1034 (1973)
10. Jeung, H., Liu, Q., Shen, H.T., Zhou, X.: A Hybrid Prediction Model for Moving Objects. In: ICDE, pp. 70–79 (2008)
11. Jeung, H., Shen, H.T., Zhou, X.: Mining trajectory patterns using hidden markov models. In: Song, I.-Y., Eder, J., Nguyen, T.M. (eds.) DaWaK 2007. LNCS, vol. 4654, pp. 470–480. Springer, Heidelberg (2007)
12. Li, Q., Zheng, Y., Xie, X., Chen, Y., Liu, W., Ma, W.Y.: Mining User Similarity Based on Location History. In: GIS (2008)
13. Liao, L., Fox, D., Kautz, H.A.: Location-Based Activity Recognition. In: NIPS (2005)
14. Liao, L., Fox, D., Kautz, H.A.: Location-Based Activity Recognition using Relational Markov Networks. In: IJCAI, pp. 773–778 (2005)
15. Lin, C.R., Chen, M.S.: On the Optimal Clustering of Sequential Data. In: SDM (2002)
16. Lo, C.H., Peng, W.C., Chen, C.W., Lin, T.Y., Lin, C.S.: CarWeb: A Traffic Data Collection Platform. In: MDM, pp. 221–222 (2008)
17. Mamoulis, N., Cao, H., Kollios, G., Hadjieleftheriou, M., Tao, Y., Cheung, D.W.: Mining, Indexing, and Querying Historical Spatiotemporal Data. In: KDD, pp. 236–245 (2004)
18. Monreale, A., Pinelli, F., Trasarti, R., Giannotti, F.: WhereNext: a Location Predictor on Trajectory Pattern Mining. In: KDD, pp. 637–646 (2009)
19. Yang, J., Hu, M.: TrajPattern: Mining sequential patterns from imprecise trajectories of mobile objects. In: Ioannidis, Y., Scholl, M.H., Schmidt, J.W., Matthes, F., Hatzopoulos, M., Böhm, K., Kemper, A., Grust, T., Böhm, C. (eds.) EDBT 2006. LNCS, vol. 3896, pp. 664–681. Springer, Heidelberg (2006)
20. Zheng, Y., Zhang, L., Xie, X., Ma, W.Y.: Mining Interesting Locations and Travel Sequences From GPS Trajectories. In: WWW, pp. 791–800 (2009)
21. Zheng, Y., Zhang, L., Xie, X., Ma, W.Y.: Collaborative Location and Activity Recommendations with GPS History Data. In: WWW, pp. 26–30 (2010)

STS: Complex Spatio-Temporal Sequence Mining in Flickr*

Chunjie Zhou and Xiaofeng Meng

School of Information, Renmin University of China, Beijing, China
{lucyzcj,xfmeng}@ruc.edu.cn

Abstract. Nowadays, due to the increasing user requirements of efficient and personalized services, a perfect travel plan is urgently needed. In this paper we propose a novel complex spatio-temporal sequence (STS) mining in Flickr, which retrieves the optimal STS in terms of distance, weight, visiting time, opening hour, scene features, etc.. For example, when a traveler arrives at a city, the system endow every scene with a weight automatically according to scene features and user's profiles. Then several interesting scenes (e.g., $o_1, o_2, o_3, o_4, o_5, o_6$) with larger weights (e.g., $w_1, w_2, w_3, w_4, w_5, w_6$) will be chosen. The goal of our work is to provide the traveler with the optimal STS, which passes through as many chosen scenes as possible with the maximum weight and the minimum distance within his travel time (e.g., one day). The difficulty of mining STS lies in the consideration of the weight of each scene, and its difference for different users, as well as the travel time limitation. In this paper, we provide two approximate algorithms: a local optimization algorithm and a global optimization algorithm. Finally, we give an experimental evaluation of the proposed algorithms using real datasets in Flickr.

Keywords: spatio-temporal, sequence, Flickr, approximate.

1 Introduction

With the rapid development of modern society, people are concentrating more on efficient and personalized services. In the tourist industry, a perfect traveling plan can help people to visit their favorite scenes as many as possible, and save a lot of time and energy. However, at present it is hard for people to make a proper and personalized traveling plan. Most of them follow other people's general travel trajectory, but do not consider their own profile and the best visiting order of scenes in this trajectory. So only after finishing their travel, do they know which scene is their favorite, which is not, and what is the perfect order of visits. Let's consider such a scenario: a person plans to travel on a holiday, but does not have a specific destination. In order to make a better plan, they scan the tourist routes on the Internet or they seek advice from travel companies. Then they choose a popular travel trajectory suggested by other people, but do not consider their own interests. As a result, this sightless travel plan may cause the following aftereffects: 1) waste a lot of time on the road among scenes; 2) waste lots of

---

* This research was partially supported by the grants from the Natural Science Foundation of China (No.60833005, 61070055, 61003205); the National High-Tech Research and Development Plan of China (No.2009AA011904).

J.X. Yu, M.H. Kim, and R. Unland (Eds.): DASFAA 2011, Part I, LNCS 6587, pp. 208–223, 2011.
© Springer-Verlag Berlin Heidelberg 2011

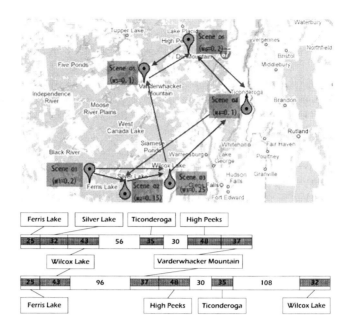

Fig. 1. Two different sequences (red, blue) including six chosen scenes with different weights on Google Maps and their timelines. Scenes of each sequence which have been visited are presented in the order of visitation. On the timelines, transitions between tourist scenes are depicted by rectangles and visiting time is given in minutes.

unnecessary money; 3) do not have enough time to visit their real favorite scenes, etc. However, with increasing interests in perfect travels and modern advanced services, more wonderful and personalized travel suggestions need to be supplied urgently.

In this paper, we propose a novel spatio-temporal sequence (STS) mining in Flickr. The goal of STS is to provide a user with the optimal STS, which minimizes the total traveling distance and maximizes the total weight within his limited travel time. In consequence a user can make a perfect and personalized travel plan based on his own profile before he starts to travel. The implementation of STS is based on two basic pieces of knowledge. 1) the user's profile. The methods of mining user's profile have been studied a lot [1]. Here we assume that every user's profile is already stored in his accompanied mobile devices. So when a user arrives at a city, the system can get his profile from his mobile devices directly. 2) scene features. The methods of mining scene features according to photos and tags in Flickr has been studied in our previous work [6]. So here we directly use the database in [6] that stores the features of each scene in each city. In this paper, we consider solutions for the STS problem as graphs where tourist scenes are nodes and paths between these scenes are edges. To the best of our knowledge, no prior work considers both point weight and edge distance together, which are inversely proportional.

Efficient STS evaluation could become a new important feature of advanced services in Flickr, and be useful for LBS (Location Based Services). The quality of these services can be greatly improved by supporting more advanced query types, like STS. An example of STS is shown in Figure 1. When a traveler arrives at a city, the system endows every scene with a weight automatically according to scene features and user's profiles. Then six interesting scenes (e.g., $o_1, o_2, o_3, o_4, o_5, o_6$) with larger weights (e.g., $w_1, w_2, w_3, w_4, w_5, w_6$) will be chosen. Given these six certain scenes, a database that stores the features of scenes will compute a proper STS (i.e., the red sequence) efficiently. The blue sequence is created by a user who follows others' general trajectory. These two different sequences are presented in Figure 1. Tourist visits cover not only the visiting time of scenes, but also the transitional cost among scenes (i.e., rectangles on the timelines)[16]. Because of the different visiting order of the scenes, the duration of these two sequences is variable, from under six hours (e.g., the red sequence) to over 9 hours (e.g., the blue sequence). Clearly, an ideal method is to propose an optimal tourist sequence that not only re-arranges the order in which these scenes are visited with the maximum value (see Section 2.1), but also should be within a limited travel time. For the red sequence, there are only two significant transitions (i.e., 56 min between Wilcox Lake and Ticonderoga, 30 min between Ticonderoga and High Peeks). The first three visited tourist scenes (Ferris Lake, Silver Lake and Wilcox Lake) and the last two ones (High Peeks and Varderwhacker) are adjacent and the tourist makes no pause between them. Some factors influencing visiting order, for example, personal preferences (i.e., scene weight) and travel time must be taken into account.

STS can be considered as a special case of the knapsack problem (KSP) which is NP-hard. The reduction from STS to KSP is straightforward. Given a set of m scenes from which we select some interesting scenes to be included in the spatio-temporal sequence in a limited travel time T_{total}. Each scene has a weight and a duration time. The objective is to choose the set of ordered scenes that optimize the STS and maximize the travel value (i.e., a minimum distance and a maximum tourist scene weight). By regarding distance as the multiplication of velocity and time, each scene of STS can be reduced to an item of KSP. There are also some differences between STS and KSP: 1) the goods in KSP is disordered, but the scenes in STS is strictly ordered; 2) KSP has only one objective function, but STS has two independent objective functions.

Contributions: This paper proposes a novel spatio-temporal sequence mining in Flickr and studies methods for solving it efficiently. Two approximate algorithms that achieve both local and global optimization are presented. In particular:

- We present a novel STS mining in Flickr, which can minimize the total traveling distance and maximize the total weight within a limited travel time. This type of mining has not been considered before.
- We give a formal definition of STS in road network. The weights of chosen scenes are specified according to personal preferences. This is more similar to the real world applications.
- We propose two types of algorithms for STS. Local optimization algorithms include approximation in terms of distance, weight and value respectively.
- We perform an extensive experimental evaluation of the proposed algorithms on real datasets in Flickr.

Paper Organization: The rest of the paper is organized as follows: Section 2 gives the problem definition and related works. The approximate algorithms are presented in Section 3. An experimental evaluation of the proposed algorithms using real datasets is presented in Section 4. Finally, we give the conclusion and future works.

2   Preliminaries

This section formally defines the STS problem and introduces the basic notation that will be used in the rest of the paper. Furthermore, a concise overview of related works is presented.

2.1   Problem Definition

Table 1 lists the main symbols we use throughout this paper. Following standard notation, we use capital letters for sets (e.g., P is a set of all scenes), and lowercase letters for vectors (e.g., o_i).

Table 1. Symbols

Symbol	Definition and Description
V	the value of the sequence $\{ o_1,...,o_k \}$
P	a set of all scenes $\{ P_1,...,P_m \}$
R	a chosen subset of scenes $\{ o_1,...,o_k \}$
Q	a travel sequence
N	the road network
T_{total}	the total travel time
$T(o_i,o_j)$	the time cost from the scene o_i to o_j
$T(o_i)$	the duration time of scene o_i
D_{dis_N}	the distance among scenes in road network
w_i	the weight of the chosen scene o_i
o_i	the ith chosen scene
α	a balance factor
m	the number of all scenes
k	the number of chosen scenes

We consider solutions for the STS problem as graphs where tourist scenes are nodes (labeled with scene's name) and paths between these scenes are edges. Given a graph $G(O,\xi)$ with n nodes $O= \{o_1,...,o_n\}$ and s edges $\xi = \{e_1,...,e_s\}$, each node in the graph has a weight denoting the interest percentage of the traveler. The value of traversing a scene sequence $(o_i,...,o_j)$ is expressed as $V(o_i,...,o_j) \geq 0$, which means the score of the sequence. As shown in Eq.(1), the value of the sequence is in proportion to the total weight of all chosen sences, but in contrast to the total distance. Here, we consider distance among scenes in road network, which is more meaningful in the real travel application scenario. Suppose that the average velocity of the traveler is υ, the distance $D_{dis_N}(o_i,o_j)$ can be denoted as $\upsilon*T(o_i,o_j)$. A balance factor α is defined between weight and distance, which may be changed in different situations.

$$V(o_i,...,o_j) = \alpha * (w_i + ... + w_j) + (1-\alpha) * \frac{1}{\sum_{k=i}^{j} D_{dis_N}(o_k, o_{k+1})}$$

$$= \alpha * \sum_{k=i}^{j} w_k + (1-\alpha) * \frac{1}{\sum_{k=i}^{j} v * T(o_k, o_{k+1})} \tag{1}$$

Given a set of m scenes $P = \{P_1,...,P_m\}$ (where $m \leq n$) and a mapping function π: $P_j \rightarrow o_i$ that maps each scene $P_j \in P$ to a node $o_i \in N$. So scenes can be regarded as special nodes, and be denoted by node symbols. In the rest of this paper, scenes and nodes will share the same symbols. The STS problem can be defined as follows:

Definition 1. *Given a set $R \subseteq P$ ($R= \{o_1, o_2, ..., o_k\}$), a source scene S and a destination scene E, identify the traveling scene sequence $Q= \{S, o_1, ..., o_k, E\}$ from S to E that visits as many scenes in R as possible (i.e., maximize the total weight of the sequence), and takes the minimum possible distance $D_{dis_N}(Q)$ (i.e., for any other feasible sequence Q' satisfying the condition $D_{dis_N}(Q) \leq D_{dis_N}(Q')$) in a limited travel time T_{total}.*

The time constraint condition is in the following, which includes not only the duration time of scenes, but also transitional cost among scenes. The duration time of scenes can be achieved by [16], which is not our focus in this paper. We mainly consider the temporal cost among scenes, namely, the distance between each couple of scenes.

$$\sum_{i=1}^{k} T(o_i) + \sum_{i=1, j=i+1}^{k} T(o_i, o_j) \leq T_{total} \tag{2}$$

2.2   Related Work

Rattenbury et al. [2] was an early attempt to discover both event and place names from Flickr geolocated textual metadata, resulting in an application [3] for geographic image retrieval, with representative and popular tags overlaid on a scalable map. Quack et al. [4] downloaded 200,000 georeferenced Flickr images from nine urban areas and clustered them using local image descriptors to discover place names and events, linking some places to their Wikipedia articles. In contrast to [4] and [5], we do not limit ourselves to geographic information of photographs since temporal information are also important for mining STS. Elsewhere [6], we detailed methods for mining the features of each scene in each city. Here we exploit these same results but shift our focus towards mining spatio-temporal sequence according to personal preferences.

Zheng et al. [7] recorded GPS tracks of 107 users during a year to determine the interestingness of tourist scenes. Cao et al. [8] presented techniques capable of extracting semantic locations from GPS data. The authors of [9] also focused on mining similar traveling sequences from multiple users' GPS logs while the authors of [10] retrieved maximum periodic patterns from spatio-temporal metadata. Girardin et al. [11] analyzed the tourist flows in the Province of Florence, Italy, based on a corpus of georeferenced Flickr photos and their results contribute to understanding how people travel.

Chen et al. [12,13] studied a problem of searching trajectories by locations, and the target was to find the K best-connected trajectories from a database such that it connected the designated locations geographically. None of these approaches considers scene features combined with user's profile, which are central pieces of our approach. Whereas [7] or [9] relied on accurate GPS traces for small scale regions and obtained from a relatively reduced number of users. Flickr data is noisy, but covers most interesting tourist regions of the world. As a result, we are able to propose itineraries in any region of the world that is sufficiently covered by Flickr data.

Visiting duration is an important characteristic of trips and it is classically estimated by domain experts [14]. The automatic extraction of visiting duration from Flickr metadata was only recently explored [15] but no separation between sightseeing and sightseeing + interior visits was proposed. Building on this latter work, Popescu et al. [16] used visual image classification to separate these two types of visits and to calculate typical visiting time of each case. In this paper, we use the method in [16] directly to get the visiting time of scenes, which is not our focus here. We mainly consider the time cost among scenes, namely, the distance between each couple of scenes.

Researches in spatial databases also address applications in spatial networks represented by graphs, instead of the traditional Euclidean space. Recent papers that extended various types of queries to spatial networks were [17]. Clustering in a road network database has been studied in [18], where a very efficient data structure was proposed based on the ideas of [19]. Li et al. [20] discussed a trip planning query in both Euclidean space and road network, which retrieved the best trip passing through at least one point from each category. However, they did not consider the point weight and the order of points. Likewise, we also study the STS problem in road network.

3   Approximation Algorithms

In this section we present two approximate algorithms for answering the spatio-temporal sequence mining.

3.1   Local Optimization Algorithms

Three local optimization algorithms in terms of distance, weight and value will be provided in the following.

Approximation in terms of distance: The most intuitive algorithm for solving STS is to form a sequence by iteratively computing the $\lceil m/2 \rceil$ nearest neighbor scenes of the current scene, comparing the value of them, choosing the scene whose value is maximum from all scenes that have not been visited yet. Then refresh the total time by adding this scene's duration time and the time cost between this scene and the last scene. If the total time is less than T_{total}, add this scene to the sequence; else, restore the total time and ignore this scene. Formally, given a partial sequence Q_k with $k<m$, Q_{k+1} is obtained by inserting the scene o_{k+1} whose value is larger than any scene in the $\lceil m/2 \rceil$ nearest neighbor of o_k. Meanwhile, this scene should not been covered yet and satisfy the time limitation. In the end, the final sequence is produced by connecting o_k to E. We call this algorithm d-LOA , which is shown in Algorithm 1.

Algorithm 1. d-LOA

Input:
    The start scene, $o = S$;
    The end scene, E;
    The set of scene IDs, $I = \{1, ..., m\}$;
    The initial spatio-temporal sequence, $Q_a = \{S\}$;
    The limited time, T_{total};
Output:
    The d-local optimal spatio-temporal sequence, Q_a;
1: $loc = S$;
2: $t = T_{total}$;
3: **while** (I *is not empty*) *and* $(t > 0)$ **do**
4:     define an array DS for storing the distances from loc to other scenes;
5:     **for** each $n \in I$ **do**
6:         $DS(loc, n) = D_{dis_N}(loc, n)$;
7:     **end for**
8:     HI = the set of $\lceil m/2 \rceil$ smallest DS scene IDs;
9:     define an array V for storing the values from loc to other scenes;
10:    **for** each $n \in HI$ **do**
11:      **if** $T(n) + T(loc, n) \le t$ **then**
12:         $V(loc, n) = \alpha * (w_{loc} + w_n) + (1 - \alpha) * \frac{1}{DS(loc, n)}$;
13:      **end if**
14:    **end for**
15:    o = the scene whose value is maximum in V;
16:    $loc = o$;
17:    pop o from I;
18:    put o to Q_a;
19:    $t = t - (T(o) + T(loc, o))$;
20: **end while**
21: $Q_a \leftarrow \{ E \}$;

Algorithm 2. w-LOA

Input:
    The start scene, $o = S$;
    The end scene, E;
    The set of scene IDs, $I = \{1, ..., m\}$;
    The initial spatio-temporal sequence, $Q_a = \{S\}$;
    The limited time, T_{total};
Output:
    The w-local optimal spatio-temporal sequence, Q_a;
1: $loc = S$;
2: $t = T_{total}$;
3: **while** (I *is not empty*) *and* $(t > 0)$ **do**
4:     **for** each $n \in I$ **do**
5:         HI = the set of $\lceil m/2 \rceil$ largest w_i scene IDs;
6:     **end for**
7:     define an array V for storing the values from loc to other scenes;
8:     **for** each $n \in HI$ **do**
9:      **if** $T(n) + T(loc, n) \le t$ **then**
10:        $V(loc, n) = \alpha * (w_{loc} + w_n) + (1 - \alpha) * \frac{1}{DS(loc, n)}$;
11:      **end if**
12:    **end for**
13:    o = the scene whose value is maximum in V;
14:    $loc = o$;
15:    pop o from I;
16:    put o to Q_a;
17:    $t = t - (T(o) + T(loc, o))$;
18: **end while**
19: $Q_a \leftarrow \{ E \}$;

Approximation in terms of weight: Another algorithm for solving STS is to form a sequence by iteratively performing the following operations. Choose the $\lceil m/2 \rceil$ maximum weight scenes, which connect to the current scene and have not been visited yet. Compare the value of these $\lceil m/2 \rceil$ scenes, and select the scene whose value is maximum. Then refresh the total time by adding this scene's duration time and the time cost between this scene and the last scene. If the total time is less than T_{total}, add this scene to the sequence; else, restore the total time and ignore this scene. We call this algorithm w-LOA, which is similar to d-LOA and shown in Algorithm 2.

Approximation in terms of value (i.e., distance and weight): A hybrid local optimization algorithm for solving STS is to form a sequence by iteratively performing the following operations. Compute the values between the current scene and every other scenes that have not been visited yet. Choose the scene whose value is maximum. Then refresh the total time by adding this scene's duration time and the time cost between this scene and the last scene. If the total time is less than T_{total}, add this scene to the sequence; else, restore the total time and ignore this scene. Formally, given a partial sequence Q_k with $k<m$, Q_{k+1} is obtained by inserting the scene o_{k+1} whose value is larger than any scene in R. Meanwhile, this scene should not been covered yet and satisfy the time limitation. In the end, the final sequence is produced by connecting o_k to E. We call this algorithm v-LOA, which is shown in Algorithm 3.

Algorithm 3. v-LOA (G, R, S, E)

Input:
    The start scene, $o = S$;
    The end scene, E;
    The set of scene IDs, $I = \{1, ..., m\}$;
    The initial spatio-temporal sequence, $Q_a = \{S\}$;
    The limited time, T_{total};
Output:
    The v-local optimal spatio-temporal sequence, Q_a;
  1: $loc = S$;
  2: $t = T_{total}$;
  3: **while** (I is not empty) and $(t > 0)$ **do**
  4:     define an array V for storing the values from loc to other scenes;
  5:     **for** each $n \in I$ **do**
  6:       **if** $T(n) + T(loc, n) \leq t$ **then**
  7:         $V(loc, n) = \alpha*(w_{loc} + w_n) + (1-\alpha)* \frac{1}{DS(loc, n)}$;
  8:       **end if**
  9:     **end for**
10:     $o =$ the scene whose value is maximum in V;
11:     $loc = o$;
12:     pop o from I;
13:     put o to Q_a;
14:     $t = t - (T(o) + T(loc, o))$;
15: **end while**
16: $Q_a \leftarrow \{ E \}$;

3.2 Global Optimization Algorithm

This section introduces a novel heuristic algorithm, called GOA. This algorithm achieves a much better result in comparison with the previous algorithms. GOA can find the optimal sequence if the heuristic function never overestimates the actual minimal value of reaching the goal. Here we select the heuristic as the value in Euclidean space, as it is always less than or equal to the actual value in road network in this scenario (see Definition 2). This can guarantee the sequence optimality in terms of road network value.

Definition 2. *For two scenes u and v ($u, v \in N$), $D_{dis_N}(u, v)$ is the road network distance, and $D_{dis_E}(u, v)$ is the Euclidean distance. Correspondingly, $V_N(u, v)$ denotes the value of the sequence from u to v in road network, while $V_E(u, v)$ means that in Euclidean space. In this paper, we care more about personal preferences, namely, we set the balance factor α larger than 0.5. So according to Eq.(1), $V_E(u, v) \le V_N(u, v)$.*

Maximum value sequence finding method. We can find the maximum value from the current scene to any scene in R using the heuristic algorithm GOA. This can be explicitly described by Theorem 1.

Theorem 1. *For an intermediary scene o along the sequence between u and v, the sequence with the maximum value $V_{NE}(u, o, v)$ is formalized by the sequence passing o. Then, the following Eq. (3) holds:*

$$V_{NE}(u, o, v) = V_N(u, o) + V_E(o, v) \tag{3}$$

Proof. Here $V_N(u, o)$ represents the value from the source scene u to the intermediary scene o, while $V_E(o, v)$ is the heuristic function that estimates the value from o to the destination scene v. According to the concept of naive heuristic algorithm, Eq. (3) holds. Then, $V_N(u, o)$ and $V_E(o, v)$ can be obtained by Lemma 1 and Lemma 2 respectively.

Lemma 1. *Assume that a source scene $u := o_0$ and a destination scene $v := o_k$. A road network traveling sequence $(o_0, o_1, \cdots, o_{k-1}, o_k)$ is a sequence of $k + 1$ interesting scenes. $V_N(o_0, o_k)$ denotes the value of the sequence from o_0 to o_k via $o_1, \cdots, o_{k-1}$:*

$$\sum_{i=1}^{k} V_N(o_{i-1}, o_i) = V_N(o_0, o_k) \tag{4}$$

Proof. The value function $V_N(o_0, o_k)$ accounts for a total value of the traveling sequence from o_0 to o_k in road network. That is, this value is the cumulative sequence value from the source scene o_0 to a destination scene o_k via as many scenes as possible from $o_1, \cdots, o_{k-1}$. So the total value $V_N(o_0, o_k)$ can be divided into $V_N(o_0, o_1) + V_N(o_1, o_2) + \cdots + V_N(o_{k-1}, o_k)$, namely, $\sum_{i=1}^{k} V_N(o_{i-1}, o_i)$.

Lemma 2. *Let node o and v be the current scene and the destination scene respectively. $h(o)$ is the heuristic estimator. Then, for the value function $V_E(o, v)$ of a heuristic value, the following Eq.(5) holds:*

$$h(o) \le V_E(o, v) \le V_N(o, v) \tag{5}$$

Proof. The heuristic estimator can find an optimal traveling sequence to a destination scene if the destination scene is reachable. Hence, according to Definition 2, the heuristic employs the value in Euclidean space as a lower bound value of a sequence from o

to v. For that reason, $h(o) \leq V_E(o, v)$ holds when $h(o)$ as the estimator is approximately equal to the value in Euclidean space. Also, for the sequence value between o and v, Eq. (5) holds by $V_E(o, v) \leq V_N(o, v)$. Because although the Euclidean distance is less than or equal to the network distance, the balance factor is larger than 0.5, as follows:

$$V_N(o, v) = \alpha * (w_o + w_{o_i} + \ldots + w_{o_j} + w_v) +$$

$$(1 - \alpha) * \frac{1}{D_{dis_N}(o, o_i) + \sum_{k=i}^{j-1} D_{dis_N}(o_k, o_{k+1}) + D_{dis_N}(o_j, v)}$$

$$V_E(o, v) = \alpha * (w_o + w_v) + (1 - \alpha) * \frac{1}{D_{dis_E}(o, v)}$$

So

$$\Delta V = V_N(o, v) - V_E(o, v) = \alpha * (w_{o_i} + \ldots + w_{o_j}) + (1 - \alpha) * \frac{1}{\Delta D_{dis}(o, v)} > 0$$

and this completes the proof.

Efficient optimal scene search. This paper employs the branch-and-bound technique [21] to search an optimal traveling sequence (i.e., a minimum distance and a maximum tourist scene weight). The technique is used to prune all of the unnecessary scenes from multiple neighbor scenes connected with a given current scene by Theorem 2. That is, to find the optimal STS whose value is maximum. This technique can select the tourist scene o_i, which has the minimum distance $D_{dis_{NE}}(u, o_i, v)$ out of the adjacent scenes $o_1, o_2, \cdots, o_i, \cdots, o_k$ emanating from u (i.e., $D_{dis_N}(u, o_1, v)$, $D_{dis_N}(u, o_2, v), \cdots, D_{dis_N}(u, o_i, v), \cdots, D_{dis_N}(u, o_k, v)$ in road network. Meanwhile, this technique can also select the tourist scene o_j, which has the maximum traveling weight $w(u, o_j, v)$ among all chosen scenes $o_1, o_2, \cdots, o_j, \cdots, o_k$. Hence, we define the optimal traveling sequence as follows:

Definition 3. *Let u and v be a source scene and a destination scene respectively. The optimal traveling sequence is a set of ordered scenes from u to v with the maximum value $V_{NE}(u, o_i, v)$ (i.e., a minimum distance $D_{dis_{NE}}(u, o_i, v)$, and a maximum weight $w(u, o_i, v)$), where o_i is a chosen scene adjacent to u.*

Theorem 2 presents how to find the optimal traveling sequence by the branch-and-bound technique and achieve $V_{NE}^*(u, o_i, v)$ from $V_{NE}(u, o_i, v)$.

Theorem 2. *Given two scenes u and v ($u, v \in N$) in road network, if there exist a set of chosen scenes $o_1, o_2, \cdots, o_i, \cdots, o_k$ connected to scenes u and v, Eq. (6) holds by Definition 3:*

$$V_{NE}^*(u, o_i, v) = \max_{1 \leq i \leq k} (V_{NE}(u, o_i, v)) \tag{6}$$

Proof. Let adjacent chosen scenes $o_i, \ldots, o_j$ be connected with a given scene u. If $V_{NE}(u, o_i, v)$ is larger than the value of any other adjacent chosen scenes, that is $V_{NE}(u, o_i, v) > V_{NE}(u, o_k, v)$, for $\forall k \in (i, j]$. Hence, $V_{NE}(u, o_i, v)$ can be the optimal traveling sequence via o_i among adjacent scenes connected to both u and v.

According to Theorem 2, the branch-and-bound technique can prone some scenes by pre-calculating values from the adjacent scenes to a destination scene.

Efficient traveling sequence finding. The result of STS query is the traveling sequence of the ordered scenes and the paths to them. Figure 2 shows an example of finding efficient STS from k scenes. First, we select a path from the source tourist scene S to each of other chosen scenes by Eq. (3). This procedure begins with the selection of the first scene with the maximum value to the source scene by Eq. (1) and then finds the path to it. In order to prevent the predetermined paths from being re-searched, we must allocate the heap for the optimal value of a scanned scene calculated by V_N (S,o_i) and the path from S to o_i. In Figure 2, we find STS starting from scanning the source scene S to scenes o_1, o_2 and o_3, choosing the scene o_2, whose value is maximum in the heap. Then calculate V_{NE} (S, o_2, o_1), $V_{NE}(S, o_2, o_3)$ and V_{NE} (S, o_2, o_4). In order to find the optimal scene, we calculate V_{NE}^* (S, o_2, o_4) by Eq. (6) and store the intermediary path into the heap. As a result, the procedure yields the sequence $S \rightarrow o_2 \rightarrow o_4$ whose value is maximum. Next, we refresh the sequence to other unvisited scenes with the maximum value to the destination scene by Eq. (3). In the same way, if the total time the sequence is less than or equal to the limited time T_{total}, we iteratively refresh the sequence to remaining scenes.

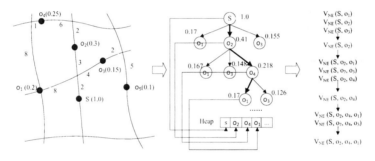

Fig. 2. Find the efficient traveling sequence

Algorithm 4 describes our GOA algorithm. At first, add the start scene to the OPEN list. Line 2 chooses a scene whose value is maximum from the OPEN list. We regard this scene as the current scene. From Line 3 to Line 6, if the destination scene is added to the CLOSE list, then the STS has been found, and the cycle stops. Else repeat the following operations from Line 7 to Line 28. Choose a scene whose value is maximum from the OPEN list. Then refresh the total time by adding the sum of this scene's duration time and the time cost between this scene and the last scene. If the total time is less than T_{total}, switch this scene to the CLOSE list; else, restore the total time and ignore this scene. For each of the other scenes adjacent to this current scene, if it is not walkable or it is in the CLOSE list, ignore it. Otherwise do the following operations. If it is not in the OPEN list, add it to the OPEN list. Regard the current scene as the parent of this scene, and calculate the value of the scene. If it is already in the OPEN list, check if there is other better path according to the value of the current sequence. If so, change the parent of the scene to the current scene, and recalculate the value of the scene.

Algorithm 4. GOA

Input:
 The start and end scene, $o = S, E$;
 The set of scenes ID, $I = \{1, ..., m\}$;
 The limited time, T_{total};
Output:
 1: The global optimal spatio-temporal sequence, Q_a;
  The array $OPEN$ which stores all chosen but not visited scenes
 2: $OPEN = [S]$;
  The array $CLOSE$ which stores all visited scenes
 3: $CLOSE = []$;
 4: **while** $OPEN$ IsNotEmpty and $T_{total} > 0$ **do**
 5: pop the first scene in $OPEN$ to o
 6: put it into $CLOSE$;
 7: **if** o equal E **then**
 8:   break;
 9: **end if**
10: M=the number of children scenes of o in I;
11: **for** each $s \in M$ **do**
12:   **if** $T(o)+T(o,s) \leq T_{total}$ **then**
13:    calculate the estimated value, $EV(o,s)=\alpha*(w_o+w_s)+(1-\alpha)*\frac{1}{D_{dis_N}(o,s)}$;
14:    **if** s is not in OPEN or CLOSE **then**
15:     pop s from I
16:     put s into $OPEN$
17:    **else if** s is in OPEN **then**
18:     **if** $EV(o,s) > EV(OPEN)$ **then**
19:      update the value of OPEN;
20:     **end if**
21:    **else**
22:     **if** $EV(o,s) < EV(CLOSE)$ **then**
23:      update the value of CLOSE;
24:      pop s from CLOSE
25:      put s into OPEN;
26:     **end if**
27:    **end if**
28:   **end if**
29: **end for**
30: put o into CLOSE;
31: sort the scenes in $OPEN$ by the EV descending
32: $T_{total} = T_{total}-T(o)+T(s))$;
33: **end while**
34: reverse $CLOSE$
35: $Q_a = CLOSE$;
36: return Q_a;

4   Experimental Evaluation

This section presents a comprehensive performance evaluation of the proposed methods for STS using Flickr datasets.

Experimental Setup. We obtained the real dataset in the city of Beijing with 286 scenes and 658 edges. In this dataset, we generated some interesting scenes according to scene features combined with user's profile. Datasets with a varying number of interesting scenes, varying balance factor, as well as varying limited total time were generated. The total number of interesting scenes is in the range $m \in [1,20]$, the balance factor is in the range $\alpha \in [0,1]$, while the limited total time is in the range $T_{total} \in [1h,8h]$, where h denotes the time granularity "hour".

Performance Results. In this part we study the performance of the proposed four algorithms. In order to prove the advantage of our algorithms, we compared them with ARM (average random method). ARM is achieved by choosing scene randomly for 30 times, and taking the average value of them.

First, we study the effects of α and T_{total} on the value of STS. Figure 3 plots the value of STS as a function of α, when T_{total} =8h. Figure 4 plots the value of STS as a function of T_{total}, when α =0.7. In both cases, GOA outperforms v-LOA, d-LOA, w-LOA and ARM obviously. The value of ARM is the lowest. With the increase of α and T_{total}, the performance of all algorithms increases. The algorithm d-LOA is greatly affected by the relative locations of scenes, because it greedily follows the nearest $\lceil m/2 \rceil$ scenes from the remaining scenes irrespective of its direction with respect to the destination scene E. With the increase of α and T_{total}, the probability that d-LOA wanders off the correct direction increases. In Figure 4 the trends of algorithm w-LOA and v-LOA are almost the same. Because when α =0.7, the distance has little effect on the value of the sequence, and the value of both algorithms is similar.

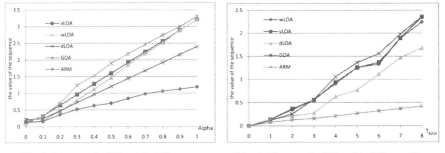

Fig. 3. The trend of V_N with different α **Fig. 4.** The trend of V_N with different T_{total}

Figure 5 plots the results of road network distance of STS as a function of T_{total}, when α =0.7. Figure 6 plots the total weight of STS as a function of T_{total}, when α =0.7. With the increase of T_{total}, both the distance and the weight of STS in all algorithms increase. The distance of algorithm d-LOA is less than the other four algorithms, because it always greedily follows the nearest $\lceil m/2 \rceil$ scenes from the remaining scenes. The weight of algorithm GOA outperforms v-LOA, d-LOA and w-LOA in Figure 6. So our four algorithms can get better results than ARM.

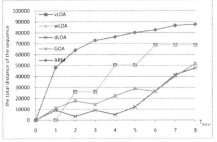

Fig. 5. The trend of D_{dis_N} with different T_{total} **Fig. 6.** The trend of w with different T_{total}

We also study the average length of STS as a function of T_{total}, when $\alpha =0.7$ in Figure 7. In general, the algorithm GOA includes more scenes than the other four ones. The reason is that GOA can get the global optimal sequence in road network. The number of scenes in ARM is the smallest. From Figure 8 we can see the value of GOA is the maximum, v-LOA and w-LOA are almost the same, which depends on the choice of α. The value of ARM is the minimum.

We examine the trend of runtime with different number of scene set in Figure 9. When the number of scene set is more than 50000, the runtime of v-LOA, d-LOA, and w-LOA increase much faster. The trend of runtime with different T_{total} is examined in Figure 10, when the number of scene set is 10000. In both cases, with the increase of scene set, the runtime of all algorithms increase. The runtime of v-LOA is the maximum, and ARM's is the minimum. The difference is the increase speed of runtime in Figure 10 is slower than that in Figure 9.

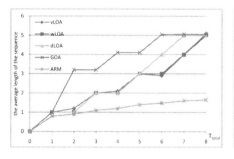
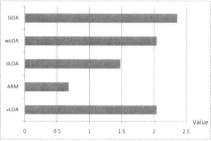

Fig. 7. The length of STS with different T_{total} **Fig. 8.** The trend of V_N of different algorithms

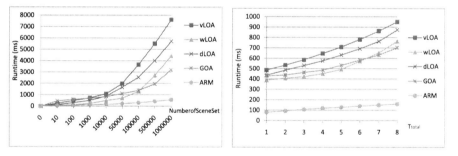

Fig. 9. Runtime with different NumSet   **Fig. 10.** The trend of runtime with different T_{total}

5   Conclusions and Future Work

The goal of this paper is to provide users with the optimal spatio-temporal sequence that passes through as many chosen scenes as possible with the maximum weight and the minimum distance within a limited travel time. We first argued that this problem is NP-hard, and gave a simple proof. Then formally defined the STS problem. We considered solutions for the STS problem as graphs where tourist scenes were nodes and paths between these scenes were edges. Two approximate algorithms: local optimization algorithms and a global optimization algorithm were provided. The experimental study using real datasets in Flickr demonstrated the effectiveness of our proposed algorithms.

Acknowledgement

We would like to thank Pengfei Dai of Beijing University of Posts and Telecommunications for his helpful comments in the experiments.

References

1. Nasraoui, O., Soliman, M., Saka, E., et al.: A Web Usage Mining Framework for Mining Evolving User Profiles in Dynamic Web Sites. IEEE Transactions on Knowledge and Data Engineering (TKDE), 202–215 (2008)
2. Rattenbury, T., Good, N., Naaman, M.: Towards Automatic Extraction of Event and Place Semantics from Flickr Tags. In: Proceedings of the 30th Annual International ACM SIGIR Conference (2007)
3. Ahern, S., Naaman, M., Nair, R., Yang, J.: World Explorer: Visualizing Aggregate Data from Unstructured Text in Georeferenced Collections. In: Proceedings of the ACM IEEE Joint Conference on Digital Libraries, JCDL (2007)
4. Quack, T., Leibe, B., van Gool, L.: World-Scale Mining of Objects and Events from Community Photo Collections. In: Proceedings of the 7th ACM International Conference on Image and Video Retrieval, CIVR (2008)
5. Crandall, D., Backstrom, L., Hutternlocher, D., Kleinberg, J.: Mapping the World's photos. In: Proceedings of the 18th International World Wide Web Conference, WWW (2009)
6. Zhou, C., Meng, X.: Complex Event Detection on Flickr. In: Proceedings of the 27th National Database Conference of China, NDBC (2010)
7. Zheng, I., Zhang, L., Xie, X., Ma, W.Y.: Mining Interesting Locations and Travel Sequences from GPS Trajectories. In: Proceedings of the 18th International World Wide Web Conference, WWW (2009)
8. Cao, X., Cong, G., Jensen, C.: Mining Significant Semantic Locations From GPS Data. In: Proceedings of the VLDB Endowment, PVLDB, vol. 3(1) (2010)
9. Gonotti, F., et al.: Trajectory Pattern Mining. In: Proceedings of the 13th ACM SIGKDD International Conference on Knowledge Discovery and Data Mining (KDD), pp. 330–339 (2007)
10. Mamoulis, N., et al.: Indexing and Quering Historical Spatiotemporal Data. In: Proceedings of the 10th ACM SIGKDD International Conference on Knowledge Discovery and Data Mining (KDD), pp. 236–245 (2004)
11. Girardin, F., Dal Fiore, F., Blat, J., Ratti, C.: Understanding of Tourist Dynamics from Explicitly Disclosed Location Information. In: Proceedings of the 4th International Symposium on LBS and Telecartography (2007)
12. Chen, Z., Shen, H.T., Zhou, X., Yu, J.X.: Monitoring Path Nearest Neighbor in Road Networks. In: Proceedings of the 35th SIGMOD International Conference on Management of Data, SIGMOD (2009)
13. Chen, Z., Shen, H.T., Zhou, X., Zheng, Y., Xie, X.: Searching Trajectories by Locations-An Efficiency Study. In: Proceedings of the 36th SIGMOD International Conference on Management of Data, SIGMOD (2010)
14. Home and Abroad, http://homeandabroad.com
15. Popescu, A., Grefenstette, G.: Deducing Trip Related Information from Flickr. In: Proceedings of the 18th International World Wide Web Conference, WWW (2009)

16. Popescu, A., Grefenstette, G., Alain, P.: Mining Tourist Information from User-Supplied Collections. In: Proceedings of The 18th ACM Conference on Information and Knowledge Management, CIKM (2009)
17. Papadias, D., Zhang, J., Mamoulis, N., Tao, Y.: Query Processing in Spatial Network Databases. In: Proceedings of 29th International Conference on Very Large Data Bases, VLDB (2003)
18. Yiu, M., Mamoulis, N.: Clustering Objects on a Spatial Network. In: Proceedings of the 30th SIGMOD International Conference on Management of Data, SIGMOD (2004)
19. Shekhar, S., Liu, D.: CCAM: A Connectivity Clustered Acccess Method for Networks and Network Computations. IEEE Transactions on Knowledge and Data Engineering (TKDE), 102–119 (1997)
20. Li, F., Cheng, D.: On trip planning queries in spatial databases. In: Anshelevich, E., Egenhofer, M.J., Hwang, J. (eds.) SSTD 2005. LNCS, vol. 3633, pp. 273–290. Springer, Heidelberg (2005)
21. Tao, Y., Papadias, D., Shen, Q.: Continuous Nearest Neighbor Search. In: Proceedings of 28th International Conference on Very Large Data Bases (VLDB), pp. 287–298 (2002)

Mining High Utility Mobile Sequential Patterns in Mobile Commerce Environments

Bai-En Shie[1], Hui-Fang Hsiao[1], Vincent S. Tseng[1], and Philip S. Yu[2]

[1] Department of Computer Science and Information Engineering,
National Cheng Kung University, Taiwan, ROC
[2] Department of Computer Science, University of Illinois at Chicago, Chicago, Illinois, USA
{brianshie,karolter1130}@gmail.com,
tsengsm@mail.ncku.edu.tw, psyu@cs.uic.edu

Abstract. Mining user behaviors in mobile environments is an emerging and important topic in data mining fields. Previous researches have combined moving paths and purchase transactions to find mobile sequential patterns. However, these patterns cannot reflect actual profits of items in transaction databases. In this work, we explore a new problem of mining high utility mobile sequential patterns by integrating mobile data mining with utility mining. To the best of our knowledge, this is the first work that combines mobility patterns with high utility patterns to find high utility mobile sequential patterns, which are mobile sequential patterns with their utilities. Two tree-based methods are proposed for mining high utility mobile sequential patterns. A series of analyses on the performance of the two algorithms are conducted through experimental evaluations. The results show that the proposed algorithms deliver better performance than the state-of-the-art one under various conditions.

Keywords: High utility mobile sequential pattern; utility mining; mobility pattern mining; mobile environment.

1 Introduction

With the rapid development of tele-communication technologies, mobile devices and wireless applications become increasingly popular. One's current position can be acquired via a mobile device with GPS service. With a series of users' moving logs, we can know the moving paths of mobile users. Besides, a greater number of people are using mobile devices to purchase mobile services online by credit cards. Combining moving logs and payment records, *mobile transaction sequences*, which are the sequences of moving paths with transactions, are obtained. Yun et al. [14] first proposed a framework for discovering *mobile sequential patterns*, i.e., the sequential patterns with their moving paths in mobile transaction sequence databases. Mobile sequential patterns can be applied in many applications, such as route planning in mobile commerce environment and maintaining website structures of online shopping websites.

However, in mobile sequential pattern mining, the importance of items is not considered. In the framework of traditional frequent pattern mining, utility mining [3, 4, 7, 8, 12, 13] is proposed for solving this problem. Instead of finding frequent patterns, utility mining discovers the patterns with high utilities, which are called *high*

J.X. Yu, M.H. Kim, and R. Unland (Eds.): DASFAA 2011, Part I, LNCS 6587, pp. 224–238, 2011.
© Springer-Verlag Berlin Heidelberg 2011

utility patterns. By utility mining, patterns with higher importance/profit/user interests can be found. For instance, the frequent patterns involving refrigerators may not be easily found from the transaction databases of hypermarkets since the frequency of purchasing refrigerators is much less than that of other items. But if we apply utility mining, the high utility patterns about refrigerators may be found since the utilities, i.e., the profits, of refrigerators are higher than that of others. Therefore, it is obvious that pushing utility mining into the framework of mobility pattern mining is an essential topic. If decision makers know which patterns are more valuable, they can choose more appropriate actions based on the useful information. Considering the utilities of items in customers' frequent purchasing patterns and moving paths is crucial in many domains, such as finding valuable patterns in mobile commerce environments, metropolitan planning and maintaining the structure and designing promotions for online shopping websites.

In view of the above issues, we aim at integrating mobility pattern mining with utility mining to find *high utility mobile sequential patterns* in this research. The proposed pattern must be not only high utility but also frequent. In other words, it is composed of both high utility purchasing pattern and frequent moving path. This is because applying only utility mining to the mobile environments is insufficient. A moving path with high utility but low frequency is unpractical. Users may be confused with a number of these redundant patterns. By this consideration, the proposed pattern is more useful than the patterns that apply only utility mining or frequent pattern mining to the mobile environments.

In this paper, we propose two tree-based methods, namely $UMSP_{DFG}$ (*mining high Utility Mobile Sequential Patterns with a tree-based Depth First Generation strategy*) and $UMSP_{BFG}$ (*mining high Utility Mobile Sequential Patterns with a tree-based Breadth First Generation strategy*). The main difference of the two algorithms is the method for generating the length 2 patterns during the mining process, which is the bottleneck of pattern mining. Both of the algorithms use a tree structure *MTS-Tree* (*Mobile Transaction Sequence Tree*) to summarize the information about locations, items, paths and utilities in mobile transaction databases. To the best of our knowledge, this is the first work that explores the integration of mobility pattern mining and utility mining. The experimental results show that $UMSP_{BFG}$ has better performance than $UMSP_{DFG}$. Moreover, the performance of the two proposed tree-based methods outperforms the compared level-wise algorithm which is improved by the state-of-the-art mobile sequential pattern algorithm [14].

Major contributions of this work are described as follows. First, this research is the first work that integrates high utility pattern mining with mobility pattern mining so as to explore the new problem of mining high utility mobile sequential patterns. Second, different methods proposed under different pattern generation strategies are proposed for solving this problem. Third, a series of detailed experiments is conducted to evaluate the performance of the proposed methods in different conditions. By the combination of high utility patterns and moving paths, highly profitable mobile sequential patterns can be found. We expect that the useful patterns can bring novel and insightful information in mobile commerce environments.

The remainder of this paper is organized as follows. We briefly review the related work in section 2. Section 3 is the problem definition of this research. In section 4, we describe the proposed algorithms. The experimental evaluation for performance study is made in section 5. The conclusions and future work are given in section 6.

2 Related Work

Extensive studies have been proposed for finding frequent patterns in transaction databases [1, 2, 5, 10]. Frequent itemset mining [1, 5] is the most popular topic among them. Apriori [1] is the pioneer for mining frequent itemsets from transaction databases by a level-wise candidate generation-and-test method. Tree-based frequent itemset mining algorithms such as FP-Growth [5] were proposed afterward. FP-Growth improves the efficiency of frequent itemset mining since it does not have to generate candidate itemsets during the mining process and it only scans the database twice. Afterwards, sequential pattern mining [2, 10] is proposed for finding customer behaviors in transaction databases. As an extension method of Apriori, AprioriAll [2] also use a level-wise framework to find sequential patterns. On the contrary, PrefixSpan [10] finds sequential patterns directly from projected databases without generating any candidate pattern. Thus, the performance can be more improved.

Mining user behaviors in mobile environments [6, 9, 11, 14] is an emerging topic in the frequent pattern mining field. SMAP-Mine [11] was first proposed for finding customers' mobile access patterns. However, in different time periods, users' popular services may be totally different. Thus, T-Map algorithm [6] was proposed to find temporal mobile access patterns in different time intervals. Although users' mobile access patterns are important, their moving paths are also essential. Therefore, Yun et al. [14] proposed a framework which combines moving paths and sequential patterns to find mobile sequential patterns. Moreover, Lu et al. [9] proposed a framework for discovering cluster-based mobile sequential patterns. The customers whose moving paths and transactions are similar will be grouped into the same clusters. By this framework, the discovered patterns may be closer to the customer behaviors in real life.

In the above researches, the profits of items are not considered yet. In transaction databases, items have different profits. Utility mining [3, 4, 7, 8, 12, 13] is proposed to conquer this problem. Among these researches, Liu et al. [8] proposed Two-Phase algorithm which utilizes the transaction-weighted downward closure property to maintain the downward closure property in the processes of utility mining. On the other hand, Ahmed et al. [3] employed a tree structure, named IHUP-Tree, to maintain essential information about utility mining. Different from Two-Phase, it avoids scanning database multiple times and generating candidate patterns. Although IHUP-Tree achieves a better performance than Two-Phase, it still produces too many high transaction weighted utilization itemsets. Therefore, Tseng et al. proposed UP-Growth [12], which applies four strategies for decreasing the estimated utilities during the mining processes. By these strategies, the number of possible high utility itemsets is effectively reduced and the performance of utility mining is further improved.

By the above literature reviews, although there are many researches about mobility pattern mining and utility mining, there is no research about the combination of the two topics. This paper is the first work which integrates the two topics to find high utility patterns with frequent moving paths in mobile environments.

3 Problem Definition

In this section, we define basic notations for mining high utility mobile sequential patterns in mobile environments in detail.

Table 1. Mobile transaction sequence database DB

SID	Mobile transaction sequence	SU
S_1	<(A; {[i_1, 2]}), (B; null), (C; {[i_2, 1]}), (D; {[i_4, 1]}), (E; null), (F; {[i_5, 2]})>	54
S_2	<(A; {[i_1, 3]}), (B; null), (C; {[i_2, 2], [i_3, 5]}), (K; null), (E; {[i_6, 10]}), (F; {[i_5, 4]}), (G; {[i_8, 2]}), (L; null), (H; {[i_7, 2]})>	132
S_3	<(A; {[i_1, 3]}), (B; null), (C; {[i_2, 1], [i_3, 5]}), (D; {[i_4, 2]}), (E; null), (F; {[i_5, 1], [i_6, 2]}), (G; null), (H; {[i_7, 1]})>	72
S_4	<(A; {[i_1, 1]}), (W; null), (C; {[i_3, 10]}), (E; null), (F; {[i_5, 1]}), (G; {[i_8, 2]}), (L; null), (H; {[i_7, 1]}), (E; {[i_9, 1]})>	59
S_5	<(A; {[i_1, 4]}), (B; null), (C; {[i_3, 10]}), (D; {[i_4, 1]}), (E; null), (F; {[i_5, 1]}), (G; null), (H; {[i_7, 2]})>	73
S_6	<(C; {[i_2, 2]}), (D; null), (E; {[i_9, 1]}), (F; {[i_5, 1]})>	31

Table 2. Utility table

Item	i_1	i_2	i_3	i_4	i_5	i_6	i_7	i_8	i_9
Utility	1	5	3	11	18	2	5	1	3

Let $L = \{l_1, l_2, ..., l_p\}$ be a set of *locations* in a mobile commerce environment and $I = \{i_1, i_2, ..., i_g\}$ be a set of *items* sold at the locations. An *itemset* is denoted as $\{i_1, i_2, ..., i_k\}$, where each item $i_v \in I$, $1 \le v \le k$ and $1 \le k \le g$. Given a *mobile transaction sequence database* D, a *mobile transaction sequence* $S = <T_1, T_2, ..., T_n>$ is a set of transactions ordered by time, where a *transaction* $T_j = (l_j; \{[i_{j_1}, q_{j_1}], [i_{j_2}, q_{j_2}], ..., [i_{j_h}, q_{j_h}]\})$ represents that a user made T_j in l_j, where $1 \le j \le n$. In T_j, the *purchased quantity* of item i_{j_p} is q_{j_p}, where $1 \le p \le h$. A *path* is denoted as $l_1 l_2 ... l_r$, where $l_j \subset L$ and $1 \le j \le r$.

Definition 1: A *loc-itemset* $<l_{loc}; \{i_1, i_2, ..., i_g\}>$ signifies the itemset $\{i_1, i_2, ..., i_g\}$ was traded in l_{loc}, where $l_{loc} \in L$ and $\{i_1, i_2, ..., i_g\} \subseteq I$. The utility of a loc-itemset $Y = <l_{loc}; \{i_1, i_2, ..., i_g\}>$ in a mobile transaction sequence database D is denoted as $u(Y)$ and defined as $\sum_{S_j : (Y \subseteq S_j) \wedge (S_j \in D)} \sum_{k=1}^{g} u(<l_{loc}; i_k>, S_j)$, where $u(<l_{loc}; i_k>, S_j)$, defined as $w(i_k) \times q_{j_k}$, is the utility of the loc-item $<l_{loc}; i_k>$ in the mobile transaction sequence S_j. $w(i_k)$ is the *unit profit* of the item i_k recorded in a *utility table*.

Take the mobile transaction sequence database in Table 1 and the utility table in Table 2 as an example, $u(<C; \{i_2, i_3\}>) = u(<C; \{i_2, i_3\}>, S_2) + u(<C; \{i_2, i_3\}>, S_3) = (5 \times 2 + 3 \times 5) + (5 \times 1 + 3 \times 5) = 25 + 20 = 45$.

Definition 2: A *loc-pattern* X is a list of loc-itemsets. It is denoted as $<l_1; \{i_{1_1}, i_{2_1}, ..., i_{g_1}\}><l_2; \{i_{1_2}, i_{2_2}, ..., i_{g_2}\}>...<l_m; \{i_{1_m}, i_{2_m}, ..., i_{g_m}\}>$. The utility of a loc-pattern X in S_j is denoted as $u(X, S_j)$ and defined as $\sum_{\forall Y \in X} u(Y, S_j)$. The utility of a loc-pattern X in D is denoted as $u(X)$ and defined as $\sum_{X \subseteq S_j) \wedge (S_j \in D)} u(X, S_j)$.

For instance, for the loc-pattern $X_1 = <A; i_1><C; \{i_2, i_3\}>$ in Table 1, $u(X_1, S_2) = u(<A; i_1>, S_2) + u(<C; \{i_2, i_3\}>, S_2) = 28$, and $u(X_1) = u(X_1, S_2) + u(X_1, S_3) = 28 + 23 = 51$.

Definition 3: A *moving pattern* is composed of a loc-pattern and a path. The utility of a moving pattern $P = <\{<l_1;\{i_{1_1},i_{2_1},....i_{g_1}\}><l_2;\{i_{1_2},i_{2_2},....i_{g_2}\}>...<l_m;\{i_{1_m},i_{2_m},....i_{g_m}\}>\}$; $l_1l_2...l_m>$, denoted as $u(P)$, is defined as the sum of utilities of the loc-patterns in P in the mobile transaction sequences which contain the path of P in D. The *support* of a moving pattern P, denoted as $sup(P)$, is defined as the number of mobile transaction sequences which contain P in D. Similarly, the support of a loc-itemset or a loc-pattern is also defined as the number of mobile transaction sequences which contain it in D.

For example, for the moving path $P_1 = <\{<A; i_1><C; \{i_2, i_3\}>\}$; ABC>, $u(P_1) = u(<A; i_1><C; \{i_2, i_3\}>, S_2) + u(<A; i_1><C; \{i_2, i_3\}>, S_3) = 28+23 = 51$, and $sup(P_1) = 2$.

Definition 4: Given a *minimum support threshold* δ and a *minimum utility threshold* ε, a moving pattern P is called a *high utility mobile sequential pattern*, abbreviated as *UMSP*, if $sup(P) \geq \delta$ and $u(P) \geq \varepsilon$. The length of a pattern is the number of loc-itemsets in this pattern. A pattern with length k is denoted as k-pattern.

For example, in Table 1, if $\delta = 2$ and $\varepsilon = 50$, the moving pattern $P_1 = <\{<A; i_1><C; \{i_2, i_3\}>\}$; ABC> is a 2-UMSP since $sup(P_1) \geq 2$ and $u(P_1) > 50$.

After addressing the problem definition of mining high utility mobile sequential patterns in mobile environments, we introduce the *sequence weighted utilization* and *sequence weighted downward closure* property (abbreviated as SWDC), which are extended from [8].

Definition 5: The *sequence utility* of mobile transaction sequence S_j is denoted as $SU(S_j)$ and defined as the sum of the utilities of all items in S_j.

For example, $SU(S_6) = u(<C; i_2>, S_6)+u(<E; i_9>, S_6)+u(<F; i_5>, S_6) = 10+3+18 = 31$.

Definition 6: The *sequence weighted utilization*, abbreviated as SWU, of a loc-itemset, a loc-pattern, or a moving pattern is defined as the sum of SU of the mobile transaction sequences which contain it in D.

For example, $SWU(<D; i_4>) = SU(S_1)+SU(S_3)+SU(S_5) = 54+72+73 = 199$; $SWU(<A; i_1><C;\{i_2,i_3\}>) = SU(S_2)+SU(S_3) = 132+72 = 204$; $SWU(<\{<A;i_1><C;\{i_2,i_3\}>\}$; ABC$)$ $= SWU(<A; i_1><C;\{i_2, i_3\}>, S_2)+SWU(<A; i_1> <C;\{i_2, i_3\}>, S_3) = 132+72 = 204$.

Definition 7: A pattern Y is called a *high sequence weighted utilization pattern*, if $sup(Y) \geq \delta$ and $SWU(Y) \geq \varepsilon$. In the following paragraphs, *high sequence weighted utilization loc-itemset*, *high sequence weighted utilization loc-pattern* and *high sequence weighted utilization mobile sequential pattern* are abbreviated as WULI, WULP and WUMSP, respectively.

Property 1. (Sequence weighted downward closure property): For any pattern P, if P is not a WUMSP, any superset of P is not a WUMSP.

Proof: Assume that there is a pattern P and P' is a superset of P. By Definition 6, $SWU(P) \geq SWU(P')$. If $SWU(P)<\varepsilon$, $SWU(P')<\varepsilon$. Similarly, by Definition 3, $sup(P) \geq sup(P')$. If $sup(P)<\delta$, $sup(P')<\delta$. By the above two conditions, we can obviously know that if P is not a WUMSP, any superset of P is not a WUMSP. ∎

Problem Statement. Given a mobile transaction sequence database, a pre-defined utility table, a minimum utility threshold and a minimum support threshold, the problem of mining high utility mobile sequential patterns from the mobile transaction sequence database is to discover all high utility mobile sequential patterns whose supports and utilities are larger than or equal to the two thresholds in this database.

4   Proposed Methods

4.1   Algorithm UMSP$_{DFG}$

The workflow of the proposed algorithm UMSP$_{DFG}$ (*high Utility Mobile Sequential Pattern mining with a tree-based Depth First Generation strategy*) is shown in Figure 1. In step 1, WULIs and a mapping table are generated. Then a *MTS-Tree (Mobile Transaction Sequence Tree)* is constructed in step 2. In step 3, WUMSPs are generated by mining the MTS-Tree with the depth first generation strategy. Finally in step 4, UMSPs are generated by checking the actual utility of WUMSPs. In this section, we describe the construction of MTS-tree first and then address the generation of WUMSP.

We first address the process of generating WULIs by an example. Take the mobile transaction sequence database in Table 1 and the utility table in Table 2 for example. Assume the minimum support threshold is 2 and the minimum utility threshold is 100. In the first step, WULIs whose supports and SWUs are larger than or equal to the two thresholds are generated by the processes similar to [8]. In this case, eight WULIs shown in Table 3 are generated. Note that they are also 1-WULPs. Then the 1-WULPs are mapped sequentially into a mapping table as shown in Table 3.

4.1.1   The Construction of MTS-Tree
The procedures of MTS-Tree construction are shown in Figure 2. The construction of MTS-Tree is completed after one scan of the original database. Without loss of generality, we give a formal definition for MTS-Tree first.

Definition 8. (MTS-Tree): In MTS-Tree, each node N includes $N._{location}$, $N._{itemset}$, $N._{SID}$ and a path table. N is represented by the form $<N._{location}$ [$N._{itemset1}$]: $N._{SID1}$; [$N._{itemset2}$]: $N._{SID2}$; ... >. $N._{location}$ records the node's location. Each node has several itemsets $N._{itemset}$ which represent the itemsets traded in the same location. For each itemset in a node, it has a string of sequence identifiers, $N._{SID}$, which records the mobile transaction sequences with the item in it. A *path table* records the paths, which are a series of locations with no item purchased from N's parent node to N, and the SIDs of the paths. Moreover, a *header table* is applied to efficiently traverse the nodes of a MTS-Tree. In a header table, each *entry* is composed of a 1-WULP, its SWU and support, and a link which points to its first occurrence in MTS-Tree.

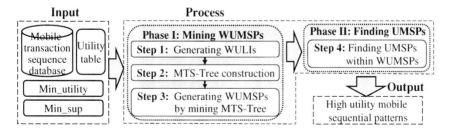

Fig. 1. The framework of the proposed algorithm UMSP$_{DFG}$

Table 3. Mapping table

1-WULP	A;i_1	C;i_2	C;i_3	D;i_4	F;i_5	G;i_8	H;i_7	C;{i_2,i_3}
After Mapping	A;t_1	C;t_2	C;t_3	D;t_4	F;t_5	G;t_6	H;t_7	C;t_8

```
Algorithm (Step 2 of UMSP_DFG: MTS-Tree construction)
Input: Mobile transaction sequence database DB, mapping table MT
Output: MTS-Tree
1. create a header table H
2. create a root R for an MTS-Tree T
3. foreach mobile transaction sequence S_i in DB do
4.    let path_start = false
5.    call InsertMTS_Tree(S_i, R, MT, sid)

Procedure InsertMTS_Tree(S_i, R, MT, sid)
1. if S_i is not NULL then
2.    divide S_i into [x|X]
      /* x: the first loc-itemset of S_i. X: the remaining list of S_i */
3.    let temppath = NULL
4.    if there is a combination y' of x exists in MT then
5.      convert x to the HTWULI y in MT
6.      if R has a child node C where C.location = y.location then
7.        if y.item exists in C.items then
8.           insert sid into C.[y.item].sid
9.        else
10.          create a new item y.item to C.items
11.          insert sid to C.[y.item].sid
12.      else
13.        create a new node C as a child node of R
14.        let C.location = y.location
15.        let C.item = y.item
16.        append sid to C.[y.item].sid
17.      update y's WULI, sup, TWU in H
18.      if temppath = NULL then
19.        append temppath and sid to C.pathtable
20.        let temppath = NULL
21.    else
22.      append x.location to temppath
23.    if X is not NULL then
24.      call InsertMTS_Tree(X, C, MT, sid)
```

Fig. 2. The procedure of MTS-Tree construction

Now we introduce the processes of the MTS-Tree construction, i.e., the second step of UMSP$_{DFG}$, by continuing the example in Section 4.1. At the beginning, the first mobile transaction sequence S_1 is read. The first transaction in S_1 is (A; {[i_1, 2]}), so we check the loc-itemset <A; i_1> in the mapping table in Table 3. After checking, the loc-itemset is converted into <A; t_1>, it is then inserted into MTS-Tree. Since there is no corresponding node in the MTS-Tree, a new node <A[t_1]: S_1> is created. The loc-itemset <A; t_1> is also inserted as an entry into the header table.

Subsequently, the second transaction (B; null) is evaluated. Since it has no purchased item, the location B is kept as a temporary path. Then the third transaction (C; {[i_2, 1]}) is checked in the mapping table and converted into <C; t_2>. Then the node <C[t_2]: S_1> is created and inserted as a child node of <A[t_1]: S_1>. Since there is a path B in the temp path, B and its SID S_1 are recorded into the path table of the node $<C[t_2]: S_1>$. Then the information in the temp path is cleared. The loc-itemset $<C; t_2>$ and its relevant information are also inserted into the header table. The remaining transactions in S_1 are inserted into the MTS-Tree sequentially by the same way.

Subsequently, the second mobile transaction sequence S_2 is read. For the first transaction (A; {[i_1, 3]}), it is converted into $<A; t_1>$. Since there is already a node $<A[t_1]: S_1>$ with the same location A in MTS-Tree, the SID S_2, SWU and support of the transaction are updated in the node and its entry in the header table, respectively. Then the location B of the second transaction (B; null) is kept into the temp path. Next, the third transaction (C; {[i_2, 2], [i_3, 5]}) is evaluated. By the mapping table, it is first converted into $<C; t_2>$, $<C; t_3>$ and $<C; t_8>$. Since there is a node $<C[t_2]: S_1>$ with location C and item t_2, $<C; t_2>$ can be just updated on the SID in the node by the processes mentioned above. On the other hand, for $<C; t_3>$ and $<C; t_8>$, their items and SIDs are stored into the node. After processing this transaction, the node becomes $<C [t_2]: S_1 S_2; [t_3]: S_2; [t_8]: S_2>$. The remaining transactions in S_2 are inserted into the MTS-Tree sequentially by the same way. After all sequences in D are inserted, we can get the MTS-Tree in Figure 3.

4.1.2   Generating WUMSPs from MTS-Tree

After constructing MTS-Tree, now we show the step 3 of UMSP$_{DFG}$. The purpose of this step is generating WUMSPs from MTS-Tree by the *depth first generation* strategy. The procedures are shown in Figure 4. First, WULPs and their *conditional MTS-Trees* are generated by tracing the links of the entries in the header table of the MTS-Tree. Then the WULPs are inserted into a *WUMSP-Tree (high sequence Weighted Utilization Mobile Sequential Pattern Tree)*, which is used for storing the WUMSPs. Then the paths of the WULPs in the WUMSP-Tree are traced in the original MTS-Tree and the WUMSPs are generated.

Definition 9. (WUMSP-Tree): In a WUMSP-Tree, each node is a WULI. For a node N, $N._{SID}$ and its path table are recorded. N is represented by the form $<N._{WULI}: N._{SID}>$. For the WULI in a node, it has a string of sequence identifiers, $N._{SID}$, which records the mobile ransaction sequences with the WULI occurring in it. A path table records the paths from the node N to root and the SIDs of the paths.

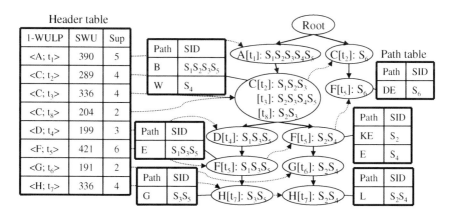

Fig. 3. An Example of MTS-Tree

```
Algorithm (Step 3 of UMSP_DFG: Generating WUMSPs)
Input: A MTS-Tree T, a header table H, a minimum utility threshold ε,
and a minimum support threshold δ
Output: A WUMSP-Tree T'
1.  Let T' be an WUMSP-Tree
2.  foreach WULI α in the bottom entry of H do
3.      trace the link of WULI α in H to get 1-WULP
4.      add 1-WULP α and sid to T'
5.      create a conditional MTS-Tree CT_α and a header table H_α
6.      call WUMSP-Mine(CT_α, H_α, α)

Procedure WUMSP-Mine(CT_α, H_α, α)
1.  foreach WULI β in H_α do
2.      if sup(β)< δ or SWU(β)< ε then
3.          delete β from CT_α and H_α
4.          if there exists an empty node X in CT_α
5.              delete X
6.              append the X's children nodes to X's parent node
7.  foreach WULI β of HT_α do
8.      add WULP βα and its sids to T'
9.      trace the paths of βα in T
10.     calculate the corresponding supports and SWUs
11.     add the paths to the path table of βα in the node of β in T'
        /* line 12-14: Path pre-checking technique*/
12.     if there exists a path in βα to form a WUMSP Y, such that
    sup(Y)≧ δ and SWU(Y)≧ ε then
13.         create a conditional MTS-Tree CT_βα and a header table H_βα
14.         call WUMSP-Mine(CT_βα, H_βα, βα)
```

Fig. 4. The procedure of generating WULPs

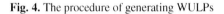
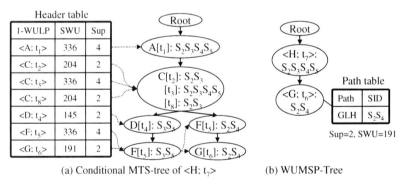

(a) Conditional MTS-tree of <H; t_7>   (b) WUMSP-Tree

Fig. 5. Conditional MTS-Tree of <H; t7> and the corresponding WUMSP-Tree

By tracing from a node to root in a WUMSP-Tree, a WULP can be derived. Furthermore, the corresponding WUMSPs of the WULP can be obtained after combining the paths in the node. Take the WUMSP-Tree in Figure 5 for example, we can get a WULP <G;t_6><H;t_7> by tracing from the node to root. Moreover, we can get a WUMSP <{<G;t_6><H;t_7>}; GLH> by combining the path GLH in the node <G;t_6> with the WULP. After tracing all nodes in WUMSP-Tree, all WUMSPs can be obtained. By storing the WUMSPs in the WUMSP-Tree, the patterns can be compressed in the tree and the memory storage can be saved.

During the processes of generating WULPs, if the length of the WULPs is larger than 1, besides the processes of tracing path, a *path pre-checking technique* will be performed to prune the moving patterns which can not fit the user-specified thresholds.

Definition 10. (Path pre-checking technique): If there exists no path in a WULP X to form a WUMP Y such that $sup(Y) \geq \delta$ and $SWU(Y) \geq \varepsilon$, X is pruned.

Path pre-checking technique is used for trimming the search space. By using this technique, the number of conditional MTS-Tree can be reduced effectively and the mining performance can be more improved.

Now we introduce the processes of the step 3, i.e., generating WUMSPs from MTS-Tree, by continuing the example in the previous section. First, the last entry $<H;t_7>$ in the header table of the MTS-Tree shown in Figure 3 is checked and a WULP $<H;t_7>$ is generated. Then $<H;t_7>$ and its SIDs are inserted as the first child node of the root of the WUMSP-Tree. Since $<H;t_7>$ is a 1-WULP, its path is not traced. Then the conditional MTS-Tree of $<H;t_7>$ shown in Figure 5 is constructed by tracing all ancestor nodes of the nodes labeled $<H;t_7>$ in MTS-Tree. The nodes labeled $<H;t_7>$ can be acquired by tracing the links from the entry of $<H;t_7>$ in header table. Note that the conditional MTS-Trees do not need any path table.

Subsequently, in the header table of the conditional MTS-Tree of $<H;t_7>$, the last entry $<G;t_6>$ is checked and a WULP $<G;t_6><H;t_7>$ is generated and inserted into the WUMSP-Tree. Since there is already a node $<H;t_7>$ in the WUMSP-Tree, we just insert $<G;t_6>$ as a child node of $<H;t_7>$. At the same time, the path of the WULP $<G;t_6><H;t_7>$ is traced in the original MTS-Tree. By the links from the entry $<H;t_7>$ in header table, we can get the node $<H[t_7]>$ with the SIDs S_2 and S_4. The node $<H[t_7]>$ is traced up until the node $<G;t_6>$ is reached to obtain the paths between the nodes. Then a path GLH is found. After tracing the path, a WUMSP $<\{<G;t_6><H;t_7>\}; GLH>$ whose support equals to 2 and SWU equals to 191 is found. By the path pre-checking technique, since its support and SWU are both no less than the two thresholds, it is kept, and the path is added into the node $<G;t_6>$ of the WUMSP-Tree. The WUMSP-Tree now is shown in Figure 5. Generating the patterns from the MTS-Tree by the above processes recursively, all WUMSPs can be generated.

4.1.3 Finding High Utility Mobile Sequential Patterns

After generating all WUMSPs, an additional database scan will be performed to find UMSPs from the set of WUMSPs. The WUMSPs whose utilities are larger than or equal to the minimum utility threshold will be regarded as UMSPs. Moreover, since the WUMSPs in WUMSP-Tree include SIDs, instead of checking all mobile transaction sequences, they will just check the specified sequences. By applying this process, the mining performance will become better.

4.2 An Improved Tree-Based Method: UMSP$_{BFG}$

In UMSP$_{DFG}$, since the number of combinations of 2-WULPs is quite large, many conditional MTS-Trees will be generated. Dealing with these conditional MTS-Trees is a hard work in the mining processes. Moreover, tracing the paths of WULPs in the processes of generating WUMSPs also consumes much time. If we can decrease the number of WULPs requiring verification, especially the large number of 2-WULPs, the performance can be more improved. Therefore, how to speed up the processes about 2-WUMSPs is a crucial problem.

To conquer this problem, we propose an improved tree-based algorithm UMSP$_{BFG}$ (*high Utility Mobile Sequential Pattern mining with a tree-based Breadth First Generation strategy*). The difference between the two algorithms is that UMSP$_{BFG}$ use a breadth first generation strategy for generating 2-WUMSPs. Within the strategy, a *possible succeeding node checking technique* is applied. By this technique, the size of the conditional MTS-Trees will be smaller, and the 2-moving patterns which cannot be 2-WUMSPs will be pruned in advance.

Instead of generating a 2-WULP by combining the last entry with the 1-WULP of a conditional MTS-Tree, in the breadth first generation strategy, 2-WULPs are generated by combining all 1-WULPs in the header table with the 1-WULP of the conditional MTS-Tree. After generating the 2-WULPs, their paths, supports and SWUs are checked in advance. The valid paths will be stored in the corresponding nodes of WUMSP-Tree. While generating 2-WULPs, UMSP$_{BFG}$ applies a *possible succeeding node checking technique* for pruning useless 2-WULPs, which is addressed as follows.

Definition 11. (Possible succeeding node checking technique): While generating 2-WULPs of a 1-WULP X in X's conditional MTS-Tree, all 1-WULPs in the header table are inserted as children nodes of X in the WUMSP-Tree in advance. If there exists no path in a WULP Y to form a WUMSP Z such that $sup(Z) \geq \delta$ and $SWU(Z) \geq \varepsilon$, Y is pruned. Furthermore, only the nodes kept in the WUMSP-Tree are able to be succeeding nodes of the WUMSP-Tree in the later mining processes.

In the following paragraphs, we use the same example as the previous section. The MTS-Tree shown in Figure 3 and the conditional MTS-Tree of <H;t$_7$> shown in Figure 5 are constructed by the same processes in previous section. <H;t$_7$> is inserted into the WUMSP-Tree as the first node. Different from UMSP$_{DFG}$, in the processes of generating 2-WULPs of UMSP$_{BFG}$, all 1-WULPs in the header table of the conditional MTS-Tree of <H;t$_7$> are inserted as children nodes of the node <H;t$_7$> in the WUMSP-Tree, that is, all 2-WULPs of the conditional MTS-Tree of <H;t$_7$> are generated in advance. The paths of the 2-WULPs are then generated by tracing the original MTS-Tree. Combining the 2-WULPs and the paths, 2-moving patterns are generated. Also, their supports and SWUs are obtained. The results are shown in Figure 6.

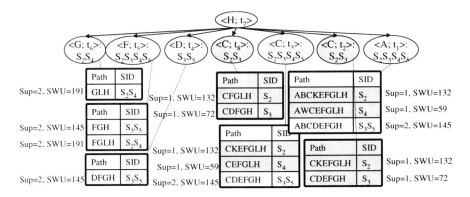

Fig. 6. An example of WUMSP-Tree generated by UMSP$_{BFG}$

By Figure 6, since the supports or SWUs of the 2-moving patterns $<\{<C;t_8><H;t_7>\}$; CFGLH>, $<\{<C;t_8><H;t_7>\}$;CDFGH>, $<\{<C;t_3><H;t_7>\}$;CKEFGLH>, $<\{<C;t_3><H;t_7>\}$;CEFGLH>, $<\{<C;t_2><H;t_7>\}$;CKEFGLH>, $<\{<C;t_2><H;t_7>\}$;CDEFGH>, $<\{<A;t_1><H;t_7>\}$;ABCKEFGLH> and $<\{<A;t_1><H;t_7>\}$;AWCEFGLH> are less than the thresholds, their relevant paths are pruned from the path tables of the WUMSP-Tree. Moreover, since there is no valid path in the nodes $<C;t_8>$ and $<C;t_2>$, the two nodes are also pruned. In Figure 6, the pruned nodes and paths in the WUMSP-Tree are labeled with grey. By the WUMSP-Tree, we can know the possible succeeding nodes of $<G;t_6>$ are $<F;t_5>$, $<D;t_4>$, $<C;t_3>$ and $<A;t_1>$.

After ascertaining which 2-moving patterns need to be pruned, the relevant nodes and entries in the conditional MTS-Tree of $<H;t_7>$ are also pruned. After this step, the mining processes proceed without the pruned nodes in both the WUMSP-Tree and the conditional MTS-Tree of $<H;t_7>$. The remaining conditional MTS-Tree is much smaller than the original one. Moreover, since the useless entries are pruned in the header table, they will never be checked in the following processes. Therefore, the search space can be further reduced and the mining performance is further improved.

5   Experimental Results

In this section, we evaluate the performance of the proposed algorithms. The experiments were performed on a 2.4 GHz Processor with 1.6 GB memory, and the operating system is Microsoft Windows Server 2003. The algorithms are implemented in Java. The default settings of the parameters are listed in Table 4. The settings of parameters related to mobile commerce environment and utility mining are similar to [14] and [8], respectively.

For comparing the performance of the proposed algorithms, we extend algorithm TJ_{PF} in [14] to form a basic algorithm for mining UMSPs which is called MSP in this paper. The processes of MSP are as follows: First, the mobile sequential patterns whose supports are no less than the minimum support threshold are generated by TJ_{PF}. Then an additional check of the actual utilities of the mobile sequential patterns is performed for finding UMSPs. In the following experiments, the performance of MSP is compared with that of the two proposed algorithms. Due to the page limit, in the experiment results, we show the number of patterns after phase I instead of the execution time of phase II since the time cost is mainly decided by the number of these patterns. The fewer patterns should be checked, the less time will be spent.

Table 4. Parameter settings

Parameter Descriptions	Default
D: Number of mobile transaction sequences	50k
P: Average length of mobile transaction sequences	20
T: Average number of items per transaction	2
N: Size of mesh network	8
n$_l$: The range of the number of items sold in each location	200
P$_b$: The probability that user makes the transaction in the location	0.5
w: Unit profit of each item	1~1000
q: Number of purchased items in transactions	1~5

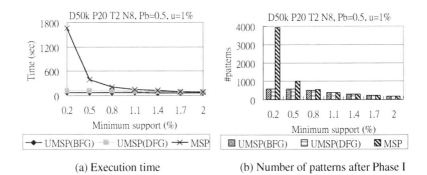

(a) Execution time   (b) Number of patterns after Phase I

Fig. 7. The performance under varied minimum support thresholds

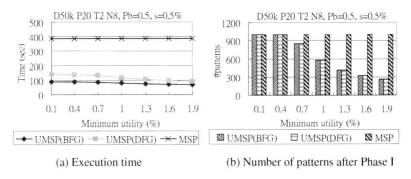

(a) Execution time   (b) Number of patterns after Phase I

Fig. 8. The performance under varied minimum utility thresholds

The first part of the experiments is the performance under various minimum support thresholds. In the experiments, the minimum utility threshold is set as 1%. The results for the execution time and the number of patterns after phase I under varied minimum support thresholds are shown in Figure 7. For the two proposed algorithms, the patterns after phase I are WUMSPs, on the other hand, for MSP, the patterns are mobile sequential patterns. In Figure 7 (a), it can be seen that MSP requires much more execution time than the other algorithms. The reason is that since MSP does not consider utility in phase I, the number of generated patterns is much larger than that of other algorithms as shown in Figure 7 (b). MSP spends much more time on processing additional patterns, so its performance is the worst. Besides, although the number of WUMSPs generated by the proposed algorithms is the same, their execution time is different. Overall, the tree-based algorithms are better than the level-wise version especially when the minimum support threshold is low.

The second part of the experiments is the performance under various minimum utility thresholds. In the experiments, the minimum support threshold is set as 0.5%. The results are shown in Figure 8. Overall, the tree-based algorithms are better than the level-wise one. Besides, since MSP does not consider utility in phase I, its execution time and number of generated patterns remain the same. On the contrary, both of the two results of the proposed two algorithms decrease with the minimum utility

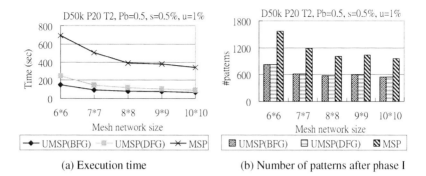

(a) Execution time   (b) Number of patterns after phase I

Fig. 9. The performance under varied mesh network size

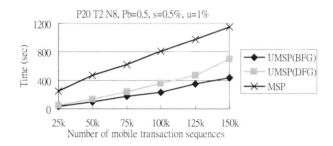

Fig. 10. The execution time under varied number of mobile transaction sequences

threshold increasing. In Figure 8 (b), when the minimum utility threshold is below 0.4%, almost no candidate can be pruned. Thus, the performance of the two algorithms is almost the same.

The third part of the experiments is the performance under varied mesh network size. The results are shown in Figure 9. By Figure 9, it can be seen that the execution time of all the three algorithms decreases with the size of mesh network. The reason is that when the size of mesh network is larger, the database will be sparser, therefore, the patterns generated in phase I will become fewer and the time cost of mining processes will be reduced.

The final part of the experiments is the performance under varied number of mobile transaction sequences. The experimental results are shown in Figure 10. In this figure, we can see that when the number of mobile transaction sequences is larger, the execution time of the algorithms increases linearly.

By the above experiments, the proposed algorithms are shown to outperform the state-of-the-art mobile sequential pattern algorithm MSP. Among the algorithms, the performance of UMSP$_{BFG}$ is the best since the MTS-tree is an efficient tree structure and the breadth first strategy effectively enhances the mining performance.

6   Conclusions

In this research, we proposed a novel data mining issue about mining high utility mobile sequential patterns in mobile commerce environments. This paper is the first research work about the combination of mobility pattern mining and utility mining. Two algorithms developed by different strategies, i.e., depth first generation and breadth first generation, are proposed for efficiently mining high utility mobile sequential patterns. The experimental results show that the proposed algorithms outperform the state-of-the-art mobile sequential pattern algorithm. For future work, additional experiments under more conditions of mobile commerce environments will be conducted for further evaluating the algorithms. Moreover, new algorithms which improve the mining performance will be designed.

Acknowledgments. This research was supported by National Science Council, Taiwan, R.O.C. under grant no. NSC99-2631-H-006-002 and NSC99-2218-E-006-001.

References

1. Agrawal, R., Srikant, R.: Fast Algorithms for Mining Association Rules. In: Proc. of the 20th Int'l. Conf. on Very Large Data Bases, pp. 487–499 (1994)
2. Agrawal, R., Srikant, R.: Mining Sequential Patterns. In: Proc. of 11th Int'l. Conf. on Data Mining, pp. 3–14 (1995)
3. Ahmed, C.F., Tanbeer, S.K., Jeong, B.-S., Lee, Y.-K.: Efficient Tree Structures for High Utility Pattern Mining in Incremental Databases. IEEE Trans. on Knowledge and Data Engineering 21(12), 1708–1721 (2009)
4. Chan, R., Yang, Q., Shen, Y.: Mining High Utility Itemsets. In: Proc. of Third IEEE Int'l Conf. on Data Mining, pp. 19–26 (November 2003)
5. Han, J., Pei, J., Yin, Y.: Mining Frequent Patterns without Candidate Generation. In: Proc. of the ACM-SIGMOD International Conference on Management of Data, pp. 1–12 (2000)
6. Lee, S.C., Paik, J., Ok, J., Song, I., Kim, U.M.: Efficient Mining of User Behaviors by Temporal Mobile Access Patterns. Int'l. Journal of Computer Science Security 7(2), 285–291 (2007)
7. Li, Y.-C., Yeh, J.-S., Chang, C.-C.: Isolated Items Discarding Strategy for Discovering High Utility Itemsets. Data & Knowledge Engineering 64(1), 198–217 (2008)
8. Liu, Y., Liao, W.-K., Choudhary, A.: A Fast High Utility Itemsets Mining Algorithm. In: Proc. of Utility-Based Data Mining (2005)
9. Lu, E.H.-C., Tseng, V.S.: Mining Cluster-based Mobile Sequential Patterns in Location-based Service Environments. In: Proc. of IEEE Int'l. Conf. on Mobile Data Management (2009)
10. Pei, J., Han, J., Mortazavi-Asl, B., Pinto, H., Chen, Q., Dayal, U., Hsu, M.C.: Mining Sequential Patterns by Pattern-Growth: The PrefixSpan Approach. IEEE Transactions on Knowledge and Data Engineering 16(10) (October 2004)
11. Tseng, V.S., Lin, W.C.: Mining Sequential Mobile Access Patterns Efficiently in Mobile Web Systems. In: Proc. of the 19th Int'l. Conf. on Advanced Information Networking and Applications, Taipei, Taiwan, pp. 867–871 (2005)
12. Tseng, V.S., Wu, C.W., Shie, B.-E., Yu, P.S.: UP-Growth: An Efficient Algorithm for High Utility Itemsets Mining. In: Proc. of the 16th ACM SIGKDD Conference on Knowledge Discovery and Data Mining (KDD 2010), Washington, DC, USA (July 2010)
13. Yen, S.-J., Lee, Y.-S.: Mining High Utility Quantitative Association Rules. In: Song, I.-Y., Eder, J., Nguyen, T.M. (eds.) DaWaK 2007. LNCS, vol. 4654, pp. 283–292. Springer, Heidelberg (2007)
14. Yun, C.-H., Chen, M.-S.: Mining Mobile Sequential Patterns in a Mobile Commerce Environment. IEEE Transactions on Systems, Man, and Cybernetics-Part C: Applications and Reviews 37(2) (2007)

Reasoning about Dynamic Delegation in Role Based Access Control Systems

Chun Ruan[1] and Vijay Varadharajan[1,2]

[1] University of Western Sydney, Penrith South DC, NSW 1797 Australia
{chun,vijay}@scm.uws.edu.au
[2] Macquarie University, North Ryde, NSW 2109 Australia
vijay@ics.mq.edu.au

Abstract. This paper proposes a logic based framework that supports dynamic delegation for role based access control systems in a decentralised environment. It allows delegation of administrative privileges for both roles and access rights between roles. We have introduced the notion of trust in delegation and have shown how extended logic programs can be used to express and reason about roles and their delegations with trust degrees, roles' privileges and their propagations, delegation depth as well as conflict resolution. Furthermore, our framework is able to enforce various role constraints such as separation of duties, role composition and cardinality constraints. The proposed framework is flexible and provides a sound basis for specifying and evaluating sophisticated role based access control policies in decentralised environments.

1 Introduction

A fundamental challenge in the development of large scale secure database systems is the design of access control and privilege management model and architecture. The dynamic aspect of privileges coupled with the fine granular nature of entities involved in large scale distributed database systems make this a significant problem, especially in the context of pervasive mobile distributed applications. At a general level, there are two types of mappings, one from requesting entity to privileges and the other from the privileges to accessed entity. Models such as Role-Based Access Control (RBAC) [3] is well-known in terms of their ability to decouple these two mappings, by having a user to role mapping and role to objects mapping. The central notion of RBAC is that access rights or privileges are associated with roles and users are assigned to appropriate roles. In distributed applications and services, we have access rights and roles are mapped to entities other than individuals. This partitioning into two mappings enables the system to manage these two mappings somewhat independently thereby helping to achieve more flexible and secure policy management. For instance, in a user based role system, if a a specific user leaves the organization and changes his/her position, only the mapping from users (subjects) to roles need to be revoked or changed (while the mapping between the role and

J.X. Yu, M.H. Kim, and R. Unland (Eds.): DASFAA 2011, Part I, LNCS 6587, pp. 239–253, 2011.
© Springer-Verlag Berlin Heidelberg 2011

the rights are not touched). On the other hand, if the access rights of the roles change, only the mapping from roles to access rights need to be changed.

When it comes to management of policies, in general, there are two types of approaches namely centralised and decentralised. Decentralised management in principle allows many administrators with different privileges and rights residing in different locations in a distributed system. A main advantage of centralised paradigm is that it enables an organization to have central control over its resources. However, when the system is large, it is often difficult, and sometimes impossible, for a single central unit to know every suitable role for a user, and every suitable access right for a role. Furthermore, large organizations may have over a thousand roles, and tens of thousands of users. A case study carried out with Dresdner Bank, a major European bank, resulted in an RBAC system that has around 40,000 users and 1300 roles [7]. In a distributed federated application environment over the Internet, where the resources are owned (and administered) by different entities, it is infeasible to envisage a centralised approach to access control management. However decentralised management with different entities, though it offers flexibility, poses several security policy and architectural challenges.

While RBAC is mainly used in a centralised management, we believe that it can be extended to a distributed environment such as web service or internet applications through proper delegation and trust management. That is, roles can be assigned to unknown entities based on the delegation and trust. For example, many newspapers such as Daily Telegraph provide free access to university students. In this case, universities can be delegated the right to assign the role "student" with full trust. This paper is concerned with dynamic delegation based access rights management in a distributed environment with multiple administrative entities. In this paper, we consider dynamic delegation of administrative privileges in the context of role based access control system. In such a situation, a major security concern is whether the organization can have some degree of control about who can access its resources; as after several steps of delegations, a user who is not trusted may get control over the system resources. Therefore how to use the delegation without compromising the security of the system is a big challenge. In this paper, we consider several ways for the delegators to control their delegations. Firstly, we allow a delegator to express their trust degrees on the delegations. For example, in a university context, the Vice-Chancellor(VC) may delegate the capability of recruiting a role "Leading Security Researcher"(LSR) to the Research Director(RD) who may further delegate to his or her Research Group Leader and an external member. The Research Director may associate different degrees of trust to these two delegations. This is especially the case when the role is fuzzy, like the above "Leading Security Researcher". Furthermore, an entity may be assigned to a role by two other entities (assigners) with different trust degrees. Hence when it comes to deciding on a user's assignment to a role, it is necessary to consider the multiple paths involved and the trust degrees associated with these paths. A role assignment can be rejected if the trust degree associated with this falls below a certain threshold. Our framework takes into account of such trust aspects in reasoning about access in decentralised systems.

Conflict resolution is another important aspect in the design of access control systems. Conflicts may arise in a number of ways involving positive and negative authorizations as well as due to dynamic delegation of privileges between roles. Solving conflicts in role based access control systems can be a complex issue that has not been well studied in the current research. A role may dynamically receive two conflicting authorizations from two different roles which are able to grant privileges. E.g. an Enrolled Student when using a website associated with a course, may receive conflicting authorizations from a Website Administrator and the Course Coordinator. Alternatively, a user may receive two conflicting authorizations from the two roles in which he/she is a member. For example, Alice is an enrolled student and a tutor. Therefore she may receive two conflicting authorizations, with one allowing her to access the students' work and the other not. Both levels of conflicts need to be properly resolved. In our framework, delegators and grantors are allowed to express trust degrees on their delegations and authorizations. Both the grantor's priority and the trust of the authorization will be taken into consideration in resolving conflicts. It will also allow us to express and enforce a range of other role requirements based on constraints such as separation of duty, role exclusion and cardinality.

Our framework is based on extended logic programs (ELP)[1]. The main contribution of this paper is the design of a logic based framework that supports dynamic delegation for role based access control systems in a decentralised environment. It has several novel features and allows for – delegation of both role administration and access right administration, trust degrees to be associated with role and access right delegations, able to specify both positive and negative authorizations to be granted to roles. This in turn enables to specify and reason about trust in delegated RBAC policies as well as resolve conflicts using trust in such policies.

The paper is organised as follows. Section 2 provides an overview of our formal dynamic delegation framework for role based access control systems. In Section 3, we define a role based authorization program (RBAP) to capture the delegations, trust degrees, access right propagations along the hierarchies of roles and objects and conflict resolution. A discussion of the features of RBAP and syntax are given in Section 3. Section 4 presents the formal semantics, and Section 5 compares our framework with previous works and discusses the future work.

2 An Overview of the Formal Framework

2.1 Administrative Privilege Delegation Correctness

In our formalization, administrative privilege delegation includes role administrative privilege delegation and access right administrative privilege delegation. We assume that System Security Officer (SSO) in an organization is the first role that can delegate. Others have to be granted the privilege to delegate.

Definition 1. *Delegation correctness: A role r can delegate other roles the priv-*
ilege to assign a role with depth d if and only if r is SSO or r has been delegated
the privilege to assign roles with delegation depth $d + 1$. A role r can delegate
other roles the privilege to grant an access right a over object o with depth d if
and only if r is SSO or r has been delegated the privilege to grant a over o with
delegation depth $d + 1$.

The delegations are dynamic in our system in several ways. Firstly, the delegations can be subjective to conditions. For example, we can say that a Head of
School is delegated the right to assign new staff if the student teacher ratio is
greater than 50:1. Hence the delegation depends on the ratio which is changing dynamically. Second, the delegation paths are changing dynamically. Since
a number of roles can be delegated the right to further delegate, or assign a role
or access rights, the delegations will be dynamically generated. Furthermore, a
delegation is allowed to be associated with a trust degree, which also affects the
delegation's evaluation. We will give more details about this later.

2.2   Role and Access Right Assignment Correctness

In this paper, we use users and entities interchangeably. The users can be individuals, agents or processes. For a particular role r, only the roles that have been
granted the role administrative privilege on r can exercise r's role assignment.
We assume SSO can assign any roles in the organization.

Definition 2. *Role assignment correctness: A role r can assign users to a role r'*
if and only if r is SSO or r has been granted the privilege to assign r'.

In our formalization, the access right administrative privilege is in terms of a
specific access right on a specific object. Thus it is possible to say that a role can
only grant read, but not write, on an object to others. For instance, a Head of
School may be able to grant "read" about the college's budget file but not write.
We also assume SSO can grant any privileges on any object in the organization.

Definition 3. *Access right assignment correctness: A role r can grant other*
roles an access right a of type $+$ or $-$ over object o if and only if r is SSO or r
has been granted the privilege to grant a over o.

2.3   Degrees of Trust and Effective Trust

In the real world, trust is often not a yes/no binary decision. People often trust
some people or something to a certain degree. This is also true when it comes
to the role or access right assignment or delegations. There is a need for people to express the degree to which they believe that someone can perform some
role/action, or someone can grant some role/action. In our formalization, we
allow a trust degree to be associated with each role/access right assignment,
and each role/access right delegation. The trust degree may come from the delegator's own knowledge about the delegate, or from the information that the

delegator obtains from reputable sources. There has been a lot research about trust evaluations in the area of trust management systems, and we will not address the issue in detail in this paper.

Consider for example the following case: (i) A says B and C can say who can assign the role R with 0.9 and 0.8 trust degrees respectively, (ii) B and C say D can assign the role R with 0.7 and 0.6 trust degrees respectively, and (iii) D assigns Alice to role R with 1. So, what should be the effective trust degree that the system adopts about assigning Alice to the role R? A statement's effective trust degree should consider all the trust degrees on a delegation path to it, and all the delegation paths to it. In our framework, for any statement, if there exists only one path to it, then the product of all the trust degrees on the path is defined as its *effective trust degree*. If there exist multiple paths to it, then the greatest value of all the paths is defined as its *effective trust degree*. In the above example, there are two paths leading to Alice's role assignment, $A- > B- > D- > Alice$ and $A- > C- > D- > Alice$, with effective trust degrees 0.9*0.7*1 and 0.8*0.6*1 respectively, so the effective trust degree for Alice's role assignment is 0.63. Effective trust degree can help to dynamically control the administrative privilege delegation and role or access right assignment. For example, if the effective trust degree falls below a certain threshold, then, the system can reject the role assignment or delegation. The trust degrees, the effective trust degrees, and the threshold can be dynamically changed, and so are the delegations and role or access right assignments. For example, a previously ineffective delegation may become effective due to the increased effective trust degree caused by an additional delegation path added to it. We believe that assigning and evaluating effective trust degrees is an easy but effective approach for delegation control.

Example 1. Let us consider the example in the university context about teaching web service. Suppose that the contents of a course website are open to all university students, and SSO has delegated the capability to assign students to the trusted universities. SSO firstly delegates the Educational Department (ED) to assign trusted university(TUni) with full trust and delegation depth of 1, so that ED can delegate once more. SSO then delegates the same right to any trusted university with a trust degree of 0.9 and delegation depth of 0 (meaning that they cannot further delegate). A university becomes trusted university if it has existed for over 10 years. Now suppose ED says UniOne is a trusted university with a trust degree of 0.9, and UniOne says Bob is a student with full trust. Then Bob will be able to access the resources on the teaching website if the role assignment threshold is set to 0.85. Now if UniOne says UniTwo is a university with a trust degree of 0.9, and UniTwo says Tom is a student with full trust, then Tom will not be able to access the contents on the website, as UniTwo is not a trusted university since its effective trust degree is only 09.*0.9=0.81 which is less than the threshold of 0.85 (suppose UniTwo has not existed for over 10 years). The situation will be represented in RBAP program in the next section.

2.4   Conflict Resolution

Two types of conflicts can arise in a role based access control system. A role may receive conflicting authorizations from multiple administrators, or an user may receive conflicting authorizations due to being a member of two different roles. For example, a staff member is allowed to borrow books from a library, but an "overdue staff" member who has overdue books is not allowed to borrow books. If a person belongs to both of the roles, then he/she will receive two conflicting authorizations. For both types of conflicts, we will use the effective trust degree to resolve the conflicts. This means that the authorization with higher effective trust degree will override. For example, by giving a higher trust degree to the negative authorization granted to "overdue staff" than that of the positive authorization granted to staff, an "overdue staff" cannot borrow books. This method would also allow the administrative privilege delegator to control their delegations flexibly. For instance, by giving a delegatee a less than one trust degree, a delegator can keep a higher priority than the delegatee in their 'can grant' delegations. Therefore, when the delegator gives an authorization a full trust degree, this authorization will not be overridden by the delegatee's authorizations. This method will help to enforce the high level policies despite the delegations.

Example 2. Consider again the example about university teaching web service. Suppose the SSO delegates the capability to grant access right on the teaching course website (CWeb) to each Course Coordinator(CC) with a trust degree of 0.9 and a delegation depth of 1. However, SSO has made a strong university wide policy that anyone who enrolled in the course can access the website's teaching materials, and all school staff can also access the website's teaching materials. This is done by granting the access to enrolled students(Enro) and school staff with a full trust degree of 1. SSO has also made a weak policy that external students can access the course website's teaching materials, and this is done by granting the access to students with a low trust degree of 0.6. Suppose SSO says Head of School(HOS) can assign the course coordinators with a trust degree of 1 and a delegation depth of 1. HOS says Helen is a course coordinator if no information shows she is on leave. If Helen wants to deny the external students to access her teaching materials, she can grant them a negative authorization with a full trust degree. In this case her grant's effective trust is 0.9*1, which is greater than SSO's 0.6, and therefore her grant will override. However, if Helen wants to deny the school staff to access in the same way, she will not be successful as her grant's effective trust degree of 0.9 (the maximum she can get) is less than the SSO's 1. This time SSO's grant will win. The situation will be represented in RBAP program in the next section.

When the two conflicting authorizations have the same effective trust, the conflicts are said to be unresolved in our framework. Although we can use the simple negative-take-precedence to solve them, we prefer to leave it to the access control mechanisms to resolve the situation.

We would like to point out that most current conflict resolution policies can be easily achieved in our framework. For example, by always giving a negative authorization a higher trust than positive one or vice versa, we can achieve *negative-take-precedence* or *positive-take-precedence* policies. By giving more specific authorizations a higher trust degree than general ones, we can achieve the *more specific-take-precedence* policy. Similarly, by always giving the strong authorizations a higher trust degree than the weak ones, we can achieve the *strong and weak policies*. By giving a trust degree of less than one to administrative privilege delegations, and giving a trust degree of one to authorizations, we can achieve the *predecessor-take-precedence* policy. This shows that our conflict resolution policy is very flexible which can meet different applications' needs.

2.5   Other Features

Various role constraints, such as strong exclusion, role cardinality, role composition requirement and role dependency, are supported in our framework. Authorization and delegation propagations along the role and object hierarchies are also supported which can greatly reduce the size of explicit role and access right delegations and assignments.

3   Role Based Authorization Programs

Role based authorization program (RBAP) is an extended logic program. Any RBAP is consisted of domain specific rules and general rules. The former is defined by users to express the desired application related security policies, while the latter is defined in this section to capture the general features stated in the last section.

3.1   Syntax of RBAP

Role based authorization program (RBAP) is a multi-sorted first order language, with seven disjoint *sorts* $\mathcal{R}, \mathcal{U}, \mathcal{O}, \mathcal{A}, \mathcal{T}, \mathcal{W}$, and $\mathcal{N}$ for role, user, object, access right, authorization type, weight and depth respectively. Variables are denoted by strings starting with lower case letters, and constants by strings starting with upper case letters. In addition, two partial orders $<_R, <_O$ are defined on sorts $\mathcal{R}$ and $\mathcal{O}$ representing the role and object hierarchies respectively. There are two authorization types denoted by $-, +$, where $-$ means *negative*, $+$ means *positive*. A negative authorization specifies that the access must be forbidden, while a positive authorization specifies that the access must be granted. $\mathcal{W}$ is a set of real number in (0,1], and $\mathcal{N}$ is a set of non-negative integers. In general, we prohibit function symbols in our language for the sake of simplicity, but allow some simple built-in arithmetic functions to be used. A *rule r* is of the form:

$$b_0 \leftarrow b_1, ..., b_k, not\ b_{k+1}, ..., not\ b_m, m >= 0$$

where $b_0, b_1, ..., b_m$ are literals, and not is the negation as failure symbol. A *Role Based Authorization Program*, RBAP, consists of a finite set of rules.

The predicate set P in RBAP consists of a set of ordinary predicates defined by users, and a set of system built-in predicate designed for user to express role assignment, role/user to privilege grant, role/access right administrative privilege delegation, and role constraints.

In our formalization, administrative privilege delegation includes role administrative privilege delegation and access right administrative privilege delegation. We assume that System Security Officer (SSO) in an organization is the first role that can delegate. Others have to be granted the privilege to delegate. For role assignment delegation, a special predicate $(g, r, w, d)canAssign(r', r'')$ is defined, where (g,r,w,d) is called the grantor of the delegation. Intuitively, it means that a user g in role r says that the role r' can not only assign users to role r'', but also further delegate this administrative privilege on r'' for the maximum delegation depth d, and g's trust degree on this delegation to r' is w. If the depth is 0, r cannot further delegate. A depth of 1 would mean that r can further delegate to some role r' with maximum depth of 0 and etc. Similarly, for privilege grant delegation, we have $(g, r, w, d)canGrant(r', a, o)$, which means that a user g in role r says that role r' can not only grant access right a on o, but also further delegate this administrative privilege to the maximum depth of d. And g's trust degree on this delegation to r' is w.

In terms of role assignment, a special predicate of the form $(g, r, w)assign(r', u)$ is defined. Intuitively, it means that a grantor g in a role r assigns user u to role r'. g's trust degree on this role assignment is w. The users can be individuals, agents or processes. Similarly, we define a predicate grant $(g, r, w)grant(r, o, t, a)$ for authorization. It means that g in role r says that role r can/cannot(depending on the type t) do access a on object o, and the grant weight is w. A role r can assign users to a role r' if and only if r is SSO $or\ r$ has been granted the privilege to assign r' (using $canAssign$). A role r can grant other roles an access right a of type $+$ or $-$ over object o if and only if r is SSO $or\ r$ has been granted the privilege to grant a over o (using $canGrant$).

For the role delegation, predicate $deleRole(u, u', R)$ means user u delegates its role R to user u' while $deleAll(u, u')$ with Type $U \times U$ means user u delegates its every role to user u'. The delegatee can perform the delegated role due to the delegation.

For the exclusion of roles, we define a predicate $exclusive(r_1, r_2, r_3, r_4)$ to represent up to 4 roles exclusion. Please note that r_2 or r_3 can be empty denoted by _ to denote 2 or 3 role exclusion.

Predicate $roleNum(r, n, m)$ is defined for the role cardinality constraint. It means that a role r should have at least n and at most m members. If there is no restriction on the minimum or maximum number, n or m is set as 0. Exact number of members is represented by setting m==n.

For the role composition constraint, we define the predicate $roleComp(r, r', n, m)$. It means that role r should contain at least n and at most m members from another role r'.

In the real world, it is often required that to be able to perform role 1, one needs to be in role 2. For example, many universities require that Unit coordinators to be full time academic staff. Role dependency requirement is thus introduced in our framework to represent this situation. For the role dependency constraint, we define the predicate $depend(r, r')$ which means that role r depends on role r'.

Example 3. We can now use RBAP program to represent Example 1.

(a_1). $(Alice, SSO, 1, 1)canAssign(TUni, Stu) \leftarrow$
(a_2). $(Alice, SSO, 1, 1)canAssign(EduD, TUni) \leftarrow$
(a_3). $(Alice, SSO, 0.9, 0)canAssign(TUni, TUni) \leftarrow$
(a_4). $(Alice, SSO, 1)assign(TUni, x) \leftarrow year(x, 10), (u, r, t)assign(Uni, x)$
(a_5). $(Mary, EduD, 1, 0.9)assign(TUni, UniOne) \leftarrow$
(a_6). $(Evan, UniOne, 1)assign(Stu, Bob) \leftarrow$
(a_7). $(Evan, UniOne, 0.9)assign(TUni, UniTwo) \leftarrow$
(a_8). $(Rose, UniTwo, 0.9, 1)assign(Stu, Tom) \leftarrow$

Example 4. The following RBAP program represents Example 2.

(b_1). $(Alice, SSO, 0.9, 1)canGrant(CC, CWeb) \leftarrow$
(b_2). $(Alice, SSO, 1)grant(Enro, CWeb, +, R) \leftarrow$
(b_2). $(Alice, SSO, 1)grant(Staff, CWeb, +, R) \leftarrow$
(b_3). $(Alice, SSO, 0.6)grant(Stu, CWeb, +, R) \leftarrow$
(b_4). $(Alice, SSO, 1, 1)canAssign(HOS, CC) \leftarrow$
(b_5). $(Alice, SSO, 0.6)grant(Stu, CWeb, +, R) \leftarrow$
(b_6). $(Jim, HOS, 1, 0)assign(Helen, CC) \leftarrow not \, onleave(Helen)$
(b_7). $(Helen, CC, 1)grant(Staff, CWeb, -, R) \leftarrow$
(b_8). $(Helen, CC, 1)grant(Enro, CWeb, -, R) \leftarrow$

3.2 Domain-Independent Rules

In this section, we define a set of domain-independent rules to formally achieve the concepts defined in the last section on role and access right administrative privilege delegation, role constraints, delegation correctness, role and privilege propagations, and conflict resolution.

Rules r_1 - r_{11} are about role delegation and assignment management.

Rules for role assignment delegation correctness

The next two rules are about delegation of role assignment. The first rule means that any delegation from SSO will be accepted, represented by predicate $canAssign1$. The second rule means that if a role has been delegated to assign roles for the maximum delegation depth d', then the role's further delegation with depth less than d' is accepted.

(r_1). $(g, SSO, w, d)canAssign1(r, r') \leftarrow (g, SSO, w, d)canAssign(r, r')$
(r_2). $(g, r_1, w, d)canAssign1(r, r_2) \leftarrow (g, r_1, w, d)canAssign(r, r_2),$
$\quad\quad (g', r_3, w_1, d')canAssign1(r_1, r_2), role(r_1, g), d' > d$

Rules for role assignment correctness

The following two rules mean any role assignment from the System Security Officer or a grantor holding the right to assign a role is accepted represented by $assign1$.

(r_3). $(g, SSO, w)assign1(r', u) \leftarrow (g, SSO, w)assign(r', u), role(SSO, g)$
(r_4). $(g, r, w)assign1(r_1, u) \leftarrow (g, r, w)assign(r_1, u),$
$\qquad (g', r_2, w_1, d)canAssign1(r, r_1), role(r, g)$

Rules for calculating the role's assigner's trust degree

A role assigner's trust degree is the production of all the trust weights along the path from the root(SSO) to it. If there are multiple paths to it, then the greatest number will be chosen as its trust degree. The following rules are used to achieve this. The first rule assigns SSO the maximum trust degree of 1. The second calculates the candidate trust based on a delegation's weight and the delegator's trust degree. The third and fourth rule check that, for any candidate trust degree x of an assigner g on a role r (represented by $trusts(g, r, x)$), if there is any other that is greater than it (represented by $existHigherTrusts(g, r, x)$). If there is no higher one, then the current one is the assigner's trust degree(represented by $trust(g, r, x)$).

(r_5). $trust(SSO, r, 1) \leftarrow$
(r_6). $trusts(r_1, r_2, w' * w) \leftarrow (g, r, w, d)canAssign1(r_1, r_2), trust(r, r_2, w')$
(r_7). $existHigherTrusts(r, r', w) \leftarrow trusts(r, r', w), trusts(r, r', w'), w' > w$
(r_8). $trust(r, r', w) \leftarrow trusts(r, r', w), not\ existHigherTrusts(r, r', w)$

Rules for role assignment acceptance

The following rule means that a role assignment from an assigner is accepted if the production of its trust degree and the assignment weight is greater than W, where W is a constant in $(0,1]$ used to express the trust degree accepted by the organization.

(r_9). $role(r, u) \leftarrow (g, r', w)assign1(r, u), trust(r', r, w'), w' * w >= W$

Rules for role delegation

The next two rules are about role delegation. The first rule means that a user u delegates its role r to another user u'. If u is a member of r, then u' will be added to the role. The second rule means that u delegates all of its roles to u'.

(r_{10}). $role(r, u') \leftarrow deleRole(u, u', r), role(r, u)$
(r_{11}). $role(r, u') \leftarrow deleAll(u, u'), role(r, u)$

Rules s_1 - s_8 are about enforcing role constraints.

Rules for enforcing the constraint about separation of duty

The next rule is about generalized static separation of duty. It means that r_1, r_2, r_3 and r_4 are required to be exclusive, and thereby a user cannot be assigned to all of them. Note that r_2 or r_3 can be empty denoted by _. In this case $s <_M$ _ is assumed to be true. On the other hand, the first two roles cannot be _. That is, at least two roles are needed to define the exclusive relationship.

(s_1). $\leftarrow exclusive(r_1, r_2, r_3, r_4), r_1 \neq \_, r_2 \neq \_,$
$role(r_1, u), role(r_2, u), role(r_3, u), role(r_4, u)$

Rules for enforcing the constraint about role dependency requirement

Role r is said to be dependent on r' if r needs to be a subset of r' The following rule enforces this requirement.

(s_2). $\leftarrow depend(r_1, r_2), role(r_1, u), notrole(r_2, u)$

Rules for enforcing the constraint about the cardinality constraint

The following rules are used to enforce the role cardinality constraints. The first rule means that it is not acceptable if there are more than m different members in r. The second and third rules mean that it is not acceptable if there are not n different members in r.

(s_3). $\leftarrow roleNum(r, n, m), role(r1, u_1), ..., role(r, u_{m+1}), m > 0, u_i \neq u_j$
$(i \neq j, i, j = 1, ..., m + 1)$
(s_4). $existRoleNum(r, n) \leftarrow role(r, u_1), ..., role(r, u_n), n > 0, u_i \neq u_j$
$(i \neq j, i, j = 1, ..., n)$
(s_5). $\leftarrow roleNum(r, n, m), notexistRoleNum(r, n)$

Rules for enforcing the constraint about role composition requirement

The following rules are used to enforce the role composition requirement. The first rule says that, if a role r should have at most m members from r' , but in fact there are $m + 1$ users who belong to both roles, then the system will fail. The second and the third rules mean that if a role r should have at least m members from r' , but the system cannot find n members from r' belonging to r, then it will fail.

(s_6). $\leftarrow roleComp(r, r', n, m), role(r, u_1), ..., role(r, u_{m+1})$
$role(r', u_1), ..., role(r', u_{m+1}), u_i \neq u_j, (i \neq j, i, j = 1, ..., m + 1, r)$
(s_7). $existRoleComp(r, r', n) \leftarrow role(r, u_1), ..., role(r, u_n),$
$role(r', u_1), ..., role(r', u_n), u_i \neq u_j, (i \neq j, i, j = 1, ..., n, r)$
(s_8). $\leftarrow roleComp(r, r', n, m), not\ existRoleComp(r, r', n)$

Rules o_1 - o_8 are about access right delegation and grant management.

Rules for administrative privilege delegation correctness

The following rules are used to guarantee that the administrative privileges are properly delegated in terms of the eligible delegators and the valid delegation depths. The first rule means the System Security Officer (SSO)'s any administrative privilege delegation is accepted (represented by predicate $canGrant1$). The second rule means if a subject has been delegated the right to grant for the maximum delegation depth d', then the subject's further administrative delegation with depth less than d' is accepted.

(o_1). $(g, SSO, w, d)canGrant1(r, o, a) \leftarrow (g, SSO, w, d)canGrant(r, o, a),$
$role(SSO, g)$

(o_2). $(g, r_2, w, d) canGrant1(r_1, o, a) \leftarrow (g, r_2, w, d) canGrant(r_1, o, a),$
$(g', r_3, w', d') canGrant1(r_2, o, a), role(r_2, g), d' > d$

Rules for administrative privilege propagation along the role hierarchy

The next rule is about administrative privilege propagations along the role hierarchy. It means that a subject's administrative privilege on some object and access right would propagate automatically to its next higher level roles represented by $<_R$ relation.

(o_3). $, (g, r'', w, d) canGrant1(r, o, a) \leftarrow (g, r'', w, d) canGrant1(r', o, a),$
$r <_R r'$

Rules for administrative privilege propagation along the object hierarchy

The next two rules are about administrative privilege propagations along the object hierarchy. It means that a subject's administrative privilege on some object and access right would propagate automatically to its next lower level objects represented by $<_O$.

(o_4). $(g, r', w, d) canGrant1(r, o, a) \leftarrow (g, r', w, d) canGrant1(r, o', a),$
$o' <_O o$

Rules for authorization correctness

The following two rules mean any grant from the System Security Officer or a grantor holding the right to grant is accepted (represented by predicate $grant1$).

(o_5). $(g, SSO, w) grant1(r, o, t, a) \leftarrow (g, SSO, w) grant(r, o, t, a),$
$role(SSO, g)$
(o_6). $(g, r_1, w) grant1(r, o, t, a) \leftarrow (g, r_1, w) grant(r, o, t, a),$
$(g', r_2, w', d) canGrant1(r_1, o, a), role(r_1, g)$

Rules for authorization propagation

The following rules are used to achieve authorization propagation along role and objects inheritance hierarchies. The first rule means any authorization given to a role would propagate to its superior roles represented by the $<_R$ relation. The second rule means any authorization on an object would propagate to its sub-objects represented by the $<_O$ relation.

(o_7). $(g, r, w) grant1(r_2, o, t, a) \leftarrow (g, r, w) grant1(r_1, o, t, a), r_1 <_R r_2$
(o_8). $(g, r, w) grant1(r_1, o', t, a) \leftarrow (g, r, w) grant1(r_1, o, t, a), o <_O o'$

Rules c_1 - c_9 are about conflict resolution for roles and users.

Rules for conflict resolution

As mentioned before, the conflict resolution is based on the priority of the grantor and the weight of the authorization. The priority of a grantor is the weighted length of the shortest delegation path from the owner to it, since there may exist multiple paths to it. In the following rules, we use predicate *priority* to represent the priority of a grantor, which is the length of the shortest path. We use *priorities* to represent all the priorities that a grantor received from their delegators, which are lengths of all possible paths to it. Predicate *existHigherPriorities* means that the corresponding *priorities* is not the highest one. It is introduced to avoid the existential quantifier to be used in (c_4), as in an extended logic program all the variables in clauses are considered to be universally quantified. For the two conflicting authorizations, we will compare the sum of the grantor's priority and weight of the authorization, and the one with smaller value will win. Predicate *overridden* is introduced to indicate that the corresponding authorization is overridden by some other authorizations. Predicate *hold* means the corresponding authorization holds as it is not overridden by any other authorizations, and its effective weight is greater than W which is a constant denoting the trust degree accepted by the organization.

(c_1). $priority(SSO, o, a, 1) \leftarrow$

(c_2). $priorities(r, o, a, w' * w) \leftarrow (r', w, d)canGrant1(r, o, a),$
$priority(r', o, a, w')$

(c_3). $existHigherPriorities(r, o, a, w) \leftarrow priorities(r, o, a, w),$
$priorities(r, o, a, w'), w' > w$

(c_4). $priority(r, o, a, w) \leftarrow priorities(r, o, a, w),$
$not\ existHigherPriorities(r, o, a, w)$

(c_5). $(g, r_1, w)overridden(r, o, t, a) \leftarrow (g, r_1, w)grant1(r, o, t, a),$
$(g', r_2, w_1)grant1(r, o, t', a),$
$priority(r_1, o, a, w_2), priority(r_2, o, a, w_3), w_1 * w_3 > w_2 * w$

(c_6). $(g, r', w * w')grant2(r, o, t, a) \leftarrow (g, r', w)grant1(r, o, t, a),$
$not\ (g, r', w)overridden(r, o, t, a), priority(r', o, a, w'), w' * w > W$

(c_7). $(g, r', w)grant3(u, o, t, a) \leftarrow (g, r', w)grant2(r, o, t, a), role(r, u)$

(c_8). $(g_1, r_1, w)overridden(u, o, t, a) \leftarrow (g_1, r_1, w_1)grant3(u, o, t, a),$
$(g_2, r_2, w_2)grant3(u, o, t', a), w_1 < w_2$

(c_9). $(g, r, w)hold(u, o, t, a) \leftarrow (g, r, w)grant3(u, o, t, a),$
$not\ (g, r, w)overridden(u, o, t, a)$

Let R denote all the general rules, i.e. $R = \{r_1, ..., r_{11}, s_1, ..., s_8, o_1, ...o_8, c_1, ..., c_9\}$. R will be combined with application dependent rules to evaluate the authorizations holding at any time (represented by predicate *hold*). Any access request will then be checked against it. For example, Example 3 and 4 will be combined with R to evaluate the authorizations holding in this application.

4   Formal Semantics of RBAP

We will adopt well known answer set semantics for RBAP because it is more suitable for our purpose of handling authorization conflicts since it provides a more flexible way to deal with incomplete and contradictory information. Let Π be a RBAP, the *Base* B_Π of Π is the set of all possible ground literals constructed from the system reserved predicates and predicates appearing in the rules of Π, the constants occurring in $\mathcal{R}, \mathcal{U}, \mathcal{O}, \mathcal{A}, \mathcal{T}, \mathcal{W}, \mathcal{N}$. A *ground instance* of r is a rule obtained from r by replacing every variable W in r by $\delta(x)$, where $\delta(x)$ is a mapping from the variables to the constants in the same sorts. Let $G(\Pi)$

denote all ground instances of the rules occurring in Π. Two ground literals are *conflicting* on subject S, object O and access right A if they are of the form $(G, R, W)hold(U, O, +, A)$ and $(G', R', W')hold(U, O, -, A)$. A subset of the Base of B_Π is *consistent* if no pair of complementary or conflicting literals is in it. An *interpretation* I is any consistent subset of the Base of B_Π.

Definition 4. *Given a RBAP Π, an interpretation for Π is any interpretation of $\Pi \cup R$. Let I be an interpretation for a RBAP $G(\Pi)$, the reduction of Π w.r.t I, denoted by Π^I, is defined as the set of rules obtained from $G(\Pi \cup R)$ by deleting (1) each rule that has a formula not L in its body with $L \in I$, and (2) all formulas of the form not L in the bodies of the remaining rules.*

Given a set R of ground rules, we denote by $pos(R)$ the positive version of R, obtained from R by considering each negative literal $\neg p(t_1, ..., t_n)$ as a positive one with predicate symbol $\neg p$.

Definition 5. *Let M be an interpretation for Π. We say that M is an answer set for Π if M is a minimal model of the positive version $pos(\Pi^M)$.*

5   Discussion and Related Work

We have developed a simple XML interface for RBAP. XML is most suitable for integration of information from various sources. This is especially useful for our scheme since there exist multiple authorities due to dynamic delegation and there is a need to exchange and integrate the security policies on the web. A visual tool interface is also developed in which one can create roles, users, objects, access rights, user to role mapping, role to access rights mapping, administrative privilege delegations etc. The XML file will be created based on it. On the other hand, Smodels based processor is used as the policy evaluation engine for RBAP. Smodels is a widely used system that implements the answer set semantics for extended logic programs. It is domain-restricted but supports extensions including built-in functions as well as cardinality and weight constraints.

ARBAC97 [4] is a model for decentralized administration of RBAC policies. In a typical ARBAC policy, roles are classified as administrative roles and normal roles. Only administrative roles can assign users and privileges. There is a single top level administrator role, called the Senior Security Officer (SSO) and a number of Junior Security Officers (JSO). The SSO partitions the organizations policy into different security domains, each of which is administered by a different JSO. To decentralise the policy details without loosing central control over broad policy, each delegation to JSO is associated with some condition. The ARBAC policy specifies, for instance, to which normal roles and under what conditions can members of a JSO role assign users. Although the framework decentralizes the administration to some degree, the scalability is still limited. Use of additional relations to limit the delegation and thereby enforcing central control is not flexible. Furthermore, negative authorizations are not discussed in ARBAC. ARBAC99 [5] and [8] improve some shortcomings of ARBAC97. But, they do not change the fundamental approach of using administrative roles, additional relations, and role ranges based on hierarchies. Our framework does not distinguish user roles and administrative roles. Delegations are conditional, and the broad policy is achieved through trust management. [2] proposes to use security analysis techniques, which views an access-control system as a state transition system, to maintain desirable security properties while delegating administrative privileges. [9] discusses how to assist a delegator in choosing the delegatee using trust-based approach. In their approach, trust is a relationship between a truster and trustee with respect to a given task and depends on several factors such as properties, experience and recommendation. The factors are quantified and used for the delegator to choose the delegatee. A permission based delegation model is presented in [10], which supports user-to-user, role-to role, and multi-depth delegations. However, how to resolve conflicts with respect to access rights in this context has not been well studied. [6] describes a way to detect conflicts, meaning the violation of specified constraints such as separation of duty constrains, caused by the role delegations. For conflict resolution, it is suggested in [6] to constrain the delegations, which in turn greatly reduces the flexibility of the model.

For our future work, we intend to study the formal model for a temporal role based access control system, where time period can be associated with roles and delegations, which will expire when the time period expires. We also intend to investigate program update including revocation of roles and access rights.

References

1. Gelfond, M., Lifschitz, V.: Classical negation in logic programs and disjunctive databases. New Generation Computing 9, 365–385 (1991)
2. Li, N., Tripunitara, M.V.: Security Analysis in Role-Based Access Control. ACM Transactions on Information and System Security 9(4), 391–420 (2006)
3. Sandhu, R.S., Coyne, E.J., Feinstein, H.L., Youman, C.E.: Role based access control models. IEEE Computer 29(2), 38–47 (1996)
4. Sandhu, R., Bhamidipati, V., Munawer, Q.: The ARBAC97 model for role-based administration of roles. ACM Trans. Inf. Syst. Secur. 2(1), 105–135 (1999)
5. Sandhu, R.S., Munawer, Q.: The ARBAC99 model for administration of roles. In: Proc of the 18th Annual Computer Security Applications Conference, pp. 229–238 (1999)
6. Schaad, A.: Conflict detection in a role-based delegation model. In: Proc. of Annual Computer Security Applications Conference (2001)
7. Schaad, A., Moffett, J., Jacob, J.: The role-based access control system of a European bank: A case study and discussion. In: Proc. of the Sixth SACMAT, pp. 3–9 (2001)
8. Oh, S., Sandhu, R.S.: A model for role admininstration using organization structure. In: Proc. of the Seventh SACMAT (2002)
9. Toahchoodee, M., Xie, X., Ray, I.: Towards trustworthy delegation in Role-Based Access Control Model. In: Samarati, P., Yung, M., Martinelli, F., Ardagna, C.A. (eds.) ISC 2009. LNCS, vol. 5735, pp. 379–394. Springer, Heidelberg (2009)
10. Zhang, X., Oh, S., Sandhu, R.: PBDM: A flexible delegation model in RBAC. In: Proc. of the 8th ACM Symposium on Access Control Models and Technologies (2003)

Robust Ranking of Uncertain Data

Da Yan and Wilfred Ng

The Hong Kong University of Science and Technology
Clear Water Bay, Hong Kong
{yanda,wilfred}@cse.ust.hk

Abstract. Numerous real-life applications are continually generating huge amounts of uncertain data (e.g., sensor or RFID readings). As a result, top-k queries that return only the k most promising probabilistic tuples become an important means to monitor and analyze such data. These "top" tuples should have both high scores in term of some ranking function, and high occurrence probability. The previous works on ranking semantics are not entirely satisfactory in the following sense: they either require user-specified parameters other than k, or cannot be evaluated efficiently in real-time scale, or even generating results violating the underlying probability model. In order to overcome all these deficiencies, we propose a new semantics called U-Pop$k$ based on a simpler but more fundamental property inherent in the underlying probability model. We then develop an efficient algorithm to evaluate U-Pop$k$. Extensive experiments confirm that U-Pop$k$ is able to ensure high ranking quality and to support efficient evaluation of top-k queries on probabilistic tuples.

1   Introduction

Many emerging applications, such as environmental surveillance and mobile object tracking, involve the generation of uncertain data which are inherently fuzzy and noisy. As a result, various probabilistic DBMSs are developed to support the storage and querying of these uncertain data [2–4]. Since precise query expressions like SQL may not be ideal to evaluate such data, top-k queries become an important means to extract information from them.

Top-k queries on deterministic data have been well studied [5, 6]. Unlike top-k queries on deterministic data, ranking probabilistic tuples requires taking both the tuple score and its occurrence probability into account, which gives rise to new challenges in defining the query semantics. Despite the many recent attempts to study top-k query semantics in the context of probabilistic relations [1, 7–10], these seemingly natural semantics lead to quite different query results.

Recently, [11] proposes a unified framework that incorporates several of the semantics and gives an approach to learn the ranking function from user preference. [12] proposes to return a number of typical top-k results to users. While these works mitigate the inconsistency of the previous semantics and provide more flexibility, they exert extra burden on users by requiring their intervention.

The work of [1] proposes five intuitive properties for top-k queries, and shows that only ExpectedRank satisfies all of them and it is considerably more efficient

J.X. Yu, M.H. Kim, and R. Unland (Eds.): DASFAA 2011, Part I, LNCS 6587, pp. 254–268, 2011.
© Springer-Verlag Berlin Heidelberg 2011

to evaluate than the previous semantics. However, ExpectedRank has two significant deficiencies: First, its results may contradict with the probability model, which will be detailed in Section 4. Second, its results deviate considerably from the results of the other semantics, which will be detailed in Section 6.1.

In this paper, we first identify a simple but fundamental property inherent in the probability model, which any robust ranking semantics for probabilistic data should satisfy. Then we describe our new semantics, U-Pop$k$, that is founded on this property to rank probabilistic tuples. It can be proved that U-Pop$k$ satisfies all the five properties of [1]. The efficiency of our evaluation algorithm and the ranking quality of U-Pop$k$ are evaluated using both real and synthetic datasets.

Compared to the state-of-the-art semantics, our proposal has many desirable features. First, U-Pop$k$ gives more reasonable results than ExpectedRank in general, with comparable evaluation cost. Second, unlike the work in [11, 12], U-Pop$k$ requires no user intervention except for the parameter k, and is thus easier to use. Finally, the evaluation of U-Pop$k$ takes considerably less time than other semantics, and thus U-Pop$k$ paves a much better way towards real time analyses.

The rest of the paper is organized as follows. Section 2 defines our probabilistic data model. We review the related work in Section 3. The robustness property and our ranking semantics are proposed in Section 4, and the corresponding algorithms are presented in Section 5. Extensive experiments are conducted in Section 6 to demonstrate the efficiency of our algorithm and the ranking quality of our semantics. Finally, we conclude the paper in Section 7.

2   Probabilistic Data Model

Among the many uncertain data models proposed in the literature, the *tuple-level probabilistic model* [1, 8–12] is one of the most important models. In this model, each tuple t is accompanied with the probability p of its occurrence. The model is able to capture the form of uncertain data that is common in many real life applications, such as sensor readings with confidence on their sensor states, and data tuples with confidence on their information sources. We adopt the tuple-level probabilistic model throughout the paper due to its popularity in real life applications.

Figure 1(a) shows our running example relation conforming to the tuple-level model, where the ranking score is defined according to the attribute "Speed", which records the car speed readings detected by different radars in a sampling moment. In this relation, a confidence field "Conf." is attached with each tuple to indicate its occurrence probability. The occurrence probability of t_1, denoted as $Pr(t_1)$, is 0.4. In contrast, the probability of the event that t_1 does not occur is given by $Pr(\neg t_1) = 1 - Pr(t_1) = 0.6$.

Note that both tuples t_2 and t_6 record the speed reading of the same car. Since a car can only have one speed in a given moment, t_2 and t_6 cannot co-exist, which we denote as $t_2 \oplus t_6$. We call such a constraint an *exclusion rule*. We have another *exclusion rule* $t_3 \oplus t_5$ defined in the relation for a similar reason. Different forms of constraints can be adopted in a probabilistic relation [2, 3]

	Radar Location	Car Make	Plate No.	Speed	Confidence
t_1	L_1	Honda	X-123	130	0.4
t_2	L_2	Toyota	Y-245	120	0.7
t_3	L_3	Mazda	W-541	110	0.6
t_4	L_4	Nissan	L-105	105	1.0
t_5	L_5	Mazda	W-541	90	0.4
t_6	L_6	Toyota	Y-245	80	0.3

Exclusion Rules: $(t_2 \oplus t_6), (t_3 \oplus t_5)$

(a)

Possible World	Probability
$PW^1=\{t_1, t_2, t_4, t_5\}$	0.112
$PW^2=\{t_1, t_2, t_3, t_4\}$	0.168
$PW^3=\{t_1, t_4, t_5, t_6\}$	0.048
$PW^4=\{t_1, t_3, t_4, t_6\}$	0.072
$PW^5=\{t_2, t_4, t_5\}$	0.168
$PW^6=\{t_2, t_3, t_4\}$	0.252
$PW^7=\{t_4, t_5, t_6\}$	0.072
$PW^8=\{t_3, t_4, t_6\}$	0.108

(b)

Fig. 1. (a)Probabilistic Relation with Exclusion Rules (b)Possible World Space

but exclusion rules are the most popular in the literature due to their simplicity and usefulness. We considered only exclusion rules in this paper.

Note that each exclusion rule corresponds to an entity (e.g. a car in our example) of the relation. Thus, each tuple appears in at most one exclusion rule. In our example, the rule $t_2 \oplus t_6$ means that the speed of the "Toyota" car takes the value 120 with probability 0.7 and 80 with probability 0.3. In general, given an exclusion rule $t_{i_1} \oplus t_{i_2} \oplus \cdots \oplus t_{i_m}$, we have (1) $Pr(t_{i_1}) + Pr(t_{i_2}) + \cdots + Pr(t_{i_m}) \leq 1$, and (2) at most one tuple in $\{t_{i_1}, t_{i_2}, \ldots, t_{i_m}\}$ can occur. There is still an event that no tuple in the rule occurs, which has probability $(1 - Pr(t_{i_1}) - Pr(t_{i_2}) - \ldots - Pr(t_{i_m}))$.

If a tuple t_i is independent of all other tuples, we say t_i itself is in a *trivial* rule. In addition, any two tuples from different rules are assumed to be independent. Thus, we have four rules in the example as follows: t_1, $t_2 \oplus t_6$, $t_3 \oplus t_5$ and t_4, where t_1 and t_4 are in trivial rules, and t_2 and t_3 are independent. This assumption simplifies the computation of the probabilities of possible worlds, which will be further elaborated when discussing the algorithm issues in Section 5.

In the tuple-level model, a possible world (PW) is a subset of the tuples in the probabilistic relation. Figure 1(b) shows the possible world space for the relation in Figure 1(a). The probability of each world is computed as the joint probability of the "occurrence events" of the tuples in the world, and the "absence events" of all the other tuples. For example, the probability of $PW^1 = \{t_1, t_2, t_4, t_5\}$ is $Pr(PW^1) = Pr(t_1) \times Pr(t_2) \times Pr(t_4) \times Pr(t_5) = 0.112$. Multiplication is used here because the occurrence events of t_1, t_2, t_4 and t_5 are independent of each other, and the absence events of t_3 and t_6 are already implied in the occurrence events of t_5 and t_2 due to the exclusion rules. Similarly, the probability of $PW^5 = \{t_2, t_4, t_5\}$ is $Pr(PW^5) = Pr(\neg t_1) \times Pr(t_2) \times Pr(t_4) \times Pr(t_5) = 0.168$.

3   Related Work

Several semantics for top-k queries on uncertain data have recently been proposed, such as U-Top$k$[8], U-kRanks[8], Global-Top$k$[9], PT-k[10] and ExpectedRank[1], all of which are defined on the possible world model. We now illustrate their semantics by performing a top-2 query on the relation in Figure 1(a).

1) U-Top$k$ returns the most probable top-k tuples that belong to a valid possible world. Consider U-Top2 and define $\langle t_i, t_j \rangle$ to be the event that t_i is ranked the first and t_j the second in a possible world. By merging the possible worlds in Figure 1(b) whose top-2 tuples are the same, we have $Pr(\langle t_1, t_2 \rangle) = Pr(PW^1) + Pr(PW^2) = 0.28$, which is the largest among all possible top-2 combinations. Therefore, the result of U-Top2 is $\langle t_1, t_2 \rangle$.

However, there can be a large number of valid possible worlds. As a result, the most probable top-k tuples that belong to a valid possible world can occur with a very small probability. [12] proposes to return c typical top-k tuple vectors in terms of the distribution of the total score of top-k tuples, from which users need to choose one, which is itself a non-trivial task for users.

2) U-kRanks returns the set of most probable top-i th tuples across all possible worlds, where $i = 1, \ldots, k$. Let us compute U-2Ranks. First, consider the most probable tuple to appear in the 1st position. Tuple t_2 appears in the 1st position with probability $Pr(PW^5) + Pr(PW^6) = 0.42$, since it appears the first only in PW^5 and PW^6. Similarly, t_1 appears in the first position with probability $Pr(PW^1) + Pr(PW^2) + Pr(PW^3) + Pr(PW^4) = 0.4$. After considering all the tuples, we can see that t_2 appears in the $1st$ position with maximum probability. Thus the first answer to U-2Ranks is t_2. The second answer to U-2Ranks should be the most probable tuple to appear in the $2nd$ position, and similarly, we find that tuple t_3 appears in the $2nd$ position with maximum probability $Pr(PW^4) + Pr(PW^6) = 0.324$. To sum up, the result of U-2Ranks is $\langle t_2, t_3 \rangle$.

Since a tuple may be the most probable tuple to appear in more than one position, the same tuple may be listed multiple times in the result of U-kRanks, which is a very unnatural answer to users.

3) PT-k returns all tuples whose probability values of being in the top-k answers in possible worlds are above a threshold; **Global-Top$k$** returns k highest-ranked tuples according to their probability of being in the top-k answers in possible worlds.

As for Global-Top2 and PT-2, we check for each tuple the probability that it is within top-2. Tuple t_2 is within top-2 in worlds PW^1, PW^2, PW^5 and PW^6, and thus with probability $Pr(PW^1) + Pr(PW^2) + Pr(PW^5) + Pr(PW^6) = 0.7$. Similarly, the probability to be within top-2 is 0.4 for t_1, 0.432 for t_3, 0.396 for t_4, 0.072 for t_5 and 0 for t_6. Global-Top2 picks the two tuples with maximum probability to be within top-2, namely t_2 and t_3. On the other hand, PT-2 picks all the tuples with probability to be within top-2 higher than a pre-specified threshold. If the threshold is set to be 0.5, then only t_2 is returned. However, if the threshold is set to be 0.3, then the result would become $\{t_2, t_1, t_3, t_4\}$.

One limitation of PT-k is that the number of returned tuples may not be k but depends on the user-specified threshold, which is difficult for a user to set.

4) ExpectedRank(k) returns k tuples whose *expected ranks* across all possible worlds are the highest. However, if a tuple does not appear in a possible world, its rank is then undetermined. To solve this, in a possible world with m tuples, ExpectedRank ranks absent tuples to be in the $(m + 1)th$ position. Thus, t_2 and t_5 are both ranked $5th$ in PW^4 shown in Figure 1(b), although t_2 have both higher score and higher occurrence probability than t_5. As a more detailed illustration, we consider tuple t_5 which is ranked $4th$ in PW^1, $5th$ in PW^2 due to its absence, $3rd$ in PW^3, $5th$ in PW^4 due to its absence, $3rd$ in PW^5, $4th$ in PW^6 due to its absence, $2nd$ in PW^7, and $4th$ in PW^8 due to its absence. Therefore the expected rank for t_5 is $0.112 \cdot 4 + 0.168 \cdot 5 + 0.048 \cdot 3 + 0.072 \cdot 5 + 0.168 \cdot 3 + 0.252 \cdot 4 + 0.072 \cdot 2 + 0.108 \cdot 4 = 3.88$. Similarly, we have the expected rank 2.8 for t_1, 2.3 for t_2, 3.02 for t_3, 2.7 for t_4, 4.1 for t_6. Therefore, the result of ExpectedRank(2) is $\{t_2, t_4\}$, since their expected ranks are the highest.

In fact, ExpectedRank defines an order on the tuples, e.g. $t_2 \succ t_4 \succ t_1 \succ t_3 \succ t_5 \succ t_6$ in the previous example, while the ranking function of Global-Top$k$ and PT-k is dependent on k, which means that a tuple in the top-k result may not appear in the top-$(k+1)$ result.

5) PRF [11] presents a unified framework that uses *parameterized ranking functions* (PRF) for ranking probabilistic data. The work proposes a basic function called PRF^e and employs a linear combination of PRF^e functions to approximate PRF by using Discrete Fourier transformation. This approach is able to achieve good performance at the expense of result quality.

PRF also gives a learning algorithm that learns the parameter from user preference. The training data come from explicit user feedback, which assumes that users know the interplay between high score and high occurrence probability of the tuples. However, users usually expect reasonable score-probability tradeoff automatically in the evaluation of such top-k queries and assume no extra effort to give explicit feedback. Another problem is that high-quality training data may not be available from casual users, even if they are willing to give feedback.

4   Robust Ranking Semantics

In this section, we first formalize a property called *top-1 robustness*, which is founded on the tuple-level probability model. Then a fundamental property of ranking deterministic data, which is called *top-stability*, is extended to rank probabilistic tuples. Top-stability enables repeated applications of *top-1 robustness*, based on which we define our new semantics.

Property 1 (Top-1 Robustness). The top-1 query on an uncertain relation D returns the tuple $t \in D$ such that $\forall t' \in D, Pr(r(t) = 1) \geq Pr(r(t') = 1)$, where $r(t)$ denotes the rank of t.

From now on, we will use "top-1 probability" to denote the probability for a tuple to be ranked top-1. Property 1 states that any top-k query semantics for probabilistic tuples should return the tuple with maximum top-1 probability when $k = 1$. Note that the semantics of U-Top1, U-1Ranks and Global-Top1 are equivalent, and they all satisfy Property 1. Although the number of tuples returned by PT-1 is determined by the threshold, the top-1 tuple defined in Property 1 must appear in the result of PT-1 if the result is not empty.

Unfortunately, ExpectedRank may violate this robustness property. Consider an example relation $\{t_1, t_2, t_3, t_4, t_5\}$ with an exclusion rule $t_1 \oplus t_2 \oplus t_3 \oplus t_5$, where tuples t_1 to t_5 are already sorted in descending order of their scores. We use p_i to denote the occurrence probability of t_i, and $p_1 = p_2 = p_3 = 0.2, p_4 = 0.45, p_5 = 0.4$. For top-1 query on this relation, t_5 should be returned since its top-1 probability $(1 - p_4)p_5 = 0.22$ is the highest. However, ExpectedRank picks t_4 whose top-1 probability $(1 - p_1 - p_2 - p_3)p_4 = 0.18$ is smaller, which contradicts Property 1 and thus the underlying probability model.

Property 2 (Top-Stability). The top-$(i + 1)th$ tuple should be the top-1st after the removal of the top-i tuples.

Property 2 is intuitive in the context of certain data. *Top-stability* implies that, in principle, we are able to adopt the following approach to obtain the top-k tuples: The top-1 tuple is repeatedly removed from the current relation until k tuples are obtained. To generalize this approach to ranking probabilistic tuples for answering a top-k query, we define a new semantics U-Pop$k$ as follows:

Definition 1 (U-Pop$k$). *Tuples are picked in order from a relation according to Property 2 until k tuples are picked, where the top-1st tuple is defined according to Property 1.*

According to Definition 1, when a tuple is picked as the result, it is removed from the relation and thus will never be considered again in later evaluation. This avoids the problem of multiple occurrences of the same tuple in the result.

Although U-Pop$k$ changes the probabilistic relation in each round of evaluation and hence the set of possible worlds, this enables the use of *top-1 robustness* to pick the result tuple in each round. If k is small compared with the size of the relation, which is not unusual in applications, the modified relation in each round is still a good approximation of the original one. Besides, by removing the "top" tuples from the relation, we only need to make comparison among the remaining tuples in the pool from which the next "top" tuple will be picked.

Our approach shares a similar spirit of the work that uses a simplification assumption to facilitate the application of a robust property. For example, a naïve Bayes classifier uses the simplification assumption that all observations are independent to the facility of the evaluation of the robust property (i.e., the Bayes' rule), and so does maximum likelihood estimation. Even in top-k queries on uncertain data, we have the independence assumption among tuples from different exclusion rules to simplify the computation of the probability of possible worlds, where the possible world model is robust. For U-Pop$k$, the simplification property is *top-stability* and the robust property is *top-1 robustness*.

Tuples are assumed to be pre-sorted in the descending order of tuple scores in all previous work, since a tuple t' with a score lower than another tuple t will be ranked lower in any possible world and thus has no influence on the probability computation for the rank of t, which is also adopted in U-Pop$k$. Recall that the tuples in Figure 1(a) are already pre-sorted according to the attribute speed. The following example illustrates how U-Pop2 works for the relation in Figure 1(a):

Example. The first step is to compute the top-1 tuple in the relation. Tuple t_1 is ranked the first with probability $Pr(t_1) = 0.4$. The probability is

$Pr(\neg t_1)Pr(t_2) = 0.42$ for t_2, since t_1 must not occur and t_2 must occur in this case, while the other tuples are immaterial. Since the probability that a tuple other than t_1 and t_2 is ranked the first is $Pr(\neg t_1)Pr(\neg t_2) = 0.18$, which is smaller than the probability for t_2 to be ranked the first, we can conclude that t_2 is ranked the first with maximum probability and is thus picked.

After removing t_2 from the relation, we have the tuples t_1, t_3, t_4, t_5, t_6 remained in the pool. Tuple t_1 is ranked the first with probability $Pr(t_1) = 0.4$. The probability is $Pr(\neg t_1)Pr(t_3) = 0.36$ for t_3. Since the probability that a tuple other than t_1 and t_3 is ranked the first is $Pr(\neg t_1)Pr(\neg t_3) = 0.24$, which is smaller than the probability for t_1 to be ranked the first, we thus conclude that t_1 is ranked the first with maximum probability and it is thus picked.

Therefore, the result for U-Pop2 on the relation in Figure 1(a) is $\langle t_2,\ t_1 \rangle$.

5   U-Pop$k$ Algorithms

In this section, we first present the algorithms to evaluate U-Pop$k$ for the case that all tuples are independent of each other (i.e., no exclusion rule is considered). Then, we extend them to handle the general case that *exclusion rules* are given.

5.1   Algorithm for Independent Tuples

We consider the special case where all tuples in a given probabilistic relation are independent of each other, and assume that tuples are pre-sorted in descending order of score.

Consider a relation having tuples $\{t_1, t_2, \ldots, t_n\}$, where t_1 to t_n are already sorted and p_i is the occurrence probability of t_i. In order to rank t_i as top-1, t_1 to t_{i-1} should not appear but t_i should appear, while all tuples after t_i (i.e., tuples with lower scores) are immaterial. Therefore, the top-1 probability of t_i is $(1 - p_1)(1 - p_2) \cdots (1 - p_{i-1})p_i$.

If we define $accum_i = (1 - p_1)(1 - p_2) \cdots (1 - p_{i-1})$ with the special case of $accum_1 = 1$, we have $accum_{i+1} = accum_i(1 - p_i)$, and the probability for t_i to be top-1 can be written as $accum_i \cdot p_i$.

Algorithm 1. Find the Top-1 Tuple

1: $accum \longleftarrow 1;\ \ max \longleftarrow -\infty;\ \ result \longleftarrow null$
2: **while** $accum > max$ **and** there are more tuples **do**
3:     {Process the next tuple t_i}
4:     $top1Prob \longleftarrow accum \cdot p_i$
5:     **if** $top1Prob > max$ **then**
6:         $max \longleftarrow top1Prob;\ \ result \longleftarrow t_i$
7:     $accum \longleftarrow accum \cdot (1 - p_i)$
8: **return** $result$

Algorithm 1 finds the top-1 tuple among a list of pre-sorted tuples. The parameter $accum$ is initialized to 1 in Line 1, and updated after each iteration

(Line 7). Lines 2–7 check the tuples one by one, and in each iteration a tuple t_i is read and its probability to be top-1 is computed as *top1Prob* (Line 4). If *top1Prob* is found to be larger than the maximum top-1 probability currently found (Line 5), i.e., *max*, it is updated and t_i is recorded (Line 6).

Note that we do not need to check all the tuples. Suppose we have checked t_i and updated *accum* to $(1 - p_1)(1 - p_2) \cdots (1 - p_i)$, where the current maximum top-1 probability is *max*. If *accum* $\leq$ *max*, then the tuple with top-1 probability equal to *max* must be the result. This is because the top-1 probability of any succeeding tuple t_j (for $j > i$) is $(1 - p_1)(1 - p_2) \cdots (1 - p_i)(1 - p_{i+1}) \cdots (1 - p_{j-1})p_j \leq (1 - p_1)(1 - p_2) \cdots (1 - p_i) \cdot 1 \cdots 1 \cdot 1 \leq$ *max*.

Intuitively, the parameter *accum* acts as an upperbound of the top-1 probabilities for the tuples to be checked, which enables early termination (Line 2).

It is straightforward to construct a naïve algorithm for U-Pop$k$ that uses Algorithm 1: All the sorted tuples are read into a memory *buffer* first. Then in each iteration, a top-1 tuple is picked from the current tuple *buffer* using Algorithm 1, removed from the *buffer* and added to the result set. This is repeated until k tuples are obtained.

However, the naïve approach is not efficient enough and can be much improved by reusing the parameters obtained from previous computation. To illustrate this, we suppose that t_3 have been checked, that the iteration stops due to *accum* $\leq$ *max*, and that t_2 is found to have maximum top-1 probability. Then after removing t_2, the top-1 probability of t_1 is still p_1, which can be reused, while that of t_3 is now $(1 - p_1)p_3$ rather than $(1 - p_1)(1 - p_2)p_3$.

In general, we can reuse the already computed top-1 probabilities of those tuples whose positions are before the picked top-1 tuple t_{top1}, since the removal of t_{top1} does not change their top-1 probabilities. Figure 2 shows an example where t_i is picked due to *accum* $\leq$ *max*, after checking t_j and updating *accum*. The top-1 probabilities for t_1 to t_{i-1} can be reused, t_i is removed from the buffer, and the top-1 probabilities for t_{i+1} to t_j should be updated (i.e., re-scanned). The update is simply to divide the original probability by $(1 - p_i)$ so as to rule out the consideration for t_i. After the update, the next iteration starts from t_{j+1}. We define $(j - i)$ to be the *rescan length* for this iteration.

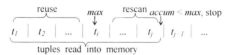

Fig. 2. Snapshot of the End of an Iteration

In order to delete the picked top-1 tuple from the memory buffer in $O(1)$ time for each iteration, we organize the buffer as a doubly-linked list and attach each tuple t in the buffer with the following three fields: (1) $t.prob$: its top-1 probability, (2) $t.id$: its index in the original sorted tuple list, and (3) $t.max$: the tuple whose top-1 probability is the maximum before t being put in the buffer.

Note that if t_i is picked as top-1, $t_i.max$ already records the tuple with maximum top-1 probability among the tuples whose positions are before t_i in the buffer (i.e., the current maximum top-1 probability for the tuples with position before t_{i+1}). While updating the top-1 probabilities from t_{i+1}, we update the current maximum top-1 probability if the updated probability is larger. Therefore, after all the updates, we get the current maximum top-1 probability for all the tuples with position before t_{j+1}. Then, the next iteration starts from t_{j+1}.

Algorithm 2 shows the process of top-1 probability adjustment between consecutive iterations discussed above, where t_i has already been identified as top-1.

Algorithm 2. Top-1 Probability Adjustment between Iterations

1: $maxTuple \longleftarrow t_i.max$
2: **if** $maxTuple == null$ **then**
3:    $max \longleftarrow -\infty$
4: **else**
5:    $max \longleftarrow maxTuple.prob$
6: **for each** tuple t after t_i in $buffer$ **do**
7:    $t.prob \longleftarrow t.prob/(1 - p_i)$; $t.max \longleftarrow maxTuple$
8:    **if** $t.prob > max$ **then**
9:       $max \longleftarrow t.prob$; $maxTuple \longleftarrow t$
10: Delete t_i from $buffer$

The variable $maxTuple$ records the tuple with the maximum top-1 probability currently, and max is its probability. Before updating from tuple t_{i+1} (Line 6), $maxTuple$ is set to $t_i.max$. However, $t_i.max$ may be $null$, because for t_1, no tuple with the maximum top-1 probability exists before it. In this case, max is set to $-\infty$. Otherwise, it is set to $maxTuple.prob$ which is the maximum top-1 probability among the tuples before t_i.

For each tuple t after t_i, $t.prob$ is divided by $(1 - p_i)$ and $t.max$ records $maxTuple$ (Line 7). If $t.prob > max$, t now has the maximum top-1 probability, and thus $maxTuple$ and max are updated accordingly (Line 9). Finally, t_i is removed from the doubly-linked list $buffer$ in $O(1)$ time in Line 10.

Algorithm 3 is a more efficient algorithm for U-Pop$k$ on independent tuples, which makes use of Algorithm 2. After the initialization in Lines 1–2, tuples are read into the memory $buffer$ one by one in Line 3. For each tuple, its fields are set in Line 4, and it is added to the end of $buffer$ (Line 5). If its top-1 probability is larger than the current maximum, the current maximum is updated accordingly (Lines 6–7). Then $accum$ is updated in Line 8 and checked in Line 9. If $accum \leq max$, the current top-1 is found to be $maxTuple$, which is thus put into the result set (Line 10). Then Algorithm 2 is called to update $buffer$ to reflect the removal of $maxTuple$ (Line 13), and so does $accum$ (Line 14).

If k results are picked, we do not need to read in more tuples (Lines 11–12). However, it is possible that all tuples are read into $buffer$ before k results are picked. Thus, after the for loop in Lines 3–14, if there are still less than k results,

Algorithm 3. U-Pop$k$ Algorithm for Independent Tuples

1: Create an empty doubly-linked list $buffer$
2: $accum \longleftarrow 1$; $resultSet \longleftarrow \phi$; $maxTuple \longleftarrow null$; $max \longleftarrow -\infty$
3: **for each** tuple t_i **do**
4:    $t_i.prob \longleftarrow accum \cdot p_i$; $t_i.max \longleftarrow maxTuple$
5:    Append t_i to the end of the doubly-linked list $buffer$
6:    **if** $t_i.prob > max$ **then**
7:       $max \longleftarrow t_i.prob$; $maxTuple \longleftarrow t_i$
8:    $accum \longleftarrow accum \cdot (1 - p_i)$
9:    **if** $accum \leq max$ **then**
10:       Put $maxTuple$ into $resultSet$
11:       **if** $|resultSet| == k$ **then**
12:          **return** $resultSet$
13:       $accum \longleftarrow accum/(1 - p_{maxTuple.id})$
14:       Call Algorithm 2 to adjust $buffer$
15: **while** $|resultSet| < k$ **and** $buffer$ is not empty **do**
16:    Put $maxTuple$ into $resultSet$
17:    **if** $|resultSet| == k$ **then**
18:       **return** $resultSet$
19:    $accum \longleftarrow accum/(1 - p_{maxTuple.id})$
20:    Call Algorithm 2 to adjust $buffer$
21: **return** $resultSet$

---

we need to pick $maxTuple$ into the result set (Line 16), and use Algorithm 2 to adjust the buffer and set the next $maxTuple$ (Line 19). This process is repeated until k results are picked, which is done by the while loop in Lines 15–20.

At most n tuples are read into $buffer$ where n is the total number of tuples in the probabilistic relation. Since all the steps in the for loop in Lines 3–14 take constant time except the adjustment in Line 13, they take $O(n)$ time in total. The adjustment of Lines 13 and 19 is executed exactly k times, since whenever a top-1 tuple is picked, Lines 14–19 or 23–28 are executed once, and thus Algorithm 2. It is straightforward to see that Algorithm 2 takes $O(L)$ where L is the *rescan length* for an iteration. The time complexity of Algorithm 3 is then $O(n + k \cdot L_{avg})$, where L_{avg} is the average *rescan length* among the k iterations.

Note that L_{avg} tends to be small since every factor of $accum$ is at most 1, which makes $accum$ smaller than max after a few tuples are read into $buffer$. For small k, it is not likely that all n tuples are read into $buffer$, but just a few top ones, the number of which is defined to be *scan depth* in [8].

The tricky case is when $p_i = 1$, where $accum$ will be updated to 0 in Line 8, and thus $accum \leq max$ and Algorithm 2 is then called in Line 13. So, t_{i+1} can never be read into $buffer$ until t_i is picked, and therefore Line 7 in Algorithm 2 is not executed and it does not cause division by 0. However, we cannot restore $accum$ using Line 14. Instead, if $p_i = 1$, we save $accum$ before executing Line 8 for later restoration when t_i is picked. The saved value is updated using Line 14, if other tuples are picked before the restoration. These details are not included in Algorithm 3 in order to make it more readable.

5.2   Algorithm for Tuples with Exclusion Rules

Now, we present the general case where each tuple is involved in an exclusion rule $t_{i_1} \oplus t_{i_2} \oplus \cdots \oplus t_{i_m} (m \geq 1)$. We assume that $t_{i_1}, t_{i_2}, \ldots, t_{i_m}$ in the rule are already pre-sorted in descending order of the tuple scores.

The upper bound of the top-1 probability for all the tuples starting from t_i is no longer $accum_i = (1 - p_1)(1 - p_2) \cdots (1 - p_{i-1})$. We denote $t_{j_1}, t_{j_2}, \ldots, t_{j_\ell}$ to be all the tuples with position before t_i and in the same exclusion rule of t_i. Then, there is a factor $(1 - p_{j_1} - p_{j_2} - \ldots - p_{j_\ell})$ in $accum_i$. Let us define $accum_{i+1} = accum_i \cdot (1 - p_{j_1} - p_{j_2} - \ldots - p_{j_\ell} - p_i)/(1 - p_{j_1} - p_{j_2} - \ldots - p_{j_\ell})$, which changes the factor of t_i's rule, in the top-1 probabilities of the tuples after t_i, from $(1 - p_{j_1} - p_{j_2} - \ldots - p_{j_\ell})$ to $(1 - p_{j_1} - p_{j_2} - \ldots - p_{j_\ell} - p_i)$. Now the top-1 probability of t_i is $accum_i \cdot p_i/(1 - p_{j_1} - p_{j_2} - \ldots - p_{j_\ell})$, where the denominator rules out the influence of the tuples exclusive with t_i.

Consider the factor in $accum_i$ that corresponds to rule $t_{j_1} \oplus t_{j_2} \oplus \cdots \oplus t_{j_m}$. If the tuples $t_{j_1}, t_{j_2}, \ldots, t_{j_\ell}$ $(\ell < m)$ are now before the current tuple considered, then the factor is $(1 - p_{j_1} - p_{j_2} - \ldots - p_{j_\ell})$. Within the same iteration, this factor can only be decreasing as we read in more tuples. For example, if we read in more tuples such that $t_{j_{\ell+1}}$ is also positioned before the current tuple, the factor then becomes $(1 - p_{j_1} - p_{j_2} - \ldots - p_{j_\ell} - p_{j_{\ell+1}})$, which is smaller.

The top-1 probability for t_i is no longer $accum_i \cdot p_i < accum_i$, but $accum_i \cdot p_i/(1 - p_{j_1} - p_{j_2} - \ldots - p_{j_\ell})$ to rule out the factor corresponding t_i's rule. Therefore, if we keep track of the factors corresponding to all the rules, where $factor_{min}$ is the smallest, the top-1 probability upper bound is **$accum_i/factor_{min}$** for all the tuples starting from t_i. This is due to two reasons. First, the top-1 probability of t_i is computed as $\frac{accum_i}{factor} \times p_i (< \frac{accum_i}{factor_{min}})$, where $factor$ is the factor corresponding to t_i's rule. Second, the factors in $accum$ that correspond to the rules can only decrease as i increases. We organize the rules in the memory by using *MinHeap* on $factor$. Thus, $factor_{min}$ can be retrieved from the top of the heap immediately, and the upper bound can be computed in $O(1)$ time. We call this MinHeap *Active Rule set* and denote it AR.

Note that we do not need to keep all the rules in AR. If all the tuples in a rule are *after* the current tuple, we do not need to fetch it into AR. Otherwise, if the rule of the current tuple is not in AR, we insert it into AR, which takes $O(\log |AR|)$ time. For each rule r in AR, we attach it with the following two fields: (1) $r.pivot$: the last tuple in rule r that is before the current tuple, and (2) $r.factor$: the current factor of r in $accum$. Suppose $r = t_{i_1} \oplus t_{i_2} \oplus \cdots \oplus t_{i_r} \oplus \cdots \oplus t_{i_m}$ and $r.pivot = t_{i_r}$, then $r.factor = (1 - p_{i_1} - p_{i_2} - \ldots - p_{i_r})$.

After checking a tuple and updating $accum$, if the upper bound is smaller than the current maximum top-1 probability, the tuple must be top-1 and thus the iteration terminates. Before next iteration, we need to update the top-1 probabilities of the tuples after the picked one. In this case, we need to update their probabilities segment by segment as illustrated in Figure 3.

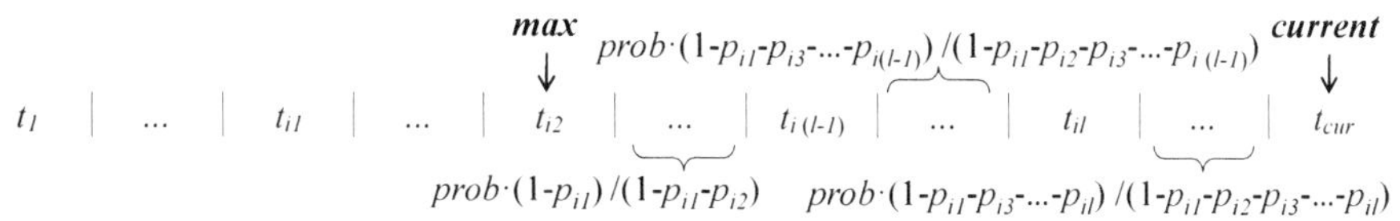

$$\overset{\textit{max}}{\downarrow} \quad prob\cdot(1\text{-}p_{i1}\text{-}p_{i3}\text{-}...\text{-}p_{i(l\text{-}1)})/(1\text{-}p_{i1}\text{-}p_{i2}\text{-}p_{i3}\text{-}...\text{-}p_{i\,(l\text{-}1)}) \quad \overset{\textit{current}}{\downarrow}$$

| t_1 | ... | t_{i1} | ... | t_{i2} | ... | $t_{i\,(l\text{-}1)}$ | ... | t_{il} | ... | t_{cur} |

$$prob\cdot(1\text{-}p_{il})/(1\text{-}p_{i1}\text{-}p_{i2}) \qquad prob\cdot(1\text{-}p_{i1}\text{-}p_{i3}\text{-}...\text{-}p_{il})/(1\text{-}p_{i1}\text{-}p_{i2}\text{-}p_{i3}\text{-}...\text{-}p_{il})$$

Fig. 3. Top-1 Probability Adjustment

In Figure 3, we assume that after processing the current tuple t_{cur}, the upper bound is smaller than tuple t_{i_2}'s top-1 probability, which is the current maximum, and thus the iteration ends. Besides, assume that $t_{i_1}, \ldots, t_{i_l}$ are the tuples in t_{i_2}'s rule and they are positioned before t_{cur}. Then the top-1 probability $prob$ of a tuple after t_{i_2} and between t_{i_h} and $t_{i_{h+1}}$ (defined to be a segment) in the *buffer* should be updated as $prob \cdot (1 - p_{i_1} - p_{i_2} - p_{i_3} - p_{i_h})/(1 - p_{i_1} - p_{i_3} - p_{i_h})$ to reflect the removal of t_{i_2} from *buffer*.

This actually changes the factor of t_{i_2}'s rule in the top-1 probabilities, which can be done by a single pass of the tuples after t_{i_2}, and a single pass over t_{i_2}'s rule. Note that the top-1 probabilities of $t_{i_1}, t_{i_3}, \ldots, t_{i_l}$ remain the same.

After the removal of the current top-1 tuple (i.e., t_{i2} in the above example) and the update of the top-1 probabilities, the tuple is also removed from its rule r. This increases $r.factor$ (i.e., by p_{i2} in the above example), and so r's position in the MinHeap AR should be adjusted, which takes $O(\log |AR|)$ time. If no tuple remains in a rule after the removal, the rule is deleted from AR.

Since the adjustment after each iteration includes $O(\log |AR|)$ time rule position adjustment in AR and a scan of the rule of the picked tuple, the total time is $O(k(\log |AR| + len_{max}))$, where len_{max} is the largest length of a rule. Besides, since each rule will be inserted into AR at most once in $O(\log |AR|)$ time, the total time is $O(|R| \log |AR|)$, where R is the rule set. So the overall time complexity is $O(n + |R| \log |AR| + k(L_{avg} + len_{max} + \log |AR|))$. We do not present the complete algorithm here due to space limitation.

6   Experiments

We conduct extensive experiments using both real and synthetic datasets on HP EliteBook with 3 GB memory, and 2.53 Hz Intel Core2 Duo CPU. The real dataset is IIP Iceberg Sightings Databases[1], which is commonly used by the related work such as [10, 11] to evaluate the result quality of ranking semantics. We study the performance of the algorithms on synthetic datasets. Our algorithms are implemented in Java.

6.1   Ranking Quality Comparison on IIP Iceberg Databases

The pre-processed verion of IIP by [10] is used to evaluate the ranking quality of different semantics. Figure 4(a) lists the occurrence probabilities of some tuples from the dataset, where the tuples are pre-sorted by scores. Figure 4(b) shows the top-10 query results of different semantics.

---

[1] `http://nsidc.org/data/g00807.html` (IIP: International Ice Patrol).

	t_1	t_2	t_3	t_4	t_5	t_6	t_7	t_8	t_9	t_{10}	t_{11}	t_{14}	t_{18}
p_i	0.8	0.8	0.8	0.6	0.8	0.8	0.4	0.15	0.8	0.7	0.8	0.6	0.8

(a) Occurrence Probabilities of the Tuples

	1	2	3	4	5	6	7	8	9	10
U-Pop$k$	t_1	t_2	t_3	t_4	t_5	t_6	t_9	t_7	t_{10}	t_{11}
ExpRank	t_1	t_2	t_3	t_5	t_6	t_9	t_{11}	t_{18}	t_{23}	t_{33}
PT-k	t_1	t_2	t_3	t_5	t_6	t_9	t_{10}	t_{11}	t_{14}	
U-Top$k$	t_1	t_2	t_3	t_4	t_5	t_6	t_7	t_9	t_{10}	t_{11}
U-kRanks	t_1	t_2	t_3	t_5	t_6	t_9	t_9	t_{11}	t_{11}	t_{18}

(b) Occurrence Probabilities of the Tuples

	U-Pop$k$	ExpRank	PT-k	U-Top$k$	U-kRanks	SUM
U-Pop$k$	0	13.5	3	1	5	22.5
ExpRank	13.5	0	13.5	12.5	2.5	42
PT-k	3	13.5	0	2	5	23.5
U-Top$k$	1	12.5	2	0	4	19.5
U-kRanks	5	2.5	5	4	0	16.5

(c) Neural Approach to Kendall's Tau Distance

	U-Pop$k$	ExpRank	PT-k	U-Top$k$	U-kRanks	SUM
U-Pop$k$	0	10	2	1	4	17
ExpRank	10	0	9	9	0	28
PT-k	2	9	0	1	3	15
U-Top$k$	1	9	1	0	3	14
U-kRanks	4	0	3	3	0	10

(d) Optimistic Approach to Kendall's Tau Distance

Fig. 4. Top-10 Results on IIP Iceberg Databases

Figure 4(b) shows that the results of U-Pop$k$, U-Top$k$ and PT-k are almost the same. U-Pop$k$ ranks t_9 before t_7, which is more reasonable, since $p_7 = 0.4$ is much smaller than $p_9 = 0.8$ and their score ranks are close. PT-k rules out t_7 but includes t_{14}. However, $p_{14} = 0.6$ is not much larger than $p_7 = 0.4$ but t_7 has a much higher score rank than t_{14}. Therefore, it is more reasonable to rank t_7 before t_{14}. In fact, the top-11th tuple for U-Pop$k$ is t_{14}. However, U-kRanks returns duplicate tuples. ExpectedRank promotes low-score tuples like t_{23} and t_{33} to the top, which is unreasonable and is also observed in [11]. This deficiency happens mainly because ExpectedRank assigns rank $(\ell + 1)$ to an absent tuple t in a world having ℓ tuples. As a result, low-score absent tuples are given relatively high ranking in those small worlds, leading to their overestimated rank. Overall, U-Pop$k$ gives the most reasonable results in this experiment.

Kendall's tau distance is extended to gauge the difference of two top-k lists in [14], which includes an optimistic approach and a neural approach. Figure 4(c) and (d) show the extended Kendall's tau distance between the top-10 lists of the different semantics in Figure 4(b), where the last column SUM is the sum of the distances in each row. We can see that ExpectedRank returns drastically different results from the other semantics, which means that ExpectedRank actually generates many unnatural rankings.

6.2   Scalability Evaluation

We find that only U-Pop$k$ and ExpectedRank are efficient enough to support real time needs, while other semantics such as PT-k are much more expensive to evaluate. Figure 5(a1)–(a2) show the results on the IIP dataset described above, where 4 seconds are consumed by PT-k even when $k = 100$ and the high probability threshold 0.5, while all the tuples can be ranked by U-Pop$k$ and ExpectedRank within 0.2 seconds. (a2) also shows that U-Pop$k$ is faster than ExpectedRank when k is within $1/10$ of the data size and never slower by a factor of 2. Overall, the efficiency of U-Pop$k$ is comparable to that of ExpectedRank.

We evaluate the scalability of U-Pop$k$ on synthetic data sets and show the important results in Figure 5. We find that the data size is not a major factor

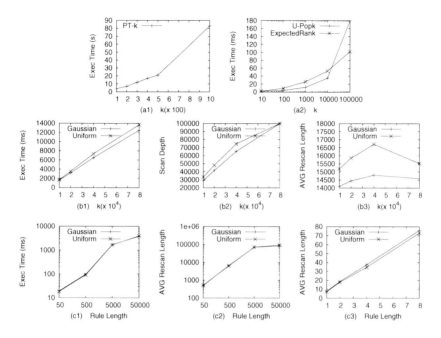

Fig. 5. Important Scalability Evaluation on Synthetic Data

on the performance, and therefore we fix it to be 100K tuples. ExpectedRank is also not shown here, since it has similar performance in Figure 5(b1). The results in each experiment are averaged over 10 randomly generated data sets, whose tuple probabilities conform to either Uniform or Gaussian distribution.

Figures 5(b1)–(b3) show the effect of k on the performance, where the rule length is set to be 1000. Figures 5(b1)–(b2) show that the execution time and scan depth are almost linear to k, while in Figure 5(b3) the average rescan length first increases and then decreases as k increases. The drop happens because almost all the tuples are read into *buffer* at the end, and therefore the number of tuples decreases for each removal of a top-1 tuple.

Figures 5(c1)–(c2) show the effect of rule length on the performance, where we set all rules to the same length. Since in real life applications it is not likely to have too many tuples in a rule, we also explore the effect of rule length for short rules (i.e. less than 10 tuples in a rule). Figure 5(c3) shows that the average rescan length increases linearly with the rule length for small rule lengths.

The time complexity of U-Pop$k$ is shown to be $O(n + |R| \log |AR| + k(L_{avg} + len_{max} + \log |AR|))$ in Section 5.2, where the $O(n)$ part is actually the *scan depth*, which is shown to be almost $O(k)$ in Figure 5(b2). Also, when the maximum rule length len_{max} is small, we have $L_{avg} = O(len_{max})$ in Figure 5(c3). Since AR is a fraction of the rule set, we have $|AR| = O(|R|)$. Thus, the time complexity can be approximated as $O(k \cdot (len_{max} + \log |R|) + |R| \log |R|)$, that is linear to k and len_{max}, and $O(|R| \log |R|)$ for the rule set R, which scales well.

7   Conclusion

We propose U-Pop$k$ as a new semantics to rank probabilistic tuples, which is based on a robust property inherent in the underlying probability model. Compared with other known ranking semantics, U-Pop$k$ is the only semantics that is able to achieve all the following desirable features: high ranking quality, fast response time, and no additional user-defined parameters other than k.

Acknowledgements. This work is partially supported by RGC GRF under grant number HKUST 618509.

References

1. Cormode, G., Li, F., Yi, K.: Semantics of ranking queries for probabilistic data and expected ranks. In: ICDE (2009)
2. Dalvi, N., Suciu, D.: Efficient query evaluation on probabilistic databases. VLDB Journal 16(4), 523–544 (2007)
3. Agrawal, P., Benjelloun, O., Das Sarma, A., Hayworth, C., Nabar, S., Sugihara, T., Widom, J.: Trio: A system for data, uncertainty, and lineage. In: VLDB (2006)
4. Antova, L., Koch, C., Olteanu, D.: From complete to incomplete information and back. In: SIGMOD (2007)
5. Fagin, R., Lotem, A., Naor, M.: Optimal aggregation algorithms for middleware. In: PODS (2001)
6. Ilyas, I.F., Beskales, G., Soliman, M.A.: Survey of top-k query processing techniques in relational database systems. In: ACM Computing Surveys (2008)
7. Re, C., Dalvi, N., Suciu, D.: Efficient top-k query evaluation on probabilistic databases. In: ICDE (2007)
8. Soliman, M.A., Ilyas, I.F., Chang, K.C.-C.: Top-k query processing in uncertain databases. In: ICDE (2007)
9. Zhang, X., Chomicki, J.: On the semantics and evaluation of top-k queries in probabilistic databases. In: DBRank (2008)
10. Hua, M., Pei, J., Zhang, W., Lin, X.: Ranking queries on uncertain data: A probabilistic threshold approach. In: SIGMOD (2008)
11. Li, J., Saha, B., Deshpande, A.: A unified approach to ranking in probabilistic databases. In: VLDB (2009)
12. Ge, T., Zdonik, S., Madden, S.: Top-k queries on uncertain data: On score distribution and typical answers. In: SIGMOD (2009)
13. Jin, C., Yi, K., Chen, L., Yu, J.X., Lin, X.: Sliding-window top-k queries on uncertain streams. In: VLDB (2008)
14. Fagin, R., Kumar, R., Sivakumar, D.: Comparing top k lists. In: SODA (2003)

Probabilistic Image Tagging with Tags Expanded By Text-Based Search

Xiaoming Zhang[1], Zi Huang[2], Heng Tao Shen[2], and Zhoujun Li[1]

[1] School of Computer, Beihang University, Beijing, 100191, China
yolixs@cse.buaa.edu.cn
[2] School of Information Technology & Electrical Engineering, University of Queensland
Room 651, Building 78, St Lucia Campus, Brisbane, Australia
{shenht,huang}@itee.uq.edu.au

Abstract. Automatic image tagging automatically assigns image with semantic keywords called tags, which significantly facilitates image search and organization. Most of present image tagging approaches assign the query image with the tags derived from the visually similar images in the training dataset only. However, their scalabilities and performances are constrained by the limitation of using the training method and the fixed size tag vocabulary. In this paper, we proposed a search based probabilistic image tagging algorithm (CTSTag), in which the initially assigned tags are mined from the content-based search result and expanded from the text-based search results. Experiments on NUS-WIDE dataset show not only the performance of the proposed algorithm but also the advantage of image retrieval using the tagging result.

Keywords: Image tagging, search based tagging, tag expansion.

1 Introduction

Image tagging is the task that assigns an image with several descriptive keywords called tags. With the tags, images can be searched like web documents. Such new characteristics of images bring new research problems to Web applications [1], [2], [3]. Recently, many automatic tagging approaches have been proposed recently [4], [5], [6], [7]. They mainly assign the query image with the tags that commonly appear in the visually similar images in the training dataset.

There are several problems existing in these automatic tagging approaches. First, these automatic tagging approaches have a high dependency on their small-scale training dataset, which restrict their effectiveness on arbitrary images. That is, only images that have visually similar images in the training dataset could be assigned with tags in a limit tag vocabulary. As a result, the novel images which have no visually similar images in the training dataset can't been assigned tags correctly, and also the fashionable tags which appear in the Web recently can't be assigned to the query images because that they aren't in the tag vocabulary. Compared with the limited number of images and tags that can be used to tag the query images, the potentially unlimited vocabulary of the web scale images and other types of web resource i.e. web documents can be utilized to tag images. By using web-scale resource, the

J.X. Yu, M.H. Kim, and R. Unland (Eds.): DASFAA 2011, Part I, LNCS 6587, pp. 269–283, 2011.
© Springer-Verlag Berlin Heidelberg 2011

assigned tags of the query image can be more complete and also can track the fashionable expression on the Web. The search based tagging is also scalable to online tagging.

The second serious problem they encounter is that the query image is only assigned with the tags which appear in its visually similar images. Since the tags of most images tend to be sparse and uncomplimentary and also the semantically similar images may not be the visually similar images, some relevant tags which don't appear in the visually similar images can't be assigned to the query image. Thus, a query image can't only be assigned with the tags derived from the visually similar images.

Finally, a tag can't be simply accepted or rejected to tag the query image. The accepted tag should be combined with a probability value which indicates its relevance to the query image, and the tag probability is helpful to image retrieval and organization. For example, for the query image of "apple" computer, tags such as "Steve Jobs" and "Stanford" and others can also be assigned to it even with small probabilities, though these tags may not appear in its visually similar images. Thus, this training based technology limit their performance and scalability.

In this paper, we propose a probabilistic image tagging algorithm (CTSTag) based on web search result mining, which assigns the query image with the tags mined from both web image and documents. The process of this algorithm is described in Fig.1. It is mainly composed of 3 steps.

1. A content-based search method is used to retrieve the K visually similar web images. Then, the most probabilistic tags are derived from the tags associated with these images. These derived tags constitute the initial tag set Q. For example, for the query image I_q about apple computer, a set of initial tags Q={"apple", "computer", "laptop",…} can be derived from the content-based search result.

2. A text-based search is used to retrieve other web images and documents using the initial relevant tag set Q as the query. As a result, other potential relevant tags which don't appear in the visually similar images can be retrieved. Based on their correlation in the text based search result, the initial tag set can be further denoised. Then it is expanded with the tags mined from the text-based search result. The probability of expanded tag is estimated by measuring the probability flow which flow into the expanded tag from the initial tag set. With the text based search and the mining of search result, the assigned tags are more complementary and can also absorb the new tags from other resources. For example, using the text based search, other web images such as apple headquarter "Silicon Valley" and document about "Steve Wozniak" the co-founder of apple computer can also be retrieved. Thus the initial tag set is expand with {"Silicon Valley", "Steve Wozniak",…}.

3. Since the initial tags and expanded tags are derived independently, we refine them consistently in the third step. The intuition is that the correlation between the initial tags and expanded tags can also be used to rank the initial tags, and all the tag which has more relation with other tags has a greater probability values.

The rest part of this paper is organized as follows. Related works are introduced in section 2. Then, the process of deriving initial relevant tags from the content-based search result is presented in section 3. The procedure of tag expansion based on text-based search result mining is introduced in section 4. In section 5, we refine all the tags using a probability propagation method. Extensive experimental results are reported in Section 6 followed by the conclusion in section 7.

2 Related Works

Image tagging has attracted more and more interests in recent years. The key problem is to decide which tag can be assigned to tag the query image. Some works use a supervised machine learning method to tag the query images. For example, Barnard et al. [17] develops a number of models for the joint distribution of image regions and words to tag the query images. Lei et al. [10] propose a novel probabilistic distance metric learning scheme which is used to retrieves k social images that share the largest visually similarity with the query image. The tags of the query image are then derived based on the tagging of the similar images. Geng et al. [18] model the concept affinity as a prior knowledge into the joint learning of multiple concept detectors, and then the concept detectors which also are classifiers are used to tag the query images. Li et al. [19] combines statistical modelling and optimization techniques to train hundreds of semantic concepts using example pictures from each concept. However, these machine learning methods try to learning a mapping between low-level visual features and high-level semantic concepts, which are not scalable to cover the potentially unlimited array of concepts existing in social tagging. Moreover, uncontrolled visual content generated by users creates a broad domain environment which has a significant diversity in visual appearance, even for the same concept.

Some other works use a mining method based on a training dataset to tag the query images. Wang et al. [4] propose a relevance model-based algorithm. Candidate tags are re-ranked by the Random Walk with Restarts (RWR) which leverages both the corpus information and the original confidence information of the tags. Finally, only the top ones are reserved as the final annotations. Siersdorfer et al. [5] propose an automatic video tagging method based on video duplicate and overlap detection, and the assigned tags are derived from the videos which are overlap with the test video. Zhou et al. [6] uses a heuristic and iterative algorithm to estimate the probabilities that words are in the caption of an image by examining its surrounding text and region matching. Words with high probabilities are selected as tags. The approach in [7] first estimates initial relevance scores for the tags based on probability density estimation, and then performs a random walk over a tag similarity graph to refine the relevance scores. There is also a learning of social tag relevance by neighbour voting [8], which learns tag relevance by accumulating votes from visual neighbours. Lei et al. [9] proposes a multi-modality recommendation model based on both tag and visual correlation of images. Rankboost algorithm is then applied to learn an optimal combination of those ranking features from different modalities.

However, all of these approaches are dependent on a small-scale high quality training set, which means that they can only tag the query images which have visually similar neighbours in the training set and only use the limit tag vocabulary. To leverage web-scale data, there are already some researches on automatic image tagging based on web search [11], [12]. X.J. Wang et al. [11] use a text-based search to find a set of semantically similar web images, and then the Search Result Clustering (SRC)[13] algorithm is used to cluster these images. Finally, the name of each cluster is used to annotate the query image. C.H. Wang et al. [12] use CBIR technology to retrieve a set of visually similar from the large-scale Web image dataset, then annotations of each web image are ranked. Finally, the candidate annotations are re-ranked using Random Walk with Restarts (RWR) and only the top ones are reserved as the final annotations. However, these approaches just use a search based method to retrieve the visually similar images from a large-scale web database instead of training dataset, and then derive tags from the feedback web images. They have no consideration of the new tags which appear in other web images or documents and are relevant to the query image.

3   Initial Tags

The initial tags are derived from the tags associated with the visually similar images directly. We use the content-based image search to retrieve K nearest neighbors. For each query image, it is desired to assign it with the most relevant tags which can describe its semantic content. We use the conditional probability that the query image generating the tag to represent the relevance, and the greater the probability is the more relevant it is. From the generation point of view, the conditional probability of tag t_i generated from the query image I_q can be estimated by the joint generating probability of every similar images of I_q. The probability $P(t_i/I_q)$ which indicate the relevance of t_i to the query image is estimated by the following formula:

$$P(t_i \mid I_q) = \frac{N_k(t_i)}{k} P_G(t_i \mid I_q) + (1 - \frac{N_k(t_i)}{k}) P_L(t_i \mid I_q)$$

where $N_k(t_i)$ represent the number of images which contain the tag t_i among the top-K similar images. This estimation includes two parts which are balanced by the parameter. The intuition behind the first part $P_G(t_i/I_q)$ is that the more number of neighbour images contain the key word and the more frequently it appears the more probably the query image is tagged with the key word. The intuition behind the second part is that the more similar the images which contain the key word are the more probably the key word is assigned. We use the following formula to estimate $P_G(t_i/I_q)$:

$$P_G(t_i \mid I_q) = \sum_{I_j \in top(k)} P(t_i \mid I_j) P(I_j \mid I_q)$$

$$P(t_i \mid I_j) = \frac{TF(t_i, I_j)}{\max_{t_k \in I_j}(TF(t_k, I_j))}, \quad P(I_j \mid I_q) \approx \frac{Sim_{visual}(I_j, I_q)}{k}$$

where $TF(t_i, I_j)$ represents the frequency of t_i in the web image of I_j, $Sim_{visual}(I_j, I_q)$ represents the visually similarity between I_j and I_q. However, the $P_G(t_i/I_q)$ will prefer the tags which appear frequently in the top-k similar images. Thus, we use the following formula to alleviate this problem:

$$P_L(t_i \mid I_q) = \frac{\sum_{I_x \in Ig(t_i)} sim_{visual}(I_x, I_q)}{\mid Ig(t_i) \mid}$$

where $I_g(t_i)$ denotes the set of images that contain tag t_i. Thus, formula (1) can alleviate the biases on either frequent tags or rare tags. Then all the tags associated with the top-k images are ranked according to their generating probability values, and the top ranked tags are selected to compose the initial relevant tag set Q.

4 Tags Expanding from Text-Based Search Result

Since there may be many relevant tags which don't appear in the visually similar images, and also there may be many new words which haven't been used to tag images. We perform a text based search to retrieve other relevant web resource by using the initial tag set as the query, and then the initial tag set can be expanded with other relevant tags mined from the search result.

4.1 Initial Tags Denoising

Because of the semantic gap problem, the content-based image search usually retrieves many noisy images. Thus it is inevitable to include many noisy tags in the initial tag set. A direct expansion of the initial tag set may include many other noisy tags. Thus, it is necessary to further denoise the initial tags set before expand it. However, only using the visually similar images is hardly to distinguish the noisy tags completely because of the semantic gap problem and the sparsity of tags. We can use the tag correlation among the feedback of the text-based search to further denoise the initial tag set.

The denoising of initial tag set is processed as following steps. First, each initial tag is regarded as a document containing the associated web pages. Then the denoising of initial tag set can be turn to removing the noisy "documents". For each initial tag t_i, a vector $<tf(t_i,D_1), tf(t_i,D_2),\dots, tf(t_i,D_n)>$ is constructed, where $tf(t_i,d_j)$ is the term frequency of tag t_i in web page d_j. The similarity between two tags can be estimated using the cosine function on the two vectors. For the noisy tags, its occurrence distribution in the feedback is different with the relevant tags and other noisy tags, which means its vector is less likely to be similar with the vectors of other tags. Thus, the more number of similar vectors a vector has the more confident that the corresponding tag is a relevant tag. We use the following formula to estimate the confidence score that a tag t_i is a relevant tag:

$$Con(t_i) = \frac{1}{|Q|}\sum_{t_j \in Q} Sim_{cosine}(\vec{t_i},\vec{t_j})$$

We use a simple denoising method which only remove the tags whose confidence score is behind a threshold value θ from the initial tags set Q. The threshold value θ is set with the average value over the initial tag set:

$$\theta = \frac{1}{|Q|}\sum_{t_i \in Q} Con(t_i)$$

4.2 Tag Expansion

Since the feedback contains many web pages, a direct tag expansion from the whole feedback will be very complexity. Thus, a ranking method is used to select a subset of more important web pages. Then features are extracted from these web pages, and the potential relevant tags are selected from these features to expand the initial tag set. The conditional probability $P(I_q/D)$, i.e., the probability of generating the query image

I_q with query Q given the observation of a web page D [15] is used to rank the feedback pages in a descending order. We use the general unigram model to formulate $P(I_q/D)$ as follows:

$$P(I_q \mid D) \approx \sum_{t_i \in Q} P(I_q \mid t_i)P(t_i \mid D) = \sum_{t_i \in Q} \frac{P(t_i \mid I_q)P(I_q)}{P(t_i)}P(t_i \mid D) \approx \sum_{t_i \in Q} P(t_i \mid I_q)P(t_i \mid D)$$

$$P(t_i \mid D) = \frac{\mid D \mid}{\mid D \mid + \mu} P_{LM}(t_i \mid D) + \frac{\mu}{\mid D \mid + \mu} P_{LM}(t_i \mid C)$$

$$P_{LM}(t_i \mid D) = \frac{tf(t_i, D)}{\mid D \mid}, \quad P_{LM}(t_i \mid C) = \frac{tf(t_i, C)}{\mid C \mid}$$

where $P_{LM}(t_i/D)$ is the maximum likelihood estimation of t_i in D, and C is the collection which is approximated by the search result in this paper and μ is the smoothing parameter. $|D|$ and $|C|$ are the length of D and the total number of words in C respectively. $tf(t_i,D)$ and $tf(t_i,C)$ are the frequency of t_i in D and C respectively.

After ranking the pages, we extract the features from the top pages. If the page contains tags, we only extract the tags as its features. Otherwise, we extract the features from the texture content of the page. The following method is used to assess the weight of features f in a page D:

$$W_D(f) = \log(\frac{tf(f,D)}{\mid D \mid} + 1) * \log(\frac{\mid Q \mid}{n_t(f,Q)} + 1)$$

where $tf(f, D)$ is the term frequency in D, and $n_t(f,Q)$ is the number of tags that f has co-occurred with, and the addition of 1 is to avoid zero or negative weights. It is similar with *TF-IDF*. But it instead the inverse document frequency with the inverse co-occurred tag frequency. The intuition is that the more tags a feature co-occurs with, the less specific and important the feature is. For each top page, we extract the feature whose weight is greater than the average weight of all the feature words in this page.

Then, we expand the initial tag with a sub set of features selected from the extracted features. Usually, the common and noisy features have similar "relations" to most of initial tags. This means that the irrelevant feature have an evenly distribution of relation among the initial tags. However, the relevant features usually have strong relation with some of the initial tags but have less relation to other initial tags, and these features are preferable to be used as the expanded tags. We use the posterior probability to represent the relation between a tag t_i and a feature f:

$$P(t_i \mid f) = \frac{C(t_i \cap f)}{\sum_{t_j \in Q} C(t_j \cap f)}$$

where $C(t_i \cap f)$ denotes the number of web pages which contain both tag t_i and feature f, and $C(\overline{t_i} \cap \overline{f})$ denotes the number of web pages which contain neither tag t_i nor feature f. Then each feature has a list of these posterior probabilities for each initial tag. Features that has a strong relation with tag t_i will has a high value of $P(t_i/f)$, and low value of $P(t_j/f)$, $\forall j \neq i$. The irrelevant features will have more evenly distributed values among the posterior probabilities. For example, the common features co-occur with most initial tags and the occurrences of other noisy features are different with the initial tags, which induces a more evenly distributed value among the posteriors. To measure the degree of the confidence that the feature is relative to the initial tags, we compute the entropy E for a feature f:

$$E(f) = -\sum_{t_i \in Q} P(t_i \mid f) \log_2 P(t_i \mid f)$$

The lower the entropy, the higher the confidence that the feature is relative to one or several initial tags is. Similarly, higher entropy feature has less confidence that it is relative to the initial tags. Thus the top features with the smallest entropy are selected as the expanded tags.

4.3 Probability Flow

Like the initial tags, the expanded tags also need a probability value to indicate its relevance to the query image. Since the expanded tags have no direct relation with the query image and only have relation with the initial tags. Thus, we estimate the probability of expanded tag based on its relation with the initial tags. We estimate it by measuring the probability flow which flow into the expanded tag from the initial tags. The probability flow reflects how strongly the probability values of the initial tags are inherited by the expanded tag. We use the formula $PF(Q \lhd t_e)$ to denote the probability flow which flows into tag t_e from tag set Q, and then the probability $P(t_e/I_q)$ of the expanded tag t_e is approximated by the probability flow:

$$P(t_e \mid I_q) \approx PF(Q \lhd t_e) = PF(\bigoplus_{t_i \in Q} t_i \lhd t_e)$$

where $\bigoplus_{t_i \in Q} t_i$ denotes the concept combination of the query tag set. It indicates that the probability value of expanded tag is the rate of probability value that flow into the expanded tag from the combination concept. The intuition behind the concept combination is that the tag which is more relevant with the expanded tag can dominate the others. We use the following heuristic method to construct the concept combination of the initial tag set:

Step 1: The vector of each tag $t_i \in Q$ is represented by $\vec{t_i} < w_{t_i}^1, w_{t_i}^2, ..., w_{t_i}^{|F|} >$, where $|F|$ is the number of pages of the search result, and $w_{t_i}^k$ is estimated as following:

$$w_{t_i}^k = \begin{cases} P(t_e \mid t_i) P(t_i \mid q) & \text{if the } k^{th} \text{ page contains tag } t_i \\ 0 & \text{else} \end{cases}$$

$$(1 \leq k \leq |F|)$$

$$P(t_e \mid t_i) = \frac{C(t_e \cap t_i)}{C(t_e \cap t_i) + C(\overline{t_e} \cap t_i)} * \frac{C(t_e \cap t_i)}{C(t_i)} + \frac{C(\overline{t_e} \cap \overline{t_i})}{C(t_e \cap \overline{t_i}) + C(\overline{t_e} \cap \overline{t_i})} * \frac{C(\overline{t_e} \cap \overline{t_i})}{C(\overline{t_i})}$$

where $C(t_i \cap t_e)$ is the co-occurrence which is the number of pages that contain both t_i and t_e, and $C(\overline{t_e} \cap \overline{t_i})$ is the number of web pages which contain neither t_i nor t_e.

Step 2: The vector of expanded tag t_e is represented by $\vec{t_e} < w^1_{t_e}, w^2_{t_e}, ..., w^{|F|}_{t_e} >$, and the weight is estimated as following:

$$w^k_{t_e} = \begin{cases} 1 & \text{if the } k^{th} \text{ page contains tag } t_e \\ 0 & \text{else} \end{cases}$$

$$(1 \leq k \leq |F|).$$

Step 3: The vector of the tag concept combination is then represented with $\overline{\oplus_{t_i \in Q} t_i} =< \oplus_{t_i \in Q} w^1_{t_i}, \oplus_{t_i \in Q} w^2_{t_i}, ..., \oplus_{t_i \in Q} w^{|F|}_{t_i} >$, where

$$\oplus_{t_i \in Q} w^k_{t_i} = P(t_e | \hat{t^k}) P(\hat{t^k} | I_q) \quad \hat{t^k} = \arg\max_{t_i \in Q} P(t_e | t_i).$$

In this vector, each element is dominated by the query tag which has the greatest generating probability. Then the probability flow from query tags Q to the expanded tag t_e is defined as the following formula:

$$PF(\oplus_{t_i \in Q} t_i \lhd t_e) = \frac{\sum_{1 \leq k \leq |F|} \oplus_{t_i \in Q} w^k_{t_i} * w^k_{t_e}}{\sum_{1 \leq k \leq |F|} w^k_{t_e}}$$

The probability flow $PF(\oplus_{t_i \in Q} t_i \lhd t_e)$ has following properties:

1. $0 \leq PF(\oplus_{t_i \in Q} t_i \lhd t_e) \leq \max_{t_i \in Q}(P(t_i | I_q))$;
2. if $t_j \in Q$, then $PF(\oplus_{t_i \in Q} t_i \lhd t_j) = P(t_j | I_q)$;

3. The tag which has a greater generating probability has a stronger influence on the probability flow, and the more frequently the tag co-occur with the query tags the more probability it inherit.

Thus, it is reasonable to estimate the probability value of expanded tag by estimating the probability flow.

5 Tag Refinement

In the above sections, we estimated the probabilities of initial tags and expanded tags independently, and it is assumed that the probability of expanded tag isn't greater than that of initial tag. Thus we refine all the tags according to their correlation with other tags in the same context. We utilize the relationship among candidate tags to boost the tags which are more relevant no matter they are initial tags or expanded tags. In order to fully utilize the probability value estimated in the former stages, we use the RWR algorithm to refine the candidate tag set based on the tag graph.

5.1 Correlation and Transition between Tags

To construct a directed tag graph, each candidate tag is considered as a vertex and each two vertexes are connected with two directed edges. The directed edge is weighted with a transition strength which indicates the probability of transition from the tail vertex to the head vertex. To estimate the transition strength, we first combine the co-occurrence and absence to estimate the correlation of two tags, and then the correlations are used to estimate the transition probability denoted by transition strength.

Usually, the co-occurrence is used to estimate the correlation of two tags in the previous works. However, these approaches don't consider the absence of tags and the hidden correlation. Thus we propose a balanced correlation measure which takes both the co-occurrence and absence to computer the correlation between two tags.

$$Cor(t_i,t_j) = \Pr(t_i,t_j)\log(\frac{C(t_i \cap t_j)}{C(t_i)+C(t_j)-C(t_i \cap t_j)}+1) + \Pr(\overline{t_i},\overline{t_j})\log(\frac{C(\overline{t_i} \cap \overline{t_j})}{C(\overline{t_i})+C(\overline{t_j})-C(\overline{t_i} \cap \overline{t_j})}+1)$$

$$\Pr(t_i,t_j) = \frac{C(t_i \cap t_j)}{|F|} \qquad \Pr(\overline{t_i},\overline{t_j}) = \frac{C(\overline{t_i} \cap \overline{t_j})}{|F|}$$,

where $\overline{t_i}$ denotes the absence of tag t_i, and $\Pr(t_i,t_j)$ is the joint probability that both t_i and t_j appear in a web page and $\Pr(\overline{t_i},\overline{t_j})$ is the joint probability of the absence of t_j and t_i. However, the correlation measured by the above-mentioned formula only considers the relation between two tags. There may be many hidden correlations which can't be discovered. This estimated correlation is also symmetric. But the transition strength between two tags is asymmetric in some time. Thus, it isn't reasonable to represent the transition strength by the correlation directly. To alleviate these problems, the transition strength between two tags is proposed to represent the probability propagation between these two tags:

$$Tran(t_i \rightarrow t_j) = \frac{\sum_{k=1}^{|\mathbb{R}|} Cor(t_i,t_k)*Cor(t_k,t_j)}{\sum_{k=1}^{|\mathbb{R}|} Cor(t_i,t_k)}$$

where $\mathbb{R}$ is the tag set which is composed of the initial tags and expanded tags. With this formula, the hidden correlation between tags can also be discovered.

5.2 Tag Refinement

Here we use the random walk with restart (RWR) [14] to propagate the probability over the tag graph in order to boost the probability of the more relevant tags. First, a transition matrix C is constructed with that each element C_{ij} be the normalized value:

$$Tran(t_i \rightarrow t_j) = \frac{Tran(t_i \rightarrow t_j)}{\sum_{i=1}^{|\mathbb{R}|} Tran(t_i \rightarrow t_j)}$$

We use $P_k(t_i|I_q)$ to denote the probability of tag t_i at the kth iterations. Then the probabilities of all tags at the kth iteration is denoted as $\overrightarrow{P_k} = [P_k(t_i|I_q)]_{|R|\times 1}$. Thus, we can formulate this process using the following formula:

$$P_k(t_i \mid I_q) = (1-\alpha)\sum_{j=1}^{|R|} C_{ij}P_{k-1}(t_j \mid I_q) + \alpha P'(t_i \mid I_q)$$

where $P'(t_i/I_q)$ is the normalized value of the initial probability of t_i estimated in the former sections, and α $(0<\alpha<1)$ is a weight parameter. The probability of each tag tends to be a fix value after a number of iterations, which result in a high probability for the tag which has strong correlation with the relevant tags.

6 Evaluations

A series of experiments are conducted on web images and documents database to evaluate the proposed algorithm CTSTag. First, we test the performance of different image tagging algorithms. Then, to show the effectiveness of the result tags on image retrieval, the retrieval performances of query by keyword based on the result tags of different image tagging algorithms are compared.

6.1 Dataset

Web-scale database. We downloaded 1 million tagged web images from Flickr using its API service. We select the most popular topics which cover the topics of the evaluation dataset also, and it is ensured that the images are evenly distributed over the different topics. The number of distinct tags per image varies from 2 to more than 100, and more than 10 million unique tags in total. Then we download about 1 million of web documents from the Internet. These documents also cover the same topics of the evaluation dataset, and each document is represented by the extracted keywords associated with their term frequencies.

Evaluation dataset. We use the NUS-WIDE dataset as the evaluation set. The number of distinct tags per image varies from 1 to more than 100, with an average value of about 30. There are 269,648 images and 425,059 unique tags in total. By removing rare tags that are used less than 10 times in the entire collection, the average number of distinct tags per images is about 15.

6.2 Image Tagging

In this section, these experiments are designed to test the effectiveness of our algorithm CTSTag used in image tagging. It is also compared with neighbor voting based image tagging algorithm (NVTag) [16] and the other search based image tagging algorithm (SBIA) [12]. For the CTSTag, 30 initial tags which are more than the average number of tags per image are derived before the denoising operation, and then we expand the initial tag set with 50 other tags. For the NVTag and SBIA, the number of result tags varies from several to the maximum number 60 based on the ground truth. We employ three standard criteria to evaluate the image tagging performance, i.e., average precision (Av_P) and average recall (Av_R). With m result tags, the precision and recall are denoted by $Av\_P@m$, $Av\_R@m$. Two parameters are evaluated first. One is the size K which indicates how many nearest neighbours are selected, and the other one is the restart parameter α. After the parameters were fixed, we compare the $Av\_P@m$ and $Av\_R@m$ of different tagging algorithms.

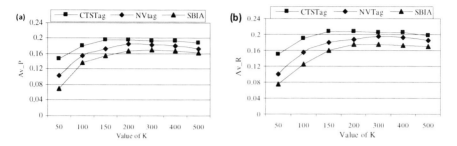

Fig. 1. a, b Average precision and recall of different value of K

By comparing the tagging performance with different K. all these algorithms retrieve K images first, and the number of tagging result is fixed to top 15 tags for all the algorithms. The average precision and recall are shown in Fig.1.a and b. It shows that the changes of these curves of different algorithms are similar though their peaks are different. The performance of NVTag is the best when K is 300, and K is 200 for SBIA to achieve its best performance. As the CTSTag select the images with smallest semantic gap, its curves are more sooth after that they achieve their peaks. The performance of CTSTag is similar when K is equal or larger than 150. Thus, in the following experiments, K is set to be 200, 300 and 150 for NVTag, SBIA and CTSTag respectively.

The other parameter to be evaluated is the restart parameter α. As the SBIA also use RWR to refine the tags, we will compare the performance of SBIA and CTSTag with α varied from 0.1 to 0.9. The number of result tags m is also set to be 15. Fig.2. a and b show the average precision and recall of both algorithms. It indicates that the change trends of both figures are similar, and both precision and recall rates reach to the lowest value when α is 0.9. Both of the two algorithms achieve their best performances when α is set to be 0.3, Therefore, in all of the following experiments, the parameter α is set to be 0.3 for SBIA and CTSTag.

Based on the aforementioned parameters evaluation, we compare the $Av\_P@m$ and $Av\_R@m$ of different algorithms in this experiment. Fig.3.a and b show the $Av\_P@m$ and $Av\_R@m$ with m varied from 1 to 15. According to these figures, CTSTag consistently outperforms both NVTag and SBIA. There are several reasons. The first one is that we reduce the effect of noisy images by selecting the images with small semantic gap. Second, the SBIA only use the TF-IDF to re-weight the tags associated with the feedback images, which rarely consider the visually similarity of images and other relation. The NVTag use the frequency of a tag minus its prior frequency to train the tags, which also consider less on the visually similarity and is not scalable. Our algorithm combines both the frequency and image visually similarity to derive the initial tags. Finally, we enhance the correlation between tags by mining the hidden correlation, and also the correlations among the expanded scale of tags can boost the tags that are more related to the query image.

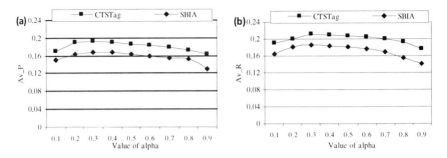

Fig. 2. a, b Average precision and recall of different value of α

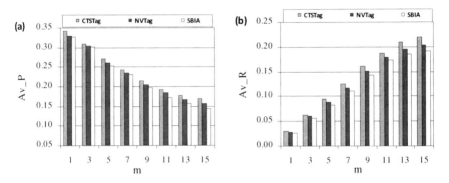

Fig. 3. a, b Average top-m precision and recall of different algorithms

6.3 Image Retrieval

In this section, we employ a general tag-based image retrieval used in existing systems such as Flickr to evaluate the effectiveness of the result tags of different tagging algorithms on image retrieval. The retrieval system indexes the result tags of the evaluation dataset, and a well-founded ranking function Okapi BM25 is used to ranking the retrieved images [20]. To evaluate the retrieval result, we use two evaluation criteria i.e. the average precision $Av\_P@m$ of top m retrieved images and the average recall $Av\_R@m$ of top m retrieved images.

Since the NVTag and SBIA only derived candidate tags from visually similar images, the tags that appear less frequently in the visually similar images are less likely to be selected. Our tagging algorithm also tag the query images with the tags that are relevant but appear less frequently or never appear in the visually similar images. To compare the effectiveness of these result tags on image retrieval, we design two experiments, i.e., one uses single-word queries, one uses double-word queries.

In the single-word queries experiment, we use the common words as the queries, i.e., {{airport}, {boat}, {beach}, {bridge}, {car}, {computer}, {dog}, {fish}, {sun}, {tree}}. In the two-word queries experiment, we combine the common word with a relatively uncommon word to set a query, i.e., {{airport girl}, {boat jumping}, {beach swimsuits}, {bridge cloud}, {car accessories}, {computer office}, {dog baseball}, {fish vocation}, {sun sports}, {tree countryside}}.

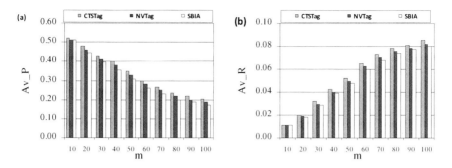

Fig. 4. a, b Average retrieval precision and recall of single-word query

Fig.4. a and b show the average retrieval performances of single-word queries experiment. Fig.5. a and b show the average retrieval performances of double-word queries experiment. Two conclusions can be drawn from these figures. First, when we set a common word as a query, the performances of image retrieval based on different result tags are similar. This is because that the common tags are preferred to be selected to tag the query images by all the three algorithms. Second, when add the single-word query with an uncommon word, the performance of image retrieval based on result tags of CTSTag are improved obviously. This because that both NVTag and SBIA tag the query image with the tags that are common in the visually similar images, and the uncommon are less likely to be selected. Since the double-word queries contain the uncommon words, many images whose result tags don't contain the uncommon tags can't be retrieved. However, the CTSTag expand the initial tag set with other relevant tags that may be uncommon or never appear in the visually similar images, and then the correlation between all the tags are used to boost the tags which may be neglected by other algorithms.

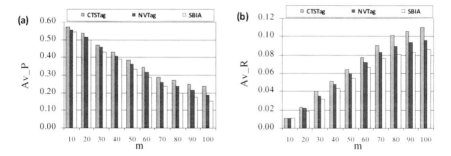

Fig. 5. a, b Average retrieval precision and recall of double-word query

7   Conclusions

In this paper, we formulate the image tagging as a search problem, and a novel probabilistic image tagging algorithm based on search result mining is proposed. First, a content-based search is used to retrieve visually similar images, and then the initial tags with their probability value are derived from the k nearest neighbour images. Second, with the initial tags as the query a text based search is used to expand the tags, and then a probability flow measuring method is proposed to estimate the probabilities of the expanded tags. Finally, a measure of tag transition strength is proposed to construct the transition matrix, and then RWR based on the transition matrix is used to refine the probability of all the tags. The search based framework make that our algorithm isn't limited to the training dataset, and it also tag image by combing web documents mining. Thus it is scalable and can use many kinds of web resource to tag image. Experimental results on NUS-WIDE dataset show not only the effectiveness of our tagging algorithm but also the effectiveness of image retrieval based on the result tags.

References

1. Joshi, D., Datta, R., Zhuang, Z., Weiss, W.P., Friedenberg, M., Li, J., Wang, J.Z.: PARAgrab: a comprehensive architecture for web image management and multimodal querying. In: Proceedings of the 32nd International Conference on Very Large Data Bases, pp. 1163–1166 (2006)
2. Zhang, B., Xiang, Q., Lu, H., Shen, J., Wang, Y.: Comprehensive query-dependent fusion using regression-on-folksonomies: a case study of multimodal music search. In: Proceedings of the 17th ACM International Conference on Multimedia, pp. 213–222 (2009)
3. Cui, B., Tung, A.K., Zhang, C., Zhao, Z.: Multiple feature fusion for social media applications. In: Proceedings of the International Conference on Management of Data, pp. 435–446 (2010)
4. Wang, C., Jing, F., Zhang, L., Zhang, H.-J.: Image annotation refinement using random walk with restarts. In: Proceedings of 14th ACM International Conference on Multimedia, pp. 647–650 (2006)
5. Siersdorfer, S., San Pedro, J., Sanderson, M.: Automatic Video Tagging using Content Redundancy. In: Proceeding of the 32nd ACM International Conference on Research and Development in Information Retrieval, pp. 16–23 (2009)
6. Zhou, X., Wang, M., Zhang, Q., Zhang, J., Shi, B.: Automatic image annotation by an iterative approach: Incorporating keyword correlations and region matching. In: Proceedings of the 6th ACM International Conference on Image and Video Retrieval, pp. 25–32 (2007)
7. Liu, D., Wang, M., Hua, X.S., Zhang, H.J.: Tag Ranking. In: Proceeding of the 18th ACM International Conference on World Wide Web, pp. 351–340 (2009)
8. Li, X.R., Snoek, C.G.M., Worring, M.: Learning tag relevance by neighbor voting for social image retrieval. In: Proceeding of 1st ACM International Conference on Multimedia Information Retrieval, pp. 30–31 (2008)
9. Lei, W., Linjun, Y., Nenghai, Y., Hua, X.S.: Learning to tag. In: Proceedings of the 18th ACM International Conference on World Wide Web, pp. 20–24 (2009)
10. Lei, W., Steven, C.H., Rong Jin, H., Jianke, Z., Nenghai, Y.: Distance Metric Learning from Uncertain Side Information with Application to Automated Photo Tagging. In: Proceeding of 17th ACM International Conference on Multimedia, pp. 15–24 (2009)
11. Wang, X.-J., Zhang, L., Li, X.R., Ma, W.-Y.: Annotating Images by Mining Image Search Results. IEEE Transactions on Pattern Analysis and Machine Intelligence 30(11), 1919–1932 (2008)

12. Wang, C., Jing, F., Zhang, L., Zhang, H.J.: Scalable search-based image annotation. Multimedia Systems 14(4), 205–220 (2008)
13. Jing, F., Wang, C., Yao, Y., Deng, K., Zhang, L., Ma, W.: IGroup: web image search results clustering. In: Proceedings of the 14th Annual ACM International Conference on Multimedia, pp. 377–384 (2006)
14. Tong, H., Faloutsos, C., Pan, J.: Fast Random Walk with Restart and Its Applications. In: Proceedings of the IEEE Sixth International Conference on Data Mining, pp. 613–622 (2006)
15. Croft, W.B., Lafferty, J.: Language Models for Information Retrieval. Kluwer Int. Series on Information Retrieval, vol. 13. Kluwer Academic Publishers, Dordrecht (2002)
16. Li, X., Snoek, C.G., Worring, M.: Learning social tag relevance by neighbor voting. IEEE Transaction on Multimedia, 1310–1322 (November 2009)
17. Barnard, K., Duygulu, P., Forsyth, D., de Freitas, N., Blei, D.M., Jordan, M.I.: Matching words and pictures. Jour. Machine Learning Research 3(6), 1107–1135 (2003)
18. Geng, B., Yang, L., Xu, C., Hua, X.: Collaborative learning for image and video annotation. In: Proceeding of the 1st ACM International Conference on Multimedia Information Retrieval, pp. 443–450 (2008)
19. Li, J., Wang, J.Z.: Real-time computerized annotation of pictures. In: Proceedings of the 14th Annual ACM International Conference on Multimedia, pp. 911–920 (2006)
20. Jones, K.S., Walker, S., Robertson, S.E.: A probabilistic model of information retrieval: development and comparative experiments - part 2. Jour. Information Processing and Management 36(6), 809–840 (2000)

Removing Uncertainties from Overlay Network

Ye Yuan[1,2], Deke Guo[3], Guoren Wang[1,2], and Lei Chen[4]

[1] College of Information Science and Engineering, Northeastern University,
Shenyang 110004, China
[2] Key Laboratory of Medical Image Computing (Northeastern University),
Ministry of Education
[3] Key lab of Information System Engineering,
School of Information System and Management,
National University of Defense Technology, Changsha 410073, China
[4] Hong Kong University of Science and Technology,
Hong Kong SAR, China

Abstract. Overlay networks are widely used for peer-to-peer (P2P) systems and data center systems (cloud system). P2P and data center systems are in face of node frequently joining, leaving and failure, which leads to topological uncertainty and data uncertainty. Topological uncertainty refers to that overlay network is incomplete, i.e., failures of node and link (between two nodes). Data uncertainty refers to data inconsistency and inaccurate data placement. Existing P2P and data center systems have these two uncertainties, and uncertainties have an impact on querying latency. In this study, therefore, we first give probabilistic lower bounds of diameter and average query distance for overlay network in face of these two uncertainties. The querying latency of existing systems cannot be better than the bounds. Also, existing systems often suffer unsuccessful queries due to uncertainties. To support an efficient and accurate query, we propose a topology constructive method and a data placement strategy for removing two uncertainties from overlay network. Also, efficient algorithms are proposed to support range queries in an overlay network. The DeBruijn graph representing overlay network is used to construct a new system, $Phoenix$, based on proposed methods. Finally, experiments show that performances of $Phoenix$ can exceed the probabilistic bounds, and they behave better than existing systems in terms of querying latency, querying costs, fault tolerance and maintenance overhead.

1   Introduction

The growing popularity of (Peer-to-Peer) P2P and data center (cloud) systems makes them very likely substrate for future large-scale information architectures [1,2,5,13,14,20,11,12]. Most P2P and data center systems are based on overlay networks. For example, for P2P systems, Chord [2], Pastry [1] are based on the Hypercube topology; Viceroy [5] and Ulysses [6] are based on the Butterfly topology; Cycloid [7] is based on the CCC topology [8]; CAN [1] is based on the d-dimensional torus topology; Koorde [4], D2B [3] are based on the deBruijn topology; Moore [18,9] and BAKE [10] are based on the Kautz topology; [19] is based on bipartite topology. For data center systems, BCube [12] and FiConn [11] are based on the Hypercube topology; VL2 [15] is based on the deBruijn-like topology.

J.X. Yu, M.H. Kim, and R. Unland (Eds.): DASFAA 2011, Part I, LNCS 6587, pp. 284–299, 2011.

However, one critical requirement of these systems is that the number of nodes must be some given values determined by the node degree and the network diameter. Hence, the corresponding approaches are often impractical when nodes frequently join, leave, and fail. The dynamic feature leads to topological uncertainty and data uncertainty of overlay network. Topological uncertainty refers to that overlay network is incomplete, i.e., failures of node and link (between two nodes). Data uncertainty refers to data inconsistency and inaccurate data placement. P2P and data center systems often suffer a large querying latency and unsuccessful queries in face of the two uncertainties. To solve the problems, there are some challenges for which we propose efficient approaches in this paper, which is summarized in next subsection.

1.1 Challenges and Our Contributions

Challenge 1: How to compute the lower bound of querying latency of overlay network in face of uncertainties? The Moore bound sets diameter lower bound for any static overlay network [22]. The Moore bound, however, cannot give a good description for overlay network in face of uncertainties. The diameters of Kautz and deBruijn graphs mostly achieve the Moore bound. However, the diameters of the systems based on them, like D2B, Koorde and FissionE, are much larger than the Moore bound. To compute lower bound, we should answer questions: how to simulate uncertainties for overlay network? how to compute probabilistic distribution of query distance?

In our approach, we define a random process that a node joins a dynamic trie tree structure. The random process can be seen as a *coupon collector's problem* [17]. Then we can compute distance distribution, from which probabilistic lower bounds of the network diameter and average query distance are derived.

Challenge 2: How to remove topological uncertainty from overlay network? Topological uncertainty refers to that overlay network is incomplete, i.e., failures of node and link (between two nodes). The main reason is that overlay network is typically dynamic with nodes frequently coming and leaving. To remove topological uncertainty, a topology constructive method should be proposed for overlay network to deal with a dynamic scenario.

In our approach, we propose a dynamic trie tree structure that can provide a flexible topological control for an overlay network in dynamic scenarios. Based on the tree structure, any overlay network can be used to construct a system that can behave good topological properties in dynamic scenarios.

Challenge 3: How to remove data uncertainty from overlay network? Data uncertainty refers to data inconsistency and inaccurate data placement. For a static overlay network, i.e., Hypercube and deBruijn, data objects are strictly mapped to corresponding nodes based on data and node identifiers. However, data objects cannot locate corresponding node identifiers in face of frequent topology changes, which can get data lost or inconsistent.

In our approach, based on dynamic trie tree structure, a data object is mapped to a node of overlay network in a fault tolerant way. Even if many nodes fail, our data placement policy guarantees that a data object is distributed to a node having the longest matching identifer with the data object's, so that the data placement policy can accurately and efficiently organize data objects and support data queries.

Challenge 4: How to support range queries in an overlay network? Range query is a universal and useful query type, and thus it is desirable to support efficient range queries in an overlay network. Our topological construction and data placement strategies guarantee an efficient exact query (routing) in an overlay network. However, it is unrealistic to transform a range query to several exact queries, since this method may incur a very large network overhead. Thus, new algorithms should be proposed to support efficient range queries in an overlay network.

In our approach, based on the trie tree, two locality-preserving data placement algorithms are proposed for one-dimensional and multi-dimensional data, so that the data whose values are close each other will obtain adjacent identifiers and data will be stored on same or neighboring peers. Therefore, the efficient routing algorithm can be used to reach a peer that covers a part of a querying range, then query is forwarded along predecessor or successor links to obtain the whole querying range.

The rest of the paper is organized as follows. The search delay is analyzed for the dynamic network, and two lower bounds are proved for the network diameter and average query distance in Section 2. The topology constructive method and data placement strategy are proposed to remove uncertainties from overlay network in Sections 3. Section 4 introduces efficient algorithms for supporting one and multiple dimensional queries. Finally, the performance of the two proposed algorithms are evaluated in Section 5 and we conclude the work in Section 6.

2   Probabilistic Lower Bounds of Query Delay of a Dynamic Overlay Network

Most existing overlay networks are in face of uncertainties. The negative effects due to possible dynamical behaviors need to be carefully considered for topology properties. In this section, we discuss the impact on the diameter and average query distance.

It is well known that Moore bound [22] is a lower bound of diameter for any static overlay network. For a static overlay network with maximum out-degree d, any node cannot reach more $d + d^2 + ... + d^h$ nodes in at most h hops. Hence, in order to reach all $n - 1$ nodes in at most h hops, the value of h should satisfy $d + d^2 + ... + d^h \geq n - 1$ which leads to the Moore bound. However, with uncertainties, the number of nodes reached from a beginning node in i hops may not be d^i nodes, for $1 \leq i \leq h$. In other words, the routing tree may not be completed in each level.

To simulate uncertainties for a overlay network, we set up a d-way trie tree. Let us consider a case that nodes join and leave randomly from a trie tree. The joining process of a node is equivalent to that of a ball with a prefix x is dropped into the root of a virtual tree and then reaches a most matched node with x of d children as the next step. The leaf at which the ball ends up is the node's parent, and shares the longest common prefix with x. The process is similar with the most existing P2P and data center systems. Node departure is an opposite process of node joining.

Now we estimate a lower bound of the diameter for an overlay network with uncertainties. Here, the diameter is defined as the largest depth of the tree.

Theorem 1. *With the probability $1 - o(1)$, the lower bound d_{max} of the diameter of an overlay network concentrates on the value d_l, where*

$$d_l = \lfloor \log_d n + \sqrt{d \log_d n} - 1.5 \rfloor \tag{1}$$

Proof. Since d_{max} is the maximum height of the tree, to compute d_{max}, we need to consider the leave process of a node. This process is constructed by dropping balls and getting rid of leaf, and it is very similar with the well know structure called the Patricia trie [16]. The tree is a collapsed version of the regular trie in which all intermediate nodes with a single child are removed. Recall that with the probability $1 - o(1)$, the height of a random Patricia trie is concentrated on the value d_l, where d_l is given by the Equation (1), we obtain the result from [16].   ∎

The diameter of an overlay network is simply the largest distance between any pair of nodes and only provides an upper bound on the delay (number of hops) experienced by users. A much more crucial metric is the average distance between any pair of nodes. We will focus on a lower bound of this metric in an overlay network.

Let h denote the minimum depth of the d-way tree. The tree is divided by two parts. First part consists of all nodes which locate between the root and $h - 1$ level. Second part includes all nodes which locate between level h to $d_{max} - 1$. Naturally, the first part is a full d-way tree with depth h. Let $d(root, x)$ (denoted by d_x) denote the distance from the root to any node x in the tree. We will discuss the distribution of d_x for the two parts separately.

To compute the distribution of the first part, we define a sequence of indicator random variables A_i, $i \geq 0$, where $A_i = 1$ if level i of the tree is full after users joined the system and $A_i = 0$ otherwise. We say that a level is full if all nodes of that level are present and nonleaf. Notice that level i can be full only if $i \leq h$ and that $A_i = 1$ implies that $A_k = 1$, for all $k < i$. It immediately follows d_x that is at least $k + 1$ if and only if all levels from 0 to k are full:

$$P(d_x \geq k + 1) = P(\bigcap [A_i = 1]) \tag{2}$$

We then have the deduction of Formula (2):

$$P(d_x \geq k + 1) = P(d_x \geq k)P(A_k|A_{k-1}) \tag{3}$$

where $P(d_x \geq 0) = 1$ and $P(A_k|A_{k-1})$ is the conditional probability of level k being full given that all previous levels $0, ..., k - 1$ are full:

$$P(A_k|A_{k-1}) = P(A_k = 1|h \geq d_x) \tag{4}$$

For the first part, we draw out the following theorem.

Theorem 2. *With the probability $1 - o(1)$, the distribution of d_x is:*

$$\exp\{-d^k \exp\{\frac{n(d-1)+1}{d^{k+1}(d-1)} - \frac{1}{d-1}\}\}$$

Proof. First notice that the first part is a d-way tree with h levels. The d-way tree built using nodes contains $\frac{n(d-1)+1}{d}$ leaves and nonleaf $\frac{n(d-1)+1-d}{d(d-1)}$ nodes. Next, examine level k of the tree and observe that all d^k possible nodes at this level must be non-leaf for level k to be fully joined. Assuming that all previous levels are full, exactly $\frac{d^{k-1}}{d-1}$ non-leaf nodes contributed to filling up levels $0, ..., k-1$ and the remaining $\frac{n(d-1)+1-d}{d(d-1)} - \frac{d^{k-1}}{d-1} = \frac{n(d+1)+1-d^{k+1}}{d(d-1)}$ non-leaf nodes had a chance to be joined. After the first levels $k-1$ have been filled up, each node at level k is hit by an incoming ball with an equal probability d^{-k}. Thus, our problem reduces to finding the probability that $u = \frac{n(d+1)+1-d^{k+1}}{d(d-1)}$ uniformly and randomly placed balls into $m = d^k$ bins manage to occupy each and every bin with at least one ball. There are many ways to solve this problem, one of which involves the application of well-known results from the coupon collector's problem [17]. We use this approach below. Define $Z(u)$ to be the random number of non-empty bins after balls u are thrown into m bins. Thus, we can write $P(A_k|A_{k-1}) = P(Z(u) = m)$. Recall that in the coupon collector's problem, u coupons are drawn uniformly randomly from a total of m different coupons. Then, the probability $Z(u) = m$ to obtain m distinct coupons at the end of the experiment is given by [17].

$$P(Z(u) = m) = \sum(-1_j)(m/j)(1 - \frac{j}{m})^u \tag{5}$$

For large u, the term $(1 - j/m)^u$ can be approximated by $e^{-uj/m}$, yielding

$$P(Z(u) = m) \approx \sum(-1_j)(m/j)e^{-uj/m} = (1 - e^{-u/m}) \tag{6}$$

Since we are only interested in the asymptotically large $m = O(log_d n)$, (6) allows a further approximation

$$P(Z(u) = m) \approx e^{-me^{-u/m}}$$
$$= \exp\{-d^k \exp\{\frac{n(d-1)+1}{d^{k+1}(d-1)} - \frac{1}{d-1}\}\} \tag{7}$$

From (5)-(7), we get the distribution of d_x:

$$P(d_x \geq k+1) = P(d_x \geq k)P(A_k|A_{k-1})$$
$$\approx \exp\{-d^k \exp\{\frac{n(d-1)+1}{d^{k+1}(d-1)} - \frac{1}{d-1}\}\} \tag{8}$$

∎

From (8), we know that

$$h = max(d_x)$$
$$= \log_d n - \log_d((1 + \varepsilon)\log n - O(\log\log n)) \tag{9}$$

with the probability at least $1 - n^{-\varepsilon}, \varepsilon \leq 1$.

For the second part of the tree, we have the following theorem from [16].

Theorem 3. *With probability $1 - o(1)$, the distribution of d_x of PATRICIA tries is:*

$$d_x \sim \sqrt{1 + d\xi\Phi'(\xi) + \xi^d\Phi''(\xi)}e^{-n\Phi(\xi)} \tag{10}$$

where $\xi = nd^{-k}$, $0 < \xi < 1$, and $\Phi(\xi) \sim \frac{1}{d}\rho_0 e^{\varphi(\log_d \xi)}\xi^{3/2}\exp(-\frac{\log^d \xi}{d\log d})$.

From the distribution of d_x for two parts of the tree, we get the lower bound of average distance for a dynamic network. We have the following theorem.

Theorem 4. *The lower bound of average distance for a dynamic network is $\log_d n + \sqrt{\log_d n/d}$ with the probability at least $1 - n^{-\varepsilon}, \varepsilon \leq 1$.*

Proof. From the above equations, we get the average distance:

$$d_{avg} = \sum_{k=0}^{h-1} k \cdot (\exp\{-d^k \exp\{\frac{n(d-1)+1}{d^{k+1}(d-1)} - \frac{1}{d-1}\}\}) +$$

$$\sum_{k=h}^{d_{max}-1} k \cdot (\sqrt{1 + d\xi\Phi'(\xi) + \xi^d\Phi''(\xi)}e^{-n\Phi(\xi)})$$

$$\approx \log_d n + \sqrt{\frac{\log_d n}{d}}$$

∎

Now we have obtained the lower bounds of the diameter and average distance for an overlay network. The search delay of any existing P2P or data center system cannot be smaller than the lower bounds.

3 Topology Constructive and Data Placement Methods of Overlay Network

Any static overlay network has desirable properties graph only if all nodes exist and are stable. Such as deBruijn and butterfly have d^k and $d^k k$ nodes respectively. This requirement, however, is impractical in face of uncertainties. To address this issue, we propose a dynamic multi-way trie tree structure to achieve the desired topology.

3.1 Dynamic Multi-way Trie Tree Structure

Definition 1. *A dynamic d-way trie tree with depth k is a rooted tree. Each node has at most d child nodes. Each edge at the same level is assigned a unique label. Each node is given a unique label. The label of a node is the concatenation of the labels along the edges on its root path. The label of each edge is assigned based on the following rules.*

- *The edge from the root node to its ith child is labeled as $x_1^i = i$ for $0 \leq i \leq d-1$. The ith child of root node is labeled as $x_1^i = i$, and is arranged from left to right. The root node does not contain any string.*
- *The edge from a node x_1 to its ith child is labeled as $x_2^i = i$ for $0 \leq i \leq d-1$. The ith child is labeled as $x_1 x_2^i$, and is arranged from left to right.*
- *The edge from a node $x_1 x_2...x_{k-1}$ to its child is labeled as x_k^i. The child of x_k^i is labeled as $x_1 x_2...x_{k-1}x_k$, and is arranged from left to right.*

– *There is a bidirectional edge between any node and its parent, and all the nodes at the same level form a ring.*

The trie tree might be imbalanced in a dynamic environment. In fact, it is impractical to keep a balanced tree in a dynamic scenario. Definition 2 gives a definition of the balanced tire tree.

Definition 2. *A d-way trie tree with depth k is balanced if all leaf nodes are at the level k. A balanced trie tree is a complete trie tree only if the parent node of any leaf node has full child, otherwise it is an incomplete trie tree.*

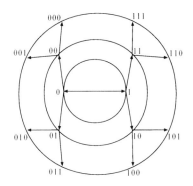

Fig. 1. Example of a trie tree

The tree shown in Fig. 1 is a complete trie tree. If the leaf node 011 fails, the tree becomes an incomplete trie tree. If nodes 010 and 011 both fail, the tree becomes an unbalanced tree.

For any level, the left-to-right traversal of nodes at that level form a total ordering, denoted as the trie ordering. We sort all children of root node in an ascending order within level 1, and then rank all the child nodes of each node at level 1 respectively. Note that the level 2 nodes inherit the order of nodes in level 1. By sorting nodes from level 1 to k recursively, we can find a trie ordering of nodes at each level. From the trie tree structure, we know each level forms a ring and we call it the trie ring. We arrange the trie ordering in each ring along the anti-clockwise direction. In the ring, each node connects its predecessor and successor.

Definition 3. *For any node x, its predecessor is the first existing node clockwise from it in a trie ring of existing nodes at the same level, and its successor is the first existing node anti-clockwise from it in the same trie ring. The concept of left adjacent node is similar to the predecessor, but the trie ring is consisted of all possible nodes not just existing nodes. So do the right adjacent node and successor node.*

For example, in Figure 1, the node 010 is the predecessor of the node 011 and its successor is the node 100. In the unbalanced tree, (the node 010 and 011 are not alive) the predecessor of the node 100 is the node 001 and 001's successor is 100.

3.2   Mapping Overlay Network to Dynamic Trie Tree

For a node $x = x_1x_2...x_k$ of an overlay network, $\sigma^i(x), 1 \leq i \leq d$, denotes a neighbor node of x. The label will be used throughput the paper. Any d-ary overlay network is mapped to the d-way trie tree as follows:

- If a node $\sigma^i(x)$ is alive in the overlay network, then it is the ith neighbor of node x;
- Otherwise, if $\sigma^i(x)$ and its trie tree predecessor y have a common prefix with length $k-1$, then node y is the ith neighbor of node x. In this case, node y also keeps the identifier $\sigma^i(x)$ to take over node $\sigma^i(x)$.
- Otherwise, if $\sigma^i(x)$ and its trie tree successor z have a common prefix with length $k-1$, then node z is the ith neighbor of node x. Similarly, node z keeps the identifier $\sigma^i(x)$.
- Otherwise, the youngest alive ancestor of $\sigma^i(x)$ is used as the ith neighbor of node x and the ancestor keeps the identifier $\sigma^i(x)$.

The mapping rule guarantees that node x of system based on any overlay network can connect to its neighbor node $\sigma^i(x)$. For example, node 010 connects to nodes 100 and 101 in Fig. 1. If node 100 fails, then node 010 connects to 101 again and node 101 keeps the identifier 100 to take over node 100. If nodes 100 and 101 both fail, node 010 connects to their parent node 10 and node 10 keeps the identifiers 100 and 101 to take over both of them.

From the mapping rule, it is easy to draw the following conclusion.

Theorem 5. *For an inner node x and leaf node y of system based on an overlay network, the former rules can make the following conclusion:*

- *For a balanced tree, x has $2d + 3$ neighbors and y has $d + 3$ neighbors.*
- *For an unbalanced tree, x has at least $2d + 1$ neighbors and y has at least $d + 1$ neighbors.*

3.3   Data Placement Rule of Overlay Network

The data placement is very important for P2P and data center systems. Normal data placement polices may not accurately and efficiently organize data and support data queries. The following scheme can assure that data objects are distributed based on the longest prefix matching policy. In addition, the scheme is fault-tolerant.

Suppose the identifier of a data object is x whose length is longer than a node identifier. The following algorithm gives the placement rule.

Algorithm 1. Placement(Data $x = x_1x_2...x_k...$)

---
if *node $x' = x_1x_2...x_k$ has appeared in the overlay network* **then**
  └ Node x' stores the data x.
else if *node x' and its predecessor y have common parent node in the trie tree* **then**
  └ Node y is the host node of data x.
else if *node x' and its successor z have common parent node in the trie tree* **then**
  └ Node z is the host node of data x.
else
  └ The youngest alive ancestor of x' is the host node of data object x.

For example, in Figure 1, data 100... is stored by node 100, and is taken over by node 101 when node 100 is not alive. If nodes 100 and 101 are both not alive, then their parent node 10 keeps the data object.

3.4 Case Study

In this section, deBruijn representing the single protocol (like Kautz, Hypercube) of static overlay network based on the trie tree and above proposed schemes are used to construct a new dynamic system $Phoenix$.

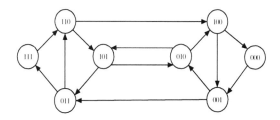

Fig. 2. Example of a deBruijn structure

The topology mapping rule of an overlay network is the most important step for constructing a dynamic system. For a deBruijn graph, $\sigma^i(x) = x_2...x_k x_{k+1}^i$, the mapping rule can be directly applied. Figure 2 shows a 3-ary deBruijn graph.

Using an example for this deBruijn graph, in Fig. 1, node 010 connects to node 100 and 101. If node 100 fails, node 101 replaces 100. As a result, node 010 connects to 101 again and node 101 stores the identifier 100. If nodes 100 and 101 both fail, node 010 connects to their parent 10 and node 10 stores the identifier 100 and 101.

The routing algorithm from peer x to peer y of the deBruijn overlay is: find the largest suffix u of x that coincides with a prefix of y, then walk towards a neighbor z of x such that its largest suffix v coincides with a prefix of y and the length of v is larger than that of u. The process continues until the route reaches y. For example, in Fig.2, the route goes from peer 010 to peer 000 along the path: 010-100-000.

$Phoenix$ can directly adopt the data placement rule to enhance the performances of their data managements.

4   Range Queries

In order to manage complex data objects and support more wide applications, $Phoenix$ should also support complex query operations besides exact-match query in a graceful fashion, for example, the one-dimensional and multi-dimensional queries. The most universal querying type is range query. Thus we focus on techniques of range query, and the techniques can be applied to other type queries after a little modification.

In the following, we will first introduce techniques for one-dimensional range queries and then for multi-dimensional range queries.

To support one-dimensional range queries of $Phoenix$, the structure of $Phoenix$ does not need to be changed. All the data is stored in leaf peers, and the inner peers act

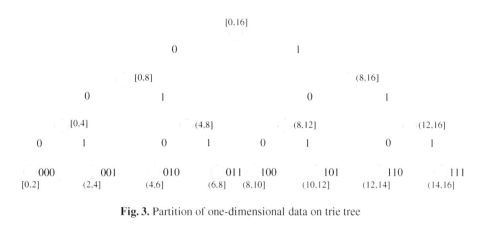

Fig. 3. Partition of one-dimensional data on trie tree

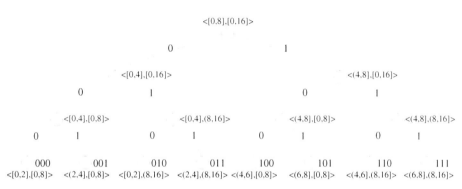

Fig. 4. Partition of multi-dimensional data on trie tree

as the routing peers. Here we first design an order-preserving data placement strategy as follows: we recursively partition the whole data space into sub-spaces in the same way to construct a complete trie tree whose height equalling that of the current trie tree. Each data finds the smallest sub-space that contains its values, and uses the identifier of that subspace as its identifier. Therefore, the data whose values are close each other will obtain adjacent identifiers in the leaf peer level, and they will be stored on same or neighboring. For instance, in Fig. 3, suppose range query, $[L, H] = [0, 16]$, based on the locality-preserving data placement algorithm, the leaf identifiers are assigned as sub-space in the way: $\{[0, 2], (2, 4], ..., (12, 14], (14, 16]\} = \{000, 001, ..., 110, 111\}$.

Any peer that issues a range query can find the identifiers of smallest sub-space that contains the whole query region. Then the query covers multiply peers and will be routed towards the peer which charges the lower bound of it region firstly. Once the peer received the query, it will first forward it to the successor peer if it cannot cover the whole region of the query. And so on and so forth, the query will be forwarded to a peer which charges the upper bound of its region. The querying method is shown in Algorithm 2. x is the requesting peer. y and z are the first and last peers that cover the whole range query. For example, in Figure 1, peer 010 issues a range query $[12, 16]$.

Based on the locality-preserving placement scheme, we find that peers 110 and 111 cover sub-spaces $\{(12, 14], (14, 16]\}$ respectively. Then the query is forwarded to peer 110 to get $(12, 14]$ firstly. Then the query reaches peer 111 for obtaining $(14, 16]$ along the successor link.

Algorithm 2. Query(x, y, z)

---

    case 1 : *Length*$(x) = $ *Length*(y) **or** *(y is a data object and x is not a prefix y)*

        if $x = y$ **then**

            └ **return** available.

        else if *Comprefix*$(x, Successor(x)) = k - 1$ **and** *Successor(x) is less than y in the ring* **then**

            └ Forward the message to node Successor(x).

        else if *Comprefix*$(x, Predecessor(x)) = k - 1$ **and** *Predecessor(x) is larger than y in the ring* **then**

            └ Forward the message to node Predecessor(x).

        else

            Node x forwards message to its neighbor $\sigma^i(x)$ which has the largest value

            Comprefix$(\sigma^i(x), y)$ among all the neighbors.

    case 2 : *Length*$(x) < $ *Length*(y) **or** *(y is a data object and x is a prefix y)*

        if x *contains* y **then**

            └ **return** available

        else if *Node x has at least one child* **then**

            └ Forward message to its child z which has the largest value of Comprefix(z, y)

        else

            Node x forwards message to its neighbor $\sigma^i(x)$ which has the largest value of

            Comprefix$(\sigma^i(x), y)$ among all the neighbors.

    case 3 : *Length*$(x) > $ *Length*(y)

        if x *has a link to its parent* **then**

            └ Node x forwards message to its parent.

        else

            Node x forwards message to its neighbors $\sigma^i(x)$ which has the largest value of

            Comprefix$(\sigma^i(x), y)$ among all the neighbors.

    Query is forwarded along the successor link until peer z for obtaining the whole range.

This method for range query can decrease the message cost caused by transferring the query to the all intersection peers, but the delay maybe be a litter larger than forwarding the query to all related peers simultaneously. It is needed to make a tradeoff between the delay and the message cost. After computing the hash values of the lower bound and upper bound, we get the common prefix b of the hash values. If they have no common prefix, we can divide the range into several (at most d) sub-regions with common prefixes and deal with each sub-region respectively. The results are covered by its offspring leaf peers. Therefore, we have another querying algorithm as follows: the query is firstly routed to peer b. Then the query is routed down links of peer b until the query reaches peers which cover the whole query region. This technique of range query is called $SPhoneix$.

The two schemes, $Phoenix$ and $SPhoenix$, for range query are flexible. If the traffic of the network is heavy, $Phoenix$ is used. Otherwise $SPhoenix$ is adopted.

For multi-dimensional queries of $Phoenix$, assume the whole data space is $< [l_0, h_0], ..., [l_i, h_i], ..., [l_{m-1}, h_{m-1}] >$. We also construct the complete trie tree as one-dimensional query. We partition the entire multi-dimensional data space $< [l_0, h_0], ..., [l_i, h_i], ..., [l_{m-1}, h_{m-1}] >$ onto the completed trie tree along dimensions $A_0, ..., A_i, ..., A_m$ in a round-robin style. Each peer in the tree represents a multi-dimensional data subspace and the root node represents the entire multi-dimensional data space. For any node B at the jth level of the trie tree that has d child peers, let i denote the value of j mod m. Then, the subspace c represented by node B is evenly divided into d pieces along the ith attribute, and each of its d child peers represents one such a piece. As a result, all leaf peer store the disconnected subspace of the entire space. Every multi-dimensional value $< a_0, ..., a_i, ..., a_{m-1} >$ can find its hash value. For example, in Figure 4, suppose multi-dimensional range query, $< [0, 8], [0, 16] >$, the leaf identifiers are assigned as sub-space in the way: $\{< [0, 2], [0, 8] >, < (2, 4],$ $[0, 8] >, ..., < (4, 6], (0, 8] >, < (6, 8], (8, 16] >\} = \{000, 001, ..., 110, 111\}$. Then queries are routed to corresponding peers to fetch desirable data. This technique of multi-dimensional query is called $MPhoenix$.

5   Performance Evaluation

We use PeerSim[1] to implement $Phoenix$, $SPhoenix$ and $MPhoenix$. PeerSim is a large-scale network simulation framework aimed at developing and testing any kind of dynamic network protocols. We evaluate networks consisting of 256 to 1 million nodes. Queries are generated randomly and uniformly from the network, and the identifiers of targeted data objects are also uniformly distributed in the key-space. The simulation environment is pentium 4 CPU 3.0 GHz, 2G memory and 160G disk. To compare the efficiency of dynamic networks, we implement Koorde based on deBruijn, and Viceroy based on Butterfly in PeerSim. To compare the efficiency of one-dimensional range queries, CAN [1] and BATON [21] are chosen to compare with $Phoenix$. To evaluate the efficiency of multi-dimensional range query, we set the dimension of data to 20. During tests, d is set to 4 for all experiments, and thus the degree of each graph is also 4.

Querying Efficiency of Dynamic Overlay Networks. In this section, we show two group of experiments so as to examine the maximum and average querying path lengths of each graph in the dynamic networks. In the first group of experiments, for different numbers of nodes, 10% nodes of the total nodes are allowed to join and (depart) in (from) the network. The maximum and average number of hops are recorded in the dynamic scenario. The diameter lower bound (Dia-bound) and average distance lower bound (Avg-bound) are also shown for the two experiments. The results are plotted in Figure 5. This figure shows that the delays of $Phoenix$ are both less than the lower bounds. $Phoenix$ has far better performance than Koorde and Viceroy whose querying efficiencies are worse than the lower bounds.

---

[1] http://www.cs.unibo.it/bison/deliverables/D07.pdf

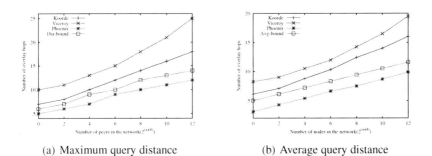

(a) Maximum query distance   (b) Average query distance

Fig. 5. Querying efficiency of dynamic overlay networks

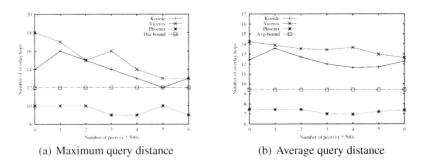

(a) Maximum query distance   (b) Average query distance

Fig. 6. Variations of hops in the presence of nodes concurrently joining and leaving

In the other group, each system initially has 64k nodes, then the same number of nodes begin concurrently join and (depart) in (from) the network. Thus the number of nodes in the network hardly changes. Figure 6 shows the variation of the lengths of query paths. As shown in the result, though the static graphs are the same, $Phoenix$ is more stable than Koorde and Viceroy when dealing with the concurrently joining and leaving operations. The experiment confirms that our constructive method guarantee that overlay networks behave stable in face of uncertainties, and the performances are not limited by the lower bound any more.

Querying Efficiency of Range Queries. This section shows evolutional results of range queries. We calculate the average number of messages and hops induced by answering one-dimensional rang query request in the network. As shown in Figure 7(a) and Figure 7(b), $Phoenix$ incurs least costs when performing a rang query request than CAN and BATON. The reason is that both CAN and BATON first identify a peer whose data is in the query region, then proceed left and/or right to cover the remainder of the query range. Though the degree of CAN is large, it still achieves the worst efficiency. In addition, it is very difficult to make a decision on the value of d in advance, because it only relies on the expected number of nodes in the system, and therefore large value of node degree is not suitable in dynamic network. Compared to other networks, $SPhoenix$ takes very small number of hops, but it takes more costs than other networks because the request can reach the leaf peers simultaneously.

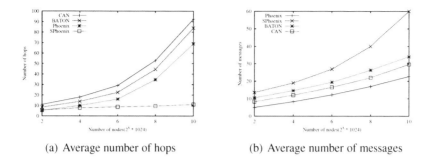

(a) Average number of hops   (b) Average number of messages

Fig. 7. Performance of one-dimensional range queries

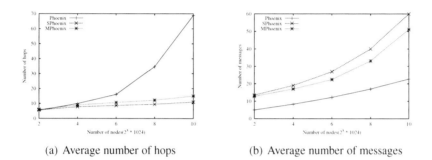

(a) Average number of hops   (b) Average number of messages

Fig. 8. Performance of multi-dimensional range queries

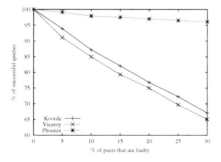

Fig. 9. Percentage of successful queries

The evaluating results of multi-dimensional range queries are shown in Figures 8(a) and 8(b). $Phoenix$ and $SPhoenix$ are also plotted in the figures. From the figures, we know that $MPhoenix$ behaves great in terms of querying hops and querying costs.

Robustness. A number of 4000 queries are uniformly generated in the network consisting of 10000 nodes. When the network becomes stable, we let a node fail with probability p ranging from 0.1 to 0.3. Figure 9 plots the percentage of successful queries when using the forwarding algorithm of corresponding protocol. For the same percentage of failed nodes, the percentages of successful queries for *Phoenix* is much higher than that of other networks, and their variations are minimal. This is because all the methods for networks are robust to guarantee the successful query. Though Viceroy and Koorde are based on the same overlay networks as *Phoenix*, they encounter more failed queries. The reason is that a message is always forwarded towards a unique neighbor that is closer to the destination and is not allowed to transfer along other paths during their querying.

6   Conclusions

In this study, we solve the problem of poor performance put on by overlay network in face of topological and data uncertainties. We first give probabilistic lower bounds of query distance in face of uncertainties. Then based on the proposed topology constructive method and data placement rule, we can remove uncertainties from overlay network in a dynamic scenario. Moreover, efficient algorithms are proposed to support range queries. Finally, extensive experiments show that the new dynamic system, *Phoenix*, derived from proposed methods, achieve an optimal querying efficiency which exceeds the lower bounds. We also have the plan of using *Phoenix* as an infrastructure to support other large-scale and distributed applications.

Acknowledgment

This research is supported by the National Science Fund for Distinguished Young Scholars (Grant No 61025007) and the National Natural Science Foundation of China (Grant Nos. 60933001 and 60803026). Deke Guo is supported in part by the Research Foundation of NUDT, the NSF China under Grant No. 60903206.

References

1. Ratnasamy, S., Francis, P., Handley, M.: A scalable content-addressable network. In: Proc. of the ACM SIGCOMM (2001)
2. Stoica, I., Morris, R., Karger, D.: Chord: a scalable peer-to-peer lookup service for internet applications. In: Proc. of the ACM SIGCOMM (2001)
3. Gauron, P.: D2B: a de bruijn based content-addressable network. Theor. Comput. Sci. (2006)
4. Kaashoek, F., Karger, D.R.: Koorde: a simple degree-optimal hash table. In: IPTPS (2003)
5. Malkhi, D.: Viceroy: a scalable and dynamic emulation of the butterfly. In: SODA (2002)
6. Xu, J., Kumar, A., Yu, X.: On the fundamental tradeoffs between routing table size and network diameter in peer-to-peer networks. IEEE J. Sel. Areas Commun. (2004)
7. Shen, H.: Cycloid: a constant-degree and lookupefficient p2p overlay network. In: IPDPS (2004)
8. Banerjee, S., Sarkar, D.: Hypercube connected rings: a scalable and fault-tolerant logical topology for optical networks. In: Computer Communications (2001)
9. Guo, D., Wu, J., Liu, Y., Jin, H., Chen, H., Chen, T.: Quasi-kautz digraphs for peer-to-peer networks. TPDS (August 24, 2010)
10. Guo, D., Liu, Y., Li, X.: BAKE: a balanced kautz tree structure for peer-to- peer networks. In: Proc. of INFOCOM, Phoenix, AZ, USA (April 2008)

11. Li, D., Guo, C., Wu, H., Zhang, Y., Lu, S.: Ficonn: using backup port for server interconnection in data centers. In: Proc. IEEE INFOCOM, Brazil (2009)
12. Guo, C., Lu, G., Li, D.: Bcube: a high performance, server-centric network architecture for modular data centers. In: Proc. SIGCOMM, Barcelona, Spain (2009)
13. He, Y., Liu, Y.: VOVO: VCR-oriented video-on-demand in large-scale peer-to-peer Networks. IEEE TPDS 20(4), 528–539 (2009)
14. Liu, Y.: A two-hop solution to solving topology mismatch. TPDS 19(11), 1591–1600 (2008)
15. Greenberg, A., Jain, N., Kandula, S., Kim, C., Lahiri, P., Maltz, D.A., Patel, P.: Vl2: A scalable and flexible data center network. In: Proc. of SIGCOMM, Barcelona, Spain (2009)
16. Szpankowski, W.: Patricia trees again revisited. Journal of the ACM 37 (1990)
17. DeGroot, M.H.: Probability and statistics. Addison-Wesley, Reading (2001)
18. Guo, D., Wu, J., Chen, H.: Moore: an extendable P2P network based on incomplete kautz digraph with constant degree. In: Proc. of 26th IEEE INFOCOM (May 2007)
19. Liu, Y., Xiao, L., Ni, L.M.: Building a scalable bipartite P2P overlay network. IEEE TPDS 18(9), 1296–1306 (2007)
20. Liu, Y., Xiao, L., Liu, X., Ni, L.M., Zhang, X.: Location awareness in unstructured peer-to-peer systems. IEEE TPDS 16(2), 163–174 (2005)
21. Jagadish, H.V., Ooi, B.C., Vu, Q.H.: BATON: A balanced tree structure for peer-to-peer networks. In: Proceedings of the 31st VLDB Conference (2005)
22. Bridges, W.G., Toueg, S.: On the impossibility of directed moore graphs. Journal of Combinatorial Theory, Series B29(3) (1980)

Probabilistic and Interactive Retrieval of Chinese Calligraphic Character Images Based on Multiple Features[*]

Yi Zhuang[1], Nan Jiang[2], Hua Hu[3], Haiyang Hu[3],
Guochang Jiang[4], and Chengxiang Yuan[1]

[1] College of Computer & Information Engineering,
Zhejiang Gongshang University, P.R. China
[2] Hangzhou No.1 People's Hospital, P.R. China
[3] School of Computer, Hangzhou Dianzi University, P.R. China
[4] The Second Institute of Oceanography, SOA, P.R. China
zhuang@zjgsu.edu.cn

Abstract. This paper proposes an efficient probabilistic indexing scheme called Probabilistic Multiple-Feature-Tree(PMF-Tree) to facilitate an interactive retrieval of Chinese calligraphic manuscript images based on multiple features such as contour points, character styles and types of character. Different from conventional character retrieval and indexing methods [18] which only adopts shape similarity as a query metric, our proposed indexing algorithm allows user to choose the above three kinds of features they prefer to as query elements. Moreover, a probabilistic modal is introduced to refine the retrieval result. Comprehensive experiments are conducted to testify the effectiveness and efficiency of our proposed retrieval and indexing methods respectively.

Keywords: Chinese calligraphic character, high-dimensional indexing, probabilistic retrieval.

1 Introduction

Chinese historical calligraphy work is a valuable part of the Chinese cultural heritage. To effectively protect these works from ruining or damage, they have been digitized to store permanently. The issue of retrieval and indexing of such digital works becomes new challenges. As shown in Fig. 1, the state-of-the-art character retrieval and indexing methods only use contour points extracted from a character as a feature in the similarity retrieval. The styles, types, even number of strokes of a character, however, have not been adopted to facilitate the character retrieval, which can be also as two effective features to prune search region in the retrieval processing.

[*] This paper is partially supported by the Program of National Natural Science Foundation of China under Grant No. 60003047, No.60873022, No.60903053; The Program of Natural Science Foundation of Zhejiang Province under Grant No. Z1100822, No.Y1080148, No.Y1090165; The Science Fund for Young Scholars of Zhejiang Gongshang University under Grant No. G09-7. The Science & Technology Planning Project of Wenzhou under Grant No. 2010G0066.

J.X. Yu, M.H. Kim, and R. Unland (Eds.): DASFAA 2011, Part I, LNCS 6587, pp. 300–310, 2011.
© Springer-Verlag Berlin Heidelberg 2011

In essence, the efficient retrieval of Chinese calligraphic characters directly relates to the category of high-dimensional data indexing with multiple features. Although considerable research efforts have been done on the high-dimensional indexing issue [10], unfortunately, the existing high-dimensional indexing methods can not be directly applied to Chinese calligraphic characters [18]. In our previous work [18], we studied the character retrieval and its indexing algorithms only based on their contour similarity measure. However, in Figs. 2 and 3, for a same character "书", there are two different styles(i.e., *Yan Ti* and *Mi Ti*, etc) and two different types (i.e., *Kai Su* and *Li Su*, etc). In most cases, people would like to get some result characters with specific style or type they prefer to, which can also be regarded as effective pruning criteria to reduce a search region.

(a). Yan Ti   (b). Mi Ti

Fig. 2. Two different styles of a same character

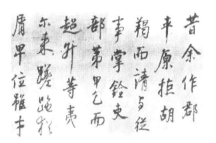

(a). Kai Su   (b). Li Su

Fig. 1. A calligraphic character images   **Fig. 3.** Two different types of a same character

In addition, due to the shape complexity of the Chinese calligraphic character, personal knowledge level and errors occurred in the feature extraction, as shown in Tables 1-3, for the character "书", the identifications of its corresponding features such as character type, style, even number of strokes are not trivial, and the precision ratio of character retrieval is not so accurate. For example, for the character in Table 1, the probabilities of the character style are *Yan Ti*, *Mi Ti* and *Liu Ti* are 10%, 85% and 5%, respectively. So the introduction of probability modal can further refine the retrieval results.

Table 1. The probability of style of"书"   **Table 2.** The probability of type of"书"

	Style	Probability
	Yan Ti	10%
	Mi Ti	85%
	Liu Ti	5%

	Type	Probability
	Kai Su	5%
	Li Su	5%
	Cao Su	85%
	Xing Su	5%

Table 3. The probability of number of strokes of"书"

	Number of strokes	Probability
	9	15%
	10	80%
	11	5%

From the above discussion, in this paper, based on the shape-similarity-based retrieval algorithm for Chinese calligraphic character in our previous work [18], we propose a probabilistic and composite high-dimensional indexing scheme based on multiple features, called PMF-Tree, which is specifically designed for indexing the large Chinese calligraphic characters. With the aid of the PMF-Tree index, a probabilistic k-nearest neighbor query of character λ_q in high-dimensional spaces is transformed into a range query in the single dimensional space.

The primary contributions of this paper are as follows:

1. We propose a novel probabilistic interactive retrieval method to effectively support the Chinese calligraphic characters retrieval by choosing multiple features of character.

2. We introduce a _Probabilistic Multiple-Feature-Tree_(PMF-Tree)-based indexing method to facilitate the interactive and efficient Chinese calligraphic characters retrieval with multiple features.

The remainder of this paper is organized as follows. In Sections 2, we provide background of our work. Then in Section 3, we propose a _Probabilistic Multiple-Feature-Tree_(PMF-Tree)- based high-dimensional indexing scheme to dramatically speed up the retrieval efficiency. In Section 4, we report the results of extensive experiments which are designed to evaluate the efficiency and effectiveness of the proposed approach. Finally, we conclude in the final section.

2 Background

Numerous promising research works have been done on the handwriting recognition [5]. For instance, a word-matching technology is used to recognize George Washington's manuscripts [3], and the historical Hebrew manuscripts were identified in [4]. However, no published research work has been done successfully on Chinese calligraphic character retrieval because it differs from other languages by its enormous numbers and complex structure of ideographs. In [6] Shi et al. have shown a content-based retrieval method for antique books, however it is unknown how well this rigid visual similarity-based method works on the Chinese calligraphic characters retrieval with different styles of handwritings in different dynasties. Belongie et al. [8] have proposed an inspirational and similar approach to ours, yet it is much more complex at least for calligraphic character retrieval. Our earlier work includes applying the Projecting method [2] and the Earth Movers' Distance (EMD) method [9] to the Chinese calligraphic character retrieval. However, these two recognition techniques are too rigid to be applied to the retrieval process. For calligraphic character retrieval, shape is a promising feature to model a character. So in this paper, we adopt a shape-similarity(SS)-based retrieval method proposed in our previous work [18] as a similarity measure between two characters.

There is a long stream of research on solving the high-dimensional indexing problems [10]. The R-tree [11] and its variants [12], etc are based on data & space partitioning, hierarchical tree index structure, however their performance deteriorates rapidly as dimensionality increases due to the "_dimensionality curse_" [10]. Another category is to represent original feature vectors using smaller, approximate representations, e.g., VA-file [13] and IQ-tree [14], etc. Although these methods accelerate the sequential scan by using data compression, they incur higher computational cost to decode the bit-string. The above two categories of indexing approaches, however, are only suitable for indexing the multi-dimensional data with the fixed dimensionality and does not fit for indexing the character features since almost every character has different number of contour points (dimensionalities). The distance-based approach (e.g., iDistance [15]) may be a promising scheme to indexing them since it does not heavily depend on the dimensionalities of the characters.

3   The PMF-Tree Index

In order to improve the probabilistic retrieval efficiency, in this section, we develop a novel probabilistic high-dimensional indexing technique, called the _Probabilistic Multiple-Feature Tree_(_PMF-Tree_ for short), to accelerate the retrieval process.

3.1   Preliminaries

The design of the _PMF-Tree_ is motivated by the following key observations. First, conventional character retrieval and indexing methods [18] are only based on similarity of two characters without considering the other factors such as number of strokes, styles, etc. So the effectiveness of these methods are not satisfactory and their query performances are not well scaled for large dataset due to the CPU-intensive distance computation in the retrieval process [18]. Second, the feature uncertainty has not been studied in the state-of-the-art character retrieval techniques, which may enable the query results to be more accurate and objective.

First we briefly introduce the notations that will be used in the rest of paper.

Table 4. Meaning of Symbols Used

Symbols	Meaning
Ω	a set of Chinese calligraphic character
λ_i	the i-th character and $\lambda_i \in \Omega$
M	the number of contour points from a character
N	the number of characters in Ω
λ_q	a query character user submits
$\Theta(\lambda_q, r)$	a query sphere with centre λ_q and radius r
$Sim(\lambda_i, \lambda_j)$	the distance between two characters defined in [18]
ε	Threshold value

As a preliminary step, before constructing the PMF-Tree, all characters in Ω are first grouped into T clusters by an AP-Cluster algorithm [16]. Each cluster is denoted as C_j, in which the centroid character(O_j) in C_j can be adaptively selected by the algorithm [16], where $j \in [1,T]$. So we can model a cluster as a tightly bounded sphere described by its _centroid_ and _radius_, which is saved in a class information file.

Definition 1 (Cluster Radius). _Given a cluster C_j, the distance between O_j and the character which has the longest distance to O_j is defined as the cluster radius of C_j, denoted as CR_j._

Given a cluster Cj, the cluster sphere of it is denoted as $\Theta(Oj,CRj)$, where Oj is the centroid character of cluster Cj, CRj is the cluster radius.

Definition 2 (Centroid Distance). *Given a character λ_i, its centroid distance is defined as the distance between λ_i and Oj, the centre of cluster that λ_i belongs to:*

$$CD(\lambda_i)=Sim(\lambda_i,O_j) \tag{1}$$

where $j\in[1,T]$, $i\in[1,\delta]$, and δ is the number of characters in C_j.

Once T clusters are obtained, then the centroid distance and the number of strokes of each character are computed. Moreover, its style and the type are identified at the same time. Finally, a uniform index key of a character is obtained, which is inserted by a B+-Tree.

As we know, for a same character, there are a number of different styles and types, respectively (see Figs. 2 and 3). To embed these two information into the unified index key that will be discussed in Section 3.2, two encoding schemes of the style and the type are needed which is shown in the following tables.

Table 5. Style of character

Style Name	Yan Ti	Liu Ti	Cai Ti	Su Ti	...
Style ID	1	2	3	4	...

Table 6. Type of character

Type Name	Li Su	Kai Su	Cao Su	...
Type ID	1	2	3	...

3.2 The Data Structure

In order to effectively prune the search region, we propose the *PMF-Tree*, an probabilistic multiple-feature indexing scheme in which the high-dimensional index for contour points is based on the iDistance[15]. As mentioned before, all characters are first grouped into T clusters using an AP-Cluster algorithm [16], then the *centroid-distance* and the number of strokes of each character are computed, the style and type of each character can be identified by user in preprocessing step. Thus the character λ_i can be modeled by a six-tuple:

$$\lambda_i::= <i, CID, CD, Style, Type, Num > \tag{2}$$

where

- i refers to the i-th character in Ω;
- CID is the ID number of the cluster λ_i belongs to;
- CD is the centroid distance of λ_i;
- $Style =\{StyID, P_s\}$, where $StyID$ is the style ID of λ_i, and $P_s=$**Prob**(the style ID of λ_i is $StyID$);
- $Type =\{TyID, P_t\}$, where $TyID$ is the type ID of λ_i, and $P_t=$**Prob**(the type ID of λ_i is $TyID$)
- $Num =\{NumS, P_n\}$, where $NumS$ is the number of strokes of λ_i, and $P_n=$**Prob**(the number of strokes of λ_i is Num)

For each character λ_i in a cluster sphere, its index key can be defined as:

$$key(\lambda_i)=CD(\lambda_i) \tag{3}$$

Since the characters are grouped into T clusters, to get a uniform index key of image in different clusters, the index key in Eq. (8) can be rewritten by Eq. (9):

$$key(\lambda_i)=CID+\frac{CD(\lambda_i)}{MAX} \tag{4}$$

where the CID is the ID number of cluster λ_i falls in.

Note that since $CD(\lambda_i)$ may be larger than one, the value of $CD(\lambda_i)$ should be normalized into the range of [0,1] by being divided a large constant MAX. Thus, it is guaranteed that the search range of centroid distance of each character can not be overlapped.

To facilitate retrieving characters via submitting an auxiliary information (e.g, the style name, type name or number of strokes) of λ_i, its index keys can be derived in Eqs. (5) and (7), respectively:

$$KEY(\lambda_i)=\alpha*StyID(\lambda_i)+P_S \tag{5}$$

$$KEY(\lambda_i)=\beta*TyID(\lambda_i)+P_T \tag{6}$$

$$KEY(\lambda_i)=\gamma*NumS(\lambda_i)+P_N \tag{7}$$

Where α, β and γ are three stretch constants which are set 10, 10^2 and 10^3 respectively.

In the above, we suppose that user submits two query elements (e.g., (a) $StyID$ and λ_i, (b) $TyID$ and λ_i, or (c) $NumS$ and λ_i). If user submits three query elements ($TyID$, $StyID$ and λ_i,), then a uniform index key for λ_i can be rewritten below:

$$KEY(\lambda_i)=\alpha*StyID(\lambda_i)+\beta*TyID(\lambda_i)+P_T*P_S \tag{8}$$

$$KEY(\lambda_i)=\alpha*StyID(\lambda_i)+\gamma*NumS(\lambda_i)+P_S*P_N \tag{9}$$

$$KEY(\lambda_i)=\beta*TyID(\lambda_i)+\gamma*NumS(\lambda_i)+P_T*P_N \tag{10}$$

Similarly, for four query elements ($TyID$, $StyID$, $NumS$ and λ_i,) submitted by a user, a uniform index key for λ_i can be derived as follows:

$$KEY(\lambda_i)=\alpha*StyID(\lambda_i)+\beta*TyID(\lambda_i)+\gamma*NumS(\lambda_i)+P_T*P_S*P_N \tag{11}$$

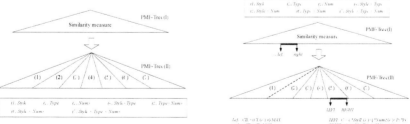

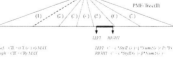

Fig. 4. The PMF-Tree index structures   **Fig. 5.** The search range in B+-Tree

Eqs. (5-11) represent the index keys of character respectively, which correspond to seven independent indexes. In order to incorporate them into an integral index, we derive a new uniform index key expression by adding seven stretch constants(i.e., C_1 to C_7), which is shown in Eq. (12):

$$KEY(\lambda_i)=\begin{cases} C_1+\alpha*StyID(\lambda_i)+P_s & \text{(a)} \\ C_2+\beta*TyID(\lambda_i)+P_t & \text{(b)} \\ C_3+\gamma*NumS(\lambda_i)+P_N & \text{(c)} \\ C_4+\alpha*StyID(\lambda_i)+\beta*TyID(\lambda_i)+P_t*P_s & \text{(d)} \\ C_5+\alpha*StyID(\lambda_i)+\gamma*NumS(\lambda_i)+P_s*P_N & \text{(e)} \\ C_6+\beta*TyID(\lambda_i)+\gamma*NumS(\lambda_i)+P_t*P_N & \text{(f)} \\ C_7+\alpha*StyID(\lambda_i)+\beta*TyID(\lambda_i)+\gamma*NumS(\lambda_i)+P_t*P_s*P_N & \text{(g)} \end{cases} \quad (12)$$

where $C_1=0$, $C_2=1\times10^4$, $C_3=1.5\times10^4$, $C_4=2\times10^4$, $C_5=2.5\times10^4$, $C_6=3\times10^4$, and $C_7=3.5\times10^4$.The above seven constants should be set large enough to stretch the value ranges of the index keys so that they do not overlap with each other.

3.3 Building PMF-Tree

For a character, its four values of *CD, NumS, styID* and *tyID* are recorded in the corresponding index key of PMF-Tree whose basic structure is the B$^+$-tree, which is shown in Fig. 4. Fig. 6 shows the detail steps of constructing a PMF-Tree. Note that the routines **TransDis**(λi) and **TransDis1**(λi) are two distance transformation function in Eq.(4) and Eq. (12) respectively, and **BInsert**(key,bt) is a B$^+$-tree insert procedure.

Algorithm 1. PMF-Tree Index Construction
Input: Ω: the character set;
Output: bt and bt': the index for PMF-Tree(I) and (II);
1.   The characters in Ω are grouped into T clusters using the AP cluster algorithm
2. $bt \leftarrow newFile()$, $bt' \leftarrow newFile()$;   * create index header file for PMF-Tree(I),(II)*
3.   **for** each character $\lambda_i \in \Omega$ **do**
4.   The CD of λ_i are computed;
5.   The style, type and stroke number of the character are identified by user with probabilities;
6. $key(\lambda_i)=$**TransDis**(λ_i);   * Function TransDis() is shown in Eq. (4) *
7. $KEY(\lambda_i)=$**TransDis1**(λ_i);   * Function TransDis1() is shown in Eq. (12) *
8.   **BInsert**$(key(\lambda_i),\ bt)$;   * insert it to B$^+$-tree *
9.   **BInsert**$(KEY(\lambda_i),\ bt')$;   * insert it to B$^+$-tree *
10. **return** bt and bt'

Fig. 6. The index construction algorithm for PMF-Tree

3.4 Probabilistic *k*-NN Search Algorithm

For n high-dimensional characters, a probabilistic k-Nearest-Neighbor(Pk-NN) search is a most frequently used search operation which retrieves the k most similar characters that are closest in distance to a given character with a probabilistic threshold. In this section, we will focus on Pk-NN searches of Chinese calligraphic character. For example, when user submits a query character "国" and a threshold ε, its type name and the style name of the result characters are *Kai Su* and *Song Ti* respectively. Then as shown in Figure 5, the retrieval process is composed of two steps: 1). Candidate characters returned by retrieving over the PMF-Tree(I) in which the range is [*left*, *right*], where $left=CID+(CD(\lambda_i)-r)/MAX$, $right=CID+CR_j/MAX$; 2). Retrieval over the PMF-Tree(II), whose range is [*LEFT*, *RIGHT*], where $LEFT=C_5+\alpha*tyID(\lambda_i)+\beta*StyID(\lambda_i)+P_t*P_s$, $RIGHT=C_5+\alpha*tyID(\lambda_i)+\beta*StyID(\lambda_i)+1$;

Algorithm 2. PkNN Search
Input: query character λ_q, k, StyID or TyID or NumS, ε
Output: query results S
1. $r\leftarrow 0$, $S\leftarrow\Phi$;   *initialization*
2.   **while** ($|S|<k$) $|S|$ *refers to the number of candidate characters in S*
3. $r\leftarrow r+\Delta r$;
4. $S\leftarrow RSearch(\lambda_q,r)$;
5.   **if** ($|S|>k$) **then**
6.   **for** $i:=1$ to $|S|-k$ **do**
7. $\lambda_{far}\leftarrow Farthest(S, \lambda_q)$;
8. $S\leftarrow S-\lambda_{far}$;
9.   **return** S;

RSearch(λ_q,r)
10. $S1\leftarrow\Phi$, $S2\leftarrow\Phi$;
11. **for** each cluster sphere $\Theta(O_j,CR_j)$ and $j \in [1, T]$
12.   **if** $\Theta(O_j,CR_j)$ **contains** $\Theta(\lambda_q,r)$ **then**
13. $S1\leftarrow S1 \cup Search(\lambda_q,r, j)$;
14.   **end loop**
15.   **else if** $\Theta(O_j, CR_j)$ **intersects** $\Theta(\lambda_q,r)$ **then**
16. $S1\leftarrow S1 \cup Search(\lambda_q,r, j)$;
17. $S5\leftarrow Search1(StyID, TyID, NumS$ and $i)$;
18.   **for** each character $\lambda_i \in S1$ **do**
19.   **if** $\lambda_i \in$ S5 **then** $S1\leftarrow S1 \cup \lambda_i$
20. **return** $S1$;   *return candidate characters*

Search(λ_q,r; i)
21. $left\leftarrow i+(CD(\lambda_q)-r)/MAX$, $right\leftarrow i+CR_j/MAX$;
22. $S3\leftarrow BRSearch[left, right]$;   *the filtering step*
23. **for** each character $\lambda_j \in S3$ **do**
24.   **if** $Sim(\lambda_q, \lambda_j)>r$ **then** $S3\leftarrow S3-\lambda_j$; // *the refinement stage*
25. **return** $S3$;   *return the candidate character set*

Search1($StyID, TyID, NumS$ and i)
26. **if** user submits a λ_q and its style **then**
27. $LEFT\leftarrow C_1+\alpha*StyID(\lambda_i)+Ps$, $RIGHT\leftarrow C_1+\alpha*StyID(\lambda_i)+1$;
28. **else** if user submits a λ_q and its type **then**
29. $LEFT\leftarrow C_2+\alpha*TyID(\lambda_i)+Pt$, $RIGHT\leftarrow C_2+\alpha*TyID(\lambda_i)+1$;
30. **else** if user submits a λ_q and the number of strokes **then**
31. $LEFT\leftarrow C_3+\alpha*NumS(\lambda_i)+Pn$, $RIGHT\leftarrow C_3+\alpha*NumS(\lambda_i)+1$;
32. **else** if user submits a λ_q, its style and the number of strokes **then**
33. $LEFT\leftarrow C_4+\alpha*StyID(\lambda_i)+\beta*TyID(\lambda_i)+Ps*P_n$, $RIGHT\leftarrow C_4+\alpha*StyID(\lambda_i)+\beta*TyID(\lambda_i)+1$;
34. **else** if user submits a λ_q, its type and the number of strokes **then**
35. $LEFT\leftarrow C_5+\alpha*StyID(\lambda_i)+\beta*NumS(\lambda_i)+P_t*P_n$, $RIGHT\leftarrow C_5+\alpha*StyID(\lambda_i)+\beta*NumS(\lambda_i)+1$;
36. **else** if user submits a λ_q, its type and style **then**
37. $LEFT\leftarrow C_6+\alpha*tyID(\lambda_i)+\beta*NumS(\lambda_i)+Pt*Ps$, $RIGHT\leftarrow C_6+\alpha*tyID(\lambda_i)+\beta*NumS(\lambda_i)+1$;
38. **else** if user submits a λ_q, its type, style and the number of strokes **then**
39. $LEFT\leftarrow C_7+\alpha*StyID(\lambda_i)+\beta*TyID(\lambda_i)+\gamma*NumS(\lambda_i)+Pt*Ps*Pn$;
40. $RIGHT\leftarrow C_7+\alpha*StyID(\lambda_i)+\beta*TyID(\lambda_i)+ \gamma*NumS(\lambda_i)+1$;
41. $S4\leftarrow BRSearch[LEFT, RIGHT]$;   *the filtering step*
42. **return** $S4$;   *return the candidate character set*

Fig. 7. Pk-NN search algorithm

Figure 7 details the whole search process. Routine **RSearch()** is the main range search function which returns the candidate characters of range search with centre λq and radius r with probability larger than ε, **Search()** and **Search1()** are the implementation of the range search. **Farthest()** returns the character which is the longest from λq in S. **BRSearch()** is a B$^+$-tree range search function.

4   Experimental Results

In this section, we present an extensive performance study to evaluate the effectiveness and efficiency of our proposed retrieval and indexing method. The Chinese Calligraphic characters image data we used are from *CADAL Project* [17] which contains a set of contour point features extracted from the 12,000 character images in which each feature point is composed of a pair of coordinates <*x axis*, *y axis*>. We implemented the shape-similarity-based retrieval approach and the PMF-Tree index in C language in which a B$^+$-tree is as the single dimensional index structure. The index page size of B$^+$-tree is set to 4096 Bytes. All the experiments are run on a Pentium IV CPU at 2.0GHz with 1G Mbytes memory. In our evaluation, we use the number of page accesses and the total response time as the performance metric.

4.1   Effectiveness of the Retrieval Method

In the first experiment, we have implemented an online interactive retrieval system for Chinese calligraphic characters to testify the effectiveness of our proposed retrieval method comparing with the conventional one [18]. As shown in the right part of Figure 8, when user submits an example Chinese calligraphic character by drawing a character "天" and the number of strokes (e.g., 4) as well, the query radius and a threshold value are set 0.8, 60% respectively, the candidate characters are quickly retrieved by the system with the aid of the PMF-Tree. The left part of the figure is the query result in which the similarity and confidence values of the answer character images are given with respective to the query one.

Figure 9 illustrates a *Recall-Precision* curve for the performance comparisons of the *shape- based* method [18] and our proposed composite search. It compares the average retrieval result (the average precision rate under the average recall rate) of 20 characters queries randomly chosen from the database. Each of them has more than 4 different calligraphic styles and types. The figure shows that the retrieval performance of the composite search is better than that of the shape-based one by a large margin.

4.2   Efficiency of PMF-Tree Index

In the following, we test the performance of our proposed indexing method—PMF-Tree under different sizes of databases and different selectivity.

4.2.1   Effect of Data Size

In this experiment, we measured the performance behavior with varying number of characters. Figure 10a shows the performance of query processing in terms of CPU cost. It is evident that PMF-Tree outperforms sequential scan significantly. The CPU cost of PMF-Tree increases slowly as the data size grows. It's worth mentioning that the CPU cost of sequential scan is ignored since the computation cost of it is very expensive. In Figure 10b, the experimental result reveals that the I/O cost of PMF-Tree is superior to sequential scan.

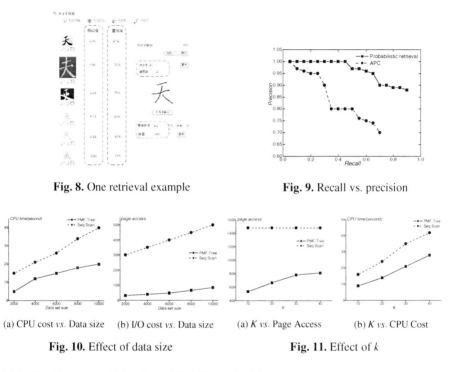

Fig. 8. One retrieval example   **Fig. 9.** Recall vs. precision

(a) CPU cost *vs.* Data size   (b) I/O cost *vs.* Data size   (a) K *vs.* Page Access   (b) K *vs.* CPU Cost

Fig. 10. Effect of data size   **Fig. 11.** Effect of k

4.2.2   Performance Behavior with k(Selectivity)

In this section, we proceed to evaluate the effect of k (selectivity) on the performance of a Pk-NN retrieval by using the PMF-Tree. Figures 11a and 11b both indicate that when k ranges from 10 to 40, the PMF-Tree is superior to other methods in terms of page access and the CPU cost. The results conform to our expectation that the search region of PMF-Tree is significantly reduced and the comparison between any two characters is a CPU-intensive task. The CPU cost of sequential scan is ignored due to the expensive computation cost of it.

5   Conclusions

In this paper, we proposed a novel probabilistic and interactive multiple-feature-based indexing scheme to support large-scale historical Chinese calligraphic character images retrieval. Two main components are included, such as 1). an effective approach to probabilistic retrieving Chinese calligraphic characters by choosing three kinds of the features is introduced; 2). a novel multiple-feature-tree(*PMF-Tree*)-based probabilistic high-dimensional indexing scheme is then proposed to boost the retrieval performance of the large Chinese calligraphic characters. The prototype retrieval system is implemented to demonstrate the applicability and effectiveness of our new approach to Chinese calligraphic character retrieval.

References

[1] Zhang, X.-Z.: Chinese Character Recognition Techniques. Tsinghua University Press, Beijing (1992)

[2] Wu, Y.-S., Ding, X.-Q.: Chinese character recognition: the principles and the implementations. High Education Press, Beijing (1992)

[3] Rath, T.M., Manmatha, R., Lavrenko, V.: A search engine for historical manuscript images. In: SIGIR, pp. 369–376 (2004)

[4] Yosef, I.B., Kedem, K., Dinstein, I., Beit-Arie, M., Engel, E.: Classification of Hebrew Calligraphic Handwriting Styles: Preliminary Results. In: DIAL 2004, pp. 299–305 (2004)

[5] Palmondon, R., Srihari, S.N.: On-Line and Off-Line hand-writing Recognition: A Comprehensive Survey. IEEE Trans. on PAMI 22(1), 63–84 (2000)

[6] Shi, B.-l., Zhang, L., Wang, Y., Chen, Z.-F.: Content Based Chinese Script Retrieval Through Visual Similarity Criteria. Chinese Journal of Software 12(9), 1336–1342 (2001)

[7] Chui, H.-l., Rangarajan, A.: A new point matching algorithm for non-rigid registration. CVIU 89(2-3), 114–141 (2003)

[8] Belongie, S., Malik, J., Puzicha, J.: Shape Matching and Object Recognition Using Shape Contexts. IEEE Trans. on PAMI 24(4), 509–522 (2002)

[9] Cohen, S., Guibas, L.: The Earth Mover's Distance under Transformation Sets. In: ICCV, Corfu, Greece, pp. 173–187 (September 1999)

[10] Böhm, C., Berchtold, S., Keim, D.: Searching in High-dimensional Spaces: Index Structures for Improving the Performance of Multimedia Databases. ACM Computing Surveys 33(3) (2001)

[11] Guttman, A.: R-tree: A dynamic index structure for spatial searching. In: SIGMOD, pp. 47–54 (1984)

[12] Berchtold, S., Keim, D.A., Kriegel, H.P.: The X-tree: An index structure for high-dimensional data. In: VLDB, pp. 28–37 (1996)

[13] Weber, R., Schek, H., Blott, S.: A quantitative analysis and performance study for similarity-search methods in high-dimensional spaces. In: VLDB, pp. 194–205 (1998)

[14] Berchtold, S., Bohm, C., Kriegel, H.P., Sander, J., Jagadish, H.V.: Independent quantization: An index compression technique for high-dimensional data spaces. In: ICDE, pp. 577–588 (2000)

[15] Jagadish, H.V., Ooi, B.C., Tan, K.L., Yu, C., Zhang, R.: iDistance: An Adaptive B+-tree Based Indexing Method for Nearest Neighbor Search. ACM Trans. on Database Systems 2(30), 364–397 (2005)

[16] Frey, B.J., Dueck, D.: Clustering by Passing Messages Between Data Points. Science 315, 972–976

[17] 2010, http://www.cadal.zju.edu.cn

[18] Zhuang, Y., Zhuang, Y.-T., Li, Q., Chen, L.: Interactive high-dimensional index for large Chinese calligraphic character databases. ACM Trans. Asian Lang. Inf. Process. 6(2) (2007)

Real-Time Diameter Monitoring for Time-Evolving Graphs

Yasuhiro Fujiwara[1], Makoto Onizuka[1], and Masaru Kitsuregawa[2]

[1] NTT Cyber Space Laboratories
[2] Institute of Industrial Science, The University of Tokyo

Abstract. The goal of this work is to identify the diameter, the maximum distance between any two nodes, of graphs that evolve over time. This problem is useful for many applications such as improving the quality of P2P networks. Our solution, G-Scale, can track the diameter of time-evolving graphs in the most efficient and correct manner. G-Scale is based on two ideas: (1) It estimates the maximal distances at any time to filter unlikely nodes that cannot be associated with the diameter, and (2) It maintains answer node pairs by exploiting the distances from a newly added node to other nodes. Our theoretical analyses show that G-Scale guarantees exactness in identifying the diameter. We perform several experiments on real and large datasets. The results show that G-Scale can detect the diameter significantly faster than existing approaches.

Keywords: Diameter, Graph mining, Time-evolving.

1 Introduction

Graphs arise naturally in a wide range of disciplines and application domains. The distances between pairs of nodes are a fundamental property in graph theory. The node-to-node distances are often studied in terms of the **diameter**, the maximum distance in a graph. However, the focus of traditional graph theory has been limited to just *static* graphs; the implicit assumption is that the number of nodes and edges never change.

Recent years have witnessed a dramatic increase in the availability of graph datasets that comprise many thousands and sometimes even millions of time-evolving nodes; this is one consequence of the widespread availability of electronic databases and the Internet. Recent studies on large-scale graphs are discovering several important principles of time-evolving graphs [12]. Thus demands for efficient approaches to the analysis of time-evolving graphs are increasing.

In this paper, we focus on the problems faced when attempting to identify the exact diameter of a graph evolving over time by the addition of nodes. In other words, the goal of this work is continuous diameter monitoring for time-evolving graphs. We propose an algorithm to solve this problem exactly in real-time. The commonly-used approach to diameter computation is based on breadth-first search, which is not practical for large-scale graphs since it requires excessive CPU time. To the best of our knowledge, this is the first study to address the diameter detection problem that guarantees exactness and achieves efficiency.

J.X. Yu, M.H. Kim, and R. Unland (Eds.): DASFAA 2011, Part I, LNCS 6587, pp. 311–325, 2011.
© Springer-Verlag Berlin Heidelberg 2011

1.1   Contributions

We propose a novel method called G-Scale that can efficiently identify the diameter of time-evolving graphs. In order to reduce monitoring cost, (1) we estimate the maximal distance to prune unlikely nodes that cannot be associated with the diameter, and (2) we maintain the answer node pairs whose distances are the diameter by exploiting distances from a newly added node to other nodes. G-Scale has the following attractive characteristics:

- **Efficient:** G-Scale is drastically faster than the existing algorithm. The existing algorithm takes $O(n^2 + nm)$ time where n and m are the number of nodes and edge, respectively, and so is prohibitively expensive for large-scale graphs.
- **Exact:** G-Scale does not sacrifice accuracy even though it prunes unlikely nodes in the monitoring process; it can exactly track the node pair that delimit the diameter of a time-evolving graph at any time.
- **Parameter-free:** Previous approximate approaches require the setting of parameters. G-Scale, however, is completely automatic; this means it does not require the user to set any parameters.

1.2   Problem Motivation

The problem tackled in this paper must be overcome to develop the following important applications. The network architecture called P2P is the basis of several distributed computing systems such as Gnutella, Seti@home, and OceanStore [2]. And content distribution is a popular P2P application on the Internet. For example, Kazaa and its variants have grown rapidly over time, over 4.5 million users share a total of 7 petabytes of data [3]. In a content distribution network, personal computers can use hop-by-hop data forwarding between themselves. An important and fundamental question is how many neighbors should a computer have, i.e., what size the routing table should be [16]. This question is important for two reasons. The number of computers in a P2P network could be extremely large, hence the complete routing table is likely to be too large to maintain. Second, because each hop in a P2P network is overhead, suppressing the query hop number by increasing table size is important in raising service efficiency.

Network diameter is a useful metric when trying to raise the search efficiency of a content distribution network, since it directly corresponds to the number of hops a query needs to travel in the worst case [9]. Moreover, by monitoring network diameter, the routing table size can be updated more effectively; if the diameter is large, the routing table size should be increased. This strategy can bound the search speed of content distribution networks.

In addition to the application presented above, robustness improvement is an important application in P2P networks of diameter monitoring. Koppula et al. showed that it can be determined which edges should be added/rewired to improve network robustness by plotting the diameters of dynamically changing graphs [8]. Furthermore, our proposed method can be used in other applications such as measuring the structural robustness of metro networks [13], monitoring the evolution of the Internet [11], and measuring citation networks size [10]. While time-evolving graphs are potentially useful in many applications, they have been difficult to utilize due to their high computational costs. However, by providing exact solutions in a highly efficient manner, G-Scale will allow many more data mining applications based on time-evolving graphs to be developed in the future.

The remainder of this paper is organized as follows. Section 2 describes related work. Section 3 overviews some of the background of this work. Section 4 introduces the main ideas of G-Scale and explains its algorithm. Section 5 gives theoretical analyses of G-Scale. Section 6 reviews the results of our experiments. Section 7 is our brief conclusion.

2   Related Work

Many papers have been published on approximation for node-to-node distances. The previous distance approximation schemes are distinguished into two types: annotation approach and embedding approach. Rattigna et al. studied two annotation schemes [17]. They randomly select nodes in a graph and divide the graph into regions that are connected, mutually exclusive, and collectively exhaustive. They give a set of annotations to every node from the regions. Distances are computed by the annotations. They demonstrated their method can compute node distances more accurately than the embedding approaches.

The Landmark approach is an embedding approach [7,15], and estimates node-to-node distance from selected nodes. The minimum distance via a landmark node is utilized as node distance in this method. Another embedding approach is Global Network Positioning which was studied by Ng et al [14]. Node distances are estimated with L_p norm between node pairs.

However, interest of these approaches lies only in the estimation; these approaches do not guarantee exactness.

3   Preliminary

In this section, we formally define some notations and introduce the background to this paper. Content distribution networks and others can be described as graph $G = (V, E)$, where V is the set of nodes, and E is the set of edges. We use n and m to denote the number of nodes and edges, respectively. That is $n = |V|$ and $m = |E|$. We define a **path** from node u to v as the sequence of nodes linked by edges, beginning with node u and ending at node v. A path from node u to v is the **shortest path** if and only if the number of nodes in the path is the smallest possible among all paths from node u to v. We use $d(u, v)$ to denote the **distance** between node u and v, and $d(u, v)$ is the number of edges in the shortest path from node u to v in the graph. By definition, $d(u, u) = 0$ for every $u \in V$, and $d(u, v) = d(v, u)$ for $u, v \in V$.

The diameter, D, is defined as the maximal distance between two arbitrary nodes as follows [5]: $D = \max(d(u,v)|u,v \in V)$. And our algorithm returns not only the diameter but the node pairs whose distances are equal to the diameter, $\mathcal{D}$. $\mathcal{D}$ is formally defined as follows: $\mathcal{D} = \{(u,v)|d(u,v) = D\}$.

The diameter of graph G can exactly be computed by the breadth-first search approach [4]. But the breadth-first search based approach generally needs $O(n^2 + nm)$ time because it computes the distances from all n nodes in a graph and $O(n+m)$ time is required for each node [6]. This incurs excessive CPU time for large-scale graphs as illustrated by the statement 'computing shortest paths among all node pairs is computationally prohibitive' made in [10]. Furthermore, the naive approach to monitoring time-evolving graphs is to perform this procedure each time a graph changes. However, considering the high frequency with which graphs evolve, a much more efficient algorithm is needed.

4   Monitoring the Diameter

In this section, we introduce the two main ideas and describe the algorithm of G-Scale. The main advantage of G-Scale is that it can efficiently and exactly solve the problem of identifying the diameter of time-evolving graphs. First we give an overview of each idea and then a full description.

4.1   Ideas Behind G-Scale

Our solution is based on the two ideas described below.

Reference node filtering. Our first idea is to prune unlikely nodes efficiently so as to reduce the high cost of the existing approach. The existing algorithm requires high computation time because it computes distances for all pairs of n nodes in the graph. Our idea is simple; instead of computing distances from all nodes, we compute the distances only from selected nodes and prune unlikely nodes. In other words, we use selected nodes to filter unlikely nodes. We refer to the selected nodes as **reference nodes**.

In the monitoring process, we select reference nodes one by one and compute the distances from the node to other nodes. In doing so, we estimate whether nearby nodes of the reference node can delineate the diameter. The time incurred to estimate node distances is $O(1)$ for each node. As a result, if the number of reference nodes is k $(k \ll n)$, $O(kn+km)$ time is required to detect the diameter, instead of the $O(n^2 + nm)$ time required by the existing algorithm solution.

This new idea has the following two major advantages. First, we can identify the diameter exactly even though we prune nearby nodes with the estimation. This means that we can safely discard unlikely nodes at low CPU cost. Note that the number of k is automatically determined. Generally, it is difficult to set parameters which would significantly impact the final result. Our approach, however, avoids user-defined parameters, and this is the second advantage.

Incremental update. Time-evolving graphs evolve by the addition of nodes over time. By a node addition, the diameter can *grow, shrink,* or be *unchanged.* We propose an algorithm that efficiently maintains the answer node pairs to detect the diameter of time-evolving graphs.

The naive method based on the above filtering approach for time-evolving graphs is to identify the diameter by setting reference nodes every time a node added. However, we ask the question, 'Can we avoid re-estimating the maximal distance every time the graph grows?'. This question can be answered by examining whether the node addition changes the answer node pairs. As described in detail later, if the diameter does not shrink with node addition, the answer node pairs can be incrementally updated by assessing *only* the distances from the added node.

This idea is especially effective for time-evolving graphs. In the case of time-evolving graphs, a small number of new nodes are continually being added to the large number of existing nodes. Therefore, there is little difference in the graphs before and after the addition of nodes, even if the new nodes arrive frequently. As a result, we can efficiently update the diameter and the answer node pairs by computing distances from the added node.

4.2   Reference Node Filtering

Our first idea involves selecting reference nodes so as to filter unlikely nodes efficiently.

Our filtering algorithm is as follows: (1) It computes the candidate distance which is expected to be diameter. (2) It selects reference nodes and estimates the maximal distances of nearby nodes to other nodes, and (3) If a node distance estimation yields a shorter distance than the candidate distance, it prunes the node since the node cannot be delineate the diameter. Accordingly, the unlikely node can be filtered quickly.

In this section, we first describe how estimate the maximal distance of a node to other nodes, and show that node distance estimation gives an upper bound for the maximal distance. We then introduce our approach of selecting the reference nodes and computing the candidate distance.

Formally, the following equation gives the maximal distance of node u: $d_{max}(u) = \max(d(u, v)|v \in V)$. We then define the estimation of the maximal distance as follows:

Definition 1 (Estimation). *For graph G, let u_r be a reference node, we define the following estimation of the maximal distance of node v, $\hat{d}_{max}(v)$, to filter unlikely nodes:*

$$\hat{d}_{max}(v) = d_{max}(u_r) + d(u_r, v) \tag{1}$$

We show the following lemma to introduce the upper bounding property of node estimation; this property enables G-Scale to identify the diameter exactly.

Lemma 1 (Upper bound). *For any node in graph G, the following inequality holds.*

$$d_{max}(v) \leq \hat{d}_{max}(v) \tag{2}$$

Algorithm 1. Filtering

Input: $G_t = (V, E)$, a time-evolving graph at time t
         $\mathcal{D}_{t-1}$, the answer node pairs of previous time tick
Output: D_t, the diameter of graph G_t
         $\mathcal{D}_t$, the answer node pairs
 1: $D_t := \max(d(u, v)|(u, v) \in \mathcal{D}_{t-1})$;
 2: $\mathcal{D}_t := \emptyset$;
 3: $V' := V$;
 4: **while** $V' \neq \emptyset$ **do**
 5:     $u_r := \operatorname{argmax}(deg(v)|v \in V')$;
 6:     compute the maximal distance $d_{max}(u_r)$;
 7:     **if** $d_{max}(u_r) = D_t$ **then**
 8:         append $\{(u_r, v)|d(u_r, v) = D_t\}$ to $\mathcal{D}_t$;
 9:     **end if**
10:     **if** $d_{max}(u_r) > D_t$ **then**
11:         $D_t := d_{max}(u_r)$;
12:         $\mathcal{D}_t := \{(u_r, v)|d(u_r, v) = D_t\}$;
13:     **end if**
14:     subtract u_r from V';
15:     subtract $\{v|\hat{d}_{max}(v) < D_t\}$ from V';
16: **end while**

Proof. Omitted for space.   □

If a node estimation yields a shorter distance than the candidate distance, the node cannot be a node of the answer node pairs. So we prune the node. Since node estimation can be computed at the cost of $O(1)$ as shown in Definition 1, we can efficiently identify the diameter by exploiting node estimation.

Selection of the reference nodes and candidate distance are very important for efficient filtering; if the maximal distance of the reference node is longer than the maximal distance of the candidate node, we cannot effectively prune unlikely nodes (see Definition 1).

We select the highest-degree nodes as the reference nodes since the maximal distances of such nodes are likely to be short than other nodes; from such nodes, all nodes can be reached in a small number hops. We utilize the answer node pairs of the previous time tick to compute the candidate distance. Since graphs are almost the same after a node addition, the prior answer node pairs are expected to remain valid. These two techniques allow us to obtain good reference nodes and candidate distances effectively as demonstrated in the experiments in Section 6.

Algorithm 1 shows the filtering algorithm that detect the diameter by the reference nodes. The number of reference nodes, k, is automatically obtained in this algorithm. In this algorithm, $deg(u)$ represents the degree of node u. The algorithm first computes the candidate distance based on the prior answer node pairs (line 1). It then selects a reference node according to degree (line 5). If the maximal distance of the reference node is equal to the candidate distance, it appends the answer node pairs (lines 7-9). If the maximal distance of the reference node is larger than the candidate distance, it sets the candidate distance and the answer node pairs (lines 10-13). It uses the candidate distance to prune the unlikely nodes in the graph. That is, if a node distance estimation is less than the candidate distance, that node cannot delineate the diameter, and so can be safely discarded (line 15). This procedure is iterated until all nodes have been processed. This implies that the number of reference nodes, k, is automatically computed. That is, this algorithm does not require any user-defined parameters.

4.3   Incremental Update

Our second idea is an incremental monitoring algorithm that efficiently maintains the answer in case a node addition; it suppresses the computation time by providing conditions that restrict the application of the filtering algorithm for node addition.

Diameter changes. In this section, we first describe the property of node distance after node addition, and then examine the conditions in which the diameter grows, shrinks, or is unchanged. We assume that one node, u_a, and its connected edges are added to a time-evolving graph at each time tick.

In we introduce below the property that underlies our update algorithm; distances of already existing node pairs can not be increased by node addition:

Lemma 2 (Distances after node addition). *Node distances at time t can not be longer than that at time $t - 1$ for all node pairs in graph G_{t-1}.*

Proof. If all shortest paths between node u and v at time t pass through the added node, the corresponding path at time $t - 1$ cannot have existed. That is, all the shortest paths at $t - 1$ must be shortened by the added node. Otherwise, there exists a shortest path between node u and v at time t that does not pass through the added node. Therefore, the same path was present at time $t - 1$. As a result, distance between node u and v is not increased by node addition.   □

After node addition, the diameter can change. We distinguish three types of changes in diameter after node addition, and we theoretically analyze the three lemmas of the changes by utilizing the above property.

The first type of change is diameter increase. The diameter increases iff the maximal distance of the added node is longer than the diameter of the previous time:

Lemma 3 (Growth in diameter). *The diameter grows at time t if and only if:*

$$d_{max}(u_a) > D_{t-1} \tag{3}$$

Proof. If $D_t > D_{t-1}$, then node u_a must delineate the diameter since the maximal distances of already existing nodes cannot be longer than D_{t-1} from Lemma 2. If $d_{max}(u_a) > D_{t-1}$, then obviously $d_{max}(u_a) = D_t$ and $D_t > D_{t-1}$.   □

The diameter shrinks iff the maximal distance of added node is shorter than the diameter at the previous time tick, and node addition invalidates all previous answer pairs:

Lemma 4 (Shrinkage in diameter). *The diameter shrinks at time t if and only if:*

$$(1) d_{max}(u_a) < D_{t-1}, and$$
$$(2) \forall (v, w) \in \mathcal{D}_{t-1}, d(v, w) < D_{t-1} \quad (4)$$

Proof. If $D_t < D_{t-1}$, then $d_{max}(u_a) < D_{t-1}$ and all distances of answer nodes pairs at the last time tick must be shorter than D_{t-1}. If (1) and (2) hold, then the diameter shrinks at time t because of Lemma 2. □

The diameter would be unchanged after node addition iff the maximal distance of the added node is equal to the diameter of the previous time, or there exists at least one node pair whose distance is equal to the diameter at the previous time tick:

Lemma 5 (Unchanged diameter). *The diameter is unchanged at time t if and only if:*

$$(1) d_{max}(u_a) = D_{t-1}, or$$
$$(2) \exists (v, w) \in \mathcal{D}_{t-1} \ s.t. \ d(v, w) = D_{t-1} \quad (5)$$

Proof. This is obvious from Lemma 3 and 4. □

Monitoring algorithm. We can efficiently maintain the answer with the incremental update approach. As the first step, we describe invalidation of answer pairs can be checked with only distances from added node, and then show our monitoring algorithm based on the incremental update approach.

We exploit the following property of the shortest path, which is shown in [4], to update the diameter and the answer node pairs:

Lemma 6 (Bellman criterion [4]). *Node u lies on a shortest path between node v and w, if and only if:*

$$d(u, v) + d(u, w) = d(v, w) \quad (6)$$

With Lemma 6, we can check whether node addition shortens distances of previous answer pairs:

Lemma 7 (Distance check). *In time-evolving graphs, the added node u_a shortens the distances of previous answer node pair (v, w) if and only if:*

$$d(v, u_a) + d(w, u_a) < D_{t-1} \quad (7)$$

Proof. If the added node u_a shortens the distances, then node u_a must lie on the shortest path between node v and w. Therefore, $D_{t-1} > d(v, w) = d(u_a, v) + d(u_a, w)$ from Lemma 6. If $d(v, u_a) + d(w, u_a) < D_{t-1}$, then the added node u_a must shorten the distance since $D_{t-1} > d(u_a, v) + d(u_a, w) \geq d(v, w)$ (see [6]). □

Lemma 7 implies that we can maintain the answer node pairs by using only the distances from the added node if the diameter grows or remains unchanged.

That is, if the added node delineates the diameter, we can compute the answer node pairs by using the distances from the added node. And if node addition shortens the distances of a previous answer node pair, that pair can be efficiently detected with from Lemma 7. If there exist no node pair whose distance is equal to the previous diameter, we detect the new diameter by the filtering algorithm.

Algorithm 2 describes our G-Scale algorithm. It first computes the maximal distance of the added node (line 1). If the maximal distance is longer than the prior time diameter, the added node must delineate the diameter (Lemma 3). It uses the distances from added node to set the diameter and the answer node pairs (lines 3-5). If the maximal distance is equal to the prior time diameter or if there exists an answer node pair whose distance is equal to the prior time diameter, the diameter remains unchanged after node addition (Lemma 5). It appends/removes the answer node pairs by distances from the added node (lines 8-13). If no node pair exists whose distance is equal to the prior time diameter after the addition, the diameter shrinks (Lemma 4). The diameter and the answer node pair are identified in Algorithm 1 (lines 15-17). Thus G-Scale limits the application of the filtering algorithm to the minimum.

Time-evolving graphs experience the addition of nodes and we assume here that a graph has only one node at $t = 1$. At $t = 1$, we set $D_{t-1} = 0$ and $\mathcal{D}_{t-1} = \emptyset$. Lifting this assumption is not difficult, and is not pursued in this paper.

Even though we assumed single node addition, we can also handle the addition of several nodes in each time tick; we simply iterate the above procedure for each additional node. If one edge is added, we assume one connected node is added to the graph. For node/edge deletion, we can detect the diameter by Algorithm 1 since such graphs do not have the property of Lemma 2. This procedure is several orders of magnitude faster than the existing approach as showed in Section 6.

We have focused on unweighted undirected graphs in this paper, but G-Scale can also handle weighted or directed graphs. For weighted graphs, we use Dijkstra's algorithm to compute distances from nodes, and bread-first search for each direction to obtain distances for directed graphs. Monitoring procedures, such as how to estimate the maximal distance from reference nodes and how to maintain the answer node pairs, are the same as those for unweighted undirected graphs.

5   Theoretical Analysis

In this section, we introduce a theoretical analysis that confirms the accuracy and complexity of G-Scale. Let k be the number of reference nodes.

5.1   Accuracy

We prove that G-Scale detects the diameter accurately (without fail) as follows:

Lemma 8 (Exact monitoring). *G-Scale guarantees the exact answer in identifying the diameter.*

Algorithm 2. G-Scale

Input: $G_t = (V, E)$, a time-evolving graph at time t
        D_{t-1}, the diameter at previous time tick
        $\mathcal{D}_{t-1}$, the answer node pairs at previous time tick
        u_a, the added node at time t
Output: D_t, the diameter of graph G_t
        $\mathcal{D}_t$, the answer node pairs
 1: compute the maximal distance $d_{max}(u_a)$;
 2: //Growth in diameter
 3: **if** $d_{max}(u_a) > D_{t-1}$ **then**
 4:    $D_t := d_{max}(u_a)$;
 5:    $\mathcal{D}_t := \{(u_a, v)|d(u_a, v) = D_t\}$;
 6: **else**
 7:    //Unchanged diameter
 8:    $D_t := D_{t-1}$;
 9:    $\mathcal{D}_t := \mathcal{D}_{t-1}$;
10:    **if** $d_{max}(u_a) = D_{t-1}$ **then**
11:       append $\{(u_a, v)|d(u_a, v) = D_{t-1}\}$ to $\mathcal{D}_t$;
12:    **end if**
13:    remove $\{(v, w)|d(v, u_a) + d(w, u_a) < D_{t-1}\}$ from $\mathcal{D}_t$;
14:    //Shrinkage in diameter
15:    **if** $\mathcal{D}_t = \emptyset$ **then**
16:       compute D_t and $\mathcal{D}_t$ by the filtering algorithm;
17:    **end if**
18: **end if**

Proof. Mathematical induction can be used to prove that G-Scale detects the diameter exactly at time $t(\geq 1)$. First, we must show that the statement is true at $t = 1$. At time $t = 1$, G-Scale detects $D_1 = 0$ and $\mathcal{D}_1 = (u_1, u_1)$ exactly since (1) it sets $D_{t-1} = 0$ and $\mathcal{D}_{t-1} = \emptyset$ and (2) $d_{max}(u_1) = 0$ (see lines 8-12 in Algorithm 2). Next, we will assume that the statement holds at $t = i$. Assuming this, we must prove that the statement holds for its successor, $t = i + 1$. If the diameter does not shrink at $t = i+1$, it detects the diameter and the answer node pairs exactly from the distances from the added node with Lemma 7. Otherwise, it finds the diameter and the answer node pairs by the filtering algorithm. The filtering algorithm discards a node if its estimated maximal distance is lower than the candidate distance, and node estimation has upper bounding property (Lemma 1). That is, a node that delineates the diameter cannot be pruned. We have now fulfilled both conditions of the principle of mathematical induction. □

5.2   Complexity

We discuss the complexity of G-Scale.

Lemma 9 (Space complexity of G-Scale). *G-Scale requires $O(n+m)$ space to compute the diameter.*

Proof. G-Scale requires $O(n+m)$ space to keep the graph. The number of answer node pairs is negligible compared to that of nodes/edges as shown in Section 6. As a result, G-Scale requires $O(n+m)$ space in diameter monitoring.   □

Lemma 10 (Time complexity of G-Scale). *G-Scale requires $O(n+m)$ time if the diameter does not shrink by node addition, otherwise it requires $O(kn+km)$ time to compute the diameter.*

Proof. To identify the diameter, G-Scale first compute the distances from the added node and examines whether the added node delimits the diameter or shortens the distances of the previous answer node pairs. It takes $O(n+m)$ time. If the diameter shrinks, G-Scale detects the diameter with the filtering algorithm which takes $O(kn+km)$ time. As a result it requires $O(n+m)$ time if the diameter does not shrink, and it takes $O(kn+km)$ time if the diameter shrinks.   □

The monitoring cost depends on the effectiveness of the filtering and incremental update techniques used by G-Scale to detect the diameter. In the next section, we confirm the effectiveness of our approach by presenting the results of extensive experiments.

6   Experimental Evaluation

We performed experiments to demonstrate G-Scale's effectiveness. We compared G-Scale to the existing common algorithm based on breadth-first search [4] and the network structure index [17]. Note that the network structure index can compute node distances quickly at the expense of exactness. Furthermore, this method requires $O(n^2)$ space and $O(n^3)$ time as described in their paper; this method has higher orders of space and time complexities than the method based on breadth-first search.

Our experiments will demonstrate that:

- Efficiency: G-Scale outperforms breadth-first search by up to 5 orders of magnitude for the real datasets tested. G-Scale is scalable to dataset size (Section 6.1).
- Effectiveness: The components of G-Scale, reference node filtering and incremental update, are effective in monitoring the diameter (Section 6.2).
- Exactness: Unlike the existing approach, which sacrifices accuracy, G-Scale can identify the diameter exactly and efficiently (Section 6.3).

We used the following three public datasets in the experiments: *Citation*, *Web*, and *P2P*. They are a U.S. patent network, web pages within 'berkely.edu' and 'stanford.edu' domain, and the Gnutella peer-to-peer file sharing network, respectively. All data can be downloaded from [1]. We extracted the largest connected component from the real data, and we added single nodes one by one in the experiments.

We evaluated the monitoring performance mainly through wall clock time. All experiments were conducted on a Linux quad 3.33 GHz Intel Xeon server with 32GB of main memory. We implemented our algorithms using GCC.

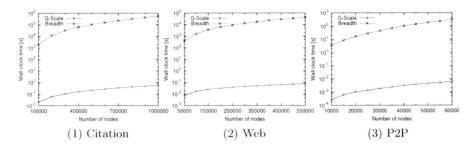

(1) Citation   (2) Web   (3) P2P

Fig. 1. Wall clock time versus number of nodes

6.1   Efficiency and Scalability

We assessed the monitoring time needed for G-Scale and breadth-first search. We conducted trials with various numbers of nodes because differences in this number are expected to have strong impact for wall clock time. Figure 1 shows the wall clock time as a function of the number of nodes to detect the diameter.

These figures show that our method is much faster than breadth-first search under all the conditions examined. Breadth-first search computes distances for all node pairs in a graph. However G-Scale requires only distances from an added node if the diameter does not shrink. Even though G-Scale computes the distances from reference nodes if the diameter shrinks, this cost has no effect on the experimental results. This is because node addition hardly changes the diameter in time-evolving graphs and the number of reference nodes, k, is very small as is shown in the next section.

6.2   Effectiveness of Each Approach

In the following experiments, we examine the effectiveness of the core techniques of G-Scale: reference node filtering and incremental update.

Reference node filtering. G-Scale prunes unlikely nodes using reference nodes and the candidate distance. As mentioned in Section 4.2, G-Scale selects the highest-degree node as a reference node and utilizes the previous answer node pairs as candidate pairs. To show the effectiveness of this idea, we removed the update approach from G-Scale to directly evaluate the filtering technique, and examined the wall clock time. In other words, we directly evaluate Algorithm 1.

Figure 2 shows the result. In this figure, G-Scale without the update technique is abbreviated to *Filtering*, and *Random* represents the results where reference nodes and the candidate distance are selected at random. The numbers of nodes in this figure are $1,000,000$ for Citation, $550,000$ for Web, and $60,000$ for P2P.

Our selection methods require less computation time than the other methods. The maximal distances of the highest-degree node are expected to be short, and the prior answer node pairs are likely to remain valid. Therefore, the filtering algorithm can efficiently detect the diameter for time-evolving graphs.

For node/edge deletion, we can detect the diameter by Algorithm 1 as described in Section 4.3. Figure 2 shows the effectiveness of this approach; it is

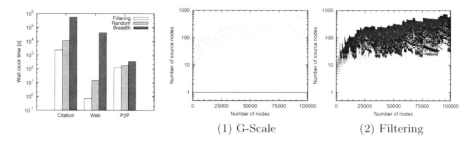

(1) G-Scale   (2) Filtering

Fig. 2. Effect of reference node filtering

Fig. 3. Number of source nodes from which distances are computed in diameter monitoring

several orders of magnitude faster than the existing approach (compare Filtering to Breadth).

Incremental update. Our update algorithm efficiently detects the diameter by reducing the application of the filtering algorithm. That is, if the diameter does not shrink, it maintains the answer node pairs by the distances from the added node. We compared the number of source nodes from which distances are computed to show the effectiveness of this approach. Note that the number of source nodes is the number of the added node plus the reference nodes. In other words, the number of source nodes is $k + 1$. Figure 3 shows the results by G-Scale and without update technique (abbreviated to *Filtering*) in monitoring time-evolving graphs. And Figure 4 shows the number of answer node pairs. We used Citation as dataset.

Figure 3 shows that the update algorithm significantly reduces the number of source nodes. As we can see from the figure, in practice, G-scale computes the distances only from the added node, while the number of reference nodes, k, is much smaller than that of graph nodes, n, with the filtering algorithm. Moreover, in most cases, the number of answer node pairs is less than 10 as shown in Figure 4; it was never more than 30 in the experiments. Therefore, it can efficiently check whether node addition invalidates the previous answer node pairs. As a result, G-scale can maintain the answer node pairs efficiently.

6.3 Exactness of the Monitoring Results

One major advantage of G-Scale is that it guarantees the exact answer, but this raises the following simple questions:

- How successful is the previous approximation approach in providing the exact answer even though it sacrifices exactness?
- Can G-Scale identify the diameter faster than the previous approximation approach that does not guarantee the exact answer?

To answer the these questions, we conducted comparative experiments using the network structure index proposed by Rattigan et al. [17]. Although they studied several estimation schemes for node distances, we compared the **distant to**

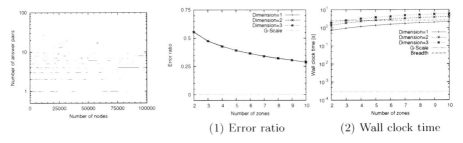

(1) Error ratio   (2) Wall clock time

Fig. 4. Number of answer pairs

Fig. 5. Comparison of G-Scale and the network structure index

zone annotation scheme to G-Scale since it outperforms the other approaches, including embedding schemes mentioned in Section 2, in all of our dataset; the same result is reported in their paper. This annotation has two parameters: **zones** and **dimensions**. Zones are divided regions of the entire graph, and dimensions are sets of zones. We measured the quality of accuracy by the error ratio, which is error value of the estimated diameter distance divided the exact diameter distance. Note that the error ratio becomes a value from 0 to 1. Figure 5 shows the error ratio and the wall clock time of the diameter detection. The dataset used was Citation where the number of nodes is $10,000$.

As we can see from the figure, the error ratio of G-Scale is 0 because it identifies the diameter accurately. However, the network structure index has much higher error ratio. And the number of dimensions has no impact on the error ratio. Therefore it is not practical to use the network structure index in identifying the diameter. This answers the first question.

Figure 5 shows that G-Scale greatly reduces the computation time while it guarantees the exact answer. Specifically, G-Scale is at least $2,400$ times faster than the network structure index in this experiment; this is our answer to the second question.

The efficiency of the network structure index depends on the parameters used; it can take much more time than breadth-first search if the parameters are wrongly chosen. Furthermore, the results show that the network structure index forces a trade-off between speed and accuracy. That is, as the number of zones and dimensions decreases, the wall clock time decreases but the error ratio increases. The network structure index is an estimation technique and so can miss the exact answer. G-Scale also estimates the maximal distances to yield efficient filtering, but unlike the network structure index, G-Scale does not discard the exact answer in the monitoring process. As a result, G-Scale is superior to the network structure index in not only accuracy, but also speed.

7   Conclusions

This paper addressed the problem of detecting the diameter of time-evolving graphs efficiently. As far as we know, this is the first study to address the diameter monitoring problem for time-evolving graphs with the guarantee of exactness. Our proposal, G-Scale, is based on two ideas: (1) It filters unlikely nodes by selecting reference nodes to estimate the maximal distances, and (2) It incrementally updates the answer node pairs by exploiting the distances from the newly added node. Our experiments show that G-Scale is significantly faster than the existing methods. Diameter monitoring is fundamental for many mining applications in various domains such as content distribution, metro networks, the Internet, and citation networks. The proposed solution allows the diameter to be detected exactly and efficiently, and helps to improve the effectiveness of future data mining applications.

References

1. http://snap.stanford.edu/data/index.html
2. Androutsellis-Theotokis, S., Spinellis, D.: A survey of peer-to-peer content distribution technologies. ACM Comput. Surv. 36(4), 335–371 (2004)
3. Bawa, M., Cooper, B.F., Crespo, A., Daswani, N., Ganesan, P., Garcia-Molina, H., Kamvar, S.D., Marti, S., Schlosser, M.T., Sun, Q., Vinograd, P., Yang, B.: Peer-to-peer research at stanford. SIGMOD Record 32(3), 23–28 (2003)
4. Brandes, U.: A faster algorithm for betweenness centrality. Journal of Mathematical Sociology 25, 163–177 (2001)
5. Brandes, U., Erlebach, T.: Network Analysis: Methodological Foundations. Springer, Heidelberg (2008)
6. Cormen, T.H., Leiserson, C.E., Rivest, R.L., Stein, C.: Introduction to Algorithms. The MIT Press, Cambridge (2009)
7. Goldberg, A.V., Harrelson, C.: Computing the shortest path: search meets graph theory. In: SODA, pp. 156–165 (2005)
8. Koppula, H.S., Puspesh, K., Ganguly, N.: Study and improvement of robustness of overlay networks (2008)
9. Kumar, A., Merugu, S., Xu, J., Yu, X.: Ulysses: A robust, low-diameter, low-latency peer-ti-peer network. In: ICNP, pp. 258–267 (2003)
10. Leskovec, J., Kleinberg, J.M., Faloutsos, C.: Graph evolution: Densification and shrinking diameters. TKDD 1(1) (2007)
11. Magoni, D., Pansiot, J.J.: Analysis of the autonomous system network topology. SIGCOMM Comput. Commun. Rev. 31(3), 26–37 (2001)
12. Newman: The structure and function of complex networks. SIREV: SIAM Review 45 (2003)
13. Ng, A.K.S., Efstathiou, J.: Structural robustness of complex networks. In: NetSci. (2006)
14. Ng, T.S.E., Zhang, H.: Predicting internet network distance with coordinates-based approaches. In: INFOCOM (2002)
15. Potamias, M., Bonchi, F., Castillo, C., Gionis, A.: Fast shortest path distance estimation in large networks. In: CIKM, pp. 867–876 (2009)
16. Ratnasamy, S., Stoica, I., Shenker, S.: Routing algorithms for dHTs: Some open questions. In: Druschel, P., Kaashoek, M.F., Rowstron, A. (eds.) IPTPS 2002. LNCS, vol. 2429, pp. 45–52. Springer, Heidelberg (2002)
17. Rattigan, M.J., Maier, M., Jensen, D.: Using structure indices for efficient approximation of network properties. In: KDD, pp. 357–366 (2006)

Handling ER-top$k$ Query on Uncertain Streams

Cheqing Jin[1], Ming Gao[2], and Aoying Zhou[1]

[1] Shanghai Key Laborary of Trustworthy Computing,
Software Engineering Institute, East China Normal University, China
{cqjin,ayzhou}@sei.ecnu.edu.cn
[2] Shanghai Key Laboratory of Intelligent Information Processing,
School of Computer Science, Fudan University, China
mgao@fudan.edu.cn

Abstract. It is critical to manage uncertain data streams nowadays because data uncertainty widely exists in many applications, such as Web and sensor networks. The goal of this paper is to handle top-k query on uncertain data streams. Since the volume of a data stream is unbounded whereas the memory resource is limited, it is challenging to devise one-pass solutions that is both time- and space efficient. We have devised two structures to handle this issue, namely *domGraph* and *probTree*. The *domGraph* stores all candidate tuples, and the *probTree* is helpful to compute the expected rank of a tuple. The analysis in theory and extensive experimental results show the effectiveness and efficiency of the proposed solution.

1 Introduction

Uncertain data management becomes more and more important in recent years since data uncertainty widely exists in lots of applications, such as financial applications, sensor networks, and so on. In general, there are two kinds of uncertainties, namely *attribute-level* uncertainty, and *existential* uncertainty that is also called as *tuple-level* uncertainty in some literatures [1]. The *attribute-level* uncertainty, commonly described by discrete probability distribution functions or probability density functions, illustrates the imprecision of a tuple's attributes. The *existential* uncertainty describes the confidence of a tuple. Recently, several prototype systems have been produced to manage uncertain data with explicit probabilistic models of uncertainty, such as MayBMS [4], MystiQ [10], and Trio [3].

For example, nowadays, radars are often used in traffic monitoring applications to detect car speeds. It is better to describe a reading record by a discrete probability distribution function rather than a single value, since the readings may have some errors caused by complicated reasons, such as nearby high voltage lines, close cars' interference, human operators mistakes, etc. Table 1 illustrates a small data set consisting of four reading records described by x-relation model that is introduced in Trio [3]. For instance, the 1st record observes a Buick car (No. Z-333) running through the monitoring area at AM 10:33 with the speed

J.X. Yu, M.H. Kim, and R. Unland (Eds.): DASFAA 2011, Part I, LNCS 6587, pp. 326–340, 2011.

estimated as 50 (miles per hour) with probability 0.6, and 70 with probability 0.4 respectively. In addition, a range is used to test the validation of a speed reading (e.g. [0, 150]). Once the reading exceeds the range, we remove this part of information from the tuple description, which makes the total confidence of a record smaller than 1. For example, the 3rd record estimates the speed of 80 with probability 0.4, and of invalidation with probability 0.6 ($= 1 - 0.4$).

Table 1. A radar reading database in x-relation model

ID	Reading Info	(Speed, prob.)
1	AM 10:33, Buick, Z-333	(50, 0.6), (70, 0.4)
2	AM 10:35, BMW, X-215	(60, 1.0)
3	AM 10:37, Benz, X-511	(80, 0.4)
4	AM 10:38, Mazda, Y-123	(20, 0.4), (30, 0.5)

The possible world semantics is widely adopted by many uncertain data models. The possible world space contains a huge number of possible world instances, each consisting of a set of certain values from uncertain tuples. A possible world instance is also affiliated with a probability value, computed by the product of all tuples within the instance and the product of the non-existing confidence of all tuples outside of the instance. Table 2 illustrates the possible world space of in total 12 possible world instances for the dataset in Table 1. Each column is a possible world instance with the probability listed below. For example, tuples t_1, t_2 and t_3 occur in w_{10} at the same time, so that the probability of this possible world is 0.016 ($= 0.4 \times 1.0 \times 0.4 \times (1 - 0.4 - 0.5)$).

Table 2. Possible worlds for Table 1

PW	w_1	w_2	w_3	w_4	w_5	w_6	w_7	w_8	w_9	w_{10}	w_{11}	w_{12}
t_1	50	70	50	70	50	70	50	70	50	70	50	70
t_2	60	60	60	60	60	60	60	60	60	60	60	60
t_3	80	80			80	80			80	80		
t_4	20	20	20	20	30	30	30	30				
Prob.	0.096	0.064	0.144	0.096	0.120	0.080	0.150	0.120	0.024	0.016	0.036	0.024

Uncertain data stream is quite popular in many fields, such as the radar readings in traffic-control applications. Each tuple arrives rapidly, and the volume of the data stream is considered unbounded. It is necessary to devise space- and time- efficient one-pass solutions to handle uncertain data streams, which are also helpful to handle traditional issues over massive data sets. Our focus in this paper is an uncertain top-k query.

A top-k query focuses on getting a small set of the most important tuples from a massive data set. Generally, a ranking function is utilized to give a score to each tuple and k tuples with maximum scores are returned as query results. Although the semantics of a top-k query is explicit for the deterministic data, several different top-k definitions are proposed for distinct purposes, whereas the ranking score could be based on attributes, confidences, or the combination of these two factors, inclusive of U-Top$k$ [18], U-kRanks [18], PT-k [12], Global-top$k$ [20], ER-top$k$ [8], c-Typical-Top$k$ [11] and so on. Cormode et al. have listed a set of properties to describe the semantics of ranking queries on uncertain data, namely exact-k, containment, unique-rank, value-invariance and stability [8]. Moreover, different from other uncertain top$k$ semantics like U-top$k$, U-kRanks, PT-k, Global-top$k$, the ER-top$k$ query satisfies all of these properties.

1.1   Our Contribution

It is trial to process a certain top-k query over high-speed data streams since we only need to maintain a buffer containing k tuples with highest scores. The lowest ranked tuple is replaced by the new tuple if its score is lower than the new tuple. However, processing an uncertain top-k query over data streams is not equally trial because the semantics of an uncertain top-k query, stemming from the integration of attribute values and the probability information, is much more complex than a certain top-k query. In this paper, we propose an efficient *exact* streaming approach to answer ER-top$k$ query. [8] has proposed A-ERank and T-ERank approaches to handle static uncertain data sets which requires all of the tuples fetched in a special order. Obviously, these approaches can't suit for the streaming environment. In our new solution, all tuples in the data stream are divided into two groups. One group contains *candidate* top-k tuples, i.e, the tuples having chance to belong to the query result, and the other contains the rest. We construct and maintain two structures, namely *domGraph* and *probTree*, to describe the two groups for efficiency.

The rest of the paper is organized as follows. We define the data models and the query in Section 2. In Section 3, we describe a novel solution to handle the ER-top$k$ query upon uncertain data streams. Some extended experiments are reported in Section 4. We review the related work In Section 5, and conclude the paper briefly in the last section.

2   Data Models and Query Definition

In this paper, we consider a discrete base domain $\mathcal{D}$, and $\perp$ a special symbol representing a value out of $\mathcal{D}$. Let S be an uncertain data stream that contains a sequence of tuples, $t_1, t_2, \cdots, t_N$. The i-th tuple in the stream, t_i, is described as a probability distribution $\{(v_{i,1}, p_{i,1}), \cdots, (v_{i,s_i}, p_{i,s_i})\}$. For each $l, 1 \leq l \leq s_i$, we have: $v_{i,l} \in \mathcal{D}$ and $p_{i,l} \in (0,1]$. For simplicity, we also assume that $v_{i,1} < v_{i,2} < \cdots < v_{i,s_i}$. In addition, $\sum_{l=1}^{s_i} p_{i,l} \leq 1$. The tuple t_i can also be treated as a random variable X_i over $\mathcal{D} \cup \perp$, such that $\forall l, \Pr[X_i = v_{i,l}] = p_{i,l}$, and $\Pr[X_i = \perp] = 1 - \sum_{l=1}^{s_i} p_{i,l}$.

This data model adapts both kinds of uncertainties. If $\forall i, \sum_{l=1}^{s_i} p_{i,l} = 1$, the stream only has *attribute-level* uncertainty. If $\forall i, s_i = 1$, the stream only has *existential* uncertainty. Otherwise, the stream contains both kinds of uncertainties.

Definition 1 (Expected Rank Top-k, ER-Top$k$ in abbr.). *[8] The ER-topk query returns k tuples with smallest values of $r(t)$, defined below.*

$$r(t) = \sum_{W \in \mathcal{W}} \Pr[W] \cdot rank_W(t) \tag{1}$$

where $\mathcal{W}$ is the possible world space, $\Pr[W]$ is the probability of a possible world instance W, and $rank_W(t)$ returns the rank of t in W, i.e, it returns the number of tuples ranked higher than t if $t \in W$, or the number of tuples in W ($|W|$) otherwise.

By the definition and linearity of expectation, the *expected rank* of the tuple t_i, $r(t_i)$, is computed as follows.

$$r(t_i) = \sum_{l=1}^{s_i} p_{i,l}(q(v_{i,l}) - q_i(v_{i,l})) + (1 - \sum_{l=1}^{s_i} p_{i,l})(E[|W|] - \sum_{l=1}^{s_i} p_{i,l}) \tag{2}$$

where $q(v)$ is the sum of probabilities of all tuples greater than v, and $q_i(v)$ is the sum of probabilities of the tuple t_i greater than v, i.e, $q_i(v) = \sum_{l,v_{i,l}>v} p_{i,l}$, and $q(v) = \sum_i q_i(v)$. Let $|W|$ denote the number of tuples in the possible world W, so that $E[|W|] = \sum_{i,l} p_{i,l}$.

Example 1. Consider the data set in Table 1. The expected size of all possible worlds $E[|W|] = \sum_{i,l} p_{i,l} = 3.3$. $r(t_1) = 0.6 \times ((1.0+0.4+0.4)-0.4)+0.4 \times (0.4-0) = 1.0$, $r(t_3) = 0.4 \times 0 + (1 - 0.6) \times (3.3 - 0.4) = 1.74$. Similarly, $r(t_2) = 0.8$, $r(t_4) = 2.4$. So, the query returns $\{t_2, t_1\}$ when $k = 2$.

3   Our Solution

In this section, we show how to compute the exact answers of ER-top$k$ over uncertain data streams. Equation (2) illustrates how to compute the expected rank of a tuple t. Moreover, it implies that the value of the expected rank may also change with time going on since the function $q(v)$ is based on all tuples till now. For example, at time 4, $r(t_2) = 0.8$, $r(t_1) = 1.0$, $r(t_2) < r(t_1)$ (in Example 1). Assume the next tuple t_5 is $\langle(65, 1.0)\rangle$. Then, $r(t_1) = r(t_2) = 1.6$. This simple example also implies that $r(t_i) > r(t_j)$ at some time point doesn't mean $r(t_i) > r(t_j)$ forever.

Fortunately, we actually find some pairs of tuples, t_i and t_j, such that $r(t_i) > r(t_j)$ or $r(t_j) < r(t_i)$ holds at any time point. For convenience, we use $t_i \prec t_j$ to denote the situation that $r(t_i) < r(t_j)$ holds forever, and $t_i \succ t_j$ if $r(t_i) > r(t_j)$ holds forever. For convenience, we also claim that t_i *dominates* t_j if $t_i \prec t_j$, or t_i is *dominated* by t_j is $t_i \succ t_j$.

Theorem 1. *Consider two tuples t_i and t_j. $t_i \prec t_j$ if and only if (i) $\forall v, q_i(v) \geq q_j(v)$, and (ii) $\exists v, q_i(v) > q_j(v)$ hold at the same time. Remember that $q_i(v) = \sum_{l,v_{i,l}>v} q_{i,l}$.*

Proof. Let $\mathcal{P}$ denote a set of points generated from t_i and t_j, $\mathcal{P} = \{\sum_{l=1}^{z_i} p_{i,l}\} \cup \{\sum_{l=1}^{z_j} p_{j,l}\} \cup \{1\}$, where $1 \le z_i \le s_i$, $1 \le z_j \le s_j$. Let m denote the distinct items in $\mathcal{P}$, i.e, $m = |\mathcal{P}| \le s_i + s_j + 1$. Without loss of generality, let $P_1, \cdots P_m$ denote m items in $\mathcal{P}$, and $P_1 \le P_2 \le \cdots P_m = 1$. Moreover, we assume the gth item and hth item in $\mathcal{P}$ satisfies: $P_g = \sum_{l=1}^{s_i} p_{i,l}$, $P_h = \sum_{l=1}^{s_j} p_{j,l}$. According to the condition (i), we have: $g \ge h$.

Let the function $v_i(z)$ be defined as $v_i(z) = v_{i,x}$, where $x = \operatorname{argmin}_y (\sum_{l=1}^{y} p_{i,l} \ge P_z)$. Equation (2) could be computed as follows.

$$r(t_i) = \sum_{l=1}^{g} \left(P_l - P_{l-1} \right) \left(q(v_i(l)) - q_i(v_i(l)) \right) + \sum_{l=g+1}^{m} \left(P_l - P_{l-1} \right) \left(E[|W|] - P_g \right) \quad (3)$$

Symmetrically, we can also compute the expected rank of t_j like Equation (3). Now, we begin to compute $r(t_i) - r(t_j)$ based on Equation (3). Since RHS of Equation (3) is the sum of m items. We check each item through three cases: $l \in [1, h]$, $l \in (h, g]$ and $l \in (g, m]$.

case 1: $l \in [1, h]$.

$$\Delta_l = \left(P_l - P_{l-1} \right) \left(\left(q(v_i(l)) - q_i(v_i(l)) \right) - \left(q(v_j(l)) - q_j(v_j(l)) \right) \right) \quad (4)$$

If $v_i(l) = v_j(l)$, then $q(v_i(l)) = q(v_j(l))$. According to condition (i), $\Delta_l \le 0$.

Otherwise, if $v_i(l) > v_j(l)$, then $q(v_j(l)) - q(v_i(l)) \ge$ the sum of probabilities that tuples' values are equal to $v_i(l)$ according to the definition of the function $q(\cdot)$. Moreover, since the tuple t_i has at least $P_l - q_i(v_i(l))$ probability to be $v_i(l)$, we have: $q(v_j(l)) - q(v_i(l)) \ge P_l - q_i(v_i(l))$. Finally, because $q_j(v_j(l)) < P_l$, we have: $\Delta_l < 0$.

It is worth noting that $v_i(l) < v_j(l)$ will never occur because it violates condition (i) otherwise.

case 2: $i \in (h, g]$.

$$\Delta_l = (P_l - P_{l-1})((q(v_i(l)) - q_i(v_i(l)) - (E[|W|] - P_h)) \quad (5)$$

This situation occurs only when $\sum_{l=1}^{s_i} p_{i,l} > \sum_{l=1}^{s_j} p_{j,l}$. Similarly, we have: $E[|W|] - q(v_i(l)) \ge P_l - q_i(v_i(l))$. Thus, $\Delta_l < 0$.

case 3: $l \in (g, m]$.

$$\Delta_l = (P_l - P_{l-1})((E[|W|] - P_g) - (E[|W|] - P_h)) \quad (6)$$

This situation occurs only when $\sum_{l=1}^{s_i} p_{i,l} < 1$ and $\sum_{l=1}^{s_j} p_{j,l} < 1$. Since $P_g > P_h$, we have $\Delta_l < 0$.

As a conclusion, $\Delta_l < 0$ under two conditions.

Finally, we show that if neither condition is satisfied, we will never have $t_i \prec t_j$. First, without condition (ii), t_i and t_j could be the same. Second, without condition (i), it means $\exists \hat{v}$, $q_i(\hat{v}) < q_j(\hat{v})$. When a new tuple, $\langle (\hat{v}, 1.0) \rangle$, arrives, the values of $r(t_i)$ and $r(t_j)$ will increase by $1 - q_i(\hat{v})$ and $1 - q_j(\hat{v})$ respectively. Obviously, $r(t_i) > r(t_j)$ will hold after inserting a number of such tuples because $1 - q_i(\hat{v}) > 1 - q_j(\hat{v})$.

Example 2. Figure 1 illustrates the functions $q_i(v)$ for all tuples in Table 1. Obviously, $t_1 \prec t_4$, $t_2 \prec t_4$. For the pair of tuples t_1 and t_2, neither $t_1 \prec t_2$ nor $t_2 \prec t_1$ holds.

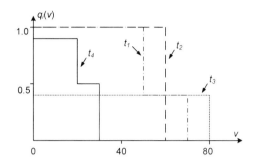

Fig. 1. The functions $q_i(v)$, for each $1 \le i \le 4$

Lemma 1. *A tuple t cannot belong to the query result if there exist at least k tuples (say, t'), $t' \prec t$.*

Proof. The correctness stems from Theorem 1.

Lemma 1 is capable of checking whether a tuple will belong to the query result potentially in future or not. Such candidate tuples must be stored in the system.

In addition, it is worth noting that a tuple cannot belong to the query result even if it is not dominated by k tuples under some situations. See the example below.

Example 3. Let's consider a situation, $k = 1$. There are three tuples, $t_1 = \langle (9, 0.6), (8, 0.2), (7, 0.1), (6, 0.1) \rangle$, $t_2 = \langle (11, 0.4), (6, 0.5), (5, 0.1) \rangle$, $t_3 = \langle (10, 0.1), (9, 0.4), (4, 0.5) \rangle$. Obviously, neither tuple dominates another tuple. But the tuple t_3 won't be output since its expected rank $r(t_3)$ will be greater than t_1 or t_2 no matter what tuples come later. In other words, We can evaluate that $r(t_1) + r(t_2) - 2r(t_3) < 0$ always holds because the function $q(\cdot)$ is monotonous (Equation (2)).

However, discovering all candidate tuples like Example 3 is quite expensive since it needs to check a huge number of tuples. Consequently, in this paper we mainly use Lemma 1 to evaluate candidates. Even though a few redundant tuples are stored, it is efficient in computing. Algorithm 1 is the main framework of our exact solution to handle data streams, which invokes `maintainDomGraph` and `maintainProbTree` repeatedly to maintain two novel structures, namely *domGraph* and *probTree*.

3.1 domGraph

The *domGraph*, with each node described in form of $(t, T_\prec, T_\succ, state, c)$, is a graph to store all candidate tuples. The entry t refers to a tuple in the stream.

Algorithm 1. processStream()

1: Empty a domGraph G and a probTree T;
2: **for each** (arriving tuple t)
3:   maintainDomGraph(t, G);
4:   maintainProbTree(t, T);

$T_{\prec}$ represents a set of tuples ranking just higher than t, i.e, (i) $\forall t' \in T_{\prec}, t' \prec t$, and (ii) any tuple in $T_{\prec}$ cannot dominate another tuple in $T_{\prec}$, i.e $\nexists t', t'' \in T_{\prec}$ such that $t' \prec t''$. $T_{\succ}$ represents a set of tuples ranking just lower than t, i.e, (i) $\forall t' \in T_{\succ}, t' \succ t$, and (ii) $\nexists t', t'' \in T_{\succ}$ that $t' \succ t''$. The entry *state* illustrates the state of the node. During processing, a node could be at one of three states: TV (To Visit), VD (Visited), or NV (No Visit). The entry c is the total number of tuples in the *domGraph* ranking higher than t.

Algorithm `maintainDomGraph` (Algorithm 2) illustrates how to maintain a *domGraph* when a new tuple t arrives. Initially, three FIFO (First In First Out) queues, Q, $Q_{\prec}$ and $Q_{\succ}$, which store nodes to be visited, nodes dominating t, and nodes dominated by t respectively are emptied at the same time. In general, $\text{pop}(Q)$ and $\text{push}(n, Q)$ are basic operators supported by any FIFO queue. The operator $\text{pop}(Q)$ returns the item at the front of a non-empty queue Q, and then remove it from Q. Otherwise, it returns $NULL$ if Q is empty. The operator $\text{push}(n, Q)$ inserts the item n at the back of the queue Q. A subroutine `set` is defined to update the *state* entry for a set of nodes. For example, at Line 1, $\text{set}(G, TV)$ means that the states of all nodes in G are set to TV. The variable b, initialized to zero, represents the number of nodes been visited.

At first, the states of all nodes in G are initialized to TV, means that these nodes are ready to be visited. All nodes that are not dominated by any other node in G are pushed into Q (at Lines 1-3). Subsequently, it begins to construct two FIFO queues, $Q_{\prec}$ and $Q_{\succ}$, by processing all nodes in Q (at Lines 4-12). The queue $Q_{\prec}$ represents all nodes that just dominates t, and $Q_{\succ}$ represents all nodes that are just dominated by t. The state of n, a node popped from Q, is updated to VD (visited), meaning that this node *has been visited*. Obviously, any node that dominates $n.t$ also dominates t if $n.t \prec t$, which means that it is unnecessary to visit these nodes in future. Under such situation, we begin to compare $n.c$ with k. It is clear that the new tuple t won't be a candidate if $n.c \geq k-1$ so that the processing for t could be terminated (Lemma 1). Otherwise, we push n into $Q_{\prec}$, and set the states of all nodes in $n.T_{\prec}$ to NV (not visit). A node with a state of NV will never be pushed into the queue Q. Subsequently, if $t \prec n.t$, we push n into $Q_{\succ}$, following which $Q_{\succ}$ is updated to make it only contains nodes directly dominated by t. It is worth noting that it is necessary to check all nodes dominating t if the tuple t is not dominated by $n.t$. In this way, the subroutine `pushDominated` (Algorithm 3) is invoked immediately to push part of nodes in $n.T_{\prec}$ into Q if satisfying following two conditions simultaneously (i) with a state of TV, and (ii) all nodes dominated by it have been visited. The condition (i) claims that a node to be pushed into Q must have not been visited. The condition (ii) shows that a node is pushed into Q after all nodes dominated by it.

Algorithm 2. maintainDomGraph(t, G)

1: Empty FIFO queues $Q, Q_\prec, Q_\succ$; $b \leftarrow 0$; set(G, TV);
2: **foreach** (node n in G)
3:   **if** ($\nexists n' \in G$, $n.t \prec n'.t$) **then** push(n, Q);
4: **while** (($n \leftarrow$ pop(Q)) $\neq$ NULL)
5:   set(n, VD); $b \leftarrow b + 1$;
6:   **if** ($n.t \prec t$) **then**
7:     **if** ($n.c \geq k - 1$) **then return**; // t isn't a candidate
8:     push($n, Q_\prec$); set($n.T_\prec$, NV);
9:   **else**
10:     **if** ($t \prec n.t$) **then**
11:       push($n, Q_\succ$); $Q_\succ \leftarrow Q_\succ - n.T_\succ$;
12:     pushDominated(n, Q);
13: create a new node $n_{new}(t, Q_\prec, Q_\succ, |G| + |Q_\prec| - b$, TV);
14: Remove old references between $n_{new}.T_\succ$ and $n_{new}.T_\prec$;
15: **foreach** (node n in G, $n.t \succ n_{new}.t$)
16:   $n.c \leftarrow n.c + 1$;
17:   **if** ($n.c \geq k - 1$) **then**
18:     Remove all nodes in $n.T_\succ$ in cascade style;

Algorithm 3. pushDominated(n, Q)

1: **for each** node n' in $n.T_\prec$ **do**
2:   **if** (n'.state = TV) **then**
3:     **if** ($\forall n'' \in n'.T_\succ$, n''.state = VD) **then** push(n', Q);

Then, it inserts a new node for t into G if necessary. The queues $Q_\prec$ and $Q_\succ$ keep all nodes dominating t and dominated by t respectively. We then compute the value of the entry c, which represents the number of nodes dominating t in G. Recall that all nodes dominating any node in $Q_\prec$ have not been visited (labeled as NV), and b represents the number of nodes been visited. So, there are $|G| + |Q_\prec| - b$ nodes in G dominating t. Subsequently, we remove all references between $n_{new}.T_\succ$ and $n_{new}.T_\prec$ to make G consistent (at Lines 13-14).

Finally, the entry c of each node dominated by t increases by 1. Additionally, if we find a node n such that $n.c \geq n - 1$, it is clear that all nodes dominated by n could be removed safely (at Lines 15-18).

Analysis. We maintain a *domGraph* for a set of candidate tuples for two reasons. First, we have tried to remove tuples which are definitely not candidates for a query. Since an arriving tuple may still contains other big attributes like text information, it will save the space consumption. Second, the *domGaph* is efficient to maintain. Without this directed acyclic graph, it is not easy to decide whether a new tuple is a candidate or not. In our algorithm, we compare the new tuple with a set of low-ranked tuples in G at first. In this way, the processing could be terminated as soon as possible (at Line 7).

Example 4. Figure 2 illustrates the evolution of a *domGraph* based on a small set of uncertain data in Figure 3. $k = 2$. Each node is affiliated with the tuple t

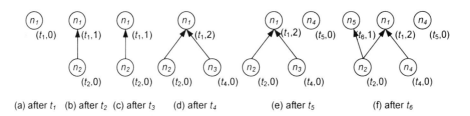

Fig. 2. The evolution of domGraph based on Figure 3

and the entry c. A directed link from n_i to n_j means that $n_i.t \prec n_j.t$. Obviously, A *domGraph* is in fact a directed acyclic graph. After time 2, both t_1 and t_2 stay in *domGraph* and $t_1 \succ t_2$. At time 3, since $t_1 \prec t_3$ and its node $n_1.c = 1 \geq k - 1$, t_3 will not be inserted into the *domGraph*. In this way, the following three tuples will be inserted into the *domGraph*.

3.2   probTree

Another indispensable task is to maintain the function $q(v)$ over all tuples in the stream. [8] provides a simple solution to handle a static data set. When a query request arrives, it begins to invoke a quick ordering algorithm to sort the data set in $O(n \log n)$ time, where n is the size of the data set. Then, the function $q(v)$ is constructed straightforwardly after conducting a linear scan upon all ordered tuples. This approach cannot suit for the streaming scenarios since it is expensive in processing a query request. Our goal is to devising a novel approach that is efficient both in tuple-maintaining and request-processing phases.

Our solution is a binary search tree that is called as *probTree* with each node in form of (v, p, l, r, par). The entry v, describing the attribute's value, is also the *key* of the tree. The entry p represents the probability sum of some tuples. The rest three entries, l, r and par, are references to its left child, right child and parent node respectively.

Algorithm `maintainProbTree` (Algorithm 4) illustrates how to maintain a *probTree* $\mathcal{T}$ continuously when a new tuple t, described as $\langle (v_1, p_1), \cdots, (v_{s_t}, p_{s_t}) \rangle$, arrives. Initially, we insert a new node of t into $\mathcal{T}$ as the root node if $\mathcal{T}$ is empty. In general, the algorithm begins to seek a target node with a *key* (or equally the entry v) equal to v_j. If such node is found, the value of its entry p increases by p_j. Otherwise, we insert a new node of (v_j, p_j) into $\mathcal{T}$. Moreover, for each node w along the path from *root* to the destine (a node with entry v equal to v_j), the entry p is updated as $w.p \leftarrow w.p + p_j$ if $w.v < v_j$.

Algorithm `getq` (Algorithm 5) illustrates how to compute the value of $q(v)$ by a probTree $\mathcal{T}$. It visits some nodes along the path from the *root* node to a destine node with v equal to $\min_{w \in \mathcal{T}, w.v > v}(w.v)$. A variable sum, representing the result value, is initialized to zero at first. For any node w in the path, the value of sum is updated as $sum \leftarrow sum + w.p$ if $w.v > v$.

The correctness of Algorithm 5 stems from the construction of a *probTree*. Let $(v_{i,l}, p_{i,l})$ be an arbitrary attribute-probability pair in the uncertain data set. Let

Algorithm 4. maintainProbTree$(t, \mathcal{T})$

1: **for** $j = 1$ **to** s_t
2:     $w \leftarrow \mathcal{T}.root$;
3:     **if** $(w = NULL)$ **then**
4:         $\mathcal{T}.root \leftarrow$ **new** $Node(v_j, p_j)$; **continue**;
5:     **while** $(w \neq NULL)$
6:         **if** $(w.v > v_j)$ **then**
7:             **if** $(w.l \neq NULL)$ **then** $w \leftarrow w.l$;
8:             **else** $w.l \leftarrow$ new$Node(v_j, p_j)$; **break**;
9:         **else if** $(w.v < v_j)$ **then**
10:           $w.p \leftarrow w.p + p_j$;
11:           **if** $(w.r \neq NULL)$ **then** $w \leftarrow w.r$;
12:           **else** $w.r \leftarrow$ new$Node(v_j, p_j)$; **break**;
13:         **else** $w.p \leftarrow w.p + p_j$; **break**;

Algorithm 5. getq$(v, \mathcal{T})$

1: $sum \leftarrow 0$;
2: $w \leftarrow \mathcal{T}.root$;
3: **while** $(w \neq NULL)$
4:     **if** $(w.v > v)$ **then**
5:         $sum \leftarrow sum + w.p$; $w \leftarrow w.l$;
6:     **else**
7:         $w \leftarrow w.r$;
8: **return** sum;

$RP(W)$ denote a tree includes the node w and its right sub-tree. For each node w in the *probTree*, the value of the entry p is the sum of probabilities of all attribute-probability pair with keys at its right side, i.e, $w.p = \sum_{v_{i,l} \in RT(w)} p_{i,l}$. In Algorithm 5, once it visits a node w such that $w.v > v$, it will choose to visit left child (at Line 7). In this way, the correctness is ensured.

Tuple ID	Attribute Value
t_1	$\langle (9, 0.3), (7, 0.7) \rangle$
t_2	$\langle (10, 0.4), (8, 0.6) \rangle$
t_3	$\langle (7, 1.0) \rangle$
t_4	$\langle (9, 0.5), (8, 0.4), (7, 0.1) \rangle$
t_5	$\langle (11, 0.3), (4, 0.7) \rangle$
t_6	$\langle (10, 0.4), (5, 0.6) \rangle$

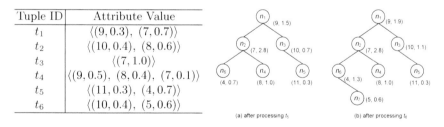

Fig. 3. A small data set   **Fig. 4.** An example of Probtree upon Figure 3

Example 5. Figure 4 illustrates the *probTree* at time 5 and 6 respectively. Each node is affiliated with information (v, p). When the tuple $t_6 \langle (10, 0.4), (5, 0.6) \rangle$ arrives, it finds the node n_3, $n_3.v = 10$. The entry $n_3.p$ is updated to 1.1 $(= 0.7 + 0.4)$. Since the parent node n_1 has $n_1.v = 9 < 10$, its entry $n_1.p$ is also updated to 1.9 $(= 1.5 + 0.4)$. Subsequently, it inserts a new node n_7 of the pair $(5, 0.6)$ into the *probTree* since no node with $v = 5$ is in the *probTree* now. Similarly, the entry p of the node n_6 is updated to 1.3 $(= 0.7 + 0.6)$ since $n_6.v < n_7.v$.

It is easy to compute the value of $q(v)$ based on a *probTree*. Assume $v = 8$. At first, it visits the *root* node n_1 to set the variable *sum* to 1.9 since $n_1.v = 9 > 8$. Next, it visits the left node, n_2, and does nothing since $n_2.v = 7 < 8$. Finally, it visits the right node, n_4, and find $n_4.v = 8$. As a result, it returns 1.9.

Theorem 2. *Let N denote the size of the data stream, s denote the maximum probability options, i.e, $s = \max_{i=1}^{N} s_i$. The size of a probTree is $O(sN)$, the per-tuple processing cost is $O(s \log(sN))$, and computing $q(\cdot)$ from a probTree costs $O(\log(sN))$.*

Proof. We assume the tuples in the stream arrive out-of-order. Obviously, the number of distinct items in the stream is $O(sN)$. The cost on inserting a tuple or computing $q(\cdot)$ is dependent on the height of the *probTree*. When each item is inserted in-order, the height is $O(sN)$ under the worst case. Contrarily, the expected height of a randomly built binary search tree on sN keys is $O(\log(sN))$ [5]. Straightforwardly, the amortized cost on inserting a tuple is $O(s \log(sN))$, and the cost on computing $q(\cdot)$ is $O(\log(sN))$.

3.3   Handle a Request

we can efficiently handle an ER-top$k$ request by using *domGraph* and *probTree*. Initially, an FIFO query Q and a result set R are emptied. Let $r_{\max}(R)$ denote the maximum expected rank of all nodes in a set R, i.e, $r_{\max}(R) = \max_{n \in R}(r(n))$. At first, all nodes not dominated by any other nodes are pushed into Q since these nodes have potentialities to be at the 1st rank. Subsequently, a node n is popped out of Q for evaluation repeatedly until Q is empty. The expected rank $r(n)$ of a node n can be computed by Equation (2). If R contains no more than k nodes, the node n is added into R immediately. Otherwise, if $r(n) < r_{\max}(R)$, we insert n into R and remove the lowest-ranked tuple in R. If the node n is pushed into R, it means that all tuples dominated by n directly are also candidate tuples. Thus, we can push each node in $n.T_{\succ}$ into Q if this node has not been pushed into Q. The program will terminate when the queue Q is empty.

4   Experiments

In this section, we present an experimental study upon synthetic and real data. All the algorithms are implemented in C++ and the experiments are performed on a system with Intel Core 2 CPU (2.4GHz) and 4G memory. Since our solution is the only method to handle ER-Top$k$ query over data stream, we will report the time- and space- efficiency below. We use two synthetic and one real 1,000,000-tuple data set in our testings, as described below.

syn-uni The *syn-uni* data set only has *existential* uncertainty. The rank of each tuple is randomly selected from 1 to 1,000,000 without replacement and the probability is uniformly distributed in $(0, 1)$.

syn-nor The *syn-nor* data set has both kinds of uncertainties. The existential
confidence of each tuple is randomly generated from a normal distribution
$N(0.6, 0.3)$ [1]. We set the maximum number of options of all tuples no more
than 10, i.e, $s_{max} = 10$. For the ith tuple, the number of attribute options
(say, s_i) is uniformly selected from $[s_{max}/2, s_{max}]$. Subsequently, we construct a normal distribution $N(\mu_i, \sigma_i)$, where μ_i is uniformly selected from
$[0, 1, 000]$ and σ_i is uniformly selected from $[0, 1]$. We randomly select s_i values from the distribution $N(\mu_i, \sigma_i)$, denoted as $v_1, \cdots, v_{s_i}$, and construct t_i
as $\langle (v_1, p_i/s_i), \cdots, (v_i, p_i/s_i) \rangle$.

IIP The (IIP) Iceberg Sightings Database [2] collects information on iceberg activity in North Atlantic near the Grand Banks of Newfoundland. Each sighting record contains the date, location, shape, size, number of days drifted,
etc. Since it also contains a confidence level attribute according to the source
of sighting, we converted the confidence levels, including R/V, VIS, RAD,
SAT-LOW, SAT-MED, SAT-HIGH and EST, to probabilities 0.8, 0.7, 0.6,
0.5, 0.4, 0.3 and 0.4 respectively. We created a data stream by repeatedly
selecting records randomly from a set of all records gathered from 1998 to
2007.

Figure 5 illustrates the space consumption of the proposed method, which mainly
contains a *domGraph* that stores all *candidate* tuples and a *probTree* for the
function $q(\cdot)$. In general, each tuple in a data set also contains other information besides the scoring attributes. For example, the IIP data set contains 19
attributes, including 2 time fields, 7 category fields and 10 numeric fields, among
which only two fields (e.g, 1 numeric field and 1 category field) are used for scoring function and confidence respectively. The full information should be stored
in the system if a tuple is probably belonging to the query result. The x-axis
in Figure 5(a) and (b), representing the size of information attributes for each
tuple, varies from 10 to 1000. The *domGraph* size is quite small, because merely
a small number of tuples are candidates. The *probTree* size is decided by the
number of distinct attribute values, independent of the information attributes.
Figure 5(c) illustrates the space consumption upon the IIP data set, where each
tuple uses 52 bytes to store the information attributes. The space consumption
is only 1% of the total data set.

Figure 6 illustrates that the per-tuple processing cost is low in general on
three uncertain data sets. When k increases, the cost will continue to grow.
Moreover, the cost for *syn-nor* is significantly higher than the other data sets,
because each tuple in *syn-nor* contains multiple attribute choices so that the
cost on maintaining *probTree* is much higher.

Figure 7 illustrates the cost on handling a request upon three data sets with
the comparison of the static method. The x-axis represents the value of parameter k, and the y-axis represents the time cost. Similarly, when k increases, the cost
will continue to grow. The cost on *syn-nor* and *IIP* is significantly greater than
syn-uni because of two reasons. First, the *syn-nor* data set has multiple choices

---

[1] $N(\mu, \sigma)$ is a normal distribution with μ as mean value and σ as standard deviation.
[2] http://nsidc.org/data/g00807.html

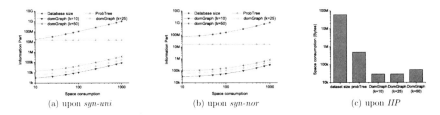

| (a) upon *syn-uni* | (b) upon *syn-nor* | (c) upon *IIP* |

Fig. 5. Space consumption upon uncertain data sets

in the scoring attribute, making it expensive to compute the rank of each candidate tuple. Second, the *domGraph* for *IIP* is more complex than *syn-uni* since it has many identical tuples. Anyway, this cost can be reduced significantly if we can scan the domGraph conveniently with the help of an additional list for candidates.

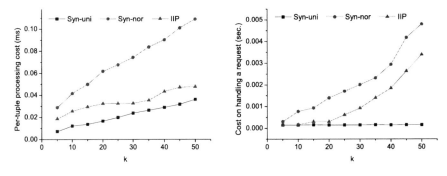

Fig. 6. Per-tuple processing cost   **Fig. 7.** Cost on handling a request

5   Related Work

Uncertain data management has attracted a lot of attentions recent years, because the uncertainty widely exists in many applications, such as Web, sensor networks, financial applications, and so on [1]. In general, there are two kinds of uncertainties, namely existential uncertainty and attribute-level uncertainty. More recently, several prototype systems have also been developed to handle such uncertain data, including MayBMS [4], MystiQ [10], and Trio [3].

Uncertain top-k query has been studied extensively in recent years. Although the semantics is explicit for the deterministic data, several different top-k definitions are proposed for distinct purposes, including U-top$k$ [18], U-$k$ranks [18], PT-k [12], global top-k [20], ER-top$k$ [8], ES-top$k$ [8], UTop-Set$k$ [17], c-Typical-Top$k$ [11] and unified top$k$ [16] queries. However, most of previous work only studies the "one-shot" top-k query over static uncertain dataset except [15] that studies how to handle sliding-window top-k queries (U-top$k$, U-kRanks, PT-k and global top$k$) on uncertain streams. In [15], some synopsis data structures are proposed to summarize tuples in the sliding-window efficient and effectively.

The focus of this paper is to study the streaming algorithm to handle top-k query. After studying main uncertain top-k query, we find that most of existing top-k semantics has the shrinkability property except the ER-topk query. In other words, for an ER-topk query, each tuple in the stream is either a candidate result tuple, or may influence the query result. So, we try to devise the exact streaming solution for ER-topk query.

Recently, there has been a lot of efforts in extending the query processing techniques on static uncertain data to uncertain data streams [2,6,7,9,13,14,19]. In [2], a method is proposed to handle clustering issue. [19], the frequent items mining is also studied.

6 Conclusion

In this paper, we aim at handling top-k queries on uncertain data streams. Since the volume of a data stream is unbounded whereas the memory resource is limited, we hope to find some heuristic rules to remove parts of redundant tuples and the rest tuples are enough for final results. Although this assumption is true for most of typical uncertain top-k semantics, we find that no tuple is redundant for the ER-top$k$ semantic. In other words, each tuple either (i) is belonging to the result sets now or later, or (ii) may influence the query results. We have devised an efficient and effective solutions for the ER-top$k$ query. A possible future work is to devising solutions for the sliding-window model.

Acknowledgement

The research of Cheqing Jin is supported by the Key Program of National Natural Science Foundation of China (Grant No. 60933001), National Natural Science Foundation of China (Grant No. 60803020), and Shanghai Leading Academic Discipline Project (Project No. B412). The research of Aoying Zhou is supported by National Science Foundation for Distinguished Young Scholars (Grant No. 60925008), and Natural Science Foundation of China (No.61021004).

References

1. Aggarwal, C.C.: Managing and mining uncertain data. Springer, Heidelberg (2009)
2. Aggarwal, C.C., Yu, P.S.: A framework for clustering uncertain data streams. In: Proc. of ICDE (2008)
3. Agrawal, P., Benjelloun, O., Sarma, A.D., Hayworth, C., Nabar, S., Sugihara, T., Widom, J.: Trio: A system for data, uncertainty, and lineage. In: Proc. of VLDB (2006)
4. Antova, L., Koch, C., Olteanu, D.: From complete to incomplete information and back. In: Proc. of SIGMOD (2007)
5. Cormen, T.H., Leiserson, C.E., Rivest, R.L., Stein, C.: Introduction to algorithms, pp. 265–268. The MIT Press, Cambridge (2001)
6. Cormode, G., Garofalakis, M.: Sketching probabilistic data streams. In: Proc. of ACM SIGMOD (2007)

7. Cormode, G., Korn, F., Tirthapura, S.: Exponentially decayed aggregates on data streams. In: Proc. of ICDE (2008)
8. Cormode, G., Li, F., Yi, K.: Semantics of ranking queries for probabilistic data and expected ranks. In: Proc. of ICDE (2009)
9. Cormode, G., Tirthapura, S., Xu, B.: Time-decaying sketches for sensor data aggregation. In: Proc. of PODC (2007)
10. Dalvi, N., Suciu, D.: Efficient query evaluation on probabilistic databases. VLDB Journal 16(4), 523–544 (2007)
11. Ge, T., Zdonik, S., Madden, S.: Top-k queries on uncertain data: On score distribution and typical answers. In: Proc. of SIGMOD (2009)
12. Hua, M., Pei, J., Zhang, W., Lin, X.: Ranking queries on uncertain data: A probabilistic threshold approach. In: Proc. of SIGMOD (2008)
13. Jayram, T., Kale, S., Vee, E.: Efficient aggregation algorithms for probabilistic data. In: Proc. of SODA (2007)
14. Jayram, T., McGregor, A., Muthukrishnan, S., Vee, E.: Estimating statistical aggregates on probabilistic data streams. In: Proc. of PODS (2007)
15. Jin, C., Yi, K., Chen, L., Yu, J.X., Lin, X.: Sliding-window top-k queries on uncertain streams. Proc. of the VLDB Endowment 1(1), 301–312 (2008)
16. Li, J., Saha, B., Deshpande, A.: A unified approach to ranking in probabilistic databases. In: Proc. of VLDB (2009)
17. Soliman, M.A., Ilyas, I.F.: Ranking with uncertain scores. In: Proc. of ICDE (2009)
18. Soliman, M.A., Ilyas, I.F., Chang, K.C.-C.: Top-k query processing in uncertain databases. In: Proc. of ICDE (2007)
19. Zhang, Q., Li, F., Yi, K.: Finding frequent items in probabilistic data. In: Proc. of SIGMOD (2008)
20. Zhang, X., Chomicki, J.: On the semantics and evaluation of top-k queries in probabilistic databases. In: Proc. of DBRank (2008)

Seamless Event and Data Stream Processing: Reconciling Windows and Consumption Modes[*]

Raman Adaikkalavan[1] and Sharma Chakravarthy[2]

[1] Computer and Information Sciences & Informatics, Indiana University South Bend
[2] ITLab & Computer Science and Engineering, The University of Texas at Arlington
raman@cs.iusb.edu, sharma@cse.uta.edu

Abstract. For a number of stream applications, synergistic integration of stream as well as event processing is becoming a necessity. However, the relationship between windows and consumption modes has not been studied in the literature. A clear understanding of this relationship is important for integrating the two synergistically as well as detecting meaningful complex events using events generated by a stream processing system. In this paper, we analyze the notion of *windows* introduced for stream processing and the notion of *consumption modes* introduced for event processing. Based on the analysis, this paper proposes several approaches for combining the two and investigates their ramifications. We present conclusions based on our analysis and an integrated architecture that currently supports one of the reconciled approaches.

1 Introduction

Event processing systems [1, 2] have been researched extensively from the situation monitoring viewpoint to detect changes in a timely manner and to take appropriate actions via active (or Event-Condition-Action) rules. *Data stream processing* systems [3–5] deal with applications that generate large amounts of data in real-time at varying input rates and to compute functions over multiple streams that satisfy quality of service (QoS) requirements. *Event stream processing* systems [3, 6–11], those integrating event and data stream processing, combine the capabilities of both models for applications that not only need to process continuous queries (CQs), but also need to detect complex events based on the events generated by CQs. Integrating both these models requires analysis of their relationships [3, 6]: inputs and outputs, consumption modes vs. windows, event operators vs. CQ operators, computation model, best effort vs. quality of service requirements, optimization vs. scheduling, buffer management vs load shedding, and rule processing management.

In complex event processing systems, *primitive* or simple events are domain-specific and are detected or generated by a system along with the time of occurrence. For example, a data stream processing system may compute the rise in temperature over a period of time using a continuous query (CQ) and trigger an event "high temperature". *Composite* or complex events are event expressions composed out of one or more primitive

---

[*] This work was supported, in part, by the NSF grant IIS 0534611. This work was supported, in part, by IU South Bend Research Grant.

J.X. Yu, M.H. Kim, and R. Unland (Eds.): DASFAA 2011, Part I, LNCS 6587, pp. 341–356, 2011.
© Springer-Verlag Berlin Heidelberg 2011

or composite events using event operators. Composite events are detected using event operator semantics and event consumption modes [1]. For example, the notion of fire may be defined as composite event when "high temperature" and "presence of smoke" occur in the vicinity of each other (spatially and temporally). *Event consumptions modes* [12–19], such as Recent, Chronicle, Continuous, and Cumulative, were defined based on the needs of various application domains and are used to detect events in a meaningful way as required by the application. Some of the *event operators* of Snoop event specification language [12, 20] are AND, OR, Not, Sequence, Plus, Periodic, and Aperiodic. The time of occurrence of a composite event depends on the detection semantics [21, 22] that can be either point-based or interval-based.

Data stream management systems support CQs that are composed using various operators. Most of them support traditional relational operators such as select, project, join, and aggregates. The data in this model can be accessed only sequentially and the data items (or tuples) are typically accessed only once. However, computations on stream processing do not preclude the use of data from a conventional DBMS. Data items that arrive as a continuous stream can be considered to be ordered by their arriving time stamp or by other attributes (e.g., sequence number in an IP header). To deal with the unbounded input size of an input stream, a *sliding window* [4, 5] is used to capture a subset of an input stream, on which an operator (e.g., join, aggregate) can compute its results. Different types of windows have been proposed in the literature, namely: tuple-based, time-based, attribute-based, partitioned and semantic.

Although there is a large body of work on stream and event processing, there is not much work in the literature on their integration. As a consequence, there is not any work that explicitly discusses the relationship between windows and consumption modes. Even the stream and event processing systems that support event operators explicitly do not indicate how multiple events of the same type are handled with respect to earlier occurrences. There is no declarative specification along the lines of consumption modes in existing event stream processing. In this paper, we investigate the reconciliation between *event consumption modes* and *windows* – specifically tuple- or time-based windows. We will discuss both consumption modes and windows in detail, and then discuss how they can be handled in a consistent manner.

2 Integrated Event Stream Processing

The integrated event stream processing architecture (from [6]) shown in Figure 1 has four stages:

Stage 1: In this stage CQs over data streams are processed. This stage processes normal CQs where it takes streams as inputs and gives computed continuous streams. In Figure 1, operators S_1, S_2, and J_1 form a CQ. CQs can generate meaningful events that can be composed further for detecting higher level events (e.g., fire as described above). Even intermediate results of stream computations may generate events. A CQ may also give rise to multiple events. In Figure 1, primitive events E_1 and E_2 generated from J_1 and J_2, respectively, are composed to form E_3.

---

[1] They are also termed as parameter contexts or simply contexts in this paper where the intention is clear.

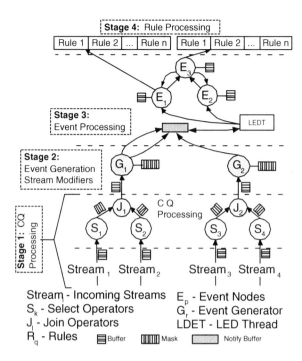

Fig. 1. Event Stream Processing Architecture

Stage 3: In this stage, events are defined and processed using the event detection graphs, and active rules are triggered. Event detection graphs (EDGs) [12, 15] can be used to detect events, where leaf nodes represent primitive events and internal nodes represent composite events.

Stage 4: Multiple rules can be associated with an event. When an event is detected in stage 3 and a rule is triggered, conditions are checked, and associated actions are carried out by this stage.

Stage 2: This is the glue that integrates the event and stream processing. In this stage, stream output is coupled with event nodes using an event generator operator. The event detector shown in stage 3 has a common event processor buffer into which all events that are raised are queued. A single queue is necessary as events are detected and raised by different components of the system (CQs in this case) and they need to be processed according to their time of occurrence. This stage adds a new event generator operator to every stream query as the root if an event is associated with that CQ. This operator can take any number of MASKS (or conditions) and for each MASK, a different event tuple/object is created and sent to the notify buffer in stage 3. As shown in Figure 1, nodes J_1 and J_2 are attached to event generator nodes G_1 and G_2, which have MASKS (or conditions on event parameters). Thus, CQ results from J_1 and J_2 are converted to events by nodes G_1 and G_2.

The seamless nature of the proposed integrated model is due to the compatibility of the chosen event processing model (i.e., an event detection graph) with the model used for stream processing. Event detection graphs correspond to operator trees and have similarities with respect to query processing whereas the other representations do not share these characteristics with query processing. The four stages shown in the architecture can be further linked in meaningful ways. For example, event streams can be fed to the stage 1 to be further processed as streams. Using the above stages as components, multiple layers of stream and event processing can be formulated based on the needs of applications. For a detailed analysis of event and stream processing, and their synergistic integration, please refer to [3, 11].

3   Event Consumption Modes

In the absence of consumption mode specification, events are detected in *unrestricted* (or with no) consumption mode. This means, once an event occurs, it cannot be discarded at all. Below we explain the detection of events in various consumption modes. For formalization and generalization, please refer to [12, 22].

The Snoop event specification language operators [12, 20] detect complex events based on event instances occurring over a time line. An event history ($E_i[H]$) maintains all event instances of a particular event type (E_i) up to a given point in time. Suppose e_1 is an event instance of type E_1, then event history $E_1[H]$ stores all the instances of the event E_1 (namely $e_1{}^j$). For each instance $e_i{}^j$, $[t_{si}, t_{ei}]$ indicates the start time (t_{si}) and end time (t_{ei}). Event instances in an event history are ordered by their end time.

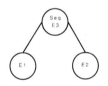

Fig. 2. Sequence Event Operator EDG

Below, we discuss the *Sequence* composite event operator that detects the sequence of two events. Consider two primitive events E_1 and E_2, and a Sequence event E_3 (i.e., E_1 SEQ E_2). Event E_3 is detected when E_2 occurs after E_1. Event detection graph of event E_3 is shown in Figure 2. With **unrestricted** (or no) consumption mode, all occurrences of event E_1 and E_2 are combined, and none of them are removed after they have taken part in an event detection. Consider the following event occurrences[2]:

$E_1[H]$: $e_1^1[10:00am]$; $e_1^2[10:03am]$; $e_1^3[10:04am]$;
$E_2[H]$: $e_2^1[10:01am]$; $e_2^2[10:02am]$; $e_2^3[10:05am]$;

In the above histories, event instance e_1^1 occurs at 10:00 am. The sequence event E_3 is detected in *unrestricted* consumption mode with the following event pairs (for example, e_2^2 is combined only with e_1^1) :

$(e_1^1, e_2^1), (e_1^1, e_2^2), (e_1^1, e_2^3), (e_1^2, e_2^3), (e_1^3, e_2^3).$

---

[2] e_i^j represents the jth occurrence of event E_i.

Even though the above event pairs are detected, the *time of occurrence* for these event pairs depend on the detection semantics. With **interval-based** semantics, event E_3 occurs with the following time of occurrences:

$(e_1^1, e_2^1)[10:00am, 10:01am]$, $(e_1^1, e_2^2)[10:00am, 10:02am]$,
$(e_1^1, e_2^3)[10:00am, 10:05am]$, $(e_1^2, e_2^3)[10:03am, 10:05am]$,
$(e_1^3, e_2^3)[10:04am, 10:05am]$.

The constituent event that starts a composite event detection is called the *initiator* event. The constituent event that detects a composite event is the *detector* event. The *terminator* event stops the future detection of a composite event by removing existing constituent event occurrences. For the Sequence event E_3, E_1 is the initiator and E_2 is the detector. There is no terminator event as no event instances are discarded without any consumption mode.

However, the number of events produced without any consumption mode will be large and not all event occurrences are be meaningful to an application. In addition, detection of these events has substantial computation and space overhead that may become a problem for situation monitoring applications that are based on stream data and their unbounded nature. In order to restrict the detection to meaningful composite events, consumptions modes, contexts, consumption conditions, or similar concepts, are supported by systems such as: ACOOD, SAMOS, Snoop/Sentinel, REACH, AMiT, SASE+, CAYUGA for restricting the unnecessary events from being detected.

Consumptions modes, based on their semantics, impose an upper bound on the number of events of the same event type that should be kept for the purpose of detecting composite events. The number of events to be kept depended solely on the consumption mode of the operator and the semantics of the operator. Below we explain four consumption modes (Recent, Chronicle, Continuous, and Cumulative) by showing the events detected for the Sequence event E_3 defined above. Consumptions modes and event operators defined in the paper are based on the Sentinel system [12]. *Note that these context definitions were proposed prior to the advent of stream data processing.* We also briefly discuss why we chose only these four modes in Section 3.5.

3.1   Recent Consumption Mode

This mode was intended for applications where *events are happening at a fast rate and multiple occurrences of the same event only refine the previous value*. Briefly, only the most recent or the latest initiator (i.e., latest instance of the initiator event) that started the detection of a composite event is used in this consumption mode. This entails that the most recent occurrence just updates (summarizes) the previous occurrence(s) of the same event type. An initiator will continue to initiate new event occurrences until a new initiator (i.e., new instance of the initiator event) or a terminator (i.e., instance of the terminator event) occurs. Binary Snoop operators (AND, OR, Sequence) use only detectors. This implies that the initiator will continue to initiate new event occurrences until a new initiator occurs. On the other hand, ternary Snoop operators (Periodic, Aperiodic) contain both detectors and terminators, which implies that the initiator will continue to initiate new event occurrences until a new initiator occurs or until a terminator occurs. Once the composite event is terminated, all the constituent event instances of that composite event will be deleted.

Example: The sequence event E_3 (defined earlier) is detected in *recent* consumption mode using interval-based semantics with the following occurrences:

$(e_1^1, e_2^1)[10 : 00am, 10 : 01am]$, $(e_1^1, e_2^2)[10 : 00am, 10 : 02am]$,
$(e_1^3, e_2^3)[10 : 04am, 10 : 05am]$.

Sequence events (e_1^1, e_2^3) [10:00am, 10:05am] and (e_1^2, e_2^3) [10:03am, 10:05am] are not part of the recent consumption mode, in comparison to the unrestricted mode, as the initiator e_1^3 replaced both e_1^1 and e_1^2 before detector e_2^3 occurred.

In general, events detected in this mode are a *subset* of the events detected using the unrestricted consumption mode. *This is true in general[3] for all Snoop operators and all modes except the cumulative mode [20, 22].*

3.2 Continuous Consumption Mode

For applications where *event detection along a moving time window is needed*, this mode can be used. Briefly, in this mode, each instance of the initiator event starts the detection of a composite event. An instance of the detector or terminator event may detect one or more occurrences of that same composite event. An instance of the initiator will be used at least once to detect that event. For binary Snoop operators, all the constituent event instances (initiator, detector and/or terminator) are deleted once the event is detected. The detector event acts as the terminator as well. For ternary Snoop operators, detector and terminator are usually different, so initiators are removed only upon termination.

Example: The sequence event E_3 is detected in *continuous* consumption mode using interval-based semantics with the following occurrences:

$(e_1^1, e_2^1)[10 : 00am, 10 : 01am]$, $(e_1^2, e_2^3)[10 : 03am, 10 : 05am]$,
$(e_1^3, e_2^3)[10 : 04am, 10 : 05am]$.

Some of the events detected are different from the ones detected in the recent mode. Event pair (e_1^1, e_2^2) [10:00am, 10:02am] is not part of the continuous consumption mode, as the detector event e_2^1 removed the occurrence of e_1^1. On the other hand, event pair (e_1^2, e_2^3) [10:03am, 10:05am] is not part of the recent context.

3.3 Chronicle Consumption Mode

This mode was proposed for applications where *there is a correspondence between different types of events and their occurrences, and this correspondence needs to be maintained*. In this mode, for a composite event occurrence, the initiator and terminator pair is unique (oldest initiator instance is paired with the oldest terminator instance). For binary Snoop operators, both the detector and terminator are the same, so once detected

---

[3] Although only the interval semantics is used in this paper, the subset relationship is true for point semantics as well.

the entire set of participating constituent events (initiator, detector and terminator) are deleted. For ternary Snoop operators, detectors and terminators are different.

Example: The event E_3 is detected in the *chronicle* consumption mode using the interval-based semantics with the following occurrences:

$$(e_1^1, e_2^1)[10:00am, 10:01am], (e_1^2, e_2^3)[10:03am, 10:05am].$$

The events detected are a *subset* of the events detected using the unrestricted consumption mode. Event pairs (e_1^1, e_2^2), (e_1^1, e_2^3), and (e_1^3, e_2^3) are not part of the chronicle consumption mode as the detector e_2^1 removed the occurrence of e_1^1 after pairing, and e_2^2 occurred before e_1^2. Event e_1^3 waits for the next occurrence of E_2 to occur.

3.4   Cumulative Consumption Mode

Applications can use this consumption mode *when multiple occurrences of the same constituent event need to be grouped and used in a meaningful way.* In this consumption mode, all occurrences of an event type are accumulated as instances of that event until the event is detected. In both binary and ternary operator, detector and terminator are same and once detected and terminated, all constituent event occurrences that were part of the detection are deleted.

Example: The event E_3 is detected in *cumulative* consumption mode using interval-based semantics with the following occurrences:

$$(e_1^1, e_2^1)\ [10:00am, 10:01am], (e_1^2, e_1^3, e_2^3)\ [10:03am, 10:04am, 10:05am].$$

Note that for this context, events detected are *not* a subset of the unrestricted consumption mode. As shown above, the second event pair includes two occurrences of E_1 as it accumulates event occurrences until a detector/terminator occurs.

3.5   Consumption Modes in Other Systems

In some of the recent systems [16–19], consumption modes have been specified as a combination of *instance selection* and *instance reuse*. The *instance selection* option specifies which event instance of a constituent event will take part in an event composition. The *instance reuse* option specifies whether an event instance is consumed after it takes part in an event detection. We will discuss both the options using AMiT [18] as it provides the more expressive version. The instance selection options (or quantifiers) supported by AMiT are first, strict first, last, strict last, each, and strict each. For example, last selects the last instance occurrence of the constituent event to take part in the composite event detection. The instance reuse option (or consumption condition) can be either *True* or *False*. If it is set to *True*, the constituent event is consumed immediately after taking part in a detection. Using both the options, events can be detected in recent mode discussed previously, for example, using *last* as the instance selection option and *True* for the instance reuse option. Both the options can be applied to each constituent event of a complex event. For example, consider a complex event E_{c3} that consists of two other complex events E_{c1} and E_{c2}. Event E_{c1} can be detected in a consumption mode (instance selection and reuse), which is different from E_{c2} and E_{c3}. In contrast, in systems such as Snoop/Sentinel, the consumption mode is applied to the entire event expression (i.e., E_{c1}, E_{c2}, and E_{c3} will be detected in one consumption mode). Though the approach of allowing consumption modes to be specified with each constituent improves the expressive power, the semantics of the complex event detected using multiple modes is rather not clear. Thus, we will discuss the reconciliation using the four consumption modes discussed previously.

4   Windows

In *stream processing*, a window [3, 5, 6, 23] is defined as a historical snapshot of a finite portion of a stream at any time point. It defines the meaningful set of data (or tuples) used by CQ operators to compute their functions. The notion of a window is defined on each input stream and does not depend upon the operator semantics. The window need not be defined only in terms of either time or physical number of tuples, although that is typical in most of the applications and supported by most systems. Windows can be also semantic, attribute-based, predicate-based, partitioned, etc. The main objective of defining a window in stream processing is to convert blocking computations (i.e., operators) into non-blocking computations to produce output in a continuous manner. The results of a continuous query form a data stream too.

Below we discuss the working of tuple- and time-based windows using a continuous query. Consider two traffic streams TS1 and TS2 with two attributes.

TS1(SourceIP, DestIP)   TS2(SourceIP, DestIP)

Consider the following tuples along with their time of occurrence:

TS1 : $(1, 2)[10 : 00$am$], (1, 3)[10 : 03$am$], (2, 5)[10 : 04$am$]$
TS2 : $(2, 3)[10 : 01$am$], (2, 4)[10 : 02$am$], (4, 2)[10 : 05$am$]$

CQ1 shown below joins streams TS1 and TS2, defined above. It selects the SourceIP when attribute DestIP's values are same. Stream operator Join performs computation on windows W1 and W2.

CQ1 :   SELECT TS1.SourceIP, TS2.SourceIP
        FROM TS1 [Window W1], TS2 [Window W2]   WHERE TS1.DestIP = TS2.DestIP

4.1   Tuple-Based Window

Given a window specification for a continuous query (CQ), each operator in the CQ performs computation over the specified window. A Tuple-based window is expressed as [Row N tuples advance M] over stream S, where $N > 0$, specifies the number of tuples to be kept in the window. When specified, tuple-based window produces a relation R with N tuples. At any time instant T, R will have N tuples of S with the maximum tuple timestamp of R is $\leq$ T. When CQ1 is executed with the input tuples shown above, the resulting tuples based on the window size are as follows:

- [Row 1 tuples advance 1] for W1 and W2: The result of the above CQ is an empty set at any time instant.
- [Row 2 advance 1] for W1 and W2: The resulting set is an empty set at timepoints 10:00, 10:01, 10:02, 10:04, and 10:05, whereas at timepoint 10:03 one tuple (1,2) is generated by joining the tuples (1,3) of TS1 and (2,3) of TS2.

- [Row Unbounded tuples] or [Row ∞] for W1 and W2: With unbounded window size the resulting set contains two tuples: First tuple (1,2) joining (1,3) of TS1 & (2,3) of TS2, and second tuple (1,4) joining (1,2) of TS1 & (4,2) of TS2.

4.2   Time-Based Window

A time-based window is similar to the tuple-based window except the fact that the size of the sliding window is specified in terms of a time range rather than tuple size. A Time-based window can be expressed by [Range N time units advance M time units], where $N \geq 0$. When N is unbounded, all tuples in the entire stream are used. As a special case, NOW denotes the window with $N = 0$. When N is 0, all the tuples (possibly empty) that occurred at that instant are used in the computation. This is particularly useful when database relations are used along with streams.

Let us consider the same CQ defined for explaining tuple-based window. The results of the CQ execution for different values of N (in seconds) are discussed below.

- [Range 1 advance 1] for W1 and W2: The result of the above CQ is an empty set at all time instants as no two tuples match within this sliding window.
- [Range 2 advance 2] for W1 and W2: The resulting set is an empty set at any time instant as the window can only hold tuples that occurred in the last two seconds.
- [Range Unbounded] or [Range ∞] for W1 and W2: With unbounded range, the resulting set contains two tuples: First tuple (1,2) that joins (1,3) from TS1 & (2,3) from TS2, and second tuple (1,4) that joins (1,2) from TS1 & (4,2) from TS2.

4.3   Summary

The above examples produce different result sets for different *window sizes*. However, the last case in both tuple- and time-based window produced the same results as none of the tuples are discarded. This is similar to the unrestricted consumption mode where no tuples are discarded either. This case, with unbounded window size, is computationally expensive, produces redundant composite events, and more importantly, results in blocking for an unbounded stream. There are multiple types of sliding windows; for brevity, in this paper, we analyze only the tuple- and time-based windows and their relationships with event consumption modes.

5   Reconciliation of Windows and Consumption Modes

When event processing and stream processing are combined end-to-end for situation monitoring applications (as shown in Figure 1), the interplay between windows and consumption modes need to be clearly understood both from the viewpoints of application specification and implementation. In this section, we will analyze this relationship or interplay between windows and consumption modes. In this paper, we only summarize the reconciliation between these two independent concepts informally and through examples in terms of implications and semantics.

The integrated end-to-end architecture is shown in Figure 1. CQs are computed using the window specification associated with a query. Primitive events are produced by the stream processing component[4] of the system and are used by the event processing component. In the integrated model, complex events can either be specified independently and associated with CQs, or can be specified along with CQs. In either case, consumption modes can be specified for events.

First, we will try to establish an intuitive difference between the concept of windows and consumption modes.

The window specification, although defined for each stream, is applied to all (blocking) operators of a CQ. Hence, for tuple- and time-based windows, the number of tuples in a window is based on a single stream and can be *assumed* to have an upper bound for practical purposes[5]. However, the concept of the consumption modes is in terms of the semantics of the mode and the semantics of the operator and hence the number of events to be maintained is *not* pre-determined. Also, the semantics of event operators solely depend on the time of occurrence of the event. In a sense, consumption modes can be viewed as time-based windows where the end time can vary with each instance of the window. Recent context is the only consumption mode where the number of events (whether primitive or complex) maintained by the operator is fixed (to one). It has been shown [12, 20] that the space needs for different contexts are different. Although one can establish the space requirements for the same set of event sequences with given rates, the actual size depends on the runtime characteristics of the events (for example, which event occurs how far apart for chronicle mode).

In contrast, other sliding windows based on partitions, predicates, and semantics have characteristics similar to that of consumption modes in that the upper bound of a window specification cannot be established (at compile time) from the definition. It is beyond the scope of this paper to analyze and reconcile between these window specifications and consumption modes.

There are multiple ways in which one can try to reconcile between these two concepts in the context of event stream processing:

1. Use windows for stream data processing and consumption modes for event processing (Independent approach),
2. Use only windows for both stream and event processing (Windows-only approach)
3. Use only consumption modes or contexts for both stream and event processing (Context-only approach).
4. Use windows along with event consumption modes during event processing and only windows for stream processing (Hybrid approach)

It is important to understand that event consumption modes are based on the semantics of the operator, whereas window specification is not dependent on the semantics of stream operators. Although, in theory, a different window can be specified for each stream (and/or operator), the same window is typically used for a query. Currently, it is not clear as to how the consumption modes can be applied to stream processing or

---

[4] Note that although stream processing is used, in this paper, any computation system can be used to generate events; then the characteristics of those systems need to be reconciled with the consumption modes.

[5] For time-based windows, the granularity of timestamp is likely to put an upper bound although in theory it may be undefined due to variable arrival rates.

even whether it is appropriate! Thus, in this paper, we will not discuss the Context-only Approach that use consumption modes for both stream data and event processing. Below, we will analyze the remaining three approaches.

5.1 Independent Approach

In this reconciliation, CQs are processed using window specifications and are passed on to the event processing subsystem where window specification is *not* used. Instead, the consumption mode specified for the complex event is used for event detection.

This approach essentially assumes that the semantics of event detection is decoupled from the CQ that generates the events and hence CQ specification is not needed here. This may or may not be the case and is likely to be application dependent. Furthermore, it is possible that the semantics of the consumption mode may conflict with the semantics of the window and hence appropriate situation monitoring is not realized.

Consider the situation where a CQ either generates at most one event per window or more than one event per window. If all the CQs that generate primitive events generate at most one event per window, then the recent context perhaps can be effectively used to make sure that the event from the next window replaces the event from the previous window. Even this may not be entirely accurate as we elaborate further later. If a window generates multiple events, then the recent context is not likely to be an appropriate one. Events from the same window replace previous events and the semantics of complex event detection may not be as intended. Using other contexts is likely to interfere with the window boundaries and pair events that occur across window boundaries. This can also happen with the recent context if some windows do not generate any events or the event generation rate is varying across CQs.

In this approach, the subset property with respect to the unrestricted context is satisfied as the events are assumed to be a stream without any window specification. In any of the contexts, it is not possible to constrain the pairing of events to the events generated by a window. On the positive side, there is no change in the event detection algorithms for this approach.

This approach may be a good starting strategy to integrate the two systems as it will not require changes to either of the system. In fact, the integrated architecture presented in Figure 1 uses this approach. Currently, we are conducting experiments to understand the effectiveness of this approach for various class of applications.

5.2 Windows-Only Approach

For this approach, we assume that events are detected applying window constraints in the unrestricted mode and not using any other consumption mode. In other words, only sliding windows (either overlapping or non-overlapping/disjoint) are used both for stream and event processing. As sliding windows are new to event detection, we discuss the effect of adding windows to replace contexts for event processing. In this paper, we consider only tuple-based windows with window size (1, n and ∞) for N. We will use events and event occurrences discussed in Section 3.

Consider queries CQ2 and CQ3 with a tuple-based window [W] that generate primitive events E_1 and E_2, respectively. Also the same window [W] is used for the detection of the Sequence composite event E_3.

$E_3 = E_1$ [Row W] SEQ E_2 [Row W]
$E_1[H]$: $e_1^1[10:00\text{am}]$; $e_1^2[10:03\text{am}]$; $e_1^3[10:04\text{am}]$;
$E_2[H]$: $e_2^1[10:01\text{am}]$; $e_2^2[10:02\text{am}]$; $e_2^3[10:05\text{am}]$;

If we impose different window sizes for the detection of composite events, we get the following composite events. Note that unrestricted mode is applied for all events within each window and events are not paired across windows.

- $W =$ [Row 1 advance 1, unrestricted mode]:
  $(e_1^1, e_2^1)[10:00\text{am}, 10:01\text{am}]$, $(e_1^1, e_2^2)[10:00\text{am}, 10:02\text{am}]$,
  $(e_1^3, e_2^3)[10:04\text{am}, 10:05\text{am}]$.
  The second window does not produce an event as the sequence semantics is not satisfied i.e., e_2^2 occurs before e_1^2.
- $W =$ [Row 2 advance 1, unrestricted mode]:
  $(e_1^1, e_2^1)[10:00\text{am}, 10:01\text{am}]$, $(e_1^1, e_2^2)[10:00\text{am}, 10:02\text{am}]$,
  $(e_1^2, e_2^3)[10:03\text{am}, 10:05\text{am}]$, $(e_1^3, e_2^3)[10:04\text{am}, 10:05\text{am}]$.
  There are other windows where no event is produced as the Sequence event semantics is not satisfied. Note that due to overlapping nature of windows, the same event may be generated more than once (in different windows).
- $W =$ [Row 2 advance 2, unrestricted mode]:
  $(e_1^1, e_2^1)[10:00\text{am}, 10:01\text{am}]$, $(e_1^1, e_2^2)[10:00\text{am}, 10:02\text{am}]$,
  $(e_1^3, e_2^3)[10:04\text{am}, 10:05\text{am}]$.
  Since the windows are disjoint, event with occurrences (e_1^2, e_2^3) [10:03am, 10:05am] is not detected in this case.
- $W =$ [Row ∞/unbounded, unrestricted mode]:
  $(e_1^1, e_2^1)[10:00\text{am}, 10:01\text{am}]$, $(e_1^1, e_2^2)[10:00\text{am}, 10:02\text{am}]$,
  $(e_1^1, e_2^3)[10:00\text{am}, 10:05\text{am}]$, $(e_1^2, e_2^3)[10:03\text{am}, 10:05\text{am}]$,
  $(e_1^3, e_2^3)[10:04\text{am}, 10:05\text{am}]$.
  Unbounded window size always produces the same set of events as the unrestricted event consumption mode (as inferred earlier) as no event is ever discarded.

The events detected are *not* the same even when the context is the same (unrestricted in this case) and the window is changing. Overlapping windows are likely to produce duplicate events whereas non-overlapping windows will not produce duplicate events.

5.3 Hybrid Approach

We discuss the effect of adding windows to event processing in addition to the consumption modes. For our running example, let us apply different windows and contexts. For an *unbounded* window, the following instances of E_3 is generated for various contexts. The unrestricted context/unbounded window consists of five events as shown above.

- $W =$ [Row ∞/unbounded, recent mode]:
  $(e_1^1, e_2^1)[10:00\text{am}, 10:01\text{am}]$, $(e_1^1, e_2^2)[10:00\text{am}, 10:02\text{am}]$,
  $(e_1^3, e_2^3)[10:04\text{am}, 10:05\text{am}]$.
- $W =$ [Row ∞/unbounded, continuous mode]:
  $(e_1^1, e_2^1)[10:00\text{am}, 10:01\text{am}]$, $(e_1^2, e_2^3)[10:03\text{am}, 10:05\text{am}]$,
  $(e_1^3, e_2^3)[10:04\text{am}, 10:05\text{am}]$.

- W = [Row ∞/unbounded, chronicle mode]:
  $(e_1^1, e_2^1)[10:00am, 10:01am]$, $(e_1^2, e_2^3)[10:03am, 10:05am]$.
- W = [Row ∞/unbounded, cumulative mode]:
  (e_1^1, e_2^1) $[10:00am, 10:01am]$, (e_1^2, e_1^3, e_2^3) $[10:03am, 10:04am, 10:05am]$.

Furthermore, we analyze the effect of imposing consumption modes in addition to the window specification. Table 1 shows the events that are detected with various consumption modes with different window sizes.

As an example, we will discuss the detection of events in Continuous mode and window [Row 2 advance 1]. Initially, windows corresponding to events E_1 and E_2 are empty. At time 10:00am event e_1^1 occurs and is added to E_1's window. At time 10:01am event e_2^1 occurs and is added to E_2's window. As event E_1's window is not empty, events in that window are checked for a sequence occurrence. Since e_1^1 has occurred at 10:00am before e_2^1 a sequence event with (e_1^1, e_2^1) is detected. After the detection, events e_1^1 and e_2^2 are deleted according to continuous mode and Sequence operator semantics. Event e_2^2 occurring at time 10:02am is deleted as there are no initiators. Event e_1^2 occurring at time 10:03am starts the next sequence event and is added to E_1's window. Event e_1^3 occurring at time 10:04am starts the next sequence event and is also added to E_1's window. Since the size of the window is 2 rows, both the events are kept in event E_1's window. When event e_2^3 occurs at 10:05am it is added to the sliding window of event E_2 and is processed. Since it satisfies the sequence event condition, two sequence events are detected with (e_1^2, e_2^3) and (e_1^3, e_2^3). After the processing all the three events are removed according to the continuous consumption mode and Sequence operator semantics. Note that not all the events detected in this approach are same as the events detected without any windows in Section 3.

Generalization. Since all modes or contexts except the cumulative context produce a subset of the composite events generated in the unrestricted context as shown in Table 1, we will generalize our solution for the Sequence operator using the unrestricted consumption mode. The formalization of interval-based Sequence event is given below.

$$I(E_1 \gg E_2, [t_{s1}, t_{e2}]) \triangleq \exists t_{e1}, t_{s2}(I(E_1, [t_{s1}, t_{e1}]) \wedge I(E_2, [t_{s2}, t_{e2}]) \\ \wedge(t_{s1} \leq t_{e1} < t_{s2} \leq t_{e2}))$$

The above formal definition is based on the history of events E_1 and E_2 (i.e., $E_1[H]$ and $E_2[H]$), respectively. At any point in time, $E_1[H]$ contains all the event occurrences of event E1 up to that point. In the windows only approach, the history of event E_1 i.e., $E_1[H]$ is partitioned. $E_1[H] = E_1^1[H] + E_1^2[H] + E_1^3[H] + ... + E_1^n[H]$. The sub-history can be disjoint or overlapping. Thus, when the same formal definition is applied over each sub-history instead of $E_1[H]$, the number of events detected will differ. This is mainly due to the removal of events by the sliding window size. Consider an event that is detected using the unrestricted mode in the *independent* approach with event e_1^1 in E_1 and e_2^3 in E_2. If the event occurrences are grouped based on the window size then e_1^1 might not combine with e_2^3 since they might be in different sub histories. The above arguments hold *True* for other operators as the above analysis is based only on the event histories and sub-histories.

Table 1. Events Detected in Hybrid Approach

Modes	Window Sizes		
	Row 1 advance 1	Row 2 advance 1	Row ∞/unbounded
Unrestricted	(e_1^1, e_2^1) [10:00,10:01am] (e_1^1, e_2^2) [10:00,10:02am] (e_1^3, e_2^3) [10:04,10:05am]	(e_1^1, e_2^1) [10:00,10:01am] (e_1^1, e_2^2) [10:00,10:02am] (e_1^2, e_2^3) [10:03,10:05am] (e_1^3, e_2^3) [10:04,10:05am]	(e_1^1, e_2^1) [10:00,10:01am] (e_1^1, e_2^2) [10:00,10:02am] (e_1^1, e_2^3) [10:00,10:05am] (e_1^2, e_2^3) [10:03,10:05am] (e_1^3, e_2^3) [10:04,10:05am]
Recent	(e_1^1, e_2^1) [10:00,10:01am] (e_1^1, e_2^2) [10:00,10:02am] (e_1^3, e_2^3) [10:04,10:05am]	(e_1^1, e_2^1) [10:00,10:01am] (e_1^1, e_2^2) [10:00,10:02am] (e_1^3, e_2^3) [10:04,10:05am]	(e_1^1, e_2^1) [10:00,10:01am] (e_1^1, e_2^2) [10:00,10:02am] (e_1^3, e_2^3) [10:04,10:05am]
Continuous	(e_1^1, e_2^1) [10:00,10:01am] (e_1^3, e_2^3) [10:04,10:05am]	(e_1^1, e_2^1) [10:00,10:01am] (e_1^2, e_2^3) [10:03,10:05am] (e_1^3, e_2^3) [10:04,10:05am]	(e_1^1, e_2^1) [10:00,10:01am] (e_1^2, e_2^3) [10:03,10:05am] (e_1^3, e_2^3) [10:04,10:05am]
Chronicle	(e_1^1, e_2^1) [10:00,10:01am] (e_1^3, e_2^3) [10:04,10:05am]	(e_1^1, e_2^1) [10:00,10:01am] (e_1^2, e_2^3) [10:03,10:05am]	(e_1^1, e_2^1) [10:00,10:01am] (e_1^2, e_2^3) [10:03,10:05am]
Cumulative	(e_1^1, e_2^1) [10:00,10:01am] (e_1^3, e_2^3) [10:04,10:05am]	(e_1^1, e_2^1) [10:00,10:01am] (e_1^2, e_1^3, e_2^3) [10:03,10:04,10:05am]	(e_1^1, e_2^1) [10:00,10:01am] (e_1^2, e_1^3, e_2^3) [10:03,10:04,10:05am]

5.4   Analysis

Below, we analyze the effect of windows and contexts on event detection. It is evident, even from the above analysis using an event operator, that the number and events detected can vary based on two *orthogonal* parameters: window and consumption mode.

It is also clear from the above discussion and examples that windows and consumption modes play different roles in the detection of events. Informally, windows (especially time-based) can be viewed as a mechanism for indicating the validity of an event and hence when an event can be dropped. This can be especially useful for the unrestricted context where there is no mechanism for dropping an event. In other words, a window can act as an *expiry mechanism* for event detection. On the other hand, a consumption mode is tied to the time of occurrence (or rate and distance of occurrence) and is based on constraints that are tied to the application semantics (e.g., sensor applications versus sliding-window applications). To some extent this is similar to the tuple-based window except that the application semantics is not perhaps dominant. The *expiry mechanism* of a consumption mode (if we want to think about it in this manner) is dynamically changing based on several variables (e.g., rate, application semantics).

From the above discussion, it is clear that the two were designed for different purposes and hence is unlikely to replace one another. A more theoretical analysis is underway to establish this more formally.

The independent approach seems to be useful where CQs and events are specified separately and independently. The window semantics seems to be not relevant or needed for the detection of events which can be guided by the specification of a consumption mode. However, if there is a semantic coupling between CQs and complex events, then this is not a good choice. The windows-only approach allows only unrestricted event consumption context for event detection. As the number of events detected is likely to be proportional (or quadratic at most) to the window size, this approach seems useful only when the window size is relatively small or when the number of events generated within a window is small. Furthermore, this approach is subsumed by the hybrid approach when the consumption mode is unrestricted.

Of all the three approaches, the hybrid approach seems most flexible and useful. This alternative can be chosen to satisfy *both* the window requirements of an application and the consumption mode requirements of an application. This approach does not subsume the independent approach as the latter assumes an unbounded window along with the specific event consumption mode.

Based on the above analysis, we postulate the following:

Postulation 1: Windows cannot be simulated by consumption modes.

Postulation 2: Consumption modes cannot be simulated by windows. The two are orthogonal and are designed for different purposes. Whether they can be combined to form a single mechanism to include components of both need to be investigated.

Postulation 3: The hybrid and *independent* approach are both needed from the expressiveness viewpoint, and to provide a flexible event stream processing system. The integrated architecture discussed in Section 2 implements the independent approach.

6   Conclusions and Future Work

In this paper, we have discussed windows and consumption modes associated with stream data and event processing, respectively. We then analyzed the impact and relationships between the concept of windows and the concept of consumption modes. We have identified three approaches for using windows and event consumption modes together in an integrated system. After studying a number of examples for analyzing: i) the number and types of events generated, ii) whether the subset property is satisfied or not, and iii) the interaction of windows with events, our conclusion is that both the hybrid approach and the independent approach are needed as they provide the maximum expressiveness and flexibility. We are currently investigating the reconciliation between other types of windows and consumption modes, and formalizing the reconciliation.

References

1. Luckham, D.C.: The Power of Events: An Introduction to Complex Event Processing in Distributed Enterprise Systems. Addison-Wesley, Boston (2001)
2. Widom, J., Ceri, S.: Active Database Systems: Triggers and Rules. Morgan Kaufmann Publishers, Inc., San Francisco (1996)
3. Chakravarthy, S., Jiang, Q.: Stream Data Processing: A Quality of Service Perspective. Advances in Database Systems, vol. 36. Springer, Heidelberg (2009)
4. Carney, D., et al.: Monitoring streams - a new class of data management applications. In: Proc. of the VLDB, pp. 215–226 (2002)

5. Arasu, A., Babu, S., Widom, J.: The CQL continuous query language: semantic foundations and query execution. VLDB Journal 15(2), 121–142 (2006)
6. Jiang, Q., Adaikkalavan, R., Chakravarthy, S.: MavEStream: Synergistic Integration of Stream and Event Processing.. In: Proc. of the International Conference on Digital Telecommunications, p. 29. IEEE Computer Society, Los Alamitos (2007)
7. Chakravarthy, S., Adaikkalavan, R.: Event and Streams: Harnessing and Unleashing Their Synergy. In: Proc. of the ACM DEBS, pp. 1–12 (July 2008)
8. Brenna, L., et al.: Cayuga: a high-performance event processing engine. In: Proc. of the ACM SIGMOD, pp. 1100–1102 (2007)
9. Wu, E., Diao, Y., Rizvi, S.: High-performance complex event processing over streams. In: Proc. of the ACM SIGMOD, pp. 407–418 (2006)
10. Rizvi, S., et al.: Events on the Edge. In: Proc. of the ACM SIGMOD, pp. 885–887 (2005)
11. Jiang, Q., Adaikkalavan, R., Chakravarthy, S.: NFM^i: An Inter-domain Network Fault Management System. In: Proc. of the ICDE, pp. 1036–1047 (2005)
12. Chakravarthy, S., Mishra, D.: Snoop: An Expressive Event Specification Language for Active Databases. DKE 14(10), 1–26 (1994)
13. Engström, H., Berndtsson, M., Lings, B.: ACOOD essentials. University of Skovde, Tech. Rep. (1997)
14. Gatziu, S., Dittrich, K.R.: Events in an Object-Oriented Database System. In: Proc. of the Int'l Workshop on Rules in Database Systems, pp. 23–39 (September 1993)
15. Branding, H., Buchmann, A.P., Kudrass, T., Zimmermann, J.: Rules in an Open System: The REACH Rule System. In: Proc. of the Int'l Workshop on Rules in Database Systems, pp. 111–126 (September 1993)
16. Zimmer, D., Unland, R.: On the semantics of complex events in active database management systems. In: Proc. of the ICDE, p. 392. IEEE Computer Society, Los Alamitos (1999)
17. Bailey, J., Mikulás, S.: Expressiveness issues and decision problems for active database event queries. In: Van den Bussche, J., Vianu, V. (eds.) ICDT 2001. LNCS, vol. 1973, pp. 68–82. Springer, Heidelberg (2000)
18. Adi, A., Etzion, O.: AMiT - The Situation Manager. VLDB Journal 13(2), 177–203 (2004)
19. Carlson, J., Lisper, B.: An event detection algebra for reactive systems. In: Proc. of the ACM International Conference on Embedded Software, pp. 147–154 (2004)
20. Chakravarthy, S., Krishnaprasad, V., Anwar, E., Kim, S.: Composite Events for Active Databases: Semantics, Contexts and Detection. In: Proc. of the VLDB, pp. 606–617 (1994)
21. Galton, A., Augusto, J.C.: Two Approaches to Event Definition. In: Hameurlain, A., Cicchetti, R., Traunmüller, R. (eds.) DEXA 2002. LNCS, vol. 2453, pp. 547–556. Springer, Heidelberg (2002)
22. Adaikkalavan, R., Chakravarthy, S.: SnoopIB: Interval-based event specification and detection for active databases. DKE 59(1), 139–165 (2006)
23. Ghanem, T.M., Aref, W.G., Elmagarmid, A.K.: Exploiting predicate-window semantics over data streams. SIGMOD Record 35(1), 3–8 (2006)

Querying Moving Objects with Uncertainty in Spatio-Temporal Databases*

Hechen Liu and Markus Schneider

Department of Computer and Information Science and Engineering
University of Florida
Gainesville, FL 32611, USA
{heliu,mschneid}@cise.ufl.edu

Abstract. Spatio-temporal uncertainty is a special feature of moving objects due to the inability of precisely capturing or predicting their continuously changing locations. Indeterminate locations of moving objects at time instants add uncertainty to their topological relationships. Spatio-temporal uncertainty is important in many applications, for example, to determine whether two moving objects could possibly meet. Previous approaches, such as the 3D cylinder model and the space-time prism model have been proposed to study the spatio-temporal uncertainty. However, topological relationships between uncertain moving objects have been rarely studied and defined formally. In this paper, we propose a model called *pendant model*, which captures the uncertainty of moving objects and represents it in a databases context. As an important part of this model, we define a concept called *spatio-temporal uncertainty predicate* (*STUP*) which expresses the development of topological relationships between moving objects with uncertainty as a binary predicate. The benefit of this approach is that the predicates can be used as selection conditions in query languages and integrated into databases. We show their use by query examples. We also give an efficient algorithm to compute an important STUP.

1 Introduction

The study of moving objects has aroused a lot of interest in many fields such as mobile networking, transportation management, weather report and forecasting, etc. Moving objects describe the continuous evolution of spatial objects over time. The feature that their locations change continuously with time makes them more complicated than static spatial objects in some aspects; one aspect refers to the topological relationships. A topological relationship, such as *meet*, *disjoint* or *inside*, characterizes the relative position between two or more spatial objects. In the spatio-temporal context, however, topological relationships between moving objects are not constant but may vary from time to time. For

---

* This work was partially supported by the National Science Foundation (NSF) under the grant number NSF-IIS-0812194 and by the National Aeronautics and Space Administration (NASA) under the grant number NASA-AIST-08-0081.

J.X. Yu, M.H. Kim, and R. Unland (Eds.): DASFAA 2011, Part I, LNCS 6587, pp. 357–371, 2011.

example, an airplane is disjoint with a hurricane at the beginning, and later it flies approaching the hurricane, and finally locates inside the hurricane. This developing relationship is named as *enter* [1]. Since we often do not have the capability of tracking and storing the continuous changes of locations all the time due to the deficiency of devices but can merely get observations of them at time instants, the movements between these time instants are always uncertain. The indeterminate locations further add uncertainty to the topological relationships between two moving objects.

Traditionally, moving objects are represented as 3D (2D+time) polylines. This representation, however, only stores the most possible trajectory of a moving object, without considering the uncertainty property. Several approaches have been proposed to handle spatio-temporal uncertainty, including the 3D cylinder model [2] and the space-time prism model [3]. However, the former assumes that the degree of uncertainty of a moving object does not change with time and thus lacks a precise representation. The latter is not able to combine both the certain part and the uncertain part of a movement. Both models rarely discuss the dynamic change of topological relationships with uncertainty.

The goal of this paper is to model and query the dynamic topological relationships of moving objects with uncertainty. This problem is solved through proposing the *pendant model* which is based on the well known space-time prism model while adds two significant advantages. First, it is an integrated and seamless model which combines both known movements and uncertain movements. Second, it formally defines spatio-temporal uncertainty predicates (STUP) that express the topological relationships between uncertain moving objects. This is important in querying moving objects with uncertainty since they can be used as selection conditions in databases. As an important part of the model, we formally define operations related to retrieving and manipulating uncertain data, and STUPs such as *possibly_meet_at*, *possibly_enter*, and *definitely_cross* that are defined on the basis of the operations. Queries related to the uncertainty in the topological relationships between moving objects can then be answered.

The paper is organized as follows: Section 2 discusses the related work on moving objects and spatio-temporal uncertainty models. Section 3 introduces our pendant model of moving objects with uncertainty. Section 4 defines spatio-temporal uncertainty predicates and shows the use of them in queries. Section 5 presents the algorithms to determine the defined spatio-temporal uncertainty predicates. Section 6 draws some conclusions and discusses future work.

2   Related Work

Several approaches have been proposed to model moving objects in spatial databases and GIS [4–6]. In some models, a moving object is represented as a polyline in the three-dimensional (2D+time) space with the assumption that the movement is a linear function between two consecutive sample points [4, 6]. The 3D polyline model, however, only yields the trajectory that the moving object may take with the highest possibility but is often not the exact route the moving object takes in reality.

An important approach that captures the uncertainty feature of movements is the *3D cylinder model* [2, 7]. A trajectory of a moving point is no longer a polyline but a cylinder in the 2D+time space. The possible location of a moving object at a time instant is within a disc representing the area of the uncertainty. The cylinder model, however, assumes that the degree of uncertainty does not change between two sample points, which is not the exact case in reality. An improved model, the space-time prism model [3] represents the uncertain movement of a moving object as a volume formed by the intersection of two half cones in the 3D space. Given the maximum speed of a moving object and two positions at the beginning and at the end of the movement, all possible trajectories between these two points are bounded within the volume of two half cones. Space-time prisms are more efficient than the cylinder models since they reduce the uncertain volume by two thirds because of the geometric properties of cones. There are many application examples that benefit from the space-time prism model [8–11]. In some GPS applications, the uncertainty is represented as as an error ellipse which is the projection of a space-time prism in 2D space [8]. The set of all possible locations is useful, for example, in determining whether an animal could have come into contact with a forest fire, or an airplane could have entered a hurricane [12]. An approach that uses the space-time prism model to provide an analytic solution to alibi queries is proposed in [13]. A most recent approach discusses the problem of efficient processing of spatio-temporal range queries for moving objects with uncertainty [14]. However, there are some remaining problems that are not solved by space-time prism-based approaches. First, some movements are more complicated since they include both known movements and unknown movements together, but this situation has not yet been included. Second, dynamic topological relationships between moving objects with uncertainty have not been discussed. We will provide solutions to the aforementioned problems in our model.

The spatio-temporal predicates (STP) model [1] represents the temporal development and change of the topological relationship between spatial objects over time. A spatio-temporal predicate between two moving objects is a temporal composition of topological predicates. For example, the predicate *Cross* is defined as $Cross := Disjoint - meet - Inside - meet - Disjoint$ where juxtaposition indicates temporal composition. However, the STP model does not deal with the uncertainty of topological relationships between moving objects.

3   Modeling the Uncertainty of Moving Objects

In this section, we introduce the pendant model which represents the uncertainty of moving objects. Section 3.1 provides the formalization of spatio-temporal uncertainty. We review the space-time prism model, and then taking it as a basis, we give the definition of a new concept called the uncertainty volume. Section 3.2 introduces the pendant model which integrates the known part of a movement and the uncertain part of movement together. Section 3.3 discusses operations on the pendant model.

3.1   The Formalization of Uncertain Movements

The pendant model is built on top of the well known space-time prism model of moving objects with uncertainty. Given the locations and time instants of the origin and the destination of a moving point and its maximum speed, all possible trajectories between these two time instants are bounded within the volume of two half cones. An advantage of using a cone to represent moving objects with uncertainty compared to a cylinder is that the former is one third of the latter in volume, which reduces the degree of uncertainty. We rewrite this relationship as a formal definition of the uncertainty volume of a moving point.

Definition 1. *Let* (x_0, y_0) *denote the origin of a moving point at time* t_0, (x_1, y_1) *denote the destination of this point at time* t_1, *where* $t_1 > t_0$, *and* v_{max} *denote the maximum velocity. Then the uncertainty volume* UV *is given as*

$$UV = \{(x, y, t) \mid$$
$$\text{(i)} \quad x, y, t \in \mathbb{R}, t_0 \le t \le t_1$$
$$\text{(ii)} \quad t_0 \le t \le (t_0 + t_1)/2 \Rightarrow \sqrt{(x - x_0)^2 + (y - y_0)^2} \le (t - t_0)v_{max}$$
$$\text{(iii)} \quad (t_0 + t_1)/2 \le t \le t_1 \Rightarrow \sqrt{(x - x_1)^2 + (y - y_1)^2} \le (t_1 - t)v_{max}\}$$

The above definition contains two different scenarios illustrated by Figure 1(a) and Figure 1(b) respectively. First, if the origin and the destination of the moving point are at the same location, the uncertain volume of the moving object is shown by two connected symmetric cones sharing a base parallel to the xy-plane, which is easy to prove. In Condition(ii), the right side of the inequality shows the maximum distance the moving point can travel up to a given time instant t, which is actually the radius of the circle representing the uncertainty area, and the left side represents the actual distance it has traveled. If we integrate all the time instants, the possible movements are bounded by the half cone with top (x_0, y_0). Condition (iii) shows that the moving point must return to its original location with the same speed bound. Thus, these two parts of movement form a symmetric double-cone. If we make a cut to the cone which is parallel to the xy-plane, we will get either a point or a circle object.

In the second scenario, the origin and the destination of the movement are different, in which case the uncertain volume is the combination of two oblique cones. The shared base of two cones is not parallel to the xy plane. The reason is that the moving point shows an apt direction from the origin to the destination, which is different from the first scenario. As Figure 1(b) illustrated, since $x_1 > x_2$, the moving point has an apt direction to the right of the plane. For the first half cone, the left most trajectory is shorter than the right most trajectory, which means that the moving point moves to left for a small distance, and suddenly changes its direction to the opposite and travels to the destination with a longer distance. Thus, the base of such an asymmetric double-cone is oblique. If we give a cut with a plane parallel to the xy plane on the oblique cone, we will get either a point or a lens object.

We further extend Definition 1 to moving regions whose areas must be considered. We assume that a moving region is represented by a polygon (a circle

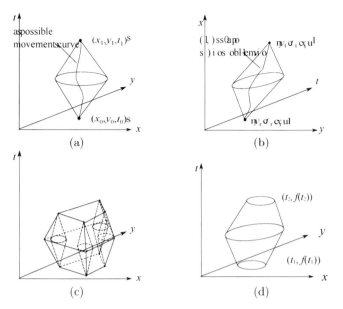

Fig. 1. Double-cone shaped volumes representing the uncertainty of a moving point with same (a) and different (b) origin and destination, uncertainty volumes of a moving polygon region (c) and a moving circle region (d)

region is a polygon with many vertices approaching infinite). Further we assume that a moving region only makes translation, i.e., there is no rotation, shrink or split of this region, then each vertices of the region is a moving point, and the uncertainty volume of each of them can be used to construct the uncertain volume of the entire moving region.

Definition 2. *Let $f(t)$ denote a moving region and let t_0 and t_1 denote two instants at the start and the end of the movement and $t_1 > t_0$, and v_{max} denote the maximum velocity. Then the uncertainty volume UV is defined as*

$$UV = \{\, (x, y, t) \,|$$
$$\text{(i)} \quad (x, y) \in f(t) \in region$$
$$\text{(ii)} \quad t_0 \le t \le (t_0 + t_1)/2 \Rightarrow \sqrt{(x - x_0)^2 + (y - y_0)^2} \le (t - t_0)v_{max}$$
$$\text{(iii)} \quad (t_0 + t_1)/2 \le t \le t_1 \Rightarrow \sqrt{(x - x_1)^2 + (y - y_1)^2} \le (t_1 - t)v_{max}\}$$

Condition (i) states that the uncertainty volume of the moving region contains all uncertainty volume of all points of the region. Condition (ii) and (iii) show that the uncertainty volume of each single point is a double-cone volume, as in Definition 1. Since the moving region only makes translation, all vertices will result the same uncertainty volume. Thus we can form the base of the entire uncertainty volume by connecting the tangent lines between bases of cones, as illustrated by Figure 1(c). In a special case, if the shape of the moving region is a circle, the uncertainty volume of it is a cone with circle shaped bases, as illustrated in Figure 1(d). We observe that Definition 1 can be treated as a special case of Definition 2 in that a moving point is a degenerated case of a moving polygon which contains only one vertex, thus we can replace (x_0, y_0) by $f(t_0)$ in the inequalities. In the rest of the paper, we will use the latter notation when treated moving points and moving regions together.

3.2 The Pendant Model as the Combination of Certain and Uncertain Movements

In some real cases, part of a movement can be exactly tracked, while part of the movement is uncertain. For example, the appearance of the signal on a radar representing the change of the position of an airplane belong to this category. A signal may exist in some intervals while it may disappear in some other intervals. The movements during periods in which the signal exists are known movements, however, the movements in the periods when the signal disappears are unknown but are interesting to us. We separate the movement of a moving object into two parts: the known part that can be represented as a function and is analogous to the string of a necklace, and the uncertain part which is analogous to a "pendant". It is allowed that in some special cases, a movement only contains known part, or only contains uncertain part, thus it will be convenient to give an integrated model that can represent all kinds of movement. In the above part, we have discussed the uncertain movement of a moving object in a single time interval. Next, we discuss how to integrate the known part of the movement. The part that is known can be represented by line segments through linear interpolation. This is different from the 3D polyline model, for the latter does not contain uncertainty parts. We say that a moving point, denoted by $(t, (x, y))$, is *linear* in $[t_0, t_1]$, if and only if

$$\forall t \in [t_0, t_1], (\frac{z - z_0}{t - t_0} = \frac{z_1 - z_0}{t_1 - t_0} \land z \in \{x, y\} \land x_1 \neq x_0, y_1 \neq y_0)$$
$$\lor (x = x_0 = x_1 \land y = y_0 = y_1)$$

Further, we say that a moving region, denoted by $f(t)$, is linear if and only if $\forall p \in f(t)$ is linear. A special case of the linear movement is that a static spatial object can be treated as a moving object whose locations remain constant over time. For example, a static point can be represented as a straight line perpendicular to the xy plane, and a static region can be represented as a cylinder volume in 2D+time space.

Definition 3. *The movement of a moving object with uncertainty, denoted by unmovement, is a function of a data type α, where $\alpha \in \{point, region\}$, defined as*

> $unmovement = \{f : time \to \alpha \mid$
> (i) $dom(f) = \cup_{i=1}^{n} [l_i, r_i], n, i \in \mathbb{N}; l_i, r_i \in time$
> (ii) $\forall 1 \leq i < n : r_i \leq l_{i+1}$
> (iii) $\forall 1 \leq i \leq n : f(t)$ is continuous at $t \in [l_i, r_i]$
> (iv) Let $v = d(f(t))/d(t)$ denote the velocity of f at t,
> $\forall 1 \leq i \leq n, \forall t \in [l_i, r_i] : v_{max_i} = max(v)\}$
> (v) $\forall 1 \leq i \leq n, \forall t \in [l_i, r_i] :$ either $(f(t), t)$ is an uncertainty volume,
>   or $f(t)$ is linear in $[l_i, r_i]$.

Definition 3 gives the formal representation of the moving object data type with uncertainty in our pendant model. Condition(i) states that an *unmovement* is a partial function defined on a union of intervals. Condition(ii) ensures that time intervals do not overlap with each other thus maximizes intervals and gives a unique representation for a known movement. Condition (iii) states that the movement is continuous in each interval thus instantaneous jump is not allowed. Condition (iv) defines v_{max_i} as the upper bound of the speed of the moving object during each interval i. Condition (v) states that within each interval of the moving object function, the movement is either a certain movement which changes linearly, or an uncertain volume of a double-cone. Thus, the uncertainty volume of a moving object is composed by the union of cones and line segments.

On the basis of Definition 3, we further define two subclasses of the *unmovement* data type, unmpoint and unmregion, representing moving point with uncertainty and moving region with uncertainty, which inherit all properties of *unmovement*.

Definition 4. *The data types moving point with uncertainty, and moving region with uncertainty are defined as follows:*

$$unmpoint \subset unmovement = \{f : time \to \alpha \,|\, \exists t \in dom(f) : f(t) \in point\}$$
$$unmregion \subset unmovement = \{f : time \to \alpha \,|\, \forall t \in dom(f) : f(t) \in region\}$$

3.3   Operations on the Pendant Model

Operations are important components in a data model. They are integrated into databases and used as tools for retrieving and manipulating data. In the rest of this section, we introduce some important operations in our pendant model. They will be useful in helping us define spatio-temporal uncertain predicates in the next section. Because of the page limitation, we only give the semantic and the explanation of each operation here.

$$
\begin{array}{lll}
construct\_segment & : \alpha \times \alpha \times interval & \to segment \\
construct\_pendant & : \alpha \times \alpha \times interval \times real & \to UV \\
construct\_movement & : segment^m \times UV^n & \to unmovement \\
at\_instance & : unmovement \times instant & \to \alpha \\
temp\_select & : unmovement \times interval & \to unmovement \\
dom & : unmovement & \to interval
\end{array}
$$

The operation *construct_segment* takes two observations and an interval as inputs, to construct the certain movement, which is the a 3D line segment between the observations. The *construct_pendant* operation takes two observations, the speed of the moving object, and the interval as inputs, and construct a double-cone volume. The *construct_movement* operation integrates known movements and movements with uncertainty together, where $m, n \in \mathbb{N}$, which means that a movement is composed by multiple known parts and uncertain parts. The *at_instance* operation takes an *unmovement* data type and an instant as inputs, and returns a spatial object representing the uncertainty of the moving object at

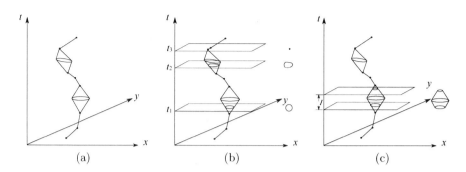

Fig. 2. An *unmpoint* data object (a); at_instance (b) and temp_select (c)

that instant, as illustrated in Figure 2(b). The *temp_select* operation is short for "temporal selection". It retrieves partial movement of the moving object during a time interval that is last for a period, as illustrated in Figure 2(c). The *dom* operator returns the life domain interval of an uncertain moving object.

4   Spatio-Temporal Predicates with Uncertainty and Queries

In this section, we discuss topological relationships between moving objects with uncertainty and show query examples. In [1], authors have defined topological relationships between moving objects over time as binary predicates, called *spatio-temporal predicates* (*STP*), the results of which are either true or false. This will make querying moving objects easier by using binary pre-defined predicates as selection conditions. Similarly, we define the dynmic topological relationships between moving objects with uncertainty as s*patio-temporal uncertain predicates* (*STUP*), which will be integrated into databases as join conditions. In Section 4.1, we formally define important STUPs on moving objects. In Section 4.2, we introduce the use of the STUP by showing query examples.

4.1   Definitions of Spatio-Temporal Uncertain Predicates

A spatio-temporal uncertain predicate (STUP) is a bool expression that is composed of topological predicates, math notations and operations on the pendant model we have defined in Section 3.3. An STUP expression may contain the distance operator *dist*, topological predicates between two regions (*disjoint, meet, overlap, covers, coveredBy, equal, inside, contains*), logic operators ($\neg, \exists, \forall, \land, \lor$), set operators ($\cap, \cup, \in, \subset$), and operations on the pendant model (*at_instance* (shortcut: @), *temp_select* (shortcut: π_t)).

The eight topological predicates describe the relationship between two regions, and they form the basis of our spatio-temporal uncertain predicates. *inside*(A, B) means that region A locates inside of region B in geometry. For simplicity, we use notation @ to represent *at_instance*, and π_t to represent

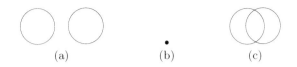

Fig. 3. Instant predicates of moving points with uncertainty: disjoint_at (a), definitely_meet_at and possibly_meet_at (c)

temp_select. We use the logic operators and set operators to connect terms and form the expressions. We first exam the topological relationship between two moving points. We name the relationship between them at a time instance as an *instant predicate*, and the relationships which lasts for a period as a *moving predicate*. The topological relationship between two static points is either *disjoint* or *meet*. However, for two moving objects, at a time instant there are three possible relationships. They can be disjoint, or certainly meet, or *possibly meet*. Thus we are able to define the following three instant predicates for moving points with uncertainty.

Definition 5. *Assume that we have two moving points* $p, q \in unmpoint$ *and a time instant* t. *Three instant predicates at* t *are defined as follows,*

$$disjoint\_at(p, q, t) \quad := \quad t \in (D(p) \cap D(q)) \wedge disjoint(@(p, t), @(q, t))$$
$$definitely\_meet\_at(p, q, t) := \quad t \in (D(p) \cap D(q)) \wedge @(p, t) \in point$$
$$\wedge dist(@(p, t), @(q, t)) = 0$$
$$possibly\_meet\_at(p, q, t) \quad := \neg disjoint\_at(p, q, t) \wedge$$
$$\neg definitely\_meet\_at(p, q, t)$$

In the above definition, two moving points are disjoint with each other at t, if the resulting regions from *at_instance* operations are disjoint. If the *at_instance* operation on two objects result two points that are of the same position, they will definitely meet at this time instance. The rest situations all belong to the *possibly_meet_at* relationship. Figure 3 illustrates the above three predicates.

After introducing the instant predicates between two moving points, now we define their moving predicates which describe topological relationships that last for a period of time. Assume that we have an interval $I \subset time$,

Definition 6. $definitely\_encounter(p, q, I) = true$, *if the following conditions hold*

    (i) $I \subset (D(p) \cap D(q)), p_I := \pi_t(p, I), q_I := \pi_t(q, I),$
    (ii) $\exists t_1, t_2, t_3 \in I \wedge t_1 < t_2 < t_3 \wedge disjoint\_at(p_I, q_I, t_1)$
        $\wedge definitely\_meet\_at(p_I, q_I, t_2) \wedge disjoint\_at(p_I, q_I, t_3)$

Definition 7. $possibly\_encounter(p, q, I) = true$, *if the following conditions hold*

    (i) $I \subset (D(p) \cap D(q)), p_I := \pi_t(p, I), q_I := \pi_t(q, I),$
    (ii) $\exists t_1, t_2, t_3 \in I \wedge t_1 < t_2 < t_3 \wedge disjoint\_at(p_I, q_I, t_1)$
        $\wedge possibly\_meet\_at(p_I, q_I, t_2) \wedge disjoint\_at(p_I, q_I, t_3)$

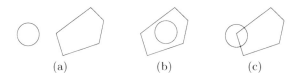

Fig. 4. Predicates between a moving point and a moving region disjoint_at (a), definitely_inside_at (b) and possibly_inside_at (c)

The *definitely_encounter* predicate describes the situation that two moving points will meet for sure during a time interval I. Similarly, *possibly_encounter* means that two moving points will meet with some possibility. Figure 5(a) illustrates this predicate.

Besides the uncertain relationship between two moving points we have introduced above, topological relationships between a moving point and a static region on the land are also of great importance. It can help, for example, detect whether an airplane has the possibility of entering a city when it disappears on the radar. Now, we define some important spatio-temporal uncertain predicates which represent topological relationships between a moving point and a static region on the land.

Definition 8. *Given the movement of a moving point $p \in unmpoint$, a static region $R \in region$, and a time instant $t \in time$, three instant predicates between them are defined as follows,*

$$
\begin{aligned}
disjoint\_at(p, R, t) &:= t \in D(p) \ \wedge \ disjoint(@(p, t), R) \\
definitely\_inside\_at(p, R, t) &:= t \in D(p) \ \wedge \ inside(@(p, t), R) \\
possibly\_inside\_at(p, R, t) &:= \neg disjoint\_at(p, R, t) \ \wedge \\
&\quad definitely\_inside\_at(p, R, t)
\end{aligned}
$$

The above definition formalize three instant predicates between a moving point and a static region. Figure 4(a)-(c) show the above three predicates, where the circle represents the uncertain region of a moving point at the given time instance and the polygon represents the static region. Based on the instant predicates we have defined, we give the following definitions of spatio-temporal uncertain predicates which represent the development of relationships between a moving point and a static region within a period.

Definition 9. *definitely_enter$(p, R, I) = true$, if the following conditions hold*

(i) $I \subset D(p), p_I := \pi_t(p, I)$
(ii) $\exists t_1, t_2 \in I \ \wedge \ t_1 < t_2 \ \wedge \ disjoint\_at(p_I, R, t_1) \ \wedge$
    $definitely\_inside\_at(p_I, R, t_2)$

Definition 10. *possibly_enter$(p, R, I) = true$, if the following conditions hold*

(i) $I \subset D(p), \ p_I := \pi_t(p, I)$
(ii) $\exists t_1, t_2 \in I \ \wedge \ t_1 < t_2 \ \wedge \ disjoint\_at(p_I, R, t_1) \ \wedge$
    $possibly\_inside\_at(p_I, R, t_2)$

Definition 11. $definitely\_cross(p, R, I) = true$, *if the following conditions hold*

(i) $I \subset D(p), p_I := \pi_t(p, I)$
(ii) $\exists t_1, t_2, t_3 \in I \ \wedge \ t_1 < t_2 < t_3 \ \wedge \ disjoint\_at(p_I, R, t_1) \ \wedge$
    $definitely\_inside\_at(p_I, R, t_2) \ \wedge \ disjoint\_at(p_I, R, t_3)$

Definition 12. $possibly\_cross(p, R, I) = true$, *if the following conditions hold*

(i) $I \subset D(p), p_I := \pi_t(p, I)$
(ii) $\exists t_1, t_2, t_3 \in I \ \wedge \ t_1 < t_2 < t_3 \ \wedge \ disjoint\_at(p_I, R, t_1) \ \wedge$
    $possibly\_inside\_at(p_I, R, t_2) \ \wedge \ disjoint\_at(p_I, R, t_3)$

Definitions 9 to 12 describe spatio-temporal uncertainty predicates between a moving point and a static region. Figure 5(b) illustrates the *possibly_cross* predicate. Similarly, we can define more predicates between an *unmpoint* object and an *unmregion* object, for example, *possibly_enter*, and *possibly_cross*. There are no *definite_enter* or *definitely_cross* relationships between an *unmpoint* object and an *unmregion* object, since there is no *definitely_inside_at* instant predicate between a moving point and a moving region with uncertainty at a time instance. Because of the page limitation, we do not give formal definitions of STUPs between an *unmpoint* and *unmregion* here.

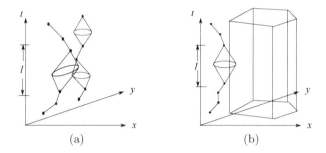

(a)   (b)

Fig. 5. Spatio-temporal predicates of possibly_encounter (a); possibly_cross (b)

4.2 Spatio-Temporal Uncertainty Queries

Now we discuss how to use STUPs we have defined in Section 4 in database queries. Current database query languages such as SQL are not able to answer temporal queries because they do not support temporal operators. This problem could be solved by implementing the STUPs as operators in queries. Thus, we are able to extend the SQL language to a more comprehensive query language, named as *spatio-temporal uncertainty language* (*STUL*). The STUL language extends SQL and supports the spatio-temporal uncertainty operations in terms of STUPs. Here we show some examples of using this language.

Assume that we have the following database schema of persons,

```
persons(id:integer, name:string, trajectory:unmpoint)
```

The query "Find all persons that may possibly become the witness of the criminal Trudy during the period from 10 am to 12 pm" can be answered by the STUL query as follows,

```
SELECT  p1.id            FROM     persons p1, persons p2
WHERE   possibly_encounter(p1.trajectory,p2.trajectory,
        10:00:00,12:00:00)       AND   p2.name='Trudy'
```

Now we give an example of SQL like query on the predicates between an uncertain moving point and a static region. Assume that we have the following schemas,

```
airplanes(id: string, flight: unmpoint)
airports(name: string, area: region)
```

The query "Find all planes that have possibly entered the Los Angeles airport from 2:00pm to 2:30pm" can be written as follows,

```
SELECT  airplanes.id         FROM     airplanes, airports
WHERE   possibly_enter(airplanes.flight, airports.area,
        14:00:00, 14:30:00) AND airports.name='LAX';
```

We are also able to query on the topological relationship between an uncertain moving point and an uncertain moving region. An example of a moving region with uncertainty is the hurricane. Assume that we have the following schema,

```
hurricanes(name: string, extent: unmregion)
```

The query "Find all planes that have possibly entered the extent of hurricane Katrina between Aug 24 to Aug 25, 2005" can be written as follows,

```
SELECT  a.id             FROM     airplanes a, hurricanes h
WHERE   possibly_enter(a.flight,h.extent,2005-08-24,2005-08-25)
        AND   h.name='Katrina'
```

5   Algorithms to Determine STUP

In this section, we give algorithms to determine the spatio-temporal uncertainty predicates. Because of the page limitation, we take the predicate between two moving points, *possibly_encounter* for example. Algorithms for other predicates will be provided in our future work. In Definition 3, we have stated that each movement is represented by a set of partial functions on a union of intervals, and each interval has its own moving pattern, i.e., either a known movement represented by a linearly function or an unknown movement represented by a double-cone volume. Thus, we are able to fragment the entire movement as a set of slices. The slice model is a discrete approach for representing moving objects, introduced in [4]. A slice is the smallest unit of evaluating the spatio-temporal

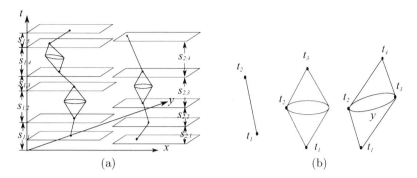

Fig. 6. Slice unit representation of moving objects with uncertainty (a), critical instants of a slice unit with different volumes

algorithm *unitIntersect* (*slice S1, slice S2*)	
1 $intersect \leftarrow false$	17 $u \leftarrow at\_instance(S1, s)$
2 $S \leftarrow empty$ // sequence of instants	18 $v \leftarrow at\_instance(S2, s)$
3 $m \leftarrow num\_of\_critical\_instants(S1)$	19 **if** $a \in point$ **and** $b \in point$
4 $n \leftarrow num\_of\_critical\_instants(S2)$	20 $intersect \leftarrow dist(u, v) = 0$
5 **while** $\{i \leq m$ and $j \leq n\}$	21 **endif**
6 **if** $time[i] < time[j]$	22 **if** $a \in point$ **and** $b \in region$
7 add $time[i]$ to S; i++	23 intersect $\leftarrow inside(u, v)$
8 **else**	24 **endif**
9 add $time[j]$ to S; j++	25 **if** $a \in region$ **and** $b \in region$
10 **endif**	26 $intersect \leftarrow overlap(u, v)$
11 **endwhile**	27 **endif**
12 add remaining instants of $S1$ to S	28 $s \leftarrow get\_next\_elem(S)$
13 add remaining instants of $S2$ to S	29 **endwhile**
14 $s \leftarrow get\_first\_elem(S)$	30 **return** $intersect$
15 **while not** $end\_of(S)$	end
16 **and** $intersect = false$	

Fig. 7. The algorithm *testUnitIntersection* to determines whether two slices intersect

uncertainty predicate, which is either a line segment or a double-cone volume. The slice unit representation of moving objects with uncertainty is illustrated in Figure 6, where moving object A is represented by $< s_{1,1}, s_{1,2}, \ldots, s_{1,5} >$ with elements ordered by time, and B is represented by $< s_{2,1}, s_{2,2}, \ldots, s_{2,4} >$. We evaluate the predicate between two entire moving objects by evaluating whether an intersection exists between a pair of slices. There are three situations: 1. Both slices are line segments; 2. one slice is a line segment while the other is a double-cone; 3. both slices are double-cone volumes. The first situation is the simplest one in that we can represent two line segments by equations in the 3D plane and compute whether they intersect at a common point, denoted by $comPoint(seg_1, seg_2)$. For situation 2 and 3, since it is cumbersome to calculate the intersection in 3D volumes, we introduce the method to test the intersection only in some time instants, which are called *critical instants*. A line segment has two critical instants: the starting time and ending time respectively. A straight double-cone volume has 3 critical instants: the instants at the bottom apex, the base and the top apex. An oblique double-cone volume has 4 critical instants: the bottom apex, the lower base point, the upper base point and the top apex. Figure 6(b) illustrates critical instants on the three types of volumes. We design an algorithm to compute whether two slices could intersect by testing whether they intersect at critical instants, shown by Figure 7. Because we only exam the intersection at a few number of critical instants, the complexity of this algorithm is constant.

The *possibly_encounter* predicate can then be determined by examining the intersection between pairs of slices. The algorithm is shown in Figure 8. Assume that the first moving object has m slices and the second has n slices, this algorithm will exam $m + n$ times of *unitIntersection*. Since the complexity of *unitIntersection* is $O(1)$, the total complexity to determine *possibly_encounter* is $O(m + n)$.

algorithm *possibly_incounter (umpoint A, umpoint B, t1, t2)*			
1	$intersect \leftarrow false$	11	$intersect \leftarrow unitIntersect(sa, sb)$
2	$A' \leftarrow temp\_select(A, t1, t2)$	12	**endif**
3	$B' \leftarrow temp\_select(B, t1, t2)$	13	**if** $endtime\_of\_sa < endtime\_of\_sb$
4	$sa \leftarrow get\_first\_elem(A')$	14	$sa = get\_next\_elem(A')$
5	$sb \leftarrow get\_first\_elem(B')$	15	**else**
6	**while not** $(end\_of(A')$ **or** $end\_of(B'))$	16	$sb = get\_next\_elem(B')$
7	**and** $intersect = false$	17	**endif**
8	**if** $sa \in segment$ **and** $sb \in segment$	18	**endwhile**
9	$intersect \leftarrow comPoints(sa, sb)$	19	**return** $intersect$
10	**else**	20	**end**

Fig. 8. The algorithm of *possibly_encounter*

6 Conclusions and Future Work

In this paper, we discuss the problem of querying topological relationships between moving objects with uncertainty. We propose the *pendant model* which properly represents the spatio-temporal uncertainty of moving objects. We define a new set of predicates called spatio-temporal uncertainty predicates (STUP) which represent moving topological relationships with uncertainty as binary predicates. The benefit of this approach is that the STUPs can be used in queries as selection conditions. We define important STUPs which describe topological relationships between two moving points, or a moving point and a static region, or a moving point and a moving region. We show query examples of how to use them and give an efficient algorithm to compute one of the most important STUPs. In our future work, we will give the algorithms for all other STUPs we have defined. We will study the topological relationships between complex regions with uncertainty as well.

References

1. Erwig, M., Schneider, M.: Spatio-temporal Predicates. IEEE Trans. on Knowledge and Data Engineering (TKDE) 14(4), 881–901 (2002)
2. Trajcevski, G., Wolfson, O., Zhang, F., Chamberlain, S.: The geometry of uncertainty in moving objects databases. In: 8th International Conference on Extending Database Technology (EDBT), pp. 233–250 (2002)
3. Hägerstrand, T.: What about people in regional science. Papers in Regional Science 24, 6–21 (1970)
4. Forlizzi, L., Güting, R.H., Nardelli, E., Schneider, M.: A data model and data structures for moving objects databases. In: SIGMOD 2000: Proceedings of the 2000 ACM SIGMOD International Conference on Management of Data, pp. 319–330 (2000)
5. Güting, R.H., Böhlen, M.H., Erwig, M., Lorentzos, C.S.J.N.A., Schneider, M., Vazirgiannis, M.: A Foundation for Representing and Querying Moving Objects. ACM Trans. on Database Systems (TODS) 25(1), 1–42 (2000)
6. Su, J., Xu, H., Ibarra, O.H.: Moving Objects: Logical Relationships and Queries. In: Jensen, C.S., Schneider, M., Seeger, B., Tsotras, V.J. (eds.) SSTD 2001. LNCS, vol. 2121, pp. 3–19. Springer, Heidelberg (2001)
7. Trajcevski, G., Wolfson, O., Hinrichs, K., Chamberlain, S.: Managing Uncertainty in Moving Objects Databases. ACM Trans. on Database Systems (TODS) 29, 463–507 (2004)
8. Pfoser, D., Jensen, C.S.: Capturing the uncertainty of moving-object representations. In: Güting, R.H., Papadias, D., Lochovsky, F.H. (eds.) SSD 1999. LNCS, vol. 1651, p. 111. Springer, Heidelberg (1999)
9. Egenhofer, M.J.: Approximations of geospatial lifelines. In: SpadaGIS, Workshop on Spatial Data and Geographic Information Systsems (2003)
10. Miller, H.J.: A measurement theory for time geography. Geographical Analysis 37, 17–45 (2005)
11. Kuijpers, B., Othman, W.: Trajectory databases: Data models, uncertainty and complete query languages. J. Comput. Syst. Sci. 76(7), 538–560 (2010)
12. Hornsby, K., Egenhofer, M.J.: Modeling moving objects over multiple granularities. Annals of Mathematics and Artificial Intelligence 36(1-2), 177–194 (2002)
13. Grimson, R., Kuijpers, B., Othman, W.: An analytic solution to the alibi query in the space-time prisms model for moving object data. International Journal of Geographical Information Science (2009)
14. Trajcevski, G., Choudhary, A., Wolfson, O., Ye, L., Li, G.: Uncertain range queries for necklaces. In: Proceedings of the 2010 Eleventh International Conference on Mobile Data Management, MDM 2010, pp. 199–208 (2010)

A Novel Hash-Based Streaming Scheme for Energy Efficient Full-Text Search in Wireless Data Broadcast[*]

Kai Yang[1], Yan Shi[1] Weili Wu[1], Xiaofeng Gao[2], and Jiaofei Zhong[1]

[1] The University of Texas at Dallas, Richardson TX 75080, USA
{kxy080020,yanshi,weiliwu,fayzhong}@utdallas.edu
[2] Georgia Gwinnett College, Lawrenceville, GA 30043
xgao@ggc.edu

Abstract. *Full-Text Search* is one of the most important and popular query types in document retrieval systems. With the development of The Fourth Generation Wireless Network (4G), *wireless data broadcast* has gained a lot of interest because of its scalability, flexibility, and energy efficiencies for wireless mobile computing. How to apply full-text search to documents transmitted through wireless communications is thus a research topic of interest. In this paper, we propose a novel data streaming scheme (named *Basic-Hash*) with hash-based indexing and inverted list techniques to facilitate energy and latency efficient full-text search in wireless data broadcast. We are the first work utilizing hash technology for this problem, which takes much less access latency and tuning time comparing to the previous literature. We further extend the proposed scheme by merging the hashed word indices in order to reduce the total access latency (named *Merged-Hash*). An information retrieval protocol is developed to cope with these two schemes. The performances of *Basic-Hash* and *Merged-Hash* are examined both theoretically and empirically. Simulation results prove their efficiencies with respect to both energy consumption and access latency.

1 Introduction

Full-text search is a popular query type that is widely used in document retrieval systems. Many commercial database systems have included full-text search as their features. For example, SQL Server 2008 provides full-text queries against character-based data in SQL Server tables [1]. Oracle Text [3] also gives powerful support for full-text search applications. Many full-text search techniques in different application areas have been proposed in literatures [2][5][6][11].

With the rapid development of mobile devices and quick rise of The Fourth Generation Wireless Network (4G), mobile communication has gained popularity due to its flexibility and convenience. Limited bandwidth in wireless communication and limited energy supply for mobile devices are the two major concerns of mobile computing. That is why *wireless data broadcast* becomes an attractive

---

[*] This work is supported by NSF grant CCF-0829993 and CCF-0514796.

J.X. Yu, M.H. Kim, and R. Unland (Eds.): DASFAA 2011, Part I, LNCS 6587, pp. 372–388, 2011.
© Springer-Verlag Berlin Heidelberg 2011

data dissemination technique for mobile communication. In a wireless data broadcast system, Base Stations (BS) broadcast public information to all mobile devices within their transmission range through broadcast channels. Mobile clients listen to the channels and retrieve information of their interest directly when they arrive. This scheme is bandwidth efficient because it utilizes most of the bandwidth as downlink and requires little uplink traffics. It is also energy efficient because receiving data costs much less energy than sending data.

Mobile devices usually have two modes: *active mode* and *doze mode*. In active mode, a device can listen, compare, and download the required data; while in doze mode, it turns off antennas and many processes to save energy. The energy consumed in active mode can be 100 times of that in doze mode [15]. In general, there are two major performance criteria for a wireless data broadcast system: *access latency* and *tuning time*. Access latency refers to the time interval between a client first tunes in the broadcast channel and it finally retrieves the data of interest, which reflects the system's time efficiency; tuning time is the total time a client remains in active mode, which indicates the system's energy efficiency.

How to apply full-text search in wireless data broadcast is an interesting but challenging topic. Since data broadcast is especially suitable for public information such as news report and traffic information, full-text search can be a very useful feature desired by mobile clients. For example, a mobile user may want to browse all news related to "FIFA", or all local traffic information that includes "accidents". Full-text search for traditional disk-storage data has been well studies [9][12][4]. However, in wireless data broadcast, the data are stored "on the air" rather than on the disk, which posts new challenges to full-text search. In disk-based storage, documents are stored in physical space, so clients can "jump" among different storage slots with little cost; while in on-air storage, documents are stored sequentially along the time line, which posts much more cost for clients to search back and forth. Traditional full-text search techniques cannot not be adopted directly because of this difference. On the other hand, since existing index techniques for wireless data broadcast [10][17][8][19][20] are mainly based on predefined structured data with key attributes, they also cannot be directly applied for full-text search which uses arbitrary words as search keys. Therefore, new design of indexing schemes are needed to facilitate full-text search in order to ensure both time efficiency and energy efficiency.

To the best of our knowledge, [7] is the only published research on full-text query processing in wireless data broadcast. They firstly utilized *inverted list* in processing full-text queries on a wireless broadcast stream, and then proposed two methods: *Inverted-List* and *Inverted-List + Index-Tree* which was extended to $(1, \alpha)$ and $(1, \alpha(1, \beta))$. They made use of an inverted list to guide full-text search and a tree-based index to locate the key word in the inverted list. However, this method is not energy efficient enough because it might take a long tuning time to locate the key word in the inverted list. It is also not latency efficient enough due to the duplication of tree-based index.

Inverted list is a mature indexing method for full-text search [18][22][14]. It is a set of word indices which guides clients to find specific documents containing a specific word. In this paper, we apply inverted list as a guide for full-text search, but implement hash function instead of searching tree as indexing method, to avoid lengthening broadcast cycle and redundant tuning time for locating target word index. Note that hash function is used to index "word indices" in an inverted list, which is the "index of indices". So the index designed in this paper is a hierarchical index scheme with two levels: (1) inverted list, the index for documents, and (2) hash function, the index for word indices in an inverted list.

Compared with tree-based indexing technique, hash-based indexing for word indices is more flexible and space efficient for full-text search in wireless data broadcast. A hash function only takes several bytes while a searching tree may take thousands of bytes depending on its design. Hash-based index is more suitable for full-text search because the nature of full-text search uses arbitrary words as search keys. Based on this idea, we propose a novel data streaming scheme named *Basic-Hash* to allocate inverted list and documents on the broadcast channel. *Basic-Hash* is further improved to another streaming scheme named *Merged-Hash*, by merging the hashed word indices to reduce access latency. A client retrieval protocol is also developed corresponding to the two schemes. We are the first work utilizing hash technology for full-text search in wireless data broadcast. We also provide detailed theoretical analysis to evaluate the performance of *Basic-Hash* and *Merged-Hash*, and then implement many numerical experiments. Simulation results prove the efficiency of these two schemes with respect to both energy and access latency.

To summarize, our main contributions include:

1. We are the first work implement inverted list and hash function for full-text search in wireless data broadcast. We propose two novel wireless broadcast streaming schemes, namely, *Basic-Hash* and *Merged-Hash*, to facilitate full-text query on broadcast documents. For each scheme, we develop algorithms for inverted index allocation, document allocation and query protocol.
2. We discuss how to turn collision issues of hash functions into advantage and utilize appropriate collisions to reduce the access latency of full-text query.
3. We analyze the performances of *Basic-Hash* and *Merged-Hash* theoretically by computing the expected access latency and tuning time for full-text queres on broadcast streams created based on these two schemes.
4. We implement simulations for the proposed systems and analyze their performances by simulation results.

The rest of the paper is organized as follows: Sec. 2 presents related works on wireless data broadcast, full-text search involving inverted list techniques, and recent research on full-text search for wireless data broadcast systems; Sec. 3 introduces the system model and some preliminaries; Sec. 4 first discusses the *Basic-Hash* broadcast streaming scheme to facilitate full-text search and then extends *Basic-Hash* to *Merged-Hash* to improve the performance; Sec. 5 theoretically analyzes the performances of *Basic-Hash* and *Merged-Hash*; Sec. 6 empirically analyzes *Basic-Hash* and *Merged-Hash* based on simulation results; and Sec. 7 concludes the paper and proposes future research directions.

2   Related Work

Wireless data broadcast has gained many attentions during the past few years. Imielinski et al. first gave an overview on wireless data broadcast systems in [10]. They also proposed a popular B^+-tree based distributed index to achieve energy efficiency. Many different index methods were proposed thereafter. Yao et al. [17] proposed an exponential index which has a linear but distributed structure to enhance error-resilience. In [21], the trade-off between confidentiality and performance of signature-based index was discussed. Hash-based index for wireless data broadcast was also proposed in [16]. All these index techniques, however, focus only on structured data with predefined key attributes. They cannot be applied directly to guide full-text queries.

Inverted list is a popular structure in document retrieval systems and a well-known technique for full-text search. Tomasic et al. [14] studied the incremental updates of inverted lists by dual-structure index. Scholer et al. [13] discussed the compression of inverted lists of document postings which contains the position and frequency of indexed terms and developed two approaches to improve the document retrieval efficiency. Zobel et al. gave a survey on inverted files for text search engines in [22]. Zhang et al. [18] studied how to process queries efficiently in distributed web search engines with optimized inverted list assignment. Most research works on inverted list are based on disk-storage documents. For on-air documents, modifications are needed to adjust to on-air storage features.

Chung et al. [7] firstly applied inverted list for full-text search in wireless data broadcast. They also combined tree-based indexing technique with inverted list for full-text query on broadcast documents. However, the construction of a searching tree and the duplication of inverted list will extend the total length of a broadcast cycle heavily, resulting additional access latency. Moreover, the average search time for a searching tree heavily relies on the depth of the tree, which is much more than the searching time of a hash function. Therefore, we replace the search tree with hash function design and construct a more efficient data streaming scheme for full-text search in wireless data broadcast.

3   Preliminary and System Model

3.1   System Model

For simplification, we only discuss the situation for one Base Station (BS) with one communication channel. The broadcast program will not update during a period of time. The BS will broadcast several documents periodically in cycle. Each document only repeats once in a broadcast cycle. Let D denote a set of t documents to be broadcast. $D = \{doc_0, doc_1, \cdots, doc_{t-1}\}$. Each doc_i will be broadcast as several buckets on a channel, each with different size. Here bucket is the smallest logical unit on a broadcast channel. Assume y_i is the size of doc_i, measured by buckets, and $Y = \{y_0, y_1, \cdots, y_{t-1}\}$.

There are altogether v non-duplicated words in D, denoted as $K = \{k_0, k_1, \cdots, k_{v-1}\}$. Let w be the length of each word measured in bytes (here we assume on average, each word has the same length).

Besides documents, we also need to insert indices to form a full broadcast cycle. As mentioned in Sec. 1, we will apply inverted list and hash function together as a searching method. A hash function will be appended to each of the bucket, while the inverted list will be split into word indices and interleave with document buckets. After all the process, we will form a whole broadcast cycle, consisting of a sequence of broadcast buckets. Each bucket will have a continuous sequence number starting from 0. Let $bcycle$ denote this bucket sequence, and $|bcycle|$ denote the whole number of buckets in one broadcast cycle.

3.2 Inverted List

To facilitate full-text search in wireless data broadcast systems, we apply inverted list technique. For full-text search, each word can be related to several documents and each document contains usually more than one word. To resolve such many-to-many relationship between documents and words, inverted list has been popularly used as an index in data retrieval systems [18][22][14].

Fig. 1. Example of an Inverted List for Broadcast System

Let I be an inverted list composed of v entries representing v non-duplicated words in D. In each entry, a word k_i is linked with a set of document address pointers which can guide clients to find documents containing this word. We name each entry a *word index*, denoted as e_i. The number of pointers in each word index depends on how many documents contain this word and is not necessarily the same. Fig. 1 is an example of an inverted list generated from 6 documents. Each pointer indicates the time offset from the index to the target document. Clients can tune off during this offset, and tune on again to save some energy.

Instead of treating an inverted list as a whole, we split it into a set of word indices, each of which is a pair of a word and a list of offsets that points to documents containing the word, that is, $I = \{e_i, i = 0, \cdots, v - 1\}$ where $e_i =< k_i : doc\_addr\_offset\_list_i >$.

For each word index $e_i \in I$, s_i denotes its number of document address offsets. Assume a is the length of one document address pointer measured in bytes, then the length of e_i can be easily computed as $w + as_i$.

3.3 Hash Function and Collisions

There are many hash functions that can hash strings into integers. In this paper, only hashing a word into an integer is not enough. We need to hash a word into a bucket on a broadcast cycle. If $|bcycle|$ denotes the length of a broadcast cycle, we should map the integer result to a sequence number between 0 and $|bcycle|$-1.

Collisions can be quantified by *collision rate* γ. In this paper, a collision happens when a word index is hashed to the same bucket as some previously hashed word indices were. In this case, the collision rate $\gamma = $ total number of collisions/v. In most hashing applications, collisions are what designers try to avoid because it increases the average cost of lookup operations. However, it is inevitable whenever mapping a large set of data into a relatively small range. While most hash function applications struggle for collision issues, our method is much more collision-tolerant. In fact, appropriate collisions are even beneficial. This is because in our method, hash scheme is only used to allocate the word indices in the inverted list and usually the sizes of most word indices are not exactly the size of a bucket. Therefore, more than one word indices hashed into the same bucket may help increasing the bucket utilization factor, which will reduce the length of a broadcast cycle and the average access latency of query processing. We will do a more detailed discussion on the collision issue in Subsec. 4.1.

3.4   Data Structure of a Bucket

A bucket is composed of two parts: *header* and *payload*. Header records basic information of a bucket such as bucket id and sequence number, while payload is the part of a bucket to store data. In our model, there are two different types of buckets: *index bucket* which stores word indices, and *document bucket* which stores documents. Index and document buckets have the same length and header structure. Hence, we can use the number of buckets to measure both Y and $|bcycle|$. If the total number of index buckets within one broadcast cycle is $|IB|$ and the total number of document buckets is $|DB|$, then we should have:

$$|bcycle| = |DB| + |IB| . \tag{1}$$

Next, we will illustrate the detailed design of index and document buckets. Assume a bucket is capable of carrying l bytes information. For both index and document buckets, the header contains the following information in Tab. 1.

Table 1. Information in a Bucket Header

Item	Description
TYPE:	whether the bucket is an index or document bucket.
LEFT:	length of unused space, measured in bytes.
END:	whether the index or document ends.
MERGEP:	time offset of merged index bucket.
SQ:	sequence number of a bucket. In this method, SQ is also the hash value.
OFFSET:	distance to the next bucket containing the same document.
HASH:	hash function to compute sequence number of target index bucket.

For a document bucket, the payload is part of a document or a complete document, depending on the document size. For an index bucket, the payload may contain a part of a long word index or several short word indices. Fig. 2 illustrates the whole view of a broadcast cycle and details of index and document buckets. Here grey block **D** denotes document buckets, while white block **I** denotes an index bucket. Each document doc_i has y_i document buckets, but they may be separated by some index bucket, not necessarily consecutively broadcasted.

Table 2. Symbol Description

Sym	Description	Sym	Description
D	document set. $D = \{doc_0, \cdots, doc_{t-1}\}$	a	document pointer size.
Y	document length. $Y = \{y_0, \cdots, y_{t-1}\}$	l	bucket size.
K	word set. $K = \{k_0, \cdots, k_{v-1}\}$	t	number of documents.
I	an inverted list. $I = \{e_0, \cdots, e_{v-1}\}$	v	number of keywords.
S	$S = \{s_0, \cdots, s_{v-1}\}$, where s_i is the	w	word size.
	number of document pointers in e_i.	γ	collision rate.
$Avg(S)$	average number of document	δ	merge rate.
	pointers in each word index.	$bcycle$	one broadcast cycle.
$\|IB\|$	number of index buckets.	$\|bcycle\|$	length of $bcycle$.
$\|DB\|$	number of document buckets.	$\|ibcycle\|$	initial broadcast cycle length.
$\|MIB\|$	number of index buckets after merging.	$Avg(AL)$	average access latency.
$\|AKE\|$	average length of the keyword entry.	$Avg(TT)$	average tuning time.

For convenience, Tab. 2 lists all symbols used in this paper. Some will be defined in the following sections.

4   Hash-Based Full-Text Search Methods

In this section, we will introduce the construction of two data streaming schemes for full-text search, which are *Basic-Hash* and *Merged-Hash*. Data streaming scheme is a preprocessing before documents are broadcasted on channels. It will interleave documents and inverted list as a whole data stream, allocate index buckets and data buckets according to the predefined hash function, and then setup corresponding address pointer and other information for clients to search words of interest and retrieve target documents.

In the following subsections, we will discuss the construction of Basic-Hash and Merged-Hash, with detailed algorithm description, examples, and scenario discussion. Finally, we propose an information retrieval protocol for mobile/wireless clients to retrieve their interest documents.

4.1   Basic-Hash Data Streaming Scheme

Full-text query processing can be achieved by adding the inverted list onto the broadcast channel. If we put the inverted list directly in front of the documents, the average tuning time can be dramatically long because the client needs to go through the inverted list one by one to find the word index of interest. This tuning time overhead can be reduced by a two-level index scheme which adds another level of index for the inverted list.

In this subsection, we propose a novel two-level index scheme for full-text search called *Basic-Hash* method. The idea is to hash all word indices in the inverted list onto the broadcast channel. The documents and word indices are interleaved with each other. We choose hashing rather than tree-based indexing as the index for the inverted list because it is faster and doesn't occupy much space. Once a client tunes in the broadcast channel, it reads the hash function in the header of a bucket and computes the offset to the target word index immediately. The tuning time to reach the word index of interest is only 2 in the ideal case. Hash function also takes little space from the header of each bucket, without occupying extra index buckets.

It takes three steps to construct a *bcycle* using *Basic-Hash* broadcast scheme:

Step 1: *Index Allocation.* Hash all word indices onto broadcast channel (Alg. 1);
Step 2: *Document Allocation.* Fill empty buckets with documents (Alg. 2).
Step 3: *Pointer setup.* Set offset information for pointers in word indices and document/index buckets.

Hash Function. Before index allocation and document allocation, we need to construct our hash function. Recall that there are t documents to broadcast, each with y_i buckets. I is split into v entries, each with a document address offset list of size s_i. Each word has w bytes, each document address pointer is a bytes, and each bucket contains l bytes (we ignore the length of the header). Initially, we do not know $|bcycle|$, so we will use an estimated $|ibcycle|$ to represent the length of a broadcast cycle. It is easy to know, $|ibcycle| = \sum_{i=0}^{t-1} y_i + \frac{1}{l}\{a \sum_{j=0}^{v-1} s_j + v \times w\}$. Then the hash function should be:

$$Hash(string) = hashCode(string) \mod |ibcycle|.$$

In the above equation, the input string will be a word (string), while the output is an integer between 0 and $|ibcycle| - 1$. This function maps a word index to a broadcast bucket with sequence number equal to hashed value.

Index Allocation. Algorithm 1 describes **Step 1** of Basic-Hash: hashing all word indices onto broadcast channel. Let **A** denote the bucket sequence array. Initially, **A** contains $|ibcycle|$ empty bucket with consecutive sequence number starting from 0. We use $A[0], \cdots, A[|ibcycle| - 1]$ to represent each bucket. The main idea of Alg. 1 is: firstly, sort I by s_i, $i = 0, \cdots, v - 1$ in increasing order as $I' = \{e'_0, \cdots, e'_{v-1}\}$. Next, hash each e'_i onto the channel in order.

The sorting process guarantees that during allocation, if more than one word index are hashed to the same bucket, the shorter word index will be assigned to the bucket first and longer word indices will be appended thereafter, which helps reducing the average tuning time for a client to find the word index of interest.

Since each bucket has l bytes capability, it may include more than one word index. Once encountering a collision, we will append e'_i right after the existing index. If this bucket is full, then find the next available bucket right forward and insert e'_i. It is also possible that there is no enough space for e'_i. In this situation, we will push other word indices in buckets with higher sequence number, and "insert" e'_i. An example is shown in Fig. 3.

Fig. 3 illuminates several scenarios for index allocation. In (a), word index e_5 should be insert into bucket $A[9]$. Since $A[9]$ is empty, we directly insert e_5 into it. If the size of e_5 is larger than $A[9]$, the rest part will be appended to $A[10]$ or more buckets. In (b), e_2, e_3, and e_5 have been allocated already, and we are going to insert e_7 to $A[7]$. Since e_2 is already allocated at $A[7]$, and there is still enough space left in $A[7]$, we append e_7 after e_2. In (c), e_1 should be inserted

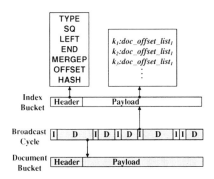

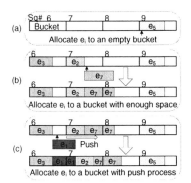

Fig. 2. Whole View of *bcycle*   **Fig. 3.** Index Allocation Scenario Analysis

---

Algorithm 1. Index Allocation For *Basic-Hash*

---

   input: I, $\mathbf{A}$;
   output: $\mathbf{A}$ filled with word indices;
1: sort I by s_i increasingly to $I' = \{e'_0, \cdots, e'_v\}$;
2: **for** $i = 0$ to $v - 1$ **do**
3:    $sq = Hash(k'_i)$;
4:    check whether $A[sq]$ is full, if yes, set $sq = sq + 1$ until $A[sq]$ is not full;
5:    insert e'_i into $A[sq]$, if there is not enough space, then push data from $A[sq + 1]$
     forward until e'_i can be successfully inserted.
6: **end for**

---

into $A[6]$. e_2, e_3, e_5, and e_7 has already been allocated onto the channel before inserting e_1. Note that $A[6]$ does not have enough space for e_5. Thus from $A[7]$, all the data entries should be moved forward until e_1 can fit into the channel. Since there is some unfilled space between e_7 and e_5, e_5 will not be influenced. Detailed description of index allocation is illustrated in Alg. 1.

Data Allocation. Alg. 2 discusses how to allocate documents after index allocation process. Since each bucket can be either an index bucket or a document bucket, we cannot append documents to these buckets which already contain indices. Therefore, starting from $A[0]$, we will scan each bucket in order, and insert documents from doc_0 to doc_{t-1} sequentially to the empty buckets. Each doc_i will take y_i buckets. We use doc_i^j to denote the jth bucket for doc_i in short.

Pointer Setup. Besides index allocation and data allocation, we need to setup the offset (address) information for pointers inside each word index, as well as **OFFSET** in each bucket header (since each document doc_i will be split into y_i buckets, and may not be consecutively allocated, we need another pointer to figure out this information). Pointers for word indices in buckets and **OFFSET** in headers can only be setup after the index and data allocation, because we did not know the locations of documents before that. To fill **OFFSET** in headers, we can scan reversely from the last bucket of the *bcycle*, record the sequence number of each doc_i^j, and then fill the offset information.

Algorithm 2. Document Allocation for *Basic-Hash*

---

    input: D, $\mathbf{A}$;
    output: a complete broadcast stream $\mathbf{A}$;
1: $sq = 0$, $j = 0$;
2: **for** $i = 0$ to $t - 1$ **do**   ▷ *Insert doc_i onto channel*
3:     **while** $j < y_i$ **do**
4:         **if** $A[sq]$ is empty **then** append doc_i^j to $A[sq]$; $j = j + 1$; $sq = sq + 1$;
5:         **else** $sq = sq + 1$; **end if**
6:     **end while**
7:     $j = 0$;   ▷ *Reset intermediate variable*
8: **end for**

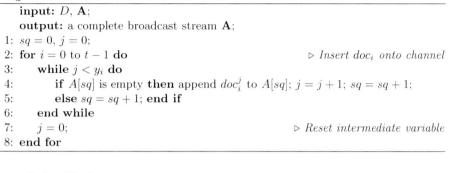

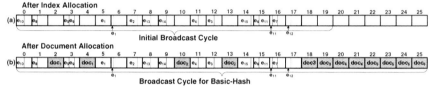

Fig. 4. Index and Document Allocation of Hash-Based Method

An Example. We apply *Basic-Hash* to the document set described in Fig. 1 as an example to demonstrate the complete streaming procedure. Assume $w = a = 4$bytes. We can then compute the length of each word index. Assume $l = 24$bytes, each of doc_1, doc_2, doc_3 and doc_4 takes 2 buckets, doc_5 takes 3 bucket, and doc_6 takes 1 bucket. Based on the above information, we can compute $|ibcycle| = 20$. Following Alg. 1, we first allocate all word indices onto the channel. Then, we use Alg. 2 to allocate 6 documents. The broadcast cycle after index and document allocations is presented in Fig. 4 respectively. Finally, after $bcycle$ is constructed, we need to fill the necessary header and pointer information to each bucket.

4.2 Merged-Hash Data Streaming Scheme

Basic-Hash method can dramatically shorten the average tuning time of the search process than the *Inverted List* method in [7]. The average access latency, however, is much longer. The reason is that in *Inverted List* method, all the word indices are combined together, and inserted into consecutive buckets. On the other hand, *Basic-Hash* method separates them and maps each index respectively into disconnect buckets, which makes $|bcycle|$ much longer. For example, if the document number is 1000, total number of words is 500, the inverted list may only occupy 100 bucket; while a non-conflict hash function maps words in different buckets from each other, which occupies 500 buckets. The $|bcycle|$ expands from 1100 to 1500, and the average access latency is thus influenced.

Merged-Hash Algorithm. *Merged-Hash* aims at reducing average access latency by reducing the number of index buckets. Compared with *Basic-Hash*, *Merged-Hash* has one more step: *Merge Word Index*. It will be performed between

Alg. 1 and 2. The purpose is to combine adjacent index fragments into one bucket to make full use of bucket space. Merge process can reduce $|bcycle|$ without increasing average tuning time.

The idea of *Merge Word Index* algorithm is: starting from the last index bucket $A[i]$, if its closest previous index $A[j]$ can be merged into $A[i]$, then append $A[j]$ to $A[i]$, and delete $A[j]$. Repeat this process until either $A[i]$ is full or its closest previous index $A[j]$ is full or cannot be merged into $A[i]$. Next, find another $A[i]$ and repeat the above process, until all index buckets have been scanned. The detailed description is showed in Alg. 3.

Algorithm 3. Merge Word Index

---
    input: A;
    output: A with merged word indices;
1: find the last non-full index $A[i]$, set $M = A[i]$;
2: **while** M is not full **do**
3:     find its closest previous index $A[j]$;
4:     **if** $A[j]$ is not full and M has enough space for $A[j]$ **then**
5:         append $A[j]$ to M, delete $A[j]$;
6:     **else** $M = A[j]$, Break; **end if**
7: **end while**
8: repeat Line 2 to Line 7 until all index buckets have been scanned.

An Example. We also use an example shown in Fig. 5(a) to illustrate Alg. 3. Merge process begins at bucket 17 with M moving backwards. We can see that word indices in bucket 16 is merged in bucket 17 because there is enough space for them to append. And we also observe that bucket 15 does not follow the step of bucket 16, because after bucket 16 is merged to bucket 17, their is not enough space any more to append the whole indices in bucket 15. For each bucket that is merged to another bucket, MERGEP in header indicates the offset between these two buckets for clients to keep track of the index. For instance, MERGEP in bucket 16's header is 1 and MERGEP in bucket 1's header is 2. Note that the merging operation is based on bucket instead of index. Hence, it is possible that an index will be split after merging. In such cases, we also need MERGEP to direct clients. For example, the second part of e_1 in bucket 6 is merged to bucket 7, so it is separated from its first part in bucket 5. With the help of MERGEP, the tuning time to read such an index only increases by 1, while merging operation dramatically reduces the $|bcycle|$.

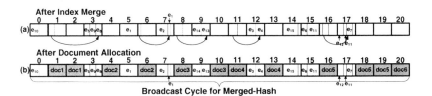

Fig. 5. Index and Document Allocation of Merged-Based Method

After *Basic-Hash*, the broadcast channel looks as Fig. 4(a), with 13 index buckets. If we merge index buckets following Alg. 3, the number of index buckets will decrease to 9, as shown in Fig. 5(a). Comparing Fig. 4(b) and 5(b), we see that $|bcycle|$ after merging is reduced to 20, which is the same as $|ibcycle|$; while $|bcycle|$ without merging is 25, which is 20% longer than $|ibcycle|$.

4.3   Information Retrieval Protocol

After document allocation, the whole *bcycle* is built. Next, we discuss information retrieval protocol. A mobile client will firstly access onto the channel, read the current bucket and get hash function. Next, it computes a sequence number hashed from target word w, and waits until this bucket appears. Then it will follow the direction of bucket pointers to find the word index containing the information of w, and read every offset inside the *doc_offset_list*. Finally, it waits according to these time offsets and download the requested documents one by one. The detailed description of this algorithm is illustrated in Alg. 4.

Algorithm 4. Information Retrieval Protocol

---
    input: keyword w;
    output: a set of documents containing w;
1: read current bucket cb, get hash function $hash(\cdot)$; compute $sq = hash(w)$;
2: **if** $A[sq]$ is not the current bucket **then** wait for the $A[sq]$; **end if**
3: read $A[sq]$, follow index pointer to find e_i with w.
4: read all the addresses of document containing w;
5: **for** each document offset **do** wait and download the document buckets; **end for**

---

5   Performance Analysis

In this section, we will give theoretical analysis for both *Basic-Hash* and *Merged-Hash* with respect to the average access latency and tuning time.

5.1   Analysis for Basic-Hash

For an inverted list, the average number of documents linked to a word is $Avg(S) = \sum_{i=0}^{v-1} s_i/v$. The average length of a word index $|AKE| = w + aAvg(S)$.

Theorem 1. *The average access latency of* Basic-Hash *is*

$$\left(\frac{1}{2} + \frac{Avg(S)}{Avg(S)+1}\right)\left(\sum_{i=0}^{t-1} y_i + \left\lceil \frac{|AKE|}{(1-\gamma)l} \right\rceil (1-\gamma)v\right). \tag{2}$$

Proof. In *Basic-Hash*, average access time (Avg(AL)) is the sum of *probe wait* and *bcast wait*, where *probe wait* denotes the latency of finding target word index bucket and *bcast wait* is the time needed to download all documents containing the requested word. If documents are uniformly distributed on the channel,

$$Avg(AL) = \frac{|bcycle|}{2} + \frac{Avg(S)|bcycle|}{Avg(S)+1}. \tag{3}$$

From Eqn. (1), we have

$$|bcycle| = \sum_{i=0}^{t-1} y_i + \left\lceil \frac{|AKE|}{(1-\gamma)l} \right\rceil v(1-\gamma) \tag{4}$$

Combining Eqn. (3) and (4), we can derive Eqn. (2).

Theorem 2. *The average tuning time of* Basic-Hash *is* $1 + \left\lceil \frac{|AKE|}{(1-\gamma)l} \right\rceil + \frac{Avg(S)}{t} \sum_{i=0}^{t-1} y_i$.

Proof. The average tuning time (Avg(TT)) includes time of 1) initial probing, 2) reading target index bucket and 3) downloading target documents. Initial probing takes time 1. After initial probe, the client computes the hashed value, dozes and tunes in the hashed bucket directly. The time needed to read the target index bucket is $\lceil |AKE|/((1-\gamma)l) \rceil$. On average, there are $Avg(S)$ documents containing a word, so the time needed to download these documents is $Avg(S) \sum_{i=0}^{t-1} y_i/t$. Summing the above three parts, we can get the conclusion.

5.2   Analysis for Merged-Hash

For *Merged-Hash* scheme, index buckets are merged according to Alg. 3 after word indices are hashed to the channel. We define $|MIB|$ to represent the total number of index buckets after merging, and merge rate $\delta = |MIB|/|IB|$ to indicate the effect of merging. δ is bounded between $[\lceil 1/|AKE| \rceil, 1]$.

Theorem 3. *The access latency of* Merged-Hash *is*

$$\left(\frac{1}{2} + \frac{Avg(S)}{Avg(S)+1} \right) \left(\sum_{i=0}^{t-1} y_i + \left\lceil \frac{|AKE|}{(1-\gamma)l} \right\rceil (1-\gamma)v\delta \right). \tag{5}$$

Proof. The access latency of *Merged-Hash* scheme is also computed as Eqn. (3). The difference is how to get $|bcycle|$. In *Merged-Hash*, the length of a *bcycle* is:

$$|bcycle| = |DB| + |MIB| = \sum_{i=0}^{t-1} y_i + \left\lceil \frac{|AKE|}{(1-\gamma)l} \right\rceil (1-\gamma)v\delta.$$

Theorem 4. *The tuning time of* Merged-Hash *is* $2 + \delta \left\lceil \frac{|AKE|}{(1-\gamma)l} \right\rceil + \frac{Avg(S)}{t} \sum_{i=0}^{t-1} y_i$.

Proof. Similar as *Basic-Hash*, the tuning time for *Merged-Hash* is also composed of three parts. If the word index of interest did not merge with any other index, the tuning time is exactly the same as in *Basic-Hash*. If the word index merged with other indices, it means the size of this index is smaller than an index bucket. So it takes 1 unit time to read. Therefore, the average tuning time to read word index is $(1-\delta) + \delta \lceil |AKE|/((1-\gamma)l) \rceil$. Combining with the tuning time for initial probing and document downloading, we can prove Thm. 4.

6   Simulation and Performance Evaluation

In this section, we will evaluate the *Basic-Hash* and *Merged-Hash* methods by simulation results. We also compare *Merged-Hash* method with *Inverted List* and *Inverted List + Tree Index* methods in [7]. The performance metrics used are average access latency (AAL) and average tuning time (ATT).

The simulation is implemented using Java 1.6.0 on an Intel(R) Xeon(R) E5520 computer with 6.00GB memory, with Windows 7 version 6.1 operating system. We simulate a base station with single broadcast channel, broadcasting a database of 10,000 documents with a dictionary of 5,000 distinct words. For each group of experiments, we generate 20,000 clients randomly tuning in the channel and compute the average of their access latency and tuning time.

6.1   Comparison between *Basic-Hash* and *Merged-Hash*

We use two experiments to compare the performance of *Basic-Hash* and *Merged-Hash*. In the first experiment, we vary the size of the word dictionary from 1,000 to 5,000, while the number of documents is fixed to 10,000. This simulates how similar a set of documents are. Documents with more similar topics may have more words in common, which results in a smaller dictionary. The content of each document is randomly generated from the dictionary. The repetitions of a word in a document is uniformly distributed between 1 and 5. The number of non-replicated words contained in a document is set between 1 and 50.

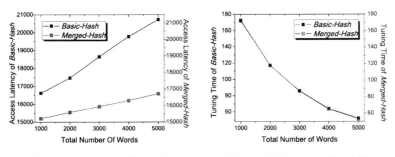

Fig. 6. AAL w.r.t. Word No.   **Fig. 7.** ATT w.r.t. Word No.

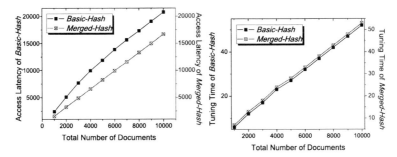

Fig. 8. AAL w.r.t. Document No.   **Fig. 9.** ATT w.r.t. Document No.

Fig. 6 shows average latency of *Basic-Hash* and *Merged-Hash* in this setting. Obviously, *Merged-Hash* has a much shorter access latency than *Basic-Hash*. This verifies our prediction that by merging hashed indices, we can reduce $|bcycle|$ and thus reduce average access latency. In fact, when the dictionary size is 5000, $|bcycle|$ of *Basic-Hash* is 14047 while tit is only 11284 for *Merged-Hash*. When the total number of words increases, the advantage of *Merged-Hash* compared with *Basic-Hash* becomes more obvious.

Fig. 7 presents the average tuning time. We can observe that no matter how the dictionary size changes, the average tuning times of *Basic-Hash* and *Merged-Hash* are very similar with each other. This is because merging word indices do not have much impact on the time for reading a word index of interest.

The second experiment is to evaluate the influence of document set size to the performances of proposed two streaming scheme. We generate D in the same way. Then, randomly choose subsets of D to form eight smaller-sized document set ranging from 1,000 to 9,000. Fig. 8 indicates that *Merged-Hash* performs better than *Basic-Hash* with respect to average access latency no matter how large the document set size is. The difference between *Merged-Hash* and *Basic-Hash* first increases as the number of documents increases, then almost remains unchanged after the document set size reaches 6000. Similar as the first experiment, whatever document set size is, the difference between average tuning times of these two streaming schemes is negligible.

6.2   Comparison with Other Methods

In [7], the authors proposed two full-text search method: *Inverted List* method (*IL*) and *Inverted List + Tree Index* method (*IL+TI*). For a fair comparison, we set the simulation environment exactly the same as in [7]. We generate 10,000 documents, each of size 1024 bytes. The contents of documents are randomly generated from 4703 distinct words. The bucket size is 1024 bytes. The repetitions of a word in a document is 1 to 5, in a uniform way. The $Avg(S)$ is 51, which is also the same as in [7]. All results are averaged based on 20,000 clients.

Table 3. Comparison of three full-text search methods

	IL	IL+TI	Merged-Hash
average access latency	14901	16323	16312
average tuning time	916	91	54

Tab. 3 compares the average access latency and tuning time of *IL*, *IL+TI* and *Merged-Hash*. Compared with *IL*, both *IL+TI* and *Merged-Hash* can dramatically reduce average tuning time by indexing the inverted list. *Merged-Hash* costs even 40.7% less average tuning time than *IL+TI*. Therefore, *Merged-Hash* is the most energy efficient scheme among three. This verifies our analysis that hashing can speed up the searching within the inverted list and consequently reduce tuning time. The average access latency of *Merged-Hash* are slightly longer (9.5%) than *IL*. The reason is that although hashing itself does not require dedicated index bucket, hashing word indices into different buckets may not make full use of the bucket capacity. Therefore, $|MIB|$ may be larger than the number of buckets needed to fill in a complete inverted list. However, *Merged-Hash* still has very similar average access latency with $IL+TI$.

7 Conclusion

In this paper, we proposed two novel wireless data broadcast streaming schemes: *Basic-Hash* and *Merged-Hash*, which provide a two-level indexing to facilitate the full-text query processing in the wireless data broadcast environment. The proposed methods utilizing hash technique to index the inverted list of document broadcasted, which itself is an index for full-text search. For each scheme, we designed detailed index allocation and document allocation algorithms, together with a corresponding querying processing protocol. The performances of these two schemes were analyzed both theoretically and empirically. Simulation results indicate that *Merged-Hash* is the most energy-efficient streaming scheme among all broadcast schemes for full-text search in existing literatures. In the future, we plan to extend *Merged-Hash* to increase the utilization ratio of index buckets in order to further reduce the access latency and tuning time of full-text query processing. We also plan to explore how to adopt other traditional full-text search methods to the wireless broadcast environment.

References

1. http://msdn.microsoft.com/en-us/library/ms142571.aspx
2. Amer-Yahia, S., Shanmugasundaram, J.: Xml full-text search: challenges and opportunities. In: VLDB 2005 (2005)
3. Asplund, M.: Building full-text search applications with oracle text, http://www.oracle.com/technology/pub/articles/asplund-textsearch.html
4. Atlam, E.S., Ghada, E.M., Fuketa, M., Morita, K., Aoe, J.: A compact memory space of dynamic full-text search using bi-gram index. In: ISCC 2004 (2004)
5. Blair, D.C., Maron, M.E.: An evaluation of retrieval effectiveness for a full-text document-retrieval system. Commun. ACM 28(3), 289–299 (1985)
6. Brown, E.W., Callan, J.P., Croft, W.B.: Fast incremental indexing for full-text information retrieval. In: VLDB 1994, pp. 192–202 (1994)
7. Chung, Y.D., Yoo, S., Kim, M.H.: Energy- and latency-efficient processing of full-text searches on a wireless broadcast stream. IEEE Trans. on Knowl. and Data Eng. 22(2), 207–218 (2010)
8. Chung, Y.C., Lin, L., Lee, C.: Scheduling non-uniform data with expected-time constraint in wireless multi-channel environments. J. Parallel Distrib. Comput. 69(3), 247–260 (2009)
9. Faloutsos, C., Christodoulakis, S.: Signature files: an access method for documents and its analytical performance evaluation. ACM Trans. Inf. Syst. 2(4), 267–288 (1984)
10. Imielinski, T., Viswanathan, S., Badrinath, B.r.: Data on air: Organization and access. IEEE Trans. on Knowl. and Data Eng. 9(3), 353–372 (1997)
11. Kim, M.S., Whang, K.Y., Lee, J.G., Lee, M.J.: Structural optimization of a full-text n-gram index using relational normalization. The VLDB Journal 17(6), 1485–1507 (2008)

12. Moffat, A., Zobel, J.: Self-indexing inverted files for fast text retrieval. ACM Trans. Inf. Syst. 14(4), 349–379 (1996)
13. Scholer, F., Williams, H.E., Yiannis, J., Zobel, J.: Compression of inverted indexes for fast query evaluation. In: SIGIR 2002, pp. 222–229 (2002)
14. Tomasic, A., García-Molina, H., Shoens, K.: Incremental updates of inverted lists for text document retrieval. SIGMOD Rec. 23(2), 289–300 (1994)
15. Viredaz, M.A., Brakmo, L.S., Hamburgen, W.R.: Energy management on handheld devices. Queue 1(7), 44–52 (2003)
16. Xu, J., Lee, W.C., Tang, X., Gao, Q., Li, S.: An error-resilient and tunable distributed indexing scheme for wireless data broadcast. IEEE Trans. on Knowl. and Data Eng. 18(3), 392–404 (2006)
17. Yao, Y., Tang, X., Lim, E.P., Sun, A.: An energy-efficient and access latency optimized indexing scheme for wireless data broadcast. IEEE Trans. on Knowl. and Data Eng. 18(8), 1111–1124 (2006)
18. Zhang, J., Suel, T.: Optimized inverted list assignment in distributed search engine architectures. In: Parallel and Distributed Processing Symposium, International, p. 41 (2007)
19. Zhang, X., Lee, W.C., Mitra, P., Zheng, B.: Processing transitive nearest-neighbor queries in multi-channel access environments. In: EDBT 2008: Proceedings of the 11th International Conference on Extending Database Technology, pp. 452–463 (2008)
20. Zheng, B., Lee, W.C., Lee, K.C., Lee, D.L., Shao, M.: A distributed spatial index for error-prone wireless data broadcast. The VLDB Journal 18(4), 959–986 (2009)
21. Zheng, B., Lee, W.C., Liu, P., Lee, D.L., Ding, X.: Tuning on-air signatures for balancing performance and confidentiality. IEEE Trans. on Knowl. and Data Eng. 21(12), 1783–1797 (2009)
22. Zobel, J., Moffat, A.: Inverted files for text search engines. ACM Comput. Surv. 38(2), 6 (2006)

Efficient Topological OLAP on Information Networks[*]

Qiang Qu, Feida Zhu, Xifeng Yan, Jiawei Han, Philip S. Yu, and Hongyan Li

{quqiang,lihy}@cis.pku.edu.cn, fdzhu@smu.edu.sg,
xyan@cs.ucsb.edu, hanj@cs.uiuc.edu, psyu@cs.uic.edu

Abstract. We propose a framework for efficient OLAP on information networks with a focus on the most interesting kind, the *topological OLAP* (called "*T-OLAP*"), which incurs topological changes in the underlying networks. T-OLAP operations generate new networks from the original ones by rolling up a subset of nodes chosen by certain constraint criteria. The key challenge is to efficiently compute measures for the newly generated networks and handle user queries with varied constraints. Two effective computational techniques, *T-Distributiveness* and *T-Monotonicity* are proposed to achieve efficient query processing and cube materialization. We also provide a T-OLAP query processing framework into which these techniques are weaved. To the best of our knowledge, this is the first work to give a framework study for topological OLAP on information networks. Experimental results demonstrate both the effectiveness and efficiency of our proposed framework.

1 Introduction

Since its introduction, OLAP (On-Line Analytical Processing) [10,2,11] has been a critical and powerful component lying at the core of the data warehouse systems. With the increasing popularity of network data, a compelling question is the following: "*Can we perform efficient OLAP analysis on* information networks?" A positive answer to this question would offer us the capability of interactive, *multi-dimensional* and *multi-level* analysis over tremendous amount of data with complicated network structure.

Example 1 (Academia Social Network Interaction). From an academic publication database like DBLP, it is possible to construct a heterogeneous information network as illustrated in Figure 1. There are four kinds of nodes each representing institutions, individuals, research papers and topics. Edges between individuals and institutions denote affiliation relationship. Edges between two individuals denote their collaboration relationship. A paper is connected to its authors, and also to its research topic. ■

OLAP operations could expose two kinds of knowledge that are hard to discover in the original network.

1. *Integrating knowledge from different parts of the network.* As an example, users could be interested in questions like "who are the leading researchers in the topic of social network?". This knowledge involves integrating information lying in two

---

[*] This work is supported by Natural Science Foundation of China (NSFC) under grant numbers: 60973002 and 60673113.

J.X. Yu, M.H. Kim, and R. Unland (Eds.): DASFAA 2011, Part I, LNCS 6587, pp. 389–403, 2011.

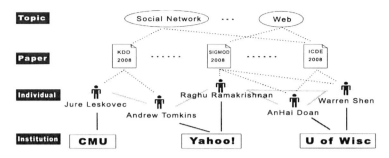

Fig. 1. A Heterogeneous Information Network

parts of the network: (1) the linkages between the individuals and papers, and (2) the linkages between the papers and the topics. As shown in the example, for the nodes representing papers, we can roll-up on them and group them by the same topics, as shown in Figure 2. As the nodes are being merged, the original edges between the papers and individuals would be aggregated accordingly, and the resulting edges would denote the authors' publication prominence in the research of every topic.

2. *Investigating knowledge embedded in different granularity levels of the network.* Besides synchronous drilling in traditional OLAP, many knowledge discovery tasks in information networks may need asynchronous drilling. For example, in Figure 3, users could be interested in the collaborative relationship between the Yahoo! Lab and related individual researchers. For instance, such analysis could show strong collaborations between AnHai Doan and researchers at both Yahoo! Lab by examining the corresponding edges. On the other hand, if the whole Wisconsin database researchers be merged into a single node, it would be hard to discover such knowledge since, collectively, there would be even stronger collaborations between Wisconsin and other universities, which may shadow Doan's link to Yahoo! Lab. Such asynchronous drilling should be guided by what can be potentially found in knowledge discovery, and thus leading to the concept of *discovery-driven OLAP*.

Based on the above motivating example, we propose a new framework for OLAP over information networks. Under this framework, we assume nodes and edges of an information network are associated with multiple hierarchical dimensions. OLAP (such as dicing and drilling) on information network takes a given network as input data and generates new networks as output. This is rather different from traditional OLAP which takes facts in the base cuboid and generates aggregate measures at high-level cuboids.

The second major difference between our OLAP model from the traditional one is the concept of asynchronous, *discovery-driven* OLAP. In the traditional data warehouse systems, drilling is performed synchronously across all the data in a cuboid. However, for OLAP in an information network, such synchronous drilling may fail to expose some subtle yet valuable information for knowledge discovery.

The information network OLAP (i.e., InfoNet OLAP) poses a major research issue: How to efficiently perform InfoNet OLAP? This paper answers this question by proposing two general properties, *T-distributiveness* and *T-monotonicity*, for efficient

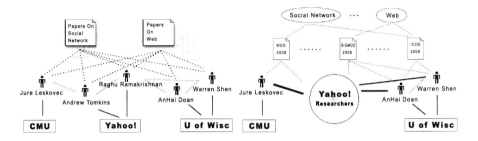

Fig. 2. Roll-up on Papers of the Same Topic

Fig. 3. Asynchronous Roll-up on Researchers to Institutions

computation of different measures in InfoNet OLAP. Our focus of this study is on efficient T-OLAP, the OLAP operations that change the topological structure (such as node merging) of the network. Moreover, we provide algorithms for computing the measures discussed in our categorization. In particular, we also examine the monotonicity property and their impact on efficient query processing. Our experiments on both real and synthetic networks demonstrate the effectiveness and efficiency of the application of our framework.

2   Problem Formulation

We study a general model of *attributed* networks, where both vertices and edges of a network G could be associated with attributes. The attributes, depending on their semantic meanings, could be either of categorical values or numeric ones. We use the DBLP co-authorship network, referred to as "DBLP network" from now on, as a running example for many illustrations in later discussions.

DBLP Network Example. In DBLP co-authorship network, each node v represents an individual author, associated with attributes: *Author Name*, *Affiliated Institution*, and *Number of Papers Published*. Each edge (u, v) between two authors u and v represents their coauthor relationship for a particular conference in a particular year, with attributes like *Conference, Year, Number of Coauthored Papers*. Evidently, there could be multiple edges between two vertices in the DBLP network if two authors have coauthored papers in different conferences. For instance, it could be found between two authors u and v edges like *(ICDE, 2007, 2)* and *(SIGMOD, 2008, 1)* and so on.

A network is *homogeneous* if every edge and vertex represents the same kind of entities and relationships, e.g. the DBLP network. Otherwise, it is *heterogeneous*.

We focus our discussion on homogeneous networks in this paper, and it should be evident that most of the results apply to heterogeneous networks as well. As a convention, the *vertex set* of a network G is denoted by $V(G)$ and the *edge set* by $E(G)$. The size of a network G is defined as the number of edges of G, written as $|E(G)|$. Let $\Sigma_V^i, 1 \leq i \leq m$ and $\Sigma_E^i, 1 \leq i \leq n$ denote the sets of valid attribute values for vertices and edges respectively.

Definition 1. [Attributed Network Model] *An attributed network is a triple* (G, L_V, L_E) *in which* $G = (V(G), E(G))$, $L_V : V(G) \mapsto \Sigma_V^1 \times \Sigma_V^2 \times \ldots \times \Sigma_V^m$ *and* $L_E \subseteq V(G) \times V(G) \times \Sigma_E^1 \times \Sigma_E^2 \times \ldots \times \Sigma_E^n$, *where* m *and* n *are numbers of attribute dimensions for vertices and edges respectively.*

In InfoNet OLAP, the underlying data for a measure is now a network, instead of isolated numerical values, thus measures could in this case take the form of a network. Given an attributed network G and a measure θ, we use $\theta(G)$ to denote the measure θ computed on the network G.

In general, given G and θ, a query in InfoNet OLAP could be represented as "SELECT G' FROM G WHERE $f_C(\theta(G'), \delta) = 1$" in which $G' \subseteq G$ and $f_C()$ is a boolean function taking as input a constraint C, a measure $\theta(G')$ computed on G' and a user-defined threshold δ, such that $f_C(\theta(G'), \delta) = 1$ if and only if $\theta(G'), \delta$ satisfies constraint C. For example, given a network G, suppose the measure θ is "diameter", then a corresponding constraint C could be "$\theta(G') <= \delta$". Then $f_C(\theta(G'), \delta) = 1, G' \subseteq G$ if and only if the diameter of G' is at most δ.

Such queries can be issued for the original data network in which every node can be considered as a data cuboid. However, for T-OLAP on InfoNet, these kind of queries could more likely be issued for some summarized network generated from the original one by merging or rolling up certain subgraphs as illustrated in Figure 2 and 3. For efficient OLAP in traditional data warehouse, data cube computation has been playing an important role with many algorithms developed. However, for InfoNet OLAP, materialization of information network "cubes" may not be realistic due to the huge number of possible flexible "cubes" that have to be precomputed, considering drilling may not even be "synchronized" (*i.e.*, rolling all the network nodes up to the same level) as one may like to perform selective drilling for effective discovery-driven OLAP. On the other hand, it is often the case that we already have some partially materialized cubes as a result of preceding queries on some summarized level. Then the central question is the following: *Can we make use of the partially materialized cubes to more efficiently answer a new coming query? If yes, how?*

3   Techniques and Framework

We propose two constraint-pushing techniques based on the unique characteristics of InfoNet OLAP, *T-Distributiveness* and *T-Monotonicity*. The framework taps the powerful techniques in traditional OLAP on data cube and extends them further into the information network setting. We use a simple motivating example to introduce the two techniques.

DBLP Query Example. Given the DBLP author network, suppose the measure θ of interest is the "total number of publications", *i.e.*, for a given node v, denoted as $\theta(v)$ its total number of publications. Depending on the level of network to which v belongs, v could represent an individual researcher, a research group, or an institution. A user could then submit queries asking to return "all researchers v such that $\theta(v) \geq \delta$".   ∎

The measure in the above example is in fact the "Degree Centrality". We use $C_D(v)$ to denote this measure, Degree Centrality, for a node v. To formally represent the concept of networks at different levels, we need a definition of *OLAP network hierarchy*

Definition 2. *[OLAP Network Hierarchy] Given a network $G(V, E)$ and a partition Π of $V(G)$ such that $\Pi_G = \{V_1, V_2, \ldots, V_m\}$, $m \leq |V(G)|$. A network G' is called a higher-level network of G if G' is obtained by merging each $V_i \in \Pi_G, 1 \leq i \leq m$ into a higher-level node v_i' and the edges accordingly. G is then called a lower-level network of G' and denoted by $G \preceq G'$. For each $v \in V(G)$, v_i' is called the higher-level node of v if $v \in V_i$, which is denoted as $v \preceq_V v_i'$.*

Notice that topological OLAP operations could be asynchronous. A higher-level network can be obtained by merging portions of a lower-level one, leaving the rest unchanged.

3.1   T-Distributiveness

Suppose we have three levels of networks where nodes represent individuals, research groups and institutions in each network respectively. Instead of individuals, users could query about the institutions with the total number of publications beyond a certain threshold δ. The straightforward way is to construct the network G'' at the institution level by merging the constituent author nodes for each institution from the original network G, and compute the measure by summing up over each. For large institutions, the computation could be costly. Now suppose we have already computed the measure for the network G' at the research group level, can we exploit this partial result to improve efficiency? It turns out we can do that in this case due to the distributiveness of this measure function. Basically, the measure value of an institution can be correctly obtained by summing up over the measure values already computed for the research groups. Consider any set of vertices $S = \{v_1, v_2, \ldots, v_k\}$ and a partition Π_S of S such that $\Pi_S = \{S_1, S_2, \ldots, S_m\}$, $m \leq k$. Each $S_i \in \Pi_S$ is merged to a new vertex v_i' and the whole set S is merged to a new vertex v'' by a T-OLAP roll-up operation. We also overload the notation to denote $\Pi_S = \{v_1', v_2', \ldots, v_m'\}$. It is easy to verify that

$$
\begin{aligned}
C_D(v'') &= \left(\sum_{v_i \in S} C_D(v_i) \right) - 2|E_S| \\
&= \sum_{1 \leq i \leq m} \left(\sum_{v_i \in S_i} C_D(v_i) - 2|E_{S_i}| \right) - 2|E_{\Pi_S}| \\
&= \left(\sum_{v_i' \in \Pi_S} C_D(v_i') \right) - 2|E_{\Pi_S}|
\end{aligned}
$$

where E_S is the set of edges with both end vertices in S. It is clear that, since addition and subtraction are commutative, distributive and associative, the result of computing by definition from the bottom-level network is the same as the result of computing from the intermediate-level one. Figure 4 is an illustration of the computation. $C_D(v'')$ is a total of $4 + 2 + 5 + 3 = 14$ from G''. We can get this measure directly from the original network G by the given formula $\sum_{v_i \in S} C_D(v_i) - 2|E_S| = (3 + 8 + 3 + 7 + 10 + 11 + 7 + 5 + 6) - 2(2 + 3 + 3 + 1 + 2 + 4 + 1 + 2 + 4 + 1) = 14$. We can also

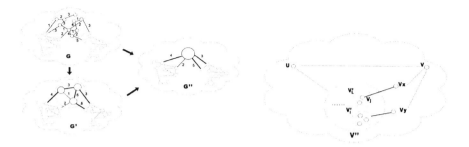

Fig. 4. T-Distributiveness for Degree Centrality   **Fig. 5.** T-Distributiveness for Shortest Path

use partial measure results computed for the intermediate network G' and compute by $\sum_{v_i' \in \Pi_S} C_D(v_i') - 2|E_{\Pi_S}| = (8+12+14) - 2(3+1+6) = 14$. The computational cost is reduced to $O(m + |E_{\Pi_S}|)$. This example shows that the computation cost is greatly reduced by taking advantage of partial measure results already computed. This kind of distributiveness of a measure function is termed *T-Distributiveness* in this topological OLAP setting.

We now give the formal definition of *T-Distributiveness*.

Definition 3. *[T-Distributiveness] Given a measure θ and three attributed networks G, G' and G'' obtained by T-OLAP operations such that $G \preceq G' \preceq G''$, suppose we have available $\theta(G)$ and $\theta(G')$, then θ is* T-Distributive *if there exists a function g such that $\theta(G'') = g(\theta(G')) = g(\theta(G))$.*

Although this example of "Degree Centrality" may seem simple, it is interesting to note that other more complicated measures, even those involving topological structures, are also T-distributive. For instance, it can be shown that the measure of "Shortest Path" is also T-distributive. Shortest path computation is a key problem underlying many centrality measures, such as *Closeness Centrality* and *Betweenness Centrality*, as well as important network measures like *Diameter*.

T-Distributiveness for Shortest Paths. It is well-known that the shortest path problem has the property of optimal substructures. In fact, shortest-path algorithms typically rely on the property that a shortest path between two vertices contains other shortest paths within it. Formally, we have the following lemma, the proof of which is omitted and readers are referred to [6].

Lemma 1. *Given an attributed network G with a weight attribute on edges given by function $w : E(G) \mapsto R$, let $p = \langle v_1, v_2, \ldots, v_k \rangle$ be a shortest path from vertex v_1 to vertex v_k and, for any i and j such that $1 \leq i \leq j \leq k$, let $p_{ij} = \langle v_i, v_{i+1}, \ldots, v_j \rangle$ be the sub-path of p from vertex v_i to vertex v_j. Then, p_{ij} is a shortest path from v_i to v_j.*

Rationale. The significance of the optimal substructure property of the shortest path problem is that it means the measure is T-distributive, thus providing an efficient way to compute the measure for T-OLAP roll-up operations.

We show our algorithm in Algorithm 2. The main algorithm is Algorithm 1 in which we show that, instead of computing from scratch from the lowest network G, we are actually able to compute the measure network $\theta(G'')$ for G'' from the measure network $\theta(G')$ already computed for an intermediate network G'.

In Algorithm 1, in Line 3, we first compute all shortest paths from the single source v'' to all other vertices. From Lines 4 to 7, we update the shortest path between each pair of vertices (u, v) by picking the smaller-weight one between the existing shortest path between them and the one which passes through the new vertex v''. In Algorithm 2, in Lines 1 and 2, we first set the shortest path weight between v'' and other vertices to be a maximum weight value. From lines 3 to 6, we calculate the shortest paths between v'' and every other vertex u by picking the one with the minimum weight among all the shortest paths between vertices in S' and u. It is easy to verify that the time complexity of computational cost of $ShortestPath\_Local$ is $O(|S'| \cdot |V(G) \setminus S|)$. The time complexity of the entire algorithm is therefore $O(|V(G)|^2)$.

The correctness of the entire algorithm can be seen from the observation that for any pair of vertices u and v, if the final shortest path $p_{u,v}$ in G'' does not pass through the new vertex v'', then it should also be the shortest path between u and v in the lower-level network G'. Therefore, the final shortest path $p_{u,v}$ in G'' must be the smaller-weight one between the existing shortest path between them in G' and the new shortest path passing through v''. By the optimal substructure property in Lemma 1, the new shortest path passing through v'' must be the union of the two shortest paths, one between u and v'', and the other between v'' and v. When computing the shortest paths between v'' and other vertices, we do not use standard single source shortest path algorithms. Instead, Algorithm $ShortestPath\_Local$ harness the T-distributiveness of the shortest path measure.

Theorem 1. *Given an attributed network G with edge weights, G'' is obtained by merging a set of vertices $S = \{v_1, v_2, \ldots, v_k\}$, $S \subseteq V(G)$ in a T-OLAP roll-up operation to a new vertex v'', and G' is obtained by partitioning S by $\Pi = \{S_1, S_2, \ldots, S_k\}$ and merging the vertices in each S_i into $v'_i \in S', 1 \le i \le k$, then given the shortest path measure network $\theta(G')$, $ShortestPath\_Local$ computes the shortest paths between v'' and all vertices in $V(G) \setminus S$.*

Proof. The proof is omitted due to the limitation of space.   ∎

Algorithm 1. ShortestPath_Main
Input: S', G and $\theta(G')$
Output: $\theta(G'')$
1: $\theta(G'') \leftarrow \theta(G')$
2: Merge S' into v'' and add v'' to G'';
3: $\theta(G'') \leftarrow ShortestPath\_Local(S', G, \theta(G''))$;
4: **for each** $u \in V(G''), u \ne v''$
5:     **for each** $v \in V(G''), v \ne v''$
6:         **if** $w(p_{uv}) > w(p_{uv'}) + w(p_{v'v})$
7:             $w(p_{uv}) \leftarrow w(p_{uv'}) + w(p_{v'v})$
8:**return** $\theta(G'')$;

Algorithm 2. ShortestPath_Local
Input: S', G and $\theta(G'')$
Output: $\theta(G'')$
1: **for each** $u \in V(G) \setminus S'$
2:     $w(p_{v''u}) \leftarrow +\infty$;
3: **for each** $u \in V(G) \setminus S'$
4:     **for each** $v \in S'$
5:         **if** $w(p_{vu}) < w(p_{v''u})$
6:             $w(p_{v''u}) \leftarrow w(p_{vu})$;
7:**return** $\theta(G'')$;

3.2   T-Monotonicity

Suppose the user queries for all pairs of collaborating researchers with the number of joint publications above a threshold δ. The observation is that the total number of publications of an institution is at least as large as that of any of its constituent individual. This simple monotone property could help prune unnecessary data search space substantially in the query processing: Given the threshold δ, any institution node pairs with its measure value less than δ could be safely dropped without expanding to explore any of its constituent nodes at lower level networks. The monotonicity of a constraint like this is termed *T-Monotonicity* in this topological OLAP setting. The definition of T-Monotonicity is as follows.

Definition 4. *[T-Monotonicity] Given a measure θ and a constraint C, let G and G' be two networks such that $G \preceq G'$, C is T-Monotone if $f_C(P_1) = 1 \rightarrow f_C(P_2) = 1$ for all $P_1 \subseteq G, P_2 \subseteq G'$ and $P_1 \preceq P_2$.* ∎

It is not just simple and common measures like the example above that are T-monotone, in fact, it can be shown that many complicated and important measures which involve network structures are also T-monotone. Interestingly, "Shortest Path" is again a good case in point.

T-Monotonicity for Shortest Paths.   For shortest path, it turns out the corresponding constraints have the property of T-monotonicity. The intuition is that when nodes from a lower-level network are merged to form nodes in a higher-level network, the shortest paths between any pair of nodes in the higher-level network *cannot* be elongated, which is proved as follows.

Theorem 2. *Given two networks G_1 and G_2 such that $G_1 \preceq G_2$, for any two nodes $u, v \in V(G_1)$, let $u', v' \in V(G_2)$ be the corresponding higher-level nodes such that $u \preceq_V u'$ and $v \preceq_V v'$. Then we have $Dist(u', v') \leq Dist(u, v)$.*

Proof. Denote $w(u, v)$ as the weight of edge (u, v). Let one of the shortest paths between u and v in G_1 be $p = \langle v_0, v_1, \ldots, v_k \rangle$ where $v_0 = u$ and $v_k = v$. Since $G_1 \preceq G_2$, there exists some i and d such that vertices $v_i, v_{i+1}, \ldots, v_{i+d}, 0 \leq i, d \leq k$ of p are merged into a single vertex w in G_2. Then the weight of the shortest path between u' and v' in G_2 will have

$$\delta(u', v') = \delta(u', w) + \delta(w, v') \leq \sum_{0 \leq j < i} w(v_j, v_{j+1}) + \sum_{i+d \leq j < k} w(v_j, v_{j+1}) \leq \delta(u, v)$$

We give a summary of some common network measures in Figure 6.

3.3   T-OLAP Query Processing Framework

Both T-distributiveness and T-monotonicity would be pushed into the framework for processing T-OLAP queries. The framework of T-OLAP query processing consists of the following stages:

Pre-computation: Given a network G and the measure θ to be computed, the query algorithm first computes the base cuboids to be materialized.

Constraints	SUM	MIN, MAX	Min Degree, Max Degree	Density	Bridging Capital	Degree Centrality	Closeness Centrality	Betweenness Centrality	Diameter	Structural Cohesion	Containment
T-Monotonicity	Yes	Yes	No	No	No	No	Yes	No	Yes	No	No
T-Distributiveness	Yes	Yes	Yes	No	Yes	Yes	Yes	Yes	Yes	No	No

Fig. 6. A General Picture of Typical InfoNet OLAP Constraints

Query Processing:

1. **Abstraction Level Processing:**
   Given the OLAP abstraction level from the user query, the algorithm locates the most immediate higher-level and lower-level networks whose corresponding cubes have been partially materialized.

2. **Measure Computation:**
   Given the constraint C from the user query, the higher-level network will be used to prune search space by applying T-monotonicity whenever available. Lower-level network will be used for more efficient measure computation for the required abstraction level by applying T-distributiveness whenever available.

4 Experimental Results

4.1 Synthetic Data

All the experiments are conducted on a Pentium(R) 3GHz with 1G RAM running Windows XP professional SP2.

T-Distributiveness. We perform experiments for two measures, *Degree Centrality* and *Closeness Centrality* on synthetic data to demonstrate the power of T-distributiveness.

Since our aim is to provide studies on measures for InfoNet OLAP in general, our synthetic data networks are not confined to specific types and statistical properties. Our synthetic data networks are generated in a random fashion such that (1) the entire network is connected, (2) the vertices have an average degree of $\hat{d}$ and (3) the edges have an average weight of $\hat{w}$.

Given a network G, users can choose a subset S of vertices to roll-up into a single vertex v' and compute the measure network for the new network G'. Such an OLAP operation is called a user OLAP request. We give a model for incoming user OLAP requests as follows. For a network G, we recursively partition G into π connected non-overlapping components of equal number of vertices, until each resulting component is of a predefined minimum number of vertices, i.e., suppose $|V(G)| = 1024$ and $\pi = 4$, we first partition G into 4 connected subgraphs each with 256 vertices, and recursively partition the 4 subgraphs. The partition process identifies a sequence T of connected subgraphs of the original network G. Now we reverse the sequence T and let the resulting sequence be T'. Consequently, observe that, for any subgraph Q in sequence T', all the subgraphs of Q appear before Q. We model the sequence of incoming user OLAP requests as the subgraph sequence T', i.e., the i-th user OLAP request would take the original network G and choose to merge the i-th subgraph in T' into a single vertex and thus obtain a higher-level network G'. The task then is to compute the measure network $\theta(G')$ for G'.

Our baseline algorithm for comparison is denoted as NaiveOLAP. For each user OLAP request, the naive algorithm would first merge the corresponding subgraph into a single vertex and then compute the measure network for the new graph directly from the original network G. Our approach, called T-distributiveOLAP (or TD-OLAP for short), would take advantage of the T-distributiveness of the measure and take the measures already computed for π lower level networks as input to compute the new measure network. In other words, if put in traditional OLAP terminology, we are considering the best scenario here in which, when computing the measure for a cuboid, all the cuboids immediately below have already been materialized.

Degree Centrality. The measure of *Degree Centrality* has the nice property of T-distributiveness. TD-OLAP could therefore make use of the measures computed for the lower-level networks and gain significant efficiency boost than the NaiveOLAP. The average vertex degree is set to $\hat{d} = 5$. The partition size π is set as 4 such that each high level vertex has 4 lower-level children vertices.

Figure 7 shows the running time comparison for the two approaches as the number of vertices for the original network increases. In this case, the original network G is recursively partitioned for a recursion depth of two with a partition size of 4. The running time is the result of summing up the computation cost for all the user OLAP requests in T'. It can be observed that with T-distributiveness the measure network computation cost increases much slower than the NaiveOLAP approach.

Figure 8 shows that, when the total number of vertices of the network G is fixed to 4096 and the average vertex degree is set to 5, how the granularity of T-OLAP operations can affect the running time of both approaches. As the number of partitions increases, the size of the set of vertices to be merged in the T-OLAP roll-up get smaller, which means the user is examining the network with a finer granularity. Since the measure of degree centrality has a small computational cost, both approaches have in this case rather slow increase in the running time. However, notice that the TD-OLAP still features a flatter growth curve compared with the NaiveOLAP approach.

Closeness Centrality. The measure of *Closeness Centrality* has the nice property of T-distributiveness. As such, TD-OLAP would use the algorithms as shown in Algorithm 1 to assemble the measures computed for the lower-level networks and save tremendous computational cost than the NaiveOLAP which simply merge subsets of vertices and run costly shortest path algorithm to compute the new measure network from scratch. In this example, the average degree is set to $\hat{d} = 5$ and the average weight on edges is set as $\hat{w} = 10$. The partition size π is set as 4 such that each high level vertex has 4 lower-level children vertices.

Figure 9 shows the running time comparison for the two approaches as the number of vertices for the original network increases. In this case, the original network G is recursively partitioned for a recursion depth of two with a partition size of 4. The running time is the result of summing up the computation cost for all the 20 user OLAP requests in T'. It is clear that, by harnessing T-distributiveness, the measure networks can be computed much more efficiently, almost in time linear to the size of the original data network, than the naive OLAP approach.

Figure 10 shows how the granularity of the T-OLAP roll-up can impact the running time for both approaches. As the number of partitions increases, the original network is partitioned into components of increasingly smaller sizes. The figure shows the average cost for computing the new measure network for one OLAP request as users choose to merge smaller set of vertices in the T-OLAP operations. The network in this case contains 1024 vertices. As shown in the figure, for **TD-OLAP**, the granularity hardly affects the computational cost since the complexity of the function to combine the measures of lower-level networks to obtain the new one is in general very low compared with the function to compute the measure itself. As the partition size only affects the number of lower-level vertices to taken into consideration, the running time therefore remains steady. On the other hand, as fewer vertices are merged with increasing number of partitions, the **NaiveOLAP** has to compute the measure network with an input network of greater size. Hence the increasing running time for the **NaiveOLAP**.

T-Monotonicity. We perform experiments on the measure of *Shortest Distance* to demonstrate the power of T-monotonicity. The number of nodes is set to 1024. The average node degree is set to 5 and the average weight on edges is set to 5. The T-OLAP scenario is the following. The user would perform T-OLAP operations on the underlying network G in the same fashion as in the experiment settings for T-distributiveness. We obtain a higher-level network G' with π partitions, each becoming a higher-level node. Then the user would present queries in an asynchronous T-OLAP manner as follows. Two partitions (nodes) of G' will be expanded into their constituent lower-level nodes while the rest partitions remain as higher-level nodes, thus generating a new network $\hat{G}_1$ such that $G \preceq \hat{G}_1 \preceq G'$. We can then choose another two partitions of G', proceed likewise and obtain another network $\hat{G}_2$. For a number of partitions π, we can obtain $\binom{\pi}{2}$ networks $\hat{G}_1, \hat{G}_2, \ldots, \hat{G}_{\binom{\pi}{2}}$ by the sequence of asynchronous T-OLAP operations. In the process, the user would query for the shortest distance for every pair of lower-level nodes u and v in $\hat{G}_i$ for $1 \leq i \leq \binom{\pi}{2}$ such that u and v are expanded out of different higher-level nodes, under the constraint that the minimum of all these shortest distances is smaller than a threshold δ. It is easy to see that the naive way would have to compute all-pair shortest distances for each $\hat{G}$ to find the minimum. Due to the T-monotonicity of shortest distance, we can prune data search space as follows. If we pre-compute the shortest distances between every pair of higher-level nodes in G', then if the shortest distance between two nodes u' and v' of G' is greater than δ, then for any pair of nodes u and v expanded out of u' and v' respectively, the shortest distance between u and v in the corresponding network $\hat{G}$ must be greater than δ. Therefore there is no need to expand u' and v' for all-pair shortest distance computation, thus reducing computational cost. Figure 11 shows how much running time we are able to save for a successful pruning by T-monotonicity as the number of partitions π increases. The curve well illustrates the cost saving which is proportional to the size of the OLAP-generated network $\hat{G}$ upon which the naive method would need to compute all-pair shortest distances. It is not monotone since the size of $\hat{G}$ first decreases and then increases as the number of partitions π increases. Figure 12 shows the situation where π is set to be 64 and the average edge weight is 500. User queries in this case also ask to return the shortest distances between all lower-level nodes for all $\hat{G}_i$ but with the constraint that the shortest distance is smaller than a threshold δ. Figure 12 shows

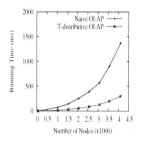

Fig. 7. Run Time Comparison

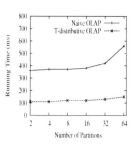

Fig. 8. T-OLAP Granularity

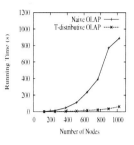

Fig. 9. Run Time Comparison

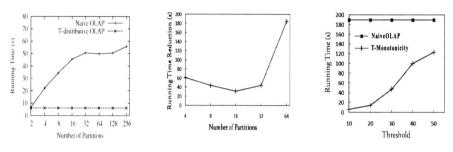

Fig. 10. T-OLAP Granularity

Fig. 11. Cost Reduction

Fig. 12. Run Time Comparison

the running time for the query processing as the user-defined threshold δ increases. It clearly shows that as δ increases, the pruning power weakens since when $\delta \to \infty$, it means all shortest distances need to be returned to the user.

4.2 Real Data

Based on the DBLP data, we can semi-automatically construct a heterogeneous network as illustrated in Figure 13. Edges between different types of entities could carry different attributes. For instance, edges connecting researchers and topics could have the relevant publication on this topic by this author; edges between two researchers could carry the publications co-authored by them; edges between a researcher and an institution could carry those researchers from the institution who have collaborations with this researcher, etc.. Wider edges indicate stronger relationships in terms of greater quantities. By performing discovery-driven, asynchronous T-OLAP operations, users would be able to examine, analyze and discover knowledge in a multi-dimensional and multi-level fashion, uncovering hidden information which is previously hard to be identified in traditional data warehouse scenario. For example, Figure 13 shows a snapshot of the network after a sequence of discovery-driven T-OLAP operations. One can easily observe that while Michael Stonebraker, Jennifer Widom and Rajeev Motwani all work on the topic of "stream data", they also have their own separate heavily-involved research topics of "C-Store", "Uncertainty" and "Web" respectively.

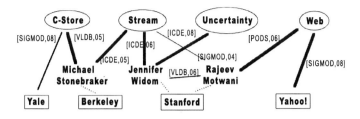

Fig. 13. A Snapshot Of A Portion Of A Real Heterogeneous Network

5   Related Work

Social network analysis, including Web community mining, has attracted much attention in recent years. Abundant literature has been dedicated to the area of social network analysis, ranging from the network property, such as power law distribution [18] and small world phenomenon [15], to the more complex network structural analysis such as [8], evolution analysis [16], and statistical learning and prediction [13]. The static behavior of large graphs and networks has been studied extensively with the derivation of power laws of in- and out- degrees, communities, and small world phenomena. This work is not to study network distribution or modeling but to examine a general analytical process, with which users can easily manipulate and explore massive information networks to uncover interesting patterns, measures, and subnetworks.

OLAP (On-Line Analytical Processing) was studied extensively by researchers in database and data mining communities [10]. Major research themes on OLAP and data cube include efficient computation of data cubes [2], iceberg cubing [7], partial materialization and constraint "pushing" [20].

Although OLAP for the traditional form of spreadsheet data has been extensively studied, there are few studies on OLAP on information networks although information networks have been emerging in many real-world applications. One interesting study that puts graphs in a multi-dimensional and multi-level OLAP framework is in [5]. However, it focuses on informational OLAP in which the rolling/drilling operations only merge multiple edges between the same pair of nodes. As there is no merging of nodes, there is no change in the underlying network structure. As such, [5] only covers a rather limited subset of all the possible OLAP operations on information networks, whereas topological OLAP (T-OLAP), the more powerful ones for knowledge discovery, has not yet been systematically explored.

InfoNet OLAP provides users with the ability to analyze the network data from any particular perspective and granularity. The T-OLAP operation of rolling-up delivers a summarized view of the underlying network. Therefore, from the perspective of generating summarized views of graph data, different aspects of the problem has been examined in one form or another such as compression, summarization, and simplification. [21,3] study the problem of compressing large graphs, especially Web graphs. Yet they only focus on how the Web link information can be efficiently stored and easily manipulated to facilitate computations like PageRank and authority vectors. [4] develops statistical summaries that analyze simple graph characteristics like degree distributions and hop-plots. While these papers studied effective summarization of graph data, they did not aim to give a comprehensive study of multi-dimensional and multi-granularity network analysis with OLAP operations.

Similar aspects have also been explored by the graphics community under the topic of graph simplification. [26,1,17], aim to condense a large network by preserving its skeleton in terms of topological features. Works on graph clustering (to partition similar nodes together), dense subgraph detection (for community discovery, link spam identification, *etc.*) and graph visualization include [19], [9,22], and [12], respectively. The visualization and summarization of cohesive subgraphs has been studied in [24]. These studies provide some kind of summaries, but the objective and results achieved are substantially different from those of this paper.

Summarizing attributed networks with OLAP-style functionalities is studied in [23]. It introduces an operation called SNAP, which merges nodes with identical labels, combines corresponding edges, and aggregates a summary graph that displays relationships for such "generalized" node groups. There have been recent works examining certain particular network measures in great detail such as shortest paths [25] and reachability [14]. However, all these work are not aimed to study measure computation in T-OLAP setting in general and offer common constraint properties for a general query processing framework.

6   Conclusion

In this paper we have performed a framework study for topological InfoNet OLAP. In particular, we propose two techniques in a constraint-pushing framework, *T-Distributiveness* and *T-Monotonicity*, to achieve efficient query processing and cube materialization. We put forward a query processing framework incorporating these two techniques. Our experiments on both real and synthetic data networks have shown the effectiveness and efficiency of the application of our techniques and framework to the measures.

References

1. Archambault, D., Munzner, T., Auber, D.: TopoLayout: Multilevel graph layout by topological features. IEEE Trans. Vis. Comput. Graph. 13(2), 305–317 (2007)
2. Beyer, K.S., Ramakrishnan, R.: Bottom-up computation of sparse and iceberg cubes. In: SIGMOD Conference, pp. 359–370 (1999)
3. Boldi, P., Vigna, S.: The WebGraph framework I: Compression techniques. In: WWW, pp. 595–602 (2004)
4. Chakrabarti, D., Faloutsos, C.: Graph mining: Laws, generators, and algorithms. ACM Comput. Surv. 38(1) (2006)
5. Chen, C., Yan, X., Zhu, F., Han, J., Yu, P.S.: Graph OLAP: Towards online analytical processing on graphs. In: Proc. 2008 Int. Conf. Data Mining (ICDM) (2008)
6. Cormen, T.H., Leiserson, C.E., Rivest, R.L., Stein, C. (eds.): Introduction to Algorithms. MIT Press, Cambridge (2001)
7. Fang, M., Shivakumar, N., Garcia-Molina, H., Motwani, R., Ullman, J.D.: Computing iceberg queries efficiently. In: VLDB, pp. 299–310 (1998)

8. Flake, G., Lawrence, S., Giles, C.L., Coetzee, F.: Self-organization and identification of web communities. IEEE Computer 35, 66–71 (2002)
9. Gibson, D., Kumar, R., Tomkins, A.: Discovering large dense subgraphs in massive graphs. In: VLDB, pp. 721–732 (2005)
10. Gray, J., Chaudhuri, S., Bosworth, A., Layman, A., Reichart, D., Venkatrao, M., Pellow, F., Pirahesh, H.: Data cube: A relational aggregation operator generalizing group-by, cross-tab, and sub totals. Data Min. Knowl. Disc. 1(1), 29–53 (1997)
11. Gupta, A., Mumick, I.S. (eds.): Materialized Views: Techniques, Implementations, and Applications. MIT Press, Cambridge (1999)
12. Herman, I., Melançon, G., Marshall, M.S.: Graph visualization and navigation in information visualization: A survey. IEEE Trans. Vis. Comput. Graph. 6(1), 24–43 (2000)
13. Jensen, D., Neville, J.: Data mining in networks. In: Papers of the Symp. Dynamic Social Network Modeling and Analysis. National Academy Press, Washington DC (2002)
14. Jin, R., Xiang, Y., Ruan, N., Wang, H.: Efficiently answering reachability queries on very large directed graphs. In: SIGMOD 2008: Proceedings of the 2008 ACM SIGMOD International Conference on Management of Data, pp. 595–608. ACM, New York (2008)
15. Kleinberg, J.M., Kumar, R., Raghavan, P., Rajagopalan, S., Tomkins, A.: The web as a graph: Measurements, models, and methods. In: Asano, T., Imai, H., Lee, D.T., Nakano, S.-i., Tokuyama, T. (eds.) COCOON 1999. LNCS, vol. 1627, pp. 1–17. Springer, Heidelberg (1999)
16. Leskovec, J., Kleinberg, J., Faloutsos, C.: Graphs over time: Densification laws, shrinking diameters and possible explanations. In: Proc. 2005 ACM SIGKDD Int. Conf. on Knowledge Discovery and Data Mining (KDD 2005), Chicago, IL, pp. 177–187 (August 2005)
17. Navlakha, S., Rastogi, R., Shrivastava, N.: Graph summarization with bounded error. In: SIGMOD Conference, pp. 419–432 (2008)
18. Newman, M.E.J.: The structure and function of complex networks. SIAM Review 45, 167–256 (2003)
19. Ng, A.Y., Jordan, M.I., Weiss, Y.: On spectral clustering: Analysis and an algorithm. In: NIPS, pp. 849–856 (2001)
20. Ng, R.T., Lakshmanan, L.V.S., Han, J., Pang, A.: Exploratory mining and pruning optimizations of constrained association rules. In: SIGMOD Conference, pp. 13–24 (1998)
21. Raghavan, S., Garcia-Molina, H.: Representing web graphs. In: ICDE, pp. 405–416 (2003)
22. Sun, J., Xie, Y., Zhang, H., Faloutsos, C.: Less is more: Sparse graph mining with compact matrix decomposition. Stat. Anal. Data Min. 1(1), 6–22 (2008)
23. Tian, Y., Hankins, R.A., Patel, J.M.: Efficient aggregation for graph summarization. In: SIGMOD Conference, pp. 567–580 (2008)
24. Wang, N., Parthasarathy, S., Tan, K.-L., Tung, A.K.H.: CSV: visualizing and mining cohesive subgraphs. In: SIGMOD Conference, pp. 445–458 (2008)
25. Wei, F.: Tedi: efficient shortest path query answering on graphs. In: SIGMOD 2010: Proceedings of the 2010 International Conference on Management of Data, pp. 99–110. ACM, New York (2010)
26. Wu, A.Y., Garland, M., Han, J.: Mining scale-free networks using geodesic clustering. In: KDD, pp. 719–724 (2004)

An Edge-Based Framework for Fast Subgraph Matching in a Large Graph

Sangjae Kim, Inchul Song, and Yoon Joon Lee

Department of Computer Science, KAIST, Republic of Korea
{sjkim,icsong}@dbserver.kaist.ac.kr, yoonjoon.lee@kaist.ac.kr

Abstract. In subgraph matching, we want to find all subgraphs of a database graph that are isomorphic to a query graph. Subgraph matching requires subgraph isomorphism testing, which is NP-complete. Recently, some techniques specifically designed for subgraph matching in a large graph have been proposed. They are based on a filtering-and-verification framework. In the filtering phase, they filter out vertices that are not qualified for subgraph isomorphism testing. In the verification phase, subgraph isomorphism testing is performed and all matched subgraphs are returned to the user. We call them a vertex-based framework in the sense that they use vertex information when filtering out unqualified vertices. Edge information, however, can also be used for efficient subgraph matching. In this paper, we propose an edge-based framework for fast subgraph matching in a large graph. By using edge connectivity information, our framework not only filters out more vertices in the filtering phase, but also avoids unnecessary edge connectivity checking operations in the verification phase. The experimental results show that our method significantly outperforms existing approaches for subgraph matching in a large graph.

1 Introduction

Since graphs are useful to represent structured, complex data, they have been used in many application areas such as Web, social networks, communication networks, bioinformatics, ontology engineering, software modeling, VLSI reverse engineering, etc. Subgraph matching, which is to find all subgraphs of a database graph that are isomorphic to a query graph, is one of the most frequently used operations in graph databases. For example, a financial crime investigator may want to find all occurrences of matches to a spurious pattern in a financial network where vertices represent account holders or banks and edges represent money transfer transactions [1]. A biologist may want to find all occurrences of matches to a specific biological pattern in protein-protein interaction networks [2] or gene regulatory networks. In addition, subgraph matching can be used for detecting and preventing some privacy attacks on anonymized social network data [3], [4].

There are two types of queries related to subgraph matching. A *subgraph containment query* finds all graphs that contain a subgraph which is isomorphic to the query graph. Much research has been done for this type of query [5],[6],[7],[8],[9]. In these works, they assume that a graph database contains many small graphs. A

J.X. Yu, M.H. Kim, and R. Unland (Eds.): DASFAA 2011, Part I, LNCS 6587, pp. 404–417, 2011.
© Springer-Verlag Berlin Heidelberg 2011

subgraph matching query finds, from a single database graph, all subgraphs that are isomorphic to the query graph. Many general purpose algorithms have been proposed for subgraph matching queries [10],[11]. Recently two techniques for subgraph matching queries in a large graph have been proposed in [12],[13]. In this paper, we focus on subgraph matching queries for a large graph.

Since subgraph matching requires subgraph isomorphism testing, which is NP-complete [14], existing methods typically use a *filtering-and-verification* framework. In the filtering phase, the vertices in the database graph are filtered out if they are not qualified as a matching vertex. This is accomplished by comparing *signatures* of vertices, which contain information about the vertices themselves and neighborhood information. After filtering, only the remaining vertices, which we call *candidate vertices* in this paper, are input to subgraph isomorphism testing. In the verification phase, subgraph isomorphism testing is performed and all subgraphs of the database graph that are isomorphic to the query graph are found and returned to the user. The main task of the verification phase is to check whether candidate vertices of different query graph vertices are properly connected to each other. Many kinds of heuristics can be employed to speed up the verification process.

Existing approaches mainly focus on reducing the input size to subgraph isomorphism testing. GADDI [12] proposed a technique for subgraph matching in a large graph based on data mining techniques. It uses *discriminative substructures*, which are small substructures in induced intersection graph between the neighborhoods of two vertices, as vertex signatures. NOVA [13] is another technique for subgraph matching in a large graph. It uses label distribution information around vertices as vertex signatures. Both of these methods are a *vertex-based framework* in the sense that they use only vertex information to filter out unqualified vertices. Edge information, however, also can be used in the filtering process. For example, by selectively checking connectivity between vertices in the database graph, we can further filter out those candidate vertices that do not have required connections to other vertices.

In this paper, we propose an edge-based framework for fast subgraph matching in a large graph. Our method follows a filtering-and-verification framework. Unlike existing vertex-based frameworks, our method uses edge connectivity information in both of the filtering and verification phases for fast subgraph matching. In the filtering phase, it filters out more candidate vertices based on edge connectivity information. Edge connectivity information is also used in the verification phase to reduce extensive connectivity checking operations between vertices. Some of the connectivity checking operations can be removed altogether. The experimental results show that our method significantly outperforms existing approaches for subgraph matching in a large graph.

The rest of this paper is organized as follows. Section 2 gives the background information and an overview of our method. Section 3 describes an index structure used in our method. The filtering and verification phases are explained in Section 4 and 5, respectively. We evaluate our method in Section 6. Section 7 discusses related work and Section 8 concludes the paper.

2   Preliminaries

In this section, we introduce the basic definitions used in the paper and give the formal problem statement. Our proposed method supports both directed and undirected graphs with labeled vertices and/or labeled edges. For ease of presentation, we assume a simple graph with labeled vertices. It is straightforward to apply our method to other types of graphs. We also assume that every query graph and database graph considered in this paper are connected, i.e., every pair of vertices is connected by a path.

Definition 1. Vertex-labeled graph. A vertex labeled graph is denoted as $G=(V, E, L, l)$, where V is the set of vertices, $E \subseteq V \times V$ is the set of edges, L is the set of vertex labels, and l is a mapping function: $V \rightarrow L$.

Definition 2. Subgraph isomorphism. Given two graphs $G = (V, E, L, l)$ and $G' = (V', E', L', l')$, G is *subgraph isomorphic* to G', if there exists an injective function f: $V \rightarrow V'$ such that

1. $\forall v \in V, l(v) = l'(f(v))$,
2. $\forall (u, v) \in E \Rightarrow (f(u), f(v)) \in E'$.

Such an injective function is called a *subgraph isomorphism mapping*.

Problem Statement. Given a query graph q and a database graph G, find all subgraph isomorphism mappings from q to G.

2.1 Filtering and Verification Framework

Our method uses the *filtering-and-verification* framework. In subgraph isomorphism testing, for each vertex v_q in the query graph q, we need to try every vertex v_G in the database graph G as a matching vertex of v_q. In the filtering phase, we filter out those vertices v_G that cannot be a matching vertex of v_q. To this end, we encode each vertex and produce its *signature*. The signature of a vertex contains information about the vertex itself and neighbor information. Unqualified vertices in the database graph are filtered out by comparing their signatures with those of vertices in the query graph. Unlike existing methods where only vertex signatures are used for filtering, our method uses edge signatures as well to filter out more vertices. The remaining vertices from the filtering phase, which we call *candidate vertices*, are used as input to subgraph isomorphism testing. In the verification phase, subgraph isomorphism testing is performed and all possible subgraph isomorphism mappings are produced and returned to the user.

2.2 Representing Vertices and Edges

Various information can be used as vertex signatures and edge signatures. For example, NOVA [13] uses a vertex label, degree, and neighbor information as vertex signatures. In order for a signature to be used for filtering out unqualified vertices, it must satisfy the *inequality property* [12], [13]. More specifically, let $sig(v)$ and $sig(u)$ be the signature of vertex v in the query graph and that of vertex u in the database graph, respectively. Then for vertex u to be a matching vertex of v, $sig(v)$ must be less than or equal to $sig(u)$, i.e., $sig(v) \leq sig(u)$. The operator $\leq$ must be properly defined for a specific signature to enforce the inequality property. Our method is designed to work with any vertex signature that satisfies the inequality property.

2.3 Vertex and Edge Signatures

In this paper, we consider two kinds of vertex signatures, namely NOVA [13] and NPV [9]. They both use a vertex label and degree, and neighbor information in their signatures. The difference between them lies in the neighbor information used. NOVA uses label distribution information around a vertex up to a user-specified distance as neighbor information. In NPV, simple paths from a vertex up to a pre-defined length are used as neighbor information. Signature comparison between two vertices $sig(v) \leq sig(u)$ is performed by checking the following conditions:

$$l(v) = l(u) \tag{1}$$
$$deg(v) \leq deg(u) \tag{2}$$
$$nInfo(v) \leq nInfo(u), \tag{3}$$

where $l(v)$ and $l(u)$ are the labels of v and u, $deg(v)$ and $deg(u)$ are the degrees of v and u, and $nInfo(v)$ and $nInfo(u)$ are the neighbor information of v and u, respectively. How to check Condition (3) is specific to each kind of signature, and information such as the number of distinct vertex labels around a vertex is commonly used in condition checking. For more details, refer to [12] or [13].

Similarly, we also define the edge signature. The edge signature for an edge contains the labels of its endpoint vertices, the sum of their degrees, and their signatures. Given an edge $e_q=(v_l,v_r)$ in the query graph q and $e_G=(u_l,u_r)$ in the database graph G, we use the following conditions to check if $sig(e_q) \leq sig(e_G)$:

$$l(v_l) = l(u_l) \ \wedge \ l(v_r) = l(u_r) \tag{4}$$
$$deg(v_l) + deg(v_r) \leq deg(u_l) + deg(u_r) \tag{5}$$
$$nInfo(v_l) \leq nInfo(u_l) \ \wedge \ nInfo(v_r) \leq nInfo(u_r) \tag{6}$$

3 Pre-processing

In this section, we describe the *Edge Index* (E-Index) and *Vertex Index* (V-Index) that are used to speed up the filtering phase. The purpose of E-Index is to find candidate edges of each query graph edge from the database graph. Given an edge (v_1, v_2) in the query graph q, an edge (u_1, u_2) in the database graph G is its candidate edge if 1) the labels of corresponding vertices are the same, i.e., Condition (4), 2) the degree sum of two vertices in edge (v_1, v_2) is less than or equal to the degree sum of two vertices in edge (u_1, u_2), i.e., Condition (5), and 3) the neighbor information of the corresponding vertices satisfies the inequality property, i.e., Condition (6). Here we need three comparisons. The E-Index is used to speed up the first two comparisons, i.e. the label comparison and degree sum comparison.

We may construct a separate index to speed up each of these two operations. For example, we construct an index whose key is a pair of vertex labels and value is the list of edges that have those vertex labels. We also construct another index whose key is a degree sum and value is the list of edges having that degree sum. We can find candidate edges by first retrieving candidate edges from each of the two indexes and then intersecting them.

Instead of having two separate indexes, the E-Index combines them and builds a two-level index to further reduce index search time. The first level index is called *Label Index* (L-Index). Its key is a pair of vertex labels and value is a pointer to a second level index. The second level index is called *Degree Index* (D-Index). Each D-Index is constructed separately for the edges having the identical vertex label pair. Given a D-Index, its key is a degree sum and value is a list of edges having that degree sum. Figure 1(a) shows the two-level structure of E-Index. Given an edge (v_1, v_2) in the query graph, we can find candidate edges as follows. First we obtain a pointer to a D-Index by querying L-Index with key ($l(v_1)$, $l(v_2)$). Then we find the candidate edges by issuing a range degree sum query over the D-Index. Both L-Index and D-Index can be easily implemented by using B+-tree index structure.

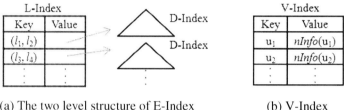

(a) The two level structure of E-Index   (b) V-Index

Fig. 1. A simple example of E-Index and V-Index

Note that we still need a signature comparison between the edge in the query graph and the candidate edges to find the final candidate edges. For this last comparison, we need to retrieve neighbor information of the vertices in the candidate edges. To efficiently retrieve neighbor information of vertices, we construct a *Vertex Index* (V-Index), whose key is a vertex identifier and value is its neighbor information. Figure 1(b) shows the structure of V-Index.

4 Filtering

In this section, we describe how to find candidate vertices from the database graph by using E-Index and V-Index described in the previous section. In the filtering phase, the main task is to find candidate vertices of each query graph vertex from the database graph. Our method has two advantages over the existing methods. First, we reduce time to retrieve candidate vertices by using E-Index, a pre-constructed index structure. Second, since E-Index stores information on vertex pairs that are directly connected to each other, we can retrieve only those candidate vertices that are directly connected to each other from E-Index. This may reduce the size of the candidate vertices.

Before we proceed, let us first give an overview of the filtering phase. In our approach, we obtain candidate vertices indirectly through candidate edges. In other words, endpoint vertices of the candidate edges will be our candidate vertices. To this end, we need to find candidate edges of the edges in the query graph. Note that we do not need to find candidate edges of every edge in the query graph. This is because we need only those edges that are enough to cover every vertex in the query graph. Here a spanning tree of the query graph is useful. Thus, we select a spanning tree of the query graph and find candidate edges of the edges in the spanning tree. Finally, we obtain candidate vertices from the candidate edges found. In what follows, we describe the filtering phase in more detail. Section 4.1 describes the spanning tree selection process and section 4.2 explains how to obtain candidate vertices.

4.1 Selecting a Spanning Tree

Given a query graph, we select a spanning tree of the query graph whose edges are to be used to find candidate edges. There may exist many different spanning trees of the query graph. Here we need a way to pick a "good" spanning tree for filtering out candidate edges. We have the following observation. Given an edge in the query graph, the number of its candidate edges can be roughly estimated by the degree sum of its two endpoint vertices. This is because candidate edges must have degree sums large than or equal to that of the query graph edge, and the larger degree sum of the query graph edge, the smaller possibility that there are many candidate edges with larger degree sums. Thus degree sums may indicate the "goodness" of the query graph edges. Based on this observation, we take the degree sum of each edge as its weight, compute the maximum cost spanning tree (we can obtain the maximum cost spanning tree by multiplying each edge weight by -1 and applying minimum spanning tree algorithms such as Kruskal's or Prim's algorithms), and use the resulting tree to retrieve candidate edges.

4.2 Discovering Candidate Vertices

After selecting a spanning tree, we find candidate vertices through candidate edges. We first obtain candidate edges of the edges in the spanning tree. We need to decide the order of visiting edges in the spanning tree. Either breadth-first search or depth-first search over the spanning tree can be used to determine the edge visiting order. During graph search, we record edge visiting order. Let the determined order be $e_1, e_2, \ldots, e_{|V(q)-1|}$, where $|V(q)|$ is the number of vertices in the query graph. Now we visit each edge in the order determined, find candidate edges, and obtain candidate vertices from the candidate edges. This step proceeds as follows. For each vertex v in the query graph, we maintain a candidate vertex set, denoted $C(v)$, and initialize it as empty. Now we visit each edge one by one. First, for the first edge $e_1 = (v_1, v_1')$ in the spanning tree, we probe L-Index by using the key $(l(v_1), l(v_1'))$ and obtain a pointer to a D-Index and then perform a range query over the D-Index to retrieve the list of candidate edges. We then compare neighbor information of the candidate edges with that of the query edge to find the final candidate edges. For each candidate edge found, we add its two endpoint vertices to the corresponding candidate vertex sets of v_1 and v_1', respectively.

For the remaining edges in the spanning tree, we process them slightly differently for better efficiency. For each edge $e_i = (v_i, v_i')$ $(i>1)$ in the spanning tree, we probe L-Index and D-Index as before and obtain a list of candidate edges. Here we make the following observation to reduce the cost of comparing neighbor information of edge e_i with the candidate edges. Given a candidate edge $e_i'=(u_i, u_i')$, if u_i is not contained in the candidate vertex set of v_i, i.e., if $u_i \notin C(v_i)$, then e_i' cannot be the final candidate edge of e_i. This is because it means u_i has been already filtered out when finding candidate edges of the query edge of the form (v, v_i). Based on this observation, we first check if $u_i \in C(v_i)$. If this is true, we proceed to check whether $nInfo(v_i') \leq nInfo(u_i')$. If this is also true, we add u_i' to $C(v_i')$.

5   Verification

In this section, we describe our verification algorithm based on depth first search with backtracking. In the verification phase, we perform subgraph isomorphism testing. To this end, we generate subgraph isomorphism mappings between the query graph and the database graph. We start with an empty mapping M and expand it incrementally by adding possible matching vertices of the vertices in the query graph. Depth first search with backtracking is used to expand M systematically. We first determine the order of visiting the vertices in the query graph. At depth d, we visit the d-th vertex in the query graph and find possible matching vertices from its corresponding candidate vertex set. If we do not find any possible matching vertex at depth d, we remove the most recently added matching vertex from M and backtrack to depth d-1. If we arrive at depth $|V(q)|$, then we have found a subgraph isomorphism mapping.

Our verification algorithm, FastMatch, uses three kinds of heuristics to speed up the verification process. Before describing FastMatch, we first explain each of these heuristics and how they may speed up the verification process. Figure 2 shows our running example that will be used in the rest of the paper. Figure 2(a), 2(b), and 2(c) show a database graph, query graph, and the spanning tree selected in the filtering phase, respectively.

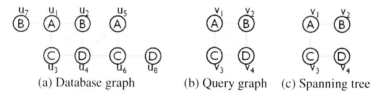

(a) Database graph   (b) Query graph   (c) Spanning tree

Fig. 2. Running example

5.1   Heuristics for Fast Verification

The first heuristic used by FastMatch is called *vertex ordering*. As mentioned above, before we start depth first search, we decide the order of visiting the vertices in the query graph. It is important to carefully decide the order of visiting the vertices in the query graph for the performance of the verification process. Given a vertex in the query graph, we can save more work if its candidate vertex size is large and it is visited more later than some vertices with smaller candidate sizes. Thus we visit the vertices of the query graph in the increasing order of their candidate sizes. This heuristic is also employed by NOVA [13]. In our method, we modify the heuristic in such a way that candidate edge information obtained from the filtering phase can be used. More specifically, we determine vertex visiting order by using the spanning tree selected in the filtering phase. We maintain a visited vertex set, denoted *Visit*. We start with a vertex with the smallest candidate size and add it to *Visit*. We choose the next vertex to visit among the vertices not in *Visit* and directly connected to any of the vertices in *Visit* on the spanning tree. The vertex with the smallest candidate size is selected as the next vertex to visit and added to *Visit*. This procedure is repeated until there is no vertex to visit.

The second heuristic is called *connection-aware forward checking*. After selecting a possible vertex mapping (v_d, u_d) between a vertex v_d in the query graph and one of its candidate vertices u_d at depth d, we check the connections between u_d and the candidate vertices of unvisited vertices v_i (i>d) that are directly connected to v_d. The candidate vertices that do not have any connection with u_d are marked as invalid. Unlike the conventional forward checking, in our method, we do not need to check the connections between u_d and the candidate vertices of v_i (i>d) if (v_d, v_i) is an edge in the spanning tree. This is because we can easily retrieve candidate vertices directly connected to u_d by using connectivity information between candidate vertices. How to do this will be described later. Note that our method eliminates the connectivity checking operations over the edges in the spanning tree altogether, which may result in a considerable performance gain.

The last heuristic is called *incompatibility learning*. When backtracking during depth first search, we lose the results of connectivity checking operations from deeper depths, which may lead to duplicated connectivity checking operations at later times. To avoid this problem, we keep track of the candidate vertices of v_i (i>d) that are invalidated by each matching vertex u_d of v_d during depth first search and use it to remove unnecessary connectivity checking operations. For ease of presentation and interest of space, we will omit the incompatibility learning heuristic in the description of our verification algorithm (Algorithm 1 in section 5.2).

5.2 The FastMatch Algorithm

The algorithm FastMatch is formally described in Algorithm 1. First it determines the order of visiting the vertices in the query graph by applying the vertex ordering heuristic, which is described in section 5.1. Then it calls RecursiveFastMatch to perform subgraph isomorphism testing.

RecursiveFastMatch performs depth first search and finds all possible subgraph isomorphism mappings. At depth d, the candidate vertices of vertex v_d are considered as possible matching vertices of v_d. By calling GetQualifedCandidateVertices, ResursiveFastMatch retrieves only those candidate vertices that are qualified as a matching vertex of v_d. The details of GetQualifiedCandidateVertices will be described later. For each qualified candidate vertex u, ResursiveFastMatch sets u as the matching vertex of v_d (line 6) and then check if we have arrived at depth |V(q)| (line 7). If this is true, we have found a subgraph isomorphism mapping. Thus we output M and backtrack. If not, we expand M further by going down to the next level. Before going down, the connection-aware forward checking heuristic is applied to filter out unqualified candidate vertices with respect to u at deeper depths (line 10). The connection-aware forward checking heuristic is described in section 5.1. Finally, RecursiveFastMatch recursively calls itself with an increased depth.

Algorithm 1. FastMatch	
Input	q: query graph, G: database graph, M: current mapping, initially empty
	d: depth, initially 1
Output	All subgraph isomorphism mappings
1.	Apply the Vertex Ordering heuristic
2.	RecursiveFastMatch(q,G,d,M)
3.	function RecursiveFastMatch(q,G,d,M):
4.	QC ← GetQualifiedCandidateVertices(v_d,M)
5.	for each u in QC do
6.	M[d] = u
7.	if d = |V(q)| then
8.	Output M and backtrack
9.	else
10.	Apply the Forward Checking heuristic
11.	RecursiveFastMatch(q,G,d+1,M)
12.	end if
13.	end for
14.	end function

5.3 The GetQualifiedCandidateVertices Function

When we arrive at depth d, only those candidate vertices that are not marked as invalid can be a qualified matching vertex of v_d. To find qualified candidate vertices efficiently, GetQualifiedCandidateVertices uses a pre-constructed data structure. For each edge (v_i,v_j) in the spanning tree selected in the filtering phase, we build a *connection map* (CM). The order of constructing CMs for the edges in the spanning tree is determined by the vertex ordering heuristic. The key of CM for (v_i,v_j) is a candidate vertex u_i of v_i and the value is the list of candidate vertices of v_j that are directly connected to u_i. The CMs for the edges in the spanning tree can be easily constructed from the candidate edges discovered in the filtering phase.

For example, Table 1 shows the candidate edges for the edges in the spanning tree in Figure 2(c). Figure 3 shows the connection maps populated from Table 1. As an example, the CM for (v_1,v_2) in Figure 3 indicates that u_2 and u_7 are the candidate vertices of v_2 that are directly connected to the candidate vertex u_1 of v_1.

Table 1. Candidate edges for the edges in the spanning tree in Figure 2(c)

Edges in the spanning tree	(v_1,v_2)	(v_1,v_3)	(v_3,v_4)
Candidate edges	(u_1,u_2)	(u_1,u_3)	(u_3,u_4)
	(u_1,u_7)	(u_5,u_6)	(u_6,u_8)

Let (v_i,v_d) (i<d) be an edge in the spanning tree of the query graph. Note that, given v_d, such an edge is uniquely determined by the vertex ordering heuristic (for more details, see section 5.1.) GetQualifiedCandidateVertices finds the candidate vertices u_d of v_d that are directly connected to u_i (obtained from $M[v_i]$) by consulting the

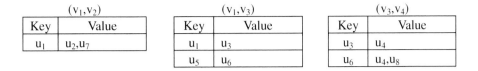

Fig. 3. Connection maps for the edges in Table 1

connection map for the edge (v_i, v_d). Additionally, it removes, from the list of candidate vertices obtained, those candidate vertices that are marked as invalid and returns the remaining candidate vertices as the qualified candidate vertices.

5.4 Improving the Connection-Aware Forward Checking Heuristic

Connection maps introduced in the previous section can also be used to enhance the connection-aware forward checking heuristic. In this section, we briefly describe how to do that. In the connection-aware forward checking heuristic, given a vertex mapping (v_d, u_d) at depth d, we check the connections between u_d and the candidate vertices u_i of unvisited vertices v_i (i>d) that are directly connected to v_d. Note that we need to consider only u_i's that are not marked as invalid. Instead of checking whether every u_i is not marked as invalid, we can retrieve valid u_i's more efficiently by using connection maps. Given an edge (v_d, v_i), if an edge (v_j, v_i) (j<d) exists in the query graph, we can retrieve u_i's that are directly connected to u_j by probing the connection map for edge (v_j, v_i) with u_j as a key. We need to check validity for only these u_i's. For the other edges, we cannot use connection maps to reduce the number of validity checking operations.

5.5 Discussion

We can estimate how many connectivity checking operations can be reduced by connection-aware forward checking as follows. Let us first calculate how many of them can be reduced at depth d over an edge (v_d, v_j) on the spanning tree T during depth first search. In the worst case, depth first search arrives at depth d for $\prod_{k=1}^{d} |C(v_k)|$ number of times, where $|C(v_k)|$ is the number of candidate vertices of v_k (this happens when all required connections between candidate vertices exist in the database graph; in such a case, the basic forward checking cannot eliminate any of the candidate vertices during depth first search). For each arrival at depth d, connection-aware forward checking eliminates $|C(v_j)|$ connectivity checking operations over edge (v_d, v_j). Thus, the number of connectivity checking operations reduced at depth d over edge (v_d, v_j) is $|C(v_j)| \times \prod_{k=1}^{d} |C(v_k)|$. Finally, the total number of connectivity checking operations reduced over all edges on the spanning tree T is:

$$\sum_{(v_d, v_j) \in E(T)} \left(|C(v_j)| \times \prod_{k=1}^{d} |C(v_k)| \right)$$

Note that the above formula computes the number of connectivity checking operations that can be reduced by connection-aware forward checking in the best case scenario. In general, it might be less than the number suggested by the formula.

6 Evaluation

In this section, we compare the performance of our method with that of NOVA [13], a state-of-the-art subgraph matching technique using synthetic data sets. Our method and NOVA have been implemented using C++ with the STL library. The experiments were conducted on a PC with a 3.0GHz quad core CPU and 4GB main memory running windows server 2003. We use two kinds of vertex signatures in our method. In the graphs that show experimental results below, "Our Method (NOVA)" represents the one that uses the vertex signature from NOVA and "Our Method (NPV)" the one that uses that from NPV [9]. In various experiments, ten database graphs are randomly generated and query graphs are selected by extracting subgraphs from the database graphs. We evaluate each method over the ten randomly generated database graphs and average the results.

6.1 Effect of the Size of the Query Graph

In this section, we evaluate our method over various query graph sizes from 10 to 50 (the size of a query graph is the number of its vertices.) Each of the randomly generated database graphs has 5,000 vertices, 80,000 edges and 20 distinct vertex labels. The average vertex degree of each query graph is set to 2.8. Figure 4 shows the experimental results. Figure 4(a) shows the query processing time and figure 4(b) shows the number of connectivity checking operations required as the query graph size increases. Because our verification algorithm reduces the number of connectivity checking operations greatly, both versions of our method show significantly better performance than that of NOVA. Our Method (NPV) outperforms Our Method (NOVA) because the vertex signature of NPV has a superior filtering power than that of NOVA.

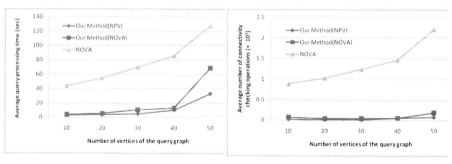

(a) Query processing time   (b) Number of connectivity checking operations

Fig. 4. Experimental results over various query graph sizes

6.2 Effect of the Average Degree of the Query Graph

In this section, we evaluate our method by varying the average degree of the query graph from 3.4 to 5.8. This time we generate ten random database graphs, each of which has 1,000 vertices, 40,000 edges and 20 distinct vertex labels. The query graph size is set to 20. Figure 5 shows the results of this experiment. Figure 5(a) shows the query processing time and figure 5(b) shows the number of connectivity checking operations as the degree of the query graph increases. As the degree of a vertex in the query graph becomes large, the number of its candidate vertices in the database graph decreases. This is because a qualified candidate vertex must have a degree larger than or equal to that of the query graph vertex. Thus as the average degree of the query graph increases, more vertices are filtered out in the filtering phase, which results in low query processing time in every method. When the average degree is small, however, many candidate vertices are used as input to subgraph isomorphism testing in the verification phase. In this case, the number of connectivity checking operations determines the performance of the verification phase. Both kinds of our method significantly outperform NOVA in such a case.

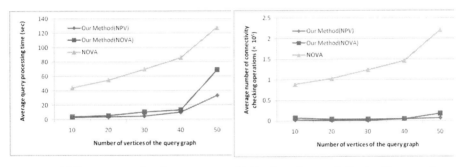

(a) Query processing time   (b) Number of connectivity checking operations

Fig. 5. Experimental results over various average query graph degrees

7   Related Work

Subgraph matching, which is to find all subgraphs of a database graph that are isomorphic to a query graph, is an important operation that has many applications over a wide range of areas. Subgraph matching requires subgraph isomorphism testing, which is NP-complete [14]. Many techniques for subgraph matching have been proposed thus far.

In graph databases, queries related to subgraph matching can be divided into two types. A subgraph containment query finds all graphs that contain a subgraph which is isomorphic to the query graph. Many techniques have been proposed for this type of query [5],[6],[7],[8],[9]. In these techniques, they assume that a graph database contains many small graphs. A *subgraph matching query* finds, from a single database graph, all subgraphs that are isomorphic to the query graph. General purpose subgraph matching algorithms are proposed in [10],[11]. Recently, techniques specifically designed for subgraphs matching queries in a large graph have been proposed in [12],[13].

Both of [12] and [13] are based on a *filtering-and-verification* framework. In the filtering phase, they compare the signatures of vertices and filter out candidate vertices if they are not qualified as a matching vertex. In the verification phase, subgraph isomorphism testing is performed and all matched subgraphs of the database graph are returned to the user. GADDI [12] uses *discriminative substructures*, which are small substructures in the intersection of the neighborhoods of two vertices, as vertex signatures. In NOVA [13], label distribution information around vertices are used as vertex signatures. Both of these methods can be classified into a *vertex-based framework* in the sense that they use only vertex information to filter out unqualified vertices. Unlike these methods, our method, which is an *edge-based framework*, uses edge connectivity information in both the filtering and verification phases for fast subgraph matching.

8   Conclusions

In this paper, we have proposed an edge-based framework for fast subgraph matching in a large graph. Our method is based on a filtering-and-verification framework. Unlike existing vertex-based frameworks, our method uses edge connectivity information in both of the filtering and verification phases for fast subgraph matching. It not only reduces the size of input to subgraph isomorphism testing, but also avoids unnecessary connectivity checking operations. We verify through experimental evaluation that our method significantly outperforms existing approaches for subgraph matching in a large graph.

Acknowledgments

This work was supported in part by Korea Institute of Science and Technology Information (KiSTi) and in part by the National Research Foundation of Korea (NRF) grant funded by the Korea government (MEST) (No. 2010-0018865).

References

1. Bröcheler, M., Pugliese, A., Subrahmanian, V.: COSI: Cloud Oriented Subgraph Identification in Massive Social Networks. In: ASONAM (2010)
2. Tian, Y., McEachin, R.C., Santos, C., States, D.J., Patel, J.M.: SAGA: a subgraph matching tool for biological graphs. Bioinformatics 23, 232–239 (2007)
3. Backstrom, L., Dwork, C., Kleinberg, J.: Wherefore art thou r3579x?: anonymized social networks, hidden patterns, and structural steganography. In: WWW (2007)
4. Cheng, J., Fu, A.W.c., Liu, J.: K-isomorphism: privacy preserving network publication against structural attacks. In: SIGMOD (2010)
5. Giugno, R., Shasha, D.: GraphGrep: A Fast and Universal Method for Querying Graphs. In: ICPR (2002)
6. Yan, X., Yu, P.S., Han, J.: Graph indexing: a frequent structure-based approach. In: SIGMOD (2004)
7. Cheng, J., Ke, Y., Ng, W., Lu, A.: Fg-index: towards verification-free query processing on graph databases. In: SIGMOD (2007)
8. Jiang, H., Wang, H., Yu, P.S., Zhou, S.: Gstring: A novel approach for efficient search in graph databases. In: ICDE (2007)

 9. Wang, C., Chen, L.: Continuous Subgraph Pattern Search over Graph Streams. In: ICDE (2009)
10. Ullmann, J.R.: An Algorithm for Subgraph Isomorphism. J. ACM 23(1), 31–42 (1976)
11. Cordella, L.P., Foggia, P., Sansone, C., Vento, M.: A (Sub)Graph Isomorphism Algorithm for Matching Large Graphs. IEEE Trans. Pattern Anal. Mach. Intell. 26, 1367–1372 (2004)
12. Zhang, S., Li, S., Yang, J.: GADDI: distance index based subgraph matching in biological networks. In: EDBT (2009)
13. Zhu, K., Zhang, Y., Lin, X., Zhu, G., Wang, W.: NOVA: A Novel and Efficient Framework for Finding Subgraph Isomorphism Mappings in Large Graphs. In: Kitagawa, H., Ishikawa, Y., Li, Q., Watanabe, C. (eds.) DASFAA 2010. LNCS, vol. 5981, pp. 140–154. Springer, Heidelberg (2010)
14. Cook, S.A.: The complexity of theorem-proving procedures. In: Proceedings of the Third Annual ACM Symposium on Theory of Computing, pp. 151–158 (1971)

Context-Sensitive Query Expansion over the Bipartite Graph Model for Web Service Search

Rong Zhang[1,2], Koji Zettsu[2], Yutaka Kidawara[2], and Yasushi Kiyoki[2,3]

[1] East China Normal University, 3663 ZhongShan Rd.(N), Shanghai 200062, China
c.zhangrong@gmail.com
[2] National Institute of Information and Communications Technology,
3-5 Hikaridai, Seika-cho, Kyoto 619-0289, Japan
{zettsu,kidawara}@nict.go.jp
[3] Keio University, 5322 Endo, Kanagawa 252-8520, Japan
kiyoki@sfc.keio.ac.jp

Abstract. As Service Oriented Architecture (SOA) matures, service consumption demand leads to an urgent requirement to service discovery. Unlike web documents, services are intended to be executed to achieve objectives and/or desired goals of users, which means to realize application requirements. This leads to the notion that service discovery should take into account the "application requirement" of service with service content (descriptions) which have been well explored. Content is defined by service developers, e.g. WSDL file and context is defined by service users, which is service usages to application requirement. We find context(application) information is more useful for query generation, especially for non-expert users. So in this paper, we propose to do context-sensitive query processing to resolve application-oriented queries for web service search engine. Context is modeled by a bipartite graph model to represent the mapping relationship between application space and service space. Application-oriented queries are resolved by query expansion based on the topic sensitive bipartite graph. The experiments verify the efficiency of our idea.

1 Introduction

Recent years have witnessed an explosive increase in online web services (WS), and tens of thousands of such services are publically accessible. Services are preferred to fulfill users' application requirements by simply assembling, e.g. composition or integration. In order to support service assembling, different tools have been designed and implemented, e.g. ActiveBpel[1], and BPMN Modeler[2], to help define services' logic collaboration graph according to application requirements. As WS consumption rises, an urgent need has arisen for designing a WS discovery mechanism that can find the suitable services to fulfill users' application requirements. Without such a mechanism, the significant amount of required manual effort will continue to bottleneck WS-based applications.

For current WS search engines, e.g. [3–5], content-based mapping is still the main technology. Generally service content includes service name, operation name, parameters and service document[1]. However, it is pointed out that the content-based mapping

---

[1] In the later, service description or service content includes all parts of such information.

J.X. Yu, M.H. Kim, and R. Unland (Eds.): DASFAA 2011, Part I, LNCS 6587, pp. 418–433, 2011.
© Springer-Verlag Berlin Heidelberg 2011

algorithm is insufficient, as service description and query are short, which makes mapping difficult between query space and service space[6]. As shown in Fig. 1, we collect services from ProgrammableWeb[7] (RESTful-based WS [2]) and acquire the term distribution, among which about 70% of services have fewer than 30 terms and 90% of services have fewer than 40 terms. In such a case, we may meet with two situations: for a specific query, the returned results will be very few because of the lack of intersection between query terms and service descriptions; for a specific query, the returned results will be too many because of the involving of popular terms.

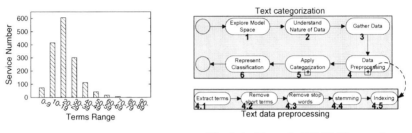

Fig. 1. Term distribution statistics for 1600 services described in Sec. 5

Fig. 2. An Example BPMN Diagram for Text Categorization

One method to improve mapping quality is to do local content analysis and perform term conceptualization[8]. For example, if it finds term a and b frequently co-occur in many content, these two terms are supposed to be semantically close (similar). Woogle[9] does the same thing to design a WS search engine. To some extend, it helps to improve system precision by clustering semantically close terms. However this local analysis method is restricted by the length of service descriptions or the overlap between service descriptions. Additionally, it tries to ask users to generate detailed queries to services (content), such as operation name, parameter name or even composite service requirement. But too many requirements on users may make the system difficult to use.

Definition 1 (Content-oriented Query). *It is the query with query terms from service description content, such as service name, operation name, or parameter. For example, in Fig. 2, for step 4.1, the query may be "termExtraction(String text)".*

Definition 2 (Application-oriented Query). *It is the query with query terms from both service content and service usage application scenario description. For example, in Fig. 2, a query may be like "term extraction for text data preprocessing" or " text data preprocessing".*

As have been said, services are defined for resembling to achieve added-value applications. For example in Fig. 2, this is a logic process procedure for text categorization application requirement drawn by BPMN modeler[2]. For this procedure, users are required to define or find the services for each (or some) step, e.g. step 4.1: to find a service to extract terms. We can not promise all users know how to compose a query

---

[2] http://en.wikipedia.org/wiki/Representational_State_Transfer

by using service name and parameters, for a specific service, e.g. for step 4.1, query may be like "termExtraction(String text)". For step 4: "data preprocessing", not all of users know that it shall be divided into 5 standard processing procedures, especially for non-expert users. But one thing is that users clearly know about their application requirement, that is in step 4, it is needed to do "data preprocessing to text". So they prefer to submit application-oriented query, defined in Def. 2, (query to application scenario) instead of content-oriented query, defined in Def. 1 (query to service content). This kinds of queries can not be resolved by content-based mapping method because of lack of application related information. On the other hand, we find service definition terms are not standardized. For example, for "termExtraction(String text)", it may be defined as "func1(String str)". Intuitively, in such a case, content-oriented query is difficult to be generated. But context-based query may be more efficient to find related services.

Definition 3 (Context). *Suppose service s has ever joined applications $\{a_j\}$. For service s, its context is represented as $A^{[s]}=\{< a_j >\}^*$, with $0 \leq |A^{[s]}|$. $a_j = \{< Des. > [s_l]^*\}$. $Des.$ and $\{s_l\}$ are a_j's application description and the involved services for a_j. For example in Fig. 2, for service in step 4.1, the context is $A^{[s_{4.1}]} = \{< text\ data\ preprocessing > [s_{4.2}, s_{4.3}, s_{4.4}, s_{4.5}]\}$. Here $s_{4.i}$ represents the service which can realize the task requirement in the corresponding step.*

In this work, we propose a novel context-sensitive WS discovery method for solving the problems mentioned above: application-oriented query processing, which may query to application scenario. Context, defined as Def. 3, displays service usages to application requirement. Based on this kind of context, we design the query expansion algorithm to solve application-oriented queries by bridging heterogeneity between query space(application space) and service space. And then even for query with the application terms, we can find services related to the application. The main contributions of our work are summarized as follows.

First, we propose to distinguish services by service usages defined as context as in Def. 3, representing the involving relationship between services and applications. As far as we know, this is the first work to introduce the real service usage information as context for WS discovery. A bipartite graph model is designed to represent context. Service content is defined by service developers, used to declare service's functionalities and context is defined by service users, used to declare how they are used. They complement each other.

Second, we propose to do topic sensitive resources organization. We learn a topic classifier by using open resources and it is used to classify services, applications and queries into different topics, e.g. "arts" or "recreation". We construct bipartite graphs for services and applications in each topic domain. We expect to resolve queries in their semantic closest topic domains.

Third, We propose to solve application-oriented queries by query expansion among different semantic space: application space and service space. We exploit the bipartite graph and extract the implicit term mapping relationship between service descriptions and application descriptions.

Fourth, we carry out a series of experiments to investigate the effects of our proposed method.

The rest of this paper is organized as follows. In Sec. 2, we describe the motivations. In Sec. 3, we propose the method to generate the topic sensitive bipartite graph from service context. In Sec. 4, we generate the semantic bridge between query space and service space. In Sec. 5, we demonstrate system performance. In Sec. 6, we discuss the related work. Finally, Sec. 7 summarizes this work.

2   Motivation

The application-oriented query is necessary. As mentioned in Sec. 1, services are preferred to accomplish users application requirements by simply assembling, e.g. composition or integration. Until now, only part of service's usage information has been taken into service discovery, especially for composite service search[10], in which the query shall point out the partner's information. As in Fig. 2, in order to find composite service for step 4.2, it shall point out the parameters requirement for step 4.1 or 4.3, which is part of context (as collaborative services) in our definition. But one of our ideas is that we want to support application-oriented query processing, e.g. query for "text preprocessing" in stead of query for each individual service like "TermExtraction(String text)" for Step 4.1. In such a case, we expect to get a set of services which are supposed to be involved for this application. This will make query generation easier, especially for non-expert users who know the application's requirements much better than detailed services requirements. On the other hand, application-oriented query processing can return users context-based related services instead of content-based relevant service. It can help to filter out useless services. Services definition are subjective. Content similar services may be totally different in functionality. For example, two services are named as "RemovingNoise", which are used for "web page clustering" and "population census" respectively. The first one may focus on removing noisy links and the second one may focus on removing persons with ages older than 150. If we can put the query together with application requirement, it will make search more accurate.

Table 1. Examples of service descriptions and involved application's representative terms. These are obtained from ProgrammableWeb[7].

Service	Service Sescription	Representative Application Terms
Yahoo Maps	It offers three ways to integrate maps into a website - a simple REST-based API for map images, an Ajax JavaScript API, and a Flash API. This summary refers to the AJAX API which is closest to Google and Microsoft models. The Yahoo Map API can also provide for integrating geo-related information from Yahoo's products, such as local listings, traffic, weather, events, and photos.	location, estate, plot, city, estate, sites, address, traffic, states, area, flight, georss, station, county, travelers, campus, restaurants, housing, rental, roads, airport
Twitter	The Twitter micro-blogging service includes an API that gives RESTful access to the Twitter database and activity streams (users, tweets, friends, followers, favorites and more), using XML and JSON formats.	social, news, photo, community, rss, event, share, messages, mobile, blogs, topic, feeds, space, website
Last.fm	The last.fm API allows for read and write access to the full slate of last.fm music data resources - albums, artists, playlists, events, users, and more.	audio, tracks, lyrics, artist

The context-sensitive query processing is feasible. As using to services, there will be a lot of service usages which are like service usage log information. It is feasible for us to collect such kinds of services usage context information, e.g. ProgrammableWeb[7]. Query log information has been successfully used for web search engine[11]. However until now, this kind of usage information has not been taken into consideration for service discovery. In Table 1. We list some popular services' descriptions and their representative context description terms. It is easy to see that, for users, application related terms are easily understood and used for query generation. But application related terms have not been covered so much by service content description terms. Generally, if service's content description terms do not exist in queries, these services will not be returned as answers to users, even though the query terms are very close to the service description terms. For example, for the commonly known "Yahoo Maps" service, with query "travel route", it will not be returned as an answer. Previous work has seldom tackled this kind of heterogenous problem between query space and service space. Generally, application related terms are more intent to tell what kind of situation it is used for instead of what it is as in Table 1. For non-experts, application-oriented query may be more useful and feasible. In such a case, correlation to heterogenous spaces is required: from the term in application space to the terms in service space.

Table 2. Content-based search for query "location code"

Service Name	Service Description	Topic
google code search	Allow client applications to search public source code for function definitions and sample code	computer
zip codes	Enable you to get the US location behind the zip code in XML format	regional
linkpoint	Provide payment processing services for those who need sophisticated payment processing options. You can control how your payments are processed via code	business

Queries have different relevances to different topic domains. Previous work only considers content-relevance. However queries or services may have different relevance to different domains, as shown in Table 2 with query "location code". We list three returned services. Here "location code" makes much more sense in "regional" topic. Then service "zip codes" may have a higher relevance for this query. Intuitively, this service shall be ranked higher to the query.

These problems motivate us to design a new method for service discovery. We collect services usages defined as **Context** as Def. 3, to solve these problems. The main idea is: Context is modeled by a bipartite graph, which represents the correlation between applications and services; we generate a topic-based classifier and then we classify context; for each topic we generate the bipartite graph; based on these graphs, we analyze the term relationships from application space to service space. The details will be introduced as followings.

In order to make it simple, in the later of our paper, "Context" is used to represent the pair of services and application; "Application" means application description; "Content" is service description.

3 Bipartite Graph Modeling to Service Context

According to Def. 3, for any service s, $A^{[s]}$ can be seen as a set of applications and the involved collaborative services [3]. We define a bipartite graph model $\mathcal{G} = (\aleph, \varepsilon)$ to represent the relationship between applications and involved services, as shown in Fig. 3. Here $\aleph = S \cup A$ with $S \cap A = \emptyset$, S is the service set and A is the application set. The set of edges is $\varepsilon \subseteq S \times A$, which represents the involving relationship between services and applications: if a service is used by an application, there is a link between them. For a single service s, its individual context is the application a_i and other services involved with a_i. For example, in Fig. 3, a sample context of service s_1 (e.g. hotel) is a_1(e.g. travel), s_2(e.g. airline), s_3(e.g. weather) and the related linkages.

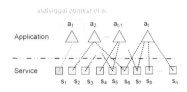

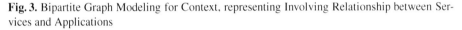

Fig. 3. Bipartite Graph Modeling for Context, representing Involving Relationship between Services and Applications

Service Context can be modeled by a bipartite graph, which catches the simple involving relationship between services and applications. However as we have mentioned in Sec. 2, generally in different topic domains, service's usefulness is different. In order to make our bipartite graph more useful, we propose to build topic-sensitive bipartite graphs that is to build the bipartite graph for each topic domain as following.

3.1 Topic Sensitive Bipartite Graph Constructing

Topic Classifier Generation. We generate a set of biasing vectors representing different topics as developed by ODP (open directory project) resources[4]. This is done offline and once. We can use a lot of resources for topic model generation. Here we prefer to use ODP which is freely available and hand constructed.

Let T_i be the set of pages in ODP topic domain c_i. Here we use the top 16 categories. We compute the topic term vector as D_i consisting of terms in T_i for topic c_i; term statistic vector D_{it} represents the number of term t occurrences in c_i.

Generally for any content composed of terms, we can calculate its topic distribution values by comparing the content vectors with these topic vectors.

Given a content string str, we can use the multinomial naive-Bayes classifier[12] to compute the probabilities for each of the ODP categories, with the parameters set to their maximum-likelihood estimates. Let str_k be the kth term in str. Then for str, we calculate str's distribution probabilities for each c_i as following:

---

[3] We just consider the existence of services to an application instead of the detailed collaboration logics as described by workflow.
[4] http://www.dmoz.org/

$$p(c_i|str) = \frac{p(c_i) \cdot p(str|c_i)}{p(str)} \propto p(c_i) \cdot \Pi_k p(str_k|c_i) \tag{1}$$

$p(str_k|c_i)$ is easily computed from term vector D_{it}. And then we take normalization for $p(c_i|str)$, $1 \leq i \leq 16$ and then $\sum_{c_i \in C} p(c_i|str) = 1$. $p(c_i)$ is uniformly valued. In this way, we can draw the topic distribution probability for a given string[13].

Topic Sensitive Bipartite Graph Generation. With topic model generated above, for each service or application o described by a string str, it is assigned a valued topic vector as $V^{[o]} = \{v_o^{c_i}\}$ with $v_o^{c_i}$ representing the tightness of current object o to topic domain c_i. We just assign the objects to the $topK$ (e.g. k=3) highly related topic domains instead of all. In order to make these topic graphs efficient and concise, we define a threshold ϱ ($\varrho \in [0, 1]$) to control the distribution scope for services and applications. The topic-sensitive graph generation algorithm is shown in Algo. 1.

---

    input : All services S, applications A, ϱ and $topK$
    output: Topic Sensitive Bipartite Graphs

1   Submit $A = \{a_i\}$ to Topic Classifier to get the topic distribution vectors;
2   **for** *each $a_i \in A$* **do**
3      $JTD^{a_i} = \emptyset$ (Join Topic Domain); $v = 0$; $K = 0$;
4      $V^{[a_i]} = \{v_{a_i}^{c_j}\}, c_j \in C$ and it is ordered descendingly;
5   **for** *each item $v_{a_i}^{c_j}$ and $K < topK$* **do**
6        $JTD^{a_i} \cup = \{c_j\}$; v+=$v_{a_i}^{c_j}$; K++;
7   **if** $v \geq \varrho$ **then**
8   break;
9   **end**
10   **end**
11 **end**
12 **for** *each topic domain $c_j \in C$* **do**
13   **for** *each $a_i \in A$* **do**
14   **if** $c_j \in JTD^{a_i}$ **then**
15        $A(c_j) \cup = a_i$; \\assign applications
16        $S(c_j) \cup = S(a_i)$; \\ assign services
17   **end**
18   **end**
19   Construct Graph for topic c_j;
20 **end**

Algorithm 1. Topic-Sensitive Bipartite Graph Constructing

First, we assign the applications A=$\{a_i\}$ rather than services to different topics C=$\{c_j\}$. The main reason is that service descriptions are less topic-sensitively distinguishable comparing to application descriptions, because service descriptions are generally used to declare service operations instead of domain-sensitive usability as in Table 1. For example, for "$YouTube$" service, based on its description, its highly related top3 topic domains are $computers$, $science$ and $business$; on the contrary, based on its applications, the highly related topic domains are $recreation$, $arts$ and $society$, which are generally acceptable. Then for each application a_i, the topic relevance vector is $V^{[a_i]} = \{v_{a_i}^{c_j}\}$ with $\sum_{c_j \in C} v_{a_i}^{c_j} = 1$ and $v_{a_i}^{c_j} = p(c_j|a_i)$ defined in Sec. 3.1.

Second, for application a_i, we select the highly related $topK$ (e.g. $topK = 3$) topic domains to join. In this algorithm we use ϱ to control application's distribution. If a_i has been related to the top K(e.g. $K = 2$) topics with $K < topK$ by the accumulated probability, calculated by $p(c_j|a_i)$, higher than ϱ (e.g. $\varrho = 99\%$), we stop a_i's distribution to other domains. For example, if application app is distributed to $recreation$, $arts$ and $society$ by probability 57%, 35% and 7% respectively, we can avoid its distribution to other topic domains with probability less than 7%, because it has a total distribution probability 99% to these three domains. If we distribute it to all other domains, it may act as noise for those domains' analysis. For each topic c_j, the selected highly related applications are $A(c_j) = \{a_i\}$. Noticed that one application may be involved in different topics. So for c_i and c_j, with $i \neq j$, $A(c_i) \cap A(c_j) \neq \emptyset$ may exist.

Third, for each topic, we collect the services used by the applications $S(c_j) \cup = S(a_i)$. $S(c_j)$ and $S(a_i)$ are services in topic c_j and used by application a_i respectively. Then the applications and services are added to the bipartite graph as introduced above: for c_j, its involved services are $S(c_j) = \{s_k\}$; by using $S(c_j)$ and $A(c_j)$, we can build the bipartite graph $g^{[c_j]}$ for topic c_j.

4   Bipartite Graph-Based Query Expansion

As mentioned in Sec. 2, if users' queries contain terms in application descriptions, which may not exist in service descriptions, this causes the mapping problem between service space and query space. The content-mapping based search will not return these services related to specific application requirements, and then it leads to low recall and precision. One popular way to solve this problem is to do query expansion for relating terms from different spaces. We exploit service usage context for such a purpose: to transform terms from query (application) space to service space.

4.1   Terms Filtering

Service or application descriptions are composed of terms. As many terms are meaningless and noisy, we take a two-step of preprocessing for filtering out these terms. First, we do stop words removing and filter out specific types of terms, as adjective, adverb, etc. Second, we do mutual information filtering[14] like Eqn. 2 and filter out terms with less information value.

$$IV(t) = p(t) \sum_{o} p(o|t) \log \frac{p(o|t)}{p(o)} \tag{2}$$

where t is a description term, and o is service or application. This calculation can remove terms with high- or low- frequency. From now on, terms we mention are the terms kept after term-filtering.

4.2 Semantics Bridging between Application (Query) Space and Service Space

Suppose two terms with t_k^a from the application description and t_j^s from the service description. Semantically, the correlation degree is calculated by $P_{c_i}(t_j^s|t_k^a)$, representing the correlation conditional probability for terms t_j^s and t_k^a, under a topic category c_i. This is the term correlation importance to a topic domain. The probability $P_{c_i}(t_j^s|t_k^a)$ is calculated as follows:

$$
\begin{aligned}
p_{c_i}(t_j^s|t_k^a) &= \frac{p_{c_i}(t_j^s, t_k^a)}{p_{c_i}(t_k^a)} = \frac{\sum_{\forall s_m \in S} p_{c_i}(t_j^s, t_k^a, s_m)}{p_{c_i}(t_k^a)} \\
&= \frac{\sum_{\forall s_m \in S} p_{c_i}(t_j^s|t_k^a, s_m) \times p_{c_i}(t_k^a, s_m)}{p_{c_i}(t_k^a)} \\
&= \frac{\sum_{\forall s_m \in S} p_{c_i}(t_j^s|s_m) \times p_{c_i}(s_m|t_k^a) \times p_{c_i}(t_k^a)}{p_{c_i}(t_k^a)} \\
&= \sum_{\forall s_m \in S} p_{c_i}(t_j^s|s_m) \times p_{c_i}(s_m|t_k^a)
\end{aligned}
\tag{3}
$$

$p_{c_i}(s_m|t_k^a)$ is the conditional probability of service s_m involved with term t_k^a in application description for topic c_i. $p_{c_i}(t_j^s|s_m)$ is the conditional probability of occurrence of t_j^s with respect to service s_m for topic c_i. Their calculation is formulated as follows:

$$
p_{c_i}(s_m|t_k^a) = \frac{f_{km}^a(t_k^a, s_m, c_i)}{f^a(t_k^a, c_i)}
\tag{4}
$$

$f_{km}^a(t_k^a, s_m, c_i)$ and $f^a(t_k^a, c_i)$ are the number of co-occurrences between application description term t_k^a and service s_m and the total number of applications involved with term t_k^a in topic c_i, respectively.

$$
p_{c_i}(t_j^s|s_m) = \frac{t_{jm}^s}{\sum_{\forall t \in s_m} t_{tm}^s}
\tag{5}
$$

t_{jm}^s and $\sum_{\forall t \in s_m} t_{tm}^s$ are the term weights calculated by TFIDF, for t_j^s and total number of term weight in service s_m, respectively.

Combining the equations in Eqn. 3, 4 and 5, we acquire the final calculation for $P_{c_i}(t_j^s|t_k^a)$ as:

$$
p_{c_i}(t_j^s|t_k^a) = \sum_{\forall s_m \in S} \frac{f_{km}^a(t_k^a, s_m, c_i)}{f^a(t_k^a, c_i)} \times \frac{t_{jm}^s}{\sum_{\forall t \in s_m} t_{tm}^s}
\tag{6}
$$

4.3 Query Expansion

Query expansion algorithm is shown in Algo. 2. Bipartite graph modeling to context provides a way to correlate services and applications. We try to create the correlations between terms in service side and terms in application side. The main differences with other work are: 1)it implements expansion between different spaces; 2) it is topic-sensitive. It means that we prefer to select most relevant topics to expand terms. The global statistical information for each related term is ranked by the topic based term relevance, as in Line 5, and we select the globally highly ranked terms as the final expansion terms.

---

input : Query q- composed of terms
output: q'- expanded query

1 Submit q to topic model;
2 Acquire q's topic distribution vector $\{v_q^{c_i}\}$;
3 **for** *each $t_k^q \in q$* **do**
4   Acquire top relevant terms $T_{t_k^q}^{c_i} = \{t_j\}$ by $p_{c_i}(t_j | t_k^q)$ in topic c_i by Eqn. 3;
5   Calculate term t_j final query relevance: $Rel_{t_j}^q = \sum_{c_i \in c} v_q^{c_i} \cdot \sum_{t_k^q \in q} p_{c_i}(t_j | t_k^q)$;
6 **end**
7 Order the selected terms to $ET = \{t_j\}$ by decreasing value of $Rel_{t_j}^q$ and select top terms as expansion terms;

Algorithm 2. Query Expansion based on Topic Sensitive Bipartite Graph Model

5 Experimental Results

5.1 Experiment Data Set

We focus on using services context to solve application-oriented queries, which does not focus on analyzing content or (parameter) structure similarity to queries, so we do not collect a great deal of well structured WSDL-based services. Instead we collect services and service-related context from ProgrammableWeb[7], which records a large number of free API services and their involved contexts. For each service, we can obtain service's description including service title, service description, and URL; for each mashup[15] application, we regard them as service contexts which have titles, descriptions, and involved service URLs. Finally, we get 1577 services and 3996 contexts. To avoid overfitting problem, we use 80% of the contexts as training data to build the semantic bridge for Eqn. 3. We use 5% of services as development data to choose query expansion depth (expansion term number) and 15% as test data.

In our experiment, we test two types of queries. One is automatically generated queries: we use (part of) the application description from development data and test data as query and take application involved services as the correct answers. For example, for query "text data preprocessing" as in Fig. 2, the possible expected results are 5 kinds of services. The other one is manually generated queries: we manually generate 24 queries from both service content description and application description. It will be introduced later.

We compare our method **OURS** with the following 4 systems, labeled as **BS**, **BS-EXP**, **Woogle-like**[9] and **APP**. **OURS** is the topic sensitive query expansion method; **BS**:baseline system implemented by content-based mapping method; **BS-EXP**: query expansion without topic classifier; **Woogle-like**:local content-based term conceptualization method to clustering similar terms. **APP**: query to application descriptions without query expansion.

5.2 Data Status

Service description term distribution status is summarized in Fig. 1 where 70% of services have fewer than 30 terms. For applications, 50% have 10-19 description terms as shown in Fig. 5. Notice that currently only 48% of services in our dataset have contexts. Fig. 4 shows applications' topic distributions based on our topic classifier, with $topK = 3$ and stop parameter $\varrho = 0.8$. The mainly involved topics are "computers", "society", "business", "arts" and "recreation". We do not emphasize on dividing the topic categories into smaller ones by diving into the hierarchical structure of ODP, which will help to get an even (better) service distribution to topic domains. However we have verified that system performance can win great increasing even with this kind of coarse resource distribution. And the means of detailing topics division will be left as future work.

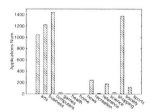 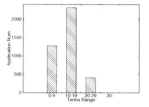

Fig. 4. Application Distribution Probability on Topics

Fig. 5. Application Des. Terms Distribution

5.3 Evaluation Metrics

We use the metrics of $P@N$, MAP and MRR to evaluate system performance [16]. For a query set Q, we calculate the average values for $P@N$, MAP and MRR. For a query q:

$P@N$: Precision (P) at top N results. $P@N = \frac{|CA_q \cap R_q|}{|R_q|}$, where CA_q is the set of tagged correct answers and R_q is the set of top N results returned by the system. In our experiment, we select N=3, 5 and 10.

MAP: Mean Average Precision. It is used to evaluate the global relevance of returned ordered results. $MAP = \frac{1}{|CA_q|} \times (\sum_{i=1}^{|CA_q|}(P@|R_{ca_i}|))$. ca_i is the ith relevant service to q in CA_q and R_{ca_i} is the set of ranked retrieval results from the top results until you get service ca_i.

Mean Reciprocal Rank is: $MRR = \frac{1}{r_q}$, r_q is the rank of the first relevant service for q. The higher the MRR value, the better the system.

5.4 Performance

Query Expansion Depth & Performance with Automatically Generated Queries.
Generally speaking, the expansion terms are not the more the better, due to the introducing of irrelevant terms. Here based on the development dataset, we choose the best expansion depth for queries those automatically generated from context descriptions. We exploit the query expansion depth (0, 3, 5, 10, 15, 20 and 25) to see the performance for P@N in Fig. 6 and MRR and MAP in Fig. 7. For these queries, expansion depth 5 can win the best performance. So we choose 5 for the next two parts of experimental usage.

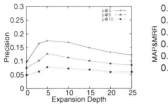

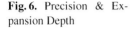

Fig. 6. Precision & Expansion Depth

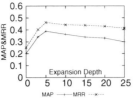

Fig. 7. MRR and MAP & Expansion Depth

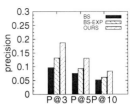

Fig. 8. Precision for Context Queries

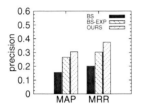

Fig. 9. MAP and MRR for Context Queries

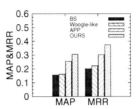

Fig. 10. MAP and MRR for Different Systems

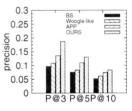

Fig. 11. P@N for Different Systems

Comparing Performance based on Different Implementations with Automatically Generated Queries. On the test dataset, we check the query performance based on our system. The results are shown in Fig. 8 and Fig. 9. Such kinds of queries are difficult for the BS system, because of few overlap between application terms and service terms. In this case, term expansion helps a lot for query processing. Both $BS - EXP$ and $OURS$ win better performance. And our method $OURS$ performs better than the non-topic system $BS - EXP$.

We also compare performance to other systems as in Fig. 10 and Fig. 11. We find $Woogle - like$ method does not help so much for query processing if the query is with application terms, because it does only local information analysis. Here even APP that is query processing on application descriptions can beat $Woogle-like$ method. $OURS$ wins best with topic sensitive query expansion.

Performance Improvement & Query Difficulty with Automatically Generated Queries. We also analyze the effectiveness of our method in helping difficult queries which have been studied in recent years [17]. In Fig. 12, we show the results. According to BS system's MAP values, we divide queries equally into five bins. The higher the MAP value means the utility of content-based mapping method for search is better. Bin 0 is assigned queries with the lowest MAP values and Bin 4 with the highest ones. "Improvement" and "Decreased" mean the improving and decreasing to "$P@5$" for these 5 bins of queries. Clearly, for difficult queries (lower MAP with BS), OURs can improve the performance. In Bin 0, $OURS$ can solve all queries with MAP values no worse than BS, e.g. 50 $improved$ vs 2 $decreased$. But for Bin 4, $OURS$ loses to BS by around 8 queries. It is verified that for application-oriented queries, our method performs better than BS system.

Table 3. Example Manually Generated Queries

ID	Short Queries	Long Queries
1	travel information browser	travel information browser for trip destinations across the globe It features videos, photos, guides and attractions, weather information and events for nearly all countries and many cities world wide
2	writers books Timeline	search the favorite writers Books And you can see results by Timeline showing a list and a publication day
3	ranking blogs	A ranking system for blogs, View the top 100 Indonesian blogs
4	talk real-time translation	talk with people all over the world by a real-time translation
5	music video rating	a grid of music videos from that band shown as thumbnails, listing their rating, view count, and the title of the video
6	US collecges information	colleges and universities in the US plotted on a Google map with street views and additional school information
7	locate zip	locate your zip automatically
8	artist mp3 information	Get lots of information about your favorite artists. View album information, find mp3 songs, lyrics, biographies, podcasts, rate artists and add comments
9	music previewing	previewing of music and has artist profiles with Music Videos from MTV
10	feeds aggregator	A feeds aggregator organized by topic where posts are auto-tagged with semantic terms
11	hotels search	Search hotels by city, check-in dates, number of guests, number of rooms, and stars
12	city event	Select a city and view an aggregated page of feeds. See what is happening in your city or a city that you are traveling or moving to

Manually Generated Queries. For previous queries, we have not ensured queries' terms overlap with service description, which is a necessary requirement for the success of the BS system. In order to make the comparison acceptable, we manually generate 24 queries for both long queries and corresponding short queries as shown in Table 3(because of lack of space, we only list 12 example queries). We promise all of the queries have terms overlap with service descriptions. We manually find the relevant services for each query by using the "Pooling" method[18], which has been frequently used in IR. The judgement pool is created as follows. We use BS, and $OURS$ to produce the $topK = 20$ candidates and then merge these results for our real candidates selection. We then select the most relevant results by our 3 evaluators from the pool. To ensure $topK = 20$ is meaningful, we select the queries which can return more than 20 relevant services by each system. Here "-short" and "-long" mean short queries and long queries, respectively.

In Fig. 13 and Fig. 14, we test the influence of expansion depth for $OURS$ to short queries and long queries. We find that query expansion has much greater influence to improve short query performance, because there is a larger change between the highest and lowest values for either precision or MAP and MRR to short queries. For example, in Fig. 13, when expansion depth is 5, the precision change for "p@3-short" is around 13.5%, but for "p@3-long", the precision change is around 10%. The main

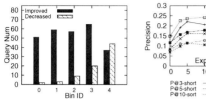

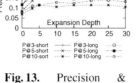

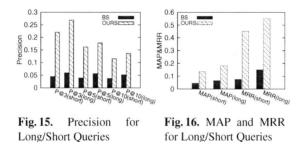

Fig. 12. Performance Improvement & Query Difficulty

Fig. 13. Precision & Expansion Depth for Long/Short Queries

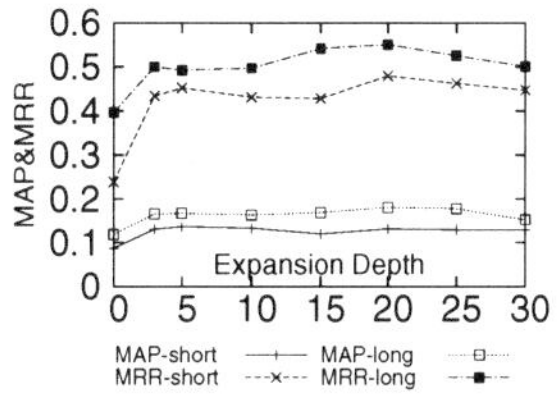

Fig. 14. MAP and MRR& Expansion Depth for Long/Short Queries

Fig. 15. Precision for Long/Short Queries

Fig. 16. MAP and MRR for Long/Short Queries

reason is that short query has less semantic information for the system to distinguish them by. But after expansion, short queries are clarified. With expansion, the precision change for the top list is obvious. In such a case, "p@3-short" wins greater improvement than "p@5-short". Long query will always have better performance than short queries because of its clearer requirement declaration.

Based on the above experiment, we take expansion depth 5 and 20 for short and long queries respectively, which are both the highest-performance parameters. We compare baseline system BS and our method for P@N, MAP and MRR in Fig. 15 and Fig. 16, respectively. For short queries, the BS system's performance is very low, because of the lack of service description and heterogeneity between query space and service space. But $OURS$ has good improvement to the BS system. Moreover, because of the low query precision of BS, current web service search engine would prefer to provide browsing functionality rather than complicated search functionality.

6 Related Work

Along with the increasing of web services, desirable web service discovery has increased its importance to win users. Content-based web service search has been found that there is noticeable amount of noise for service search, because of service description is short or not enough to distinguish themselves from others. Much effort has been put on solving this problem by detecting or extending service self static description. Until now, we summarize the work into 3 groups.

1. Content matching. It is supposed to compare service's functional descriptions or functional attributes or parameters with queries to check whether advertisements support the required type of services. And it checks the functionality capabilities of web

services in terms of input and output, as used by woogle[9]. For service search and similarity calculation, it tries to include as more information as it can to characterize services. It is challenged by the length of service descriptions, the variation of parameter names and the scale of service repository. Generally if the overlap among services are few, woogle-like method may not win good performance improvement.

2. Ontology/semantics mapping [19–21]. These approaches require clear and formal semantic annotations. But as we know most of the services those are active on WWW do not contain so many ontology annotations. One of big challenges is the definition to ontology. However, in order to make the work successful, it shall not contain so much semantic constraints which bound the activity of users and developers.

3. Context matching [10, 22–24]. Recently context-aware approaches have been proposed to enhance web service discovery and composition. [24] proposes a context-related service discovery technique for mobile environments by defining a set of attributes to service. The search is still based on tradition content-mapping search mechanism and context attributes act as filters. [10] suggests to define the context from two aspects: client-related context and provider-related context. It prefers to absorbing all the information related to service activity as the context, which makes context complicated and difficult to follow. By the way, real experiment experience has not been with the work. Context is well used in web search [25], but we deal with different spaces.

7 Conclusion and Future Work

The mapping problem between query space and service space has caused low search precision and it affects the development and popularity of Web Service Search Engines, resulting in some of them only providing service browsing functionality, instead of search. One kind of queries is application-oriented queries, which has been checked to be useful and necessary especially for non-expert users. This paper proposes to do query expansion based on service usage context to solve this kind of query. We generate the topic sensitive bipartite graph model to represent service context. By exploiting context, we establish term correlation between service space and application space. And it is proved to be an efficient way for application-oriented query processing. The experimental results show that our approach outperforms other systems.

Our work is different with but complementary to previous works. Service local properties are well exploited by great effort, e.g. Woogle[9]. Rather than taking context as local properties for results filtering[24], we define context as service usages to applications. In the future, we want to combine the local content analysis with our context data analysis; we want to detail the topic classification to improve the performance.

References

1. ActiveVOS: Activebpel engine,
   http://www.activevos.com/community-open-source.php
2. BPMNModeler, http://www.eclipse.org/bpmn/
3. BindingPoint, http://www.bindingpoint.com/
4. WebServiceList, http://www.webservicelist.com/

5. Salcentral, `http://www.salcentral.com/`
6. Fan, J., Kambhampati, S.: A snapshot of public web services. Journal of the ACM SIGMOD RECORD (2005)
7. mashups, `http://www.programmableweb.com/`
8. Xu, J., Croft, W.: Improving the effectiveness of information retrieval with local context analysis. ACM Trans. Information Systems 18(1), 79–112 (2000)
9. Dong, X., Halevy, A., et al.: Similarity search for web services. In: Proc. VLDB, pp. 372–383 (2004)
10. Medjahed, B., Atif, Y.: Context-based matching for web service composition. Distributed and Parallel Databases 21(1), 5–37 (2007)
11. Vectomova, O., Wang, Y.: A study of the effect of term proximity on query expansion. Journal of Information Science 32(4), 324–333 (2006)
12. Mitchell, T.: Machine Learning. McGraw-Hill, Boston (1997)
13. Haveliwala, T.H.: Topic-sensitive pagerank. In: Proc. WWW (2002)
14. Hsu, W.H., Chang, S.F.: Topic tracking across broadcast news videos with visual duplicates and semantic concepts. In: Proc. ICIP, pp. 141–144 (2006)
15. Mashup, `http://en.wikipedia.org/wiki/Mashup`
16. Wikipedia: Information retrieval, `http://en.wikipedia.org/wiki/Information_retrieval`
17. Yom-Tov, E., Fine, S., et al.: Learning to estimate query difficulty: including applications to missing content detection and distributed information retrieval. In: Proc. SIGIR, pp. 512–519 (2005)
18. Voorhees, E., Harman, D.: Overview of the sixth text retrieval conference (TREC-6). Information Processing & Management 36(1), 3–35 (2000)
19. Selvi, S., Balachandar, R.A., et al.: Semantic discovery of grid services using functionality based matchmaking algorithm. In: Proc. WI, pp. 170–173 (2006)
20. Paolucci, M., Kawamura, T., et al.: Semantic matching of web services capabilities. In: Horrocks, I., Hendler, J. (eds.) ISWC 2002. LNCS, vol. 2342, p. 333. Springer, Heidelberg (2002)
21. Klusch, M., Fries, B., et al.: Owls-mx: Hybrid semantic web service retrieval. In: Proc. 1st Intl. AAAI Fall Symposium on Agents and the Semantic Web (2005)
22. Morris, M.R., Teevan, J., et al.: Enhancing collaborative web search with personalization: groupization, smart splitting, and group hit-highlighting. In: Proc. CSCW, pp. 481–484 (2008)
23. Wong, J., Hong, J.I.: Making mashups with marmite: Towards end-user programming for the web. In: Proc. CHI, pp. 1435–1444 (2007)
24. Lee, C., Helal, S.: Context attributes: an approach to enable context- awareness for service discovery. In: Proc. SAINT, pp. 22–30 (2003)
25. Cao, H., Jiang, D., et al.: Context-aware query suggestion by mining click-through and session data. In: Proc. KDD, pp. 875–883 (2008)

BMC: An Efficient Method
to Evaluate Probabilistic Reachability Queries

Ke Zhu, Wenjie Zhang, Gaoping Zhu, Ying Zhang, and Xuemin Lin

University of New South Wales, Sydney, NSW, Australia
{kez,zhangw,gzhu,yingz,lxue}@cse.unsw.edu.au

Abstract. Reachability query is a fundamental problem in graph databases. It answers whether or not there exists a path between a source vertex and a destination vertex and is widely used in various applications including road networks, social networks, world wide web and bioinformatics. In some emerging applications, uncertainties may be inherent in the graphs. For instance, each edge in a graph could be associated with a probability to appear. In this paper, we study the reachability problem over such uncertain graphs in a threshold fashion, namely, to determine if a source vertex could reach a destination vertex with probabilty larger than a user specified probability value t. Finding reachability on uncertain graphs has been proved to be NP-Hard. We first propose novel and effective bounding techniques to obtain the upper bound of reachability probability between the source and destination. If the upper bound fails to prune the query, efficient dynamic Monte Carlo simulation technqiues will be applied to answer the probabilitistic reachability query with an accuracy guarantee. Extensive experiments over real and synthetic datasets are conducted to demonstrate the efficiency and effectiveness of our techniques.

1 Introduction

In many real world applications, complicatedly structured data could be represented by graphs. These applications include Bioinformatics, Social Networks, World Wide Web, etc. Reachability query is one of the fundamental graph problems. A reachability query answers whether a vertex u could reach another vertex v in a graph. Database community has put considerable efforts into studying the reachability problem, for example, [7], [5], [1], [14], [9], [2], [8], [4], [3], etc.

All of the above works focus on the applications where edges between two vertices exist for certain. However, in many novel applications, such an assumption may not capture the precise semantics and thus the results produced are also imprecise.

Example 1: In Protein-Protein interaction networks, an edge between two proteins means they have been observed to interact with each other in some experiments. However, not all interactions can be consistantly observed in every experiment. Therefore, it is more accurate to assign probabilities to edges to represent the confidence on the relationship. In this application, biologists may want to query whether a particular protein is related to another protein through a series of interactions.

J.X. Yu, M.H. Kim, and R. Unland (Eds.): DASFAA 2011, Part I, LNCS 6587, pp. 434–449, 2011.
© Springer-Verlag Berlin Heidelberg 2011

Example 2: Social Network Analysis has recently gained great research attention with the emergence of large-scale social networks like LinkedIn, Facebook, Twitter and MySpace. In these social networks, connections between entities/individuals(vertices) may not be completely precise due to various reasons including errors incurred in data collection process, privacy protection, complexed semantics, disguised information, etc([22]).

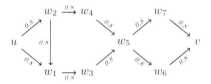

Fig. 1. A running example

In above applications, an edge connecting two vertices is associated with a probability value indicating the confidence of its existence. Reachability queries over this kind of uncertain graphs are thus called Probabilistic Reachability Queries. The Probabilistic Reachability problem is intrinsically difficult. As a running example in Fig. 1, this graph consists of only 11 edges. To accurately answer the Probabilistic Reachability from u to v, we need to enumerate up to 2^{11} possible instances of the uncertain graph. For each of these instances, we need to check whether u can reach v, and then aggregate the probabilities of the instances in which u can reach v. In [10], Valiant has proved this problem is NP-Hard.

Monte Carlo simulation provides an approximate solution to this problem. A considerable number of articles([15],[16],[17],[18],[19], etc) studied how to use Monte Carlo simulation to solve the probabilistic reachability problem. The focus of those studies are on utilizing different sampling plans to reduce sampling error. Due to the dramatical increase of the scale of graphs and the large number of iterations required by Monte Carlo simulation to guarantee the accuracy, the computational cost of traditional Monte Carlo method is still considerably expensive.

In this paper, we propose a more efficient dynamic Monte Carlo method to approximate the answer. This dymanic Monte Carlo method will only simulate necessary part of the graph and will share most of the overlapping cost between different iterations. In addition to that, we also propose an index which can assist in calculating upper bound of probabilistic reachability. Queries pruned by the bound do no need to be approximated by the Morte Carlo method which is relatively more expensive. The main contributions of the paper are:

1. To the best of our knowledge, we are the first to address the efficiency issues of Probabilistic Reachability Queryby using indexing techniques. We formally define Probabilistic Reachability Queryusing *Possible World* semantics.
2. We propose an efficient dynamic Monte Carlo algorithm to calculate approximate result. In addition, we also give a theoretical accuracy guarantee for the Monte Carlo method.
3. We propose an index which efficiently calculates the upper bound of Probabilistic Reachability Queries.
4. We perform extensive experiments on real datasets and synthetic datasets to demonstrate the efficiency of our proposed method.

Table 1. Notations

$u \rightsquigarrow v$, $u \not\rightsquigarrow v$	u can reaches v; u cannot reach v
$p(e)$	the probability that edge e will exist
$R_{u,v}$	the probability that u could reach v
$R^U_{u,v}$	the upper bound probability that u could reach v
ω, Ω	a possible world and the set of all possible worlds respectively
p_ω	the probability of a possible world
$s(u,v)$	the shortest distance between u and v
$Prob(Event)$	the probability that an $Event$ will occur

The whole paper is organized as follows: Section 2 will introduce the background knowledge of this problem. Section 3 will briefly outline our Bound and Monte Carlo(BMC) framework. Section 4 will propose a novel bound-based scheme to address the problem. Section 5 will introduce our dynamic Monte Carlo method. Section 6 will demonstrate and analyze the experiments. Section 7 will introduce related works and Section 8 concludes the paper.

2   Background

2.1   Problem Definition

In this paper, we study the reachability problem in graphs in which each edge is associated with an existence probability and we call such graphs *Uncertain Graphs*

Definition 1 (Uncertain Graph). *An uncertain graph is defined as $G = (V, E, P_E)$ where V is the set of vertices, E is the set of edges and $P_E : E \rightarrow (0, 1]$ is the edge probability function. We use $p(e)$ to denote the probability that e exists where $e \in E$.*

The probability that vertex u could reach v can be calculated by summing the probability of all possible combinations of the edge states. Each of the combination corresponds to a *Possible World* in the *Possible World* semantics. We use $R_{u,v}$ to represent the probability of being reachable and $\bar{R}_{u,v}$ otherwise.

Definition 2 (Possible World). *Let $x_e = 1$ if e exists and $x_e = 0$ if otherwise. We call ω a possible world where $\omega = \{x_e | e \in E\}$.*

We use Ω to denote the set of all possible worlds of an uncertain graph and let $r_{u,v}(\omega) = 1$ if u could reach v in ω or $r_{u,v}(\omega) = 0$ when otherwise. We also use p_ω to represent the probability for ω to occur. $p_\omega = \prod_{e \in E} h(e)$ where:

$$h(e) = \begin{cases} p(e) \text{ if } x_e = 1 \\ 1\text{-} p(e) \text{ if } x_e = 0 \end{cases}$$

Definition 3 (Probabilistic Reachability). *The probabilistic reachability between two vertices u and v, $R_{u,v}$, is the sum of probability of all possible worlds in which u can reach v. That is, $R_{u,v} = \sum_{\omega \in \Omega} r_{u,v}(\omega) \cdot p_\omega$.*

Definition 4 (Probabilistic Reachability Queries). *Given a large uncertain database graph G, two vertices u, v, where $u, v \in V$, and a threshold t where $1 \geq t > 0$, the database outputs true if $R_{u,v} \geq t$ or false if $R_{u,v} < t$. We call this type of queries* Probabilistic Reachability Queries.

2.2 Preliminaries

Naive Enumeration. Without any pruning strategy, we need to enumerate every possible world $\omega \in \Omega$ and to increment the probability of success or failure till t or $1 - t$ is reached. Algorithm **NaiveEnumerate** outlines the naive enumeration process.

Procedure. `NaiveEnumerate`(G, u, v, t)

1 **begin**
2     $fail, operate = 0$;
3     $\Omega = $ all possible worlds of G ;
4   **foreach** ω *in* Ω **do**
5   **if** u *can reach* v *in* ω **then**
6           $\lfloor$ $operate = operate + p_\omega$;
7   **else**
8           $\lfloor$ $fail = fail + p_\omega$;
9   **if** $fail > 1 - t$ **then**
10          $\lfloor$ **return** $false$;
11   **if** $operate \geq t$ **then**
12          $\lfloor$ **return** $false$;
13 **end**

Monte Carlo Sampling. The complexity of caculating probablistic reachability has been proved to be NP-Hard[10]. The cost grows exponentially as size of graphs grows. Monte Carlo sampling method is generally a widely accepted method of approximating the result. Briefly, it has three steps:

1. Randomly and independently determine a state for every edge in the graph according to the operational probability of each edge. A sample graph consists of all edges with a *exist* state.
2. Test the the reachability for this sample graph.
3. Repeat the above step 1 and 2 for k iterations.

The approximate probabilistic reachability is: $R'_{u,v} = \frac{\sum_{\omega \in \Omega_k} r_{u,v}(\omega)}{k}$ where Ω_k is a set of k sampled states.

3 Framework

As finding the probablistic reachability is infeasible when data graphs are large, we propose a framework integrating an effective bounding-pruning technique and an efficient Monte Carlo sampling technique. Intuitively, if $R_{u,v}$ is small and the threshold is large, it is possible to obtain an upper bound, $R^U_{u,v}$, to immediately

reject the query. If the bounding technique fails to prune the query, then Monte
Carlo sampling technique will be applied to produce an approximate answer. We
observe that a major portion of the traditional Monte Carlo simulation can be
shared, thus we propose a more efficient Dynamic Monte Carlo method to approximate the result. The following sumarizes the major steps in our framework.

1. We create an index on the database graph so that the upper bound of the exact reachability could be calculated efficiently. We will attempt to prune the
   query by calculating the upper bound of the probablistic reachability. This
   technique is to be detailed in **Section 4**.
2. If the upper bound cannot prune the query, we will sample a number of possible worlds and use our proposed Dynamic Monte Carlo simulation to estimate
   the reachability. This technique is to be detailed in **Section 5**.

4   Upper Bound Index

Naive Enumeration is impractical to answer Probabilistic Reachability Queries.
However, Probabilistic Reachability Queries only need to answer whether the
reachability is above a threshold t, rather than the exact probabilistic reachability. With the help of indices, we can efficiently calculate an effective upper
bound of the reachability between the source and the destination. If the upper
bound can be used to prune the query before Monte Carlo Simulation, we can
avoid the relatively more expensive sampling and reachability testing.

As mentioned previously, it is infeasible to enumerate every single possible
world when a graph is large. The following observations inspired us to propose
an efficient upper bound index:

Observation 1: Many real-world graphs are sparse and a local neighbourhood
graph surrounding a vertex is usually small in sparse graphs. It is affordable to
enumerate all possible worlds in a small neighbourhood graph.
Observation 2: The local neighbourhood structure surrounding a vertex can
usually provide an upper bound of its ability to reach(or to be reached by) other
vertices.

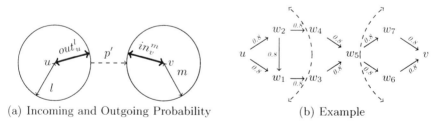

(a) Incoming and Outgoing Probability   (b) Example

Fig. 2. Upper Bound Calculation

As shown in Fig. 2 (a), a vertex u has to reach at least one vertex outside
the circle of radius l before it can reach vertex v if l is less than the unweighted
shortest distance between u and v. We will use $s(u, v)$ to denote the shortest
distance between u and v. Similarly, a vertex v has to be reached by at least

one vertex outside the circle of radius m before it can be reached by v if m is less than $s(u,v)$. The probabilistic reachability is bounded from above by the *outgoing probability* and the *incoming probability*, which are defined as below.

Definition 5 (Outgoing and Incoming Probability). *The outgoing probability of vertex u, $out^k{}_u$, represent the probability that u could reach at least one vertex w where $s(u,w) \geq k$. Similarly, the incoming probability of vertex v, $in^k{}_v$, means the probability that at least one vertex w could reach v where $s(w,v) \geq k$.*

Table 2. u's Outgoing Probability

Possible World ω	p_ω	Possible World ω	p_ω
$\{uw_1, \overline{uw_2}, w_2w_1, w_2w_4, w_1w_3\}$	0.08192	$\{uw_1, \overline{uw_2}, \overline{w_2w_1}, w_2w_4, w_1w_3\}$	0.02048
$\{uw_1, \overline{uw_2}, w_2w_1, \overline{w_2w_4}, w_1w_3\}$	0.02048	$\{uw_1, \overline{uw_2}, \overline{w_2w_1}, \overline{w_2w_4}, w_1w_3\}$	0.00512
$\{\overline{uw_1}, uw_2, w_2w_1, w_2w_4, w_1w_3\}$	0.08192	$\{\overline{uw_1}, uw_2, w_2w_1, \overline{w_2w_4}, w_1w_3\}$	0.02048
$\{\overline{uw_1}, uw_2, w_2w_1, w_2w_4, \overline{w_1w_3}\}$	0.02048	$\{\overline{uw_1}, uw_2, \overline{w_2w_1}, w_2w_4, \overline{w_1w_3}\}$	0.00512
$\{\overline{uw_1}, uw_2, \overline{w_2w_1}, w_2w_4, w_1w_3\}$	0.02048	$\{uw_1, uw_2, w_2w_1, w_2w_4, w_1w_3\}$	0.32768
$\{uw_1, uw_2, w_2w_1, \overline{w_2w_4}, w_1w_3\}$	0.08192	$\{uw_1, uw_2, w_2w_1, w_2w_4, \overline{w_1w_3}\}$	0.08192
$\{uw_1, uw_2, \overline{w_2w_1}, w_2w_4, w_1w_3\}$	0.08192	$\{uw_1, uw_2, \overline{w_2w_1}, \overline{w_2w_4}, w_1w_3\}$	0.02048
$\{uw_1, uw_2, \overline{w_2w_1}, w_2w_4, \overline{w_1w_3}\}$	0.02048		

In Table 2, we give an example of how to calculate the outgoing probability of u when the radius is 2 for the graph in Fig. 2 (b). In this example, we list all possible worlds in which u can reach at least one of w_3 and w_4. Each possible world is represented by a list of edge states. For example, $\{uw_1, \overline{uw_2}\}$ represents a possible world in which the edge between u and w_1 exists, and the edge between u and w_2 does not exist. The outgoing reachability is the aggregated probability of all listed possible worlds. Please note that we do not need to enumerate edges which are further than the specified radius because they have no effect on the outgoing reachability.

Generally speaking, as the outgoing(or incoming) edges get denser, it is more likely to reach(or to be reached by) other vertices. We can index the outgoing and incoming probability for every vertex for a specific small radius. In addition to the incoming and the outgoing probability, every u-v cut will bound the reachability between u and v from above.

Definition 6 (u-v Cut and Non-overlapping u-v Cut Set). *A u-v cut is a set of edges which will make v unreachable from u if all edges in the set are missing. A non-overlapping set of u-v cuts $C^{l,m}_{u,v}$ is a set of u-v cuts such that $\forall c_1, c_2 \in C^{l,m}_{u,v}, c_1 \cap c_2 = \emptyset$ and $\forall c \in C^{l,m}_{u,v}, \forall (x,y) \in c, s(u,y) > l \wedge s(x,v) > m$.*

Let us denote a cut with c, the event that all edges in c are missing with $Cut(c)$ and the probability for this event to occur with $Prob(Cut(c))$. For any l and m where $l + m \leq s(u,v)$ and we are given $C^{l,m}_{u,v}$, u could not reach v if any one of the following conditions is true: 1). $\forall x \in V, \neg(s(u,x) \geq l \wedge u \rightsquigarrow x)$. 2). $\exists c \in C^{l,m}_{u,v}$, $Cut(c)$ occurs. 3). $\forall x \in V, \neg(s(x,v) \geq m \wedge x \rightsquigarrow v)$

Theorem 1. *For any u and v, if we are given out^l_u, in^m_v, $C^{l,m}_{u,v}$ where $l + m \leq s(u,v)$, l ,$m \geq 0$:*

$$R_{u,v} \leq out_u^l \cdot in_v^m \cdot \prod_{c \in C_{u,v}^{l,m}} 1 - Prob(Cut(c)). \tag{1}$$

Proof. We will prove the above theorem by proving the following three conditions are necessary conditions and they are independent to each other.

1. If $\forall x \in V, \neg(s(u,x) \geq l \wedge u \rightsquigarrow x)$, then $(u \not\rightsquigarrow v)$ because $s(u,v) \geq l$. Therefore if $u \rightsquigarrow v$, then $\exists x \in V$, such that $s(u,x) \geq l \wedge u \rightsquigarrow x$. Let us call this event E_1.
2. If $\forall x \in V, \neg(s(x,v) \geq m \wedge x \rightsquigarrow v)$, then $(u \not\rightsquigarrow v)$ because $s(u,v) \geq m$. Therefore if $u \rightsquigarrow v$, then $\exists x \in V$, such that $s(x,v) \geq m \wedge x \rightsquigarrow v$. Let us call this event E_2.
3. $\exists c \in C_{u,v}^{l,m}$, $Cut(c)$ occurs, then $(u \not\rightsquigarrow v)$ by definition of cut. Therefore if $u \rightsquigarrow v$, $\forall c \in C_{u,v}^{l,m}$, $Cut(c)$ cannot occur. Let us call this event E_3.

Since E_1, E_2, E_3 are necessary conditions for the event $u \rightsquigarrow v$, therefore $R_{u,v} \leq Prob(E_1 \cap E_2 \cap E_3)$. $Prob(E_1) = out_u^l$ and $Prob(E_2) = in_v^m$. E_1 could only overlap with E_2 if there exists an edge (x,y) such that $s(u,x) \leq l-1 \wedge s(y,v) \leq m-1$. There will exist a path starting from u and *arriving* at v via the edge (x,y) with length $s(u,x) + 1 + s(y,v)$ which is less or equal than $l + m - 1$. However, this is less than the shortest distance between u and v because $s(u,v) \geq l+m$ is given. This is a contradiction. Therefore E_1 cannot overlap with E_2. E_1 and E_2 also cannot overlap with E_3 because $\forall c \in C_{u,v}^{l,m}, \forall(x,y) \in c, s(u,y) > l \wedge s(x,v) > m$. Informally, any edges included in $C_{u,v}^{l,m}$, must be outside of the circle centred at u with radius l, and be outside of the circle centred at v with radius m.   □

Given l and m, there are different choices of $C_{u,v}^{l,m}$. In this paper, we will simply define $C_{u,v}^{l,m}$ as $C_{u,v}^{l,m} = \{c_i | i \in \mathbb{I} \wedge l \leq i < s(u,v) - m\}$ where $c_i = \{(x,y) | s(u,x) = i \wedge s(u,y) = i+1 \wedge (x,y) \in E(g)\}$.

In the running example in Fig. 2 (b), the source and the destination is u and v respectively. The numbers on the edges are the existence probability. Suppose we have already indexed the outgoing/incoming probability up to the radius 2 and the shortest distance. The outgoing probability of radius 2 for u is the total proability of the possible worlds in which u can reach at least one of w_3 and w_4 whose shortest distance from u is 2. We initialize the upper bound as $out_u^2 \times in_v^2$ and this upper bound can be further reduced by independent u-v cuts. In this case, $\{\{w_3 w_5, w_4 w_5\}\}$ is the only cut to be considered. Algorithm **Upper** formalizes the above statements.

Procedure. Upper(g, u, v)

---

1 **begin**
2   choose the l and m such that $l + m < s(u,v)$ and $out_u^l \times in_v^m$ is minimum ;
3      $upper = out_u^l \times in_v^m$;
4   **for** $i = l$ **to** $s(u,v) - m - 1$ **do**
5          $c_i = \{(x,y) | s(u,x) = i \wedge s(u,y) = i+1 \wedge (x,y) \in E(g)\}$;
6          $upper = upper \times (1 - \prod_{e \in c} p(e))$;
7   **return** $upper$;
8 **end**

---

5   Dynamic Monte Carlo Simulation

Traditionally, Monte Carlo simulation has been widely accepted as one of the efficient methods to answer reachability problems in uncertain graphs due to the NP-hard nature of this problem. In this section, we will propose a dynamic Monte Carlo simulation method which integrates sample generation and reachability test to maximize the computational sharing.

There are two major costs in Monte Carlo method:

1. **Generating Sample:** running a sample pool of size k requires to generate k samples, for each sample, every edge is to be assigned *exist* or *not exist* according to the existence probability. This costs $O(k|E|)$ of time.
2. **Checking Sample Reachability:** For each sample, we need to check the reachability between u and ve. This operation costs $O(|E|)$ of time for each iteration.

We have two observations in regarding to these two costs.

Observation 3: When some edges are missing, the presence of some other edges are no longer relevant. For example, in the example Fig. 2 (b), the states of other edges will no longer affect the reachability between u and v if uw_1 and uw_2 are missing.

Observation 4: Many samples share a significant portion of *existing* or *missing* edges, the reachability checking cost could be shared among them.

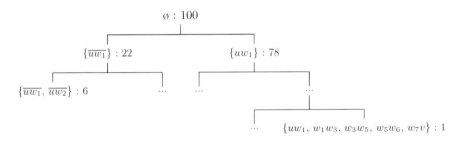

Fig. 3. A Dynamic Monte Carlo Sampling Example

Procedure. DynamicMontecarlo(g, s, t, p, k)

   Input: g, u, v, t, k

1 $succ$, $fail = 0$;
2 $succ\_threshold = kt$;
3 $fail\_threshold = k(1 - t)$;
4 $visited = \{s\}$;
5 $expand =$ outgoing edges of s ;
6 TestSample($visited$, $expand$, k, t) ;
7 **if** $succ \geq kp$ **then**
8   **return** *True*
9 **return** *False*

Procedure. TestSample(*visited, expand, n, t*)

    Input: *visited*: visted vertices, *expand*: edges
               that can expand, *n*: number of samples in this group, *v*: destination

1   **if** $expand = \emptyset$ **then**
2 | $fail = fail + n$;
3 | **return**

4   **if** $u \in visited$ **then**
5 | $succ = succ + n$;
6 | **return**

7   **if** $succ \geq succ\_threshold$ *Or* $fail > fail\_threshold$ **then**
8 | **return**

9   $e = expand.pop\_back$;
10   $k1, k2 = 0$;
11   $visited2 = visited \bigcup v$ where v is the new vertex brought in by e;
12   $expand2 = expand \bigcup v's$ outgoing edges which at least one end not in $visited2$;
13   **foreach** $i = 0$ *to* n **do**
14 | r = random number from 0 to 1 ;
15 | **if** $r > p(e)$ **then**
16 | | $k2 \mathrel{+}= 1$;
17 | **else**
18 | | $k1 \mathrel{+}= 1$;

19   TestSample(*visited, expand, k1, t*) ;
20   **if** $succ \geq kp$ *Or* $fail > k(1 - p)$ **then**
21 | **return**
22   TestSample(*visited2, expand2, k2, t*) ;

In our dynamic Monte Carlo method, starting with the source vertex u, we say u is already *reached*. An edge e is *expandable* if it starts from a *reached* vertex. We randomly pick an expandable edge e, then sample the existence of e for k iterations. The next step is to divide the samples into two groups, one group with e existing and another with e not existing. In the group with e existing, we can reach a new vertex w, and more edges become *expandable*. For both groups, we repeat the process of picking a random *expandable* edge, sampling its existence, and dividing the group into smaller batches. If a group contains no more *expandable* edges, the whole group cannot reach v. On the other hand, if v is contained in a group's *reached* vertices, then the whole group can reach v.

In the running example Fig. 2(b), we assume the number of samples to draw is 100. In the first step, we simultaneously poll the states of uw_1 for 100 samples, the result is shown in Fig. 3, uw_1 is *missing* in 22 of the 100 samples and *exists* in the rest. If uw_1 is failed, the next step is to poll on the other possible outgoing edge, uw_2, 6 of the 22 are failed in this case, and a u-v cut is formed in these 6 samples and the states of other edges are no longer relevent. If uw_1 is operational, we have more choices on which outgoing edges to poll next. If the destination can be reached in any step, we can conclude the corresponding sub-batch of samples are $u - v$ reachable. For example, the rightmost leaf node in Fig. 3.

Based on above observations, we present our Monte Carlo algorithm in Algorithm **DynamicMontecarlo** and Algorithm **TestSample**. In **DynamicMontecarlo**, we initializes a few global variables and invoke **TestSample** at line 6.

Then it checks whether the number of reachable samples is greater than kp. In **TestSample**, we firstly check whether there exist any more edges to expand. We could determine the whole group fails the reachability test if there is no more edges to expand. At line 9, we will randomly pick one edge to expand. From line 11 to line 12, we set up two groups which each represents the sample group in which the chosen edge is missing or is present, respectively. At line 13 to line 18, we will split all samples into these two group. At the end, it will recursively invoke TestSample. Line 7 to line 8, and line 20 to line 21 will check whether the current number of reachable or unreachable samples are enough to accept or reject the query.

Accuracy Guarantee: Let $R_{u,v}$ be the Probabilistic Reachability between u and v, the variance of the expected value, $E(R_{u,v})$ sampled by the Monte Carlo method is as following([16]:

$$\sigma^2(E(R_{u,v})) = \frac{R_{u,v} - R_{u,v}^2}{k} \tag{2}$$

As we introduce a threshold t into the Probabilistic Reachability Query, the result approximated by the Monte Carlo is correct as long as $R_{u,v} - E(R_{u,v}) \leq R_{u,v} - t$ when a query is rejected or $E(R_{u,v} - R_{u,v}) \leq t - R_{u,v}$.

Theorem 2 (Cantelli's Inequality [21]). *Suppose that r is a random variable with mean $E(r)$ and variance $\sigma^2(E(r))$, $Prob(r - E(r) \geq a) \leq \delta(a, \sigma(E(r))$ for any $a \geq 0$, where $Prob(r - E(r) \geq a)$ denotes the probability of $r - E(r) \geq a$, and $\delta(x, y)$ is defined as:*

$$\delta(x,y) = \begin{cases} 1 & \text{if } x = 1 \text{ and } y = 0 \\ 0 & \text{if } x = 0 \text{ and } y = 0 \\ \frac{1}{1+\frac{x^2}{y^2}} & \text{else} \end{cases}$$

Theorem 3. *The probability that the Monte Carlo method returns a false positive(or false negative) answer to a Probabilistic Reachability Query is less or equal than* $\frac{R_{u,v} - R_{u,v}^2}{(k-1)R_{u,v}^2 - (2kt-1)R_{u,v} + kt^2}$, *if we assume the exact probabilistic reachability values and the threshold follow uniform distribution.*

Proof. By using the Theorem 2, the probability that the Monte Carlo method returns a false negative answer:

$$Prob(R_{u,v} - E(R_{u,v}) \geq R_{u,v} - t) \leq \delta(R_{u,v} - t, \sigma(E(R_{u,v}))$$

$$= \frac{1}{1 + \frac{(R_{u,v}-t)^2}{\sigma(E(R_{u,v}))^2}}$$

$$= \frac{R_{u,v} - R_{u,v}^2}{(k-1)R_{u,v}^2 - (2kt-1)R_{u,v} + kt^2}$$

Similarly, the probability that the Monte Carlo method returns a false positive answer can be deduced and the details are omitted. □

6 Experiment

We have performed extensive experiment to demonstrate our approach(Bounds and Dynamic Monte Carlo, or BMC) significantly outperforms plain Monte Carlo(PMC) simulation and Naive Enumeration. Note that, to the best of our knowledge, there are no other existing techniques aiming at efficiently support Probabilistic Reachabilityover large scale datasets. In the experiment, we used both real datasets and synthetic datasets to evaluate the performance. All experiments are conducted on a PC with 2.4GHz 4-core cpu, and 4GB main memory running Linux. Programs are implemented with C++. Every experiment is run against a group of 1000 randomly generated queries, and the average response time is taken as the result.

6.1 Real Dataset

In the experiments, we use 3 real datasets, Anthra, Xmark, and Reactome. All of these datasets were used by Jin in [14]. Anthra is a metabolic pathway from EcoCyc[1]. It contains 13736 vertices and 17307 edges. Xmark is a XML document containing 6483 vertices and 7654 edges. Reactome is a metabolic network with 3678 vertices and 14447 edges. We uniformly assign each edge a probability between 0 to 1. The index construction time for Anthra, Xmark, and Reactome is 11, 9, 16 seconds respectively. The index size is 1.5MB, 700KB, 120KB respectively. Please also note that the index size and construction time do not include the shortest path index since this depends on the technique chosen, which is not the scope of this paper.

As expected, Naive Enumeration cannot complete any experiment within 6 hours. This is because, out of the 1000 queries in each experiment, Naive Enumeration always freezes on at least one of them. The reason is that the cost of Naive Enumeration is almost the same as calculating the reachability if the probabilistic reachability between two vertices is close to the threshold. If the number of edges involved in the calculation is large, the enumeration cost is unaffordably expensive. This result shows that Naive Enumeration is not pratical in solving Probabilistic Reachability Queries. As a result, we will not include the experiment result for Naive Enumeration explicitly.

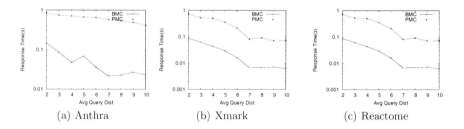

(a) Anthra   (b) Xmark   (c) Reactome

Fig. 4. Query Distance vs Response Time, *threshold* = 0.6

---

[1] www.ecocyc.org

The first group of experiments studies how distance between query vertices affects the performance. The threshold is set at 0.6 and 1000 queries are issued. The result is shown in Fig. 4. The results show that BMC performs approximately one order of magnitude faster than PMC. An interesting observation is that as the distance increases, BMC and PMC both perform faster. In our analysis of this phenomenon, we have two observations: 1) It is generally much more time consuming to prove the destination is reachable than to prove it is not reachable in a possible world. Similarly, it is generally more time consuming to accept a query than to reject a query; 2) When the difference between the threshold and the probabilistic reachability is large, the Monte Carlo simulation requires less samples to answer the query, and also there is a better chance for the upper bound to be able to answer the query. These two observations can explain the above phenomenon. When the distance is small, generally speaking, the probabilistic reachability is higher, and thus the possibility for a state to be reachable is also higher. As the distance increases, the probabilistic reachability drops drastically. It means the average difference between probabilistic reachability and the threshold becomes larger. It also means the possibility for a state to be reachable becomes lower. We can also notice when the distance increases, the gap between BMC and PMC expands. This indicates the upper bound have played a more important role in this scenario.

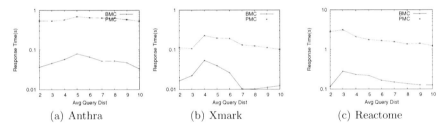

(a) Anthra   (b) Xmark   (c) Reactome

Fig. 5. Query Distance vs Response Time, $threshold = 0.1$

In order to confirm our analysis, we repeat the experiment with a threshold of 0.1. The result is presented in Fig. 5. In this case, when the distance is small, BMC and PMC are both very efficient. This is because the average probabilistic reachability is much higher than 0.1. As the distance increases, the probabilistic reachability slowly drops towards the threshold and then keeps going lower. This represents the peak points of the three result graphs.

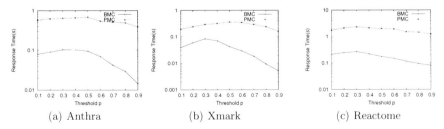

(a) Anthra   (b) Xmark   (c) Reactome

Fig. 6. Threshold vs Response Time, $QueryDistance = 0.6$

In next experiment, we fix the query distance to 5 and vary the threshold from 0.1 to 0.9. The result is shown in Fig. 6. We have observed that as the threshold increases, the response time generally increases slightly to a peak then drops drastically. This is because when the threshold is small, the probabilistic reachability is generally considerably higher than the threshold. As the threshold increases, it approaches the probabilistic reachability, thus the response time increases. In addition, as the threshold increases, more queries are rejected. To some extents, this effect offsets the increase caused by smaller gap between the threshold and the probabilistic reachability. This explains why the increase from 0.1 to the peak is moderate, as well as why the response time decreases drastically beyond the peak point.

The Anthra dataset has a similar density (average vertex degree) to Xmark but approximately 2 times the number of vertices. Reactome is much dense than both of Anthra and Xmark. As we can see from all of the above experiments, Anthra and Xmark have similar response time whereas Reactome is much slower than them. This suggests that the graph size will have limited effect on response time whereas the density plays a major role.

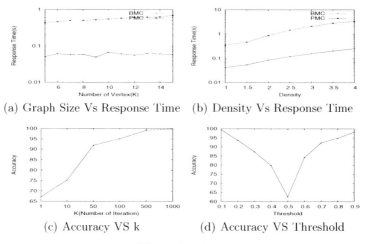

(a) Graph Size Vs Response Time   (b) Density Vs Response Time

(c) Accuracy VS k   (d) Accuracy VS Threshold

Fig. 7. Synthetic Data

6.2   Synthetic Dataset

In the first experiment, we generate graphs with 5000 to 15000 vertices with a fixed density of 1.5. The query distance is fixed at 5 and the threshold is set to 0.6. The result is shown in Fig. 7(a). We found that the increase of graph size has limited affect on BMC . This is because when the distance is fixed, the number of edges which can significantly affect the probabilistic reachability is somewhat unchanged. However, PMC's response time increases slightly as the graph size increases this is because PMC needs to draw a complete sample before testing the reachability.

In the second experiment, we generate graphs with density from 1 to 4 with a fixed graph size of 10000 vertices. The rest of the set up remains the same. The result is shown in Fig. 7(b). This shows the density will impact the response time significantly. However, BMC 's response time increases 5 times from density of 1 to density of 4, whereas PMC's response time increases 10 times. This is because a portion of queries can be rejected by upper bounds. This portion of cost is affected less by the density.

6.3   Accuracy

Since Probabilistic Reachabilityis NP-hard, it is impractical to obtain probabilistic reachability precisely over large scale datasets. Thus, in this set of experiments, we use a small synthetic graph with 10 vertices to demonstrate the accuracy of the Monte Carlo simulation method. For the first experiment, we pick a pair of nodes whose reachability is approximately 0.7, and fix the threshold to be 0.85. We then vary the number of iteration k and the result is shown in Fig. 7 (c). We notice that the increase of k can dramatically increase the accuracy initially, and the increase diminishes when k grows larger. In the second part, we pick a pair of nodes whose reachability is approximately 0.5 and fix the number of iteration to be 100. We then vary the threshold from 0.1 to 0.9. The result is shown in Fig. 7 (d). In this case, the accuracy drops dramatically when the threshold approach 0.5 and again increases when the threshold moves further away from 0.5. This is because the Monte Carlo simulation will perform the worst when the threshold is very close to the probabilistic reachability.

7   Related Work

The Probabilitic Reachability problem has been studied in a number of papers from the 1970s on small scaled graphs, for example, [10], [13]. Valiant([10]) proved it is NP-hard in 1979. As Monte Carlo becomes the widely accepted method to approximate the answer, there are many studies([15], [16], [17], [18], [19]) to propose different sampling plans to reduce the estimation error.

There has been considerable effort put on the the certain reachability problem. A group of techniques([7], [5], etc) named chain decomposition, proposed to speed up online calculation of certain reachability by decomposing graphs into chains. Agrawal *et. al.* shows that using trees instead of chains is more efficient([1]). Based on the tree cover strategy, a few variants were proposed to improve Agrawal *et. al.*'s work. For example, Path-Tree([14]), Dual-Labeling([9]), Label+SSPI([2]), GRIPP([8]), etc. In [4], Cohen *et. al.* proposed a technique called 2-Hop. 2-Hop indexes each vertex with an in-set and an out-set which are used to infer the reachability between any two vertices. However, finding an optimal 2-Hop cover requires $O(n^4)$ time complexity. In order to improve the index building process, Cheng *et. al.* proposed an approximation 2-Hop cover([3]).

The techniques and applications of Uncertain Graphs have been studied in a number of recent papers, including mining frequent subpatterns([24], [25]), finding top-k maximal cliques([26]), etc.

8   Conclusion

In this paper, we study the problem of Probabilistic Reachability Queriesand proposed effective and efficient techniques to solve this problem. To the best of our knowledge, we are the first to efficiently support Probabilistic Reachability Queries over large scale graphs using indexing techniques. We propose an index structure which assists in calculation of upper bound of probabilistic reachability efficiently. Should the bounds fail to answer a query, a dynamic Monte Carlo method is proposed to output an approximate answer. Through comprehensive experiments, we demonstrate that our solution is one order magnitude faster than the most widely accepted plain Monte Carlo simulation.

Acknowledgement

Xuemin Lin is supported by ARC Discovery Projects DP0881035, DP0987557, and DP110102937.

References

1. Agrawal, R., Borgida, A., Jagadish, H.V.: Efficient management of transitive relationships in large data and knowledge bases. In: SIGMOD, pp. 253–262 (1989)
2. Chen, L., Gupta, A., Kurul, M.E.: Stack-based algorithms for pattern matching on dags. In: VLDB, pp. 493–504 (2005)
3. Cheng, J., Yu, J.X., Lin, X., Wang, H., Yu, P.S.: Fast computation of reachability labeling for large graphs. In: Ioannidis, Y., Scholl, M.H., Schmidt, J.W., Matthes, F., Hatzopoulos, M., Böhm, K., Kemper, A., Grust, T., Böhm, C. (eds.) EDBT 2006. LNCS, vol. 3896, pp. 961–979. Springer, Heidelberg (2006)
4. Cohen, E., Halperin, E., Kaplan, H., Zwick, U.: Reachability and distance queries via 2-hop labels. In: Proceedings of the 13th Annual ACM-SIAM Symposium on Discrete Algorithms, pp. 937–946 (2002)
5. Jagadish, H.V.: A compression technique to materialize transitive closure. ACM Trans. Database Syst. 15(4), 558–598 (1990)
6. Schenkel, R., Theobald, A., Weikum, G.: HOPI: An efficient connection index for complex XML document collections. In: Hwang, J., Christodoulakis, S., Plexousakis, D., Christophides, V., Koubarakis, M., Böhm, K. (eds.) EDBT 2004. LNCS, vol. 2992, pp. 237–255. Springer, Heidelberg (2004)
7. Simon, K.: An improved algorithm for transitive closure on acyclic digraphs. Theor. Comput. Sci. 58(1-3), 325–346 (1988)
8. Tribl, S., Leser, U.: Fast and practical indexing and querying of very large graphs. In: SIGMOD 2007: Proceedings of the 2007 ACM SIGMOD International Conference on Management of Data, pp. 845–846 (2007)
9. Wang, H., He, H., Yang, J., Yu, P.S., Yu, J.X.: Dual labeling: Answering graph reachability queries in constant time. In: ICDE, p. 75 (2006)
10. Valiant, L.G.: The complexity of enumeration and reliability problems. SIAM J. Compt. 8, 410–421 (1979)
11. Jiang, B., Pei, J., Lin, X., Cheung, D.W., Han, J.: Mining preferences from superior and inferior examples. In: KDD, pp. 390–398 (2008)
12. Provan, J.S., Ball, M.O.: Computing Network Reliability in Time Polynomial in the Number of Cuts. Operations Research, Reliability and Maintainability 32(3), 516–526 (1984)
13. Shier, D.R., Liu, N.: Bounding the Reliability of Networks. The Journal of the Operational Research Society, Mathematical Programming in Honour of Ailsa Land 43(5), 539–548 (1992)
14. Jin, R., Xiang, Y., Ruan, N., Wang, H.: Efficiently Answering Reachability Queries on Very Large Directed Graphs. In: SIGMOD (2008)

15. Easton, M.C., Wong, C.K.: Sequential Destruction Method for Monte Carlo Evaluation of System Reliability. IEEE, Reliability 29, 191–209 (1980)
16. Fishman, G.S.: A Monte Carlo Sampling Plan for Estimating Network Reliability. Operational Research 34(4), 581–594 (1986)
17. Karp, R., Luby, M.G.: A New Monte Carlo Method for Estimating the Failure Probability of An N-component System. In: Computer Science Division. University of California, Berkley (1983)
18. Okamoto, M.: Some Inequalities Relating To the Partial Sum of Binomial Probabilities. Annals Inst. Statistical Mathematics 10, 29–35 (1958)
19. Fishman, G.S.: A Comparison of Four Monte Carlo Methods for Estimating the Probability of s-t Connectedness. IEEE, Trans. Reliability 35(2) (1986)
20. Chan, E.P., Lim, H.: Optimization and Evaluation of Shortest Path Queries. VLDB Journal 16(3), 343–369 (2007)
21. Meester, R.: A Natural Introduction to Probability Theory (2004)
22. Adar, E., Ré, C.: Managing Uncertainty in Social Networks. Data Engineering Bulletin 30(2), 23–31 (2007)
23. Zou, Z., Gao, H., Li, J.: Discovering Frequent Subgraphs over Uncertain Graph Databases under Probablistic Semantics. In: KDD (2010)
24. Zou, Z., Li, J., Gao, H., Zhang, S.: Mining Frequent Subgraph Patterns from Uncertain Graph Data. TKDE 22(9), 1203–1218 (2010)
25. Zou, Z., Gao, H., Li, J.: Discovering Frequent Subgraphs over Uncertain Graph Databases under Probabilistic Semantics. In: SIGKDD, pp. 633–642 (2010)
26. Zou, Z., Li, J., Gao, H., Zhang, S.: Finding Top-k Maximal Cliques in an Uncertain Graph. In: ICDE, pp. 649–652 (2010)

Improving XML Data Quality with Functional Dependencies*

Zijing Tan and Liyong Zhang

School of Computer Science,
Fudan University, Shanghai, China
{zjtan,09210240049}@fudan.edu.cn

Abstract. We study the problem of repairing XML functional dependency violations by making the smallest value modifications in terms of repair cost. Our cost model assigns a weight to each leaf node in the XML document, and the cost of a repair is measured by the total weight of the modified nodes. We show that it is beyond reach in practice to find optimum repairs: this problem is already NP-complete for a setting with a fixed DTD, a fixed set of functional dependencies, and equal weights for all the nodes in the XML document. To this end we provide an efficient two-step heuristic method to repair XML functional dependency violations. First, the initial violations are captured and fixed by leveraging the conflict hypergraph. Second, the remaining conflicts are resolved by modifying the violating nodes and their related nodes called determinants, in a way that guarantees no new violations. The experimental results demonstrate that our algorithm scales well and is effective in improving data quality.

1 Introduction

Integrity constraints are used to define the criteria that data should satisfy. However, data in real world is typically dirty; we often encounter data sets that violate the predefined set of constraints and hence are inconsistent. Recently there has been an increasing interest in the study of automatically repairing relational databases. One basic problem, known as the *optimum repair computing*, tries to find a repair that satisfies the given constraints and has the minimum cost among all repairs for a given cost model. In most versions of this problem, to compute a repair with a minimum cost is at least NP-complete. That is, we have no methods to find a repair in polynomial-time, which is the most accurate and closest to the original data. Thus, some heuristic or approximation algorithms, *e.g.,* [5,8,13,14], are presented to solve the problem.

Most of the prior works on optimum repair computing are discussed for the relational model, however, there is no reason to believe that data quality is any better for the Web data. The prevalent use of the Web has made it possible to extract and integrate data from diverse sources, but it has also increased the risks

---

* This work is supported by NSFC under Grant No. 60603043.

J.X. Yu, M.H. Kim, and R. Unland (Eds.): DASFAA 2011, Part I, LNCS 6587, pp. 450–465, 2011.

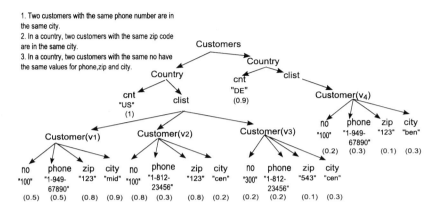

Fig. 1. Example: An XML document violating constraints

of creating and propagating dirty data. XML is fast emerging as the dominant standard for representing and exchanging data on the Web, which motivates us to study the problem of repairing inconsistent XML data.

Example 1. Fig. 1 gives an XML document about the customer information, which is the integration result of different data sources. Customers are grouped by their country, and each country is identified by its name (cnt). For each customer (named with a variable), we have values for no, phone, zip and city, respectively. We also associate each leaf node with a weight (in bracket), to reflect our confidence on the data accuracy of each node.

The company has some integrity constraints defined on the data, as shown in Fig. 1. Constraint 1 asserts that the phone number determines the city in the whole document. According to constraint 2, the zip code determines the city only in a country. Constraint 3 defines the no as a key of Customer, *i.e.*, it determines all the child nodes of a customer, but the key is only valid in a country. As discussed by prior works [3,4], one needs not only *absolute* constraints that hold on the entire document, but also *relative* ones that hold only on certain sub-documents, to cope with the hierarchical structure of XML data. Here constraint 1 is an absolute constraint, and constraints 2 and 3 are relative constraints that are only valid in a Country. All the constraints in Fig. 1 are in the form of functional dependencies (FDs); we focus on this type of constraint in this paper.

The given XML document violates all the constraints. Customers v_1 and v_4 have the same phone number but are in different cities, which violates constraint 1. Customers v_1 and v_2 violate constraint 2, for they have the same zip code but are in different cities. A constraint violation with respect to constraint 3 is that customers v_1 and v_2 have the same no, but their phones and cities are different. Note that customers v_2 and v_4 do not violate constraint 2; the two customers are in different countries, thus constraint 2 is not enforced on them.   □

Contributions. The XML optimum repair problem is studied in this paper: given an XML document that violates a set of FDs, our goal is to apply the lowest cost changes to the document, such that the FDs will be satisfied.

(1) We provide a repair framework for XML. The XML data repairing is necessarily harder to deal with than the relational counterpart. In the framework, (a) we use FDs to describe data consistency, as they are the most commonly discussed and used constraints in real applications. (b) we support modification of node value as the repair primitive; value modification is more flexible, as the deletion and insertion of nodes may be restricted by the XML schema definition. (c) we assign a weight to each leaf node in the XML document, and the cost of a repair is measured by the total weight of the modified nodes.

(2) We establish the lower and upper bounds for the optimum repair computing problem for XML. We show that our model is non-trivial; it is NP-complete to find the optimum repair. The complexity bounds are rather robust: the lower bound is already NP-complete in the size of the data, even with a fixed DTD, a fixed set of FDs, and equal weights for all the nodes in the XML document.

(3) We develop an efficient algorithm to repair inconsistent XML document. The key difficulty in fixing FD violations is that repairing one FD can break another, and most simple heuristics could fail to terminate in the presence of complex, inter-related dependencies. We first construct the conflict hypergraph to capture the initial FD violations, and fix them by modifying the values of all the nodes on a vertex cover of the conflict hypergraph. We then resolve the remaining violations by modifying not only the violating nodes, but also their *core determinants*, to guarantee that no new conflicts will be introduced.

(4) We provide a comprehensive evaluation of the algorithms presented in the paper. We evaluate the accuracy and scalability of our methods with synthetic XMark data, and the experimental results demonstrate that our repairing algorithm scales well and is effective in improving data quality.

Related Work. There are many research studies about how to get consistent information from inconsistent relational databases, following two basic ideas [1]. *Repairing* a database is to find another consistent database that minimally differs from the original one. *Consistent query answering* for a given query is to find the answer that is common in every possible repairs of the original database. Two recent surveys [7,9] provide an overview of this field.

Repair is used as an auxiliary concept to define consistent answers in most of the settings of consistent query answering, and is not actually computed. In the recently proposed optimum repair computing problem, the minimum repairs according to some criteria are found as the data cleaning result. Database repairs are computed for FDs and inclusion dependencies [5], and conditional functional dependencies [8] using heuristic methods. Approximation algorithms to find the minimum repair are given for local denial constraints [14] and FDs [13]. A recent work [6] proposes a cleaning approach that is not limited to finding a single repair; it samples from the space of possible repairs. The study of data quality has mostly focused on the relational data [9], although the data quality on the web can be even worse. In this paper, the XML optimum repair computing problem for FD violations is discussed. The data model, the constraint model and the update operations for XML are far more complicated than the relational counterparts, thus, this problem is necessarily more difficult. For instance, it is known that to cope with the hierarchical structure of XML data, one needs relative dependencies that hold only on certain sub-documents. As a result, the algorithm for XML FD repairing must handle relative FDs effectively, *i.e.*, the inference rules for XML FDs need to be incorporated in the process of repair. The XML FD and related inference are themselves much harder than the relational ones. Relational data can be expressed in XML format, our method can therefore be applied to relations as well by minor modifications.

[15] considers the problem of resolving the inconsistency of merged XML documents w.r.t. a set of FDs. As opposed to this work, [15] discusses how to put conflicting information from merged documents together in a concise merged format, by introducing new element node types. [11] considers inconsistent XML data w.r.t. a set of FDs, where the notion of repair is used as an auxiliary concept: consistent query answers are computed by removing those unreliable nodes identified by repairs. [12] investigates the existence of repairs w.r.t. a set of integrity constraints and a DTD. If the existence of a repair is decidable, then the complexity of providing consistent answers to a query is characterized. In this paper, we study the optimum repair computing problem based on repair cost, which clearly differs from the prior works.

2   Preliminaries

In this section, we review some related preliminary definitions, and revisit the definition of XML functional dependency.

A DTD [10] is (Ele, P, r) , where (a) Ele is a finite set of *element types*; (b) r is a distinguished type in Ele, called the *root type*; (c) $P(A)$ is the *production* of A in Ele; it is a regular expression: $\alpha ::= str \mid \epsilon \mid B_1 + ... + B_n \mid B_1, ..., B_n \mid B^*$. Here str denotes PCDATA, ϵ is the empty word, B is a type in Ele(called a *child* of A), and "+", "," and "*" denote *disjunction*, *concatenation* and the *Kleene star*, respectively. We call a child B of A a *singleton* if each A has exactly one B, *i.e.*, B occurs once and is not in the form of B^* or $B + B_1$ in $P(A)$.

An XML document T *conforms to* a DTD D, if (a) there is a unique *root* node in T labeled with r; (b) each node in T is labeled with an Ele type A, called an *A element* node; (c) an *A element* node is a *leaf* node carrying a string value if $P(A) := str$, otherwise it has a list of element node children whose labels are in the regular language defined by $P(A)$. We use $lab(v)$ and $val(v)$ to denote the label and the value of a node v. Given two nodes v and v', if $lab(v) = lab(v')$ and $val(v) = val(v')$, we say v and v' are of *value equality*, denoted $v \equiv v'$.

A path is a sequence of element types, with the form $P := \epsilon \mid e/P$. ϵ represents the empty path, $e \in Ele$, and "/" denotes concatenation of two paths. A path $e_1/ \ldots /e_n$ is called (a) a root path if $e_1 = r$, (b) a *singleton path* if each e_i

($i \in [1, n]$) is a singleton, and (c) a *leaf path* if $P(e_n):=str$. R is called a *prefix* of P if $P = R/O$ (O can be ϵ). If R is a prefix of both P and Q, R is called a common prefix of P and Q. If a node v_2 is *reachable* from node v_1 following path P, we alternatively say that v_2 is *qualified* by P from v_1, and that v_2 is a child node of v_1, which *matches* P. We write $\{v[\![P]\!]\}$ for the set of nodes that can be reached from v following P. In particular, when there is only one node in $\{v[\![P]\!]\}$, we use $v[\![P]\!]$ to denote this node. If v is the *root* node, we write $\{[\![P]\!]\}$ for $\{v[\![P]\!]\}$.

Example 2. Based on the XML document shown in Fig. 1, we present some examples to illustrate the notation. $\{[\![Customers/Country/clist/Customer]\!]\}$ is a set composed of the four Customer nodes, *i.e.*, $\{v_i \ (i \in [1, 4])\}$. $\{v_4[\![Zip]\!]\}$ contains the Zip node with the value "123". There is only one node in $\{v_4[\![Zip]\!]\}$, so we can write $val(v_4[\![Zip]\!]) = $ "123".   □

We give the definition of XML functional dependency following and extending the way to define XML keys [4].

Definition 1. *With a given DTD D, a functional dependency (FD) is of the form* $\sigma = (P, P', (P_1, \ldots, P_n \rightarrow P_{n+1}, \ldots, P_{n+m}))$. *Here P is a root path, or* $P = \epsilon$. *Each* P_i *(*$i \in [1, n+m]$*) is a singleton leaf path, and there is no non-empty common prefix for* $P_1, \ldots, P_{n+m}$. *Given an XML document T conforming to D, we say T satisfies* σ*: iff* $\forall v \in \{[\![P]\!]\}$, $\forall v_1, v_2 \in \{v[\![P']\!]\}$, *if* $v_1[\![P_i]\!] \equiv v_2[\![P_i]\!]$ *for all* $i \in [1, n]$, *then* $v_1[\![P_{n+j}]\!] \equiv v_2[\![P_{n+j}]\!]$ *for all* $j \in [1, m]$.

In the definition, we call P the context path, P' the target path, P_1, ..., P_{n+m} value paths, $v \in \{[\![P]\!]\}$ context nodes, $v_1, v_2 \in \{v[\![P']\!]\}$ target nodes, and $v_1[\![P_i]\!]$, $v_2[\![P_i]\!]$ ($i \in [1, n+m]$) value nodes, respectively.

Example 3. We give the formal definitions of the FDs in Fig. 1.
(1) For customers, phone determines city inside the document.
    $\sigma_1 = (\epsilon, Customers/Country/clist/Customer, (phone \rightarrow city))$
(2) For customers of a given country, zip determines city.
    $\sigma_2 = (Customers/Country, clist/Customer, (zip \rightarrow city))$
(3) For customers of a given country, no determines phone, zip and city.
    $\sigma_3 = (Customers/Country, clist/Customer, (no \rightarrow phone, zip, city))$ □

Remark. With a given DTD D and a set Σ of FDs, there always exists an XML document T that conforms to D and satisfies Σ. This is because an XML document will satisfy all the FDs if (a) it has distinct values for every nodes, or (b) it has a same value for all the nodes. In light of this, node value modifications suffice to repair any inconsistent XML document T w.r.t. FD violations.

XML FDs are studied in [3,18], with most of the key concepts introduced. For example, *logical implication* and *inference rules* are discussed for XML FDs. Below we use Σ^+ to denote the *closure* of Σ, composed of all the FDs logical implied by Σ. Some of the inference rules are extensions of the Armstrong's axioms [2]. As an example, $(P, P', (P_1, \ldots, P_k, \ldots, P_n \rightarrow P_k))$ is a *trivial* FD. It is also easy to see that $(P, P', (P_1 \rightarrow P_2))$ and $(P, P', (P_2 \rightarrow P_3))$ logical implies $(P, P', (P_1 \rightarrow P_3))$, as extension of the transitivity. W.l.o.g., we consider FDs of the form $(P, P', (P_1, \ldots, P_n \rightarrow P_{n+1}))$ in the rest of the paper.

Some inference rules are required to handle the hierarchical structure of XML data. If we have a FD $(Q, Q', (P_1, \ldots, P_n \rightarrow P_{n+1}))$, and know $P/P' = Q/Q'$, and $P=Q/R$ for some non-empty path R, we can infer a new FD $(P, P', (P_1, \ldots, P_n \rightarrow P_{n+1}))$. Intuitively, if a FD holds inside a context qualified by Q, it will certainly hold inside a smaller context qualified by P. Consider σ_1 in our running example. Since phone determines city inside the document, it must also determine city inside a given country. We therefore have a new FD $(Customers/Country, clist/Customer,(phone \rightarrow city))$, which is logical implied by σ_1.

3   Problem Formulation

Cost Model. A weight in the range $[0, 1]$ is associated with each leaf node v, denoted by $w(v)$ (see the weights given in Fig. 1). The weight of a leaf node may be assigned by a user to reflect his confidence on the accuracy of the data, or be automatically generated by some statistical methods. We assume that a leaf node with a small weight implies less reliability than a node with a large weight.

Repairing Primitive. For a FD $\sigma=(P, P', (P_1, \ldots, P_n \rightarrow P_{n+1}))$, in the subtree rooted at a node in $\{[\![P]\!]\}$, consider two nodes v_1 and v_2 matching path P'. If their child nodes qualified by paths P_i have equal values for all $i \in [1, n]$, and their child nodes qualified by path P_{n+1} have different values, then v_1 and v_2 violate σ. The violation is resolved as follows: (a) change the value of the node qualified by P_{n+1} from v_2 to the value of v_1's child node that matches P_{n+1}(or reversely), or (b) choose an arbitrary P_i ($i \in [1, n]$), and introduce a fresh new value outside the *active domain* to the node qualified by P_i from v_1 (or v_2).

We give some definitions on the domain of an XML document T. The *current domain* of T is composed of all the current node values in T. The *active domain* of T is composed of all the values that have occurred in T, *i.e.*, the *active domain* also contains those "past" values that are changed to current values by modifications. We assume that a procedure *gen_new_value* is available, which will generate a fresh new value outside the *active domain* each time when called.

Remark. As remarked earlier, it suffices to repair any inconsistent XML document using only rule (a) or rule (b) of the repairing primitive. We find that rule (a) can preserve more constant values in the current domain and rule (b) can prevent the excessive "equivalence propagation" incurred by rule (a). In light of this, the combination of the two rules helps us find a good quality repair.

Repairs. With a given XML document T and a set Σ of FDs, a repair of T is an XML document T_R such that (a) T is changed to T_R using node value modifications, and (b) T_R satisfies Σ.

We assume that a distinct identifier is assigned to each node, which is not subject to the update. The identifier plays an auxiliary role; we use v_R to denote the node in T_R that has the same identifier as the node v in T.

Definition 2. *Given an inconsistent XML document T violating a set Σ of FDs, we define the distance between T and a repair T_R of T as:*

$$\triangle(T, T_R) = \sum_{v \ in \ T} w(v) \times dist(v, v_R),$$

where $dist(v, v_R)=1$ if $val(v) \neq val(v_R)$, otherwise $dist(v, v_R)=0$.
   We define $cost(T_R)=\triangle(T, T_R)$, and say that T_R is an optimum repair of T, if T_R has the minimum cost among all repairs of T. We denote the optimum repair of T as $T_R{}^{opt}$.

Example 4. Consider the document in Fig. 1. One repair is got by changing the value of $v_1[\![city]\!]$ to "cen", and introducing two different fresh new values to nodes $v_2[\![no]\!]$ and $v_4[\![phone]\!]$, respectively. This repair has a cost $0.9+0.8+0.3=2$. We find another repair as follows. We first modify the values of nodes $v_2[\![city]\!]$ and $v_4[\![city]\!]$ to "mid", and change the value of node $v_1[\![no]\!]$ to a fresh new value. Note that the current document is not yet a repair. Our modifications have caused a new violation: customers v_2 and v_3 have the same phone number, but are now in different cities. We then change the value of $v_3[\![city]\!]$ to "mid". The new XML document is a repair, with a cost $0.2+0.3+0.5+0.3=1.3$.   □

For an inconsistent XML document T, we are interested in finding a minimum cost repair T_R of T. The provided framework is nontrivial; it is NP-complete to find the optimum repair following it.

Theorem 1. *The optimum repair computing problem is NP-complete. It is already NP-complete in the size of the XML document, even with a fixed DTD, a fixed set Σ of FDs, and equal weights for all the nodes in the XML document.*   □

4   Fixing Initial Conflicts Based on Hypergraph

We employ hypergraph as a tool to model the conflicts in an inconsistent XML document w.r.t. FDs. Formally, a hypergraph g is a pair $g = (V, E)$; V is a set of elements, called nodes or vertices, and E is a set of non-empty subsets of V called hyperedges. We use the version of weighted hypergraph to carry our cost model. The following definition constructs the initial conflict hypergraph, which extends the notions in [13] to XML with relative FDs.

Definition 3. *Given an XML document T, a set Σ of FDs, the initial conflict hypergraph of T w.r.t. Σ is defined to be $g = (V,E)$. V is the set of value nodes w.r.t. Σ in T. Each node $v \in V$ is assigned a weight $w(v)$, the same as the weight of v in T. E is a set of non-empty subsets of V, constructed as follows:*

(1) for each FD $(P, P', (P_1, \ldots, P_n \rightarrow P_{n+1})) \in \Sigma$, $\forall u \in \{\llbracket P \rrbracket\}, \forall u_1, u_2 \in \{u \llbracket P' \rrbracket\}$, if $u_1 \llbracket P_i \rrbracket \equiv u_2 \llbracket P_i \rrbracket$ for all $i \in [1, n]$, and $u_1 \llbracket P_{n+1} \rrbracket \not\equiv u_2 \llbracket P_{n+1} \rrbracket$, nodes $u_1 \llbracket P_i \rrbracket$, $u_2 \llbracket P_i \rrbracket (i \in [1, n+1])$ form a hyperedge in E;

(2) for two FDs $(P, P', (P_1, \ldots, P_n \rightarrow P_{n+1}))$, $(Q, Q', (Q_1, \ldots, Q_m \rightarrow P_{n+1})) \in \Sigma$, where $P/P' = Q/Q'$. W.l.o.g., we assume that Q is a prefix of P, that is, $P = Q/R$ for some path R. $\forall u \in \{\llbracket P \rrbracket\}, \forall u_1, u_2, u_3 \in \{u \llbracket P' \rrbracket\}$, if

    (a) $u_1 \llbracket P_i \rrbracket \equiv u_2 \llbracket P_i \rrbracket$ for all $i \in [1, n]$,

    (b) $u_1 \llbracket Q_j \rrbracket \equiv u_3 \llbracket Q_j \rrbracket$ for all $j \in [1, m]$, and

    (c) $u_2 \llbracket P_{n+1} \rrbracket \not\equiv u_3 \llbracket P_{n+1} \rrbracket$,

nodes $u_1 \llbracket P_i \rrbracket$, $u_2 \llbracket P_i \rrbracket (i \in [1, n])$, $u_1 \llbracket Q_j \rrbracket, u_3 \llbracket Q_j \rrbracket (j \in [1, m])$, $u_2 \llbracket P_{n+1} \rrbracket$, $u_3 \llbracket P_{n+1} \rrbracket$ form a hyperedge in E;

(3) for two FDs $(P, P', (P_1, \ldots, P_n \rightarrow P_{n+1}))$, $(Q, Q', (P_{n+1}, \ldots, P_{n+m} \rightarrow P_{n+m+1})) \in \Sigma$, where $P/P' = Q/Q'$. W.l.o.g., we assume that Q is a prefix of P, that is, $P = Q/R$ for some path R. $\forall u \in \{\llbracket P \rrbracket\}, \forall u_1, u_2, u_3 \in \{u \llbracket P' \rrbracket\}$, if

    (a) $u_1 \llbracket P_i \rrbracket \equiv u_2 \llbracket P_i \rrbracket$ for all $i \in [1, n]$,

    (b) $u_2 \llbracket P_{n+1} \rrbracket \equiv u_3 \llbracket P_{n+1} \rrbracket$, and $u_1 \llbracket P_{n+1+j} \rrbracket \equiv u_3 \llbracket P_{n+1+j} \rrbracket$ for all $j \in [1, m-1]$,

    (c) $u_1 \llbracket P_{n+m+1} \rrbracket \not\equiv u_3 \llbracket P_{n+m+1} \rrbracket$,

nodes $u_1 \llbracket P_i \rrbracket (i \in [1, n] \cup [n+2, n+m+1])$, $u_2 \llbracket P_j \rrbracket (j \in [1, n+1])$, $u_3 \llbracket P_k \rrbracket (k \in [n+1, n+m+1])$ form a hyperedge in E.

Intuitively, each hyperedge in the initial conflict hypergraph indicates a set of value nodes violating FDs. In rule 2 and rule 3 of Definition 3, if $P = Q/R$ for some non-empty path R, then FDs defined in a large context are considered when generating hyperedges for a small context. This is necessary to handle the hierarchical structure of XML document.

Example 5. We give some hyperedges (not all hyperedges) for the XML document in Fig. 1. Following rule 1, the set of nodes $\{v_1 \llbracket phone \rrbracket, v_1 \llbracket city \rrbracket, v_4 \llbracket phone \rrbracket, v_4 \llbracket city \rrbracket\}$ forms a hyperedge. According to rule 2, the set of nodes $\{v_1 \llbracket zip \rrbracket, v_1 \llbracket city \rrbracket, v_2 \llbracket phone \rrbracket, v_2 \llbracket zip \rrbracket, v_3 \llbracket phone \rrbracket, v_3 \llbracket city \rrbracket\}$ forms a hyperedge. If rule 3 is applied, then the set of nodes $\{v_1 \llbracket no \rrbracket, v_1 \llbracket city \rrbracket, v_2 \llbracket no \rrbracket, v_2 \llbracket phone \rrbracket, v_3 \llbracket phone \rrbracket, v_3 \llbracket city \rrbracket\}$ forms a hyperedge. It can be verified that to find a repair for this XML document, in each hyperedge we must change the value of at least one node. □

Let $g = (V, E)$ be the initial conflict hypergraph, where each hyperedge indicates a set of value nodes violating FDs. In any repair of the original document, the value of at least one node should be modified for each hyperedge. This motivates us to find a set $S \subseteq V$, such that for all edges $e \in E$, $S \cap e \neq \phi$. Note that S is known as a vertex cover (VC) for g. Moreover, recall that each node in the hypergraph is associated with a weight. If we define the weight of a vertex cover S to be the total weight of all vertices in S, then to find a low cost repair actually leads to the well-known problem of weighted vertex cover for hypergraph [17].

The vertex cover for hypergraph is well studied (also NP-complete), we can therefore use a known approximation algorithm [17]. If the size of each hyperedge is bounded by a constant d, finding a minimum weighted vertex cover has a d-approximation algorithm. In the initial conflict hypergraph, the size of each

---

Algorithm 1. *Fix-Initial-Conflicts*

---

    input : *An XML document T, a set Σ of FDs.*
    output: *A modified document T.*

1 Create the initial conflict hypergraph g of T w.r.t. Σ;
2 Use a known algorithm to find an approximation VC for the minimum
    weighted vertex cover of g;
3 *remaining* := VC;
4 **while** *There are two target nodes $v_1, v_2 \in T$ violating a FD $\sigma \in \Sigma$, and*
    $v_1 \llbracket P_{n+1} \rrbracket$ or $v_2 \llbracket P_{n+1} \rrbracket$ is the only node in VC from the set of nodes $\{v_1 \llbracket P_i \rrbracket$
    $(i \in [1, n+1])\} \cup \{v_2 \llbracket P_i \rrbracket (i \in [1, n+1])\}$. (W.l.o.g., assume the violation is as
    follows: $\sigma = (P, P', (P_1, \ldots, P_n \to P_{n+1}))$, $v \in \{\llbracket P \rrbracket\}$, $v_1, v_2 \in \{v \llbracket P' \rrbracket\}$, $v_1 \llbracket P_i \rrbracket$
    $\equiv v_2 \llbracket P_i \rrbracket$ for all $i \in [1, n]$, and $v_1 \llbracket P_{n+1} \rrbracket \not\equiv v_2 \llbracket P_{n+1} \rrbracket$.) **do**
5 | $val(v_1 \llbracket P_{n+1} \rrbracket) := val(v_2 \llbracket P_{n+1} \rrbracket)$;
         // W.l.o.g., we assume $v_1 \llbracket P_{n+1} \rrbracket$ is in VC
6 | *remaining* := *remaining* $\setminus \{v_1 \llbracket P_{n+1} \rrbracket\}$;
7 **endw**
8 **foreach** node $u \in$ *remaining* **do** $val(u) := gen\_new\_value()$;
    // Introduce fresh new values to all the remaining nodes in VC

---

hyperedge, *i.e.*, the number of nodes in each hyperedge, is determined by the number of paths involved in the violated FDs. Let MP be the maximum number of paths involved in a FD. Each hyperedge constructed based on rule 1 of Definition 3 can have at most $2 \times MP$ nodes. Rule 2 and rule 3 can generate hyperedges with at most $4 \times MP$–2 nodes. Therefore, $d \leq 4 \times MP$–2 for our hyperedges.

We use hypergraph to capture and fix initial FD violations, shown in Algorithm 1. (a) We compute the initial conflict hypergraph and find an approximation VC for the minimum weighted vertex cover (lines 1-2). (b) We save the unmodified nodes from VC in a set *remaining* (line 3). (c) We assign a value from the current domain to each node u in VC, if u is a value node on the right side of a violated FD σ, and u is the only value node w.r.t. this violation in VC (lines 4-7). More specifically, for $\sigma = (P, P', (P_1, \ldots, P_n \to P_{n+1}))$, if two target nodes v_1 and v_2 violate σ, and $u = v_1 \llbracket P_{n+1} \rrbracket$ is the only node in VC from the set of nodes $\{v_1 \llbracket P_i \rrbracket, v_2 \llbracket P_i \rrbracket \ (i \in [1, n+1])\}$, then we change the value of u to be equal to the value of $v_2 \llbracket P_{n+1} \rrbracket$ (line 5). Once a node is modified, we remove it from the set *remaining* (line 6). (d) For all the remaining nodes, we give them fresh new values outside the active domain by calling procedure *gen_new_value* (line 8). Note that each node in VC is modified exactly once, either in (c) or (d).

5   Resolving Violations Thoroughly

The major challenge in repairing FD violations is the interplay among FDs: the resolution of initial conflicts may introduce new violations. In this section, based on the document after the first step, we carry out modifications in such a way that no new violations will be introduced. After this step, we can guarantee that the resulting document is consistent with the FDs, to be a repair of the original document. Before we explain how this is achieved, we first give some notations.

Definition 4. *Given an XML document T, a set Σ of FDs and a node u in T, we say a set of nodes $\{u_1, u_2, \ldots, u_n\}$ is a σ-determinant of u, if there exists a nontrivial FD $\sigma = (P, P', (P_1, \ldots, P_n \to P_{n+1}))$ logical implied by Σ, such that $\exists v \in \{[\![P]\!]\}, \exists v_1 \in \{v[\![P']\!]\}, v_1[\![P_i]\!] = u_i$ for $i \in [1, n]$, and $v_1[\![P_{n+1}]\!] = u$.*

We say that a set C_u of nodes is a core determinant of u, if (a) for every nontrivial FD σ implied by Σ, and every set W that is a σ-determinant of u, $C_u \cap W \neq \phi$; and (b) for any proper subset C_u' of C_u, there exists some nontrivial FD σ implied by Σ, and a set W that is a σ-determinant of u, $C_u' \cap W = \phi$.

We use $CoreDeter_\Sigma(u)$ to denote the set of core determinants of u w.r.t. Σ.

Intuitively speaking, a core determinant of u is a minimum set of nodes, which intersects with every σ-determinant of u for every nontrivial $\sigma \in \Sigma^+$.

Example 6. Consider the XML document and FDs given in Fig. 1. $\{v_1[\![phone]\!]\}$, $\{v_1[\![zip]\!]\}$ and $\{v_1[\![no]\!]\}$ are *determinants* of $v_1[\![city]\!]$ w.r.t. σ_1, σ_2 and σ_3 respectively; and the set $\{v_1[\![phone]\!], v_1[\![zip]\!], v_1[\![no]\!]\}$ is the *core determinant* of $v_1[\![city]\!]$. $\{v_2[\![no]\!]\}$ is the *determinant* of $v_2[\![phone]\!]$ w.r.t. σ_3, and is the *core determinant* of $v_2[\![phone]\!]$. □

A core determinant of u is computed as follows. (a) We find all the nontrivial FDs in Σ^+ with u as a value node on the right side. The logical implication of FDs is required to compute Σ^+; it is with respect to the number of FDs in Σ, and therefore is acceptable when dealing with Σ of a reasonable size. With a given FD σ and a node u in the XML document T, it takes linear time in the size of T ($|T|$) to check whether u is a value node of σ w.r.t. the right side path. (b) We compute a minimum cover of the paths on the left side of these FDs, *i.e.*, a minimum set of paths that intersects with at least one left side path for each FD. This is irrelevant of $|T|$. (c) We get a core determinant of u by evaluating the paths in the cover on T, with a time linear in $|T|$.

We can resolve the remaining violations by doing a limited number of value modifications, based on the following finding.

Theorem 2. *For every node $w \in C_u$, where $C_u \in CoreDeter_\Sigma(u)$ is a core determinant of u, if there is a σ-determinant Y of w for some nontrivial FD σ, then there exists a core determinant $C_w \in CoreDeter_\Sigma(w)$, such that $(C_u \cup \{u\}) \supseteq C_w$.*

We provide a method to do value modifications without incurring new conflicts in Algorithm 2. We continue to select remaining violating value node u (line 2) and all the nodes in a core determinant C_u of u (line 3), and then modify the values of all the selected nodes to fresh new values (line 4), until all the violations are resolved. The termination of Algorithm 2 is guaranteed by the fact that no new violations will be introduced, since (a) we have changed values of all the nodes in C_u to distinct fresh new values, and C_u intersects with every σ-determinant of u; distinct values for the left side value nodes of σ guarantee no violations, and (b) according to Theorem 2, for every node $w \in C_u$, $C_u \cup \{u\}$ contains a core determinant of w. Thus, we have also assigned fresh new values to every nodes in a core determinant of w; this will guarantee no new violations.

Algorithm 2. *Resolve-Remaining-Violations*

---

    input : *An XML document T, a set Σ of FDs.*
    output: *A modified document T, with all the violations fixed.*

1 **while** *there are FD violations in T w.r.t. Σ* **do**
2   pick a violating value node u from T w.r.t. Σ;
3   let C_u be a core determinant of u;
4   **foreach** *node $w \in (C_u \cup \{u\})$* **do** $val(w) := gen\_new\_value()$;
         `// it guarantees that no new violations will be introduced`
5 **endw**

We give our repair computing method. Given an inconsistent XML document T and a set Σ of FDs, we compute a repair of T w.r.t. Σ in two steps. (a) Algorithm 1 is used to fix the initial violations in T. (b) Algorithm 2 is applied to the result of Algorithm 1, to resolve all the remaining conflicts. The correctness and complexity of the method is stated as follows.

Theorem 3. *With a given set Σ of FDs, the provided method computes a repair of an XML document T w.r.t. Σ in polynomial-time in the size of the XML document T ($|T|$).* □

6 Implementation

In this section, we present several techniques to improve our algorithms both in the repair quality and in the running time.

(1) We employ an algorithm to find an approximation of the optimum vertex cover of the initial conflict hypergraph. Since the size of each hyperedge is bounded by a constant d only determined by the set of FDs, we use the layer algorithm [17] for a d-approximation of the optimum cover.

(2) In Algorithm 2, we continually select violating nodes, and introduce fresh new values to each selected node and all the nodes in one of its core determinants. We use a greedy method to select the most cost-effective node v from the vertex cover VC at each step: we select the node v with the smallest $\frac{w(v)}{n_v}$, where n_v is the number of violations involving v.

(3) We find that FD validations on the given XML document are the most time-consuming procedures in our repairing method. In the construction of the initial conflict hypergraph, with a FD $\sigma = (P, P', (P_1, \ldots, P_n \rightarrow P_{n+1}))$, the validation against σ is optimized as follows. (a) Under a context node, the target nodes of σ are first clustered using a hash function on the values of their child nodes matching paths $P_1, \ldots, P_n$. More specifically, for a context node $v \in \{\llbracket p \rrbracket\}$, $\forall v_1, v_2 \in \{v\llbracket P' \rrbracket\}$, if $v_1\llbracket P_i \rrbracket \equiv v_2\llbracket P_i \rrbracket$ for all $i \in [1, n]$, then the hash function will put v_1 and v_2 in the same cluster. (b) Validations against σ are only done for the target nodes in the same cluster.

FD validation is also required in the second step of our repairing approach, because new conflicts may occur. We do the validation incrementally. (a) The target nodes whose child value nodes are modified in Algorithm 1 are identified, and are re-clustered using the same hash function if the values of any child nodes matching $P_1, \ldots, P_n$ are modified. (b) For any comparison in a cluster, at least one target node in the comparison should be an identified node in step (a).

7   Experimental Study

We next present an experimental study of our XML repairing algorithms. Using both real-life and synthetic XMark [16] data, we conduct two sets of experiments to evaluate (a) the effectiveness of our algorithm in terms of repair quality, and (b) the efficiency and scalability of our algorithm in terms of running time.

7.1   Experimental Setting

(1) Real-life data. We scrape book information from three web sites: Google Books, Amazon and Book Depository. By adjusting the searching keyword, we get datasets of different sizes. For each book, we create FDs from the isbn node to other child nodes of a book. There are many natural discrepancies between the three sites. We assume that all the isbns are trusted, and books from different sites are the same book if they have the same isbn. To resolve conflicts among the three sites, we choose the value that the majority of the sites agree on; if no such value exists, we pick the value from web sites in the order of Google Books, Amazon and Book Depository. Note that with these rules, a deterministic "best" merging result is available (though it is not seen by our algorithm, of course). We denote this result as T, and refer to the original data taken from the web as T_{dirty} because of its inconsistency. We then assign weights to the leaf nodes in T_{dirty} as follows. (a) Google Books, Amazon and Book Depository are associated with a weight 0.5, 0.4, 0.3, respectively. (b) Leaf nodes in T_{dirty} are assigned the same weight as their original site, except the isbn nodes. (c) All the isbn nodes are given a higher weight than other nodes.

(2) Synthetic data. We create XMark instances with different sizes, and consider FDs expressing data semantics. The initial generated data is correct, referred to as T. We then introduce noise to leaf nodes involved in FDs; with probability p (the noise ratio) the value of a leaf node is replaced with another value, which is guaranteed to cause a conflict. This inconsistent data set is referred to as T_{dirty}. We call a node v "dirty" if the value of v is modified in T_{dirty}, otherwise we say that v is "clean". We randomly assign a weight $w(v)$ in $[0, a]$ for each dirty leaf node v, and randomly select a weight $w(v)$ in $[b, 1]$ for each clean leaf node v. Based on the assumption that a clean node usually has a slightly higher weight than a dirty node, we set $a = 0.4$ and $b = 0.6$ in the experiments.

(3) Measuring repair quality. The repair found by the algorithm satisfies all the FDs, but still may contain two types of errors: the noises that are not fixed, and the new noises introduced in the repairing process. We extend the notions of *precision* and *recall* [8] to measure repair quality, in terms of correctness and completeness, respectively. The *precision* is computed as the ratio of the

number of correctly repaired nodes to the number of nodes modified by the repairing algorithm, and *recall* is the ratio of the number of correctly repaired nodes to the number of total error nodes. The total error nodes are the "dirty" nodes in T_{dirty}, and the correctly repaired nodes in the repair T_{rep} found by the algorithm are the modified nodes in T_{rep}, which are "dirty" in T_{dirty}.

We have implemented all the algorithms in Java. All the experiments are run on a PC with Intel Pentium Dual CPU E2140, 1GB RAM and Ubuntu Linux 9.10. Each experiment is run five times and the average reading is recorded.

7.2 Experimental Results

Exp-1: Effectiveness in terms of repair quality. We evaluate the quality of the repair that our repairing algorithm computes. The basic settings of synthetic data in these experiments are an XMark instance of 30M size ($|T| = 30M$), 20% noise ratio ($p = 20\%$) and a set Σ of 5 FDs ($|\Sigma| = 5$). In each of Fig. 2(a), Fig. 2(b) and Fig. 2(c), one parameter varies. The results demonstrate that our repairing algorithm consistently computes a repair with good quality, *i.e.*, with a *precision* value $> 80\%$ and an even higher *recall* value. The values of *recall* are high, which means that our algorithm can repair most of the introduced errors. As indicated by the values of *precision*, some new errors may be introduced by our algorithm.

Fig. 2(b) shows that the *precision* and *recall* decrease slightly with the increase of p, as expected. In Fig. 2(c), we find that the increase of $|\Sigma|$ also has a negative impact on the repair quality. The reason is that our algorithm tends to modify nodes involved in more FDs for a low cost repair, especially when the interplay among FDs becomes complicated. However, these nodes may be "clean" even if they are involved in many FDs. The repair quality improves when we assign higher weights to those "clean" nodes by adjusting the weight bound b in the settings (not shown in the figures).

We also evaluate our algorithm on real-life data, reported in Fig. 2(d). We can see that the repair quality on real-life data is even better than that on the synthetic data, since (a) the FDs on real-life data are in the form of keys, the interactions of FDs are therefore relatively simple, and (b) the isbn nodes are associated with a high weight, and hence, they are rarely modified in the repair.

Exp-2: Efficiency and Scalability. We evaluate the efficiency and scalability of our algorithm on synthetic data, in the same setting as in Exp-1. In Fig. 2(e), we fix p, $|\Sigma|$, and vary $|T|$. From the results we can see that the algorithm scales well with $|T|$. To further analyze the results, we give the details of time composition in Fig. 2(f). The time to cover and modify the initial conflict hypergraph and the time to modify values based on core determinants are omitted, because they are trivial in the overall time. We find the following. (a) The overall time is governed by the time to construct the initial conflict hypergraph. (b) The hash based clustering technique helps the algorithm conduct FD validations, and it is

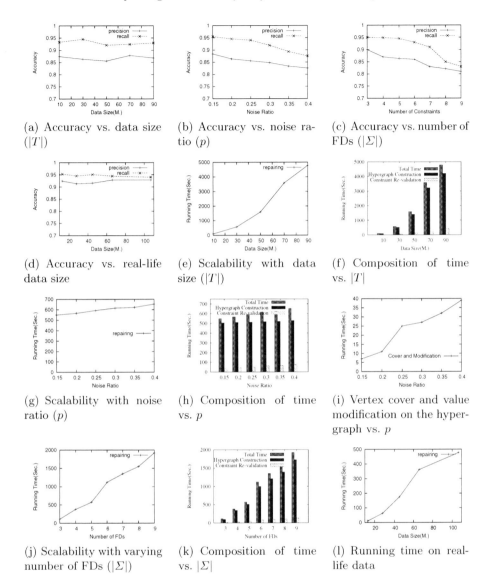

Fig. 2. Experimental Results

critical to the scalability of the algorithm with $|T|$. (c) The re-validation significantly takes less time than the initial one, which confirms that our optimization based on incremental FD validation is efficient.

By varying p, Fig. 2(g) and 2(h) show the overall time and detailed time composition, respectively. The results tell us that the algorithm is insensitive to the increase of p. To better explain the results, we give the time of vertex cover and value modification on the hypergraph in Fig. 2(i). The time increases as expected because the conflict hypergraph becomes larger with increasing p, but this time is still not very significant compared with the overall time. As remarked earlier, the time of conflict hypergraph construction is the dominating factor of the running time. The most expensive part of hypergraph construction is the initial constraint validation, which is not sensitive to p. In light of this, the algorithm scales well with p.

Fig. 2(j) and Fig. 2(k) show the impact of $|\Sigma|$. As expected, it takes more time for our algorithm when new FDs are added in the experiments, and we can see that the algorithm scales well with the increase of $|\Sigma|$.

Finally, the results for real-life data are reported in Fig. 2(l). Note that all the FDs considered on real-life data are defined from isbn to other nodes. Therefore, all of the hyperedges are generated by rule 1 of Definition 3. From the results we find that the construction of the initial conflict hypergraph becomes far more efficient if only this type of hyperedge exists.

8   Conclusions

We have proposed a framework for repairing XML data, and established the complexity bounds on the optimum repair computing problem. We have also developed techniques to compute repairs, and experimentally verified their efficiency and scalability. We are currently investigating the problem by incorporating more types of constraints, enriching the model with different repair primitives, and studying the optimization techniques to improve our method.

References

1. Arenas, M., Bertossi, L.E., Chomicki, J.: Consistent query answers in inconsistent databases. In: PODS, pp. 68–79 (1999)
2. Abiteboul, S., Hull, R., Vianu, V.: Foundations of databases. Addison-Wesley, Reading (1995)
3. Arenas, M., Libkin, L.: A normal form for XML documents. TODS 29(1), 195–232 (2004)
4. Buneman, P., Davidson, S., Fan, W., Hara, C., Tan, W.: Keys for XML. In: WWW, pp. 201–210 (2001)
5. Bohannon, P., Fan, W., Flaster, M., Rastogi, R.: A cost based model and effective heuristic for repairing constraints by value modification. In: SIGMOD, pp. 143–154 (2005)
6. Beskales, G., Ilyas, I., Golab, L.: Sampling the repairs of functional dependency violations under dard constraints. In: VLDB (2010)
7. Chomicki, J.: Consistent query answering: Five easy pieces. In: Schwentick, T., Suciu, D. (eds.) ICDT 2007. LNCS, vol. 4353, pp. 1–17. Springer, Heidelberg (2006)
8. Cong, G., Fan, W., Geerts, F., Jia, X., Ma, S.: Improving data quality: Consistency and accuracy. In: VLDB, pp. 315–326 (2007)
9. Fan, W.: Dependencies revisited for improving data quality. In: PODS, pp. 159–170 (2008)
10. Fan, W., Bohannon, P.: Information preserving XML schema embedding. TODS 33(1) (2008)

11. Flesca, S., Furfaro, F., Greco, S., Zumpano, E.: Repairs and consistent answers for XML data with functional dependencies. In: Bellahsène, Z., Chaudhri, A.B., Rahm, E., Rys, M., Unland, R. (eds.) XSym 2003. LNCS, vol. 2824, pp. 238–253. Springer, Heidelberg (2003)
12. Flesca, S., Furfaro, F., Greco, S., Zumpano, E.: Querying and repairing inconsistent XML data. In: Ngu, A.H.H., Kitsuregawa, M., Neuhold, E.J., Chung, J.-Y., Sheng, Q.Z. (eds.) WISE 2005. LNCS, vol. 3806, pp. 175–188. Springer, Heidelberg (2005)
13. Kolahi, S., Lakshmanan, L.: On approximating optimum repairs for functional dependency violations. In: ICDT, pp. 53–62 (2009)
14. Lopatenko, A., Bravo, L.: Efficient approximation algorithms for repairing inconsistent databases. In: ICDE, pp. 216–225 (2007)
15. Ng, W.: Repairing inconsistent merged XML data. In: Mařík, V., Štěpánková, O., Retschitzegger, W. (eds.) DEXA 2003. LNCS, vol. 2736, pp. 244–255. Springer, Heidelberg (2003)
16. Schmidt, A., Waas, F., Kersten, M., Carey, M., Manolescu, I., Busse, R.: XMark: A benchmark for XML data management. In: VLDB, pp. 974–985 (2002)
17. Vazirani, V.V.: Approximation algorithms. Springer, Heidelberg (2001)
18. Vincent, M., Liu, J., Liu, C.: Strong functional dependencies and their application to normal forms in XML. TODS 29(3), 445–462 (2004)

Identifying Relevant Matches with NOT Semantics over XML Documents

Rung-Ren Lin[1], Ya-Hui Chang[4,*], and Kun-Mao Chao[1,2,3]

[1] Department of Computer Science and Information Engineering
[2] Graduate Institute of Biomedical Electronics and Bioinformatics
[3] Graduate Institute of Networking and Multimedia,
National Taiwan University, Taipei, Taiwan
{r91054,kmchao}@csie.ntu.edu.tw
[4] Department of Computer Science and Engineering
National Taiwan Ocean University, Keelung, Taiwan
yahui@ntou.edu.tw

Abstract. Keyword search over XML documents has been widely studied in recent years. It allows users to retrieve relevant data from XML documents without learning complicated query languages. SLCA (smallest lowest common ancestor)-based keyword search is a common mechanism to locate the desirable LCAs for the given query keywords, but the conventional SLCA-based keyword search is for AND-only semantics. In this paper, we extend the SLCA keyword search to a more general case, where the keyword query could be an arbitrary combination of AND, OR, and NOT operators. We further define the query result based on the *monotonicity* and *consistency* properties, and propose an efficient algorithm to figure out the SLCAs and the relevant matches. Since the keyword query becomes more complex, we also discuss the variations of the monotonicity and consistency properties in our framework. Finally, the experimental results show that the proposed algorithm runs efficiently and gives reasonable query results by measuring the processing time, scalability, precision, and recall.

Keywords: keyword search, XML, Smallest LCA, NOT semantics.

1 Introduction

XML is a standard format for presenting and exchanging the information in the World-Wide-Web, and thus the requirement of getting information from XML documents is raised. Keyword search is a widely-used approach to retrieve meaningful information from XML documents, which does not require the knowledge of the XML document specification and the XML query language [14]. Many keyword search mechanisms are based on the lowest common ancestor (LCA) concept to identify the desirable subtrees. The LCA of a set of query keywords refers to the node that contains every keyword at least once in the subtree rooted

---

* Corresponding author.

J.X. Yu, M.H. Kim, and R. Unland (Eds.): DASFAA 2011, Part I, LNCS 6587, pp. 466–480, 2011.
© Springer-Verlag Berlin Heidelberg 2011

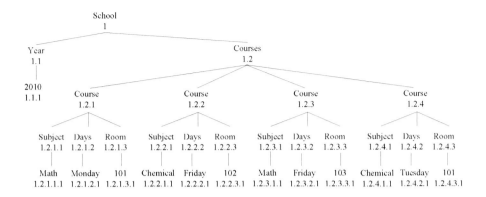

Fig. 1. A sample XML tree

Table 1. Sample keyword queries with NOT operators

Q_1 Subject $\wedge$ 101 $\wedge$!Monday
Q_2 2010 $\wedge$ Subject $\wedge$ 101 $\wedge$!Monday
Q_3 Course $\wedge$ (101 $\vee$ 103) $\wedge$ (! Friday $\vee$!Math)

at this node. The SLCA keyword search was further proposed by Wu and Pa-pakonstantinou [10]. A node n is said to be an SLCA node if: (i) n is an LCA node, and (ii) none of n's children are also the LCA nodes. Take Figure 1 as an example. It displays a sample XML tree, which describes the school's course information in year 2010, and each node is assigned a unique Dewey number. The SLCAs of querying ($Subject, Friday$) will be nodes 1.2.2 and 1.2.3. However, the original SLCA-based keyword search is for AND-only semantics. Sun *et al.* [8] then provided a solution to process more general keyword search queries which support any combination of AND and OR boolean operators. For instance, the keyword query could be ($Subject \wedge (Monday \vee Friday)$), where "$\wedge$" and "$\vee$" represent AND and OR operators respectively. Since its semantics is to search the subjects which are held on Monday or Friday, the SLCAs are now nodes 1.2.1, 1.2.2, and 1.2.3.

Liu and Chen [7] then proposed a concept *contributor* to determine the *relevant matches* in the subtrees rooted at SLCAs. That is, only "important" nodes, instead of the whole subtree, will be returned. In addition, their output also satisfies both *monotonicity* and *consistency* properties, which capture a reasonable connection between the new query result and the original query result after the data or the query updates.

In this paper, we address the issue of identifying the relevant matches where the keywords can be combined by AND, OR, and NOT operators. In AND-only semantics, the node is qualified to be the LCA node if it contains every query keywords under its subtree. For the query keyword with the NOT operator, it is intuitive that the LCA node should *not* contain such keyword under its subtree. For instance, consider query Q_1 in Table 1, where the NOT operator is marked as exclamation "!" in the prefix of the keyword. It asks for the subjects that are held in room 101 but not on Monday. We can see that totally two courses, nodes 1.2.1 and 1.2.4, are held in room 101. However, since the course of node 1.2.1 is held on Monday, the desirable answer would be node 1.2.4. Nevertheless, based on the intuitive definition of LCA, there may not exist any node to return in some cases, even though there are desirable answers. Consider another example query Q_2, whose semantics is similar to that of Q_1 but limited to year 2010. Observe that only node 1 (*School*) contains the first three keywords under its subtree, but it is not a qualified LCA since it also contains the undesirable keyword *Monday*. Hence, the meaningful information, that is, the subtree rooted at node 1.2.4, will not be returned. We further propose the definition of the *masked* LCA to overcome this problem, and will have more details in Section 2.

In summary, the main technical contributions of this paper are as follows:

- We first define the SLCA and relevant match for the keyword query with an arbitrary combination of AND, OR, and NOT operators. In addition, we propose the idea of the masked LCA to process special cases such as Q_2.
- We also give an efficient algorithm RELMN (RELevant Matches with NOT semantics) to figure out the SLCA nodes and the relevant matches corresponding to our definitions.
- We discuss the variations of the monotonicity and consistency properties since these properties naturally become more complex for the keyword query with NOT semantics.
- We have performed an empirical evaluation of the query result by measuring the precision, recall, processing time, and scalability. The experiments show that our algorithm works efficiently and gives reasonable query results.

The remainder of this paper is organized as follows. In Section 2, we deliver the definitions of the SLCA nodes and the relevant matches for the query with NOT semantics. Section 3 presents the RELMN algorithm that figures out all of the relevant matches with NOT semantics. The variations of monotonicity and consistency properties in our framework are discussed in Section 4. In Section 5, we show the experimental studies with the metrics precision, recall, processing time, and scalability. Related works and conclusions are given in Section 6 and Section 7 respectively.

2   Basic Definitions

In this section, we define the SLCA nodes and the relevant matches for the keyword query with NOT semantics. In our approach, the keyword query should be first converted into CNF (conjunctive normal form). Hence, the clauses in this paper are connected with the AND operators, and literals in the same clause are supposed to be connected with the OR operators. Besides, the literal refers to a given keyword, and is either positive or negative. The negative literal is represented with the exclamation mark "!" in the prefix of the keyword.

2.1   SLCA Nodes with NOT Semantics

As mentioned before, in AND-only semantics, a node is said to be an LCA node of a query if it contains every keyword at least once under its subtree. Therefore, it is reasonable and intuitive that the negative literals are "not" expected to be the descendants of the LCA node. Namely, the positive literal is logically true to node n if such keyword is contained in the subtree rooted at n, and the negative literal is logically true to node n if n contains "no" such keyword under its subtree. We formally define the concept of "satisfy" in the following:

Definition 1. [Satisfy]: Given a query $Q = \{c_1 \wedge c_2... \wedge c_w\}$, c_i (which consists of several positive and negative literals) is logically true to node n if n contains at least one positive literal of c_i or does not contain at least one of the negative literals of c_i under its subtree. We also say that "node n *satisfies* clause c_i".   □

Based on this definition, n is the LCA node of Q if n satisfies every clause of Q, which is also stated as "n satisfies query Q". Recall that there may not exist any node to be the LCA for some queries such as Q_2. To overcome this problem, we further define the masked LCA (mLCA) as the relaxation response. The mLCA of the query is a node that satisfies every *pure* positive clause, where a pure positive clause refers to the clause that has no negative literals. By this relaxed definition, a node whose descendants represent both the desirable information, such as the subtree rooted at node 1.2.4, and the undesirable information, such as the subtree rooted at node 1.2.1, will be kept for further processing. Moreover, the masked SLCA (mSLCA) is an mLCA, and none of its descendants are also mLCAs. Finally, we propose the following definition:

Definition 2. [SLCA]: A node is said to be an SLCA of query Q if: (i) n satisfies Q, and (ii) none of n's descendants satisfy Q. If no such node exists, the mSLCA nodes are considered to be the SLCA nodes in our framework.   □

2.2   Relevant Matches with NOT Semantics

Liu and Chen [7] defined the relevant matches under the subtree rooted at SLCA nodes for the keyword query with AND-only semantics. We first describe their original definitions and then define the relevant matches in our framework.

A node is a *match* if its tag name or the content contains a query keyword. The *descendant matches* of a node n, denoted as $n.dMatch$, is a set of query keywords, each of which has at least one match under the subtree rooted at n. For simplicity, $n.dMatch$ could also be seen as a bit array of size w, where w is the number of keywords. Namely, the jth bit of $n.dMatch$ is set to 1 if n contains the jth keyword under its subtree. In addition, a node n_1 will be considered as a *contributor* if (i) n_1 is itself or the descendant of a given SLCA, and (ii) n_1 does not have a sibling n_2 such that $n_2.dMatch \supset n_1.dMatch$ (pruning rule). The main concept of the contributor is that if n_2 contains richer information under its subtree than n_1 does, n_1 should be *pruned* by n_2. Finally, node n is considered *relevant* if every node on the path from n up to an SLCA t is a contributor, and the *query result* for a query over the XML tree is defined as (t, M), where M is the set of relevant matches (including value children, if any) under the subtree rooted at an SLCA t.

Example 1. Consider $Q = (2010) \wedge (Math) \wedge (Friday)$. Node 1 (*School*) is the only SLCA node. Besides, the $dMatch$ value of nodes 1.2.1, 1.2.2, and 1.2.3 are "010", "100", and "110" respectively. Note that the first keyword (2010) is mapped to the right-most bit, and so on. By the definitions mentioned above, nodes 1.2.1 and 1.2.2 are pruned by node 1.2.3[1], and those descendants of the pruned nodes are also skipped. Therefore, the relevant matches is the node list [1, 1.1, 1.1.1, 1.2, 1.2.3, 1.2.3.1, 1.2.3.1.1, 1.2.3.2, and 1.2.3.2.1]. □

In our framework, we modify the pruning rule of the contributor as follows:

Definition 3. [Pruning Rules]: Let $n.C$ be the set of clauses that is satisfied by node n, and $n.P$ ($n.N$) be the set of positive (negative) literals that is contained under the subtree rooted at n. Node n_2 will be able to prune node n_1 if n_2 satisfies the following three rules: (1) $n_2.C \supset n_1.C$, (2) $n_2.P \supseteq n_1.P$, and (3) $n_2.N \subseteq n_1.N$ (referred to as Rules 1, 2, and 3). □

Obviously, Rule 1 inherits from the original pruning rule. Rules 2 and 3 are added because if n_1 has more positive literals than n_2 or has less negative literals than n_2, it might contain desirable information to the users. Moreover, note that except for the pruning rules, the definitions of *contributor*, *relevant match* and *query result* in our framework are identical to the original ones.

3   The RELMN Algorithm

We propose an efficient algorithm RELMN to search the SLCAs and the relevant matches for the keyword query with an arbitrary combination of AND, OR, and NOT operators. We first give some more definitions, and then illustrate the algorithm in detail.

3.1   Definitions

The *match* in our framework refers to a node whose tag name or value equals to a positive literal or a negative literal. The *match tree* of a node n, denoted as $mTree(n)$, consists of the nodes along the path from each match (including its value child, if any) up to the node n.

Given a query Q without any pure positive clause, all the leaf nodes which do not match any negative literal would satisfy Q, and thus will be returned according to our definition. For instance, given a query $(!Monday) \wedge (!101)$, all the leaf nodes except for nodes 1.2.1.2.1, 1.2.1.3.1, and 1.2.4.3.1 are the relevant matches. On the contrary, if Q has at least one pure positive clause, the SLCA nodes are guaranteed to be in $mTree(root)$, where $root$ is the root of the whole

---

[1] Given two nodes n_1 and n_2, n_2 prunes n_1 if (i) $n_2.dMatch > n_1.dMatch$ and (ii) bitwise $AND(n_1.dMatch, n_2.dMatch) = n_1.dMatch$, which can be done in $O(1)$.

XML tree. Since the query without any pure positive clause is easier to deal with, we assume there exists at least one pure positive clause in this paper. Therefore, we construct $mTree(root)$ first, because it helps to figure out the relevant matches efficiently.

We then discuss how to achieve the pruning rules defined in Definition 3. As shown in Example 1, the bit array can efficiently implement the pruning rule. We therefore continue to adopt this data structure. For each node $n \in mTree(root)$, we use bit arrays *C-array*, *P-array*, and *N-array* to implement $n.C$, $n.P$, and $n.N$. The C-array of n is a bit array of length w bits, where w is the number of clauses. The ith bit of C-array is set to 1 if n satisfies the ith clause of the keyword query. The P-array of n records the occurrences of the positive literals in $mTree(n)$, and its length equals to the number of distinct positive literals. The jth bit of the P-array is finally set to 1 if n contains the jth positive literal under its subtree. The N-array of n is similar to the P-array, but takes negative literals into consideration only.

Example 2. Consider Q_3 in Table 1. The positive literals are {*Course, 101, 103*}, and the negative literals are {*!Friday, !Math*}. Thus, the bit numbers of P-array and N-array are 3 and 2 respectively. Recall that the literals are numbered from left to right, and the first bit is the right-most bit in this paper. Hence, the P-arrays of nodes 1.2.1 and 1.2.3 are "011" and "101", and their N-arrays are "10" and "11", respectively. Moreover, node 1.2.3 does not satisfy the third clause of Q_3 because both *Friday* and *Math* appear in its subtree, and thus its C-array is "011" while the C-array of node 1.2.1 is "111".   □

To produce the final correct C-arrays, each node is associated with one more bit array *B-array*. Given a keyword query of w clauses, every node needs w *blocks*, and each block corresponds to a specific clause. The block is a bit array and is composed of several (or zero) "negative bits" and one (or zero) "positive bit". Suppose block *blk* corresponds to clause c. It will consist of k negative bit(s) if c has k negative literal(s). On the other hand, if c has at least one positive literal, *blk* also needs one positive bit. Otherwise, *blk* needs no positive bit. Specifically, every negative literal corresponds to a specific negative bit, but all the positive literals in the same clause share a common positive bit. Such design corresponds to the concept of satisfaction given in Definition 1. The B-array for each node will consist of the blocks of all the clauses, and all the negative bits are initialized as 1 and the positive bit is initialized as 0. We then give an example to show the initial state of B-array.

Continue Example 2. The first clause c_1 (*Course*) of Q_3 has only one positive literal, and thus its corresponding block is initialized as "0". The second clause c_2 has two positive literals. Since all the positive literals share the common positive bit, its corresponding block is also initialized as "0". The third clause c_3 has exactly two negative literals and has no positive literal, and thus its corresponding block is initialized as "11". Besides, the blocks are displayed from right to left in the figure. Hence, the B-array of each node in $mTree(root)$, as shown in Figure 2 (a), is initialized as "11 0 0".

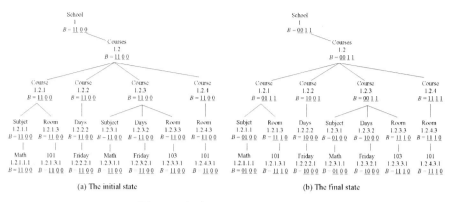

(a) The initial state   (b) The final state

Fig. 2. The B-arrays of query Q_3

3.2   Algorithm

The pseudo code of Algorithm RELMN is shown in Figure 3. Given a keyword query with an arbitrary combination of AND, OR, and NOT operators, we convert the keyword query into CNF in line 1. To initialize the B-array, P-array, and N-array for each node in $mTree(root)$ (in line 2), we first retrieve the match lists from the B-tree index, where the key of the B-tree index is the query keyword and its associated value is all the matches sorted by their Dewey numbers of this query keyword. Then we construct and initialize every node along the path from each match up to the root. Next, we iteratively process the literals of each clause in lines 3 to 8. If the literal is positive, we call Procedure $SetPositive$; otherwise, we call Procedure $SetNegative$.

In Procedure $SetPositive$, let l be the currently active literal, and m be the match of l. Besides, suppose l corresponds to the ith bit of B-array and the jth bit of P-array. For each node n from m up to $root$, if the ith bit of $n.B$ and the jth bit of $n.P$ are both 1 (in line 17), it implies that another match of l has already visited node n (including n's ancestors). Therefore, we can quit the $SetPositive$ procedure. On the contrary, if line 17 is not active, we then set the ith bit of B-array and the jth bit of P-array to 1. Procedure $SetNegative$ is quite analogous to $SetPositive$. The primary difference is to set the ith bit of B-array to 0 at each iteration. Furthermore, we only need to check $n.B[i]$ (in line 27) to determine whether n is visited or not, because every negative literal has its own negative bit in B-array. The next example shows how these two procedures work.

Example 3. Consider the second literal "103" in the second clause of Q_3. It is a positive literal and has only one match (node 1.2.3.3.1). Since it corresponds to the second bit of B-array and the third bit of P-array, these two bits of each node in the node list [1.2.3.3.1, 1.2.3.3, 1.2.3, 1.2, 1] are set to 1. Take another literal as an example. Consider "$!Friday$" in the third clause of Q_3, which corresponds to the third bit of B-array and the first bit of N-array. "Friday" has two matches 1.2.2.2.1 and 1.2.3.2.1. In the loop of the former match, we iteratively set the third bit of B-array of each node in the node list [1.2.2.2.1, 1.2.2.2, 1.2.2, 1.2, 1]

Input: A query input *query* with arbitrary combination of AND, OR, and NOT operators
Output: All of the relevant matches
Global Variable: *Result* ← empty

RELMN(query)
1: Q ← convert *query* into CNF
2: initialize the *mTree(root)* of Q
3: **for** each clause c of Q **do**
4:     **for** each literal l in c **do**
/* suppose l maps to i^{th} bit of B-array */
5:         **if** l is positive **then**
/* suppose l maps to j^{th} bit of P-array */
6:             *SetPositive(i, j)*
7:         **else**
/* suppose l maps to j^{th} bit of N-array */
8:             *SetNegative(i, j)*
9: convert B-array to C-array for each node
10: *FindSLCA(root)*
11: **if** *Result* = *empty* **then**
12:     *FindMaskedSLCA(root)*
13: **for** each node $n \in$ *Result*
14:     *SearchRelevant(n)*

SetPositive(i, j)
15: **for** each match m of l **do**
16:     **for** each node n from m up to *root* **do**
17:         **if** $n.B[i] = 1$ **and** $n.P[j] = 1$ **then**
18:             return
19:         set the ith bit of $n.B$ to 1
20:         set the jth bit of $n.P$ to 1

SetNegative(i, j)
21: **for** each match m of l **do**
22:     **for** each node n from m up to *root* **do**
23:         **if** $n.N[j] = 0$ **then**
24:             return
25:         set the ith bit of $n.B$ to 0
26:         set the jth bit of $n.N$ to 1

FindSLCA(n)
27: *slca* ← 0
28: **for** each child *nc* of n **do**
29:     *slca* ← *slca* + *FindSLCA(nc)*
30: **if** n satisfies every clause **then**
31:     **if** *slca* = 0 **then**
32:         *Result* = *Result* $\cup$ $\{n\}$
33:     return 1 + *slca*
34: return *slca*

FindMaskedSLCA(n)
35: *mslca* ← 0
36: **for** each child *nc* of n **do**
37:     *mslca* ← *mslca* + *FindMasked-SLCA(nc)*
38: **if** n satisfies every pure positive clause **then**
39:     **if** *mslca* = 0 **then**
40:         *Result* = *Result* $\cup$ $\{n\}$
41:     return 1 + *mslca*
42: return *mslca*

SearchRelevant(n)
43: **if** *IsContributor(n)* = *true* **then**
44:     output n /* n is relevant */
45:     **for** each child *nc* of n **do**
46:         *SearchRelevant(nc)*

IsContributor(n)
47: **if** n is the SLCA node **then**
48:     return *ture*
49: **for** each sibling *ns* of n **do**
50:     **if** $ns.C \supset n.C$ and $ns.P \supseteq n.P$
        and $ns.N \subseteq n.N$ **then**
51:         return *false*
52: return *true*

Fig. 3. The RELMN algorithm

to 0, and set the first bit of N-array of each node to 1. For the latter match, each node in the node list [1.2.3.2.1, 1.2.3.2, 1.2.3] is treated as the previous ones. And we can quit Procedure *SetNegative* since the first bit of N-array of node 1.2 is set to 1.   ☐

Once the processing of all matches of a given clause c is done, node $n \in mTree(root)$ satisfies c if and only if the bits of *n.blk* are not all zero, where *n.blk* is the corresponding block of c in node n. Therefore, the final state of the B-array is converted into C-array for each node when the setting of all the literals are done (in line 9). As Figure 2 (b) indicates, only nodes 1.2.1 and 1.2.4 satisfy all three clauses, and clearly they are the SLCA nodes of query Q_3 since they do not have any descendant nodes that also satisfy every clause. We then call *FindSLCA*, which traverses the nodes in postorder, to find the SLCA nodes (in line 10). If line 31 of $FindSLCA$ is active, it indicates that none of n's descendants satisfy the query. Therefore, n is an SLCA node. Furthermore, if variable $Result$ is empty in line 11, meaning that none of the nodes satisfy the keyword query, we further search mSLCA nodes by Procedure $FindMaskedSLCA$. The main difference is that it considers pure positive clauses only (in line 38). At the final step, Procedure $SearchRelevant(n)$ traverses the tree in preorder and identifies all the relevant matches of $mTree(n)$ by Procedure $IsContributor$. If n is an SLCA node, it is guaranteed to be a contributor since it has no sibling (in line 47). If there exists n's sibling n_s which satisfies the pruning rules (in line 50), n will be pruned by n_s.

3.3   Time Complexity

Now we analyze the time complexity of RELMN. Given a query $Q = c_1 \wedge c_2 ... \wedge c_w$, let $|M_i|$ denote the number of total matches of c_i, where $1 \leq i \leq w$. Besides, let d be the maximum depth of the XML tree and $|M| = \Sigma_{i=1}^{w} |M_i|$.

The initialization of $mTree(root)$ takes $O(d|M|)$ time since the arrays of each node could be initialized in constant time. Besides, the setting of ith bit and j^{th} bit in Procedures $SetPositive$ and $SetNegative$ can be done in constant time since it is determined when we parse and analyze the query keywords. Hence, it takes $O(d|M|)$ time to complete lines 3 to 8. In line 9, the conversion takes totally $O(d|M| \cdot w)$ time, because it takes $O(|w|)$ time to convert B-array into C-array for each node. The $FindSLCA$ and $FindMaskedLCA$ procedures visit every node n in $mTree(root)$ exactly once, and determine whether n is the SLCA node or not in constant time when C-arrays are constructed. Clearly, both of these procedures take $O(d|M|)$ time. The number of the nodes visited by Procedure $SearchRelevant$ is bounded in $O(d|M|)$ too. However, the $IsContributor$ procedure takes $O(b)$ time to check each of n's sibling where b is the number of n's siblings. Let b_i denote the number of siblings of node n_i in $mTree(root)$, and suppose the number of the total nodes of $mTree(root)$ is z. Procedure $SearchRelevant$ takes $O(\Sigma_{i=1}^{z} b_i)$ time to visit every node and determine whether it is a contributor or not. In total, the time complexity of RELMN is $O(d|M| \cdot w) + O(\Sigma_{i=1}^{z} b_i)$.

4   Properties of Monotonicity and Consistency

We discuss the variations of the monotonicity and consistency properties under NOT semantics in this section. In short, the monotonicity property describes how $|SLCA|$ varies with respect to data updates and query updates, and the consistency property describes how the relevant matches vary when the data or the query updates. Our pruning rules will make the query results satisfy these properties presented in this section, but the proofs are omitted due to space limitation. In the following, we deliver the modified properties and the examples.

Property 1. [Data Monotonicity]: After a new node n is added to the XML tree, $|SLCA|$ should be monotonically increasing if n matches a given positive literal. On the contrary, if n matches a given negative literal, $|SLCA|$ should be monotonically decreasing.

For example, given query $Q = (Friday) \wedge (Math)$. If we add a new node 1.2.2.1.2 (*Math*) as the sibling of node 1.2.2.1.1, then $|SLCA|$ changes from 1 to 2. The next example which concerns negative literals is $Q' = (Days) \wedge (Math) \wedge$ (!102). If we add a new node 1.2.1.3.2 (*102*) as the sibling of node 1.2.1.3.1, then $|SLCA|$ changes from 2 to 1 after the node insertion.   □

Property 2. [Query Monotonicity]: If we add a new literal, either positive or negative, into an existing clause, $|SLCA|$ should be monotonically increasing. In the opposite, if we add a new literal as a new clause, $|SLCA|$ should be monotonically decreasing.

Consider query $Q = (Days) \wedge (Math) \wedge$ (!103). Node 1.2.1 is the only SLCA node. If we add "Chemical" into the second clause, $Q' = (Days) \wedge (Math \vee Chemical) \wedge$ (!103), $|SLCA|$ becomes 3 since nodes 1.2.2 and 1.2.4 become new SLCA nodes. Consider query $Q_* = (101 \vee 102) \wedge (Chemical)$. If we add a new clause "Friday" as the third clause, $Q'_* = (101 \vee 102) \wedge (Chemical) \wedge (Friday)$, $|SLCA|$ changes from 2 to 1 since node 1.2.4 is no longer an SLCA node.   □

Property 3. [Data Consistency]: Let R denote the set of query results of a given query over the XML tree, and R' denote the set of query results after data updates or query updates. *Delta result tree* $\delta(R, R')$ is defined as $R' - R$ which results $\delta(R, R') \cap R = \emptyset$ and $\delta(R, R') \cup R = R'$. Suppose $\delta(R, R')$ is rooted at n_1, and n_2 prunes n_1 before the insertion of node n. If n is the descendant of n_1, then n must match a given positive literal. If n is the descendant of n_2, then n must match a given negative literal.

Consider query $Q = (2010) \wedge (Subject) \wedge (102) \wedge (!Friday)$. Node 1.2.3, which is originally pruned by node 1.2.1, becomes a contributor if we add a new node 1.2.3.3.2 (*102*) as the sibling of node 1.2.3.3.1, because the newly added node makes node 1.2.3 break Rules 1 and 2. Another example is that if we add a new node 1.2.1.2.2 (*Friday*) as the sibling of node 1.2.1.2.1 instead of adding node 1.2.3.3.2 (*102*), it would make node 1.2.1 satisfy fewer clauses. Therefore, node 1.2.3 is no longer pruned by node 1.2.1 since Rule 1 is broken.   □

Property 4. [Query Consistency]: Suppose the definitions of the nodes n_1 and n_2 are the same as those in Property 3. The property of query consistency in our framework could be classified into two cases, too. In the first case, if we add a new positive literal l into the keyword query, there must exist a match of l in the subtree rooted at n_1. In the second case, if the newly added literal l is negative, $mTree(n_2)$ must contain a match of such negative literal.

For example, consider $Q = (2010) \wedge (Friday) \wedge (Math)$. We know that node 1.2.3 may prune node 1.2.1. If we add "Monday" into the second clause, that is, $Q' = (2010) \wedge (Friday \vee Monday) \wedge (Math)$, node 1.2.1 becomes a contributor since Rules 1 and 2 are both broken. Consider another query, $Q'' = (2010) \wedge (Friday) \wedge (Math) \wedge$ (!103), where we add "!103" as a new clause instead of adding "Monday" into the second clause. Node 1.2.1 may not be pruned by node 1.2.3 since Rules 1 and 3 are both broken.   □

Table 2. Test queries for Reed College and Baseball

Reed College	Baseball
QR_1 title ∧ (M ∨ W) ∧ 01:00PM	QB_1 Indians ∧ Relief Pitcher ∧ surname
QR_2 room ∧ Love in Shakespeare ∧ !02:40PM	QB_2 (Indians ∨ Tigers) ∧ starting pitcher ∧ surname
QR_3 subj ∧ 04:00PM ∧ !T ∧ !W	QB_3 player ∧ !starting pitcher ∧ !relief pitcher ∧ home_runs
QR_4 subj ∧ !T	QB_4 Yankees ∧ !starting pitcher ∧ !relief pitcher ∧ home_runs
QR_5 subj ∧ room ∧ 301 ∧ !M	QB_5 player ∧ home_runs ∧ 35
QR_6 title ∧ F ∧ (ANTH ∨ BIOL) ∧ !09:00AM	QB_6 !Yankees ∧ second base ∧ home_runs
QR_7 instructor ∧ CHEM ∧ !M ∧ 01:10PM	QB_7 West ∧ !Mariners ∧ Trevor ∧ position
QR_8 instructor ∧ CHEM ∧ M ∧ 01:10PM	QB_8 National ∧ !West ∧ team_name

5   Experimental Evaluation

We have performed several experiments to evaluate the effectiveness and the efficiency of the proposed algorithm. The measurements include precision, recall, processing time, and scalability. The RELMN algorithm is implemented in C++ with the environment of Visual Studio 6.0 on Windows XP. We applied four data sets to perform the experiments. They are DBLP, Treebank_e, Reed College, and Baseball, and the data sizes are 127MB, 82MB, 277KB, 1.01MB, respectively. The first three data sets can be downloaded from [12], and the last one (Baseball) is available in [11]. In addition, another B-tree index is built to answer the tag name or value by a specific Dewey number. That is, the key is the Dewey number of a given node, and the associated value is its tag name or value. Both indices (the first B-tree index is introduced in Section 3.2) are implemented based on the Oracle Berkeley DB.

5.1   Precision and Recall

The test queries used for the experiments discussed in this subsection are listed in Table 2. The precision and recall of the Reed College data set is displayed in Figure 4. The query could consist of any arbitrary combination of AND, OR, and NOT operators. Observe that the proposed algorithm usually provides reasonable query results. Take query QR_3 for example. We want to search the subject of the courses which start at 04:00PM but not on Tuesday or Wednesday. There are totally eight courses that start at 04:00PM. Two of them are on Tuesday and three of them are on Wednesday. Since Algorithm RELMN returns the subject of the remaining three courses, the precision and recall of QR_3 are both 100%. On the other hand, the semantics of QR_4 is to search the subjects that are not held on Tuesday. Refer to Figure 6 (a), which shows the portion of the Reed College XML tree. Observe that every node with "subj" is the SLCA node of QR_4. Because each of them has only one value child that describes the name of the subject, and thus satisfy "!T", the precision is therefore imperfect. This situation could be improved by changing the query. Let $QR_4' = (subj) ∧ (days) ∧ (!T)$, which has an extra clause "days" than QR_4. Though semantics of QR_4' stay the same, it gets 100% precision. Figure 5 displays the precision and recall of the Baseball data set. Recall that if none of the nodes satisfy the keyword query, the mSLCA nodes would be considered as the SLCA nodes. QB_4 is one of such cases. Its semantics is to search the home runs of non-pitcher players in Yankees team, and it gets perfect precision and

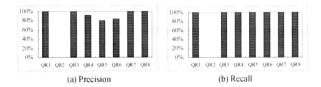

(a) Precision   (b) Recall

Fig. 4. Precision and recall of the Reed College data set

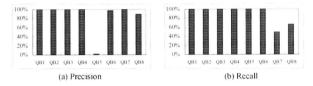

(a) Precision   (b) Recall

Fig. 5. Precision and recall of the Baseball data set

recall by RELMN. On the other hand, the precision and recall are also affected by ambiguous keywords. For example, the semantics of QB_5 is to find the players who hit 35 home runs in 1998. However, "35" is also an ambiguous keyword since many entries (walks, errors, etc) may also be 35. Hence, QB_5 has a low precision. Besides, QB_8 also gets low precision and recall since "West" could be the player's surname and the league's division name.

Though our approach answers desirable query results for most cases except for ambiguous keywords, it still has room to be improved. Consider query QR_2. Its semantics is to ask the classroom of the course with the title "Love in Shakespeare" which does not start from 2:40PM. There is only one course titled "Love in Shakespeare" in the whole XML tree, and the course starts from 2:40PM. Namely, the reasonable answer should be empty. But in our approach, the subtree rooted at the course of "Love in Shakespeare" will be returned, and thus we get poor precision and recall for QR_2. Consider query QB_6, its semantics is to search the home runs of the second base player except for Yankees team. The portion of the Baseball XML tree is shown in Figure 6 (b). We can see that the node of Yankee's second base player satisfies QB_6 which is not a desirable node to the semantics of QB_6. However, such node is still being returned in our approach, and thus causes an imperfect precision.

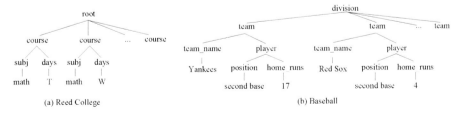

(a) Reed College   (b) Baseball

Fig. 6. Portions of the XML trees

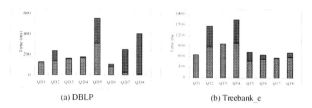

(a) DBLP   (b) Treebank_e

Fig. 7. Processing time on the DBLP and Treebank_e data sets

Table 3. Statistics of test queries of DBLP and Treebank_e

	DBLP					Treebank_e			
	Min fre.	Max fre.	Total fre.	Output size		Min fre.	Max fre.	Total fre.	Output size
QD_1	984	112K	137K	2.0K	QT_1	74	187K	236K	0.7K
QD_2	984	112K	151K	21.6K	QT_2	74	187K	321K	92K
QD_3	984	112K	161K	0.3K	QT_3	74	187K	350K	12
QD_4	984	112K	168K	1.8K	QT_4	74	187K	363K	101K
QD_5	112K	114K	338K	172K	QT_5	306	154K	202K	82K
QD_6	17.4K	20.5K	57K	19K	QT_6	306	154K	219K	47K
QD_7	5.9K	6.4K	18K	0.6K	QT_7	306	154K	227K	2.1K
QD_8	1.9K	2.0K	5.9K	6.2K	QT_8	306	154K	233K	47K

5.2   Processing Time and Scalability

The processing time was measured on a 1.67GHz dual-core CPU with 1.5GB RAM, and the cache size of Oracle Berkeley DB was set to 1.0 GB. Besides, each query was executed three times, and we computed the average processing time of the last two times. The experiments are based on the DBLP and Treebank_e data sets. Each processing time is divided into two parts. The first part (lower part) represents the cost of constructing the match tree and searching the SLCA nodes, and the second part (upper part) represents the cost of determining the contributors and outputting the query results. Table 3 displays the statistics of the test queries, in which "total frequency" denotes the sum of the numbers of all matches, and "output size" denotes the number of all relevant matches. For the first four input queries (QD_1 to QD_4) of the DBLP data set, we kept the minimum and maximum frequencies of literals constant, and varied the number of clauses (3 to 6). Each clause has exactly one literal with random positive or negative. For QD_5 to QD_8, the number of clauses is fixed as 3, and the minimum and maximum frequencies are close. As in the previous scenario, each clause has exactly one positive or negative literal.

The processing time of QD_1 to QD_8 is shown in Figure 7 (a). We can see that the cost of the first part is roughly in linear proportion to the total frequencies of all literals. The cost of the second part mainly depends on the number of relevant matches. Another factor to influence the cost of the second part is the determination of contributors. For example, the numbers of relevant matches of QD_7 and QD_8 are not very large, but the costs of their second parts are especially high. The reason is that each of them has only one SLCA node, and the number of the children rooted at the SLCA node is pretty large. Therefore, to determine the contributors, each child of the SLCA node has to check all of its siblings. The processing time of the Treebank_e data set is shown in Figure 7 (b). We selected the Treebank_e data set because it has an extraordinary depth which is

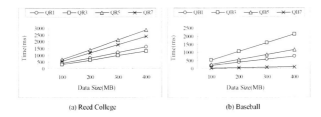

(a) Reed College   (b) Baseball

Fig. 8. Scalability of the Reed College and Baseball data sets

much larger than general data sets. The scenario of QT_1 to QT_4 is quite similar to that of QD_1 to QD_4. Namely, the numbers of the clauses are from 3 to 6, and each clause has one positive or negative literal. The numbers of clauses of QT_5 to QT_8 are from 2 to 5, where each clause has exactly two literals, and the literals are randomly set to positive or negative too. Overall, the analysis of processing time is similar to that of the DBLP data set.

Finally, the scalability of RELMN was tested with different data sizes. We selected the Reed College (277K) and Baseball (1.01M) data sets as the data sources. We then duplicated the content and gave a unique tag name as the root of the whole contents. The test queries are specified in Table 2, and the result is depicted in Figure 8. We can see that the processing time is broadly in linear proportion to the data size for each of the query input, which is satisfiable.

6   Related Works

The primary task of keyword search over XML documents is to figure out meaningful subtrees, each of which contains the query keywords at least once. The LCA-based strategy is commonly used to define the query result [2][4][9][10]. Since the nodes under the subtree rooted at a given LCA may not all be meaningful, some researchers further defined the relevant nodes related to the LCA [1][3][5][6][7]. In addition, because of the widespread use of XML data, the INitiative for the Evaluation of XML retrieval (INEX) [13] provides a coordinated effort to promote the evaluation of content-based XML retrieval.

We next discuss the performance and time complexity of some representative algorithms in the following. Given a keyword query $Q = \{k_1, k_2, ..., k_w\}$ with w keywords, let M_i denote the sorted match list of keyword k_i, and $|M_i|$ denote the frequency of k_i. Without loss of generality, suppose $|M_1|$ is the minimum one and $|M_w|$ is the maximum one. The ILE (Indexed Lookup Eager) algorithm, which was proposed by Wu and Papakonstantinou [10], runs in $O(wdM_1 \log M_w)$ time to determine the SLCA nodes, where d is the maximum depth of the XML tree. They also gave another algorithm SE (Scan Eager) which runs in $O(\Sigma_{i=1}^{w} M_i d)$ time. The ILE algorithm performs better when $|M_w|$ is much larger than $|M_1|$ since it adopts a binary search strategy to quickly locate the right-most and left-most nodes of the given node. The SE algorithm scans all the matches of keywords exactly once, so it is a good choice when the maximum frequency and the minimum frequency of the keywords are close.

7   Conclusions

The contribution of this paper is to propose an extension of searching relevant matches with NOT semantics. Our algorithm RELMN is mainly based on detecting the positive and negative literals under the subtree rooted at SLCA nodes. We also discuss the variant properties of monotonicity and consistency since they become more complex if OR and NOT operators are allowed. In addition, the experimental results show that the proposed algorithm runs efficiently and gives reasonable query results by measuring the processing time, scalability, precision, and recall. As part of our future work, we intend to improve the effectiveness for keyword query with NOT semantics, such as the QR_2, QB_6, and QB_7 cases in Table 2.

Acknowledgements. Ya-Hui Chang was supported in part by the NSC grant 97-2221-E-019-028- from the National Science Council, Taiwan. Kun-Mao Chao was supported in part by the NSC grants 97-2221-E-002-097-MY3 and 98-2221-E-002-081-MY3 from the National Science Council, Taiwan.

References

1. Cohen, S., Mamou, J., Kanza, Y., Sagiv, Y.: XSEarch: A Semantic Search Engine for XML. In: VLDB, pp. 45–56 (2003)
2. Guo, L., Shao, F., Botev, C., Shanmugasundaram, J.: XRANK: Ranked Keyword Search over XML Documents. In: SIGMOD, pp. 16–27 (2003)
3. Hristidis, V., Koudas, N., Papakonstantinou, Y., Srivastava, D.: Keyword Proximity Search in XML Trees. IEEE TKDE, 525–539 (2006)
4. Kong, L., Gilleron, R., Lema, A.: Retrieving Meaningful Relaxed Tightest Fragments for XML Keyword Search. In: EDBT, pp. 815–826 (2009)
5. Li, G., Feng, J., Wang, J., Zhou, L.: Effective Keyword Search for Valuable LCAs over XML Documents. In: CIKM, pp. 31–40 (2007)
6. Lin, R.-R., Chang, Y.-H., Chao, K.-M.: Faster Algorithms for Searching Relevant Matches in XML Databases. In: Bringas, P.G., Hameurlain, A., Quirchmayr, G. (eds.) DEXA 2010. LNCS, vol. 6261, pp. 290–297. Springer, Heidelberg (2010)
7. Liu, Z., Chen, Y.: Reasoning and Identifying Relevant Matches for XML Keyword Search. In: VLDB, pp. 921–932 (2008)
8. Sun, C., Chan, C.-Y., Goenka, A.K.: Multiway SLCA-based Keyword Search in XML Data. In: WWW, pp. 1043–1052 (2007)
9. Xu, Y., Papakonstantinou, Y.: Efficient LCA Based Keyword Search in XML Data. In: EDBT, pp. 535–546 (2008)
10. Xu, Y., Papakonstantinou, Y.: Efficient Keyword Search for Smallest LCAs in XML Database. In: SIGMOD, pp. 527–538 (2005)
11. http://www.cafeconleche.org/books/biblegold/examples/
12. http://www.cs.washington.edu/research/xmldatasets/
13. http://www.inex.otago.ac.nz/
14. http://www.w3.org/TR/xpath-full-text-10/

Evaluating Contained Rewritings for XPath Queries on Materialized Views

Rui Zhou[1], Chengfei Liu[1], Jianxin Li[1], Junhu Wang[2], and Jixue Liu[3]

[1] Swinburne University of Technology, Melbourne, Australia
{rzhou,cliu,jianxinli}@swin.edu.au
[2] Griffith University, Gold Coast, Australia
J.Wang@griffith.edu.au
[3] University of South Australia, Adelaide, Australia
Jixue.Liu@unisa.edu.au

Abstract. In this paper, we study the problem how to efficiently evaluate a set of contained rewritings on materialized views. Previous works focused on how to find a set of contained rewritings given a view and a query, but did not address how to evaluate the rewritings on materialized views. To evaluate a potential exponential number of contained rewritings, we design two algorithms, a basic algorithm and an optimized algorithm. Both algorithms are built on the observation that the exponential number of contained rewritings are actually composed by a linear number of component patterns. In the optimized algorithm, we further design four important pruning rules and several heuristic rules that can effectively reduce the number of component patterns we need to evaluate. The experiments demonstrate the efficiency of our algorithms.

1   Introduction

Answering queries using views refers to using previously defined and materialized views to answer new queries in order to save the cost of accessing the large underlying database. With the prevalence of XML technologies, answering XML queries using XML views has caught the attention of both researchers and system designers, and is believed as a promising technique in web and XML database applications. Since XPath serves as the core sub-language of the major XML languages such as XQuery, XSLT, we focus on answering XPath queries using XPath views.

There are two rewriting schemes to answer XPath queries with materialized XPath views: equivalent rewriting [1] and maximal contained rewriting [2]. We explain them formally as follows: given an XPath query q, an XPath view v, and an XML database T representing a large XML document, let the materialized view of v on T be $T_v = v(T)$, to find an equivalent rewriting for q using v is to find a *compensation pattern* q_c such that $q_c(T_v) = q(T)$. Meanwhile, the pattern produced by merging the root node of q_c with the answer node of v, denoted as $q_c \oplus v$, is called an equivalent rewriting of q. Fig. 1(a) shows an example. v (/a/b) is a view, q_1 (/a/b/c) is a query, q_c (b/c) is the compensation query, and ER is the

J.X. Yu, M.H. Kim, and R. Unland (Eds.): DASFAA 2011, Part I, LNCS 6587, pp. 481–495, 2011.
© Springer-Verlag Berlin Heidelberg 2011

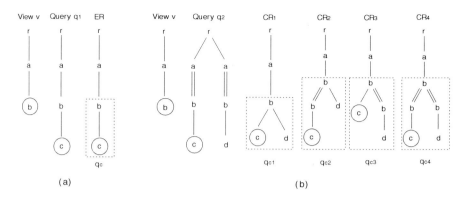

Fig. 1. Equivalent rewriting and Contained rewriting

equivalent rewriting of q_1 using v. In some cases (like data integration scenarios), a view may not answer a query completely, i.e. there does not exist an equivalent rewriting for a query using a view, but it is still reasonable to efficiently give users part of the query answers with the view. To this end, maximal contained rewriting (MCR) is proposed to give the best effort to answer a query using a view. Different from equivalent rewriting, maximal contained rewriting usually requires to find a set of compensation patterns $Q_c = \{q_{c_1}, \ldots, q_{c_i}\}$ satisfying $Q_c(T_v) \subseteq q(T)$, where $Q_c(T_v) = q_{c_1}(T_v) \cup \ldots \cup q_{c_i}(T_v)$, and there does not exist another compensation pattern set Q'_c such that $Q_c(T_v) \subset Q'_c(T_v) \subseteq q(T)$. Each $q_{c_j} \oplus v$ ($j \in [1, i]$) is called a contained rewriting (CR) of q, and it can be produced by merging the root node of q_{c_j} with the answer node of v, similar to constructing an equivalent rewriting. Fig. 1(b) shows an example for a maximal contained rewriting. q_2 and v are the query and the view, $\{q_{c_1}, q_{c_2}, q_{c_3}, q_{c_4}\}$ are the compensation patterns, and $\{CR_1, CR_2, CR_3, CR_4\}$ are the CRs.

The previous work [2] has examined, given a query q, a view v, how to find an MCR $\{q_{c_1} \oplus v, \cdots, q_{c_i} \oplus v\}$ for q using v. However, in many applications, such as query caching, to find the MCRs is not the final goal, we need to evaluate the MCRs on the materialized views. Back to our problem, we need to evaluate the compensation patterns $\{q_{c_1}, \cdots, q_{c_i}\}$ on the materialized view of v, T_v. Unfortunately, the set of compensation patterns of an MCR can be exponential, because one compensation pattern corresponds to one contained rewriting, and the number of contained rewritings may be exponential. As one can observe from Fig. 1(b), the maximal contained rewriting consists of an exponential number of contained rewritings due to the combination of patterns from different paths. Although we can speed up the evaluation by discarding some redundant CRs (if there are any) from an MCR set, in the worst case, we still have an exponential number of CRs in an MCR, which means all the CRs are irredundant (like the example we have given in Fig. 1). Here, a CR is called redundant with respect to an MCR implying that answers produced by the redundant CR can be covered by other CRs belonging to the MCR set. To tackle the challenge of evaluating an exponential number of compensation patterns, an efficient scheme is highly sought after.

To solve the problem, rather than naturally transform it into a multiple-query optimization problem [3], we examine and utilize the unique feature of this problem. Our observation is that the exponential number of compensation patterns are composed by a linear number of subpatterns, i.e. a linear number of subpatterns are shared among those exponential number of compensation patterns. Therefore, taking advantage of the special structural characteristics, we can solve the problem by evaluating only a linear number of patterns in the worst case. Moreover, to do the evaluation more wisely, we develop a set of rules and heuristics that can effectively exclude many useless patterns. We highlight our contributions as follows:

– We are the first to investigate the problem of evaluating contained rewritings on materialized views by considering the particular feature of the problem: only a linear number of patterns need to be evaluated.
– We have proposed two algorithms, basic algorithm and optimized algorithm, and both algorithms can be built inside an existing query evaluation engine or in a middleware outside the query evaluation engine. This provides enough flexibility to software developers.
– We have conducted extensive experiments to show the efficiency of our algorithms. The optimized algorithm is very promising, benefiting from its pruning rules and heuristics.

The rest of this paper is organized as follows. In Section 2, we will give some notations and background knowledge. Then we give a basic algorithm in Section 3. Important optimization techiques are unfolded in Section 4. We report experiment results in Section 5. Related work and conclusions are in Section 6 and Section 7.

2   Preliminaries

We will first introduce XPath tree patterns to represent XPath queries, and then introduce how to find contained rewritings (compensation patterns) using pattern mapping from the query pattern to the view pattern.

2.1   XPath Tree Pattern

We consider a subset of XPath language featuring child axes ($/$), descendant axes ($//$), branches ($[\,]$), denoted as $XP^{\{/,//,[]\}}$. An XPath query q can be expressed as a tree pattern (N_q, E_q, r_q, d_q), where

– N_q is the node set, and for any node $n \in N_q$, n has a label in an infinite alphabet Σ, denoted as $label(n)$;
– E_q is the edge set, and $\forall e \in E_q, type(e) \in \{/,//\}$. We use the term "pc-edge"("ad-edge") to represent the type of an edge, "$/$"("$//$");

- r_q is the root node of the query; (If q starts with "/" or "//", we add a virtual root node with a unique label r, so that every query corresponds to a unique tree pattern.)
- d_q is the answer (also called distinguished or return) node of the query, identified with a circle;

Similarly, an XPath view v can be expressed as a 4-tuple, (N_v, E_v, r_v, d_v). We also define the following functions: (i) $pc(n_1, n_2)$ holds if n_1 is the parent node of n_2; (ii) $ad(n_1, n_2)$ holds if n_1 is an ancestor node of n_2.

2.2   Useful Embedding

To rewrite an XPath query using a view, is to find the set of conditions that are not satisfied on the view query, but may be satisfied under the distinguished node of the view. The solution is to find an *embedding* from the query to the view, and see whether the embedding is valid to produce a contained rewriting. A valid embedding is called *useful embedding*. Embedding and useful embedding are defined below.

Given a query $q = (N_q, E_q, r_q, d_q)$ and a view $v = (N_v, E_v, r_v, d_v)$, an *embedding* is a mapping $e : N_q' \to N_v'$, where $N_q' \subseteq N_q$, $N_v' \subseteq N_v$, satisfying:

- root preserving: $e(r_q) = r_v$;
- label preserving: $\forall n \in N_q'$, $e(n) \in N_v' \land label(n) = label(e(n))$;
- structure preserving: $\forall (n_1, n_2) \in E_q$ and $n_1, n_2 \in N_q'$, if (n_1, n_2) is a pc-edge, $pc(e(n_1), e(n_2))$ holds in v; otherwise $ad(e(n_1), e(n_2))$ holds in v;
- e is upward closed: if node n in N_q is defined by e (means $n \in N_q'$), all ancestors of n in N_q are defined by e. Namely, $n \in N_q' \Rightarrow \forall n', n' \in N_q$ and $ad(n', n)$ holds in q, we have $n' \in N_q'$.

Not every node in N_q needs to be defined by e, and here N_q' is the set of nodes that are defined by e, $N_q' \subseteq N_q$. An embedding implies that part of the conditions of query q have been satisfied in the view v, so if the left conditions of q (unembedded parts) are possible to be satisfied under the distinguished node of the view, we will be able to use the view v to answer the query q. Such embedding is called a useful embedding. Before giving the formal definition of useful embedding, we give the definition of *anchor node* first. Given an embedding e, for each unfully embedded path $path_i$ of q, the last node on $path_i$ that can be embedded on or above d_v is called an *anchor* node of embedding e with respect to path $path_i$. An embedding is called a useful embedding, if both of the following conditions hold:

1. every anchor node n_a satisfies either of (a), (b):
   - (a) $e(n_a) = d_v$;
   - (b) $ad(e(n_a), d_v)$ holds in v and let n_c be the child node of n_a on $path_i$ in q, (n_a, n_c) is an ad-edge.
2. for the anchor node n_a on the distinguished path of q, either $n_a = d_q$ or $ad(n_a, d_q)$ holds in q.

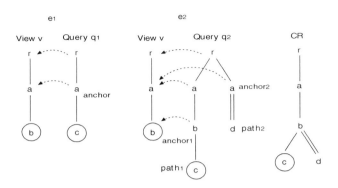

Fig. 2. An example of useful embedding

The idea behind condition (1) is: unfulfilled conditions in q have the possibility to be satisfied under d_v in the view, which requires that an anchor node either maps to d_v (condition (a)) or the anchor node maps to an ancestor of d_v and the anchor connects its descendant with "//" on the corresponding path (condition (b)). Condition (2) implies: the return node of q should not be mapped onto a node above d_v. For ease of understanding, we give an example to illustrate embedding, useful embedding and anchor nodes. In Fig 2, e_1 is an embedding, but not a useful embedding, because its anchor node, the a-node, does not satisfy either condition (1)(a) or (1)(b). While e_2 is a useful embedding, both condition (1) and (2) are satisfied by e_2's anchor nodes. To be specific, on path $path_1$, the anchor node b-node maps onto the distinguished node of the view v (satisfying condition (1)(a)); on path $path_2$, the anchor node a-node maps onto a node above the distinguished node of the view, and a-node connects its successor d-node on $path_2$ with "//" (satisfying condition (1)(b)); the anchor node b-node on the distinguished path $path_1$ is above the distinguished node in the query q (satisfying condition (2)). As a result, e_2 is a useful embedding. The corresponding CR produced by e_2 is also given on the rightmost side of Fig. 2. In the CR, $b[//d]/c$ is called a compensation pattern to the view. In the following, we will use *clip-away tree* (CAT) to represent compensation pattern in alignment with a pioneer work [2].

We define another concept, *component pattern*: given a useful embedding, let a_n be an anchor node in q with respect to some path and let a_c be the immediate successor of a_n on that path, we call such a constructed pattern a component pattern: the pattern has a root node with $label(d_v)$, and the root node connects $q_{sub}(a_c)$ with the same type of edge as edge (a_n, a_c), here $q_{sub}(a_c)$ is the subpattern in q rooted at node a_c. A component pattern may be a *predicate component pattern* (not containing node d_q) or a *distinguished component pattern* (containing node d_q). In the CR in Fig. 2, $b[//d]$ is a predicate component pattern, b/c is the distinguished component pattern. We will use p to denote a component pattern, and $P = \{p_1, \cdots, p_n\}$ to denote a CAT, since a CAT usually contains several component patterns fusing at the root.

3   Basic Algorithm

We regard a materialized XPath view as a set of subtrees, whose roots have the same label as the distinguished node of the view, obtained by matching the view pattern to the base document. To evaluate a CAT on these trees, equals to match the CAT pattern on each tree and combine the results. For simplicity, we only consider one tree of the materialized view in this paper, for computation on other trees are the same. After obtaining the CATs using useful embedding, the problem turns to be: given a set of CAT queries (representing the compensation patterns) and a materialized data tree T_v, how to efficiently evaluate the CAT queries over the tree T_v. From the previous study, we know an MCR may consist of an exponential number of CR queries, corresponding to an exponential number of CATs, bounded by k^l, where l is the number of paths in q (also the number of leaf nodes), k is the maximum number of component patterns that are residing on one particular path of q[1]. The naive method is to evaluate the exponential number of CATs one by one, and union the final results. As a result, we need to evaluate $O(k^l)$ CAT patterns.

However, if we consider the characteristics of the CATs, we are able to produce the same results by only evaluating up to $min\{|N_q| - 1, kl\}$ component patterns. The observation is that a CAT is composed by a number of component patterns fusing at their roots, and one component pattern may be shared by a few different CATs. To evaluate the CATs one by one on a materialized view will result in repeated computation of the same component patterns. Although the number of CATs is exponential, the number of component patterns is polynomial. We now explain the bound $min\{|N_q| - 1, kl\}$ for the number of component patterns: (1) Notice that component patterns share the same labeled root node, and distinguish each other starting from the single child of the root, eg. in Fig. 2, component patterns $b[//d]$ and b/c share the same b-node, and distinguish each other at the d-node and c-node. This observation implies that each node in N_q (the node set of the query) except r_q determines one possible component pattern, and there are at most $|N_q| - 1$ component patterns. Meanwhile, kl is also the bound for the number of component patterns, because there are at most k component patterns on one path of query q, and there are l paths in the query q. In consequence, the maximum number of component patterns $min\{|N_q| - 1, kl\}$ could be far less than the maximum number of CATs k^l.

The idea of our basic algorithm, shown in Algorithm 1, is to break down the CATs into component patterns, and evaluate the component patterns. As a result, each component pattern is examined only once. We now introduce the basic algorithm, and then develop some optimization techniques on the basic algorithm in Section 4. We first evaluate all the predicate component patterns on the materialized view T_v, and then, for each CAT P, if all the predicate component patterns of P are satisfied on T_v, we evaluate the distinguished component pattern of P on T_v and add the result into the final result set R. The basic

---

[1] By a close examination, one can further find that k is bounded by the number of nodes residing on the distinguished path from r_v to d_v in the view v.

Algorithm 1. Basic Algorithm

Input: all CATs of rewriting q using v, a materialized view T_v
Output: the union of the result of evaluating all CATs on T_v

1: union predicate component patterns of each CAT;
2: evaluate all predicate component patterns on T_v;
3: $R := \phi$;
4: **for** each CAT P such that $\forall p_i \in P$, p_i is predicate component pattern and
    $p_i(T_v) = true$ **do**
5:     evaluate the distinguished component pattern $\hat{p} \in P$;
6:     $R := R \cup \hat{p}(T_v)$;
7: **end for**
8: return R;

---

algorithm can be regarded as an optimization of evaluating multiple CATs on the materialized view by taking advantage of the special feature of CATs.

4 Optimizing Techniques

In this section, we first introduce four pruning rules and three heuristic rules to optimize the basic algorithm, and explain the rationale behind these rules. Then we give the outline of an optimized algorithm to illustrate how to effectively combine these rules together.

4.1 Pruning Rules

In the Basic Algorithm, every component pattern is evaluated against the materialized view T_v. In fact, some component patterns may not need to be evaluated. We now introduce several rules to prune them. In the following, when we say component pattern p is satisfied on T_v, we mean: $p(T_v) = true$ if p is a predicate component pattern; or $p(T_v) \neq \phi$ if p is a distinguished component pattern.

- **Rule 1:** If one component pattern p is not satisfied in the materialized view, those CATs that contain p as a component pattern do not need to be evaluated.

Example 1. In Fig. 3, if component pattern p_3 is not satisfied[2], we only need to evaluate component patterns in CAT P_3 and CAT P_4, since p_3 appears in CAT P_1 and CAT P_2.

A CAT can be regarded as a conjunction of its component patterns, and each component pattern is a condition. Any unsatisfied condition will prevent the CAT from producing answers, no matter whether other component patterns are satisfied or not.

---

[2] Each component pattern should start with a c-node root, while we didn't include the c-node root into those dashed rectangles for the sake of the neatness of Fig. 3.

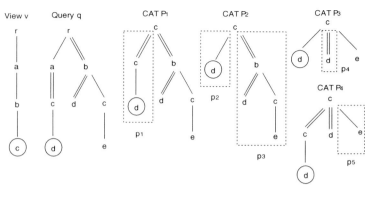

Fig. 3. CATs of the IMCR

Pruning Rule 2 is based on the following proposition:

Proposition 1. *Let CAT_1 and CAT_2 be two CATs having the same distinguished component pattern, then, for any data tree T, either $CAT_1(T) = CAT_2(T)$ or one of $CAT_1(T)$ and $CAT_2(T)$ is empty.*

Proof. The result of evaluating a CAT on a tree T is the same as the result of evaluating the distinguished component pattern of the CAT on T, when all predicate component patters of the CAT are satisfied on T, because the predicate component patterns only serve as conditions imposed on T. Therefore, if another CAT Q has the same distinguished component pattern as CAT P, then no matter what predicate component patterns P and Q possess, if both P and Q have answers on T, their answer sets will be the same.   □

– **Rule 2:** If the answers of one CAT P are produced, other CATs having the same distinguished component pattern as P do not need to be evaluated.

Example 2. In Fig. 3, if CAT $P_1 = \{p_1, p_3\}$ has been evaluated (implying $p_3(T_v) = true$), CAT $P_4 = \{p_1, p_4, p_5\}$ can be discarded, since both CATs share the same distinguished component pattern p_1, no matter whether $p_1(T_v)$ is empty or not.

There are two other optimizing methods by taking advantage of component pattern containment. This type of optimization is based on the fact that checking pattern containment is usually more light-weighted than evaluating a pattern on a materialized view, because view size is usually much larger than pattern size. Here, we use $c(p)$ to denote the set of component patterns that contain p, and $\bar{c}(p)$ to denote the set of component patterns that are contained in p. In other words, $\forall p_i(p_i \in c(p) \rightarrow p \subseteq p_i)$ and $\forall p_i(p_i \in \bar{c}(p) \rightarrow p_i \subseteq p)$ hold. Containment relationship between a predicate component pattern and a distinguished component pattern can also be determined after we transform the distinguished component pattern into a boolean pattern by adding a uniquely-labeled child node to the distinguished node and making the distinguished node no more distinguished [4].

– **Rule 3:** If a component pattern p is satisfied on the materialized view, then any pattern in $c(p)$ will be satisfied on the view.

Example 3. In Fig. 3, if p_1 or p_2 or p_3 is satisfied, p_4 can be induced to be satisfied.

– **Rule 4:** If a component pattern p is not satisfied on the materialized view, then any component pattern in $\bar{c}(p)$ cannot be satisfied on the view.

Example 4. In Fig. 3, if p_4 is not satisfied, we know none of p_1, p_2 and p_3 could be satisfied, then all CATs P_1, P_2, P_3, P_4 do not have answers on the materialized view.

All of the four rules introduced above try every means to prune some patterns, and hence we can evaluate as fewer component patterns as possible. For instance, if an unsatisfied pattern is selected and computed early, the CATs containing this unsatisfied pattern will be eliminated early. Obviously, there is an optimal order to schedule these component patterns, but it is unlikely for us to find this order without knowing in advance whether a component pattern is satisfied in the materialized view or not. To this end, we design some heuristics to find a reasonably good evaluation order.

4.2   Heuristic Rules

In this section, we introduce a few heuristics to determine the order of evaluating the component patterns of the input CATs. We will first list the heuristics, and discuss the rationale behind the heuristics afterwards.

1. **To evaluate frequently shared component patterns first.** This rule can be applied to both distinguished component patterns and predicate component patterns. If the pattern is satisfied in the view, Rule 2 and Rule 3 can be applied to prune other component patterns, otherwise Rule 1 and Rule 4 can be used to prune other component patterns. Frequently shared patterns contribute to more CATs than rare patterns, and thus are worth to be evaluated first.
2. **To investigate component patterns belonging to the same CAT first.** The idea of this heuristic is try to produce answers early. Once the component patterns in one CAT are fully evaluated, we can use Rule 2 to prune other CATs sharing the same distinguished component pattern. This heuristic is specially driven by Rule 2.
3. **To group the CATs by their distinguished component patterns.** For the CATs in each group, since they share the same distinguished component pattern, it is sufficient to evaluate only one of them. We can start with the CAT with the least number of component patterns or apply the above two heuristics for evaluating this subset of CATs.

All the above heuristics are designed to maximize the effect of applying the pruning rules in Section 4.1. Heuristic 1 is more akin to Rule 3 and Rule 4, and it also has substantial pruning power if the inspected pattern is not satisfied in the materialized view. In the best case, heuristic 1 could remove a maximum number of CATs in one step. The reason is that it always picks the most shared component pattern, and hence if the picked pattern cannot match the materialized view, all the CATs containing that pattern can be discarded. Heuristic 2 implies an eager strategy to find some answers as early as possible by inspecting component patterns in the same CAT. It is akin to Rule 2, because, once a CAT is found able to produce answers on the materialized view, other CATs containing the same distinguished component pattern can be disregarded. Heuristic 3 also corresponds to Rule 2. After grouping the CATs by their distinguished component patterns, we can apply other heuristic rules within each group. Once a CAT in a certain group is found to be able to produce some results, other CATs in the same group will be freed of examination.

4.3 Optimized Algorithm

We want to stress that the heuristics in Section 4.2 are orthogonal to the pruning rules in Section 4.1. Any heuristic to determine an order of evaluating the component patterns can be integrated into the algorithm shown in Algorithm 2.

We now go through Algorithm 2 step by step. In the beginning, the result set R is set to ϕ. Then, each component pattern is evaluated according to an order determined by heuristic rules from line 2 to line 14. In each loop, line 3 evaluates the component pattern p on the materialized view T_v. If $p(T_v)$ satisfies, we use Rule 3 to find out other satisfied component patterns by comparing pattern containment in line 5. In line 6, if p is a distinguished component pattern, and all other component patterns of a CAT containing p have been satisfied already, $p(T_v)$ will be added into R. This means we have found some answers. All other CATs containing p as the distinguished component pattern will be eliminated using pruning Rule 2 at line 9. On the other hand, if $p(T_v)$ does not satisfy, we first find out other unsatisfied patterns in line 11, and then use Rule 1 to prune a number of unsatisfied CATs (line 12). In the end (line 15), we return the final answer result R.

4.4 Discussion

We provide some discussion from the engineering point of view for implementing the basic algorithm and the optimized algorithm. Both algorithms can be built on top of an existing query evaluation engine, recall that evaluating a component pattern is an abstracted procedure in both algorithms. Therefore, we can implement the algorithms as a query optimizer in a middleware, which interacts with a query engine by feeding component patterns into it and receiving the corresponding results from it. In such case, it does not matter what type of query evaluation method the engine uses, whether the materialized view is indexed or not. On the other hand, we can also build the algorithms inside a query evaluation engine. The engine itself will contain functions of finding contained rewritings and evaluating selected patterns over the materialized views. Such choices provide enough flexibility to software engineers.

Algorithm 2. Optimized algorithm with pruning rules and heuristics

---

Input: all CATs of rewriting q using v, a materialized view T_v
Output: R, the answers of evaluating all CATs on T_v

1: final result $R := \phi$;
2: **for** each component pattern p chosen by some heuristic **do**
3:     evaluate p on T_v;
4:     **if** $p(T_v)$ satisfies or produces some answers **then**
5:         use Rule 3 to find other component patterns that are also satisfied;
6:         **if** p is a distinguished component pattern $\wedge$ all the predicate component patterns of a CAT containing p as the distinguished component pattern are satisfied **then**
7:             $R := R \cup p(T_v)$;
8:         **end if**
9:         use Rule 2 to prune other CATs;
10:    **else**
11:        use Rule 4 to find other component patterns that are not satisfied;
12:        use Rule 1 to prune other CATs;
13:    **end if**
14: **end for**
15: return R;

---

5   Experiments

We build a prototype system MCRE (MCR Evaluator) to evaluate a generated MCR on materialized views. Our experiments are conducted on a PC with Pentium(R) 4 3GHz CPU and 1G memory.

View and Query Generation. Due to the challenge of collecting view and query specimens deployed in real applications (also mentioned in [5]), we generated views and queries synthetically. In order to cover a variety of cases, the parameters can be tuned within a wide range. To make the generated patterns reasonable, we enforce the generated views and queries to conform to a given DTD, though the query evaluation can be done without knowing the DTD. The view patterns are generated in a top-down manner. For each node in the view query, its children are selected with four parameters: (1) a child node is selected with probability α_1; (2) the edge connecting the child to itself is labeled as $//$ with probability α_2; (3) a descendant node is selected with probability α_3 directly connecting to its parent; (4) the maximum fanout f is fixed and set within a limit. We do not generate value predicates in the pattern, because checking value predicate can be easily integrated into the system.

The queries are generated based on the views to ensure some rewritings exist. The generation is performed in a bottom-up manner. For each view pattern, (1) a node is deleted with probability β_1. After deleting the node, if the node is a internal node, we should connect the node's parent to its children with $//$. Note that we never delete the view root; (2) a pc-edge is replaced by an ad-edge with probability β_2; (3) some new nodes are added under a node with probabilities

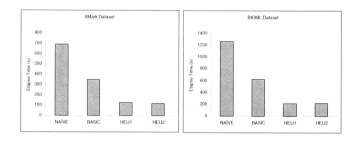

Fig. 4. Average Case Study

similar to the view generation part. We do not set a limit to the fanout of query patterns, because deleting an internal node may increase the out degree of its parent (if the internal node has multiple children), and thus increase the fanout of the result pattern.

Datasets. We test our algorithms on two datasets, XMark[3] and BIOML[4]. The former is widely used in the literature, and the latter is famous for its recursive feature, and is ideal to build materialized views with. BIOML DTD is tailored with only "chromosome" and its descendant elements. XML data is generated with IBM XML Generator[5]. We use eXist[6] database as the underlying engine to store and query the documents. Each materialized view is generated by evaluating the view pattern on documents and saving back into eXist database.

5.1   Average Case Study

In this study, we investigate the performance of four different algorithms to evaluate the CATs of the MCRs on materialized views. We use the same set of views and queries generated above. 50 views are materialized, and a set of 20 queries are rewritten and evaluated on each materialized view. In the NAIVE algorithm, each CAT of an MCR is evaluated on the materialized views. In BASIC algorithm, only component patterns are evaluated on the materialized views, but all the component patterns are computed. In the optimized algorithms, we use the proposed four rules to prune unnecessary component patterns, and also use heuristic information to schedule the evaluation order. Specifically speaking, in HEU1, we use the first two heuristics with heuristic 2 prior to heuristic 1. In HEU2, we first group the CATs by their distinguished component patterns (heuristic 3), and then apply heuristic 2 and 1.

The result is shown in Fig. 4, BASIC algorithm takes almost half time of the NAIVE algorithm, because redundant CATs are pruned in advance, and

---

[3] XMark An XML Benchmark Project, http://www.xml-benchmark.org/

[4] BIOpolymer Markup Language, http://xml.coverpages.org/bioml.html

[5] IBM XML Generator, http://www.alphaworks.ibm.com/tech/xmlgenerator

[6] eXist Open Source Native XML Database, http://exist.sourceforge.net/

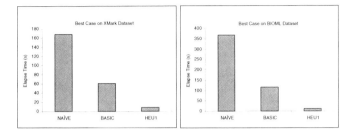

Fig. 5. Best Case Study

pruning these CATs is not expensive. Furthermore, HEU1 and HEU2 perform even better, which demonstrates our heuristic methods are very effective and encouraging. It is not obvious to find a better one between HEU1 and HEU2. Although HEU2 seems to provide more effective pruning heuristics, it also suffers in updating component pattern statistics for each distinguished pattern group, and may spare some time on maintaining the auxiliary information.

5.2   Best and Worst Case Study

In the above experiments, our aim is to test the performance of NAIVE, BASIC, HEU1 and HEU2 algorithms in the average case, where a number of views and a number of queries are randomly generated to capture all pattern types as various as possible. It is reasonable that our heuristic performs best in the average case. One may wonder how far our heuristic algorithms can achieve and what is the worst performance our heuristic methods will reach. We examine the best and worst cases by manually designing two queries, because randomly generated queries are not that extreme.

For the best case, the query is designed to have four paths, and on each path there are three component patterns, two out of which are irredundant. And hence there are $81=3^4$ CR CATs, $16=2^4$ irredundant CR CATs, 8 irredundant component pattern. The query time is shown in Fig. 5. On BIOML dataset, the BASIC algorithm beats NAIVE in two-thirds time, and HEU1 needs only 10% query time of the BASIC algorithm. Similar observation is obtained on XMark dataset, but the result is not that dramatic.

For the worst case part, we designed a query which does not produce redundant CRs. Every component pattern produced from the query is not contained in its pals. Therefore, in the evaluation process, every component pattern is evaluated on the materialized view, with no one can be pruned. In Fig. 6, HEU1 performs almost the same as BASIC, because both of them have evaluated all component patterns. The part of updating heuristic statistics in HEU1 does not apparently degrade, though it may take some extra time. Both BASIC and HEU1 are a little costly than NAIVE. The reason may be that NAIVE has evaluated less number of patterns, although each larger pattern has a larger size.

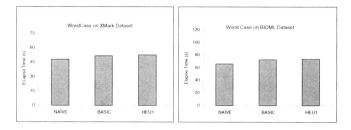

Fig. 6. Worst Case Study

6   Related Work

Answering queries using views has been studied for a long time. Halevy [6] has done a survey on this problem over relational database. In the XML context, equivalent rewriting and contained rewriting, have been studied for XPath [7, 8, 1, 9], XQuery [5], and tree patterns [2, 10]. The works [7, 8] propose to use materialized views to speed up query evaluation in the query caching scenario, where to find an equivalent rewriting for a query with given views is the key subtask. Xu and Ozsoyoglu [1] have provided a theoretical study on equivalent rewriting based on query containment [4] and query minimization [11]. Afrati et al. [9] have extended Xu and Ozsoyoglu's result in fragment $XP^{\{[],*,//\}}$. They have discovered a coNP-complete upper bound for some sub-fragments of $XP^{\{[],*,//\}}$. Onose et al. [5] have investigated the equivalent rewriting of XQuery queries using XQuery views. While XQuery queries are more expressive, the shortcomings of using them as views are also noted in [5]. Tang et al. [12] studied the materialized view selection problem, which is to select fewer materialized views to answer a query equivalently. Zhou et al. [13] also studied view selection for answering contained rewritings. Wu et al. [14] and Chen. et al [15] proposed two different view implementations, and how to use the materialized views to evaluate queries. View definition nodes are also materialized in [14, 15], while our work does not have this requirement. Lakshmanan et al. [2] addressed how to find contained rewritings, but did not cover how to efficiently evaluate the compensation patterns over materialized views. A recent work [3] discussed how to perform the evaluation. It is based on an approach to examine each path pattern and then verify the rewriting, and each path pattern needs to be built into an automata. While, in our approach, we use useful embedding to find the compensation patterns, and try to evaluate as few component patterns as possible. Although useful embedding is not sufficient to find all the compensation patterns for $XP^{\{[],*,//\}}$, other complete methods exist, despite in coNP-complete. The proposed algorithms still work if we can find all the compensation patterns.

7   Conclusions

In this paper, we have proposed two algorithms, basic algorithm and optimized algorithm, to evaluate contained rewritings on materialized views. Both algorithms are built on the observation that an exponential number of rewritings in fact share a linear number of component patterns. In consequence, the idea of our algorithms is to evaluate the component patterns, rather than to evaluate the whole compensation patterns for each contained rewriting. We have also designed four important pruning rules and several heuristic rules to effectively reduce the number of component patterns that need to be evaluated. Experiments show that the optimized algorithm is advantageous in most cases.

Acknowledgments. Rui Zhou, Chengfei Liu, Jianxin Li and Jixue Liu are supported by the Australian Research Council Discovery Grant DP0878405, and Junhu Wang is supported by the Australian Research Council Discovery Grant DP1093404.

References

1. Xu, W., Özsoyoglu, Z.M.: Rewriting XPath queries using materialized views. In: VLDB, pp. 121–132 (2005)
2. Lakshmanan, L.V.S., Wang, H., Zhao, Z.J.: Answering tree pattern queries using views. In: VLDB, pp. 571–582 (2006)
3. Gao, J., Lu, J., Wang, T., Yang, D.: Efficient evaluation of query rewriting plan over materialized xml view. Journal of Systems and Software 83(6), 1029–1038 (2010)
4. Miklau, G., Suciu, D.: Containment and equivalence for a fragment of XPath. J. ACM 51(1), 2–45 (2004)
5. Onose, N., Deutsch, A., Papakonstantinou, Y., Curtmola, E.: Rewriting nested XML queries using nested views. In: SIGMOD Conference, pp. 443–454 (2006)
6. Halevy, A.Y.: Answering queries using views: A survey. VLDB J. 10(4), 270–294 (2001)
7. Balmin, A., Özcan, F., Beyer, K.S., Cochrane, R., Pirahesh, H.: A framework for using materialized XPath views in XML query processing. In: VLDB, pp. 60–71 (2004)
8. Mandhani, B., Suciu, D.: Query caching and view selection for XML databases. In: VLDB, pp. 469–480 (2005)
9. Afrati, F.N., Chirkova, R., Gergatsoulis, M., Kimelfeld, B., Pavlaki, V., Sagiv, Y.: On rewriting xpath queries using views. In: EDBT, pp. 168–179 (2009)
10. Arion, A., Benzaken, V., Manolescu, I., Papakonstantinou, Y.: Structured materialized views for XML queries. In: VLDB, pp. 87–98 (2007)
11. Amer-Yahia, S., Cho, S., Lakshmanan, L.V.S., Srivastava, D.: Tree pattern query minimization. The VLDB Journal 11(4), 315–331 (2002)
12. Tang, N., Yu, J.X., Özsu, M.T., Choi, B., Wong, K.F.: Multiple materialized view selection for xpath query rewriting. In: ICDE, pp. 873–882 (2008)
13. Zhou, R., Liu, C., Li, J.-x., Wang, J.: Filtering techniques for rewriting xPath queries using views. In: Bailey, J., Maier, D., Schewe, K.-D., Thalheim, B., Wang, X.S. (eds.) WISE 2008. LNCS, vol. 5175, pp. 307–320. Springer, Heidelberg (2008)
14. Wu, X., Theodoratos, D., Wang, W.H.: Answering xml queries using materialized views revisited. In: CIKM, pp. 475–484 (2009)
15. Chen, D., Chan, C.Y.: Viewjoin: Efficient view-based evaluation of tree pattern queries. In: ICDE, pp. 816–827 (2010)

XStreamCluster: An Efficient Algorithm for Streaming XML Data Clustering*

Odysseas Papapetrou[1] and Ling Chen[2]

[1] L3S Research Center, University of Hannover, Germany
papapetrou@L3S.de
[2] QCIS, University of Technology Sydney, Australia
ling.chen@uts.edu.au

Abstract. XML clustering finds many applications, ranging from storage to query processing. However, existing clustering algorithms focus on static XML collections, whereas modern information systems frequently deal with streaming XML data that needs to be processed online. Streaming XML clustering is a challenging task because of the high *computational* and *space* efficiency requirements implicated for online approaches. In this paper we propose *XStreamCluster*, which addresses the two challenges using a two-layered optimization. The bottom layer employs Bloom filters to encode the XML documents, providing a space-efficient solution to memory usage. The top layer is based on Locality Sensitive Hashing and contributes to the computational efficiency. The theoretical analysis shows that the approximate solution of XStreamCluster generates similarly good clusters as the exact solution, with high probability. The experimental results demonstrate that XStreamCluster improves both memory efficiency and computational time by at least an order of magnitude without affecting clustering quality, compared to its variants and a baseline approach.

1 Introduction

In the past few years we have seen a growing interest in processing streaming XML data, motivated by emerging applications such as management of complex event streams, monitoring the messages exchanged by web-services, and publish/subscribe services for RSS feeds [1]. Various research activities have been triggered accordingly, including query evaluation over streaming XML data [2], summarization of XML streams [1] as well as classification of XML tree streams [3]. However, to the best of our knowledge, there exists no work on clustering *streaming* XML data, albeit extensive research has been carried out toward clustering *static* XML collections [4–7].

Streaming XML clustering is important and useful in many applications. For example, it enables the building of individual indices for each of the clusters, which in turn improves the efficiency of query execution over XML streams. The

---

* This work is partially supported by the FP7 EU Project GLOCAL (contract no. 248984).

J.X. Yu, M.H. Kim, and R. Unland (Eds.): DASFAA 2011, Part I, LNCS 6587, pp. 496–510, 2011.
© Springer-Verlag Berlin Heidelberg 2011

problem is different than the one of clustering static XML collections, due to the typical high *computation* and *space* efficiency requirements of online approaches. As we explain later, existing approaches for clustering static XML collections do not meet these requirements, and therefore cannot be applied to streaming data. Therefore, this work designs an online approach for clustering of streaming XML.

More specifically, we focus on clustering of streaming XML documents based on the structural similarity of the documents in terms of the common edges shared by their XML graphs. As discussed in [5], this kind of clustering is particularly important for XML databases, as it yields clusters supporting efficient XML query processing. First, clusters not containing the query can be efficiently filtered out, thereby eliminating a large portion of the candidate documents inexpensively. Second, each cluster of documents can be indexed more efficiently in secondary memory, due to the structural similarity of the documents.

In the context of structural XML clustering, an XML document can be represented as a set of edges $(e_1, e_2, e_3, \ldots)$. This representation makes the problem similar to clustering of streaming categorical data. However, existing approaches for clustering streaming categorical data are also not sufficient for streaming XML. Although most of them are designed with special concerns on computational efficiency, according to [8] they are not sufficiently efficient in terms of memory, especially when clustering *massive-domain data* where the possible domain values are so large that the intermediate cluster statistics cannot be maintained easily. Therefore, considering XML streams that encode massive-domain data, it is critical to design online XML clustering approaches which are both time and space efficient.

In the massive domains case, the edges are drawn from a universe of millions of possibilities. Therefore, maintaining the cluster statistics for all clusters in main memory becomes challenging. Recently, an approximate algorithm was proposed which uses compact sketches for maintaining cluster statistics [8]. The promising results delivered by this approximate solution, motivated us to apply an even more compact sketching technique based on Bloom filters to encode the intermediate cluster statistics. In addition, considering XML streams consisting of heterogeneous documents where a large number of clusters is created, we reduce the number of required comparisons between each newly incoming document and all existing clusters using another approximation technique based on Locality Sensitive Hashing (LSH).

Precisely, we propose XStreamCluster, an effective algorithm which employs two optimization strategies to improve time and space efficiency respectively. At the top level, LSH is used to quickly detect the few candidate clusters for the new document out of all clusters. This first step reduces drastically the required document-cluster comparisons, improving the time efficiency of the algorithm. At the bottom level, Bloom filters are employed to encode the intermediate cluster statistics, contributing to the space efficiency. Although the two levels introduce a small probability of errors, our theoretical analysis shows that XStreamCluster provides similar results to an exact solution with very high probability.

2   Related Work

As an important data mining technique, clustering has been widely studied by different communities. Detailed surveys can be found in [9, 10]. In the scenario of streaming data, the problem of clustering has also been addressed before, e.g., [11–13]. Considering the large volume of incoming data, computational efficiency is one of the most critical issues addressed by these works. Recently, the issue of space efficiency in clustering massive-domain streaming data was stressed [8], and an approximate solution based on the *count-min sketch* technique was proposed. Our algorithm also provides an approximate solution by using compact sketches to maintain intermediate cluster information in main memory. However, since we measure the similarity between XML documents in terms of number of shared edges (rather than the frequency of shared edges), our algorithm utilizes the more compact structure of Bloom filters, further reducing the memory requirements.

Clustering of *static* XML documents also attracted a lot of attention. Based on the adopted similarity/distance measure, existing static XML clustering approaches can be broadly divided into the following categories: *structure-based approaches*, *content-based approaches*, and *hybrid approaches*.

Early structure-based approaches usually represent XML documents as tree structures. The *edit distance* is then used to measure the distance between two XML trees, based on a set of edit operations such as inserting, deleting, and relabeling a node or a subtree [4, 6]. However, computing tree edit distances requires quadratic time complexity, making it impractical for clustering of XML streams. Differently, Lian et al. [5] proposed to represent XML documents as a set of parent-child edges from the corresponding structure graphs, which enables the efficient calculation of structural similarity of XML documents. Therefore, in our work, we employ the distance measure defined in [5] to design our online clustering algorithm for streaming XML data.

An important efficiency consideration for the existing structure-based algorithms is the large number of incurred document-cluster comparisons, as shown in Section 4. Particularly for the case of massive domain data, a large number of clusters is expected, aggravating the costs substantially. These costs become prohibitively expensive for clustering of *streaming* XML, rendering the existing algorithms unsuitable. Therefore, in this work we also include a probabilistic method which reduces drastically the comparisons between the incoming document and existing clusters, allowing XStreamCluster to handle streams.

Content-based XML clustering approaches are mainly used for clustering text-centric XML documents. Vector space models have been widely used to represent XML documents [14, 15]. Recently, there were also a few hybrid approaches which cluster XML documents by considering both structures and contents. For example, Doucet and Lehtonen [16] extract bags-of-words and bags-of-tags from the content and structure of XML documents as feature vectors. The type of clustering performed is thus substantially a textual clustering, while the results were shown to be better than those of other competing methods in the INEX 2006 contest. Although our work focuses on structural similarity, the proposed

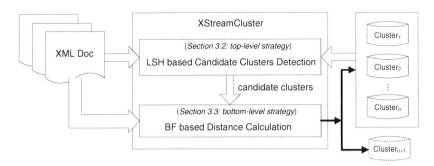

Fig. 1. The framework of XStreamCluster

algorithm can be extended to cluster XML documents based on content, as well as the the combination of structure and content, once XML documents are represented as vector space models. For example, we can use *counting Bloom filters* to encode the feature vectors of XML documents and clusters.

3   Streaming XML Clustering with XStreamCluster

We start by introducing the framework of our algorithm, and the preliminaries concerning the XML representation and distance measure. We then elaborate on the two optimization strategies. XStreamCluster (Fig. 1) clusters streaming XML documents at a single pass. When a new XML document arrives, instead of comparing it against all existing clusters, the top-level strategy - the LSH-based Candidate Cluster Detection - efficiently selects a few candidate clusters which are most similar to the new document. The algorithm then proceeds to compute the distance between the new document and each of the candidate clusters. To reduce memory requirements, the bottom-level strategy - the Bloom filter based Distance Calculation - computes the distance between the XML document and each of the candidate clusters based on their Bloom filter representations. Finally, a decision is made to either assign the new document to one of the existing clusters, if their distance is sufficiently low, or to initialize a new cluster for the current document.

3.1   Preliminaries

As discussed in Section 2, existing work on clustering static XML documents adopted various similarity/distance measures, ranging from structure-based measures to content-based measures. In our work, we focus on clustering XML documents based on their structure.

In order to define the structural distance between two XML documents, the documents are first represented as structure graphs, or *s-graphs* [5].

Definition 1. (*Structure Graph*) *Given a set of XML documents C, the structure graph of C, $sg(C) = (N, E)$, is a directed graph such that N is the set of*

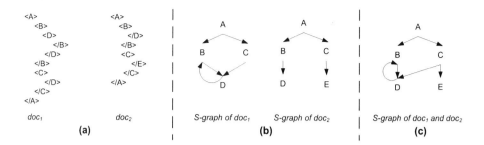

Fig. 2. The S-Graph Representation of XML documents

all the elements and attributes in the documents in C and $(a, b) \in E$ if and only if a is a parent element of element b or b is an attribute of element a in some document in C.

For example, Fig. 2(b) shows the s-graphs of the two XML documents of Fig. 2(a). Given the s-graphs of two XML documents, a revised Jaccard coefficient metric is used to measure their distance.

Definition 2. (XML Distance) *For two XML documents d_1 and d_2, the distance between them is defined as $dist(d_1, d_2) = 1 - \frac{|sg(d_1) \cap sg(d_2)|}{\max\{|sg(d_1)|, |sg(d_2)|\}}$ where $|sg(d_i)|$ is the number of edges in $sg(d_i)$ and $sg(d_1) \cap sg(d_2)$ is the set of common edges of $sg(d_1)$ and $sg(d_2)$.*

As an example, consider the two XML documents and their s-graphs in Fig. 2 (a) and (b). Since $|sg(d_1) \cap sg(d_2)| = 3$ and $\max\{|sg(d_1)|, |sg(d_2)|\} = 5$, the distance between the two documents is $1 - 3/5 = 0.4$. As stated in [5], this distance metric enables generating clusters to support efficient query answering. Note however that XStreamCluster can be adapted to other distance measures which might be more appropriate for other domains, as outlined in the technical report [17].

3.2 LSH-Based Candidate Clusters Detection

Traditional single-pass clustering algorithms need to compare each incoming document against all existing clusters, to find out the cluster with the minimum distance. However, considering XML datasets with heterogeneous structures, there may exist a large number of clusters, requiring a huge amount of time for comparing each document with all existing clusters. XStreamCluster addresses this issue by reducing the number of required document-cluster comparisons drastically. This reduction is based on an inverted index of clusters, built using Locality Sensitive Hashing (LSH).

The main idea behind LSH is to hash points from a high-dimensional space such that nearby points have the same hash values, and dissimilar points have different hash values. LSH is probabilistic, that is, two similar points will end up with the same hash value with a high probability p_1, whereas two dissimilar points will have the same hash value with a very low probability p_2. Central to

LSH is the notion of *locality sensitive hash families*, i.e., an ordered collection of hash functions, formally defined as (r_1, r_2, p_1, p_2)-sensitive hash families [18].

Definition 3. *Let $\mathcal{S}$ denote a set of points, and $Dist(\cdot, \cdot)$ denote a distance function between points from $\mathcal{S}$. A family of hash functions $\mathcal{H}$ is called (r_1, r_2, p_1, p_2)-sensitive, where $r_1 \leq r_2$ and $p_1 \geq p_2$, if for any two points $p, q \in \mathcal{S}$ and for any $h_i \in \mathcal{H}$:*
- *if $Dist(p, q) \leq r_1$ then $Pr[h_i(p) = h_i(q)] \geq p_1$*
- *if $Dist(p, q) \geq r_2$ then $Pr[h_i(p) = h_i(q)] \leq p_2$*

For the case where the points of $\mathcal{S}$ are sets of elements (e.g., s-graphs are sets of edges), and $Dist(\cdot, \cdot)$ denotes the Jaccard coefficient, a suitable locality sensitive hash family implementation is *minwise independent permutations* [19]. Particularly, when hashing is conducted using minwise independent permutations, the probability that two points have the same hash value is $Pr[h_i(p) = h_i(q)] = 1 - Dist(p, q)$. For the case that $Dist(p, q) \leq r_1$, $Pr[h_i(p) = h_i(q)] \geq 1 - r_1$.

XStreamCluster employs LSH for efficiently detecting the candidate clusters for each document. Let $\mathcal{H}$ denote a locality sensitive hash family, based on minwise independent permutations. L hash tables are constructed, each corresponding to a composite hash function $g_i(\cdot)$, for $i = 1 \ldots L$. These hash functions $g_1(\cdot), g_2(\cdot), \ldots, g_L(\cdot)$ are obtained by merging k hash functions chosen randomly from $\mathcal{H}$, i.e., for a point $p : g_i(p) = [h_{i1}(p) \oplus h_{i2}(p) \ldots \oplus h_{ik}(p)]$. Each cluster s-graph is hashed to all L hash tables, using the corresponding hash functions. When XStreamCluster reads a new document d, it computes $g_1(sg(d)), \ldots, g_L(sg(d))$ and finds from the corresponding hash tables all clusters that collide with d in at least one hash table. These clusters, denoted as $\mathcal{C}(d)$, are returned as the candidate clusters for the document.

There is a latent difference between our approach for constructing the LSH inverted index of clusters and previous LSH algorithms, e.g., [18]. Previous approaches construct $g_i(p)$ by mapping each of $h_{ij}(p)$ to a single bit, for all $j : [1 \ldots k]$, and concatenating the results to a binary string of k bits. Due to this mapping to bits, the probability that two points $h_{ij}(p)$ and $h_{ij}(q)$ will map into the same bit value is at least 0.5, independent of their distance $Dist(p, q)$. Therefore, the probability for false positives is high. Previous works compensate for this issue by increasing the number of hash functions k, and thereby increasing the number of bits in each hash key $g(\cdot)$. But increasing the number of hash functions has a negative effect on computational complexity, which we want to avoid for the streaming data scenario. To this end, instead of mapping each of $h_{ij}(p)$ to a single bit, we represent the value of $h_{ij}(p)$ in the binary numeral system. We then generate $g_i(p)$ using the logic operation of *exclusive or* (denoted with XOR) on the set of $h_{ij}(p)$ values. Our theoretical analysis shows that the LSH based candidate cluster detection strategy retrieves the optimal cluster for each document with high probability.

Theorem 1. *The optimal cluster C_{opt} for document d will be included in $\mathcal{C}(d)$ with a probability $Pr \geq 1 - (1 - (1 - \delta)^k)^L$, where δ denotes the maximum acceptable distance between a document d and a cluster C for assigning d to C.*

Proof of the theorem can be found in the technical report [17].

For initializing the LSH inverted index, XStreamCluster needs to set the values of δ, k and L. The value of δ corresponds to the maximum acceptable distance between a document and the cluster for assigning the document to that cluster. Therefore, it depends on the requirements of the particular application, as well as the characteristics of the data. Nevertheless, as we show in the experimental evaluation, XStreamCluster offers substantial performance benefits for a wide range of δ. Note that δ is expressed using the standard Jaccard coefficient. Since the interesting measure for our work is the revised Jaccard coefficient, proposed in [5], we compute δ as follows $\delta \leq (1-\delta')/(1+\delta')$, where δ' is the same threshold expressed using the revised Jaccard coefficient. In order to set the values of L and k, the user first decides on the probability pr that a lookup in the LSH inverted index will return the optimal cluster for a document. Then, the values for L and k can be selected appropriately by considering Theorem 1. For example, let $\delta = 0.1$, and $pr \geq 0.95$. Then, according to Theorem 1: $1-(1-(1-0.9^k))^L \geq 0.95$. If we create $L = 10$ hashtables, then setting k as any value no greater than 12.8 should satisfy the required probability. However, the lower the value of k, the more candidate clusters will be returned, which incurs more time to filter false positive candidates. Consequently, we can set $k = 12$ hash functions for each hash table, which satisfies the probability requirements and minimizes the false positives.

After assigning the new document d to a cluster C, we need to update the L hash keys of C in the LSH hash tables. Normally, we would need to recompute these keys from scratch, which requires additional computation. Minwise hashing allows us to compute the updated hash values for the cluster C, denoted with $h'_{ij}(C)$, by using the values of $h_{ij}(d)$ and the current values of $h_{ij}(C)$ as follows: $h'_{ij}(C) = \min(h_{ij}(d), h_{ij}(C))$. The updated values of the $g_1(sg(C)), \ldots, g_L(sg(C))$ can then be computed accordingly.

3.3   Bottom-Level Strategy: Bloom Filter Based Distance Calculation

After the top-level strategy detects a set of candidate clusters, we need to compute the distance between the new document and each candidate cluster, for finding the nearest one. As mentioned, for space efficiency XStreamCluster encodes s-graphs with Bloom filters. We now describe this encoding, and show how the distance between two s-graphs can be computed from their Bloom filters.

A Bloom filter is a space-efficient encoding of a set $S = \{e_1, e_2, \ldots, e_n\}$ of n elements from a universe U. It consists of an array of m bits and a family of λ pairwise independent hash functions $F = \{f_1, f_2, \ldots, f_\lambda\}$, where each function hashes elements of U to one of the m array positions. The m bits are initially set to 0. An element e_i is inserted into the Bloom filter by setting the positions of the bit array returned by $f_j(e_i)$ to 1, for $j = 1, 2, \ldots, \lambda$. To encode an s-graph with a Bloom filter, we hash all s-graph edges in an empty Bloom filter with a predefined length m and λ hash functions.

Recall from Section 3.1 that the XML distance between a document d and a cluster C is $dist(d, C) = 1 - \frac{|sg(d) \cap sg(C)|}{max\{|sg(d)|, |sg(C)|\}}$. Therefore, we need to estimate the values of $|sg(d)|$, $|sg(C)|$, and $|sg(d) \cap sg(C)|$ from the Bloom filter representations of $sg(d)$ and $sg(C)$. With BF_x we denote the Bloom filter encoding of $sg(x)$, where x denotes a document or a cluster. Let m and λ denote the length and number of hash functions of BF_x, and t_x be the number of true bits in BF_x. We can estimate $|sg(x)|$ and $|sg(x) \cap sg(y)|$ as follows:

$$E(|sg(x)|) = \frac{\ln(1 - t_x/m)}{\lambda \ln(1 - 1/m)} \tag{1}$$

$$E(|sg(x) \cap sg(y)|) = 1 - \frac{\ln \left(m - \frac{m t_\wedge - t_x t_y}{m - t_x - t_y + t_\wedge} \right) - \ln(m)}{\lambda \ln(1 - 1/m)} \tag{2}$$

where $t_\wedge$ denotes the number of true bits in the Bloom filter produced by merging BF_x and BF_y with bitwise-AND. Proofs are given in [20].

Notice that the distances calculated using the estimated values of $|sg(d)|$, $|sg(C)|$, and $|sg(d) \cap sg(C)|$, may deviate slightly from the actual distance values. These deviations do not necessarily lead to a wrong assignment, as long as the nearest cluster (the one with the smallest distance) is correctly identified using the estimated values. A wrong assignment occurs only when the nearest cluster is not identified. Some of the wrong assignments have negligible effects, e.g., when the difference of the distances between the document and the two clusters is very small; others may have a significant negative effect, e.g., when the assigned cluster is significantly worse than the optimal one. We are interested in the latter case, which we refer to as significantly wrong assignments, and analyze the probability of such errors.

Given a document d, the optimal cluster C_{opt} for d, and a suboptimal cluster C_{sub}, we define the assignment of d to cluster C_{sub} as a *significantly wrong assignment* if $dist(sg(d), sg(C_{sub})) - dist(sg(d), sg(C_{opt})) > \Delta$, where Δ is a user-chosen threshold. Since d was assigned to C_{sub} instead of C_{opt}, the estimated distance of d with C_{sub}, denoted as $\overline{dist}(sg(d), sg(C_{sub}))$, was smaller than the corresponding distance for C_{opt}. Therefore, we aim to find the probability $Pr[dist(sg(d), sg(C_{sub})) - dist(sg(d), sg(C_{opt})) > \Delta]$, given that $\overline{dist}(sg(d), sg(C_{sub})) < \overline{dist}(sg(d), sg(C_{opt}))$.

We use the following notations. $|ovl(d, C)|$ and $|\overline{ovl}(d, C)|$ denote the actual and expected overlap cardinalities (computed with Eqn. 2) of the sets $sg(d)$ and $sg(C)$. With t_d and t_C we denote the number of true bits in the Bloom filter of $sg(d)$ and $sg(C)$, whereas $t_\wedge$ denotes the number of true bits in the Bloom filter produced by merging the two Bloom filters with bitwise-AND. With $S(t_d, t_C, x)$ we denote the expected value of $t_\wedge$, given that $|ovl(d, C)| = x$. As shown in [20], $S(t_d, t_C, x)$ can be computed as follows: $S(t_d, t_C, x) = \frac{t_d t_C + m(1 - (1 - 1/m)^{\lambda x})(m - t_d - t_C)}{m(1 - 1/m)^{\lambda x}}$.

In particular we study the worst-case scenario, where the expected cardinalities of the overlap of the two clusters ($|\overline{ovl}(d, C_{sub})|$ and $|\overline{ovl}(d, C_{opt})|$) get the minimum possible value, given that the two clusters are candidates for

Algorithm 1. XStreamCluster

INPUT: XML Stream $\mathcal{D}$, dist. threshold δ	10: $\mathcal{C}(d) = \mathcal{C}(d)/C$
OUTPUT: Set of clusters $\mathcal{C}$	11: **end if**
	12: **end for**
1: Initialize L hash tables $ht_1, \ldots, ht_L$, corre-	13: **if** $|\mathcal{C}(d)| \neq 0$ **then**
sponding to $g_1, \cdots, g_L$	14: Assign d to cluster C = argmin
2: $\mathcal{C} \leftarrow \{\}$	dist(d,C)
3: **for** each document d from $\mathcal{D}$ **do**	15: **else**
4: **for** each ht_i, $i : [1 \ldots L]$ **do**	16: Initialize a new cluster C with d, $\mathcal{C}$ =
5: $\mathcal{C}(d) = \mathcal{C}(d) \cup ht_i.get(g_i(sg(d)))$	$\mathcal{C} \cup \{C\}$
6: **end for**	17: **end if**
7: Hash $sg(d)$ in BF_d	18: Update Bloom filter of C and L hashta-
8: **for** each cluster $C \in \mathcal{C}(d)$ **do**	bles
9: **if** $\overline{dist}(d, C) \geq \delta$ **then**	19: **end for**

the document. This value, denoted with $\overline{minOvl}$, is determined from the parameter δ as follows: $\delta = 1 - \overline{minOvl}/card \Rightarrow \overline{minOvl} = card(1 - \delta)$, with $card = \min(|sg(C_{opt})|, |sg(C_{sub})|)$. This is without loss of generality, because the accuracy of the estimations further increases when the overlap increases [20]. Furthermore, for simplification, we assume $|sg(C_{opt})|$ and $|sg(C_{sub})|$ are known, and that $|sg(d)| < |sg(C_{opt})|$ and $|sg(d)| < |sg(C_{sub})|$. We relax these assumptions later. For the theorem we use as shortcuts $t_{mopt} = S(t_d, t_{C_{opt}}, \overline{minOvl})$ and $t_{msub} = S(t_d, t_{C_{sub}}, \overline{minOvl})$.

Theorem 2. *The probability of a significantly wrong assignment* $Pr[dist(sg(d), sg(C_{sub})) - dist(sg(d), sg(C_{opt})) > \Delta]$ *is at most* $1 - (1 - (\frac{t_l}{t_{msub}-1})^{t_{msub}-1} \times e^{t_{msub}-1-t_l}) \times (1 - e^{-\frac{(t_{mopt}+1-t_r)^2}{2t_r}})$, *where* $t_l = S(t_d, t_{C_{sub}}, \overline{minOvl} - \frac{\Delta'}{2|sg(C_{opt})|})$, $t_r = S(t_d, t_{C_{opt}}, \overline{minOvl} + \frac{\Delta'}{2|sg(C_{sub})|})$, *and* $\Delta' = \Delta \times |sg(C_{opt})| \times |sg(C_{sub})| - \overline{minOvl} \times (|sg(C_{sub})| - |sg(C_{opt})|)$.

Proof of the theorem can be found in the technical report[17].

As an example, consider the case when $\delta = \Delta = 0.2$, $m = 4096$, $\lambda = 2$, and $|sg(C_{sub})| = |sg(C_{opt})| = 1000$. Then, according to Theorem 2, the probability of a significantly wrong assignment is less than 0.025. We can further reduce this error probability by increasing the Bloom filter length. For example, for $m = 8192$ the probability is reduced to less than 0.002, and for $m = 10000$, the probability becomes less than 7×10^{-4}.

In Theorem 2, for simplification we assume that $|sg(C_{opt})|$ and $|sg(C_{sub})|$ are given. In practice, we can closely approximate both cardinalities using Eqn. 1. In addition, we can obtain probabilistic lower and upper bounds for $|sg(C_{opt})|$ and $|sg(C_{sub})|$, as described in [20], and use these to derive the worst-case values (i.e., the ones that minimize $\overline{minOvl}$, and maximize the probability of a significantly wrong assignment). Integrating these probabilistic guarantees in the analysis of Theorem 2 is part of our future work.

The Bloom filter encoding also allows an efficient updating of the s-graph representations of a cluster when a new document is assigned to it. As explained in [20], the bitwise-OR operation of two Bloom filters equals to the creation of a new Bloom filter of the union of two sets. We can therefore simply merge the corresponding Bloom filters of the document and the cluster with bitwise-OR, rather than generating the new Bloom filter for the updated cluster from scratch. The full algorithm of XStreamCluster is illustrated in Algorithm 1.

4   Experimental Evaluation

XStreamCluster was evaluated in terms of efficiency, scalability, and clustering quality, using streams of up to 1 million XML documents. Our simulations were carried out in a single dedicated Intel Xeon 3.6Ghz core.

Datasets. We conducted experiments with two streams. The first (STREAM1) was generated using a set of 250 synthetic DTDs. To verify the applicability of the experimental results for real DTDs, the second stream (STREAM2) was created following a set of 22 real, publicly available DTDs. In particular, for STREAM1 we first generated x DTDs, out of which we created y different XML documents with XML Generator [21]. For generating each document, we randomly selected one of the available DTDs as an input for the XML Generator. The values of x and y varied for each experiment, with a maximum of 250 and 1 million respectively. The resulting documents were fed to the stream in a random order. For generating STREAM2 we followed the same procedure, but using a set of real, frequently used DTDs. The full list of the used DTDs and the DTD generator are available online, in http://www.l3s.de/~papapetrou/dtdgen.html.

Algorithms and Methodology. To evaluate in depth the contribution of each of the strategies to the algorithm's efficiency and effectiveness, we have compared three different variants of XStreamCluster. Furthermore, XStreamCluster was compared with the existing static algorithm which employs s-graphs for representing and comparing clusters and documents [5], called S-GRACE. In particular, we implemented and evaluated the following algorithms:

S-GRACE: We adapted S-GRACE [5] to streaming data. This required the following extensions: (a) changing the clustering algorithm from ROCK to K-Means, and (b) representing the s-graphs as extensible bit arrays, instead of bit arrays of fixed sizes. Note that S-GRACE achieves the same quality as comparing each document to all clusters without any optimization.

XStreamBF: XStreamCluster with only the bottom-level strategy in place (i.e., encoding of s-graphs as Bloom filters; documents were compared to all clusters).

XStreamLSH: XStreamCluster with only the top-level strategy (i.e., indexing clusters using LSH; s-graphs were represented as extensible bit arrays).

XStreamCluster: The algorithm as presented in this paper.

To evaluate efficiency and scalability, we measured the average time and memory required for clustering streams of up to 1 million documents. Quality was evaluated using the standard measure of normalized mutual information. In the following, we report average measures after 4 executions of each experiment. We present results with Bloom filters of 1024 bits, with 2 hash functions. The

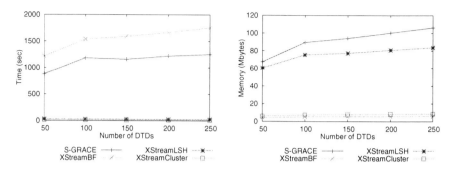

Fig. 3. (a) Time requirements, and, (b) Memory requirements, for clustering STREAM1 with respect to the number of DTDs

LSH index was configured for satisfying a correctness probability of 0.9. Due to page constraints, we present detailed results for STREAM1, and summarize the results for STREAM2, noting the differences.

4.1   Efficiency

With respect to efficiency, we compared the memory and execution time of each algorithm for clustering the two streams. To ensure that time measures were not affected by other activities unrelated to the clustering algorithm, e.g., network latency, we excluded the time spent in reading the stream.

For the first experiment, we studied how the efficiency of the algorithms changes with respect to the diversity of the stream. We controlled the diversity of the stream by choosing the number of DTDs out of which STREAM1 was generated. Figures 3 (a) and (b) plot the time and memory requirements of the four algorithms for clustering different instances of STREAM1, each generated by a different number of DTDs. The distance threshold for this experiment was set to 0.1, and the number of documents in the stream was set to $100k$. We see that XStreamCluster clearly outperforms S-GRACE in terms of speed; it requires up to two orders of magnitude less time for clustering the same stream. XStreamLSH presents the same speed improvement. The efficiency of both algorithms is due to the top-level strategy for candidate clusters detection, which drastically reduces the cluster-document comparisons. XStreamBF does not present this speed improvement as it does not employ an LSH inverted index.

We also observe that the speed improvement of XStreamLSH and XStream-Cluster is more apparent for higher number of DTDs. This is because more DTDs lead to more clusters. For the S-GRACE and XStreamBF algorithms, more clusters lead to longer bit arrays, thereby requiring more time to cluster a document. Furthermore, more clusters lead to an increase in the cluster-document comparisons since each document needs to be compared to all clusters. This is not the case for XStreamLSH and XStreamCluster though, which pre-filter the candidate clusters for each document by using the top-level strategy. Therefore, the execution time of XStreamLSH and XStreamCluster is almost constant.

With respect to memory requirements (Figure 3(b)) we see that XStream-Cluster and XStreamBF require at least one order of magnitude less memory compared to S-GRACE. The difference is again particularly visible for a higher number of DTDs, which results to a higher number of clusters. The huge memory savings are due to the Bloom filter encodings employed by the two algorithms.

The experiment was also conducted using STREAM2. Since STREAM2 was generated from a limited number of DTDs, instead of varying the number of DTDs we varied the value of the distance threshold δ, which also had an influence on the number of clusters: reducing the δ value resulted to more clusters. Table 1 presents example results, for $\delta = 0.1$ and 0.2. As expected, reducing the δ value leads to an increase in memory and computational cost for S-GRACE and XStreamBF, due to the increase in the number of clusters. On the other hand, the speed of XStreamCluster and XStreamLSH actually increases by reducing the distance threshold, because the LSH index is initialized with less hash tables and hash functions.

4.2 Scalability

To verify the scalability of XStreamCluster, we compared it against S-GRACE and its variants for clustering streams of different sizes, reaching up to 1 million documents. In particular, we generated instances of STREAM1 and STREAM2 with 1 million documents, and used all four algorithms to cluster them. During clustering, we monitored memory and execution time every $100k$ documents. The experiment was repeated for various configurations. Due to space limitations we report only the results for STREAM1 corresponding to 100 DTDs where δ is set to 0.1. The results for other settings lead to the same conclusions.

Figures 4(a) and 4(b) present the execution time and memory usage with respect to number of documents. With respect to execution time, we see that S-GRACE and XStreamBF fail to scale. Their execution time increases exponentially with the number of documents, because of the increase in the number of clusters. On the other hand, XStreamCluster and XStreamLSH have a linear scale-up with respect to the number of documents, i.e., the cost for clustering each document remains constant with the number of clusters. This is achieved due to the efficient filtering of clusters with the LSH-based candidate cluster detection strategy.

With respect to memory requirements (Fig. 4(b)), all algorithms scale linearly with the number of documents, but the algorithms that use Bloom filters

Table 1. Example results for STREAM2

Algorithm	Time (sec)		Memory (Mbytes)		NMI	
Distance thres.	0.1	0.2	0.1	0.2	0.1	0.2
S-GRACE	61	35	23.1	14	0.72	0.76
XStreamBF	363	253	1.7	1.1	0.72	0.76
XStreamLSH	8	11	19.1	13.8	0.716	0.755
XStreamCluster	4	7	2.5	3.1	0.715	0.755

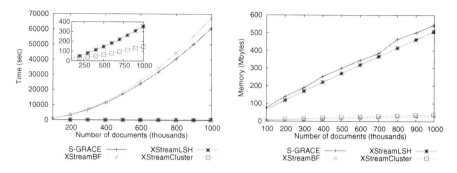

Fig. 4. (a) Execution time, and (b) Memory requirements, for clustering STREAM1 with respect to number of documents

require an order of magnitude less memory. Interestingly, for clustering 1 million documents, XStreamCluster requires only 39 Mbytes memory, which is an affordable amount for any off-the-shelf PC. Therefore, XStreamCluster can keep all its memory structures in fast main memory, instead of resorting to the slower, secondary storage. Keeping as many data structures as possible in main memory is very important for algorithms working with streams, because of their high efficiency requirements.

4.3   Clustering Quality

We evaluated the clustering quality of XStreamCluster by using the standard measure of Normalized Mutual Information (NMI). NMI reflects how close the clustering result approaches an optimal classification – a ground truth – which is usually constructed by humans [22]. An NMI of 0 indicates a random assignment of documents to clusters, whereas an NMI of 1 denotes a clustering which perfectly resembles the optimal classification. For our datasets, the optimal classification was defined by the DTD of each document: two XML files were considered to belong to the same class when generated from the same DTD.

Figure 5(a) presents NMI with respect to distance threshold δ for STREAM1, which consists of $100k$ documents generated from 100 different DTDs. We see that all XStreamCluster variants achieve a clustering quality nearly equal to S-GRACE. Quality of XStreamBF is practically equal to quality of S-GRACE, which means that introducing the Bloom filters as cluster representations does not result to quality reduction. XStreamCluster and XStreamLSH have a small difference compared to S-GRACE, which is due to the aggressive filtering of clusters that takes place during clustering at the top-level strategy. This difference is negligible, especially for small distance threshold values.

We further studied how the diversity of the stream influences the algorithms' quality, by repeating the experiment using streams generated from a different number of DTDs. Figure 5(b) shows the NMI with respect to the number

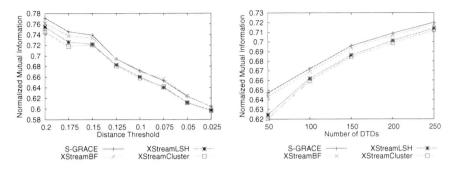

Fig. 5. Normalized Mutual Information for STREAM1. Varying (a) the distance threshold, and, (b) the number of DTDs.

of DTDs used for generating STREAM1. We see that XStreamCluster variants again achieve a quality almost equal to S-GRACE. The difference between XStreamCluster and S-GRACE reduces with an increase in the number of DTDs, and becomes negligible for the streams generated from more than 100 DTDs.

As shown in the last column of Table 1, the same outcome was observed on the experiments with STREAM2. XStreamBF produced an equivalent solution to S-GRACE, whereas XStreamLSH and XStreamCluster approximated closely the optimal quality. The difference between the approximate solution produced by XStreamCluster and the exact solution produced by S-GRACE was less than 0.01 in terms of NMI, in all experiments.

Summarizing, XStreamCluster achieves good clustering of XML documents requiring at least an order of magnitude less cost compared to S-GRACE, with respect to both execution time and memory. The experimental results show that it is especially suited for clustering large and diverse streams, both with respect to quality and efficiency. Owing to the low memory and time requirements, it is easily deployable in standard off-the-shelf PCs and scales to huge XML streams.

5 Conclusions

We presented XStreamCluster, the first algorithm that addresses clustering of *streaming XML documents*. The algorithm combines two optimization strategies, Bloom filters for reducing the memory requirements, and Locality Sensitive Hashing to reduce significantly the cost of clustering. We provided theoretical analysis showing that XStreamCluster provides an approximately similar quality of clustering as exact solutions do. Our experimental results also confirmed the efficiency and effectiveness contributed by the two strategies of XStreamCluster. For future work, we plan to extend this work by considering additional distance measures, including the ones which combine both content similarity and structure similarity.

References

1. Mayorga, V., Polyzotis, N.: Sketch-based summarization of ordered XML streams. In: Proc. of ICDE (2009)
2. Josifovski, V., Fontoura, M., Barta., A.: Querying XML streams. VLDB Journal 14(2) (2005)
3. Bifet, A., Gavald, R.: Adaptive XML tree classification on evolving data streams. In: Buntine, W., Grobelnik, M., Mladenić, D., Shawe-Taylor, J. (eds.) ECML PKDD 2009. LNCS, vol. 5781, pp. 147–162. Springer, Heidelberg (2009)
4. Dalamagas, T., Cheng, T., Winkel, K.J., Sellis, T.: A methodology for clustering XML documents by structure. Inf. Syst. 31(3), 187–228 (2006)
5. Lian, W., Cheung, D.W.L., Mamoulis, N., Yiu, S.M.: An efficient and scalable algorithm for clustering XML documents by structure. IEEE TKDE 16(1), 82–96 (2004)
6. Nierman, A., Jagadish, H.V.: Evaluatating structural similarity in XML documents. In: Proc. of ACM SIGMOD WebDB Workshop, pp. 61–66 (2002)
7. Tagarelli, A., Greco, S.: Toward semantic XML clustering. In: Proc. SDM (2006)
8. Aggarwal, C.C.: A framework for clustering massive-domain data streams. In: Proc. of IEEE ICDE (2009)
9. Jain, A.K., Dubes, R.C.: Algorithms for clustering data. Prentice-Hall, Englewood Cliffs (1988)
10. Kaufman, L., Rousseuw, P.: Finding groups in data - An introduction to cluster analysis. Wiley, Chichester (1990)
11. Aggarwal, C.C., Han, J., Wang, J., Yu, P.S.: A framework for clustering evolving data streams. In: Proc. of VLDB (2003)
12. Guha, S., Mishra, N., Motwani, R., O'Callaghan, L.: Clustering data streams. In: Proc. of IEEE FOCS (2000)
13. O'Callaghan, L., Mishra, N., Meyerson, A., Guha, S., Motwani, R.: Streaming-data algorithms for high-quality clustering. In: Proc. of ICDE (2002)
14. Candillier, L., Tellier, I., Torre, F.: Transforming XML trees for efficient classification and clustering. In: Fuhr, N., Lalmas, M., Malik, S., Kazai, G. (eds.) INEX 2005. LNCS, vol. 3977, pp. 469–480. Springer, Heidelberg (2006)
15. Doucet, A., Ahonen Myka, H.: Naive clustering of a large XML document collection. In: INEX, pp. 81–87 (2002)
16. Doucet, A., Lehtonen, M.: Unsupervised classification of text-centric XML document collections. In: Fuhr, N., Lalmas, M., Trotman, A. (eds.) INEX 2006. LNCS, vol. 4518, pp. 497–509. Springer, Heidelberg (2007)
17. Papapetrou, O., Chen, L.: XStreamCluster: an Efficient Algorithm for Streaming XML data Clustering. Technical report (2010),
   http://www.l3s.de/~papapetrou/publications/XStreamCluster-long.pdf
18. Gionis, A., Indyk, P., Motwani, R.: Similarity search in high dimensions via hashing. In: Proc. of VLDB (1999)
19. Broder, A.Z., Charikar, M., Frieze, A.M., Mitzenmacher, M.: Min-wise independent permutations. In: Proc. of STOC 1998, pp. 327–336. ACM, New York (1998)
20. Papapetrou, O., Siberski, W., Nejdl, W.: Cardinality estimation and dynamic length adaptation for bloom filters. DAPD 28(1) (2010)
21. Diaz, A.L., Lovell, D.: XML generator (1999),
   http://www.alphaworks.ibm.com/tech/xmlgenerator
22. Christopher, D., Manning, P.R., Schtze, H.: Introduction to Information Retrieval. Cambridge University Press, Cambridge (2008)

Efficient Evaluation of NOT-Twig Queries in Tree-Unaware Relational Databases

Kheng Hong Soh[2] and Sourav S. Bhowmick[1,2]

[1] Singapore-MIT Alliance, Nanyang Technological University, Singapore
[2] School of Computer Engineering, Nanyang Technological University, Singapore
assourav@ntu.edu.sg

Abstract. Despite a large body of work on XML query processing in relational environment, systematic study of NOT-*twig queries* has received little attention in the literature. Such queries contain not-predicates and are useful for many real-world applications. In this paper, we present an efficient strategy to evaluate NOT-twig queries on top of a dewey-based *tree-unaware* system called SUCXENT++ [11]. We extend the encoding scheme of SUCXENT++ by adding two new labels, namely AncestorValue and AncestorDeweyGroup, that enable us to *directly* filter out elements satisfying a not-predicate by comparing their *ancestor group identifiers*. In this approach, a set of elements under the same common ancestor at a specific level in the XML tree is assigned *same ancestor group identifier*. Based on this encoding scheme, we propose a novel SQL translation algorithm for NOT-twig query evaluation. Real and synthetic datasets are employed to demonstrate the superiority of our approach over industrial-strength RDBMS and native XML databases.

1 Introduction

Querying XML data over relational framework has gained popularity due to its stability, efficiency, expressiveness, and its wide spread usage in the commercial world. On the one hand, there has been a host of work, *c.f.*, [3], on enabling relational databases to be *tree-aware* by invading the database kernel to support XML. On the other hand, some completely jettison the invasive approach and resort to a *tree-unaware* approach, *c.f.*, [4, 7, 11, 13, 14], where the database kernel is not modified to support XML queries.

Generally, the tree-unaware approach reuses existing code, has a lower cost of implementation, and is more portable since it can be implemented on top of off-the-shelf RDBMSs. This has triggered recent efforts to explore how far we can push the idea of using mature tree-unaware RDBMS technology to design and build a relational XQuery processor [4, 5, 7]. Particularly, a wealth of existing literature has extensively studied evaluation of various navigational axes in XPath expressions and optimization techniques in a tree-unaware environment [4, 5, 7, 11, 13, 14]. *However, to the best of our knowledge, no systematic study has been carried out in efficiently evaluating NOT-twig queries in this relational environment.* Such queries contain not-predicates and are useful for many real-world applications. For example, the query /catalog/book[not(review) and

J.X. Yu, M.H. Kim, and R. Unland (Eds.): DASFAA 2011, Part I, LNCS 6587, pp. 511–527, 2011.
© Springer-Verlag Berlin Heidelberg 2011

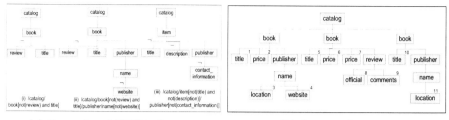

(a) Examples of NOT-twig query   (b) An XML document

Fig. 1. Examples of NOT-twig queries and XML document

Id	Size	No. of Attributes	No. of Nodes	Total	Max Depth
DC10	10.3MB	15,000	152,673	167,673	8
DC100	103.3MB	150,000	1,520,322	1,670,322	8
DC1000	1033.3MB	1,500,000	15,215,868	16,715,868	8

(a) XBENCH data sets

Id	Size	No. of Attributes	No. of Nodes	Total	Max Depth
U28	28MB	771,880	422,972	1,194,852	6
U284	284MB	7,791,620	4230,003	12,021,623	6
U2843	2843MB	77,977,270	42,421,745	120,399,015	6

(b) UniProt data sets

Id	Data Source	NOT-twig Query
Q1	XBench	/catalog/item/authors/author[not(biography)]/name
Q2	UniProt	/uniprot/entry[not(geneLocation/name) and not(comment/location)]
Q3	UniProt	/uniprot/entry[not(geneLocation) and not(protein/domain)]/comment[not(note) and not(event)]

(c) NOT-twig queries

Data Set	Q1		Data Set	Q2		Q3	
	XSysA	XSysB		XSysA	XSysB	XSysA	XSysB
DC10	1,439	334	U28	20,399	2,457	5,899	1,977
DC100	2,332	2,823	U284	207,982	24,429	47,386	20,819
DC1000	11,298	57,309	U2843	-	-	-	-

(d) XBench data set   (e) UniProt data set

Fig. 2. Data sets and query evaluation times (in msec.)

title] retrieves all books that have a title but no reviews (Figure 1(a)(i)). Figures 1(b)(ii) and 1(c)(iii) show graphical representations of two more NOT-twig queries.

At first glance, it may seem that such lack of study may be primarily due to the fact that we can efficiently evaluate these NOT-twig queries by leveraging on the XML query processor of an existing industrial-strength RDBMS and relying on its query optimization capabilities. However, our initial investigation showed that fast evaluation of NOT-twigs still remains a bottleneck in several industrial-strength RDBMSs. To get a better understanding of this problem, we experimented with the XBench DCSD [15] and UNIPROT (downloaded from www.expasy.ch/sprot/) data sets shown in Figures 2(a) and 2(b) and queries $Q_1 - Q_3$ in Figure 2(c). We fix the result size of Q_1 to be 500. Figures 2(d) and 2(e) report the query evaluation times in two commercial-strength RDBMSs. Note that due to legal restrictions, these systems are anonymously identified as *XSysA* and *XSysB* in the sequel. Observe that the evaluation cost can be expensive as it can take up to 208 seconds to evaluate these queries. Also, both these commercial systems do not support processing of XML documents having size greater than 2GB (U2843 data set). *Is it possible to design a tree-unaware scheme that can address this performance bottleneck?* In this paper, we demonstrate that novel techniques built on top of an industrial-strength RDBMS can make up for a large part of the limitation. We show that the above queries can be evaluated in *a second or less* on smaller data sets and *less than* 13s for *Q2* on U284 data sets.

We built our proposed NOT-twig evaluation technique on top of dewey-based SUCXENT++ system [2, 11], a tree-unaware approach designed primarily for read-mostly workloads. As SUCXENT++ is designed primarily for fast evaluation of normal path and twig queries, it does not support efficient evaluation of NOT-twig queries. Hence, in Section 3 we extend SUCXENT++ encoding scheme by adding two new labels, namely AncestorValue and AncestorDeweyGroup, to each level and leaf elements, respectively. These labels enable us to efficiently group a set of elements under the same common ancestor at a specific level with the *same ancestor group identifier*. As we shall see later, this will allow us to efficiently filter out elements satisfying a not-predicate by comparing their *ancestor group identifiers*.

Based on the *extended* encoding scheme, we propose a novel SQL translation algorithm for NOT-twig evaluation (Section 4). In our approach, we use the AncestorDeweyGroup and AncestorValue labels to evaluate *all* paths in a NOT-twig query. In Section 5, we demonstrate with exhaustive experiments that the proposed approach is significantly faster than XML supports of *XSysA* and *XSysB* (highest observed factor being 40 times).

Our proposed approach differs from existing efforts in evaluating NOT-twigs using structural join algorithms [1, 8, 10, 16] in the following ways. Firstly, we take relational-based approach instead of native strategy used in aforementioned approaches. Secondly, our encoding scheme is different from the above approaches. In [16], region encoding scheme is employed to label the elements whereas a pair of *(path-id, node id)* [9] is used in [10]. In contrast, we use a dewey-based scheme where only the leaf elements and the levels of the XML tree are explicitly encoded. Thirdly, these existing approaches typically report query performance on documents smaller than 150MB and containing at most 2.5 million nodes. In contrast, we explore the scalability of our approach for larger XML documents (2.8GB size) having more than 120 million nodes.

2   Preliminaries

XML Data Model. We model XML documents as ordered trees. In our model we ignore comments, attributes, processing instructions, and namespaces. Queries in XML query languages make use of twig patterns to match relevant portions of data in an XML database. A twig pattern can be represented as a tree containing all the nodes in the query. A node m_i in the pattern may be an element tag, a text value or a wildcard "*". We distinguish between query and data nodes by using the term "node" to refer to a query node and the term "element" to refer to a data element in a document. Each node m_i and its parent (or ancestor) m_j are connected by an edge, denoted as $edge(m_i, m_j)$.

A twig query contains a collection of *rooted path* patterns. A *rooted path* pattern (RP) is a path from the root to a node in the twig. Each rooted path represents a sequence of nodes having parent-child (PC) or ancestor-descendant (AD) edges. We classify the rooted paths into two types: *root-to-leaf* and *root-to-internal* paths. A *root-to-leaf* path is a RP from the root to a leaf node in the query. In contrast, a RP ending at a non-leaf node is called a *root-to-internal* path. If the number of children of a node in the twig query is more than one, then we call this node a NCA (nearest common ancestor) *node*. Otherwise, when the node has only one child, it is a *non*-NCA *node*. The level of the NCA node is called NCA-*level*.

In this paper, we focus on twig queries with not-predicates. We refer to such queries as NOT-*twig queries*. The twig pattern edges of a NOT-twig query can be classified into one of the following two types. (a) *Positive edge:* This corresponds to an $edge(m_i, m_j)$ without not-predicate in the query expression. It is represented as "|" or "||" in a twig pattern for PC or AD edges, respectively. Node m_j is called the *positive* PC *(resp.* AD*) child of m_i. A rooted path that contains only positive children is called a *normal* rooted path. (b) *Negative edge:* This corresponds to an $edge(m_i, m_j)$ with not-predicate and is represented as "|¬" or "||¬" in the twig for PC or AD edges, respectively. In this case, node m_j is called the *negative* PC *(resp.* AD*) child of m_i. A rooted path pattern that contains a negative child is called a *negative* rooted path. For example, consider the NOT-twig query in Figure 1(a)(ii). $edge(\texttt{book},\texttt{title})$ and $edge(\texttt{book},\texttt{publisher})$ are positive edges whereas $edge(\texttt{book},\texttt{review})$ and $edge(\texttt{name},\texttt{website})$ are negative edges. Node **book** has three children, in which **title** and **publisher** are positive PC children and node **review** is a negative PC child. The RP **catalog/book/review** is a negative RP as it contains the negative PC child **review**. On the other hand, **catalog/book/title** is a normal RP.

NOT-Twig Pattern Matching. Given a NOT-twig query Q, a query node n, and an XML tree D, an element e_n (with the tag n) in D satisfies the subquery rooted at n of Q iff: (1) n is a leaf node of Q; or (2) For each child node n_c of n in Q: (a) If n_c is a positive PC (resp. AD) child of n, then there is an element e_{n_c} in D such that e_{n_c} is a child (resp. descendant) element of e_n and satisfies the sub-query rooted at n_c in D. (b) If n_c is a negative PC (resp. AD) child of n, then there does not exists any element e_{n_c} in D such that e_{n_c} is a child (resp. descendant) element of e_n and satisfies the sub-query rooted at n_c in D.

3   Encoding Scheme

In this section, we first briefly describe the encoding scheme of SUCXENT++ [2, 11] and highlight its limitations in efficiently processing NOT-twig queries. Then, we present how it can be extended to efficiently support queries with not-predicates.

3.1   SUCXENT++ Schema and Its Limitations

In SUCXENT++, each level ℓ of an XML tree is associated with an attribute called RValue (denoted as R_ℓ). Each leaf element n is associated with four attributes,

Document

DocID	Name
1	catalog.xml

DocumentRValue

DocID	Level	RValue
1	1	29
1	2	4
1	3	2
1	4	1

PathValue

DocID	Leaf Order	Brach Order	PathID	Dewey Order Sum	Sibling Sum	Leaf Value
1	1	0	1	0	0	....
1	2	2	2	7	0	....
1	3	2	3	14	0	....
1	4	4	4	15	0	....
1	5	1	1	57	57	....
1	6	2	2	64	57	....
1	7	2	2	71	64	......
1	8	2	5	78	57	....
1	9	3	6	81	57	......
1	10	1	1	114	114	......
1	11	2	3	121	114	....

Path

PathID	PathExp
1	.catalog#.book#.title#
2	.catalog#.book#.price#
3	.catalog#.book#.publisher#.name#.location#
4	.catalog#.book#.publisher#.name#.website#
5	.catalog#.book#.review#.official#
6	.catalog#.book#.review#.comments#

Fig. 3. Storage of a shredded XML document

namely LeafOrder, BranchOrder, DeweyOrderSum, and SiblingSum. Each non-leaf element n' is *implicitly* assigned the DeweyOrderSum of the first descendant leaf element. Here we briefly define the relevant attributes necessary to understand this paper. The reader may refer to [2, 11] for details related to their roles in XML query processing.

The schema of SUCXENT++ [2, 11] is as follows: (a) Document(<u>DocID</u>, Name), (b) Path(<u>PathId</u>, PathExp), (c) PathValue(<u>DocID</u>, <u>DeweyOrderSum</u>, PathId, BranchOrder, LeafOrder, SiblingSum, LeafValue), and (d) DocumentRValue(<u>DocID</u>, <u>Level</u>, RValue). Document stores the document identifier DocID and the name Name of a given input XML document D. Each distinct root-to-leaf path appearing in D, namely PathExp, is associated with an identifier PathId and stored in Path table. Essentially each path is a concatenation of the labels of the elements in the path from the root to the leaf. An example of the Path table containing the root-to-leaf paths of Figure 1(b) is shown in Figure 3. Note that '#' is used as a delimiter of steps in the paths instead of '/' for reasons described in [14].

For each leaf element n in D, a tuple in the PathValue table is created to store the LeafOrder, BranchOrder, DeweyOrderSum, and SiblingSum values of n. The data value of n is stored in LeafValue. Given two leaf elements n_1 and n_2, n_1.LeafOrder $<$ n_2.LeafOrder *iff* n_1 precedes n_2 in document order. LeafOrder of the first leaf element in D is 1 and n_2.LeafOrder $=$ n_1.LeafOrder$+1$ *iff* n_1 is a leaf element immediately preceding n_2. For example, the superscript of each leaf element in Figure 1(b) denotes its LeafOrder value.

Given two leaf elements n_1 and n_2 where n_1.LeafOrder$+1$ $=$ n_2.LeafOrder, n_2.BranchOrder is the level of the nearest common ancestor (NCA) of n_1 and n_2. For example, the BranchOrder of the `location` leaf element with LeafOrder value 3 in Figure 1(b) is 2 as the NCA of this element and the preceding `price` element is at the second level. Note that the BranchOrder of the first leaf element is 0.

Next we define RValue. We begin by introducing the notion of *maximal k-consecutive leaf-node list*. Consider a list of consecutive leaf element $\mathcal{S}$: $[n_1, n_2, n_3, \ldots, n_r]$ in D. Let $k \in [1, L_{max}]$ where L_{max} is the largest level of D. Then, $\mathcal{S}$ is called a *k-consecutive leaf-node list* of D *iff* $\forall 0 < i \leq r$ n_i.BranchOrder $\geq k$. $\mathcal{S}$ is called a *maximal k-consecutive leaf-node list*, denoted as M_k, if there does not exist a k-consecutive leaf-node list $\mathcal{S}'$ such that $|\mathcal{S}| < |\mathcal{S}'|$. For example, M_2 in Figure 1(b) contains four leaf elements as $|\mathcal{S}| = 4$ for M_2.

The RValue of level ℓ, denoted as R_ℓ, is defined as follows: (i) If $\ell = L_{max} - 1$ then $R_\ell = 1$; (ii) If $0 < \ell < L_{max} - 1$ then $R_\ell = 2R_{\ell+1} \times |M_{\ell+1}| + 1$. For example, consider Figure 1(b). Here $L_{max} = 5$. The values of $|M_2|$, $|M_3|$, and $|M_4|$ are 4, 1, and 1, respectively. Then, $R_4 = 1$, $R_3 = 2 \times 1 \times |M_4| + 1 = 3$, $R_2 = 2 \times 3 \times |M_3| + 1 = 7$, and $R_1 = 2 \times 7 \times |M_2| + 1 = 57$. In order to facilitate evaluation of XPath queries, the RValue attribute in DocumentRValue stores $\frac{R_\ell - 1}{2} + 1$ instead of R_ℓ (denoted as R'_ℓ). For instance, in Figure 3 the RValue of level 1 is stored as 29 instead of 57.

DeweyOrderSum is used to encode an element's order information together with its ancestors' order information using a single value. Let $parent(w)$ denote the parent of an element w. Consider a leaf element n at level ℓ in D. Then, for $1 < k \leq \ell$, $\mathsf{Ord}(n, k) = i$ *iff* (i) there exists an element a at level k which is either an ancestor of n or n itself; and (ii) a is the i-th child of $parent(a)$. For example, consider the rightmost leaf element in Figure 1(b) (denoted as d). $\mathsf{Ord}(d, 2) = 3$ as the rightmost book element in the second level is an ancestor of d as well as the third child of the root. Similarly, $\mathsf{Ord}(d, 3) = 2$.

Then DeweyOrderSum of n, $n.$DeweyOrderSum, is defined as $\sum_{j=2}^{\ell} \Phi(j)$ where $\Phi(j) = [\mathsf{Ord}(n, j) \text{-} 1] \times R_{j-1}$. The DeweyOrderSum of the first leaf element is 0. Reconsider the rightmost leaf element again. It has a Dewey path "1.3.2.1.1". DeweyOrderSum of this element is: $n.$DeweyOrderSum $= (\mathsf{Ord}(n, 2) - 1) \times R_1 + (\mathsf{Ord}(n, 3) - 1) \times R_2 + (\mathsf{Ord}(n, 4) - 1) \times R_3 + (\mathsf{Ord}(n, 5) - 1) \times R_4 = 2 \times 57 + 1 \times 7 + 0 \times 3 + 0 \times 1 = 121$. The DeweyOrderSum of remaining elements are shown in the DeweyOrderSum attribute of the PathValue table in Figure 3.

Limitations of SUCXENT++. DeweyOrderSum and RValue attributes are designed primarily to evaluate normal twig queries. Consequently, they are unable to *directly* filter out elements satisfying negative RPs without having to first evaluate the rooted paths as normal RPs and then use the intermediate results to filter out irrelevant elements (see details in [12]). For instance, for the query in Figure 1(a)(i), DeweyOrderSum and RValue attributes fail to reveal those `title` and `review` elements that *do not* share the same common book ancestors without exhaustively comparing them. Furthermore, they do not always support efficient evaluation of descendant (ancestor) axis. In the subsequent sections, we shall present a novel technique that addresses these limitations.

3.2   AncestorValue Attribute

We now elaborate on the extension of the encoding scheme of SUCXENT++. Due to space constraints, the proofs of lemmas and theorem presented in the sequel are given in [12]. Each level ℓ of an XML tree is added an attribute called AncestorValue along with its existing RValue. Each leaf element n is added an attribute called AncestorDeweyGroup. These attributes are materialized in the DocumentRValue and PathValue tables, respectively. As we shall see later, our proposed strategy aims to group a set of leaf elements under the same common ancestor at level ℓ with the *same ancestor group identifier*. AncestorDeweyGroup and AncestorValue attributes will be used to compute these identifiers.

AncestorValue, similar to RValue, is used for encoding the level of the NCA of any pairs of leaf elements.

Definition 1. *[AncestorValue] Let L_{max} be the maximum level of an XML tree. Then the **AncestorValue** of level ℓ for $0 < \ell < L_{max}$, denoted as A_ℓ, is defined as follows: (a) If $\ell = L_{max} - 1$, then $A_\ell = 1$; (b) If $0 < \ell < L_{max} - 1$, then $A_\ell = A_{\ell+1} \times (|M_{\ell+1}| + 1)$.*

For example, reconsider the XML tree in Figure 1(b). Here $L_{max} = 5$, $|M_4| = 1$, $|M_3| = 1$, and $|M_2| = 4$. Hence, $A_4 = 1$, $A_3 = 1 \times (1+1) = 2$, $A_2 = 2 \times (1+1) = 4$, and $A_1 = 4 \times (4 + 1) = 20$.

Lemma 1. *Let ℓ be a level in an XML tree where $0 < \ell < L_{max}$. Then, A_ℓ is divisible by all $A_{\ell+m}$ where $0 < m < (L_{max} - \ell)$.*

Consider the previous example. Let $\ell = 2$. Then, $0 < m < 3$. Hence based on the above lemma, $A_2/A_3 = 4/2 = 2$ and $A_2/A_4 = 4/1 = 4$. Note that existing RValue do not have such divisibility property [12].

3.3 AncestorDeweyGroup Attribute

The AncestorDeweyGroup attribute, similar to DeweyOrderSum, is used to encode an element's order information using a single value. The only difference between AncestorDeweyGroup and DeweyOrderSum is that the former uses each level's AncestorValue whereas the latter uses the RValue of each level.

Definition 2. *[AncestorDeweyGroup] Consider a leaf element n at level ℓ in an XML document. Then, for $1 < k \le \ell$, $\mathsf{Ord}(n, k) = i$ iff (i) there exists an element a at level k which is either an ancestor of n or n itself; and (ii) a is the i-th child of $parent(a)$. Then **AncestorDeweyGroup** of n, n.AncestorDeweyGroup, is defined as $\sum_{j=2}^{\ell} \Omega(j)$ where $\Omega(j) = [\mathsf{Ord}(n,j)\text{-}1] \times A_{j-1}$.*

For example, reconsider the last leaf element in Figure 1(b) with Dewey value "1.3.2.1.1". AncestorDeweyGroup of this element is: n.AncestorDeweyGroup $= (Ord(n, 2) - 1) \times A_1 + (Ord(n, 3) - 1) \times A_2 + (Ord(n, 4) - 1) \times A_3 + (Ord(n, 5) - 1) \times A_4 = 2 \times 20 + 1 \times 4 + 0 \times 2 + 0 \times 1 = 44$. The AncestorDeweyGroup values of remaining leaf elements in Figure 1(b) are (in document order): 0, 4, 8, 9, 20, 24, 28, 32, 34, and 40.

4 Ancestor Group-Based Approach

We begin by formally introducing the notion of *ancestor group identifier*. Then, we present how such identifiers can be used for evaluating NOT-twig queries.

4.1   Ancestor Group Identifier

Informally, given an internal element n at level $\ell > 1$ of an XML tree, a unique ancestor group identifier with respect to ℓ is assigned to all the descendant leaf element(s) of n. It is computed using AncestorDeweyGroup values of the leaf elements and the AncestorValue of level of n.

Definition 3. *[Ancestor Group Identifier] Let n_i be a leaf element in the* XML *tree D. Let n_a be an ancestor element of n_i at level $\ell > 1$. Then* **Ancestor Group Identifier** *of n_i w.r.t n_a at level ℓ is defined as* $\mathcal{G}_i^\ell = \left\lfloor \dfrac{n_i.\mathsf{AncestorDeweyGroup}}{A_{\ell-1}} \right\rfloor$.

For example, consider the leaf elements n_1, n_2, n_3, and n_4 (we denote a leaf element as n_i where i is its LeafOrder value) in Figure 1(b). The AncestorDeweyGroup values of these elements are 0, 4, 8, and 9, respectively. Also, $A_1 = 20$ and $A_2 = 4$. If we consider the first book element at level 2 as the ancestor element of these elements, then $\mathcal{G}_1^2 = \left\lfloor \frac{0}{A_{2-1}} \right\rfloor = 0$, $\mathcal{G}_2^2 = \left\lfloor \frac{4}{A_{2-1}} \right\rfloor = 4/20 = 0$, $\mathcal{G}_3^2 = \left\lfloor \frac{8}{20} \right\rfloor = 0$, and $\mathcal{G}_4^2 = \left\lfloor \frac{9}{20} \right\rfloor = 0$. However, if we consider the `publisher` element at level 3 as ancestor element, then $\mathcal{G}_3^3 = \left\lfloor \frac{8}{4} \right\rfloor = 2$, and $\mathcal{G}_4^3 = \left\lfloor \frac{9}{4} \right\rfloor = 2$. Note that we do not define ancestor group identifier with respect to the root element ($\ell = 1$) because it is a trivial case as all leaf elements in the document shall have same identifier values.

Ancestor group identifiers of non-leaf elements: Observe that in the above definition only the leaf elements have *explicit* ancestor group identifiers. We assign the ancestor group identifiers to the internal elements implicitly. The basic idea is as follows. Let n_c be the NCA at level ℓ of two leaf elements n_i and n_j with ancestor group identifiers equal to $\mathcal{G}^\ell$. Then, the ancestor group identifiers of all non-leaf elements in the subtree rooted at n_c is $\mathcal{G}^\ell$. For example, reconsider the first book element at level 2 as the root of the subtree. Then, the ancestor group identifiers of the `publisher` and `name` elements are 0. Note that these identifiers are not stored explicitly as they can be computed from AncestorDeweyGroup and AncestorValue values.

Role of ancestor group identifiers to evaluate descendant axis. Observe that a key property of the ancestor group identifier is that all descendants of an ancestor element at a specific level must have *same* identifiers. We can exploit this feature to efficiently evaluate descendant axis. Given a query a//b, let n_a and n_b be elements of types a and b, respectively. Then, whether n_b is a descendant of n_a can be determined using the above definition as all descendants of n_a must have *same* ancestor group identifiers. As we shall see later, *this equality property is also important for our* NOT-*twig evaluation strategy.*

Remark. Due to the lack of divisibility property of RValue (Lemma 1), it cannot be used along with the DeweyOrderSum to correctly compute the *ancestor group identifiers* of elements. Consequently, they are not particularly suitable for efficient evaluation of NOT-twig queries. Due to the space limitations, these issues are elaborated in [12].

4.2   Computation of Common Ancestors

Lemma 2. *Let n_i and n_j be two leaf elements in D at level ℓ_1 and ℓ_2, respectively. Let $\ell < \ell_1$ and $\ell < \ell_2$. (a) If $\mathcal{G}_i^\ell \neq \mathcal{G}_j^\ell$ then n_i and n_j do not have a common ancestor at level ℓ. (b) If $\mathcal{G}_i^\ell = \mathcal{G}_j^\ell$ then n_i and n_j must have a common ancestor at level ℓ.*

Example 1. Consider the leaf elements n_1, n_2, n_5, and n_6 in Figure 1(b). The AncestorDeweyGroup values of these elements are 0, 4, 20, and 24, respectively. Also, $A_1 = 20$. Then, with respect to level 2 $\mathcal{G}_1^2 = \lfloor \frac{0}{20} \rfloor = 0$, $\mathcal{G}_2^2 = \lfloor \frac{4}{20} \rfloor = 0$, $\mathcal{G}_5^2 = \lfloor \frac{20}{20} \rfloor = 1$, and $\mathcal{G}_6^2 = \lfloor \frac{24}{20} \rfloor = 1$. Based on Lemma 2, since $\mathcal{G}_1^2 \neq \mathcal{G}_5^2$ then n_1 and n_5 does not have a common ancestor at level 2. Similarly, (n_1, n_6), (n_2, n_5), and (n_2, n_6) do not have common ancestors at the second level.

Since $\mathcal{G}_1^2 = \mathcal{G}_2^2$, n_1 and n_2 must have a common ancestor at level 2 (the first book element in Figure 1(b)). ∎

Observe that by using Lemma 2 we can filter out leaf elements that belong to the same common ancestor directly for negative rooted paths.

Theorem 1. *Let r_k and r_m be two RPs in a query Q on D. Let N_k and N_m be the sets of leaf elements that match r_k and r_m, respectively in D. Let $n_i \in N_k$ and $n_j \in N_m$. For $\ell > 1$, n_i must have the same ancestor as n_j at level ℓ iff $\mathcal{G}_i^\ell = \mathcal{G}_j^\ell$.*

Note that Lemma 2 and Theorem 1 can also be used for internal elements since ancestor group identifier of an internal element of a subtree rooted at the NCA is identical to that of any leaf element in the subtree (Section 4.1). Also, it immediately follows from the above theorem that n_i needs to be filtered out if r_m is a negative RP in Q. Note that we ignore the trivial case of $\ell = 1$ [12].

Example 2. Assume that the `price` and `location` elements in Figure 1(b) match a normal and a negative RPs, respectively in a NOT-twig query. Hence, we want to filter out all leaf elements having the same ancestor as `location` at level 2. Let $n_i \in N_{price}$ and $n_j \in N_{location}$ where N_{price} and $N_{location}$ are sets of leaf elements satisfying the normal and negative RPs, respectively. Here $N_{price} = \{n_2, n_6, n_7\}$, $N_{location} = \{n_3, n_{11}\}$, and $A_{2-1} = 20$. The AncestorDeweyGroup values of n_2, n_6, and n_7 are 4, 24, and 28, respectively. Similarly, AncestorDeweyGroup values of n_3 and n_{11} are 8 and 44, respectively. Then, $\mathcal{G}_2^2 = \mathcal{G}_3^2 = 0$, $\mathcal{G}_6^2 = \mathcal{G}_7^2 = 1$, and $\mathcal{G}_{11}^2 = 2$. Consequently, based on Theorem 1 n_2 has to be filtered out as n_2 share the same ancestor as n_3 (at level 2) which matches the negative RP. ∎

4.3   Evaluation of NOT-Twig Queries

We now discuss in detail how ancestor group identifiers are exploited for evaluating NOT-twig queries. As our focus is on not-predicates, for simplicity we assume that $edge(m_i, m_j)$ in a query is PC edge. Note that the proposed technique can easily support AD edges as discussed in Section 4.1.

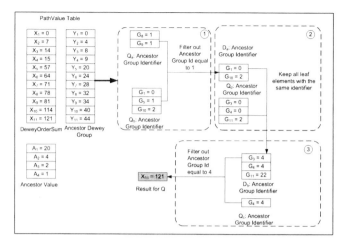

Fig. 4. Overview of NOT-twig evaluation

Consider the evaluation of the query Q in Figure 1(a)(ii) on the XML document in Figure 1(b). Figure 4 depicts a step-by-step evaluation of Q. In this example, we consider the fragment of the PathValue table in Figure 3 for illustration. Note that for clarity, in Figure 4 we only show DeweyOrderSums and AncestorDeweyGroups in the PathValue table. The DeweyOrderSum and AncestorDeweyGroup of each leaf element are denoted as X_i and Y_i, respectively, where i is the LeafOrder value of the element. First, Q is decomposed into the following normal rooted path patterns (without not-predicates). These paths are extracted from Q in left-to-right order and consists of all root-to-leaf paths in Q and the rightmost root-to-internal path representing the path after removing all qualifiers (Q_a: `/catalog/book/review`, Q_b: `/catalog/book/title`, Q_c: `/catalog/book/publisher/name/website`, Q_d: `/catalog/book/publisher/name`).

Evaluation order of RPs. If RPs are evaluated sequentially in left-to-right order ignoring the presence of negative RPs, then it will produce incorrect answers. Hence, we follow the following order. If the rooted path (say r) being evaluated is a negative RP then it is not evaluated immediately. On the other hand, if r is a normal RP, then it is evaluated immediately. First, elements matching r is evaluated with those that match the first preceding normal RP (if exists). Next, the elements will be evaluated with previously encountered negative RPs (if any) to filter out irrelevant elements. For example, in the aforementioned query Q_a is not immediately evaluated as it is a negative RP. Next, the normal RP Q_b is encountered. Since there does not exist any normal RP preceding Q_b, it is evaluated along with the negative RP Q_a. Next, the evaluation of the negative RP Q_c is skipped and normal RP Q_d is encountered. Since Q_b is the first preceding normal RP, Q_d is evaluated along with Q_b. Lastly, Q_d is evaluated in conjunction with the previously recorded negative RP Q_c. Hence, the order of evaluation of the above query is: Q_a and Q_b (results are represented as D_a), Q_d and D_a (results are represented as D_b), and D_b and Q_c.

Evaluation of RPs. In Step 1, the negative RP Q_a and normal RP Q_b are evaluated. Note that the NCA level of these RPs is 2. Since Q_a is a negative RP, all elements that satisfy Q_b but not Q_a are required. Therefore, we can directly select these elements using Theorem 1 for level 2. All elements in the results of Q_b that share same ancestor group identifiers with the results of Q_a are removed. Since $\mathcal{G}_8^2 = \mathcal{G}_9^2 = \mathcal{G}_5^2 = 1$, n_5 will be removed. Therefore, this step returns elements n_1 and n_{10} (denoted as D_a) as their ancestor group identifiers are not equal to 1. In Step 2, we compute the ancestor group identifiers of all elements satisfying D_a and Q_d and retrieve those elements that share the same identifiers. This results in the leaf elements n_3, n_4, and n_{11} (D_b). Finally, we process the previous negative RP Q_c. We now retrieve all leaf elements in D_b that are missing in Q_c using Theorem 1. Here $\ell = 4$ (**name** element). Observe that for D_b, $\mathcal{G}_3^4 = \mathcal{G}_4^4 = 4$ and $\mathcal{G}_{11}^4 = 22$. For Q_c, $\mathcal{G}_4^4 = 4$. Since ancestor group identifier of n_4 satisfying Q_c is identical to those of n_3 and n_4, we remove n_3 and n_4 from D_b (Step 3). Since there are no more rooted paths, the final result is n_{11}.

4.4   SQL Translation Algorithm

The Query Decomposition Phase. First, given a NOT-twig query Q, the SQL translation algorithm decomposes Q into a list of normal and negative rooted paths T. It extracts from Q the root-to-leaf paths and rightmost root-to-internal path (in absence of qualifiers), and store them into a list T in the following order. First, all root-to-leaf paths are inserted according to the left-to-right order of Q. Next, the root-to-internal path is added in T. The list also stores predicate information. We assume that T has a *size* method which returns the total number of RPs in T and a *countNotPred* method which returns the total number of negative RPs.

The SQL Generation Phase. This phase generates the SQL query S_{not} for retrieving elements that satisfy Q. This query only retrieves the LeafOrder values of the matching elements. The algorithm is shown in Algorithm 1. Given a set of rooted paths T of Q, the *generateSQLforNot* procedure outputs a SQL statement consisting of three clauses: *select_sql*, *from_sql*, and *where_sql*. In the sequel we assume that a clause has an *add()* method which encapsulates some simple string manipulations and simple joins for constructing valid SQL statements. Also, the *NCAlevel()* function computes the level of an NCA in Q. We preprocess the PathId and RValue to reduce the number of joins.

For each rooted path $r_i \in T$, the procedure first checks if it is a negative RP. Recall that a negative RP is not evaluated immediately. Specifically, all *consecutive* negative RPs are recorded (using the counter *cntNotPred*) until the next normal RP is encountered (Lines 03–04). When a normal RP r_i is encountered, it checks if it is a root-to-leaf path (Line 08). If it is then the algorithm generates the SQL fragment that retrieves the representative leaf elements by using instances of r_i's PathId and BranchOrder values (Line 09). Next, the algorithm generates statement for NCA computation of normal RPs in the following ways.

Algorithm 1. Algorithm *generateSQLforNot*.

Input: A list of normal and negative RPs T
Output: SQL query S_{not}

1 Initialize $cntNotPred = 0$;
2 **for** $(i = 1$ to $T.size())$ **do**
3 | **if** *(rooted path r_i is negative RP)* **then**
4 | | $cntNotPred++$;
5 | **else**
6 | from_sql.**add**("PathValue AS V_i");
7 | where_sql.**add**("V_i.PathId IN r_i.getPathId()");
8 | **if** $(i < T.size())$ **then**
9 | | where_sql.**add**("V_i.BranchOrder $< r_i$.**level**()");
10 | **if** $(i > 1$ and $cntNotPred = 0)$ **then**
11 | | where_sql.**add**("V_i.AncestorDeweyGroup/AncestorValue $(r_{i-1}$.**NCAlevel**() - 1)
           = V_{i-1}.AncestorDeweyGroup/AncestorValue $(r_{i-1}$.**NCAlevel**() - 1");
12 | **else**
13 | | set $x = cntNotPred$;
14 | | **while** $(x > 0)$ **do**
15 | | | where_sql.**add**("V_i.AncestorDeweyGroup/ AncestorValue
           $(r_{i-x}$.**NCAlevel**()-1) NOT IN (");
16 | | | where_sql.**add**(select_sql.**add**("V_{i-x}.AncestorDeweyGroup/
           AncestorValue$(r_{i-x}$.**NCAlevel**()-1)"));
17 | | | where_sql.**add**(from_sql.**add**("PathValue AS V_{i-x}"));
18 | | | where_sql.**add**(where_sql.**add**("V_{i-x}.PathId IN r_{i-x}.**getPathId**())"));
19 | | | where_sql.**add**(where_sql.**add**("V_{i-x}.BranchOrder $<r_i$.**level**())"));
20 | | | x- - ;
21 | | **if** $(i - cntNotPred > 1)$ **then**
22 | | | where_sql.**add**("V_i.AncestorDeweyGroup/ AncestorValue
           $(r_{i-cntNotPred-1}$.**NCAlevel**()-1) =
           $V_{i-cntNotPred-1}$.AncestorDeweyGroup/
           AncestorValue$(r_{i-cntNotPred-1}$.**NCAlevel**()-1)");
23 | | set $cntNotPred = 0$;

24 select_sql.**add**("DISTINCT V_i.DocID, V_i.LeafOrder");
25 **return** $S_{not} = $ select_sql $+$ from_sql $+$ where_sql;

- r_i *is the first* RP *in* T: Let r_1 $(i = 1)$ be a normal RP in T (without not-predicate). In this case, r_1 does not have any preceding RP. Then, r_1 will not be evaluated immediately (conditions in Lines 10 and 21 are not satisfied) as a pair of RPs is required for NCA evaluation (Theorem 1).
- r_i *is not the first* RP *in* T *and* $i > 1$: In this case, the algorithm may have encountered a normal RP r_j earlier $(j < i)$. Hence, if $countNotPred = 0$ it will execute Line 11 to generate the SQL statement to retrieve pairs of leaf elements that have NCA at the specified NCA level (based on Theorem 1). Otherwise, if $countNotPred > 0$ then the condition in Line 21 is true. Consequently, Line 22 will be used to generate the SQL fragment for NCA evaluation.

For all consecutive negative RPs, the procedure directly evaluates them using ancestor group identifiers (Lines 14-20). Specifically, Line 16 returns the ancestor group identifiers and Line 15 filters out elements based on Lemma 2. Note that the counter $cntNotPred$ will be reset to 0 whenever the procedure encounters a normal RP (Line 23).

The Final SQL Generation Phase. Finally, in this phase the final SQL query S for retrieving entire subtrees that match Q is generated. This procedure is

Algorithm 2. Algorithm *finalSQLGen*

Input: SQL query S_{not}, number of RPS x, number of negative RPS y
Output: SQL query S

1 order_sql.**add**("DocID, LeafOrder") ;
2 select_sql.**add**("V_{x+1}.LeafValue, ... V_{x+1}.LeafOrder");
3 from_sql.**add**("(" + S_{not} + ") AS V_x INNER JOIN PathValue V_{x+1} ON V_{x+1}.DocID = V_x.DocID AND V_{x+1}.LeafOrder = V_x.LeafOrder");
4 where_sql.**add**("V_{x+1}.PathID IN r_x.**getPathID()**");
5 **if** $(x - y > 1)$ **then**
6 | option_sql.**add**("FORCE ORDER, ORDER GROUP");
7 **else**
8 | option_sql.**add**("ORDER GROUP");
9 **return** $S = select\_sql + from\_sql + where\_sql + order\_sql + option\_sql$;

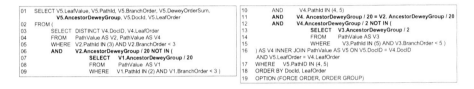

```
01  SELECT V5.LeafValue, V5.PathId, V5.BranchOrder, V5.DeweyOrderSum,     10        AND       V4.PathId IN (4, 5)
        V5.AncestorDeweyGroup, V5.DocId, V5.LeafOrder                     11        AND       V4. AncestorDeweyGroup / 20 = V2. AncestorDeweyGroup / 20
02  FROM (                                                                12        AND       V4.AncestorDeweyGroup / 2 NOT IN (
03      SELECT   DISTINCT V4.DocId, V4.LeafOrder                          13            SELECT    V3.AncestorDeweyGroup / 2
04      FROM     PathValue AS V2, PathValue AS V4                         14            FROM      PathValue AS V3
05      WHERE    V2.PathId IN (3) AND V2.BranchOrder < 3                  15            WHERE     V3.PathId IN (5) AND V3.BranchOrder < 5 )
06      AND      V2.AncestorDeweyGroup / 20 NOT IN (                      16    ) AS V4 INNER JOIN PathValue AS V5 ON V5.DocId = V4.DocId
07              SELECT    V1.AncestorDeweyGroup / 20                            AND V5.LeafOrder = V4.LeafOrder
08              FROM      PathValue  AS V1                                17  WHERE    V5.PathId IN (4, 5)
09              WHERE     V1.PathId IN (2) AND V1.BranchOrder < 3 )       18  ORDER BY DocId, LeafOrder
                                                                         19  OPTION (FORCE ORDER, ORDER GROUP)
```

Fig. 5. Translated SQL query

outlined in Algorithm 2 and contains five clauses: $select\_sql$, $from\_sql$, $where\_sql$, $order\_sql$, and $option\_sql$. It includes an addition instance of PathValue V_{x+1} which uses the same path in the PathValue table V_x representing the rightmost root-to-internal path in S_{not} (Line 04). V_{x+1} is joined on DocID and LeafOrder attributes with V_x to retrieve entire subtrees of matched elements (Line 03). Since the results must be in document order, the tuples are sorted according to DocID and LeafOrder attributes using the $order\_sql$ clause (Line 01). Lastly, the option clause ($option\_sql$) is used to enforce the **distinct** and **order by** operations to use sort operator using the ORDER GROUP query hint (Lines 05 - 08). Also, if there exists at least one normal root-to-leaf path in Q then FORCE ORDER hint is used to enforce a "left-to-right" join order on the translated SQL query (Line 06). The performance benefits of join order enforcement is highlighted in [4, 7, 11]. Note that the translated SQL has at least one instance of PathValue table representing the normal root-to-internal path. Further, if all root-to-leaf paths in Q are negative RPS, then join order enforcement is discarded as these paths will be evaluated by subqueries (generated by Lines 15–19 in Algorithm 1).

Reconsider the query Q in Section 4.3. The list of root-to-leaf and root-to-internal paths T is: $[r_1 = Q_a, r_2 = Q_b, r_3 = Q_c, r_4 = Q_d]$. The translated SQL is shown in Figure 5. The reader may refer to [12] for details related to this example.

5   Performance Study

In this section, we present the experiments conducted to evaluate the performance of our proposed approach and report some of the results obtained. A

(a) XBench query sets

Id	Query
Q1	/catalog/item/publisher/contact_information[not(website) and phone_number]
Q2	/catalog/item/publisher[not(contact_information/website) and contact_information/phone_number]/name
Q3	/catalog/item[not(subject) and not(description)]/title
Q4	/catalog/item/authors/author[not(biography)]/name
Q5	/catalog/item[not(description) and not(subject)]/authors/author[not(biography)]/contact_information[not(email_address)]/mailing_address
Q6	/catalog/item[pricing/quantity_in_stock]/authors/author[not(biography) and contact_information/mailing_address/name_of_country]/name[not(middle_name)]/first_name
Q7	/catalog/item[date_of_release]/pricing[not(when_is_available)]/cost
Q8	/catalog/item[not(media) and not(pricing/quantity_in_stock)]/publisher[not(contact_information/FAX_number)]
Q9	/catalog/item[title and not(subject) and not(publisher/contact_information/website)]/authors/author[biography]/date_of_birth
Q10	/catalog/item[not(media) and not(subject)]/publisher[contact_information/phone_number]/name
Q11	/catalog/item[not(description)]/authors/author[not(contact_information/email_address)]/biography
Q12	/catalog/item[not(media) and not(attributes)]/publisher/contact_information[not(website) and not(FAX_number)]

(b) UniProt query sets

Id	Query	No. of matching subtrees		
		U28	U283	U2843
UQ1	/uniprot/entry[not(geneLocation/name) and not(comment/location)]	3212	35277	354813
UQ2	/uniprot/entry[not(organismHost) and not(evidence) and geneLocation]	157	1622	16392
UQ3	/uniprot/entry[not(gene)]/protein[not(component) and not(domain)]	189	1626	16479
UQ4	/uniprot/entry[not(geneLocation) and not(protein/domain)]/comment[not(note) and not(event)]	13835	140196	1403859
UQ5	//comment[not(event)]/following::text	14620	145799	1461240
UQ6	//comment[not(note) and not(event)]/preceding::text	14620	145799	1461240
UQ7	/uniprot/entry/comment[not(note) and not(event)]/preceding::component	160	1576	15599
UQ8	/uniprot/entry[not(geneLocation)]/protein/following::component	158	1574	15597
UQ9	/uniprot/entry/protein[not(component)]//domain//name	185	1433	13914

Fig. 6. Query sets

more detailed results is available in [12]. Prototype for our ancestor group-based approach (denoted by AG-SX) was implemented by extending SUCXENT++ using Java JDK 1.6. The experiments were conducted on an Intel Pentium IV 3GHz machine running on Windows XP Service Pack 2 with 2GB RAM. The RDBMS used was MS SQL Server 2008 Developer Edition.

We are not aware of any existing tree-unaware approaches that have undertaken a systematic study to evaluate NOT-twig queries. Hence, we compared our approach to the native XML supports of *XSysA* and *XSysB* (Recall from Section 1). For all these approaches appropriate indexes were created. Prior to our experiments, we ensure that statistics had been collected. The bufferpool of the RDBMS was cleared before each run. The queries in AG-SX were executed in the *reconstruct* mode [13] where not only the internal elements are selected, but also all descendants of those elements. Each query was executed 6 times and the results from the first run were always discarded. All rows were fetched from the answer set; however, they were not sent to output. Note that we did not select $TwigStackList\neg$ [16] and $NJoin$ [10] as we were unable to get the implementation from the authors. However, an intuitive comparison with these approaches is discussed later.

Datasets. We use XBench DCSD [15] shown in Figure 2(a) as synthetic data set. We also modified the data set so that we can control the number of subtrees (denoted as K) that matches a NOT-twig query and the number of instances of the rooted paths in the XML document. We set $K \in \{100, 500\}$. We use the UNIPROT dataset shown in Figure 2(b) as real-world data set. Since the original UNIPROT data is 2.8GB in size (denoted as U2843), we also truncated this document into smaller XML documents of sizes 28MB and 284MB (denoted as U28 and U284, respectively) to study scalability of various systems.

Querysets. Figure 6 depicts the benchmark queries. As our primary objective is to assess the performance of not-predicates evaluation, we choose two categories of queries. In the first category ($Q1 - Q12$ and $UQ1 - UQ4$), we fix the XPath axis in the twigs to *child* and generate queries by varying the number of

Id	DC10			DC100			DC1000		
	AG-SX	XSysA	XSysB	AG-SX	XSysA	XSysB	AG-SX	XSysA	XSysB
Q1	111	346	205	377	588	1,761	3,022	3,303	48,288
Q2	107	475	197	183	806	2,075	1,273	3,359	47,208
Q3	249	595	178	517	952	1,712	1,363	5,394	42,980
Q4	223	574	331	495	1,585	3,082	3,385	12,348	59,087
Q5	362	2,501	200	925	3,055	1,564	6,804	10,879	44,289
Q6	256	3,038	631	624	3,950	5,889	4,205	14,648	93,219
Q7	166	1,119	306	376	1,678	2,789	1,742	6,283	49,235
Q8	279	1,228	290	684	1,620	2,235	4,816	7,209	47,209
Q9	179	1,436	324	631	2,240	2,822	5,313	9,331	63,980
Q10	135	1,642	230	268	2,357	1,962	1,669	8,910	50,842
Q11	188	1,008	206	318	1,559	1,708	2,709	9,748	60,109
Q12	382	972	231	842	1,366	2,026	4,997	6,921	46,924

(a) K=100

Id	DC10			DC100			DC1000		
	AG-SX	XSysA	XSysB	AG-SX	XSysA	XSysB	AG-SX	XSysA	XSysB
Q1	124	883	249	452	1,165	1,796	3,148	4,156	47,556
Q2	104	964	248	209	1,325	1,843	1,378	4,244	45,871
Q3	232	1,576	213	392	1,732	1,601	1,401	5,643	45,209
Q4	184	1,439	334	475	2,332	2,823	3,790	11,298	57,309
Q5	346	3,351	296	908	4,000	1,791	7,206	12,134	42,109
Q6	269	4,022	686	641	4,712	5,767	4,543	16,180	94,770
Q7	162	1,967	321	410	2,832	2,777	1,806	7,282	50,259
Q8	297	2,247	241	727	2,691	2,163	4,705	8,937	46,295
Q9	171	2,175	479	655	3,369	2,694	4,953	10,153	63,820
Q10	154	2,386	280	275	3,431	2,024	1,641	9,975	51,872
Q11	212	1,719	288	331	2,593	2,001	2,711	10,887	59,952
Q12	404	1,877	270	867	2,490	2,013	5,009	7,424	49,825

(b) K= 500

Id	U28			U284			U2843
	AG-SX	XSysA	XSysB	AG-SX	XSysA	XSysB	AG-SX
UQ1	1,530	20,399	2,457	12,809	207,982	24,429	128,208
UQ2	1,130	1,541	450	21,740	11,279	4,391	86,816
UQ3	494	2,299	591	485	15,038	5,342	3,302
UQ4	505	5,899	1,977	1,210	47,386	20,819	6,681
UQ5	1,604	NS	NS	8,878	NS	NS	86,555
UQ6	1,829	NS	NS	12,488	NS	NS	113,547
UQ7	815	NS	NS	3,615	NS	NS	31,310
UQ8	659	NS	NS	2,896	NS	NS	26,265
UQ9	2,040	4,965	3,988	9,420	65,624	34,716	124,900

(c) Query performance on UniProt data sets

Fig. 7. Query evaluation times of AG-SX, *XSysA*, and *XSysB* (in msec.)

normal and negative rooted paths, number of NCA nodes, and structure of twigs. For instance, $Q3 - Q5$, $Q8$, $Q11$, $Q12$, $UQ1$, $UQ3$, and $UQ4$ are queries with purely not-predicates while the remaining queries contain a mixture of normal and negative rooted paths. The number of instances of root-to-leaf paths that matches the query set varies between 150 and 2,035,889. In the second category ($UQ5$-$UQ9$), we include different XPath axes (e.g., descendant, following, preceding) in the NOT-twigs to study the performance of these queries in the presence of such axes.

5.1   Query Evaluation Times

Figure 7 depicts the NOT-twig query evaluation times. As *XSysA* and *XSysB* are unable to handle XML documents having size larger than 2GB, no query evaluation times are reported for these approaches on U2843 data set. Also, as AG-SX is orders of magnitude faster than the not-predicate evaluation approach on the original SUCXENT++ (see [12]), we only report query evaluation times of AG-SX. The symbol NS in Figure 7 denotes that the query is not currently supported in the current version of a particular system.

We observe that AG-SX significantly outperforms both *XSysA* and *XSysB* for majority of the queries (highest observed factors being 37 and 40, respectively). As the data size increases, the performance gap between AG-SX and these approaches increases. Particularly, we noticed that except for $Q5$, our proposed approach is at least 9 times faster than *XSysB* for all values of K. For the real-world data sets (U28 and U284), AG-SX is faster than *XSysA* and *XSysB* for 90% and 80% of the benchmark queries, respectively. In summary, AG-SX outperforms *XSysA* and *XSysB* primarily due to the effectiveness of the former approach to generate a relatively simple SQL statement, which exploits ancestor group identifiers to efficiently evaluate common ancestors and not-predicates using the equality property (Theorem 1). Also, interestingly *XSysA* is less efficient than *XSysB* for smaller data sets (DC10 and DC100). However, it is faster than *XSysB* for DC1000.

Comparison with $TwigStackList\neg$ **[16] and** $NJoin$ **[10]:** Based on the results reported in [10, 16] we can make the following observations. For a data set of size 100MB and less than 2.5 million nodes, the average running time of benchmark NOT-twig queries using $TwigStackList\neg$ is $15 - 30s$ [16] whereas majority of our queries on similar data sets take less than a second to retrieve results. In [10], it is shown that $NJoin$ is 2-3 faster than $TwigStackList\neg$ for simple NOT-twig queries. Based on this observation, we expect AG-SX to outperform these approaches.

6　Conclusions and Future Work

In this paper, we present an efficient strategy to evaluate NOT-twig queries in a tree-unaware relational environment. We extended the encoding scheme of dewey-based SUCXENT++ [11] by adding two new labels, namely AncestorValue and AncestorDeweyGroup, that enable us to efficiently filter out elements satisfying a not-predicate by comparing their ancestor group identifiers. We proposed a novel NOT-twig query evaluation algorithm that reduce useless structural comparisons by exploiting these labels. Our results showed that the our proposed approach have superior performance compared to existing state-of-the-art tree-unaware and native approaches. In future, we plan to investigate if some of the optimization techniques proposed in [4] (e.g., choosing right join algorithms, eliminating redundant ordering (if any)) are beneficial for evaluating NOT-twig queries in our proposed framework.

References

1. Al-Khalifa, A., Jagadish, H.V.: Multi-level Operator Combination in XML Query Processing. In: ACM CIKM (2002)
2. Bhowmick, S.S., Leonardi, E., Sun, H.: Efficient Evaluation of High-Selective XML Twig Patterns with Parent Child Edges in Tree-Unaware RDBMS. In: ACM CIKM (2007)
3. Boncz, P., Grust, T., et al.: MonetDB/XQuery: A Fast XQuery Processor Powered by a Relational Engine. In: SIGMOD (2006)
4. Georgiadis, H., Vassalos, V.: Xpath on Steroids: Exploiting Relational Engines for Xpath Performance. In: SIGMOD (2007)
5. Georgiadis, H., et al.: Cost-based Plan Selection for XPath. In: SIGMOD (2009)
6. Gou, G., Chirkova, R.: Efficiently Querying Large XML Data Repositories: A Survey. IEEE TKDE 19(10) (2007)
7. Grust, T., et al.: Why Off-the-Shelf RDBMSs are Better at XPath Than You Might Expect. In: SIGMOD (2007)
8. Jiao, E., Ling, T.-W., Chan, C.-Y.: PathStack: A Holistic Path Join Algorithm for Path Query with Not-Predicates on XML Data. In: Zhou, L.-z., Ooi, B.-C., Meng, X. (eds.) DASFAA 2005. LNCS, vol. 3453, pp. 113–124. Springer, Heidelberg (2005)
9. Li, H., Li Lee, M., Hsu, W.: A Path-Based Labeling Scheme for Efficient Structural Join. In: Bressan, S., Ceri, S., Hunt, E., Ives, Z.G., Bellahsène, Z., Rys, M., Unland, R. (eds.) XSym 2005. LNCS, vol. 3671, pp. 34–48. Springer, Heidelberg (2005)

10. Li, H., Lee, M.-L., et al.: A Path-Based Approach for Efficient Structural Join with Not-Predicates. In: Kotagiri, R., Radha Krishna, P., Mohania, M., Nantajeewarawat, E. (eds.) DASFAA 2007. LNCS, vol. 4443, pp. 31–42. Springer, Heidelberg (2007)
11. Seah, B.-S., Widjanarko, K.G., et al.: Efficient Support for Ordered XPath Processing in Tree-Unaware Commercial Relational Databases. In: Kotagiri, R., Radha Krishna, P., Mohania, M., Nantajeewarawat, E. (eds.) DASFAA 2007. LNCS, vol. 4443, pp. 793–806. Springer, Heidelberg (2007)
12. Soh, K.-H., Bhowmick, S.S.: Efficient Evaluation of not-Twig Queries in A Tree-Unaware RDBMS. Technical Report (December 2009), `http://www.cais.ntu.edu.sg/~assourav/TechReports/NotTwig-TR.pdf`
13. Tatarinov, I., Viglas, S., et al.: Storing and Querying Ordered XML Using a Relational Database System. In: SIGMOD (2002)
14. Yoshikawa, M., et al.: XRel: A Path-based Approach to Storage and Retrieval of XML documents Using Relational Databases. ACM TOIT 1(1) (2001)
15. Yao, B., Tamer Özsu, M., Khandelwal, N.: XBench: Benchmark and Performance Testing of XML DBMSs. In: ICDE (2004)
16. Yu, T., Ling, T.-W., Lu, J.: TwigStackList¬: A Holistic Twig Join Algorithm for Twig Query with Not-Predicates on XML Data. In: Li Lee, M., Tan, K.-L., Wuwongse, V. (eds.) DASFAA 2006. LNCS, vol. 3882, pp. 249–263. Springer, Heidelberg (2006)

A Hybrid Algorithm for Finding Top-k Twig Answers in Probabilistic XML

Bo Ning[1] and Chengfei Liu[2]

[1] Dalian Maritime University, Dalian, China
`ningbo@dlmu.edu.cn`
[2] Swinburne University of Technology, Melbourne, Australia
`CLiu@groupwise.swin.edu.au`

Abstract. Uncertainty is inherently ubiquitous in data of real applications, and those uncertain data can be naturally represented by the XML. Matching twig pattern against XML data is a core problem, and on the background of probabilistic XML, each twig answer has a probabilistic value because of the uncertainty of data. The twig answers that have small probabilistic values are useless to the users, and the users only want to get the answers with the largest k probabilistic values. In this paper, we address the problem of finding twig answers with top-k probabilistic values against probabilistic XML documents directly. To cope with this problem, we propose a hybrid algorithm which takes both the probability value constraint and structural relationship constraint into account. The main idea of the algorithm is that the element with larger path probability value will more likely contribute to the twig answers with larger twig probability values, and at the same time lots of useless answers that do not satisfy the structural constraint can be filtered. Therefore the proposed algorithm can avoid lots of intermediate results, and find the top-k answers quickly. Experiments have been conducted to study the performance of the algorithm.

1 Introduction

Nowadays, uncertainty is inherently ubiquitous in data of real applications. For instance, in sensor applications, sensors produce uncertain data since readings of sensors are inherently imprecise. In scientific research, error-prone experimental machinery, polluted samples, and simple human error bring the uncertainty to experimental data. Therefore uncertain data management is becoming a critical issue. The current relational database technologies can not deal with this problem very well, because to store imprecise information in structured data format can lead to high complexity of space and processing time. While the XML data is a natural representation of uncertain data due to its flexible characteristics. XML has hierarchical structure, therefore the probability values can be assigned to elements and subtrees, dependency and independency of elements can be expressed. In addition, XML supports incomplete information gracefully. The data models for representing uncertainty in XML have been studied in [1–6].

J.X. Yu, M.H. Kim, and R. Unland (Eds.): DASFAA 2011, Part I, LNCS 6587, pp. 528–542, 2011.

As to the query processing on probabilistic XML, the queries on the probabilistic XML are often in the form of twig patterns. Compared with the query on ordinary XML, the matched answers are associated with the probabilistic values when querying probabilistic data. Therefore the answers as well as the probability values need to be returned. Many kinds of twig queries with different semantics were proposed, and their evaluations were studied in [7]. It is obvious that the answers with small probabilistic values are useless to users who submit the queries, and it makes sense to only return the twig answers with top-k probabilistic values. There are also some other works on querying uncertain data. On uncertain relational data, matching the twig answers with probability values above a threshold was investigated in [8], and query ranking was studied in [9, 11, 12]. The paper [10] studied the query ranking in probabilistic XML by possible world model, and a dynamic programming approach was deployed that extends the approach in [9] to deal with the containment relationships in probabilistic XML, and ranks the results by the interplay between score and uncertainty. Those works are based on the documents generated from probabilistic XML or the relational data model in which the possible worlds are stored.

It is more flexible if the twig answers with top-k probabilistic values can be matched against the probabilistic XML directly. The algorithm *ProTJFast* and *PTopKTwig*[13] belong to this catalog. In those algorithms, by the use of a novel encoding scheme and the effective use of lower bounds, elements or paths with small probabilities can be filtered. Matching a twig query against an ordinary XML document only needs the answer to satisfy the structural relationship constraint, while finding top-k twig answers against probabilistic XML also needs the answers to satisfy the constraint that the probabilistic values of twig answers are k largest ones. The algorithm *ProTJFast* uses element streams ordered by document order (pre-order) as input, and the process of algorithm follows the document order. This may cause the constraint of probabilistic values not met as soon as possible. The algorithm *PTopKTwig* is based on the element streams ordered by path probabilistic value, and does not consider the structural constraint too much, therefore, to satisfy the structural constraint, there are lots of times of detection whether the elements of leaf nodes in query can be matched to be an twig answer. Although the use of enhanced lower bound makes algorithm *PTopKTwig* efficient, there are still lots of useless path answers included in the candidate twig answers that will not contribute to the final top-k answers.

In this paper, we also address the problem of efficiently finding top-k twig answers against probabilistic XML directly. Our algorithm takes both of the structural constraint and probabilistic value constraint into account, and can find the k twig answers which satisfy the structural relationships and their probabilistic values are largest as early as possible. In our algorithm, the intermediate path answers which do not satisfy the structural constraint and probabilistic value constraint can be filtered rapidly. Also we improve the encoding scheme that makes the process of calculating the probabilistic values more efficiently.

The rest of this paper is organized as follows. Section 2 introduces the background and relate work including the data model twig answers and encoding scheme of probabilistic XML. In Section 3, we improve the encoding scheme by redesigning the float vector. In Section 4, we present a hybrid algorithm *HyTop-KTwig* for matching twig answers with top-k probabilities. Section 5 shows our experimental results. Conclusions are included in Section 6.

2   Preliminaries

2.1   Probabilistic XML Model

Nierman et al. proposed the Probabilistic Tree Data Base(ProTDB) [4] to manage uncertain data represented in XML. Actually it belongs to the catalog of $PrXML^{\{ind,\ mux\}}$ model [6], in which the independent distribution and mutually-exclusive distribution are considered.

A probabilistic XML document T_P defines a probability distribution over an XML tree $T(V, E)$ and it can be regarded as a weighted XML tree $T_P(V_P, E_P)$. In T_P, $V_P = V_D \cup V$, where V is a set of ordinary elements that appear in T, and V_D is a set of distribution nodes, including independent nodes and mutually-exclusive nodes (*ind* and *mux* for short). An ordinary element, $u \in V_P$, may have different types of distribution nodes as its child elements in T_P that specify the probability distributions over its child elements in T. E_P is a set of edges, and an edge which starts from a distribution node can be associated a positive probability value as weight.

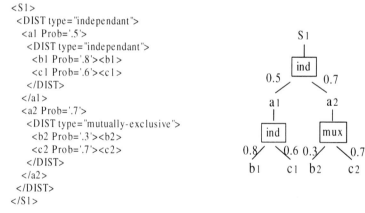

```
<S1>
  <DIST type="independant">
    <a1 Prob='.5'>
      <DIST type="independant">
        <b1 Prob='.8'><b1>
        <c1 Prob='.6'><c1>
      </DIST>
    </a1>
    <a2 Prob='.7'>
      <DIST type="mutually-exclusive">
        <b2 Prob='.3'><b2>
        <c2 Prob='.7'><c2>
      </DIST>
    </a2>
  </DIST>
</S1>
```

(a) Fragment of probabilistic XML   (b) Tree model of probabilistic XML

Fig. 1. Example of probabilistic XML

Figure 1 (a) shows an fragment of probabilistic XML document. By using the tag *DIST*, *Prob* and *VAL*, the XML has the ability to express the probabilistic distributions and the probabilistic values of elements. Figure 1 (b) is the tree model for the probabilistic XML fragment, which contains *ind* and *mux* nodes. The element a_1 has an *ind* node as its child, which specifies that its twig child nodes, b_1 and c_1 are independent. The probabilities of having b_1 and c_1 are 0.8 and 0.6, as indicated in the incoming edges to b_1 and c_1 respectively. The element a_2 has a mux node as its child, which specifies that b_2 and c_2 cannot appear as a_2's child at the same time. Because of the mutually-exclusive distribution, the sum of probability values of b_2 and c_2 cannot be larger than one.

2.2   Twig Query and Answers

A twig query q is an XPath query with predicates, and it can be modeled as a small tree $T_q(V_q, E_q)$, where V_q is a set of nodes representing types(tag name) and E_q is a set of edges. There are two kinds of edges in E_q including parent-child edge (PC for short) and ancestor-descendant edges (AD for short). Usually, AD edge corresponds to the descendant axis in XPath, and PC edge corresponds to the child axis. The answer of a twig query is a set of tuples, in which there are elements from the probabilistic XML, and those elements match the nodes in q and satisfy all the structural relationships specified in q.

However, to find the answers of a twig query q against a probabilistic XML document, not only the structural relationships specified in q have to be satisfied, but also the types of distribution nodes and weights of edges from distribution nodes have to be considered. Actually, a mux distribution node u_{mux} can be regarded as constraint to restrict two elements under different child elements of u_{mux} so that they do not contribute to the same result. In contrast, an ind distribution node does not affect the existence of an answer, but determine the probability value of the answer.

Given an answer expressed by a tuple $t = (e_1, e_2, ..., e_n)$, there exist a subtree $T_s(V_s, E_s)$ of T_P, which contains all those elements. The probability of t can be computed by the following equation.

$$prob(t) = prob(T_s) = \Pi_{e_i \in E_s} prob(e_i);$$

2.3   Encoding Scheme

For encoding an ordinary XML, the encoding scheme should support the structural relationships and keep the document order, because matching a twig query against ordinary XML document only needs the answer to satisfy the structural relationships constraint. However the encoding scheme for ordinary XML can not meet the requirements of probabilistic XML, as there are new characteristics of probabilistic XML including the distribution nodes and the probabilistic values. Therefore encoding scheme for probabilistic XML should contain the information of probabilistic values and provides the ability for matching twig answers that satisfy the probabilistic value constraint.

Region-based encoding[14, 15] and prefix encoding are two kinds of encoding schemes for ordinary XML documents. Both encoding schemes support the structural relationships and keep the document order, and these two requirements are essential for evaluating queries against ordinary XML documents. As to the aspect of encoding elements in probabilistic XML, a new requirement needs to be met, that is how to record the probability values of elements on different levels. Depending on different kinds of processing, the probability value of the current element which is under a distribution node in PXML (node-prob for short) needs to be recorded, and so does the probability value of the path from the root to the current element (path-prob for short). Ning et al.[13] conclude that both node-prob and path-prob in twig pattern matching against PXML are needed, and property for supporting ancestor vision and ancestor probability vision are also needed. The prefix encoding scheme naturally have the properties above, therefore it is better to encode PXML elements by a prefix encoding scheme.

For efficiently processing twig matching against ordinary XML, Lu et al. proposed a prefix encoding scheme named extended Dewey [16]. Extended Dewey is a kind of Dewey encoding, which uses the modulus operation to create a mapping from an integer to an element name, so that given a sequence of integers, it can be converted into the sequence of element names. This characteristic provides extended Dewey encoding the tag name vision of ancestor, that makes the evaluation of twig join efficiently.

For the purpose of supporting twig pattern matching against probabilistic XML, Ning et al.[13] extend Lu's encoding scheme[16] by adding the properties of the *probability vision* and the *ancestor probability vision*, and propose a new encoding scheme called *PEDewey*. Compared with Extended Dewey, *PEDewey* takes the distribution node into account and assigns a float vector to each element, which records the probabilistic value information.

3   Improvement of the *PEDewey* Encoding

In this paper, we improve the float vector part of *PEDewey* for the efficient calculation of twig answers' probabilistic values.

In *PEDewey*, an additional float vector is assigned to each element. The length of the vector is equal to that of a normal Dewey encoding, and each component holds the node probability value of the element. Given the vector v and each component $node_i$ in v, the path probability value can be calculated by the following equation:

$$prob(path) = \Pi_{node_i \in v} prob(node_i);$$

We can see from the equation that there are lots of multiplication operations for calculating a path probability value, and it is not efficient, so we improve the float vector by recording the natural logarithm of probability value in each component. After that, the path probability value can be calculated by equation:

$$prob_{ln}(path) = \Sigma_{node_i \in v} prob_{ln}(node_i);$$

During the processing of finding top-k twig answers, the probability values are in the form of natural logarithm, and when the final answers are found, we calculate the probability value by the equation $prob(t) = e^{prob_{ln}(t)}$. Notice that the components for elements of ordinary, *ind* and *mux* are all assigned to 0 which is the natural logarithm of 1.

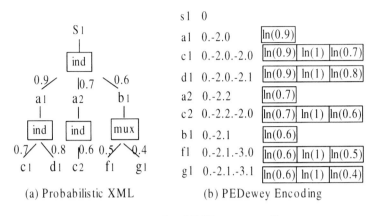

Fig. 2. Example of *PEDewey* encoding

We redefine the operations on the float vector. (1) Given element e, function $pathProb_{ln}(e)$ returns the path-prob of element e in natural logarithm form, which is calculated by adding the node-prob values of all ancestors of e in the float vector (2) Given element e and its ancestor e_a, function $ancPathProb_{ln}(e, e_a)$ returns the path-prob of e_a in natural logarithm form by adding those components from the root to e_a in e's float vector. (3) Given element e and its ancestor e_a, function $leafPathProb_{ln}(e, e_a)$ returns the path-prob of the path from e_a to e in form of natural logarithm by adding those components from e_a to e in e's float vector. (4) Given elements e_i and e_j, function $twigProb_{ln}(e_i, e_j)$ returns the probability of the twig whose leaves are e_i and e_j. Assume e_i and e_j have common prefix e_c, and the probability of twig answer containing e_i and e_j is:

$$twigProb_{ln}(e_i, e_j) = pathProb_{ln}(e_i) + pathProb_{ln}(e_j) - ancPathProb_{ln}(e_i, e_c);$$

For the PXML shown in Figure 2(a), the encodings of its elements are shown in Figure 2(b). For example, $pathProb_{ln}(c_1)$=-0.462 (0+(-0.10536)+0+(-0.35667)), $ancPathProb_{ln}(c_1, a_1)$= -0.10536 (0+(-0.10536)+0), $leafPathProb_{ln}$ (c_1, a_1) = -0.35667(0+(-0.35667)) and $twigProb_{ln}(c_1, d_1)$=$pathProb_{ln}(c_1)$+$pathProb_{ln}(d_1)$ - $ancPathProb_{ln}(c_1, a_1)$ = -0.6852, where a_1 is the common prefix of c_1 and d_1. At last we can get the probability value of this twig answer is 0.504 ($e^{-0.6852}$).

4   *HyTopKTwig*: A Hybrid Algorithm

In this part, we propose an algorithm for finding top-k twig answers against the probabilistic XML directly. Firstly, we analyze the characteristics of the problem that finding top-k twig answers, then we propose the algorithm. At last, we discuss the correctness of our algorithm.

4.1   Analysis of the Problem

When matching twig pattern against the ordinary XML documents, the only consideration is the structural relationships of query. That means the answers only need to satisfy the structural relationships of twig pattern. However the problem we are going to deal with is to find not only the answers that match the twig pattern, but also the k answers that have the largest probability values among all the twig answers. Therefore how to find top-k answers quickly without large amount of useless intermediate results is the challenge of the problem.

Most of the algorithms for matching twig pattern against the ordinary XML use elements streams ordered by the document order as the input data. Although those algorithms can be easily adjusted to solve the top-k answers problem against probabilistic XML, the efficiency is very low. That is because all the twig answers need to be found out no matter how small their probabilistic values are. Many useless intermediate results are computed, and the algorithm *ProTJFast* is in this case. The algorithm *ProTJFast* uses the elements streams ordered by document order, and the document order is good for matching the structural relationships, and makes the twig matching algorithm holistic. However, at the background of our problem, the document order limits the efficiency.

Intuitively the element with larger path probability value will more likely contribute to the twig answers with larger twig probability values. Based on this idea, algorithm using the elements streams ordered by probability values is proposed, for example the algorithm *PTopKTwig*, to deal with the top-k matching of twig queries against probabilistic XML. The algorithm *PTopKTwig* mainly takes the probability value order into account, and ignores the characteristics of the structural relationships constraints. It needs to compare whether the two leaf elements of leaf nodes in query can be joined to contribute to a final answers lots of times. Although the use of enhanced lower bound makes algorithm *PTopKTwig* efficient, without the documents order, the merge-joins can not be performed, that leads to the low efficiency because lots of comparisons can not be avoided.

From above we can see that there are structural relationships constraint and largest probabilistic values constraint in our problem, and it is a tradeoff between finding top-k probability values and satisfying the structural relationships. So we intend to design a hybrid algorithm which takes both constraints into account, and in our algorithm, the intermediate path answers which do not satisfy the structural constraint and probabilistic value constraint can be filtered rapidly.

4.2   Data Structures and Notations

Firstly we give the definition of skeleton pattern, which is the key point to balance the tradeoff between finding top-k probability values and satisfying the structural relationships.

Definition 1. *The skeleton pattern s is a subtree of twig pattern q, and it can be obtained by deleting all the subtrees of twig pattern which do not contain any branching node.*

For example, the skeleton pattern of A[//E]//B[//D]//C is A//B.

In our algorithm, we associate each leaf node f in a twig pattern q with a stream T_f, which contains encoding of all elements that match the leaf node f.

The elements in the stream are sorted by their path-prob values. It is very fast to sort those elements by using the float vector in our encodings. Notice that in our encoding scheme, the component of float vector is in the form of natural logarithm, so the order of those natural logarithm floating numbers should be ascending, so that the order of real probability values is descending. We maintain *cursorList*, a list pointing to the head elements of all leaf node streams. Using the function cursor(f), we can get the position of the head element in T_f.

A list L_c for keeping top-k candidates is allocated for q. A set S_b is associated with skeleton pattern of query q. Each element cached in S_b are the skeleton part of the candidate answers. Initially set S_b is empty. For each element stream, we assign a signature list. The signature for an element represents whether the current element is a descendant of any skeleton result in set S_b. Initially, all the signatures are zero.

4.3   Algorithm *HyTopKTwig*

There are three phases in main algorithm of *HyTopKTwig*(Algorithm 1). The first phase (Lines 2-6) is to find the initial answers so that we can have k skeleton results in S_b. In the second phase(Lines 7-10), we use the signature list to identify the descendants of skeleton results in set S_b in corresponding element streams of leaf node in query q. The function of signature list is to filter the elements that can not contribute to the final answers. So in third phase(Line 11), we run algorithm *ProTJFast* against the document ordered element streams where useless elements have been filtered in phase two.

In the first phase, we intend to find the answers whose probability values are as large as possible, so we use the element streams ordered by probability value. The processing of this phase is just like that of algorithm *PTopKTwig*. Algorithm 1 firstly proceeds in the probability order of all the leaf nodes in query q, by calling the function *getNextP()*. This function returns the tag name of the leaf node stream which has the biggest probability value in its head element among all leaf node streams. So that, each processed element will not be processed repetitively. After function *getNextP()* returns a tag q_{act}, we may find the twig answers which the head element in stream $T_{q_{act}}$ contributes to, by invoking function *matchTwig()*. In function *matchTwig()*, the twig answers containing $e_{q_{act}}$ can all be found. During the process of finding other elements that contribute to the twig answers with $e_{q_{act}}$, there is no duplicated computation of comparing the prefixes, due to the order of probability values and the use of *cursorList*. The *cursorList* records the head elements in respective streams which is next to be processed. The elements before the head elements have been compared with elements in other streams, and the twig answers that these elements might contribute to have been considered. Therefore we only compare $e_{q_{act}}$ with the elements after the head elements in the related streams (Lines 3-4 in Algorithm 2). Once a twig answer are found, we add the skeleton result of this answer to the set S_b, until the size of the S_b equals to k (Line 5 in Algorithm 1). So we can see that the task of phase 1 is to find k skeleton results.

Algorithm 1. $HyTopkTwig(q)$

Data: Twig query q, and streams T_f of the leaf node in q.
Result: The matchings of twig pattern q with top-k probabilities.

```
 1 begin
 2 │   while Sizeof(S_b) < k ∧ ¬ end(q) do
 3 │   │    q_act=getNextP(q);
 4 │   │    tempTwigResults=matchTwig(q_act,q);
 5 │   │    addSkeletonResults(S_b,tempTwigResults);
 6 │   │    advanceCursor(cursor(q_act));
 7 │   foreach q_i ∈ leafNodes(q) do
 8 │   │   foreach e_j ∈ T_{q_i} do
 9 │   │   │   if ∃s_k ∈ S_b, s_k is the prefix of e_j then
10 │   │   │   │   e_j.signature = 1;
11 │   ProTJFastBySignature(q);
12 end
13 Function end(q)
14 begin
15 │   Return ∀f ∈ leafNodes(q) → eof(T_f);
16 end
17 Function getNextP(n)
18 begin
19 │   foreach q_i ∈ leafNodes(q) do
20 │   │   e_i = get(T_{q_i});
21 │   max = maxarg_i(e_i);
22 │   return n_max
23 end
24 Function ProTJFastBySignature(q)
25 begin
26 │   Sort the elements whose signature equals to 1 in respective element streams
   │   by document order.
27 │   By using algorithm ProTJFast, output the k answers with largest
   │   probability values.
28 end
```

In the second phase of Algorithm 1(Lines 7-10), we update the signature lists of element streams by assigning the signature of element below any skeleton result in set S_b to 1. As such, we can get a subtree of original probabilistic XML document by the signature lists. We can prove that the final top-k twig answers against the original XML can be found in the subtree.

In the last phase of Algorithm 1(Line 11), we firstly sort the elements whose signatures equal to 1 in respective element streams by document order. And then we can perform the algorithm *ProTJFast* to output the k answers with largest probability values. Notice that if the k of top-k query is small, we can use algorithm *TJFast* at the third phase directly, because the number of elements under the k skeleton results is also small, therefore there is no need to use the more complex algorithm *ProTJFast*.

Algorithm 2. $matchTwig(q_{act}, q)$

---

1 **begin**
2   **for** *any tags pair $[T_{q_a}, T_{q_b}]$ $(q_a, q_a \in leafNodes(q) \wedge q_a, q_b \neq q_{act})$ **do***
3   Advance head element in T_{q_a} to the position of cursor(q_a);
4   Advance head element in T_{q_b} to the position of cursor(q_b);
5   **while** $\neg$ $end(q)$ **do**
6   **if** *elements e_{q_a}, e_{q_a} match the common path pattern with $e_{q_{act}}$ in query q, and the common prefix of e_{q_a}, e_{q_a} match the common path pattern which is from the root to the branching node q_{bran} of q_a and q_b in query q, and the common prefix is not a element of mux node.* **then**
7   add e_{q_a}, e_{q_a} to the set of intermediate results.

8   return twig answers from the intermediate set.
9 **end**

For the twig query (a) in Figure 3: S[//C]//D against probabilistic XML (b), assume that the answers for top-2 probabilities are required. In the first phase of Algorithm 1, streams T_C and T_D are scanned, and the elements in streams are sorted by path-prob values shown in Figure 3 (c). The processing order of elements in streams are marked by dotted arrow line in Figure 3(c), which is obtained by invoking the $getNextP()$ function. Firstly, $getNextP()$ returns c_3 because its probability value is largest among the elements in respective streams, and c_3 start to join the element in T_D, and find a match (c_3, d_3), next the Algorithm 1 add the skeleton result s_2 of (c_3, d_3) to set S_b. Then $getNextP()$ returns the element c_1, and an answer (c_1, d_1) is found, also the skeleton result s_1 is added to set s_b. At this moment, the size of set s_b equals to the value of k, so the first phase ends. In the second phase, Algorithm 1 marks the signatures of elements that are the descendants of skeleton results in set S_b, so the signatures of elements c_1, d_1, c_2, d_2, c_3 and d_3 are updated. In the last phase, the elements of respective element streams are ordered by document order (In Figure (d)), and then the Algorithm $ProTJFast$ can be performed on them. So the final top-2 twig answers is (c_1, d_1) with probability value 0.576 and (c_2, d_2) with probability value 0.512. Notice that, the temporal results (c_3, d_3) in phase one is not the final answer, by the skeleton result s_2 generated from temporal results (c_3, d_3), we can bring the elements c_2 and d_2 to the final phase, and finally they can be merge-joined to be a final answer.

4.4 Analysis of Algorithm

We can see that in the first phase of Algorithm $HyTopKTwig$, the element streams are ordered by probability value, so that we can find the answers whose probability value are as potentially large as possible, while in the last phase, we use Algorithm $ProTJFast$ against the element streams ordered by document order, therefore Algorithm $HyTopKTwig$ is a hybrid algorithm that takes both probability value constraint and structure constraint into account.

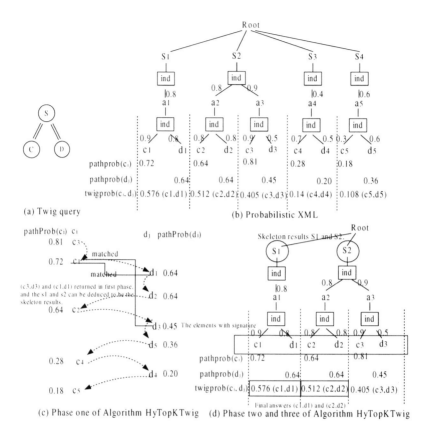

Fig. 3. Example of HyTopKTwig

However, are those temporary results in phase one the final answers? In Algorithm *PTopKTwig*, an enhanced lower bound is proposed to get the final answer quickly and to ensure the correctness of algorithm. The *HyTopKTwig* algorithm also needs to ensure the correctness and trys to find other candidate answers which may be the final answers, so we use the skeleton results to bound the structural region of those candidate answers. The correctness of Algorithm *HyTopKTwig* is proved below:

Theorem 1. *Given a twig query q and an probabilistic XML database PD, Algorithm* HyTopKTwig *correctly returns all the answers for q with top-k probabilities on PD.*

Proof. If the k skeleton results in set S_b can bound the final answers, then we can prove the correctness of algorithm *HyTopKTwig*, so we change the problem to prove that there is no element, which is not the descendant of any skeleton results in set S_b, can contribute to the final answers.

Assuming that there is an element e which is not the descendant of any skeleton results in set S_b, and can match a twig answer with element e_x whose probability value is larger than the k-th twig answer (merge-joined by ek_1 and ek_2). So we can get the inequations:

$$pathProb(ek_1) * preProb(ek_2) < pathProb(e_x) * preProb(e) \qquad (1)$$

equation (1) can be deduced to equation (2)

$$pathProb(e_x) > pathProb(ek_1) * preProb(ek_2)/preProb(e) \qquad (2)$$

In phase one, once the function $getNextP()$ returns a tag t, algorithm will regard the tag t as main path and find matching with other element streams ordered by predicate probability value, so preProb(ek_2)/preProb(e) must be larger than one because the path answer with larger predicate probability value has been scanned, we can conclude that pathProb(e_x) is larger than pathProb(ek_1), this is contradictory with that the processed main path has larger path probability value than the unprocessed ones. Or if the pathProb(e_x) is really larger than pathProb(ek_1), it means element e_x has been processed before ek_1, and the skeleton result of e_x has existed in set S_b. Because the element e can be merge-joined with e_x, the element e must be the descendant of skeleton result of e_x. It means element e will be dealt with in phase three of algorithm $HyTopKTwig$. Therefore we can conclude that there is no element which is not the descendant of any skeleton results in set S_b, and can match a twig answer whose probability value is larger than the k-th twig answer. So the correctness of algorithm $HyTopKTwig$ is proved.

5   Experiments

5.1   Experimental Setup

The algorithms $HyTopKTwig$, $ProTJFast$ and $PTopKTwig$ were implemented in JDK 1.4. Both real-world data set DBLP and synthetic data are used to test the performance of algorithm above, and the synthetic data set was generated by IBM XML generator and a synthetic DTD. We made the corresponding probabilistic XML documents from ordinary ones by inserting distribution nodes and assigning probability values to the child elements of distribution nodes. Table 1 lists the used queries. We take the metrics elapsed time and processed element rate $rate_{proc} = num_{proc}/num_{all}$ to compare the performance among those algorithms, where num_{proc} is the number of processed elements, and num_{all} is the number of all elements.

5.2   Performance Study

Influence of Document Size   We evaluated Q_1 against the DBLP data set of different sizes, ranging from 20MB to 110MB, and the answers with top-20 probability values were returned. From Figure 4, we can see that the elapsed time

Table 1. Queries

ID	queries
Q_1	dblp//article[//author]//title
Q_2	S//[//B]//A
Q_3	S//[//B][//C]//A
Q_4	S//[//B][//C][//D]//A
Q_5	S//[//B][//C][//D][//E]//A

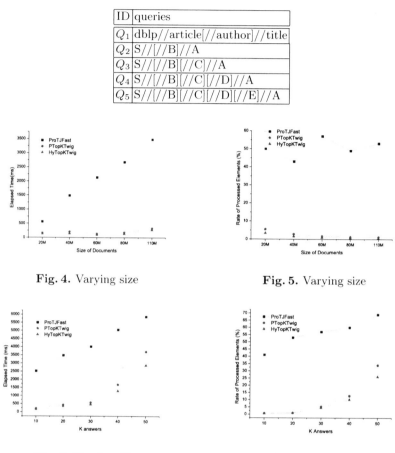

Fig. 4. Varying size   **Fig. 5.** Varying size

Fig. 6. Varying K **Fig. 7.** Varying K

of *ProTJFast* is linear to the size of documents, while varying size of documents has almost no impact on *HyTopKTwig* and *PTopKTwig*, and the algorithm *HyTopKTwig* performs better than *PTopKTwig*.

Influence of Number of Answers. Query Q_1 was evaluated against the DBLP data set. From Figures 6 and 7, we can see that by varying k from 10 to 50, the elapsed time and the rate of processed elements of those algorithms increase, the algorithm *HyTopKTwig* performs best, and algorithm *ProTJFast* performs worst. When k is small, the performance of *HyTopKTwig* is much better than *ProTJFast* and is better than *PTopKTwig*.

Influence of Multiple Predicates. To test the influence of multiple predicates, the queries Q_2 to Q_5 were evaluated on the synthetic data set. By varying the fan-out of query from 2 to 5, from Figures 8 and 9, we can see that the

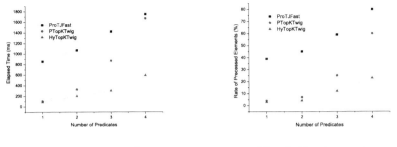

Fig. 8. Varying *pred*   **Fig. 9.** Varying *pred*

elapsed times of algorithm *PTopKTwig* increases rapidly. The situation is similar in Figure 9 when testing the rate of processed elements. The increasing speed of algorithm *HyTopKTwig* is steady and is slower than the other two, especially when the query's fan-out is large.

6   Conclusions

In this paper, we addressed the problem of finding top-k matching of a twig pattern against probabilistic XML data. Firstly, we improved the float vector part of *PEDewey* encoding, then we proposed a hybrid algorithm named *HyTopK-Twig* that has three phases. The element streams in first phase are ordered by probabilistic value, and the element streams in third phase are ordered by document order, therefore the algorithm *HyTopKTwig* considers both the probability value constraint and structural relationship constraint. Finally we presented experimental results on a range of real and synthetic data.

Acknowledgement. This work is supported by the Australian Research Council Discovery Project (Grant No. DP0878405, DP110102407), and the Fundamental Research Funds for the Central Universities of China (Grant No. 2009QN030).

References

1. Abiteboul, S., Senellart, P.: Querying and updating probabilistic information in XML. In: Ioannidis, Y., Scholl, M.H., Schmidt, J.W., Matthes, F., Hatzopoulos, M., Böhm, K., Kemper, A., Grust, T., Böhm, C. (eds.) EDBT 2006. LNCS, vol. 3896, pp. 1059–1068. Springer, Heidelberg (2006)
2. Hung, E., Getoor, L., Subrahmanian, V.S.: Probabilistic interval XML. In: Calvanese, D., Lenzerini, M., Motwani, R. (eds.) ICDT 2003. LNCS, vol. 2572, pp. 358–374. Springer, Heidelberg (2002)
3. Hung, E., Getoor, L., Subrahmanian, V.S.: PXML: A probabilistic semistructured data model and algebra. In: Proceeding of ICDE, pp. 467–478 (2003)
4. Nierman, A., Jagasish, H.V.: ProTDB: Probabilistic data in XML. In: Proceeding of VLDB, pp. 646–657 (2002)

5. Senellart, P., Abiteboul, S.: On the complexity of managing probabilistic XML data. In: Proceeding of PODS, pp. 283–292 (2007)
6. Kimelfeld, B., Kosharovsky, Y., Sagiv, Y.: Query efficiency in probabilistic XML models. In: Proceeding of SIGMOD, pp. 701–714 (2008)
7. Kimelfeld, B., Sagiv, Y.: Matching twigs in probabilistic XML. In: Proceeding of VLDB, pp. 27–38 (2007)
8. Hua, M., Pei, J., Zhang, W., Lin, X.: Ranking queries on uncertain data: A probabilistic threshold approach. In: Proceeding of SIGMOD, pp. 673–686 (2008)
9. Hua, M., Pei, J., Zhang, W., Lin, X.: Efficiently answering probabilistic threshold top-k queries on uncertain data. In: Proceeding of ICDE, pp. 1403–1405 (2008)
10. Chang, L., Yu, J.X., Qin, L.: Query Ranking in Probabilistic XML Data. In: Proceeding of EDBT, pp. 156–167 (2009)
11. Yi, K., Li, F., Kollios, G., Srivastava, D.: Efficient processing of top-k queries in uncertain databases. In: Proceeding of ICDE, pp. 1406–1408 (2008)
12. Yi, K., Li, F., Kollios, G., Srivastava, D.: Efficient processing of top-k queries in uncertain databases with x-relations. TKDE 20(12), 1669–1682 (2008)
13. Ning, B., Liu, C., Yu, J.X., Wang, G.: Matching Top-k Answers of Twig Patterns in Probabilistic XML. In: Kitagawa, H., Ishikawa, Y., Li, Q., Watanabe, C. (eds.) DASFAA 2010. LNCS, vol. 5981, pp. 125–139. Springer, Heidelberg (2010)
14. Grust, T.: Accelerating XPath Location Steps. In: Proceeding of SIGMOD, pp. 109–120 (2002)
15. Zhang, C., Naughton, J., DeWitt, D., Luo, Q., Lohman, G.: On Supporting Containment Queries in Relational Database Management Systems. In: Proceeding of SIGMOD, pp. 425–436 (2001)
16. Lu, J., Ling, T.W., Chan, C.-Y.: Ting Chen. From region encoding to extended dewey: On efficient processing of XML twig pattern matching. In: Proceeding of VLDB, pp. 193–204 (2005)

Optimizing Incremental Maintenance of Minimal Bisimulation of Cyclic Graphs

Jintian Deng[1], Byron Choi[1], Jianliang Xu[1], and Sourav S. Bhowmick[2]

[1] Hong Kong Baptist University, China
{jtdeng,bchoi,xujl}@comp.hkbu.edu.hk
[2] Nanyang Technological University, Singapore
assourav@ntu.edu.sg

Abstract. Graph-structured databases have numerous recent applications including the Semantic Web, biological databases and XML, among many others. In this paper, we study the maintenance problem of a popular structural index, namely *bisimulation*, of a possibly *cyclic* data graph. In comparison, previous work mainly focuses on acyclic graphs. In the context of database applications, it is natural to compute minimal bisimulation with merging algorithms. First, we propose a maintenance algorithm for a minimal bisimulation of a cyclic graph, in the style of merging. Second, to prune the computation on non-bisimilar SCCs, we propose a feature-based optimization. The features are designed to be constructed and used more efficiently than bisimulation minimization. Third, we conduct an experimental study that verifies the effectiveness and efficiency of our algorithm. Our features-based optimization pruned 50% (on average) unnecessary bisimulation computation. Our experiment verifies that we extend the current state-of-the-art with a capability on handling cyclic graphs in maintenance of minimal bisimulation.

1 Introduction

Graph-structured databases have a wide range of recent applications, *e.g.*, the Semantic Web, biological databases, XML and network topologies. To optimize the query evaluation in graph-structured databases, indexes have been proposed to summarize the paths of a data graph. In particular, many indexing techniques, *e.g.*, [3,4,11,17,19,23], have been derived from the notion of *bisimulation* equivalence. In addition to indexing, bisimulation has been adopted for selectivity estimation [14,20] and schemas for semi-structured data [2].

To illustrate the applications of bisimulation in graph-structured databases, we present a simplified sketch of a popular graph used in XML research, shown in the left hand side of Figure 1, namely XMark. XMark is a synthetic auction dataset: open_auction contains an author, a seller and a list of bidders, whose information is stored in persons; person in turn watches a few open_auctions. To model the bidding and watching relationships, open_auctions reference persons and vice versa. The references are encoded by IDREFs and represented by the dotted arrows in the figure. Two nodes in a data graph are bisimilar if they

J.X. Yu, M.H. Kim, and R. Unland (Eds.): DASFAA 2011, Part I, LNCS 6587, pp. 543–557, 2011.

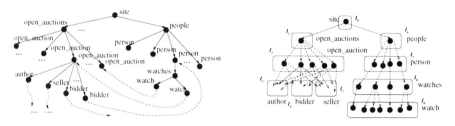

Fig. 1. Illustration – A sketch of XMark and its bisimulation

have the same set of incoming paths. A sketch of the bisimulation graph of XMark is shown in the right hand side of Figure 1. In the bisimulation graph, bisimilar data nodes are placed in a partition, denoted as I_i. Consider a query q /site//open_auction//seller that selects all sellers of open_auctions. We can evaluate q on the partitions and simply retrieve the data nodes in I_3. Therefore, it is crucial to minimize the bisimulation graph for efficient index to reduce I/O.

In practice, many data graphs are cyclic (e.g., [1]) and subject to updates. Therefore, different from other applications of bisimulation, its maintenance problem is much more important in database applications [12,21]. Furthermore, previous work [12,21] on maintenance of bisimulation of graphs mainly focuses on *directed acyclic* graphs. In contrast, this paper focuses on the maintenance problem of bisimulation of possibly cyclic graphs.

In this paper, we take the first step to systematically and comprehensively investigate *incremental maintenance of minimal bisimulation of cyclic graphs*. There are two key challenges in the maintenance problem. Firstly, merging algorithms for bisimulation as opposed to partition refinement are more natural for incremental maintenance of bisimulation. However, it is known [12] that merging algorithms fail to determine the minimum bisimulation of cyclic graphs. The main reason is that nodes of SCCs must be considered *together*, which is not the case in merging algorithms.

The first contribution is a maintenance algorithm for minimal bisimulation of cyclic graphs (Section 5), in the style of merging algorithms. Our algorithm consists of a split and a merge phase. In the split phase, we split and mark the index updated nodes (*i.e.*, the equivalence partitions) into a correct but non-minimal bisimulation. In the merge phase, we apply a (partial) bisimulation minimization algorithm on the marked index nodes. Our algorithm has an explicit handling of bisimulation between SCCs, when compared to previous work. As such, our algorithm *always* produces smaller (if not the same) bisimulation graphs when compared to previous work. In case of acyclic graphs, our algorithm and the previous work will produce the same bisimulation.

The second contribution is on a feature-based optimization for determining bisimulation between two SCCs (Section 6). On one hand, the computation of bisimulation between two SCCs can be costly. On the other hand, there may not be many bisimilar SCCs, in practice. We aim at deriving structural features from SCCs such that two SCCs are bisimilar *only if* they have the same or bisimilar features. Specifically, we explore label- or edge-based, path-based, tree-based and circuit-based features. With these, the merging algorithm has a more global information of SCCs and may prune computation on non-bisimilar SCCs early.

Third, we conduct an experimental study that verifies the effectiveness and efficiency of our algorithm. In particular, our feature-based optimization prunes an average of 50% unnecessary bisimulation computation. The results validate the practical feasibility of extending the current state-of-the-art with a capability of maintaining minimal bisimulation on cyclic graphs.

2   Related Work

Previous work on maintaining bisimulation can be categorized into two: *merging* and *partition-refinement* algorithms. There have been two previous merging algorithms [12,21] for incremental maintenance of bisimulation of cyclic graphs. The algorithm proposed in [12] contains a split and a merge phase. Upon an update on the data graph, the bisimulation graph is split to a correct but non-minimal bisimulation of the updated graph. Next, the bisimulation graph is minimized in the merge phase. For acyclic graphs, [12] produces the minimum bisimulation of the updated graph. If the graph is cyclic, [12] returns a minimal bisimulation only. Thus, to support cyclic graphs, the minimum bisimulation is occasionally re-computed from scratch. [21] proposes a split-merge-split algorithm with a rank flag for SCCs, which is originally proposed in [6]. [21] also returns a minimal bisimulation in response to an update of a cyclic graph. However, there is neither experimental evaluation [21] nor implementation for us to perform comparisons. A difference between our work and the previous work is that we introduce explicit handling of SCCs and propose features to optimize bisimulation maintenance.

A recent partition-refinement algorithm [10] can be considered as a variant of Paige and Tarjan's algorithm [18] – a construction algorithm for the minimum bisimulation. The algorithm proposes its own split to handle edge changes. It has been extended to support maintenance of k-bisimulation. Their experiment shows that [10] produces a bisimulation that is always within 5% of the minimum bisimulation. It is shown, through a later experiment, that [12] may produce even smaller bisimulations, which we compared via experiments in Section 7.

Bisimulation (relation) [16] has its root at symbolic model checking, state transition systems and concurrency theories. In a nutshell, two state transition systems are bisimilar if and only if they *behave* the same from an observer's point of view. Bisimulation minimization has been extensively studied through experiments in [7], in the context of modeling checking. A conclusion of [7] is that minimization may not be worthwhile for model checking as it may easily be more costly than checking invariance properties of systems. In comparison, when bisimulation is used as an index structure for query processing, bisimulation minimization and therefore its maintenance are far more important.

3   Background

This section presents the background and the notations of this paper.

Definition 3.1. *A* graph-structured database *(or* data graph*) is a rooted directed labeled graph* $G(V, E, r, \rho, \Sigma)$, *where* V *is a set of nodes and* $E: V \times V$ *is a set of edges,* $r \in V$ *is a root node and* $\rho : V \to \Sigma$ *is a function that maps a vertex to a label, and* Σ *is a finite set of labels.*

For clarity, we may often denote a data graph as $G(V, E)$ when r, ρ and Σ are irrelevant to our discussions. Since our work focuses on cyclic graphs, we recall some relevant definitions below.

Cyclic graphs. A *strongly connected component* (SCC) in a graph $G(V, E)$ is a subgraph $G'(V', E')$ whose nodes are a subset of nodes $V' \subseteq V$ where the nodes in V' can reach each other. The SCCs of a graph can be determined by classical graph contraction algorithms, *e.g.*, Gabow's algorithm, in $O(|V|+|E|)$, where each SCC is reduced to a supernode. The resulting graph is a *directed acyclic graph* DAG, which is often called the *reduced graph*. In subsequent discussions, we use SCCs to refer to non-trivial SCCs (SCCs with more than one node) only. In the definition below, we highlight two special kinds of nodes in SCCs, namely, exit and entry nodes.

Definition 3.2. *A node* n *of an* SCC $G'(V', E')$ *of a graph* $G(V, E)$ *is an* exit *node if there exists an edge* (n, n_1) *where* $n \in V'$ *and* $n_1 \notin V'$. *Similarly,* n *is an* entry *node if there exists an edge* (n_0, n) *where* $n_0 \notin V'$ *and* $n \in V'$.

Bisimulation. Next, we recall the relevant definitions of bisimulation.

Definition 3.3. *Given two graphs* $G_1(V_1, E_1, r_1, \rho_1)$ *and* $G_2(V_2, E_2, r_2, \rho_2)$, *an* upward bisimulation $\sim$ *is a binary relation between* V_1 *and* V_2:

$$\forall v_1 \in V_1, v_2 \in V_2 . v_1 \sim v_2 \to$$
$$\forall (v_1', v_1) \in E_1 \exists (v_2', v_2) \in E_2 . v_1' \sim v_2' \wedge \rho_1(v_1') = \rho_2(v_2') \wedge$$
$$\forall (v_2'', v_2) \in E_2 \exists (v_1'', v_1) \in E_1 . v_1'' \sim v_2'' \wedge \rho_1(v_1'') = \rho_2(v_2'').$$

Two graphs G_1 *and* G_2 *are* upward bisimilar *if an upward bisimulation* $\sim$ *can be established between* G_1 *and* G_2.

Examples of bisimilar nodes can be found in Figure 1, where the bisimilar nodes are placed in the same rounded rectangle. Definition 3.3 presents upward bisimulation in the sense that two nodes can be bisimilar only if their parents are bisimilar. The definition can be paraphrased in terms of paths, which is often convenient to simplify our discussions[1] .

Proposition 3.1: *Two nodes are upward bisimilar if and only if the incoming path set of the two nodes are the same.*   □

A set of bisimilar nodes is often referred to as an *equivalence partition* of nodes, or simply partitions. Hence, a bisimulation of a graph can be described as a partition graph. In the context of indexing, the partitions are sometimes referred

---

[1] We should remark that there have been other notions of bisimulation, such as downward bisimulation and k-bisimulation, that have been applied in indexing/selectivity estimation but have not been the focus of this paper. Our techniques can be extended to support them with minor modifications.

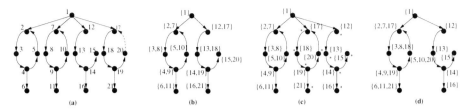

Fig. 2. (a) A cyclic data graph; (b) the minimal bisimulation graph; (c) the split bisimulation graph; and (d) an updated minimal bisimulation graph

to as *index nodes*, or simply *Inodes*, whereas the nodes of the data graph are referred to as *data nodes*, or simply *nodes*.

In this work, we consider the notion of bisimulation minimality defined in Definition 3.5. First, we recall the notion of *stability*.

Definition 3.4. *Given two partitions of nodes X and I, X is stable with respect to I if either (i) X is contained in the children of the nodes in the partition I or (ii) X and the children of the nodes in I are disjoint.*

Definition 3.5. *Given a bisimulation B of a graph G, B is minimal if for any two partitions I, $J \in B$, either (i) the nodes in I and J have different labels, or (ii) merging I and J results in some partition $K \in B$ unstable.*

Definition 3.6. *A bisimulation B of a graph G is the minimum bisimulation if B contains the minimum number of partitions, among all bisimulations of G. According to [12], the minimum bisimulation of a graph is unique.*

Bisimulation minimization. Next, we illustrate the intuitions of merging algorithm for bisimulation minimization with a brief example shown in Figure 2. Assume the nodes of the data graph shown in Figure 2(a) have the same label. The node id is shown next to each node. We use {} to denote an Inode. A merging algorithm initially places each node in a single partition. Assume that the algorithm merges pairs of partitions top-down, which attempts to merge Nodes 2 and 7. However, the algorithm has not yet determined Nodes 5 and 10. Hence, the algorithm terminates and fails to return the minimum bisimulation shown in Figure 2(b), unless it memorizes the SCCs containing Nodes 2 and 7 together.

4   Bisimulation of Cyclic Graphs

This section presents a minimization algorithm for bisimulation of cyclic graphs, shown in Figure 3, which is a component of the maintenance algorithm. We focus on the logic of handling SCCs during the minimization.

The algorithm can be divided into two parts. First, Lines 01-06, if n_1 and n_2 are not both in some SCCs, we compute bisimulation between n_1 and n_2 in the style of any merging algorithm. We assume the existence of a procedure next_nodes_top_order(G) of a node n which returns the next n's child in topological order in G. Then, we recursively invoke bisimilar_cyclic.

Procedure `bisimilar_cyclic`
Input: Nodes n_1 and n_2 where $\rho(n_1) = \rho(n_2)$; B, the current bisimulation
Output: An updated bisimulation relation B'

01 **if** n_1 and n_2 are not both in some SCC
02   **if** $\forall p_1 \in n_1.\text{parent } \exists p_2 \in n_2.\text{parent s.t. } p_1 \sim p_2$ **then**
03   add (n_1, n_2) to B
04   **for all** c_1 in $n_1.\text{next\_nodes\_top\_order}(G_1)$
05   **for all** c_2 in $n_2.\text{next\_nodes\_top\_order}(G_2)$
06                 $B = \text{bisimilar\_cyclic}(c_1, c_2, B)$

07 **else**   /* check bisimulation of the two SCCs */
08   assume n_1 and n_2 are in SCCs S_1 and S_2, respectively
        if $\text{feature\_pruning}(S_1, S_2)$ **return** B /* Sec. 6*/
09   clone S_1 to S_1';   create an artificial node n_1' for n_1
10   **for all** $(n, n_1) \in S_1'.E$
11   replace (n, n_1) with $(n, n_1') \in S_1'$
12   clone S_2 to S_2';   create an artificial node n_2' for n_2
13   **for all** $(n, n_2) \in S_2'.E$
14   replace (n, n_2) with $(n, n_2') \in S_2'$

15   clone B to B';   add (n_1, n_2) to B' /* assume $n_1 \sim n_2$ */
16   **for all** c_1 in $n_1.\text{next\_nodes\_top\_order}(S_1')$
17   **for all** c_2 in $n_2.\text{next\_nodes\_top\_order}(S_2')$
18             $B' = \text{bisimilar\_cyclic}(c_1, c_2, B')$
19   **if** (n_1', n_2') in B' **then** $B = B \cup B'$ /* $S_1 \sim S_2$ */
20 **return** B

Fig. 3. Bisimulation minimization of cyclic graphs

Second, if both n_1 and n_2 are in some SCCs, Lines 07-20 check if S_1 *and* S_2, as opposed to simply n_1 and n_2, can be bisimilar. We prune non-bisimilar SCCs by using the feature-based optimization presented in Section 6, in Line 08. For presentation clarity, we assume that n_1 and n_2 are in two different SCCs. Then, we break the SCCs and check bisimulation recursively, in Lines 09-15. The main idea is illustrated with Figure 4. Specifically, we redirect the incoming edges of n_1 in SCC (Lines 09-11) to an artificial node n_1'. Similarly, we redirect the incoming edges of n_2 to n_2' (Lines 12-14). We clone the current bisimulation relation determined thus far (Line 15). Assuming that n_1 and n_2 are bisimilar, we check the possible bisimulation between the children of n_1 and n_2 by calling `bisimilar_cyclic` recursively (Lines 16-18). If we can construct a possible bisimulation between n_1' and n_2' (Line 19), then S_1 and S_2 are bisimilar.

The main idea of `bisimilar_cyclic` on handling SCCs is that `bisimilar_cyclic` explicitly breaks a cycle, whereas previous work *overlooks* cycles. `bisimilar_cyclic` may be recursively called due to nested SCCs (Line 18). Without breaking a cycle in each call, `bisimilar_cyclic` may not terminate and the feature-based optimization (Line 07) may always derive features of the "topmost" SCC.

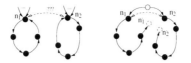

Fig. 4. Breaking one cycle in an SCC in `bisimilar_cyclic`

Analysis. For presentation clarity, `bisimilar_cyclic` did not incorporate with classical indexing techniques. `bisimilar_cyclic` runs in $O(|E|^2)$ due to the for loops at Lines 04-06 and Lines 16-18, assuming that `feature_pruning` can be performed more efficiently than $O(|E^2|)$.

5   Maintenance of Bisimulation

In this section, we present the overall maintenance algorithm. For simplicity, we present an edge insertion algorithm `insert` in Figure 5. Edge deletions are discussed at the end of this section. Our algorithm consists of a split phase and a merge phase. In the following, we focus on the split phase, as the merge phase is essentially `bisimilar_cyclic`.

The split phase. The split phase is presented in Lines 05-20. We maintain two variables to record two kinds of nodes that are needed to be split. More specifically, we use S to record the nodes of *SCCs* needed to be split and Q to record the *nodes* that are not in any SCCs but needed to be split. In the split phase, we mark the affected Inodes, which will be examined in the merge phase.

   Suppose the insertion makes the Inode of n_2 unstable. To initialize S (Line 03), we set S to the Inode of n_2 and n_2, *i.e.*, $\{(I_{n_2}, n_2)\}$, if n_2 is in an SCC. Otherwise, S is empty. Similarly, we initialize Q to I_{n_2} if n_2 is not in any SCC and Q is empty otherwise (Line 04). Next, we split the Inodes in S and Q recursively until they are empty (Line 05).

(1) We process the nodes in S as follows (Lines 06-12): We select a node n from S and retrieve its Inode I_n. We split n from I_n as the SCC of n is potentially non-bisimilar to the SCC of other nodes in I_n (Line 09). We mark the split Inodes so that they will be checked in the merge phase (Line 10). In Lines 11-12, we insert the children of the split Inode to S and Q, similar to Lines 03-04.

(2) The handling of Q is shown in Lines 13-20. We select an Inode I_n from Q (Line 14). If I_n is not stable, we split I_n into a set of stable Inodes $\mathcal{I}$, as in the pervious work [12] for acyclic graphs (Lines 15-16). We mark Inodes in $\mathcal{I}$ in Line 18. In Lines 19-20, we update the affected nodes S and Q, similar to Lines 03-04.

   The split phase essentially traverses the bisimulation graph B and SCCs in the data graph to spilt and collect the Inodes that are affected by the update. SCCs themselves may be affected by an update. In Line 21, we call Gabow's algorithm to update SCC information of a graph, which is needed in the merge phase.

The merge phase. The merge phase can be done by applying the minimization algorithm presented in Section 4 (Figure 3). An optimization is that we apply merging on only the Inodes that are marked in the split phase.

Procedure insert

Input: an insertion of an edge (n_1, n_2) to a graph G; its minimal bisimulation B

Output: An updated graph G' and its updated minimal bisimulation B'

01 $G' =$ insert (n_1, n_2) into G

02 **if** n_2 is new

    then create a new Inode I_{n_2}; insert I_{n_2} into B; mark I_{n_2}

    else if I_{n_2} is not stable

03        $S = \{(I_{n_2}, n_2) \mid n_2$ is in an SCC$\}$

04        $Q = \{I_{n_2} \mid n_2$ is not in any SCC$\}$

05 **while** $Q \neq \emptyset$ or $S \neq \emptyset$

06   **if** $S \neq \emptyset$ **then**   /* split the relevant SCC */

07   pick a node (I_n, n) from S; remove (I_n, n) from S

08   **while** I_n is not stable or a singleton

09   split I_n into $I_1 = I_n - \{n\}$ and $I_2 = \{n\}$

10   mark I_1 and I_2

11        $S = S \cup \{(I_{n_s}, n_s) \mid n_s$ is n_i's child, $n_i \in I_2$ and n_s in the SCC of $n\}$

12        $Q = Q \cup \{I_{n_q} \mid n_q$ is a child of n_i, $n_i \in I_2$ and n_q not in any SCCs$\}$

13   **if** $Q \neq \emptyset$ **then**   /* split nodes not related to SCCs */

14   pick a node $I_n \in Q$; remove I_n from Q

15   **if** I_n is not stable or a singleton

16   split I_n into a stable set $\mathcal{I}$ /* [12] */

17   **for each** I in $\mathcal{I}$

18   mark I

19        $S = S \cup \{(I_{n_s}, n_s) \mid n_s$ is n_i's child, $n_i \in I$ and n_s in the SCC of $n\}$

20        $Q = Q \cup \{ I_{n_q} \mid n_q \in$ child of n_i, $n_i \in I$ and n_q not in any SCCs$\}$

21 Gabow(G') /* update the SCC information in G' */

22 $(G', B') =$ bisimilar_cyclic_marked(G, B) /* merging the marked Inodes */

23 **return** (G', B')

Fig. 5. Insertion for minimal bisimulation of cyclic graphs

Example 5.1. We illustrate Algorithm insert with an example. Reconsider the cyclic data graph that is shown in Figure 2(a). Its minimal bisimulation is shown in Figure 2(b). Assume that we insert an edge (20,17) into the data graph. Algorithm insert initially puts {12,17} into Q (Line 04). Then, in Line 16, Node 17 is split from {12,17}. The split Inodes are marked, with a "*" sign in the figure. The split phase proceeds recursively and finally produces the graph in Figure 2(c). Then, we update the SCC information of the data graph. By bisimilar_cyclic_marked, we obtain the bisimulation at Figure 2(d).

While the previous work [12] produces the same split graph (Figure 2(c)), it returns the bisimulation in Figure 2(c), due to the lack of the handling on SCCs. Subsequently, any subgraphs that are connected to the SCC (Nodes 17-20), *e.g.*, Node 21, are not merged, as the SCCs are not merged.

Analysis. The recursive procedure in Lines 05-20 traverses the graph $O(|E|)$. With optimization in [18], stablizing a set can be done in $O(log(|V|))$. Hence.

the split phase runs in $O(|E|log(|V|))$. Gabow's algorithm in Line 21 runs in $O(|V| + |E|)$. The merge phase with optimization runs in $O(|E|^2)$. Thus, the overall runtime of Algorithm `insert` is $O(|E|^2)$.

Edge deletions. While our discussions focused on insertions, our technique can be generalized to support edge deletions with the following modifications. (i) In Line 01, we delete the edge from the data graph. (ii) If n_2 is connected after the deletion, we check the stability of I_{n_2} in Line 02, initialize S and Q and then invoke the split phase as before.

6   Feature-Based Optimization

The maintenance algorithm presented in Section 5 involves splitting the updated bisimulation into a non-minimal bisimulation followed by bisimulation minimization. As discussed, determining if two SCCs are bisimilar can be computationally costly, $O(|E|^2)$. In practice, SCCs may often be non-bisimilar. This motivates us to optimize the minimization of cyclic graphs by proposing features to prune computations on non-bisimilar SCCs. The main idea is to derive features of SCCs such that two SCCs can be bisimilar *only if* their features are the same or bisimilar. Ideally, the features are discriminative enough and can be efficiently constructed and used. Furthermore, features may be efficiently maintainable so that they are constructed once and maintained with the bisimulation.

6.1   Properties of Bisimulation of Cyclic Graphs

Prior to the discussions on features, we list some properties of bisimulation of cyclic graphs. These properties show that a number of classic properties of graphs are not suitable for our feature-based optimization. Due to space constraints, we omitted the proofs, which are established by simple proof by contradictions [5].

Property 1. Two SCCs with the same cycle height may not be bisimilar. Two SCCs with different cycle heights can be bisimilar.

Property 2. Two SCCs with the same number of simple cycles may not be bisimilar. Two bisimilar SCCs may have different number of simple cycles.

Property 3. Two SCCs with different numbers of entry nodes can be bisimilar.

The design of features exploits the following proposition on bisimulation of SCCs. The intuition is that as long as we find a node in a SCC that is not bisimilar to any node in another SCC, the two SCCs will not be bisimilar.

Proposition 6.2: *An SCC $G_1(V_1, E_1)$ is not bisimilar to another SCC $G_2(V_2, E_2)$ if and only if there is a node v in V_1 such that it is not bisimilar to any node in V_2.*   □

6.2 Features of SCCs

Merging algorithms for bisimulation minimization are iterative in nature. The current merging step of a SCC may not have sufficient information for determining bisimulation between SCCs. Hence, we propose some features that give merging algorithms some "lookahead" of SCCs to check Proposition 6.2.

1. Label-based or edge-based features. We begin with the label-based and edge-based features, which are straightforward, and have many alterative implementations. For example, we may use all label and edge types that appeared in an SCC as an SCC feature. Two bisimilar graphs must contain the same type of labels and edges. In our experiments, we found that the incoming label or edge sets of an entry node are relatively concise and effective in distinguishing non-bisimilar SCCs. For example, in Figure 1, the incoming label set of the entry node open_auction is {open_auction, watch} and that of the entry node watches is {person, bidder}. The construction and maintenance of such labels can be efficiently supported by hashtables.

2. Path-based features. Regarding path-based features, one may be tempted to use all simple paths in an SCC. However, determining all simple paths of a cyclic graph is a problem in PSPACE [15].

Proposition 6.3: *Two SCCs are bisimilar only if they have the same set of simple path(s) from their entry node(s).*   □

Next, the longest paths of a cyclic graph are not appropriate for our problem either, as they cannot be determined in PTIME.

In this work, we propose to use the set of incoming paths with a length at most k (or simply k-paths) as a feature of the entry nodes, where k is a user parameter. The value of k may be increased when maintenance of bisimulation spends substantial time on bisimulation computation. From Proposition 3.1, two bisimilar graphs must have the same set of k-paths. Contrarily, two graphs with different sets of k-paths are non-bisimilar. Hence, k-paths can be used as a feature. It is straightforward that k-paths can be efficiently constructed. However, since k-paths is local, k-paths may not contain a node that is not bisimilar to any nodes in any other SCCs. Another simple remark is that a node in an SCC may appear in a k-path set multiple times.

3. Feature of canonical spanning tree. We further explore more complex structural features of SCCs. First, we define the weight used in determining the canonical spanning tree. The *weight* of an edge (n_1, n_2) is *directly proportional* to the count of $(\rho(n_1), \rho(n_2))$-edges in the graph. We exploit a popular trick to perturb the edge's weight such that each kind of edges has a unique weight.

Given the weight defined above, we can compute a minimum spanning tree, in the style of a greedy breath first traversal in $O(|V|+|E|)$. As the weight is defined to be directly proportional to the edge count, a minimum spanning contains more infrequent edge kinds of a graph. However, minimum spanning trees of a *directed* graph are often difficult to maintain. In comparison, maintenance of spanning trees of an undirected graph is much simpler, *e.g.*, in amortized time $O(|V|^{1/3} log(|V|))$ [9]. Hence, we perform some simple tricks on the data graph

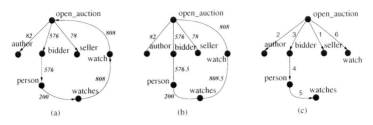

Fig. 6. The construction of the canonical spanning tree from a simplified open_auction

when constructing the spanning tree. First, we ignore the direction of the edges. Second, we adopt Prim's algorithm to construct the minimum spanning tree of the undirected graph. From the root of the minimum spanning tree, we derive the edge direction, which gives us the *canonical spanning tree*. Note that the edge direction is simply needed for checking bisimulation between canonical spanning trees and the direction of the edges in the canonical spanning tree may differ from that of the edges in the original graph.

Proposition 6.4: *Two* SCCs *are bisimilar only if their minimum canonical spanning trees returned by Prim's algorithm are bisimilar.*   □

It should be remarked that SCCs are often nested. In the worst case, the total size of the spanning trees of all possible entry nodes of an SCC is $O((|V| + |E|)^2)$. In addition, computing bisimulation between large canonical spanning trees can be costly. Therefore, we introduce a termination condition to the Prim's algorithm – we do not expand the spanning tree further from a node n when there is an ancestor of n having the same label as n. The total size of the canonical spanning trees is then $O(|V| \times |E|)$.

Example 6.2. We illustrate the construction of a canonical spanning discussed above with an example shown in Figure 6. Figure 6(a) shows a simplified SCC of open_auction from XMark with a scaling factor 0.1. The count of each edge type is shown on the edge. We perturb the weight to make each weight in the SCC unique. We ignore the direction of the edges, shown in Figure 6(b). Then, it is straightforward to compute the spanning tree (shown in Figure 6(c), where the number on an edge shows the order of the edge returned by Prim's algorithm). Finally, the direction of the edges are derived from the root of the spanning tree open_auction.

4. Circuit-based features. While the time for checking bisimulation between minimum spanning trees is very close to that between SCCs, one may be tempted to explore structural features further. Here, we illustrate that complicated structural features can lead to inefficient maintenance. For example, circuit bases contain much more structural information than spanning trees. It has been shown that the minimum circuit bases of directed graphs is unique [8]. However, determining the circuit bases is $O(|V|^3)$. It is therefore more efficient to simply compute the bisimulation of two SCCs than using the feature of circuit bases.

Proposition 6.5: *Two* SCCs *are bisimilar if their circuit bases are bisimilar.*   □

6.3   Offline versus Online Feature Construction

Since the proposed features can be constructed relatively efficiently, they may be constructed and used during bisimulation computation, *i.e.*, runtime. Then, during runtime, we may incorporate the features with not only the labels but also the partial bisimulation constructed so far. Specifically, some nodes in SCCs have been associated with Inode. The ids of Inodes together with the labels, as opposed to the labels alone, are used in online feature construction.

In comparison, the features may be built offline and maintained with each update of the graph. However, given a cyclic graph, we may determine features for each entry node, in the worst case, to build all possible features offline. However, this size requirement may sometimes be prohibitive, in practice.

7   Experimental Evaluation

This section presents an experimental study that verifies the efficiency of our algorithms. Our implementation is written in JDK building on top of Ke *et al.* [12]. It is available at http://code.google.com/p/minimal-bisimulation-cyclic-graphs/. The experiments were run on a laptop computer with a dual CPU at 2.0 GHz and 2GB RAM running Ubuntu hardy.

Datasets. We used both synthetic and real-life graph data to test various aspects of our algorithms. (i) XMark is a synthetic XML dataset provided by the XMark Benchmark Projects [22]. The cycles in XMark is essentially composed by IDREFs of open_auction to person and vice versa. We ran Gabow's algorithm on XMark. We note that there are few very large SCCs. It is easy to verify that very few, or none, of the SCCs are bisimilar. Hence, we randomly decompose SCCs into smaller SCCs as follows: We define a parameter s to set the average number of open_auction nodes and another parameter r to define the ratio between open_auction and person nodes in an SCC. For example, when s and r are set to 10 and 1.2, respectively, an SCC contains approximately 10 open_auctions and 12 persons. In our experiment, the dataset generated directly from XMark is referred to Large and the decomposed Large is refered to Cyclic.

In the experiment on Algorithm insert, we generated a dataset Base to test the performance difference between insert and Ke *et al.* The performance difference can hardly be shown because Large only contains few large SCCs and Cyclic contains numerous random non-bisimilar SCCs. Therefore, we constructed Base by connecting two XMark graphs with the same scaling factor (s.f.) and removing a number of edges from the graph. When the edges are inserted by Algorithm insert, the bisimilar SCCs will be recovered and merged.

We tested insert over real-life data Cite. Cite is a citation graph extracted from papers on high energy physics [13]. It covers those papers in the period from Jan. 1993 to Apr. 2003 and contains 35k papers in total. Cite represents each paper as a data node and a citation in paper i to paper j as an edge. We removed self-citing edges, for simplicity. Cite is highly cyclic. Similar to Cyclic, we removed citation edges randomly and used them for insertions.

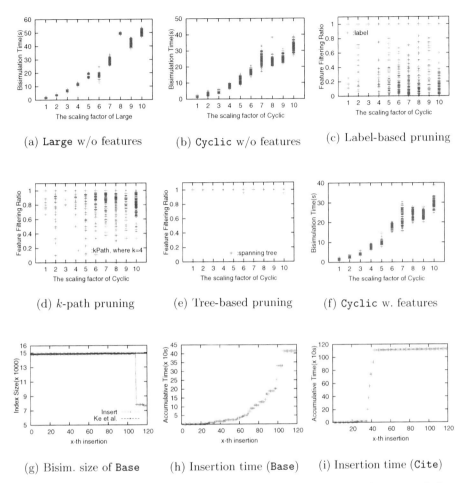

(a) `Large` w/o features (b) `Cyclic` w/o features (c) Label-based pruning

(d) k-path pruning (e) Tree-based pruning (f) `Cyclic` w. features

(g) Bisim. size of `Base` (h) Insertion time (`Base`) (i) Insertion time (`Cite`)

Fig. 7. Performance results on `bisimilar_cyclic` with and without feature optimization on `Large` and `Cyclic` and `insert` performance on `Base` and `Cite`

Performance analysis. To test the runtime of `bisimilar_cyclic` with feature-based optimization, we ran 100 random `Large` and `Cyclic` for each s.f. ranging from 0.01 to 0.1 (*i.e.*, 17k nodes to 168k nodes). Figures 7(a) and 7(b) show that the runtimes are roughly linear to s.f.. At the same s.f., the runtimes for `Large` are longer than those for `Cyclic`. The reason is that `Cyclic` contains are more smaller random SCCs, which are often non-bisimilar, and `bisimilar_cyclic` identifies them relatively earlier. In comparison, `bisimilar_cyclic` in `Large` may spend more time in checking sub-SCCs inside a large SCC.

Next, we verified the effectiveness of the features by using each feature on 100 `Cyclic` graphs for each s.f.. The features were *computed in runtime* and k in the path-based feature is 4. We skipped the edge-based feature as its performance is similar to the label-based feature, in `Cyclic`. The pruning of each feature is plotted in Figures 7(c), 7(d) and 7(e). The y-axis is the percentage of pruned non-bisimilar SCCs. In all, the label-based, path-based and canonical-tree feature pruned (on average) 14%, 62% and 73%, respectively. Figure 7(f) shows the runtime of bisimilar_cyclic with all feature optimization. On average, it is 4% faster than the one without optimization (Figure 7(b)). We remark that on average, 7.7% of the runtime was due to *online* feature construction.

Lastly, we conducted an experiment on Algorithm insert over Base and Cite. The results are shown in Figures 7(g),(h) and (i). Figure 7(g) shows the size of the minimal bisimulation produced by insert and Ke *et al.* [12]. We did not show the minimum bisimulation as insert always produces a bisimulation that is within 2% of the minimum. Initially, both insert and [12] are very close. After some number of insertions, the two bisimilar SCCs in the Base were recovered. We ran this experiment multiple times and found that the drop occurs randomly between 100th and 120th insertion. As illustrated in Figure 7(g), the difference in the size of bisimulation returned by insert and [12] depends on the number and the size of bisimilar SCCs in a graph. In this particular graph, insert returns a bisimulation graph that is 100% smaller than that by [12].

The accumulative runtime of insert over Base is shown in Figure 7(h). The accumulative runtime increases as we insert more edges into Base. After some insertions, insert ran slower because the two SCCs in Base became similar. bisimilar_cyclic checked many nodes before it declared the SCCs were not bisimilar. The runtime of [12] is close to 0s as it does not process SCCs, as the minimal bisimulation remains the same.

The accumulative runtime of insert over Cite is shown in Figure 7(i). As expected, the runtime increases as more edges are inserted. Between the 30th and 40th insertion, the largest SCC in Cite was involved and insert ran slower. In most of the cases, the runtime of insert is close to 0s when it did not process the SCCs. The average runtime for one insertion is around 10s. However, there is no bisimilar SCCs in Cite and insert and [12] returned the same bisimulation.

8   Conclusions

In this paper, we studied the optimization in maintaining the minimal bisimulation of cyclic graphs. Our first contribution is a bisimulation minimization algorithm that *explicitly* handles SCCs and a maintenance algorithm for minimal bisimulation of cyclic graphs. Second, we propose a feature-based optimization to avoid computing non-bisimilar SCCs. Third, we presented an experiment to verify the effectiveness and efficiency of our algorithms. Our experimental results show that the features can prune unnecessary bisimulation computation and our maintenance algorithm can return smaller bisimulation graphs than previous work, depending on the size and number of bisimilar SCCs in the data graph. As for future work, we plan to refine the selection of discriminative features to further reduce the maintenance time. We are studying on maintenance algorithms that can produce either the minimum bisimulation or one whose size is bounded by a theoretical guarantee.

References

1. Batagelj, V., Mrvar, A.: Pajek datasets,
   http://vlado.fmf.uni-lj.si/pub/networks/data/
2. Buneman, P., Davidson, S.B., Fernandez, M.F., Suciu, D.: Adding structure to un-structured data. In: Afrati, F.N., Kolaitis, P.G. (eds.) ICDT 1997. LNCS, vol. 1186, Springer, Heidelberg (1996)
3. Buneman, P., Grohe, M., Koch, C.: Path queries on compressed XML. In: VLDB (2003)
4. Chen, Q., Lim, A., Ong, K.W.: D(k)-index: an adaptive structural summary for graph-structured data. In: SIGMOD (2003)
5. Deng, J., Choi, B., Xu, J., Bhowmick, S.S.: Optimizing incremental mainte-nance of minimal bisimulation of cyclic graphs. Technical report, HKBU (2010), http://www.comp.hkbu.edu/~jtdeng/techreport.pdf
6. Dovier, A., Piazza, C., Policriti, A.: An efficient algorithm for computing bisimu-lation equivalence. Theor. Comput. Sci. 311(1-3), 221–256 (2004)
7. Fisler, K., Vardi, M.Y.: Bisimulation minimization and symbolic model checking. Form. Methods Syst. Des. 21(1), 39–78 (2002)
8. Gleiss, P.M., Leydold, J., Stadler, P.F.: Circuit bases of strongly connected di-graphs. Working Papers 01-10-056, Santa Fe Institute (2001), http://ideas.repec.org/p/wop/safiwp/01-10-056.html
9. Henzinger, M.R., King, V.: Maintaining minimum spanning trees in dynamic graphs. In: Degano, P., Gorrieri, R., Marchetti-Spaccamela, A. (eds.) ICALP 1997. LNCS, vol. 1256. Springer, Heidelberg (1997)
10. Kaushik, R., Bohannon, P., Naughton, J.F., Shenoy, P.: Updates for structure indexes. In: VLDB (2002)
11. Kaushik, R., Shenoy, P., Bohannon, P., Gudes, E.: Exploiting local similarity for indexing paths in graph-structured data. In: ICDE (2002)
12. Ke, Y., Hao, H., Ioana, S., Jun, Y.: Incremental maintenance of XML structural indexes. In: SIGMOD (2004)
13. Leskovec, J.: Stanford large network dataset collection, http://snap.stanford.edu/data
14. Li, H., Lee, M.L., Hsu, W., Cong, G.: An estimation system for XPath expressions. In: ICDE (2006)
15. Mendelzon, A.O., Wood, P.T.: Finding regular simple paths in graph databases. In: VLDB (1989)
16. Milner, R.: Communication and Concurrency. Prentice Hall, Englewood Cliffs (1989)
17. Milo, T., Suciu, D.: Index structures for path expressions. In: Beeri, C., Bruneman, P. (eds.) ICDT 1999. LNCS, vol. 1540, pp. 277–295. Springer, Heidelberg (1999)
18. Paige, R., Tarjan, R.E.: Three partition refinement algorithms. SIAM J. Com-put. 16(6), 973–989 (1987)
19. Polyzotis, N., Garofalakis, M.: XCluster synopses for structured XML content. In: ICDE (2006)
20. Polyzotis, N., Garofalakis, M.: XSketch synopses for XML data graphs. ACM Trans. Database Syst. 31(3) (2006)
21. Saha, D.: An incremental bisimulation algorithm. In: Arvind, V., Prasad, S. (eds.) FSTTCS 2007. LNCS, vol. 4855, pp. 204–215. Springer, Heidelberg (2007)
22. Schmidt, A., Waas, F., Kersten, M., Carey, M.J., Manolescu, I., Busse, R.: XMark: A benchmark for XML data management. In: VLDB (2002)
23. Spiegel, J., Polyzotis, N.: Graph-based synopses for relational selectivity estima-tion. In: SIGMOD (2006)

Social Based Layouts for the Increase of Locality in Graph Operations[⋆]

Arnau Prat-Pérez, David Dominguez-Sal, and Josep L. Larriba-Pey

DAMA-UPC, Departament d'Arquitectura de Computadors
Universitat Politècnica de Catalunya,
Campus Nord, C/Jordi Girona 1-3, 08034 Barcelona, (Catalonia, Spain)
{aprat,ddomings,larri}@ac.upc.edu

Abstract. Graphs provide a natural data representation for analyzing the relationships among entities in many application areas. Since the analysis algorithms perform memory intensive operations, it is important that the graph layout is adapted to take advantage of the memory hierarchy.

Here, we propose layout strategies based on community detection to improve the in-memory data locality of generic graph algorithms. We conclude that the detection of communities in a graph provides a layout strategy that improves the performance of graph algorithms consistently over other state of the art strategies.

Keywords: graph mining, performance, community detection.

1   Introduction

The number of application and research areas where data can be intuitively cast into relationship networks (i.e. graphs) is huge [2]. Just to cite a few examples, Internet is represented by the web sites and the links among these sites [24], social networks are represented by the individuals and their friendship or professional connections [24], protein interaction networks are represented by the proteins and how they are linked to perform particular biological functions [7] and bibliographic networks represent how the authors are linked by their co-authorship relationships [17]. In all these cases, the graphs are large and querying them requires significant computational effort to meet specific time restrictions.

Graph based applications query their data periodically in order to extract information about the relationships among nodes, or about their topology. For instance, computing routes in navigation systems [11], getting information from recommendation systems [6], analyzing the security of computer networks [33], visualizing proteins [7], designing drugs [20] or analyzing the relevance of particular users in social networks [21] are applications where computing connected

---

[⋆] The members of DAMA-UPC thank the Ministry of Science and Innovation of Spain and Generalitat de Catalunya, for grant numbers TIN2009-14560-C03-03 and GRC-1087 respectively.

J.X. Yu, M.H. Kim, and R. Unland (Eds.): DASFAA 2011, Part I, LNCS 6587, pp. 558–569, 2011.

components, looking for the minimum distance between pairs of nodes and computing cycles, forests, minimum spanning trees and centrality have been proven to be very important. All those operations require graph traversals, which are typically variants of Breadth First Search (BFS) and Depth First Search (DFS) and can be accelerated if the graph layout exploits the cache architecture.[1]

There has been an effort to improve the locality for some specific graph operations by means of designing compact structures for storing the graph [27] or by changing the way that specific algorithms access the data set [32, 36]. However, the issue of creating generic strategies that benefit different types of graph algorithms is still to be investigated. Leskovec et al. showed that typical large graphs coming not only from social relations but other fields such as citations networks, web graphs, authorship relations or data file sharing in peer to peer networks, have their nodes clustered into communities [24]. We have used this result to derive cache-aware graph layouts, which target social networks, but which in fact are flexible to target many other data sets.

This paper has three main contributions. The first one is the proposal of a new method for laying out a graph in order to improve data locality. These method is called COM, and is based on the conjecture that the nodes that belong to a community are spatially related and are likely to be traversed together. However, community detection techniques are difficult to scale to huge graphs because of their computational complexity. Thus, our second contribution is the proposal of COM(x). COM(x), which is based on COM, restricts the search of communities to sets of "x" nodes that are connected by BFS, reducing the cost of laying out the data while improving the performance of traversals. Finally, the third contribution of this paper is to provide a comparison of state of the art techniques for laying out social graphs. These strategies are BFSL [1] (from Breadth First Search Layout), a sparse matrix reordering strategy, i.e. the Cuthill-McKee reordering scheme [9] (referred to as CUTHILL from now on), Multilevel Spectral Bisection [3] (referred to as SPECTRAL), and GPART [18].

We conclude that laying out the graphs following the community structure (COM) not only allows for better performance of traversals than the state of the art strategies due to higher hit rates in the L1 and L2 caches, but also reduces the sparseness of the graph and leads to more compact memory layouts.

This paper is structured as follows. First, we present the related work in Section 2. Then, we propose communities as a data layout and its variations in Section 3. In Section 4, we describe the computational environment and the characteristics of the benchmarks generated. Section 5 describes the experiments we have performed to prove the goodness of community based layouts, and in Section 6 we perform a profiling of BFS traversal to understand the impact of the layouts on the memory hierarchy. Finally, in Section 7, we summarize our conclusions and give directions for future work.

---

[1] In this paper, we focus on the BFS. For the DFS there is an extended version of this work in [34].

2   Related Work

The memory hierarchy of the computer has a very important impact in applications that work with large data sets. The research literature has published many cache-aware solutions for different data structures such as lists [5], heaps [37], etc. However, most solutions for graphs are oriented towards data structures for storing them in external memory [36], assume a certain simple distribution (such as a uniform distribution, which is not typical for social networks) of the nodes and edges in the graph [4] or optimize certain expensive computing graph algorithms [32]. Our approach contrasts with them because we target irregular graphs that emerge naturally in social networks and we focus on the data layout but not on the final algorithm.

Given that graphs are typically represented as matrices, the first proposals of graph reorganization techniques correspond to matrix manipulation, which exchange the rows and columns of the matrix in order to move all the non-zero values in the matrix towards the diagonal. Social graphs correspond to very sparse matrices, in other words, they contain more zeroes than connections in a matrix representation of nodes and edges, thus the exchange of rows allow for increasing the density of non-zeroes in certain parts of the matrix. This diagonalization operation is known as the the bandwidth minimization problem (BMP), which is NP-complete [31]. The most popular solution to this problem was presented by Cuthill and McKee in [9], which performs a BFS traversal with a heuristic to select the nodes in decreasing order of degree. Other approaches for matrix manipulation, such as the algorithm proposed by Gibbs et al. [15], achieve faster execution time than CUTHILL but not better quality. Alternative approaches to matrix reorganization are spectral methods [35] such as Multilevel Spectral Bisection (SPECTRAL), which compute certain eigenvectors of the matrix that indicate the closeness between nodes.

Regarding the optimization of large graphs, Al-Furaih and Ranka proposed several reorganization methods to obtain better memory performance for particle interaction problems arising from physics in [1]. Among these methods, they concluded that layouts based on BFS (BFSL from now on) performed the best, with a low preprocessing overhead. Some other hierarchical methods which are used to reshape graphs are METIS [22], and GPART [18], which are based on locating graph partitions.

Finally, we find a survey of graph algorithms that encourage algorithm access patterns with a high locality in [36]. This approach is different from ours, where we do a generic optimization of the graph data structure based on the data relations, not on the particular algorithm. We believe that the combination of our graph layout with cache conscious algorithms would increase the performance.

3   Community Based Data Layouts

The analysis real world graphs (which are typically neither uniform nor completely random) has found that graphs are characterized by probability distribution laws that model the inhomogeneities of the graph, and the overall organization of the edges among nodes [23]. The distribution of edges reveal the presence of communities [13], which are groups of nodes with high densities of edges and low densities of edges between nodes of other groups.

Communities are groups of vertices which probably share common properties and/or play similar roles within the graph. Communities may correspond to groups of pages of the World Wide Web dealing with related topics [14], functional modules such as cycles and pathways in metabolic networks [30], groups of related individuals in social networks [16], etc. The vertices within a community are highly connected, so the probability for an edge to exist between pairs of vertices that belong to a community is high.

The rationale behind the use of communities to layout the data of a graph is as follows. In general, graph traversals such as BFS and DFS, visit the graphs in a way where topologically close nodes are visited in nearby iterations. Thus, intuitively, if we put those nodes that belong to the same community close in memory, we will achieve higher spatial locality, because those nodes will be topologically close.

Although the problem of community detection is intuitively clear, there is not a standard formal definition for communities. Thus, several methods have been proposed [13,10,19,29]. We will base our work on the community detection method for large networks proposed by Clauset et al. in [8], because it has been classified as one of the most efficient methods for detecting non overlapping communities on large networks. In the next section, we explain the algorithm that we use to layout the graph using communities.

3.1 Community Layout - COM

We propose Community Layout (COM) as an algorithm to arrange the layout of graph data, following the topology of the communities. In order to detect the communities, we take the community detection procedure proposed by Clauset et al., which is widely used in the literature [8, 26]. Clauset et al. propose a greedy algorithm, which uses a metric called modularity. Modularity measures the quality of a partition of the network into non overlapping communities. At the beginning of the algorithm, each node is a community. At each step, the algorithm merges the two most related communities until the modularity does not increase. In other words, for every pair of communities, the algorithm searches for those that improve the modularity maximally if merged. The algorithm is executed until no improvement on the modularity can be reached.

After the communities have been detected, COM arranges the layout of the graph. For each community found, COM labels all the nodes with consecutive node identifiers and stores them contiguously in memory.

3.2 Truncated Community Layout - COM(x)

Since applying COM over huge graphs may be very time consuming due to the complexity of the community extracting algorithm ($O(md\ log(n))$, where m is the number of edges, n the number of nodes and d is the depth of the "dendogram" describing the network's community structure), we propose a faster alternative that we call Truncated Community layout COM(x).

Although some communities in the graph are very large, there are typically many smaller communities. Since it is not necessary to consider the whole graph topology to locate such small communities. COM(x) simplifies the community detection problem, exploring the graphs in chunks of x connected nodes. COM(x) is an iterative algorithm that, at each step, selects an unvisited node n from the graph and builds a subgraph following a BFS traversal of unvisited nodes starting from n. Once COM(x) finds x connected unvisited nodes (or the connected component has no more nodes to visit), it applies COM on this subset of nodes to build the layout of the graph. This procedure is repeated until all the nodes in the graph have been visited once.

In general, COM(x) detects a larger number of communities than COM. For instance, if a community is larger than x or if a community is not fully included in the subset of x nodes, COM(x) takes them as two separate communities. Thus, if our conjecture that the arrangement of the graph in communities improves the data locality is true, then the layout generated by COM(x) will not be as efficient as for COM. Nevertheless, we will see in the experimental section that for most of the graphs, COM(x) achieves speedups not very far from COM with a significantly shorter preprocessing time.

4   Experimental Setup

In this section, we present the experimental setup used to test how the graph layout affects the performance of the application. We implement the graphs with the aid of DEX, which is a very compact and efficient graph representation library [25]. DEX stores the adjacency lists of nodes and edges in bitmaps, which are more cache-friendly and more suitable for large graphs than the standard adjacency lists. DEX stores the adjacency lists in a B-Tree that maps the ids of the nodes to compressed bitmaps, which are a compact implementation of an adjacency list. BFS is implemented using a queue and a visited node vector which helps in the backtracking procedure.

The computer used to execute the experiments has the following characteristics. It has an Intel Xeon processor at 2.83 Ghz, with 32 KB of L1 cache for instructions and 32 KB for data, 2x6 MB of L2 shared cache and 64 GB of main memory. The algorithms compared are the following:

1. BFSL [1] , CUTHILL [9], SPECTRAL [3] and GPART [18], which correspond to the state of the art.
2. BFSL(x) which we propose as the simplest way to find community like structures. It consists in performing several BFS traversals of a graph. Each traversal starts from a different node selected at random. Once x nodes from the graph have been visited, the traversal stops and a new one is started. The process continues until all the nodes of the graph have been visited.
3. COM and COM(x) which are described in section 3.

Table 1. Different graphs created for the experiments

Num. nodes	Max. degree	Avg. degree	Min. community size	Max. community size
100K	500	40	20	500
500K	100	8	4	100
500K	500	40	20	500
500K	1000	80	40	1000
500K	2000	160	80	2000
1,000K	500	40	20	500

4.1   Social Network Generation

We generated several graphs with the aid of the graph generator proposed by Lancichinetti et al. [23]. This tool generates graphs that have the characteristics of social network graphs and allows us to compare the performance of the strategies for different graph sizes and densities. The seven parameters used to configure the graphs are the following: the number of nodes, the maximum degree, the average degree, the minimum community size, the maximum community size, the degree distribution exponent, the community size distribution exponent and the mixing factor.

Table 1 shows the different graphs we have created for the experiments and the values for the parameters used to generate them. We have used six different graphs in our experiments. Four of them have the same size, 500K nodes, and we increase the number of edges by increasing the maximum and average degree, and the minimum and maximum community sizes. The degree distribution exponent was set to 2, the community size distribution exponent was set to 1 and the mixing factor which was set to 0.2. All of them are set by default in the software to make the graphs adapt to the characteristics of social data graphs [23].

In the experiments we refer to the layout output by the graph generator as RANDOM. It becomes the baseline of our testing workbench. Additionally, the graph generator provides information about the communities created, i.e. nodes that belong to the same community. We use this information to know the communities created by the graph generator for the 1,000K node graph, to avoid using the community detection algorithm, which is computationally expensive.

5   Experiments

5.1   Comparison of Layout Methods

In this section, we compare the different methods previously described, for the social network with 500K nodes and an average of 160 nodes per edge. Each observation is obtained as follows: we execute ten series of ten executions of the BFS traversal algorithm for a fixed graph layout, starting each series from a different node selected at random. Our reported observation is the average execution time of the 100 measurements. The results for the execution of the DFS traversal follow a similar trend as the BFS, and can be found in [34].

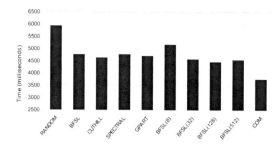

Fig. 1. Average execution time of BFS for different layouts. The graph has 500K nodes and 160 edges per node on average.

In Figure 5.1(a), we show the average execution time for the previously described BFS traversal experiment. The leftmost bar in each plot corresponds to the original graph with the nodes laid out in the order given by the network data generator. We observe that all the reorganization techniques under test provide a significant reduction of the execution time because of a better spatial locality.

Among all the techniques that we are testing, COM is the best layout in terms of performance. It improves the performance of the second best algorithm by 18% and over the basic layout by 58%. Regarding the previously published techniques, BFSL, CUTHILL, SPECTRAL and GPART, we observe that reduce the execution time by a similar amount of time, approximately 28%, which is not as good as for COM.

CUTHILL reorganizes the data in the matrix in blocks of nodes that have a high connectivity, and therefore the execution time is reduced. On the other hand, BFSL is effective because it groups the nodes that are accessed in sequence in a BFS traversal. However, this locality is better for the first nodes of the traversal than for the rest. If the BFS traversal started from a different node than the one selected to build the layout, then the nodes would be accessed in a very different order, and thus it would not get such a good locality. In order to deal with this problem, BFSL(x) clusters the groups of nodes with a depth limit. BFSL(x) detects groups of x nodes with spatial locality. This intuition is confirmed by Figure 5.1, which shows that BFSL(128) is better than the BFSL. We note that BFSL(x) resembles a simple community detection algorithm of fixed size, and thus it seems natural that COM, which is more precise setting the communities, performs better.

5.2 Scalability Analysis

The scalability of the algorithms is very important because we aim at arranging layouts for arbitrarily large or dense graphs. In this experiment, we test the speedup trends of the different layout techniques with respect to the size of the graph. We vary the two dimensions of the graph size: the number of edges, and the number of nodes.

Figure 2 shows the performance of BFS for graphs with a variable number of edges. According to previous studies, typical graphs have an average number of

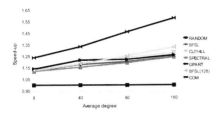

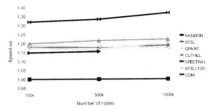

Fig. 2. Speedup of BFS for graphs with different edge densities. The number of nodes is fixed to 500K.

Fig. 3. Speedup of BFS for graphs with different number of nodes. The average number of edges per node is 40.

edges that range from a few units to a few hundred units [24, 12]. We observe that all the techniques improve their locality on denser graphs because graphs with more edges have matrices where more nodes can be clustered together. We see that COM is the algorithm with the best performance for all the tested edge densities. The detection of communities is flexible to locate groups of nodes that are very connected with respect to the density of the rest of the graph, which is independent of the average edge degree.

Figure 3 shows the evolution of the performance for varying numbers of nodes using BFS. We observe that state of the art approaches are stable with an approximate speedup of 1.20. However, COM is able to obtain speedups of at least 1.30, and up to 1.37. This happens because if the number of nodes is larger but the average number of edges is constant, then the probability that two random nodes are connected is smaller, and thus the graph is sparser. COM detects these clusters of nodes and groups them in nearby regions of memory. For the rest of layouts, the arrangement improves the speedup over RANDOM but below COM.

Overall, BFSL, BFSL(x), SPECTRAL and GPART did not prove better than CUTHILL. So, in the following sections, we will use CUTHILL as the baseline.

5.3 Community Size Discussion

In our previous experiments, we showed that COM is the best algorithm in terms of performance and scalability. However, as already mentioned in Section 3, the detection of communities is an expensive computing operation. Thus, we propose COM(x), which is discussed in this section.

Figure 4 depicts the preprocessing time to generate the layout of a 100K and 1,000K node graph in logarithmic scale. We were not able to compute the communities for the largest graph because it exceed three computing days. For this particular configuration, we used the communities provided by the graph generator for estimating the speedup of COM. In Figure 4, we see that COM(x) reduces the search space to find the community to a subset of x nodes, which is very effective to reduce the preprocessing time. For example, COM(2048) has a preprocessing time comparable to CUTHILL.

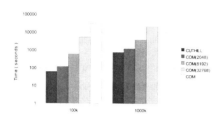

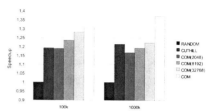

Fig. 4. Preprocessing time for arranging the graph layout for 100K and 1000K nodes and 160 edges per node

Fig. 5. Speed up of BFS for different layouts for 100K and 1000K nodes and 160 edges per node

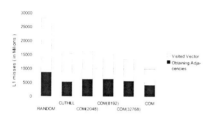

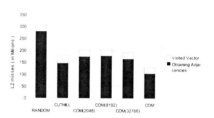

Fig. 6. Number of L1 misses of BFS for 500K nodes and 160 edges per node

Fig. 7. Number of L2 misses of BFS for 500K nodes and 160 edges per node

In Figure 5, we show the speedup of the BFS traversal for the graphs of 100K and 1,000K nodes. We observe that CUTHILL and COM(2048) behave similarly with respect to the traversal speedup. Nevertheless, the values between 2048 and the full graph show a progressive increase in the speedup for COM(x) which is far above CUTHILL. Large values of x provide better overall performance because larger subsets of nodes are more likely to contain the large communities.

Given that for very large graphs the larger communities tend to increase too, it is necessary to set large values of x in order to detect these communities. Although the computational time for COM(2048) is still comparable to CUTHILL, we obtained a performance speedup slightly worse for this configuration. Nevertheless, if we set the limit higher to ease the location of large communities, such as COM(32768), the speedup is over CUTHILL. All in all, COM(x) is an efficient approach that addapts to the necessities of the graph application by adjusting the value of the parameter x.

6   Profiling

In this section, we evaluate the behavior of the architecture at hand with a profile of the execution of BFS. We repeated the experiments reported in Section 5.1 with the Oprofile daemon activated [28]. We measure the main factors that determine the memory performance of an application, which are the number of misses in the memory hierarchy of the processor (the number of misses in the L1 and L2 caches). We report our observations in Figures 6-7. For each of the executions, we divided the measure into the accesses to the two main data structures of the traversals: (a) the boolean vector that indicate whether a node is visited or not and (b) the adjacency lists.

Regarding the cache hierarchy, in Figure 6 we observe that the number of misses in the L1 cache diminishes with any rearrangement of the graph under study. However, we observe that the reduction is not homogeneous between the accesses to the two data structures. We observe that the number of misses to the vector is up to one order of magnitude larger than to the adjacency lists. Each time a node is visited, the traversal checks whether its neighbors have been visited or not in order to continue the exploration. Since COM arranges the nodes by dense regions and there are more edges inside the community than outside, then it is more likely that the neighboring nodes in the graph lay contiguous in memory. We also find that the number of misses to the adjacency lists by COM is smaller and thus, performs better. Although the volume of cache misses is smaller than for the visited vector, we find it is less relevant fo the execution time.

When we turn to the analysis of the L2, in Figure 7 we observe that the behavior is different. The fraction of misses to the boolean vector is one order of magnitude smaller than to the adjacency lists. This is because the boolean vector fits in the L2 cache but not in the L1. Therefore, the data structure to optimize is the adjacency lists in contrast to the boolean vector for L1. Nevertheless, COM is also the algorithm that reduces more the misses to the adjacency lists.

The profiling of the traversals demonstrates that our performance improvement comes from a smaller number of misses in when accessing both caches, specially for L1 which produces the largest benefit.

7   Conclusions and Future Work

The research described in this paper has the objective to improve the performance of graph algorithms by improving the spatial locality of the in-memory graph representation. We have departed from the conjecture that the nodes that belong to a community are spatially related and have a significant importance in shaping the traversals of graph algorithms. Our first important result shows that layouts based on communities like the one we propose, COM, improves the performance of common graph algorithms more significantly than other state of the art layouts because of a better usage of the cache hierarchy. Moreover, for graph traversals, COM works better in laying out the graphs than the same traversal algorithms used as a layout strategy.

Our second important result is related to the cost of community detection in large graphs. Given that this is a very expensive procedure, we propose truncated approaches to layout the graph based on community detection. This new technique, called COM(x), is able to preprocess data at a comparable speed to previous state of the art proposals and gets results comparable to COM. Furthermore, the quality of the layout is proportional to the value of the parameter "x" while the preprocessing speed is inversely proportional. This behaviour allows the user to adjust the value of the parameter "x" according to his necessities.

The future of our research will go towards the adaptation of community based techniques to environments where graphs have additional information besides the graph structure. Our main focus goes towards the reorganization of attributed and labeled graphs (i.e. typed graphs) where the nodes and edges have associated information that can be taken into account for improving the speed and quality to find communities in the graph.

References

1. Al-Furaih, I., Ranka, S.: Memory hierarchy management for iterative graph structures. In: IPPS/SPDP, pp. 298–302 (1998)
2. Angles, R., Gutiérrez, C.: Survey of graph database models. ACM Comput. Surv. 40(1), 1–39 (2008)
3. Barnard, S., Simon, H.: Fast multilevel implementation of recursive spectral bisection for partitioning unstructured problems. CPE 6(2), 101–117 (1994)
4. Baswana, S., Sen, S.: Planar graph blocking for external searching. Algorithmica 34(3), 298–308 (2002)
5. Bender, M.A., Cole, R., Demaine, E.D., Farach-Colton, M.: Scanning and traversing: Maintaining data for traversals in a memory hierarchy. In: Möhring, R.H., Raman, R. (eds.) ESA 2002. LNCS, vol. 2461, pp. 139–151. Springer, Heidelberg (2002)
6. Mirza, B.J., Keller, B.J., Ramakrishnan, N.: Studying recommendation algorithms by graph analysis. JIIS 20(2), 131–160 (2003)
7. Brown, K., Otasek, D., Ali, M., McGuffin, M., Xie, W., Devani, B., van Toch, I.L., Jurisica, I.: Navigator: Network analysis, visualization and graphing toronto. Bioinformatics 25(24), 3327–3329 (2009)
8. Clauset, A., Newman, M., Moore, C.: Finding community structure in very large networks. Physical Review E 70(6) (2004)
9. Cuthill, E., McKee, J.: Reducing the bandwidth of sparse symmetric matrices. In: Proceedings of the 1969 24th National Conference, pp. 157–172. ACM, New York (1969)
10. Dourisboure, Y., Geraci, F., Pellegrini, M.: Extraction and classification of dense communities in the web. In: WWW, pp. 461–470 (2007)
11. Duckham, M., Kulik, L.: Simplest Paths: Automated Route Selection for Navigation. In: Kuhn, W., Worboys, M.F., Timpf, S. (eds.) COSIT 2003. LNCS, vol. 2825, pp. 169–185. Springer, Heidelberg (2003)
12. Facebook: Press room - statistics,
http://www.facebook.com/press/info.php?statistics
(Last retrieved in January 2010)
13. Flake, G., Lawrence, S., Giles, C.: Efficient identification of web communities. In: KDD, pp. 150–160 (2000)
14. Flake, G., Lawrence, S., Giles, C., Coetzee, F.: Self-organization and identification of web communities. IEEE Computer 35(3), 66–71 (2002)
15. Gibbs, N., Poole, J., Stockmeyer, P.: An algorithm for reducing the bandwidth and profile of a sparse matrix. SIAM Journal on Numerical Analysis 13(2), 236–250 (1976)
16. Girvan, M., Newman, M.E.: Community structure in social and biological networks. PNAS 99(12), 7821–7826 (2002)

17. Gómez-Villamor, S., Soldevila-Miranda, G., Giménez-Vañó, A., Martínez-Bazan, N., Muntés-Mulero, V., Larriba-Pey, J.: Bibex: a bibliographic exploration tool based on the dex graph query engine. In: EDBT, pp. 735–739 (2008)

18. Han, H., Tseng, C.: Exploiting locality for irregular scientific codes. TPDS, 606–618 (2006)

19. Ino, H., Kudo, M., Nakamura, A.: Partitioning of web graphs by community topology. In: WWW, pp. 661–669 (2005)

20. Ivanciuc, O., Balaban, A.: Graph theory in chemistry. In: The Encyclopedia of Computational Chemistry, pp. 1169–1190 (1998)

21. Musiał, K., Kazienko, P., Bródka, P.: User position measures in social networks. In: SNA-KDD, pp. 1–9. ACM, New York (2009)

22. Karypis, G., Kumar, V.: METIS: Unstructured graph partitioning and sparse matrix ordering system, vol. 2. The University of Minnesota (1995)

23. Lancichinetti, A., Fortunato, S., Radicchi, F.: Benchmark graphs for testing community detection algorithms. PRE 78(4) (2008)

24. Leskovec, J., Lang, K., Dasgupta, A., Mahoney, M.: Statistical properties of community structure in large social and information networks. In: World Wide Web Conference, pp. 695–704 (2008)

25. Martínez-Bazan, N., Muntés-Mulero, V., Gómez-Villamor, S., Nin, J., Sánchez-Martínez, M., Larriba-Pey, J.: Dex: high-performance exploration on large graphs for information retrieval. In: CIKM, pp. 573–582 (2007)

26. Newman, M., Girvan, M.: Finding and evaluating community structure in networks. Physical Review E 69(2), 026113 (2004)

27. Niewiadomski, R., Amaral, J.N., Holte, R.: A performance study of data layout techniques for improving data locality in refinement-based pathfinding. JEA 9, 1–2 (2004)

28. Oprofile: Oprofile documentation, http://oprofile.sourceforge.net/docs/ (Last retrieved in January 2010)

29. Padrol-Sureda, A., Perarnau-Llobet, G., Pfeifle, J., Muntés-Mulero, V.: Overlapping community search for social networks. In: ICDE, pp. 992–995 (2010)

30. Palla, G., Derényi, I., Farkas, I., Vicsek, T.: Uncovering the overlapping community structure of complex networks in nature and society. Nature 435(7043), 814–818 (2005)

31. Papadimitriou, C.: The np-completeness of the bandwidth minimization problem. Computing 16(3), 263–270 (1976)

32. Park, J., Penner, M., Prasanna, V.: Optimizing graph algorithms for improved cache performance. IEEE TPDS 15(9), 769–782 (2004)

33. Phillips, C., Swiler, L.: A graph-based system for network-vulnerability analysis. In: NSPW, pp. 71–79 (1998)

34. Prat-Pérez, A.: Master thesis: Social based layouts for the increase of locality in graph operations (2010), http://www.dama.upc.edu

35. Barnard, S.T., Pothen, A., Simon, H.D.: A spectral algorithm for envelope reduction of sparse matrices. In: SC, pp. 493–502 (1993)

36. Vitter, J.: Algorithms and data structures for external memory. FTTCS 2(4), 305–474 (2006)

37. Wilson, P., Lam, M., Moher, T.: Effective "static-graph" reorganization to improve locality in garbage-collected systems. In: PLDI, pp. 177–191 (1991)

Generating Random Graphic Sequences[*]

Xuesong Lu and Stéphane Bressan

School of Computing,
National University of Singapore
{xuesong,steph}@nus.edu.sg

Abstract. The graphs that arise from concrete applications seem to correspond to models with prescribed degree sequences. We present two algorithms for the uniform random generation of graphic sequences. We prove their correctness. We empirically evaluate their performance. To our knowledge these algorithms are the first non trivial algorithms proposed for this task. The algorithms that we propose are Markov chain Monte Carlo algorithms. Our contribution is the original design of the Markov chain and the empirical evaluation of mixing time.

1 Introduction

In this paper we are interested in the random generation of graphic sequences. The problem is trivial if one wishes to generate graphic sequences according to the underlying graph distribution. The problem is particularly difficult if one wishes to generate graphic sequences uniformly at random.

There is evidence that useful graphs in most applications domains follow a prescribed degree sequence or degree law (typically the power law). Numerous algorithms have been developed that generate graphs from prescribed degree sequences [1–3] and laws [4] or that evaluate their structure and dynamics [5]. The latter example mines and evaluates the interestingness of motifs in all kinds of graphs such as transcription networks, ecological food webs and neuron synaptic connection networks.

It is therefore necessary to provide algorithms that generate graphic sequences (realizable degree sequences), in particular uniformly at random, in order to evaluate these algorithms. Using random graphs or graphic sequence with an underlying distribution of random graphs would neglect rare but possibly significant graphic sequences.

The rest of the paper is organized as follows. Section 2 introduces preliminary definitions and results as well overviews the relevant related work. Section 3 presents two algorithms for the uniform generation of random graphic sequences. Section 4 presents an empirical evaluation of the effectiveness and efficiency of the algorithms proposed. Finally, we conclude in Section 5.

Further discussions, results and detailed proofs are available in [28].

---

[*] This research is supported by NUS grant R-252-000-328-112.

J.X. Yu, M.H. Kim, and R. Unland (Eds.): DASFAA 2011, Part I, LNCS 6587, pp. 570–579, 2011.
© Springer-Verlag Berlin Heidelberg 2011

2   Background and Related Work

2.1   Degree Sequences

Without loss of generality and for the sake of simplicity of exposition we consider *simple graphs*, that is undirected graphs without self-loops and multiple edges. The degree sequence of a graph [26] is the non-increasing sequence of natural numbers corresponding to the degrees of vertices in G.

For a variety of applications, authors have considers the realizability, construction, enumeration and, less so, counting and generation of graphs with prescribed degree sequences. The list of applications is long and increasing. For instance, the authors of [14] assess the interest of data mining results by comparing them with the results obtained from mining random graphs with the same degree sequence. Interesting patterns and rules are those who are specific to the original graph rather than those frequently appearing in graphs with the same degree sequence. The authors of [10] propose a notion of k-degree anonymity and anonymization algorithms for privacy preservation in graphs in general and in social networks in particular. In this approach, identity disclosure and its prevention depend and rely on degree sequence.

One particularly interesting problem is the random generation of graphs with prescribed degree sequences. The authors of [1] present and compare three mainstream algorithms: a naïve configuration algorithm that is painstakingly matching stubs under the prescribed sequence constraints, a local optimization algorithm and a Markov Chain Monte Carlo algorithm called the switching algorithm. Recently, the authors of [3] have proposed a polynomial time algorithm that avoids the drawbacks of backtracking and uncontrolled rejection of the three approaches above. The exact problem statement may vary depending on the nature of the graph and on additional constraints. For instance, the authors of [2] consider the uniform generation of random simple connected graphs with a prescribed degree sequence.

Other applications need to determine the structure and dynamics of graphs with prescribed degree sequences. For instance, the authors of [5] mine and evaluate the significance of motifs in transcription networks, ecological food webs and neuron synaptic connection networks. The authors of [6] investigate the number of vertices and the number of cycles in the largest component of random graphs with a given degree sequence. Connected components [8] and Hamilton cycles [7] are also investigated in the graphs with prescribed degree sequences.

2.2   Graphic Sequences

Therefore it is important to consider, upstream from the problems of graphs with prescribed degree sequences, the realizability, construction, enumeration, counting and generation of degree sequences themselves.

However, not every non-increasing sequence of natural numbers is a degree sequence. Non-increasing sequence of natural numbers that is the degree sequence of a graph is called a *graphic sequence* [26] or a *realizable degree sequence*.

The realizability of degree sequences is first investigated by Havel [17]. Hakimi [16] then complements the work and obtains a sufficient and necessary condition for a degree sequence to be graphic. The authors of [22] shows the equivalence of seven previously proposed necessary and sufficient criteria for a sequence of integers to be graphic.

The original construction problem is trivial since a sequence of zeros is graphic. We shall use this trivial graphic sequence as a starting state in the Markov Chain Monte Carlo method that we devise in this paper.

As remarked by the Ruskey of [9] the enumeration of graphic sequences has been largely overlooked. They propose an algorithm that enumerates graphic sequences with prescribed length. The algorithm leverages Havel and Hakimi condition [17]. The authors conjecture and verify experimentally that their algorithm is running in constant amortized time, i.e. in time proportional to the size of the output. Nevertheless its worst case complexity remains exponential (see [23] and the following discussion about counting.)

Barnes and Savage propose in [24] an algorithm for the enumeration of graphic sequences with prescribed sum.

There is no known analytical formula for the counting of graphic sequences with prescribed length. Burns [23] gives a upper bound of $4^n/(\log n)^C \sqrt{n}$ and a lower bound of $4^n/Cn$.

Other authors have addressed related but different problems. Barnes, in [19], proposes a recurrence and a polynomial-time algorithm for counting graphic sequences with prescribed sum. Stanley, in [20] and Peled, in [21], count the number of ordered graphic sequences (as opposed to graphic sequences which are multi-sets) as a (exponential) function of the number of odd cycles in the corresponding forest.

While the construction problem, without or with further constraints, is rather simple, the counting problem seems difficult. This apparent paradox makes the problem of counting, and the problem of generating graphic sequences uniformly at random, interesting and challenging. To our knowledge, we are the first to address the random generation of graphic sequences with prescribed length.

2.3   Random Walks on Markov Chains

Markov chain Monte Carlo (MCMC) algorithms are random walks on Markov chains. They stem from the Monte Carlo methods used in applied statistics where they were used for simulation and sampling [11]. Sinclair, in his monograph [12], has formalized and popularized their use for random generation and counting.

An MCMC algorithm (see [12]) builds and randomly walks on a Markov chain whose states correspond to the objects being sampled. If the Markov chain is carefully designed, if it is finite and ergodic (irreducible and aperiodic), then it has a stationary distribution. Namely, the stationary probability vector of the Markov chain is $\pi = \pi P$, where P is the transition matrix of the Markov chain. For instance, the stationary distribution is uniform if the Markov chain's graph is regular. It is not necessary that all the states of the Markov chain correspond to object being sampled as a rejection mechanism can filter out those undesirable objects. The mixing time of the Markov chain is the number of steps t required to reach the stationary distribution. A sufficiently long random walk, longer than the mixing time of the chain, will reach states at random according approximately to the stationary distribution.

For a given generation problem the challenge is to devise a rapidly mixing ergodic Markov chain whose states are the objects to be generated with the desired stationary distribution. For example, the authors of [1] and those of [14], among others, use MCMC algorithms to generate random graphs with prescribed degree sequences. The authors of [15] use it to sample graph patterns. As their Markov chain is not a regular graph, they modify the weights of certain edges, as suggested by Sinclair in [12], to obtained the wished stationary distribution. In this paper, we devise MCMC algorithms to generate graphic sequences with prescribed length and graphic sequences with prescribed length and sum. Our contribution is the design of the corresponding Markov Chains and the empirical evaluation of the (rapid) mixing times.

3   Generating Random Graphic Sequences

We now present algorithms for the random generation of graphic sequences. It is straightfoward to generate random graphic sequences with prescribed length, and prescribed length and sum, according to the underlying distribution of graphs. This is done by generating the corresponding random graph (see [13]) and observing its degree sequence. However generating graphic sequences uniformly at random is more challenging. We propose two algorithms to generate uniformly at random graphic sequences with prescribed length, and prescribed length and sum, respectively. We contribute the first non trivial algorithms for these tasks and study their effectiveness and efficiency. We also discuss practical optimization of the algorithms.

3.1   Uniformly Random Graphic Sequence with Prescribed Length

We consider graphic sequences with prescribed length uniformly at random. We call this model $\mathcal{D}_u(n)$.

Here a naïve algorithm to generate sequences in the model $\mathcal{D}_u(n)$ consists in enumerating the different graphic sequences (for instance using the algorithm of [9]) and then choosing one at random among the different ones. This approach runs in exponential time.

A Markov chain Monte Carlo approach may be able to give us acceptable approximations (almost uniform) algorithms in polynomial time. The two issues at hand are the construction of the Markov chain and the evaluation of its mixing time.

We can now present the construction of the Markov chain for the algorithm $D_u(n)$ that we advocate. We consider a chain that contains both graphic and non-graphic sequences. The initial state[1] is the zero sequence as it is graphic. At

---

[1] The initial state could be any other state and we use the words "state" and "sequence" indistinctively.

each transition, we increment or decrement by 1 one element in the sequence. We only consider a new state if its elements are in the $[0, n-1]$ interval. We only consider the new state if it is in non-increasing order. We do not consider states that are non-graphic (This is tested using one of the available necessary and sufficient conditions [16–18].) and whose sum is even. However the Markov chain contains both graphic and non-graphic sequences. We call this Markov chain MC, We call P its transition matrix.

We consider the Markov chain MC' with transition matrix P^2. We cannot directly construct and walk on MC'. Rather we will walk on MC and consider even numbers of steps.

We remark that two adjacent states in MC cannot be both graphic or both non-graphic. Therefore MC has the interesting property that states that correspond to graphic sequences can only be reached in an even number of steps (from a graphic state). This may look as if compromised the aperiodicity of the Markov chain but it does not as far as we are concerned since we will walk even numbers of steps and are only interested in graphic states.

Furthermore, because we have not included in MC the non-graphic states whose sum is even, only graphic states can be reached in an even number of steps (from a graphic state). We show in [28] that MC' is irreducible and aperiodic, therefore ergodic, and that all graphic states are included in MC'.

Unfortunately different states have different stationary probabilities. It is possible to modify the weight to make the stationary distribution uniform. We use a technique suggested by Sinclair in [12] and used by the authors of [15], that consists in allocating appropriate weights to the transitions. The result that authorizes and justifies these changes is given by Cover and Thomas in [25]. Namely the stationary probability of each state is proportional to the sum of the weights of its incident transitions, with the following transition matrix P.

$$p_{(i,j)} = \begin{cases} \frac{w_{(i,j)}}{\sum_{l \in adj(i)} w_{(i,l)}} & \text{if } j \in adj(i) \\ 0 & \text{if } j \notin adj(i), \end{cases} \tag{1}$$

where $w_{(i,j)}$ is the weight of edge corresponding to the transition from state i to j.

We show in [28] that the stationary probability of graphic state in MC' is proportional to the total weight incident to that state and the stationary probability for a non-graphic state is zero for the transition Matrix P^2.

For state i and j where i and j are adjacent, the weight $w(i,j)$ is assigned as per Equation 2 where d_i is the degree of state i. Remember that i and j cannot be both graphic or both non-graphic.

$$w(i,j) = \begin{cases} \frac{1}{d_i} & \text{if } i \text{ is a graphic state and } j \text{ is a non-graphic state} \\ \frac{1}{d_j} & \text{if } i \text{ is a non-graphic state and } j \text{ is a graphic state.} \end{cases} \tag{2}$$

Note that $w(i,j) = w(j,i)$. Consequently, the sum of weights of every graphic state is 1. We have shown that the stationary distribution of graphic states is uniform.

In summary, $D_u(n)$ is an algorithm that builds and randomly walks in the Markov chain MC with transition matrix P, yet outputs only evenly (or oddly, depending on the initial state) reachable states. It simulates the ergodic Markov chain MC' with transition matrix P^2 which has a uniform stationary distribution thanks to a modification of the weights in MC. The algorithm is illustrated in Algorithm 1. The parity of t depends on the initial state.

Algorithm 1. Algorithm $D_u(n)$

    Input: n : the length of the sequence; t : the steps of random walk
    Output: DS : a graphic sequence generated from uniform distribution

1 *Start from any initial state S_0 of non-increasing sequence;*
2 $i = 1$;
3 **while** $i \leq t$ **do**
4 | *Compute locally the transition matrix P;*
5 | *Transfer from state S_{i-1} to S_i according to P;*
6 | $i = i + 1$;
7 **end**
8 $DS = S_t$;

3.2 Uniformly Random Graphic Sequence with Prescribed Length and Sum

We consider random graphic sequences with prescribed length and sum uniformly at random. We call this model $\mathcal{D}_u(n, s)$.

We propose an algorithm, which we call $D_u(n, s)$. It is also a Markov chain Monte Carlo algorithm. The Markov chain in which $D_u(n, s)$ is walking is similar to the one in which $D_u(n)$ is walking. However $D_u(n, s)$ considers less states than $D_u(n)$ since the graphic sequences of the model $\mathcal{D}_u(n, s)$ are only those with prescribed sum s. The states in the Markov chain in which $D_u(n, s)$ is walking are restricted to those that correspond to sequences with sum s , for graphic states and $s - 1$ and $s + 1$, for non-graphic states. We do not illustrate the algorithm as it is a simple variation of Algorithm 1.

3.3 Practical Optimization for $D_u(n)$

A practical optimization for $D_u(n)$ is based on the observation that complement graphs have related graphic sequences. We define the *complement sequence* of a graphic sequence $(d_1, d_2, \ldots, d_n)$ as $(n - 1 - d_n, n - 1 - d_{n-1}, \ldots, n - 1 - d_1)$. In [28] we prove that the complement sequence of a graphic sequence is graphic. The proof uses Erdös-Gallai Lemma [18].

The number of states that Algorithm 1 visits can then be reduced to half. The Markov chain only includes states that correspond to sequences with sum of degrees less than or equal to $n(n - 1)/2$. For every random sequence generated, we then further toss a coin to decide whether we output the sequence or its complement (or keep the sequence or discard it for those sequences who equal their complement.)

4   Performance Evaluation

In this section, we empirically evaluate the effectiveness and efficiency of the proposed algorithms. We implement the algorithms and run the experiments on a Microsoft Windows 7 machine with an Intel Core 2 Quad 2.83G CPU and 3GB memory. All the algorithms are implemented using Visual C++ 9.0.

The effectiveness and efficiency of $D_u(n)$ and $D_u(n, s)$ are evaluated by measuring the fitness of a sample generated by the algorithm to the uniform distribution after each transition. We use for that purpose the standard deviation. We have verified and confirmed the results and the conclusions below with other tests of fitness such as χ^2 and Jensen-Shannon divergence. We evaluate the efficiency, that is the number of steps, needed to achieve a desired effectiveness, measured by the test of fitness.

4.1   Performance of $D_u(n)$

We vary the length n of the sequences from 3 to 11 for $D_u(n)$. We measure the standard deviation for number of steps in MC varying from $n \times (n + 1)$ to $10 \times n \times (n + 1)$ (there are half this number of steps in MC') in increments of $n \times (n + 1)$.

Figure 1 illustrates the standard deviation for varying number of steps for $D_u(n)$. For each n the number of generated graphic sequences is $k \times d(n)$ where $d(n)$ is the number of distinct graphic sequences of length n and $k = 100$ in the experiments. The numbers of distinct graphic sequences of some lengths can be found in A004251 of Sloane and Plouffe's encyclopedia [27]. We observe from

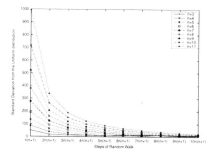

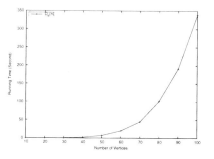

Fig. 1. Standard Deviation from Uniform for varying number of steps for $D_u(n)$ for different n

Fig. 2. Running time of $D_u(n)$

Figure 1 that the algorithm is quickly reaching a minimum standard deviation (which is not 0 because of the parameter k). We can also see that this empirical mixing time increases with n. Empirically, we conservatively estimate the mixing time to be n^3 number of steps. Figure 2 shows the average running time for the generation of 10 graphic sequences of $D_u(n)$ for n varying from 10 to 100 in increments of 10. The curve is expected to be of the order of n^3 by construction.

It however reveals the actual value of a high constant which is the cost of one step in the Markov chain. We further notice that this constant depends on n. Therefore the actual complexity of the algorithm should combine the number of steps with the processing at each step. We can generate a graphic sequence of 100 elements uniformly at random in 340 seconds.

The mixing time and the running time can be effectively divided by 2 using the practical optimization that we have proposed. Figure 3 and Figure 4 show the mixing and running performance of $D_u(n)$ with the practical optimization. We can generate a graphic sequence of 100 elements uniformly at random in 160 seconds.

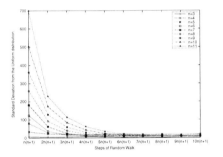

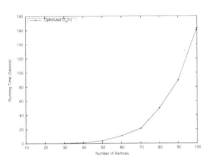

Fig. 3. Standard Deviation from Uniform for varying number of steps for $D_u(n)$ with the practical optimization for different n

Fig. 4. Running time of $D_u(n)$ with the practical optimization

4.2   Performance of $D_u(n, s)$

We vary the length n of the sequences from 3 to 13 for $D_u(n, s)$. We fix the value of s to $2 \times \lceil \frac{n \times (n-1)}{4} \rceil$. With this value of s, the number of distinct graphic sequences is the largest for length n. We measure the standard deviation for number of steps in MC varying from 2 to 20 in increments of 2.

Figure 5 illustrates the standard deviation for varying number of steps for $D_u(n, s)$. For each n the number of generated graphic sequences is $k \times d(n, s)$ ($k = 100$ in the experiments) where $d(n, s)$ is the number of distinct graphic sequences of length n and sum $s = 2 \times \lceil \frac{n \times (n-1)}{4} \rceil$. The standard deviation for $n = 3$ and $s = 4$ is always 0 because the graphical sequence can only be $(2, 1, 1)$. The standard deviations for lower values of n and s are not stable because of the small numbers of distinct graphic sequences. Empirically, we conservatively estimate the mixing time to be n number of steps. Figure 6 shows the average running time for the generation of 10 graphic sequences of $D_u(n, s)$ for n varying from 1000 to 10000 in increments of 1000. The curves are expected to be of the order of n by construction. Similarly to $D_u(n)$, they reveal the cost of each step in the Markov chain. We can generate a graphic sequence of 10000 elements with sum $49,995,000$ uniformly at random in 4200 seconds.

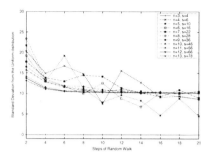
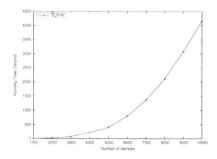

Fig. 5. Standard deviation from Uniform for varying number of steps for $D_u(n,s)$ for different n and $s = 2 \times \lceil \frac{n \times (n-1)}{4} \rceil$

Fig. 6. Running time of $D_u(n,s)$, for n varies in $\{1000, 2000, \ldots, 10000\}$

5   Conclusion

We have presented two new algorithms for the uniform random generation of graphic sequences. We have proved their correctness. We have empirically evaluated their performance. To our knowledge these algorithms are the first non trivial algorithms proposed for this task. There is strong evidence that the problem, if not #P-complete, is intrinsically difficult. The algorithms that we propose are Markov chain Monte Carlo algorithms. Our contribution is the original design of the Markov chain and the empirical evaluation of its mixing time. Nevertheless the practical problem of generating graphic sequences uniformly at random remains open for large lengths. We are currently investigating alternative approaches such as local optimization approaches.

References

1. Milo, R., Kashtan, N., Itzkovitz, S., Newman, M.E.J., Alon, U.: On the uniform generation of random graphs with prescribed degree sequences. Arxiv preprint cond-mat/0312028 (2003)
2. Viger, F., Latapy, M.: Efficient and simple generation of random simple connected graphs with prescribed degree sequence. In: Wang, L. (ed.) COCOON 2005. LNCS, vol. 3595, pp. 440–449. Springer, Heidelberg (2005)
3. Genio, C.I.D., Kim, H., Toroczkai, Z., Bassler, K.E.: Efficient and exact sampling of simple graphs with given arbitrary degree sequence. CoRR abs, 1002.2975 (2010)
4. Gkantsidis, C., Mihail, M., Zegura, E.: The Markov chain simulation method for generating connected power law random graphs. In: Proc. 5th Workshop on Algorithm Engineering and Experiments (ALENEX). SIAM, Philadelphia (2003)
5. Milo, R., Shen-Orr, S., Itzkovitz, S., Kashtan, N., Chklovskii, D., Alon, U.: Network motifs: Simple building blocks of complex networks. Science 298, 824–827 (2002)
6. Molloy, M., Reed, B.: A critical point for random graphs with a given degree sequence. Random Structures and Algorithms 6, 161–179 (1995)
7. Cooper, C., Frieze, A., Krivelevich, M.: Hamilton cycles in random graphs with a fixed degree sequence. SIAM J. Discrete Math. 24(2), 558–569 (2010)

8. Chung, F., Lu, L.: Connected components in random graphs with given expected degree sequences. Annals of Combinatorics 6(2), 125–145 (2002)
9. Ruskey, F., Eades, P., Cohen, B., Scott, A.: Alley CATs in search of good homes. Congressus Numerantium 102, 97–110 (1994)
10. Liu, K., Terzi, E.: Towards identity anonymization on graphs. In: SIGMOD (2008)
11. Hastings, W.K.: Monte Carlo sampling methods using Markov chains and their applications. Biometrika 57(1), 97–109 (1970)
12. Sinclair, A.: Algorithms for Random Generation and Counting: A Markov Chain Approach. Progress in Theoretical Computer Science. Birkhauser, Boston (1992)
13. Nobari, S., Lu, X., Karras, P., Bressan, S.: Fast Random Graph Generation. Accepted and to appear in EDBT (2011)
14. Gionis, A., Mannila, H., Mielikäinen, T., Tsaparas, P.: Assessing data mining results via swap randomization. In: SIGKDD, pp. 167–176 (2006)
15. Hasan, M.A., Zaki, M.: Musk: Uniform sampling of k maximal patterns. In: SIAM Data Mining (2009)
16. Hakimi, S.: On the realizability of a set of integers as degrees of the vertices of a graph. SIAM J. Applied Math. 10, 496–506 (1962)
17. Havel, V.: A remark on the existence of finite graphs (Hungarian). Časopis Pěst. Mat. 80, 477–480 (1955)
18. Erdös, P., Gallai, T.: Graphs with prescribed degrees of vertices. Mat. Lapok (1960)
19. Barnes, T.: A recurrence for counting graphical partitions. The Electronic Journal of Combinatorics 2(1) (1995)
20. Stanley, R.P.: A zonotope associated with graphical degree sequences (manuscript) (1989); appeared in Applied Geometry and Discrete Math. Dimacs Series. AMS, Providence (1991)
21. Peled, U.N., Srinivasan, M.K.: The polytope of degree sequences. Linear Algebra Appl. 114, 349–377 (1989)
22. Sierksma, G., Hoogeveen, H.: Seven criteria for integer sequences being graphic. Journal of Graph Theory 15(2), 223–231 (1991)
23. Burns, J.M.: The number of degree sequences of graphs. Phd thesis. Massachusetts Institute of Technology (2007)
24. Barnes, T.M., Savege, C.D.: Efficient generation of graphical partitions. Discrete Applied Mathematics 78, 17–26 (1997)
25. Cover, T.M., Thomas, J.A.: Elements of information theory. Chapter 4.3
26. Skiena, S.: Implementing discrete mathematics: combinatorics andgraph theory with mathematica. Addison-Wesley, MA (1990)
27. Sloane, N., Plouffe, S.: The Encyclopedia of Integer Sequences. Academic Press, London (1995)
28. Lu, X., Bressan, S.: Generating Random Graphic Sequences. School of Computing, National University of Singapore, Technical Report TRA1/11 (January 2011)

Author Index